AF817132

commutative ring theory and applications

PURE AND APPLIED MATHEMATICS

A Program of Monographs, Textbooks, and Lecture Notes

LECTURE NOTES IN PURE AND APPLIED MATHEMATICS

1. *N. Jacobson,* Exceptional Lie Algebras
2. *L.-Å. Lindahl and F. Poulsen,* Thin Sets in Harmonic Analysis
3. *I. Satake,* Classification Theory of Semi-Simple Algebraic Groups
4. *F. Hirzebruch et al., Differentiable Manifolds and Quadratic Forms*
5. *I. Chavel,* Riemannian Symmetric Spaces of Rank One
6. *R. B. Burckel,* Characterization of C(X) Among Its Subalgebras
7. *B. R. McDonald et al.,* Ring Theory
8. *Y.-T. Siu,* Techniques of Extension on Analytic Objects
9. *S. R. Caradus et al.,* Calkin Algebras and Algebras of Operators on Banach Spaces
10. *E. O. Roxin et al.,* Differential Games and Control Theory
11. *M. Orzech and C. Small,* The Brauer Group of Commutative Rings
12. *S. Thomier,* Topology and Its Applications
13. *J. M. Lopez and K. A. Ross,* Sidon Sets
14. *W. W. Comfort and S. Negrepontis,* Continuous Pseudometrics
15. *K. McKennon and J. M. Robertson,* Locally Convex Spaces
16. *M. Carmeli and S. Malin,* Representations of the Rotation and Lorentz Groups
17. *G. B. Seligman,* Rational Methods in Lie Algebras
18. *D. G. de Figueiredo,* Functional Analysis
19. *L. Cesari et al.,* Nonlinear Functional Analysis and Differential Equations
20. *J. J. Schäffer,* Geometry of Spheres in Normed Spaces
21. *K. Yano and M. Kon,* Anti-Invariant Submanifolds
22. *W. V. Vasconcelos,* The Rings of Dimension Two
23. *R. E. Chandler,* Hausdorff Compactifications
24. *S. P. Franklin and B. V. S. Thomas,* Topology
25. *S. K. Jain,* Ring Theory
26. *B. R. McDonald and R. A. Morris,* Ring Theory II
27. *R. B. Mura and A. Rhemtulla,* Orderable Groups
28. *J. R. Graef,* Stability of Dynamical Systems
29. *H.-C. Wang,* Homogeneous Branch Algebras
30. *E. O. Roxin et al.,* Differential Games and Control Theory II
31. *R. D. Porter,* Introduction to Fibre Bundles
32. *M. Altman,* Contractors and Contractor Directions Theory and Applications
33. *J. S. Golan,* Decomposition and Dimension in Module Categories
34. *G. Fairweather,* Finite Element Galerkin Methods for Differential Equations
35. *J. D. Sally,* Numbers of Generators of Ideals in Local Rings
36. *S. S. Miller,* Complex Analysis
37. *R. Gordon,* Representation Theory of Algebras
38. *M. Goto and F. D. Grosshans,* Semisimple Lie Algebras
39. *A. I. Arruda et al.,* Mathematical Logic
40. *F. Van Oystaeyen,* Ring Theory
41. *F. Van Oystaeyen and A. Verschoren,* Reflectors and Localization
42. *M. Satyanarayana,* Positively Ordered Semigroups
43. *D. L Russell,* Mathematics of Finite-Dimensional Control Systems
44. *P.-T. Liu and E. Roxin,* Differential Games and Control Theory III
45. *A. Geramita and J. Seberry,* Orthogonal Designs
46. *J. Cigler, V. Losert, and P. Michor,* Banach Modules and Functors on Categories of Banach Spaces
47. *P.-T. Liu and J. G. Sutinen,* Control Theory in Mathematical Economics
48. *C. Byrnes,* Partial Differential Equations and Geometry
49. *G. Klambauer,* Problems and Propositions in Analysis
50. *J. Knopfmacher,* Analytic Arithmetic of Algebraic Function Fields
51. *F. Van Oystaeyen,* Ring Theory
52. *B. Kadem,* Binary Time Series
53. *J. Barros-Neto and R. A. Artino,* Hypoelliptic Boundary-Value Problems
54. *R. L. Sternberg et al.,* Nonlinear Partial Differential Equations in Engineering and Applied Science
55. *B. R. McDonald,* Ring Theory and Algebra III
56. *J. S. Golan,* Structure Sheaves Over a Noncommutative Ring
57. *T. V. Narayana et al.,* Combinatorics, Representation Theory and Statistical Methods in Groups
58. *T. A. Burton,* Modeling and Differential Equations in Biology
59. *K. H. Kim and F. W. Roush,* Introduction to Mathematical Consensus Theory

60. *J. Banas and K. Goebel,* Measures of Noncompactness in Banach Spaces
61. *O. A. Nielson,* Direct Integral Theory
62. *J. E. Smith et al.,* Ordered Groups
63. *J. Cronin,* Mathematics of Cell Electrophysiology
64. *J. W. Brewer,* Power Series Over Commutative Rings
65. *P. K. Kamthan and M. Gupta,* Sequence Spaces and Series
66. *T. G. McLaughlin,* Regressive Sets and the Theory of Isols
67. *T. L. Herdman et al.,* Integral and Functional Differential Equations
68. *R. Draper,* Commutative Algebra
69. *W. G. McKay and J. Patera,* Tables of Dimensions, Indices, and Branching Rules for Representations of Simple Lie Algebras
70. *R. L. Devaney and Z. H. Nitecki,* Classical Mechanics and Dynamical Systems
71. *J. Van Geel,* Places and Valuations in Noncommutative Ring Theory
72. *C. Faith,* Injective Modules and Injective Quotient Rings
73. *A. Fiacco,* Mathematical Programming with Data Perturbations I
74. *P. Schultz et al.,* Algebraic Structures and Applications
75. *L Bican et al.,* Rings, Modules, and Preradicals
76. *D. C. Kay and M. Breen,* Convexity and Related Combinatorial Geometry
77. *P. Fletcher and W. F. Lindgren,* Quasi-Uniform Spaces
78. *C.-C. Yang,* Factorization Theory of Meromorphic Functions
79. *O. Taussky,* Ternary Quadratic Forms and Norms
80. *S. P. Singh and J. H. Burry,* Nonlinear Analysis and Applications
81. *K. B. Hannsgen et al.,* Volterra and Functional Differential Equations
82. *N. L. Johnson et al.,* Finite Geometries
83. *G. I. Zapata,* Functional Analysis, Holomorphy, and Approximation Theory
84. *S. Greco and G. Valla,* Commutative Algebra
85. *A. V. Fiacco,* Mathematical Programming with Data Perturbations II
86. *J.-B. Hiriart-Urruty et al.,* Optimization
87. *A. Figa Talamanca and M. A. Picardello,* Harmonic Analysis on Free Groups
88. *M. Harada,* Factor Categories with Applications to Direct Decomposition of Modules
89. *V. I. Istrătescu,* Strict Convexity and Complex Strict Convexity
90. *V. Lakshmikantham,* Trends in Theory and Practice of Nonlinear Differential Equations
91. *H. L. Manocha and J. B. Srivastava,* Algebra and Its Applications
92. *D. V. Chudnovsky and G. V. Chudnovsky,* Classical and Quantum Models and Arithmetic Problems
93. *J. W. Longley,* Least Squares Computations Using Orthogonalization Methods
94. *L. P. de Alcantara,* Mathematical Logic and Formal Systems
95. *C. E. Aull,* Rings of Continuous Functions
96. *R. Chuaqui,* Analysis, Geometry, and Probability
97. *L. Fuchs and L. Salce,* Modules Over Valuation Domains
98. *P. Fischer and W. R. Smith,* Chaos, Fractals, and Dynamics
99. *W. B. Powell and C. Tsinakis,* Ordered Algebraic Structures
100. *G. M. Rassias and T. M. Rassias,* Differential Geometry, Calculus of Variations, and Their Applications
101. *R.-E. Hoffmann and K. H. Hofmann,* Continuous Lattices and Their Applications
102. *J. H. Lightbourne III and S. M. Rankin III,* Physical Mathematics and Nonlinear Partial Differential Equations
103. *C. A. Baker and L. M. Batten,* Finite Geometrics
104. *J. W. Brewer et al.,* Linear Systems Over Commutative Rings
105. *C. McCrory and T. Shifrin,* Geometry and Topology
106. *D. W. Kueke et al.,* Mathematical Logic and Theoretical Computer Science
107. *B.-L. Lin and S. Simons,* Nonlinear and Convex Analysis
108. *S. J. Lee,* Operator Methods for Optimal Control Problems
109. *V. Lakshmikantham,* Nonlinear Analysis and Applications
110. *S. F. McCormick,* Multigrid Methods
111. *M. C. Tangora,* Computers in Algebra
112. *D. V. Chudnovsky and G. V. Chudnovsky,* Search Theory
113. *D. V. Chudnovsky and R. D. Jenks,* Computer Algebra
114. *M. C. Tangora,* Computers in Geometry and Topology
115. *P. Nelson et al.,* Transport Theory, Invariant Imbedding, and Integral Equations
116. *P. Clément et al.,* Semigroup Theory and Applications
117. *J. Vinuesa,* Orthogonal Polynomials and Their Applications
118. *C. M. Dafermos et al.,* Differential Equations
119. *E. O. Roxin,* Modern Optimal Control
120. *J. C. Díaz,* Mathematics for Large Scale Computing

121. *P. S. Milojević* Nonlinear Functional Analysis
122. *C. Sadosky,* Analysis and Partial Differential Equations
123. *R. M. Shortt,* General Topology and Applications
124. *R. Wong,* Asymptotic and Computational Analysis
125. *D. V. Chudnovsky and R. D. Jenks,* Computers in Mathematics
126. *W. D. Wallis et al.,* Combinatorial Designs and Applications
127. *S. Elaydi,* Differential Equations
128. *G. Chen et al.,* Distributed Parameter Control Systems
129. *W. N. Everitt,* Inequalities
130. *H. G. Kaper and M. Garbey,* Asymptotic Analysis and the Numerical Solution of Partial Differential Equations
131. *O. Arino et al.,* Mathematical Population Dynamics
132. *S. Coen,* Geometry and Complex Variables
133. *J. A. Goldstein et al.,* Differential Equations with Applications in Biology, Physics, and Engineering
134. *S. J. Andima et al.,* General Topology and Applications
135. *P Clément et al.,* Semigroup Theory and Evolution Equations
136. *K. Jarosz*, Function Spaces
137. *J. M. Bayod et al.*, *p*-adic Functional Analysis
138. *G. A. Anastassiou*, Approximation Theory
139. *R. S. Rees*, Graphs, Matrices, and Designs
140. *G. Abrams et al.*, Methods in Module Theory
141. *G. L. Mullen and P. J.-S. Shiue*, Finite Fields, Coding Theory, and Advances in Communications and Computing
142. *M. C. Joshi and A. V. Balakrishnan*, Mathematical Theory of Control
143. *G. Komatsu and Y. Sakane*, Complex Geometry
144. *I. J. Bakelman*, Geometric Analysis and Nonlinear Partial Differential Equations
145. *T. Mabuchi and S. Mukai*, Einstein Metrics and Yang–Mills Connections
146. *L. Fuchs and R. Göbel*, Abelian Groups
147. *A. D. Pollington and W. Moran*, Number Theory with an Emphasis on the Markoff Spectrum
148. *G. Dore et al.*, Differential Equations in Banach Spaces
149. *T. West*, Continuum Theory and Dynamical Systems
150. *K. D. Bierstedt et al.*, Functional Analysis
151. *K. G. Fischer et al.*, Computational Algebra
152. *K. D. Elworthy et al.*, Differential Equations, Dynamical Systems, and Control Science
153. *P.-J. Cahen, et al.*, Commutative Ring Theory
154. *S. C. Cooper and W. J. Thron*, Continued Fractions and Orthogonal Functions
155. *P. Clément and G. Lumer*, Evolution Equations, Control Theory, and Biomathematics
156. *M. Gyllenberg and L. Persson*, Analysis, Algebra, and Computers in Mathematical Research
157. *W. O. Bray et al.,* Fourier Analysis
158. *J. Bergen and S. Montgomery*, Advances in Hopf Algebras
159. *A. R. Magid*, Rings, Extensions, and Cohomology
160. *N. H. Pavel*, Optimal Control of Differential Equations
161. *M. Ikawa*, Spectral and Scattering Theory
162. *X. Liu and D. Siegel*, Comparison Methods and Stability Theory
163. *J.-P. Zolésio*, Boundary Control and Variation
164. *M. Křížek et al.*, Finite Element Methods
165. *G. Da Prato and L. Tubaro*, Control of Partial Differential Equations
166. *E. Ballico*, Projective Geometry with Applications
167. *M. Costabel et al.*, Boundary Value Problems and Integral Equations in Nonsmooth Domains
168. *G. Ferreyra, G. R. Goldstein, and F. Neubrander*, Evolution Equations
169. *S. Huggett*, Twistor Theory
170. *H. Cook et al.*, Continua
171. *D. F. Anderson and D. E. Dobbs*, Zero-Dimensional Commutative Rings
172. *K. Jarosz,* Function Spaces
173. *V. Ancona et al.*, Complex Analysis and Geometry
174. *E. Casas*, Control of Partial Differential Equations and Applications
175. *N. Kalton et al.*, Interaction Between Functional Analysis, Harmonic Analysis, and Probability
176. *Z. Deng et al.*, Differential Equations and Control Theory
177. *P. Marcellini et al.* Partial Differential Equations and Applications
178. *A. Kartsatos*, Theory and Applications of Nonlinear Operators of Accretive and Monotone Type
179. *M. Maruyama*, Moduli of Vector Bundles
180. *A. Ursini and P. Agliano*, Logic and Algebra
181. *X. H. Cao et al.,* Rings, Groups, and Algebras
182. *D. Arnold and R. M. Rangaswamy,* Abelian Groups and Modules
183. *S. R. Chakravarthy and A. S. Alfa*, Matrix-Analytic Methods in Stochastic Models

184. *J. E. Andersen et al.*, Geometry and Physics
185. *P.-J. Cahen et al.*, Commutative Ring Theory
186. *J. A. Goldstein et al.*, Stochastic Processes and Functional Analysis
187. *A. Sorbi*, Complexity, Logic, and Recursion Theory
188. *G. Da Prato and J.-P. Zolésio*, Partial Differential Equation Methods in Control and Shape Analysis
189. *D. D. Anderson*, Factorization in Integral Domains
190. *N. L. Johnson*, Mostly Finite Geometries
191. *D. Hinton and P. W. Schaefer*, Spectral Theory and Computational Methods of Sturm–Liouville Problems
192. *W. H. Schikhof et al.*, *p*-adic Functional Analysis
193. *S. Sertöz*, Algebraic Geometry
194. *G. Caristi and E. Mitidieri*, Reaction Diffusion Systems
195. *A. V. Fiacco*, Mathematical Programming with Data Perturbations
196. *M. Křížek et al.*, Finite Element Methods: Superconvergence, Post-Processing, and A Posteriori Estimates
197. *S. Caenepeel and A. Verschoren*, Rings, Hopf Algebras, and Brauer Groups
198. *V. Drensky et al.*, Methods in Ring Theory
199. *W. B. Jones and A. Sri Ranga*, Orthogonal Functions, Moment Theory, and Continued Fractions
200. *P. E. Newstead*, Algebraic Geometry
201. *D. Dikranjan and L. Salce*, Abelian Groups, Module Theory, and Topology
202. *Z. Chen et al.*, Advances in Computational Mathematics
203. *X. Caicedo and C. H. Montenegro*, Models, Algebras, and Proofs
204. *C. Y. Yıldırım and S. A. Stepanov*, Number Theory and Its Applications
205. *D. E. Dobbs et al.*, Advances in Commutative Ring Theory
206. *F. Van Oystaeyen*, Commutative Algebra and Algebraic Geometry
207. *J. Kakol et al.*, *p*-adic Functional Analysis
208. *M. Boulagouaz and J.-P. Tignol*, Algebra and Number Theory
209. *S. Caenepeel and F. Van Oystaeyen*, Hopf Algebras and Quantum Groups
210. *F. Van Oystaeyen and M. Saorin*, Interactions Between Ring Theory and Representations of Algebras
211. *R. Costa et al.*, Nonassociative Algebra and Its Applications
212. *T.-X. He*, Wavelet Analysis and Multiresolution Methods
213. *H. Hudzik and L. Skrzypczak*, Function Spaces: The Fifth Conference
214. *J. Kajiwara et al.*, Finite or Infinite Dimensional Complex Analysis
215. *G. Lumer and L. Weis*, Evolution Equations and Their Applications in Physical and Life Sciences
216. *J. Cagnol et al.*, Shape Optimization and Optimal Design
217. *J. Herzog and G. Restuccia*, Geometric and Combinatorial Aspects of Commutative Algebra
218. *G. Chen et al.*, Control of Nonlinear Distributed Parameter Systems
219. *F. Ali Mehmeti et al.*, Partial Differential Equations on Multistructures
220. *D. D. Anderson and I. J. Papick*, Ideal Theoretic Methods in Commutative Algebra
221. *Á. Granja et al.*, Ring Theory and Algebraic Geometry
222. *A. K. Katsaras et al.*, *p*-adic Functional Analysis
223. *R. Salvi*, The Navier-Stokes Equations
224. *F. U. Coelho and H. A. Merklen*, Representations of Algebras
225. *S. Aizicovici and N. H. Pavel*, Differential Equations and Control Theory
226. *G. Lyubeznik*, Local Cohomology and Its Applications
227. *G. Da Prato and L. Tubaro*, Stochastic Partial Differential Equations and Applications
228. *W. A. Carnielli et al.*, Paraconsistency
229. *A. Benkirane and A. Touzani*, Partial Differential Equations
230. *A. Illanes et al.*, Continuum Theory
231. *M. Fontana et al.*, Commutative Ring Theory and Applications

Additional Volumes in Preparation

commutative ring theory and applications

proceedings of the fourth international conference

edited by

Marco Fontana
Università degli Studi Roma Tre
Rome, Italy

Salah-Eddine Kabbaj
King Fahd University of Petroleum and Minerals
Dharan, Saudi Arabia

Sylvia Wiegand
University of Nebraska—Lincoln
Lincoln, Nebraska, U.S.A.

MARCEL DEKKER
MARCEL DEKKER, INC. NEW YORK • BASEL

Library of Congress Cataloging-in-Publication Data
A catalog record for this book is available from the Library of Congress.

ISBN: 0-8247-0855-5

This book is printed on acid-free paper.

Headquarters
Marcel Dekker, Inc.
270 Madison Avenue, New York, NY 10016
tel: 212-696-9000; fax: 212-685-4540

Eastern Hemisphere Distribution
Marcel Dekker AG
Hutgasse 4, Postfach 812, CH-4001 Basel, Switzerland
tel: 41-61-260-6300; fax: 41-61-260-6333

World Wide Web
http://www.dekker.com

The publisher offers discounts on this book when ordered in bulk quantities. For more information, write to Special Sales/Professional Marketing at the headquarters address above.

Current printing (last digit):
10 9 8 7 6 5 4 3 2 1

PRINTED IN THE UNITED STATES OF AMERICA

Preface

This volume draws on the contributors' talks at the Fourth International Conference on Commutative Algebra held in Fez, Morocco. The goal of this conference was to present recent progress and new trends in the growing area of commutative algebra, with primary emphasis on commutative ring theory and its applications. The conference also facilitated a fruitful interaction among the participants, whose various mathematical interests shared the same (commutative) algebraic roots.

The book consists of 34 chapters which, while written as separate articles, provide nonetheless a comprehensive report on questions and problems of contemporary interest. Some articles are surveys of their subject, while others present a narrower, in-depth view. All the manuscripts were subject to a strict refereeing process.

This volume encompasses wide-ranging topics in commutative ring theory (along with connections to algebraic number theory, algebraic geometry, homological algebra, and model-theoretic algebra). The topics covered include: algebroid curves, arithmetic rings, chain conditions, class groups, constructions of examples, divisibility and factorization, linear Diophantine equations, the going-down and going-up properties, graded modules and analytic spread, Gröbner bases and computational methods, homological aspects of commutative rings, ideal and module systems, integer-valued polynomials, integral dependence, Krull domains and generalizations, local cohomology, prime spectra and dimension theory, polynomial rings, power series rings, pullbacks, tight closure, ultraproducts, and zero-divisors.

Graduate students and established commutative algebraists will find the book a valuable and reliable source, as will researchers in many other branches of mathematics.

The conference was organized by the University of Fez with the scientific collaboration of the Università degli Studi "Roma Tre," Italy, and the University of Nebraska, U.S.A. Financial support was provided by the Commutative Algebra and Homological Aspects Laboratory, the Faculty of Sciences "Dhar Al-Mehraz," the International Mathematical Union (CDE), the "Espace Sciences & Vie" Association, and the Università degli Studi "Roma Tre."

We wish to express our gratitude to the local organizing committee, especially Professors A. Benkirane, Chairman of the Department of Mathematics, R. Ameziane Hassani, and A. Touzani, as well as to Professor M. H. Kadri and Mr. M. A. Chad, Dean and Secretary-General, respectively, of the Faculty of Sciences "Dhar Al-Mehraz" at Fez. Special thanks are due to Mr. A. Bennani and Mrs. T. Ibn Abdelmoula for their constant help with conference arrangements. The efforts of the contributors and the referees are greatly appreciated; without their work this volume would never have been produced. Last, we thank the editorial staff at Marcel Dekker, Inc., in particular, Maria Allegra and Ana Pacheco, for their patience, hard work, and assistance with this volume.

Marco Fontana
Salah-Eddine Kabbaj
Sylvia Wiegand

Contents

Contributors

D. F. Anderson, Department of Mathematics, The University of Tennessee, Knoxville, Tennessee 37996-1300 USA Email: anderson@math.utk.edu

A. Badawi, Department of Mathematics, Birzeit University, P.O. Box 14, Birzeit, West Bank, Palestine. Email: abring@birzeit.edu

A. Badra, Département de Mathématiques, Université Blaise Pascal, 63177 Aubière, France. Email: badra@ucfma.univ-bpclermont.fr

V. Barucci, Dipartimento di Matematica, Università di Roma "La Sapienza", 00185 Roma, Italy. Email: barucci@mat.uniroma1.it

A. Benhissi, Département de Mathématiques, Faculté des Sciences, B.P. 5000, Monastir, Tunisia. Email: ali.benhissi@fsm.rnu.tn

A. Bouanane, Department of Mathematics, University of Tetouan, P.O. Box 2121, Tetouan, Morocco. Email: abouanane@hotmail.com

J. Brewer, Department of Mathematics, Florida Atlantic University, Boca Raton, FL 33431-6498, USA. Email: brewer@fau.edu

G. W. Chang, Department of Mathematics, Inha University, Incheon, 402-751 Korea

J.-L. Chabert, Département de Mathématiques, Université de Picardie, 80039 Amiens, France. Email: jlchaber@worldnet.fr

T. Coquand, Dept. of Computer Science, Chalmers University, Eklandagatan 86, 412 96 Goteborg, Sweden Email: coquand@cs.chalmers.se

M. D'Anna, Dipartimento di Matematica, Università di Catania, 95125 Catania, Italy. Email: mdanna@dmi.unict.it

D. Delfino, Department of Mathematics, Rutgers University, 110 Frelinghuysen Rd, Piscataway, New Jersey 08854-8019, USA. Email: ddelfino1@ yahoo.com

Y. Diagana, Département de Mathématiques, Université d'Abobo-Adjamé, 03 B.P. 1644, Abidjan, Côte d'Ivoire. Email: y_diagana@yahoo.com

H. Dichi, Département de Mathématiques, Université Blaise Pascal, 63177 Aubière, France. E-mail: henri.dichi@math.univ-bpclermont.fr

D. Dobbs, Department of Mathematics, University of Tennessee, Knoxville, TN 37996-1300, USA. Email: dobbs@math.utk.edu

R. Eddahabi, Department of Mathematics, University of Tetouan, P.O. Box 2121, Tetouan, Morocco.

S. El Baghdadi, Department of Mathematics, Faculté des Sciences et Techniques, P.O. Box 523, Beni Mellal, Morocco. Email: said_el1@yahoo.com

M. Fontana, Dipartimento di Matematica, Università degli Studi "Roma Tre", Largo S.L. Murialdo, 1, 00146 Roma, Italy. Email: fontana@mat.uniroma3.it

R. Fröberg, Department of Mathematics, Stockholm University 106 91 Stockholm, Sweden.
Email: ralff@matematik.su.se

S. Gabelli, Dipartimento di Matematica, Università degli Studi "Roma Tre", Largo S.L. Murialdo, 1, 00146 Roma, Italy. Email: gabelli@mat.uniroma3.it

S. Glaz, Department of Mathematics, University of Connecticut, Storrs, CT 06269-0001, USA. Email: glaz@uconnvm.uconn.edu

F. Halter-Koch, Institute of Mathematics, Graz University, A-8010 Graz, Austria. Email: franz.halterkoch@kfunigraz.ac.at

W. Hassler, Institut fü r Mathematik Karl-Franzens-Universitä t Graz Heinrichstr. 36, A-8010 Graz, Austria. Email: wolfgang.hassler@kfunigraz.ac.at

W. Heinzer, Department of Mathematics, Purdue University, West Lafayette, IN 47907-1968, USA. Email: heinzer@math.purdue.edu

A. Hetzel, Department of Mathematics, University of Tennessee, Knoxville, TN 37996-1300, USA. Email: ahetzel@math.utk.edu

M. Hochster, Department of Mathematics, University of Michigan, Ann Arbor, MI 48109-1109, USA. Email: hochster@math.lsa.umich.edu

C. Huneke, Department of Mathematics, University of Kansas, Lawrence, KS 66045, USA. Email: huneke@math.ukans.edu

S. Kabbaj, Department of Mathematics, KFUPM, P.O. Box 5046, Dhahran 31261, Saudi Arabia and University of Fez, Fez, Morocco. Email: kabbaj@kfupm.edu.sa

M. Khalis, Department of Mathematics, Faculté des Sciences et Techniques, P.O. Box 577, Settat, Morocco. Email: mohammed_khalis@hotmail.com

L. Klingler, Department of Mathematics, Florida Atlantic University, Boca Raton, FL 33431-6498, USA. Email: klingler@fau.edu

F. Kourki, Department of Mathematics, University of Tetouan, P.O. Box 2121, Tetouan, Morocco. Email: kourki@yahoo.com

Y. Lequain, IMPA, 22.460-320, Rio de Janeiro, Brazil. Email: ylequain@impa.br

H. Lombardi, Equipe de Mathématiques, UMR CNRS 6623, UFR des Sciences et Techniques, Université de Franche-Comté, 25030 Besancon cedex, France.
Email: lombardi@math.univfcomte.fr

T. G. Lucas, Department of Mathematics, University of North Carolina, Charlotte, NC 28223, USA. Email: tglucas@email.uncc.edu

N. Mahdou, Department of Mathematics, FST Fez-Saïss, University of Fez, P.O. Box 2202, Fez, Morocco. Email: mahdou@hotmail.com

A. Mimouni, Department of Mathematics, University of Fez, P.O. Box 1796, Fez, Morocco. Email: a_mimouni@hotmail.com

B. Olberding, Department of Mathematics, University of Louisiana, Monroe LA 71209, USA. E-mail: maolberding@ulm.edu

J. Park, Department of Mathematics, Inha University, Incheon 402-751, Korea Email: jnpark@math.inha.ac.kr

G. Picavet, Département de Mathématiques, Université Blaise Pascal, 63177 Aubière, France E-mail: gabriel.picavet@math.univ-bpclermont.fr

M. Picavet L'Hermitte, Département de Mathématiques, Université Blaise Pascal, 63177 Aubière, France. E-mail: martine.picavet@math.univ-bpclermont.fr

C. Quitté, Laboratoire de Mathématiques, SP2MI, Boulevard 3, Teleport 2, BP 179, 86960 Futuroscope Cedex, France. Email: quitte@mathlabo.univ-poitiers.fr

C. Rotthaus, Department of Mathematics, Michigan State University, East Lansing, MI 48824-0001, USA. Email: rotthaus@math.msu.edu

A. S. Saydam, Department of Mathematics, University of Louisiana, Monroe, LA 71209, USA. Email: masaydam@ulm.edu

I. Swanson, Department of Mathematical Sciences, New Mexico State University - Las Cruces, New Mexico 88003-8001, USA. Email: iswanson@nmsu.Edu

F. Tartarone, Dipartimento di Matematica, Università degli Studi "Roma Tre", Largo S. L. Murialdo, 1, 00146 Roma, Italy. Email: tfrance@mat.uniroma3.it

A. Taylor, Department of Mathematics, Rutgers University, 110 Frelinghuysen Rd, Piscataway, New Jersey 08854-8019, USA. Email: ataylor@math.rutgers.edu

W. V. Vasconcelos, Department of Mathematics, Rutgers University, 110 Frelinghuysen Rd, Piscataway, New Jersey 08854-8019, USA. Email: vasconce@math.rutgers.edu

R. H. Villarreal, Departamento de Matemáticas, Centro de Investigación y de Estudios Avanzados del IPN, Apartado Postal 14–740, 07000, México City, D.F. MEXICO. Email: vila@math.cinvestav.mx

N. Weininger, Department of Mathematics, Rutgers University, 110 Frelinghuysen Rd, Piscataway, New Jersey 08854-8019, USA. Email: nweining@math.rutgers.edu

R. Wiegand, Department of Mathematics, University of Nebraska, Lincoln, NE 68588-0323, USA. Email: rwiegand@math.unl.edu

S. Wiegand, Department of Mathematics, University of Nebraska, Lincoln, NE 68588-0323, USA. Email: swiegand@math.unl.edu

M. Zafrullah, Department of Mathematics, University of Arkansas, Fayetteville AR 72701, USA. Email: mzafrullah@usa.net

$D[X^2, X^3]$ Over an Integral Domain D

DAVID F. ANDERSON Department of Mathematics, The University of Tennessee, Knoxville, Tennessee 37996-1300.

GYU WHAN CHANG Department of Mathematics, Inha University, Incheon, Korea 402-751

JEANAM PARK Department of Mathematics, Inha University, Incheon, Korea 402-751

1. INTRODUCTION

Let D be an integral domain with identity and quotient field K. In this paper, we study the ring $D[X^2, X^3] = D + X^2D[X] \subset D[X]$, and we compare its behavior to its polynomial overring $D[X]$. Of course, $D[X^2, X^3]$ is never integrally closed (or seminormal, root closed, etc.); so in this paper, we are mainly interested in ring-theoretic properties that do not involve "closedness" conditions. Quite often $D[X^2, X^3]$ satisfies a given ring-theoretic property if and only if $D[X]$ satisfies that property. However, in Section 3, the characteristic of D plays an important role. The ring $K[X^2, X^3]$ has proved useful in constructing examples concerning the Picard group (see Theorem 3.4) and nonunique factorization (see [10]). This paper gives several other cases where the ring $D[X^2, X^3]$ can be used to construct interesting, elementary examples (for instance, see Section 3). Many of the results in this paper generalize to monoid domains; we leave this to a future paper.

We first recall some of the properties we will investigate in this paper. An integral domain D is said to be a *weakly factorial domain* (WFD) [4] if each nonzero nonunit of D is a product of primary elements. Following [9], D is called a *generalized weakly factorial domain* (GWFD) if each nonzero prime ideal of D contains a primary element. Clearly, WFDs are GWFDs; however, if D is a Dedekind domain with nonzero torsion divisor class group, then R is a GWFD, but not a WFD (cf. [7, p. 912], [9, Proposition 3.1]). Following [5], D is called a *weakly Krull domain* if $D = \cap_{P \in X^{(1)}(D)} D_P$ and the intersection has finite character, where $X^{(1)}(D)$ is the set of height-one prime ideals of D. A Krull domain is weakly Krull, and a Noetherian domain is weakly Krull if and only if every grade-one prime ideal has height one. In [7, Theorem], it was shown that D is a WFD if and only if D is a weakly Krull domain and $Cl_t(D) = 0$. A Krull domain D is called *almost factorial* if $Cl(D)$ is torsion. As in [5], we say that an integral domain D is an *almost weakly factorial domain (AWFD)* if for each nonzero nonunit $x \in D$, there is an integer $n = n(x) \geq 1$ such that x^n is a product of primary elements. Thus an AWFD is a GWFD. It was shown in [5, Theorem 3.4] that D is an AWFD if and only if D is a weakly Krull domain and $Cl_t(D)$ is torsion. We say that an integral domain D is an *almost GCD-domain (AGCD-domain)* if for all nonzero $x, y \in D$, there exists an integer $n = n(x, y) \geq 1$ such that $(x^n, y^n)_v$ is principal. In [6, Theorem 3.4], it was proved that $Cl_t(D)$ is torsion when D is an AGCD-domain.

Throughout this paper, D denotes an integral domain with quotient field K, $Spec(D)$ its set of prime ideals, and $X^{(1)}(D)$ its set of height-one prime ideals. For $f \in K[X]$, let A_f be the fractional ideal of D generated by the coefficients of f. Recall that for a nonzero fractional ideal A of D, we have $A^{-1} = \{x \in K | xA \subseteq D\}$, $A_v = (A^{-1})^{-1}$, and $A_t = \cup\{(a_1, ..., a_n)_v | 0 \neq (a_1, ..., a_n) \subseteq A\}$. A nonzero fractional ideal A of D is called a *divisorial ideal* (resp., *t-ideal*) if $A_v = A$ (resp., $A_t = A$). We say that D has *t-dimension one*, written t-dim$D = 1$, if each prime t-ideal of D has height one (note that a height-one prime ideal is necessarily a t-ideal). A weakly Krull domain D has t-dim$D = 1$ [5, Lemma 2.1]. An integral ideal of D is said to be a maximal t-ideal if it is maximal with respect to being a t-ideal, and a maximal t-ideal is necessarily a prime ideal.

A nonzero fractional ideal A of D is said to be *t-invertible* if there exists a fractional ideal B of D with $(AB)_t = D$, and in this case we can take $B = A^{-1}$. It is well known that if A is a t-invertible t-ideal, then $A = J_v$ for some finitely generated subideal J of A. The set of t-invertible t-ideals of D forms an abelian group under the t-product $A * B = (AB)_t$. The *t-class group* of D is $Cl_t(D)$ - the group of t-invertible fractional t-ideals of D modulo its subgroup of principal fractional ideals. For D a Krull domain, $Cl_t(D) = Cl(D)$, the divisor class group; while for D a Prüfer domain or one-dimensional integral domain, $Cl_t(D) = C(D) = Pic(D)$, the ideal class group (or Picard group). For a recent survey article on the t-class group, see [8].

2. THE RING $D[X^2, X^3]$

In this section, we study the ring $D[X^2, X^3] = D + X^2D[X]$ and prove some analogs of the polynomial ring $D[X]$. Our first goal is to show that $D[X^2, X^3]$ is a UMT-domain if and only if D is a UMT-domain. The next lemma also holds for monoid domains (cf. [11, Lemma 2.3]).

LEMMA 2.1. *Let I be a nonzero fractional ideal of D. Then*

(1) $(ID[X^2, X^3])^{-1} = I^{-1}D[X^2, X^3]$.

(2) $(ID[X^2, X^3])_v = I_vD[X^2, X^3]$.

(3) $(ID[X^2, X^3])_t = I_tD[X^2, X^3]$.

Proof. (1) It is clear that $I^{-1}D[X^2, X^3] \subseteq (ID[X^2, X^3])^{-1}$. Note that since $I(ID[X^2, X^3])^{-1} \subseteq D[X^2, X^3] \subseteq K[X^2, X^3]$, we have $(ID[X^2, X^3])^{-1} \subseteq K[X^2, X^3]$. If $f \in (ID[X^2, X^3])^{-1}$, then $A_fI \subseteq D$, and hence $A_f \subseteq I^{-1}$. So $f \in A_fD[X^2, X^3] \subseteq I^{-1}D[X^2, X^3]$. Therefore, $(ID[X^2, X^3])^{-1} = I^{-1}D[X^2, X^3]$.

(2) $(ID[X^2, X^3])_v = ((ID[X^2, X^3])^{-1})^{-1} = (I^{-1}D[X^2, X^3])^{-1} = I_vD[X^2, X^3]$ by (1).

(3) It is clear that if $f_1, f_2, \ldots, f_k \in ID[X^2, X^3]$, then

$$(f_1, \ldots, f_k)_v \subseteq ((A_{f_1}, \ldots, A_{f_k})D[X^2, X^3])_v$$

$$= (A_{f_1}, ..., A_{f_k})_v D[X^2, X^3] \subseteq I_t D[X^2, X^3].$$

So $(ID[X^2, X^3])_t \subseteq I_t D[X^2, X^3]$. For the converse, let J be a nonzero finitely generated subideal of I. Then $J_v D[X^2, X^3] = (JD[X^2, X^3])_v \subseteq (ID[X^2, X^3])_t$ by (2). Thus $I_t D[X^2, X^3] \subseteq (ID[X^2, X^3])_t$, and hence $(ID[X^2, X^3])_t = I_t D[X^2, X^3]$. □

LEMMA 2.2. *(cf. [18, Proposition 1.1]) Let Q be a maximal t-ideal of $D[X^2, X^3]$ such that $Q \cap D \neq 0$. Then $Q = (Q \cap D)[X^2, X^3]$. In particular, $Q \cap D$ is a maximal t-ideal of D.*

Proof. It suffices to show that $c(Q)[X^2, X^3] \subseteq Q$, where $c(Q)$ is the ideal of D generated by the coefficients of all the polynomials in Q. If $c(Q) \nsubseteq Q$, then $Q \subsetneq c(Q)[X^2, X^3]$. Since Q is a maximal t-ideal, we have $c(Q)_t[X^2, X^3] = (c(Q)[X^2, X^3])_t = D[X^2, X^3]$. So $c(Q)_t = D$; whence there is a polynomial $f \in Q$ such that $(A_f)_v = D$. Let $0 \neq a \in Q \cap D$.

We claim that $(a, f)^{-1} = D[X^2, X^3]$. First note that $(a, f)^{-1} \subseteq K[X^2, X^3]$ because for $g \in (a, f)^{-1}$, $ag \in D[X^2, X^3] \subseteq K[X^2, X^3]$. Next, if $g \in (a, f)^{-1}$, then there is an integer $m \geq 1$ such that $A_f^{m+1} A_g = A_f^m A_{fg}$ [16, Theorem 28.1]. Thus $(A_f^{m+1} A_g)_v = (A_f^m A_{fg})_v$ and $A_g \subseteq (A_g)_t = ((A_f^{m+1})_v A_g)_v = (A_f^{m+1} A_g)_v = (A_f^m A_{fg})_v = ((A_f^m)_v A_{fg})_v = (A_{fg})_v \subseteq D$. Hence $g \in A_g[X^2, X^3] \subseteq D[X^2, X^3]$. Thus $(a, f)^{-1} = D[X^2, X^3]$, and hence $(a, f)_v = D[X^2, X^3]$, which is a contradiction since Q is a t-ideal. Therefore $c(Q)[X^2, X^3] = Q$, and hence $Q = (Q \cap D)[X^2, X^3]$. □

As in [18], D is called a *UMT-domain* if every upper to zero (a nonzero prime ideal of $D[X]$ which contracts to zero in D) of $D[X]$ is a maximal t-ideal. Recall that $D[X]$ is a UMT-domain if and only if D is a UMT-domain [14, Theorem 3.4]. Thus, as a consequence of our next result, $D[X^2, X^3]$ is a UMT-domain if and only if $D[X]$ is a UMT-domain.

THEOREM 2.3. *$D[X^2, X^3]$ is a UMT-domain if and only if D is a UMT-domain.*

Proof. (⇒) Suppose that $D[X^2, X^3]$ is a UMT-domain. Let P be a maximal t-ideal of D. Then $PD[X^2, X^3]$ is a maximal t-ideal of $D[X^2, X^3]$ by Lemma 2.2. Also, note that $D[X^2, X^3]_{PD[X^2, X^3]} = D[X]_{P[X]}$. Since $D[X^2, X^3]$ is a UMT-domain, $D[X]_{P[X]}$ is a t-linkative UMT-domain [14, Theorem 1.5], and hence D_P

is a t-linkative UMT-domain (see the proof of [14, Theorem 2.4]). Thus D is a UMT-domain [14, Theorem 1.5].

($\Leftarrow$) Suppose that D is a UMT-domain. To show that $D[X^2, X^3]$ is a UMT-domain, it is enough to show that if Q is a maximal t-ideal of $D[X^2, X^3]$, then the integral closure of $D[X^2, X^3]_Q$ is a Prüfer domain [14, Theorem 1.5].

Let Q be a maximal t-ideal of $D[X^2, X^3]$ and let $Q \cap D = P$. If $P \neq 0$, then $Q = P[X^2, X^3]$ by Lemma 2.2. Moreover, since $X^2 \notin P[X^2, X^3]$ we have $D[X^2, X^3]_Q = D[X]_{P[X]}$. Thus the integral closure of $D[X^2, X^3]_Q$ is a Prüfer domain by [14, Theorem 1.5] (note that $D[X]$ is a UMT-domain [14, Theorem 2.4] and $P[X]$ is a prime t-ideal of $D[X]$). If $P = 0$, then $D[X^2, X^3]_Q = K[X^2, X^3]_{QK[X^2,X^3]}$, and hence $D[X^2, X^3]_Q$ is a one-dimensional Noetherian domain. Thus the integral closure of $D[X^2, X^3]_Q$ is a Dedekind domain (cf. [22, Theorem 33.10]), and hence a Prüfer domain. □

LEMMA 2.4. *If Q is a prime ideal of $D[X^2, X^3]$, then there is a unique prime ideal of $D[X]$ lying over Q. Thus the natural map Spec$(D[X]) \to$ Spec$(D[X^2, X^3])$, given by $P \to P \cap D[X^2, X^3]$, is an order-preserving bijection.*

Proof. Let Q be a prime ideal of $D[X^2, X^3]$, $P = Q \cap D$, and $S = \{X^n | n = 0, 2, 3, \ldots\}$.

Case 1. $P = 0$. If $QD[X^2, X^3]_S = D[X^2, X^3]_S$, then $Q = XD[X] \cap D[X^2, X^3]$ and $XD[X]$ is the unique prime ideal of $D[X]$ lying over Q. Assume that $QD[X^2, X^3]_S \subsetneq D[X]_S$. Note that $D[X^2, X^3]_S = D[X]_S = D[X, X^{-1}]$. So $QD[X^2, X^3]_S \cap D[X]$ is the unique prime ideal of $D[X]$ lying over Q.

Case 2. $P \neq 0$. If $Q = P[X^2, X^3]$, then $P[X]$ is the unique prime ideal of $D[X]$ lying over Q. Assume that $P[X^2, X^3] \subsetneq Q$. Note that $D[X^2, X^3]/P[X^2, X^3] \cong (D/P)[X^2, X^3]$, $D[X]/P[X] \cong (D/P)[X]$, and $(Q/P[X^2, X^3]) \cap (D/P)[X^2, X^3] = 0$. Thus there is a unique prime ideal of $(D/P)[X]$ lying over $Q/P[X^2, X^3]$ by Case 1. Since every prime ideal of $D[X]$ lying over Q contains $P[X]$, there is a unique prime ideal of $D[X]$ lying over Q. □

We next show that the bijection in Lemma 2.4 preserves t-ideals.

THEOREM 2.5. *Let Q be a prime ideal of $D[X]$ and let $Q' = Q \cap D[X^2, X^3]$. Then Q' is a prime t-ideal of $D[X^2, X^3]$ if and only if Q is a prime t-ideal of $D[X]$.*

Proof. Let $P = Q \cap D = Q' \cap D$ and $S = \{X^n | n = 0, 2, 3, \dots\}$.

Case 1. $P = 0$. Then htQ' = ht$Q = 1$ by Lemma 2.4. Thus Q and Q' are prime t-ideals of $D[X]$ and $D[X^2, X^3]$, respectively.

Case 2. $P \neq 0$. Then $Q = P[X]$ if and only if $Q' = P[X^2, X^3]$ (Lemma 2.4). Thus, by Lemma 2.1, Q is a prime t-ideal of $D[X]$ if and only if P is a prime t-ideal of D, if and only if Q' is a prime t-ideal of $D[X^2, X^3]$.

Case 3. $P \neq 0$. Then $P[X] \subsetneq Q$ if and only if $P[X^2, X^3] \subsetneq Q'$. Note that if either Q or Q' is a t-ideal, then $X \notin Q$. For if $0 \neq a \in P$, then $((a, X)D[X])_v = D[X]$ and $((a, X^2)D[X^2, X^3])_v = D[X^2, X^3]$. Note that $D[X^2, X^3]_S = D[X]_S$, $Q'D[X^2, X^3]_S = QD[X]_S$, $Q = QD[X]_S \cap D[X]$, and $Q' = Q'D[X^2, X^3]_S \cap D[X^2, X^3]$. Thus it suffices to show that if either Q or Q' is a t-ideal, then $Q'D[X^2, X^3]_S = QD[X]_S$ is a t-ideal of $D[X]_S$ by [19, Lemma 3.17].

Let A be a fractional ideal of $D[X]$ such that $A \cap D \neq 0$. We claim that $(AD[X]_S)^{-1} = A^{-1}D[X]_S$. It is clear that $A^{-1}D[X]_S \subseteq (AD[X]_S)^{-1}$. For the converse, let $u \in (AD[X]_S)^{-1}$. Then $uA \subseteq u(AD[X]_S) \subseteq D[X]_S$. Since $A \cap D \neq 0$, $u \in K[X]_S$. Thus $u = \frac{g}{X^m}$ for some $g \in K[X]$ and integer $m \geq 0$. For any $f \in A$, since $uf = (\frac{g}{X^m})f \in D[X]_S$, $fgX^n \in D[X]$ for some integer $n \geq 0$, and hence $fg \in D[X]$. Thus $g \in A^{-1}$ and $u = \frac{g}{X^m} \in A^{-1}D[X]_S$. Hence $(AD[X]_S)^{-1} \subseteq A^{-1}D[X]_S$, and thus $(AD[X]_S)^{-1} = A^{-1}D[X]_S$. A similar argument shows that if A is a fractional ideal of $D[X^2, X^3]$ with $A \cap D \neq 0$, then $(AD[X]_S)^{-1} = A^{-1}D[X]_S$.

Suppose that Q is a t-ideal of $D[X]$ and let B be a finitely generated subideal of Q. Note that $P \neq 0$, and for any $a \in Q$, (B, a) is also a finitely generated subideal of Q and $(BD[X]_S)_v \subseteq ((B, a)D[X]_S)_v$. So we may assume that $B \cap D \neq 0$. By the previous paragraph, we have that $(BD[X]_S)_v = B_vD[X]_S$. Thus $QD[X]_S$ is a t-ideal. Similarly, we have that if Q' is a t-ideal of $D[X^2, X^3]$, then $Q'D[X^2, X^3]_S$ is a t-ideal. Therefore, the proof is completed. □

Recall that D is a *Mori domain* if it satisfies the ascending chain condition on integral divisorial ideals. The class of Mori domains includes Noetherian domains and Krull domains, and is closed under finite intersections. Recall that $D[X]$ is a Mori domain if D is an integrally closed Mori domain [24]. However, an example is given in [25] of a Mori domain D for which $D[X]$ is not a Mori domain. We say that an integral domain D satisfies the *Principal Ideal Theorem (PIT)* if each prime ideal of D which is minimal over a nonzero principal ideal has height one. It follows from [12, Proposition 3.1(b)] that an integral domain D satisfies PIT if and only if each nonzero prime ideal of D is a union of height-one prime ideals. An integral domain D is called an *S-domain* if $htP[X] = 1$ for each prime ideal P of D with $htP = 1$ [20]. Note that $D[X]$ is an S-domain for any integral domain D [2, Theorem 3.2]. Also, note that if $D[X]$ satisfies PIT, then D satisfies PIT and D is an S-domain [12, Proposition 6.1]; but, D satisfies PIT does not imply that $D[X]$ satisfies PIT [12, Remark 6.2]. However, if D is integrally closed, then $D[X]$ satisfies PIT if and only if D satisfies PIT and D is an S-domain [13, Theorem 4].

We next show that $D[X^2, X^3]$ satisfies any of the above three properties if and only if $D[X]$ does.

THEOREM 2.6. *Let D be an integral domain. Then*

(1) *$D[X^2, X^3]$ is an S-domain.*

(2) *$D[X^2, X^3]$ satisfies PIT if and only if $D[X]$ satisfies PIT.*

(3) *$D[X^2, X^3]$ is a Mori domain if and only if $D[X]$ is a Mori domain.*

Proof. (1) Since $D[X]$ is integral over $D[X^2, X^3]$ (or by Lemma 2.4) and $D[X]$ is an S-domain, $D[X^2, X^3]$ is also an S-domain.

(2)($\Rightarrow$) Suppose that $D[X^2, X^3]$ satisfies PIT. Let Q be a prime ideal of $D[X]$ and $P = Q \cap D[X^2, X^3]$. We need to show that $Q = \cup Q_\alpha$, where $\{Q_\alpha\}$ is the set of height-one prime ideals of $D[X]$ contained in Q. Since $D[X^2, X^3]$ satisfies PIT, $P = \cup(Q_\alpha \cap D[X^2, X^3])$ by Lemma 2.4. If $X \in Q$, then $Q = (Q \cap D, X)$. Thus $P = (Q \cap D, X^2, X^3)$. Hence $Q = (Q \cap D, X) \subseteq \cup Q_\alpha$; so $Q = \cup Q_\alpha$. If $X \notin Q$, then $fX^2 \in P$ for any $f \in Q$. Then $fX^2 \in Q_\alpha$ for some Q_α, and hence $f \in Q_\alpha$; so $Q = \cup Q_\alpha$. Thus $D[X]$ satisfies PIT. ($\Leftarrow$) Suppose that $D[X]$ satisfies PIT. Let P be a prime ideal of $D[X^2, X^3]$. By Lemma 2.4, $P = Q \cap D[X^2, X^3]$ for some prime ideal Q of $D[X]$. Since $D[X]$ satisfies PIT, Q is a union of height-one prime ideals of $D[X]$. Since each height-one prime ideals of $D[X]$ contracts to a height-one prime ideal of $D[X^2, X^3]$ by Lemma 2.4, P is thus a union of height-one prime ideals. Hence $D[X^2, X^3]$ satisfies PIT.

(3)($\Rightarrow$) Suppose that $D[X^2, X^3]$ is a Mori domain. Let $S = \{X^n | n = 0, 2, 3, \dots\}$.

Then $D[X] = K[X] \cap D[X^2, X^3]_S$ and $D[X^2, X^3]_S$ is a Mori domain [23, Corollary 3]. Thus $D[X]$ is also a Mori domain. ($\Leftarrow$) This follows since $D[X^2, X^3] = D[X] \cap K[X^2, X^3]$ and $K[X^2, X^3]$ is a one-dimensional Noetherian domain (and hence a Mori domain). $\square$

Our next result is the $D[X^2, X^3]$ analog of [3, Proposition 4.11] that $D[X]$ is a weakly Krull domain if and only if D is a weakly Krull UMT-domain.

PROPOSITION 2.7. *(cf. [3, Proposition 4.11]) $D[X^2, X^3]$ is a weakly Krull domain if and only if D is a weakly Krull UMT-domain.*

Proof. ($\Rightarrow$) Suppose that $D[X^2, X^3]$ is a weakly Krull domain, and hence $D[X^2, X^3]$ has $t-$dimension one. Let P be a prime t-ideal of D. Then $PD[X^2, X^3]$ is a prime t-ideal of $D[X^2, X^3]$, and hence $\text{ht}P = \text{ht}P[DX^2, X^3] = 1$; whence $t\text{-dim}D = 1$. Moreover, if $0 \neq a \in D$, then the number of height-one prime ideals of $D[X^2, X^3]$ that contain a is finite. Hence $\{P \in X^1(D) | a \in P\}$ is finite, and thus D is weakly Krull.

Let $P \in X^1(D)$. Then $\text{ht}P[X^2, X^3] = 1$, and hence $\text{ht}P[X] = 1$, which implies that D is a UMT-domain because $t\text{-dim}D = 1$.

($\Leftarrow$) Suppose that D is a weakly Krull UMT-domain, and let $P \in X^1(D)$. Then $\text{ht}(P[X^2, X^3]) = 1$ by Lemma 2.4. Thus $t\text{-dim}(D[X^2, X^3]) = 1$ by Lemma 2.2 (note that $t\text{-dim}D = 1$ since D is weakly Krull). Hence by [17, Proposition 4] or [19, Proposition 2.8],

$$D[X^2, X^3] = \cap_{Q \in X^1(D[X^2, X^3])} D[X^2, X^3]_Q.$$

Let $0 \neq f \in D[X^2, X^3]$, $A = \{PD[X^2, X^3] | P \in X^1(D) \text{ and } f \in PD[X^2, X^3]\}$, and $B = \{Q \in X^1(D[X^2, X^3]) | Q \cap D = 0 \text{ and } f \in Q\}$. Since D is weakly Krull, A is finite. Moreover, since $K[X^2, X^3]$ is a one-dimensional Noetherian domain, B is also finite. Therefore, $D[X^2, X^3]$ is weakly Krull. $\square$

The final result of this section is the $D[X^2, X^3]$ analog of [24, Lemme 1].

PROPOSITION 2.8. *Let D be integrally closed and $0 \neq f \in K[X^2, X^3]$. Then*

(1) $fK[X^2, X^3] \cap D[X^2, X^3] = fA_f^{-1}[X^2, X^3]$.

(2) $fK[X] \cap D[X^2, X^3] = \begin{cases} fA_f^{-1}[X^2, X^3], & \text{if } f(0) \neq 0 \\ fA_f^{-1}[X], & \text{if } f(0) = 0. \end{cases}$

Proof. (1) Let $fg \in fK[X^2, X^3] \cap D[X^2, X^3]$. Then $A_f A_g \subseteq (A_f A_g)_v = (A_{fg})_v \subseteq D$ because D is integrally closed (cf. [16, Proposition 34.8]). Thus $g \in A_f^{-1}[X^2, X^3]$ and $fK[X^2, X^3] \cap D[X^2, X^3] \subseteq fA_f^{-1}[X^2, X^3]$. The converse is clear.

(2) Case 1. $f(0) = 0$. Let $fg \in fK[X] \cap D[X^2, X^3]$, where $g \in K[X]$. Since D is integrally closed, $A_f A_g \subseteq (A_f A_g)_v = (A_{fg})_v \subseteq D$ (cf. [16, Proposition 34.8]). Thus $g \in (A_f)^{-1}[X]$, and hence $fK[X] \cap D[X^2, X^3] \subseteq fA_f^{-1}[X]$. Moreover, since $f(0) = 0$, we have $fh \in D[X^2, X^3]$ for any $h \in (A_f)^{-1}[X]$. Therefore, $fK[X] \cap D[X^2, X^3] = fA_f^{-1}[X]$ for any $h \in (A_f)^{-1}[X]$.

Case 2. $f(0) \neq 0$. Since $f(0) \neq 0$, $fg \notin K[X^2, X^3]$ for any $g \in K[X] - K[X^2, X^3]$, which implies that the proof is identical to the proof of Case 1. □

3. GENERALIZED WEAKLY FACTORIAL DOMAINS

One of the purposes of this section is to find equivalent conditions for $D[X^2, X^3]$, over an almost factorial domain D, to be a GWFD. The other is to study the t-class group $Cl_t(D[X^2, X^3])$. Recall that a GWFD is weakly Krull and has t−dimension one [9, Corollary 2.3], and that an almost factorial domain is a Krull domain with torsion divisor class group.

THEOREM 3.1. *The following statements are equivalent for an almost factorial domain D.*

(1) *$D[X^2, X^3]$ is an AGCD-domain.*
(2) *$D[X^2, X^3]$ is an AWFD.*
(3) *$D[X^2, X^3]$ is a GWFD.*
(4) *$charD = p \neq 0$.*

Proof. (1) ⇒ (2): Recall that a Krull domain is a weakly Krull UMT-domain. Thus $D[X^2, X^3]$ is a weakly Krull domain by Proposition 2.7. Also, note that an AGCD-domain has torsion t-class group. Hence $D[X^2, X^3]$ is an AWFD.

(2) ⇒ (3): Let Q be a nonzero prime ideal of $D[X^2, X^3]$ and let $0 \neq f \in Q$. By the definition of an AWFD, there is an integer $n \geq 1$ such that f^n is a product of primary elements. Thus Q contains a nonzero primary element of $D[X^2, X^3]$. Therefore, $D[X^2, X^3]$ is a GWFD.

(3) $\Rightarrow$ (4): Since $D[X^2,X^3]$ is a GWFD, $D[X^2,X^3]_{D-\{0\}} = K[X^2,X^3]$ is also a GWFD by [9, Remark 2.5(4)]. Moreover, since char$D$ = char$K$, it suffices to show that char$K \neq 0$.

Suppose that char$K = 0$, and let $Q = (1+X)K[X] \cap K[X^2,X^3]$. Since $K[X^2,X^3]$ is a GWFD, there is a primary element $f \in Q$ such that $Q = \sqrt{fK[X^2,X^3]}$. Let $S = \{X^n | n = 0,2,3,\dots\}$. Then $K[X^2,X^3]_S = K[X]_S = K[X,X^{-1}]$. Note that $K[X]_S$ is a PID and $QK[X]_S = (1+X)K[X]_S$. Thus $fK[X]_S = (1+X)^n K[X]_S$ for some integer $n \geq 1$, and hence $f = \frac{u(1+X)^n}{X^m}$ for some integer m and $0 \neq u \in K$.

If $m \geq 0$, then $X^m f = u(1+X)^n$, and hence $m = 0$. Thus $f = u(1+X)^n$, and $u(1+X)^n \in K[X^2,X^3] \Leftrightarrow nuX \in K[X^2,X^3] \Leftrightarrow X \in K[X^2,X^3]$ (note that char$K = 0$), a contradiction. Hence $m < 0$ and $f = u(1+X)^n X^{-m} \in Q \cap (XK[X] \cap K[X^2,X^3])$, which contradicts that f is primary. Thus char$K \neq 0$.

(4) $\Rightarrow$ (1): Let $0 \neq f,g \in D[X^2,X^3]$. Then there is an integer $k \geq 1$ and $h \in K[X]$ such that $(((f,g)D[X])^k)_v = ((f^k,g^k)D[X])_v = hD[X]$ (note that $D[X]$ is a Krull domain with torsion divisor class group)[6, Lemma 3.3]. Thus $f^k = hf_1$ and $g^k = hg_1$ for some $f_1,g_1 \in D[X]$, and $((f_1,g_1)D[X])_v = D[X]$. Since char$D = p \neq 0$, $f_1^p, g_1^p \in D[X^2,X^3]$.

Assume that $(f_1^p,g_1^p)_v \subsetneq D[X^2,X^3]$. Then there is a height-one prime ideal Q of $D[X^2,X^3]$ such that $(f_1^p,g_1^p)_v \subseteq Q$ (note that t-dim$(D[X^2,X^3]) = 1$ since $D[X^2,X^3]$ is weakly Krull). Since $D[X]$ is integral over $D[X^2,X^3]$ (or by Lemma 2.4), there is a height-one prime ideal Q' of $D[X]$ such that $Q' \cap D[X^2,X^3] = Q$. Thus

$$D[X] \supsetneq Q' = Q'_t \supseteq ((f_1^p,g_1^p)D[X])_v = (((f_1,g_1)D[X])^p)_v$$
$$= ((((f_1,g_1)D[X])_v)^p)_v = (D[X]^p)_v = D[X],$$

a contradiction. Hence $(f_1^p,g_1^p)_v = D[X^2,X^3]$. Therefore,

$$(f^{kp},g^{kp})_v = ((hf_1)^p,(hg_1)^p)_v = h^p(f_1^p,g_1^p)_v = h^pD[X^2,X^3].$$

Thus $D[X^2,X^3]$ is an AGCD-domain. □

COROLLARY 3.2. *The following statements are equivalent for a field K.*

(1) *$K[X^2,X^3]$ is an AGCD-domain.*

(2) *$K[X^2,X^3]$ is an AWFD.*

(3) $K[X^2, X^3]$ *is a GWFD.*

(4) *char*$K = p \neq 0$.

Our next result generalizes Theorem 3.1.

THEOREM 3.3. *(cf. [9, Theorem 3.3]) Let D be an integrally closed domain with char$D = p \neq 0$. Then the following statements are equivalent.*

(1) $D[X^2, X^3]$ *is an AWFD.*

(2) $D[X^2, X^3]$ *is a GWFD.*

(3) $D[X]$ *is an AWFD.*

(4) $D[X]$ *is a GWFD.*

(5) D *is a generalized weakly factorial AGCD-domain.*

(6) D *is an almost weakly factorial AGCD-domain.*

(7) D *is a weakly Krull AGCD-domain.*

Proof. (1) $\Rightarrow$ (2) : This follows from the definitions.

(2) $\Rightarrow$ (4) : By [9, Theorem 2.2], it suffices to show that if Q is a maximal t-ideal of $D[X]$, then $Q = \sqrt{fD[X]}$ for some $f \in D[X]$ because t-dim$D[X]$ = t-dim$D[X^2, X^3] = 1$. Let $P = Q \cap D$. If $P \neq 0$, then $Q = P[X]$ and $P[X^2, X^3] = \sqrt{aD[X^2, X^3]}$ for some $a \in P$ (note that $P[X^2, X^3]$ is a height-one prime ideal). Thus $P[X] = \sqrt{aD[X]}$.

Assume that $P = 0$, and let $Q \cap D[X^2, X^3] = \sqrt{fD[X^2, X^3]}$. Note that if $g \in D[X]$, then $g^p \in D[X^2, X^3]$ because char$D = p \neq 0$. Thus $Q = \sqrt{fD[X]}$.

(3) $\Rightarrow$ (1) : Recall that $D[X]$ is a weakly Krull domain $\Leftrightarrow$ D is a weakly Krull UMT-domain $\Leftrightarrow$ $D[X^2, X^3]$ is a weakly Krull domain. Hence it suffices to show that $Cl_t(D[X^2, X^3])$ is torsion.

Let Q be a t-invertible t-ideal of $D[X^2, X^3]$. Then $((QD[X])(Q^{-1}D[X]))_t = (QQ^{-1}D[X])_t \subseteq D[X]$. Since Q is t-invertible, QQ^{-1} is not contained in any height-one prime ideal of $D[X^2, X^3]$ (note that t-dim$D[X] = t$-dim$D[X^2, X^3] = 1$). Thus $(QD[X])(Q^{-1}D[X])$ is not contained in any height-one prime ideal of $D[X]$, and hence $((QD[X])(Q^{-1}D[X]))_t = D[X]$. Since $D[X]$ is an AWFD and thus has torsion t-class group, there is an integer $n \geq 1$ and an $f \in D[X]$ such that $((QD[X])^n)_v = (Q^nD[X])_v = fD[X]$. Since Q is a finite type t-ideal, by the same argument as in the proof of (4) $\Rightarrow$ (1) in Theorem 3.1, we have that $(Q^{pn})_v = f^p D[X^2, X^3]$. Therefore, $D[X^2, X^3]$ is an AWFD.

(3) $\Leftrightarrow$ (4) $\Leftrightarrow$ (5) $\Leftrightarrow$ (6) $\Leftrightarrow$ (7): These implications are in [9, Theorem 3.3]. □

We close this paper with a discussion of $Cl_t(D[X^2, X^3])$. Recall that an integral domain D with quotient field K is *seminormal* if whenever $x^2, x^3 \in D$ for some $x \in K$, then $x \in D$; and that $Pic(D[X]) = Pic(D)$ if and only if D is seminormal. Using the Mayer-Vietoris exact sequence for (U, Pic) (cf. [21, pp. 39-40]), one may show that $Pic(D[X^2, X^3]) = Pic(D) \oplus D$ (as additive abelian groups) when D is seminormal. Also, $Cl_t(D[X]) = Cl_t(D)$ if and only if D is integrally closed [15, Theorem 3.6]. In analogy with the Picard group case, we ask if $Cl_t(D[X^2, X^3]) = Cl_t(D) \oplus K$ (as additive abelian groups) when D is integrally closed. Our final theorem shows that this does hold in the special case when D is a GCD-domain since then $Cl_t(D) = 0$. For example, letting $D = \mathbb{Z}$, we have $Pic(\mathbb{Z}[X^2, X^3]) = \mathbb{Z}$ and $Cl_t(\mathbb{Z}[X^2, X^3]) = \mathbb{Q}$.

THEOREM 3.4. *Let D be a GCD-domain with quotient field K. Then, as additive abelian groups,*

(1) $Pic(D[X^2, X^3]) = D$.
(2) $Cl_t(D[X^2, X^3]) = K$.

Proof. (1) This follows using the Mayer-Vietoris exact sequence for (U, Pic).

(2) Let $S = D - \{0\}$. Then S is an *lcm* splitting set in $D[X^2, X^3]$. Thus

$$Cl_t(D[X^2, X^3]) \cong Cl_t(D[X^2, X^3]_S) = Cl_t(K[X^2, X^3]) = Pic(K[X^2, X^3]) = K$$

by [1, Theorem 4.1]. □

QUESTION 3.5. Compute $Cl_t(D[X^2, X^3])$ for an arbitrary integral domain D with quotient field K. In particular, does $Cl_t(D[X^2, X^3]) = Cl_t(D) \oplus K$ (as additive abelian groups) when D is integrally closed?

ACKNOWLEDGMENTS

The third author's work was supported grant No R02-2000-00016 from the Basic Science Research Program of the Korea Science & Engineering Foundation.

REFERENCES

1. D.D. Anderson, D.F. Anderson, and M. Zafrullah, *Splitting the t-class group*, J. Pure and Appl. Algebra 74 (1991), 17-37.
2. D.D. Anderson, D.F. Anderson, and M. Zafrullah, *Rings between* $D[X]$ *and* $K[X]$, Houston J. Math. 17 (1991), 109-129.
3. D.D. Anderson, E. Houston, and M. Zafrullah, *t-linked extensions, the t-class group, and Nagata's theorem*, J. Pure and Appl. Algebra 86 (1993), 109-124.
4. D.D. Anderson and L. Mahaney, *On primary factorizations*, J. Pure and Appl. Algebra 54 (1988), 141-154.
5. D.D. Anderson, J. L. Mott, and M. Zafrullah, *Finite character representations for integral domains*, Boll. Un. Mat. Ital. B(7) 6 (1992), 613-630.
6. D.D. Anderson and M. Zafrullah, *Almost Bézout domains*, J. Algebra 142 (1991), 285-309.
7. D.D. Anderson and M. Zafrullah, *Weakly factorial domains and groups of divisibility*, Proc. Amer. Math. Soc. 109 (1990), 907-913.
8. D.F. Anderson, *The class group and local class group of an integral domain*, in Non-Noetherian Ring Theory, Math. Appl., Kluwer Acad. Publ., Dordrecht 520 (2000), 33-55
9. D.F. Anderson, G.W. Chang, and J. Park, *Generalized weakly factorial domains*, Houston J. Math., to appear.
10. D.F. Anderson, S. Chapman, F. Inman, and W.W. Smith, *Factorization in* $K[X^2, X^3]$, Arch. Math., 61(1993), 521-528.
11. S. El Baghdadi, L. Izelgue, and S. Kabbaj, *On the class group of a graded domain*, preprint.
12. V. Barucci, D.F. Anderson, and D. Dobbs, *Coherent Mori domains and the Principal Ideal Theorem*, Comm. Algebra 15 (1987), 1119-1156.
13. G.W. Chang, *On the Principal Ideal Theorem*, Bull. Korean Math. Soc. 36 (1999), 655-660.
14. M. Fontana, S. Gabelli, and E. Houston, *UMT-domains and domains with Prüfer integral closure*, Comm. Algebra 26(4) (1998), 1017-1039.

15. S. Gabelli, *On divisorial ideals in polynomial rings over Mori domains*, Comm. Algebra 15 (1987), 2349–2370

16. R. Gilmer, *Multiplicative Ideal Theory*, Marcel Dekker, New York, 1972.

17. M. Griffin, *Some results on v-multiplication rings*, Canad. J. Math., 19 (1967), 710-722.

18. E. Houston and M. Zafrullah, *On t-invertibility II*, Comm. Algebra 17(8) (1989), 1955-1969

19. B.G. Kang, *Prüfer v-multiplication domains and the ring* $R[X]_{N_v}$, J. Algebra 123 (1989), 151-170.

20. I. Kaplansky, *Commutative Rings*, revised edition, Univ. of Chicago Press, 1974.

21. T.Y. Lam, *Lectures on Modules and Rings*, Springer, New York, 1999.

22. M. Nagata, *Local Rings*, Interscience, New York, 1962.

23. J. Querre, *Sur une propriété des anneaux de Krull*, Bull. Sci. Math. 99(2) (1971), 341-354.

24. J. Querre, *Idéaux divisoriels d'un anneaux de polynômes*, J. Algebra 64 (1980), 270-284

25. M. Roitman, *On polynomial extensions of Mori domains over countable fields*, J. Pure and Appl. Algebra 64 (1990), 315-328.

On the Complete Integral Closure of Rings that Admit a ϕ-Strongly Prime Ideal

AYMAN BADAWI, Department of Mathematics, Birzeit University, Box 14, Birzeit, West Bank, Palestine, via Israel. E-Mail: abring@birzeit.edu

ABSTRACT:

Let R be a commutative ring with 1 and $T(R)$ be its total quotient ring such that $Nil(R)$ (the set of all nilpotent elements of R) is a divided prime ideal of R. Then R is called a *ϕ-chained ring (ϕ-CR)* if for every $x, y \in R \setminus Nil(R)$, either $x \mid y$ or $y \mid x$. A prime ideal P of R is said to be a ϕ-strongly prime ideal if for every $a, b \in R \setminus Nil(R)$, either $a \mid b$ or $aP \subset bP$. In this paper, we show that if R admits a regular ϕ-strongly prime ideal, then either R does not admit a minimal regular prime ideal and $c(R)$ (the complete integral closure of R inside $T(R)$) $=$ $T(R)$ is a ϕ-CR or R admits a minimal regular prime ideal Q and $c(R) = (Q : Q)$ is a ϕ-CR with maximal ideal Q. We also prove that the complete integral closure of a conducive domain is a valuation domain.

1 INTRODUCTION

We assume throughout that all rings are commutative with $1 \neq 0$. We begin by recalling some background material. As in [17], an integral domain R, with quotient field K, is called a *pseudo-valuation domain (PVD)* in case each prime ideal P of R is *strongly prime*, in the sense that $xy \in P, x \in K, y \in K$ implies that either $x \in P$ or $y \in P$. In [4], Anderson, Dobbs and the author generalized the study of pseudo-valuation domains to the context of arbitrary rings (possibly with nonzero zerodivisors). Recall from [4] that a prime ideal P of R is said to be *strongly prime (in R)* if aP and bR are comparable (under inclusion) for all $a, b \in R$. A ring R is called a *pseudo-valuation ring (PVR)* if each prime ideal of R is strongly prime. A PVR is necessarily quasilocal [4, Lemma 1(b)]; a chained ring is a PVR [4, Corollary 4]; and an integral domain is a PVR if and only if it is a PVD (cf. [1, Proposition 3.1], [2, Proposition 4.2], and [6, Proposition 3]). Recall from [7] and [14] that a prime ideal P of R is called *divided* if it is comparable (under inclusion) to every ideal of R. A ring R is called a *divided ring* if every prime ideal of R is divided.

In [8], the author gave another generalization of PVDs to the context of arbitrary rings (possibly with nonzero zerodivisors). As in [8], for a ring R with total quotient ring $T(R)$ such that $Nil(R)$ (the set of all nilpotent elements of R) is a divided prime ideal of R, let $\phi : T(R) \longrightarrow K := R_{Nil(R)}$ such that $\phi(a/b) = a/b$ for every $a \in R$ and every $b \in R \setminus Z(R)$. Then ϕ is a ring homomorphism from $T(R)$ into K, and ϕ restricted to R is also a ring homomorphism from R into K given by $\phi(x) = x/1$ for every $x \in R$. A prime ideal Q of $\phi(R)$ is called a *K-strongly prime* ideal if $xy \in Q$, $x \in K, y \in K$ implies that either $x \in Q$ or $y \in Q$. If each prime ideal of $\phi(R)$ is K-strongly prime, then $\phi(R)$ is called a *K-pseudo-valuation ring (K-PVR)*. A prime ideal P of R is called a *ϕ-strongly prime* ideal if $\phi(P)$ is a K-strongly prime ideal of $\phi(R)$. If a ϕ-strongly prime ideal P of R contains a nonzerodivisor, then we say that P is a regular ϕ-strongly prime ideal. If each prime ideal of R is ϕ-strongly prime, then R is called a *ϕ-pseudo-valuation ring ($\phi - PVR$)*. For an equivalent characterization of a ϕ-PVR, see Proposition 1.1(7). It was shown in [9, Theorem 2.6] that for each $n \geq 0$ there is a ϕ-PVR of Krull dimension n that is not a PVR. Also, recall from [10], that a ring R is called a *ϕ-chained ring (ϕ-CR)* if $Nil(R)$ is a divided prime ideal of R and for every $x \in R_{Nil(R)} \setminus \phi(R)$, we have $x^{-1} \in \phi(R)$. For an equivalent characterization of a ϕ-CR, see Proposition 1.1(9). A ϕ-CR is a divided ring [10, Corollary 3.3(2)], and hence is quasilocal. It was shown in [10, Theorem 2.7] that for each $n \geq 0$ there is a ϕ-CR of Krull dimension n that is not a chained ring.

Suppose that $Nil(R)$ is a divided prime ideal of a commutative ring R such that R admits a regular ϕ-strongly prime. In this paper, we show that $c(R)$ (the complete integral closure of R inside $T(R)$) is a ϕ-chained ring. In fact, we will show that either $c(R) = T(R)$ or $c(R) = (Q : Q) = \{x \in T(R) : xQ \subset Q\}$ for some minimal regular ϕ-strongly prime ideal Q of R.

In the following proposition, we summarize some basic properties of PVRs, ϕ-PVRs, and ϕ-CRs.

PROPOSITION 1.1. *1. An integral domain is a PVR if and only if it is a ϕ-PVR if and only if it is a PVD([1, Proposition 3.1], [2, Proposition 4.2], [6, Proposition 3], and [8]).*

2. *A PVR is a divided ring [4, Lemma 1], and hence is quasilocal.*

3. *A ring R is a PVR if and only if for every $a, b \in R$, either $a \mid b$ in R or $b \mid ac$ in R for each nonunit c in R [4, Theorem 5].*

4. *If R is a PVR, then $Nil(R)$ and $Z(R)$ are divided prime ideals of R ([4], [8]).*

5. *A PVR is a ϕ-PVR [8, Corollary 7(3)].*

6. *If P is a ϕ-strongly prime ideal of R, then P is a divided prime. In particular, if R is a ϕ-PVR, then R is a divided ring [8, Proposition 4], and hence is quasilocal.*

7. *Suppose that $Nil(R)$ is a divided prime ideal of R. Then a prime ideal P of R is ϕ-strongly prime if and only if for every $a, b \in R \setminus Nil(R)$, either $a \mid b$ in R or $aP \subset bP$. In particular, a ring R is a ϕ-PVR if and only if for every $a, b \in R \setminus Nil(R)$, either $a \mid b$ in R or $b \mid ac$ in R for every nonunit $c \in R$ [8, Corollary 7].*

8. *Suppose that $Nil(R)$ is a divided prime ideal of R. If P is a ϕ-strongly prime ideal of R and Q is a prime ideal of R contained in P, then Q is a ϕ-strongly prime ideal of R [8, Proposition 5].*

9. *Suppose that $Nil(R)$ is a divided prime ideal of R. Then a ring R is a ϕ-CR if and only if for every $a, b \in R \setminus Nil(R)$, either $a \mid b$ in R or $b \mid a$ in R [10, Proposition 2.3].*

10. *A ϕ-CR is a ϕ-PVR [10, Corollary 2.3].* □

2 The COMPLETE INTEGRAL CLOSURE OF RINGS THAT ADMIT A REGULAR ϕ-STRONGLY PRIME IDEAL

Throughout this section, $Nil(R)$ denotes the set of all nilpotent elements of R, $Z(R)$ denotes the set of all zerodivisor elements of R, and $c(R)$ denotes the complete integral closure of R inside $T(R)$. The following two lemmas are needed in the proof of Proposition 2.3.

LEMMA 2.1. *Suppose $Nil(R)$ is a divided prime ideal of R and P is a regular ϕ-strongly prime ideal of R. If s is a regular element of R and $z \in Z(R)$, then $s \mid z$ in R. In particular, $Z(R) \subset P$.*

Proof: Let s be a regular element of P and $z \in Z(R)$. Suppose that $s \nmid z$ in R. Then $sP \subset zP$ by Proposition 1.1(7). Since $s \in P$, we have $z \mid s^2$ in R, which is impossible. Hence, $s \mid z$ in R. Thus, $Z(R) \subset P$. Now, suppose that s is a regular element of $R \setminus P$. Since P is divided by Proposition 1.1(6), we conclude that $P \subset (s)$. Hence, since $Z(R) \subset P$, we conclude that $s \mid z$ in R. □

LEMMA 2.2. *Suppose that $Nil(R)$ is a divided prime ideal of R and P is a regular ϕ-strongly prime ideal of R. Then $x^{-1}P \subset P$ for each $x \in T(R) \setminus R$. In particular, if $x \in T(R) \setminus R$, then x is a unit of $T(R)$.*

Proof: First, observe that $Z(R) \subset P$ by Lemma 2.1. Now, let $x = a/b \in T(R) \setminus R$ for some $a \in R$ and for some $b \in R \setminus Z(R)$. Since $b \nmid a$ in R, $Z(R) \subset P$, and P is divided, we conclude that $a \in R \setminus Z(R)$. Hence, $x^{-1} \in T(R)$. Thus, since $b \nmid a$ in R, we have $bP \subset aP$ by Proposition1.1(7). Thus $x^{-1}P = \frac{b}{a}P \subset P$. □

In light of the Lemmas 2.1 and 2.2, we have the following proposition.

PROPOSITION 2.3. *Suppose that $Nil(R)$ is a divided prime ideal of R and P is a regular prime ideal of R. Then the following statements are equivalent:*

1. *P is a ϕ-strongly prime ideal of R.*

2. *$(P : P)$ is a ϕ-CR with maximal ideal P.*

Proof: (1) $\Longrightarrow$ (2). First, we show that P is the maximal ideal of $(P : P)$. Let $s \in R \setminus P$. Then s is a regular element of R (because P is a divided regular prime ideal of R, and therefore $Z(R) \subset P$). Hence $1/s \in (P : P)$. Thus, s is a unit of $(P : P)$. Hence, P is the maximal ideal of $(P : P)$. Now, we show that $(P : P)$ is a ϕ-CR. Since $Nil(R)$ is a divided prime ideal of R, $Nil((P : P)) = Nil(R)$. Let $x, y \in (P : P) \setminus Nil(R)$ and suppose that $x \nmid y$ in $(P : P)$. Then $x = a/s$, $y = b/s$

for some $a, b \in R \setminus Nil(R)$, and some $s \in R \setminus Z(R)$. Since $x \nmid y$ in $(P : P)$, it is impossible that a be a regular element of R and $b \in Z(R)$. Thus, we consider three cases. Case 1: suppose that $a \in Z(R)$ and $b \in R \setminus Z(R)$. Then $b \mid a$ in R by Lemma 2.1. Hence, $y \mid x$ in $(P : P)$. Case 2: suppose that $a, b \in R \setminus Z(R)$. Since $x \nmid y$ in $(P : P)$, we conclude that $w = y/x \in T(R) \setminus R$. Hence, $w^{-1}P = \frac{x}{y}P \subset P$ by Lemma 2.2. Hence, $y \mid x$ in $(P : P)$. Case 3: suppose that $a, b \in Z(R)$. Since $x \nmid y$ in $(P : P)$, we conclude that $a \nmid b$ in R. Thus, $aP \subset bP$ by Proposition 1.1(7). Let h be a regular element of P. Then $ah = bc$ for some $c \in P$. Suppose that $h \mid c$ in R. Then $b \mid a$ in R. Hence, $y \mid x$ in $(P : P)$. Thus, suppose that $h \nmid c$ in R. Then, c is a regular element of P. Hence, $f = c/h \in T(R) \setminus R$. Thus, $f^{-1}P = \frac{h}{c}P \subset P$ by Lemma 2.2. Hence, $f^{-1} \in (P : P)$. Thus, $ah = bc$ implies that $xf^{-1} = y$. Hence, $x \mid y$ in $(P : P)$, a contradiction. Thus, $h \mid c$ in R, and therefore $y \mid x$ in $(P : P)$. Hence, $(P : P)$ is a ϕ-CR by Proposition 1.1(9). **(2) $\Longrightarrow$ (1).** This is clear by Proposition 1.1(10). □

PROPOSITION 2.4. *Suppose that $Nil(R)$ is a divided prime ideal of R and P is a regular ϕ-strongly prime ideal of R. Then $Q = \cap_{i=1}^{\infty}(s^i)$ is a prime ideal of R for every regular element s of P.*

Proof: Suppose that $xy \in Q$ for some $x, y \in R$. Since $Z(R) \subset (s^i)$ for each $i \geq 1$ by Lemma 2.1, we conclude that $Z(R) \subset Q$. Hence, we may assume that neither $x \in Z(R)$ nor $y \in Z(R)$. Thus, assume that $x \notin Q$. Then $s^n \nmid x$ for some $n \geq 1$. Hence, $s^nP \subset xP$ by Proposition 1.1(7). In particular, since $s^n \in P$, we have $s^{2n} \subset xP$. Hence, we have $xy \in (s^{2n+i}) \subset xs^iP \subset (xs^i)$ for every $i \geq 1$. Thus, $y \in (s^i)$ for every $i \geq 1$. Hence, $y \in Q$. □

PROPOSITION 2.5. *Let P be a regular prime ideal of R. Then $(P : P) \subset c(R)$.*

Proof: Let $x \in (P : P)$, and let s be a regular element of P. Then $sx^n \in P$ for every $n \geq 1$. Hence, x is an almost integral element of R. Thus, $x \in c(R)$. □

PROPOSITION 2.6. *Suppose that $Nil(R)$ is a divided prime ideal of R and P is a regular ϕ-strongly prime ideal of R. Then $T(R)$ is a ϕ-CR.*

Proof: First, observe that $Nil(T(R)) = Nil(R)$. Hence, it suffices to show that if $a, b \in R \setminus Nil(R)$, then either $a \mid b$ in $T(R)$ or $b \mid a$ in $T(R)$. Hence, let $a, b \in R \setminus Nil(R)$. Suppose that $a \nmid b$ in $T(R)$. Then $a \nmid b$ in R. Hence, $aP \subset bP$ by Proposition 1.1(7). Thus, let s be a regular element of P. Then $as = bc$ for some $c \in P$. Thus, $a = b\frac{c}{s}$. Hence, $b \mid a$ in $T(R)$. □

Now, we state our main result in this section

THEOREM 2.7. *Suppose that $Nil(R)$ is a divided prime ideal of R and P is a regular ϕ-strongly prime ideal of R. Then exactly one of the following statements must hold:*

1. *R does not admit a minimal regular prime ideal and $c(R) = T(R)$ is a ϕ-CR.*
2. *R admits a minimal regular prime ideal Q and $c(R) = (Q : Q)$ is a ϕ-CR with maximal ideal Q.*

Proof: **(1)**. Suppose that R does not admit a minimal regular prime ideal. We will show that $1/s \in c(R)$ for every regular element $s \in R$. Hence, let s be a regular element of R. Suppose that $s \in R \setminus P$. Then $1/s \in (P : P)$ because P is a divided prime ideal of R by Proposition 1.1(6). Hence $1/s \in (P : P) \subset c(R)$ by Proposition 2.5. Thus, suppose that $s \in P$. We will show that there is regular prime ideal $H \subset P$ such that $s \notin H$. Deny. Let $F = \{D : D$ is a regular prime ideal of R and $D \subset P\}$ and $N = \cap_{D \in F} D$. Then, $s \in N$. Now, by Proposition 1.1(8) and (6), we conclude that the prime ideals in the set F are linearly ordered. Hence, N is a minimal regular prime ideal of R, which is a contradiction. Thus, there is a regular prime ideal $H \subset P$ such that $s \notin H$. Hence, once again $1/s \in (H : H) \subset c(R)$ by Proposition 2.5. Thus, $c(R) = T(R)$. Now, $T(R)$ is a ϕ-CR by Proposition 2.6.

(2). Suppose that Q is a minimal regular prime ideal of R. First, observe that $Q \subset P$ by Proposition 1.1(6). Thus, Q is a minimal ϕ-strongly prime ideal of R by Proposition 1.1(8). Now, $(Q : Q) \subset c(R)$ by Proposition 2.5. We will show that $c(R) \subset (Q : Q)$. Suppose there is an $x \in c(R) \setminus R$. Then x is a unit of $T(R)$ by Lemma 2.2. We consider three cases. Case 1: suppose that $x^{-1} \in T(R) \setminus R$. Then $xQ \subset Q$ by Lemma 2.2. Hence, $x \in (Q : Q)$. Case 2: suppose that $x^{-1} \in R \setminus Q$. Then $Q \subset (x^{-1})$ by Proposition 1.1(6). Thus, $x \in (Q : Q)$. Case 3: suppose that $x^{-1} \in Q$. This case can not happen, for if $x^{-1} \in Q$, then $D = \cap_{i=1}^{\infty}(x^{-1})^i$ contains a regular element of R because $x \in c(R)$. But D is a prime ideal of R by Proposition 2.4. Hence, D is a regular prime ideal of R that is properly contained in Q. A contradiction, since Q is a minimal regular prime ideal of R. Hence, $c(R) = (Q : Q)$. Now, $c(R) = (Q : Q)$ is a ϕ-CR by Proposition 2.3. □

Suppose that $Nil(R)$ is a divided prime ideal of R and $P \neq Nil(R)$ is a ϕ-strongly prime ideal of R. Then observe that $Nil(\phi(R))$ is a divided prime ideal of $\phi(R)$ and $\phi(P)$ is a regular K-strongly prime ideal of $\phi(R)$ (recall that $K = R_{Nil(R)}$). Now, since $\phi(R)_{Nil(\phi(R))} = K_{Nil(R)}$, we may think of $\phi(P)$ as a ϕ-strongly prime ideal of $\phi(R)$. In light of this argument and Theorem 2.7, we have the following corollary.

COROLLARY 2.8. *Suppose that $Nil(R)$ is a divided prime ideal of R and $P \neq Nil(R)$ is a ϕ-strongly prime ideal of R. Then exactly one of the following statements must hold:*

1. *$\phi(R)$ does not admit a minimal regular prime ideal and $c(\phi(R)) = T(\phi(R)) = K_{Nil(R)}$ is a K-CR.*
2. *$\phi(R)$ admits a minimal regular prime ideal Q and $c(\phi(R)) = (Q : Q)$ is a K-CR.* □

COROLLARY 2.9. *Suppose that R admits a regular strongly prime ideal. Then exactly one of the statements in Theorem 2.7 must hold.* □

COROLLARY 2.10. *Suppose that an integral domain R admits a nonzero strongly prime ideal of R. Then exactly one of the statements in Theorem 2.7 must hold (observe that in this case $c(R)$ is a valuation domain).* □

COROLLARY 2.11. *Suppose that $Nil(R)$ is a divided prime ideal of R and P is a regular ϕ-strongly prime ideal of R. If P contains a finite number, say n, of regular* *prime ideals of R, $P_1 \subset P_2 \subset \cdots \subset P_{n-1} \subset P_n = P$, then $c(R) = (P_1 : P_1)$.* □

Let $J(R)$ denotes the Jacobson radical ideal of R. We have the following result.

COROLLARY 2.12. *Suppose that R is a Prüfer domain such that $J(R)$ contains a nonzero prime ideal of R. Then exactly one of the statements in Theorem 2.7 must hold (once again, observe that in this case $c(R)$ is a valuation domain).*

Proof: Let P be a nonzero prime ideal of R such that $P \subset J(R)$. Then P is a strongly prime ideal by [11, Proposition 1.3, and the proof of Theorem 4.3]. Hence, the claim is now clear. □

It is well-known [17, Proposition 3.2] that if R is a Noetherian pseudo-valuation domain (which is not a field), then R has Krull dimension one. The following is an alternative proof of this fact.

PROPOSITION 2.13. *([17, Proposition 3.2]). If R is a Noetherian pseudo-valuation domain (which is not a field), then R has Krull dimension one.*

Proof: Deny. Let M be the maximal ideal of R. Then there is a nonzero prime ideal P of R such that $P \subset M$ and $M \neq P$. Hence, there is an element $m \in M \setminus P$. Since P is divided, we have $P \subset (m)$. Thus, $1/m \in c(R)$. Since R is Noetherian, $1/m$ is also integral over R, which is impossible. Hence, R has Krull dimension one. □

3 THE COMPLETE INTEGRAL CLOSURE OF CONDUCIVE DOMAINS

Throughout this section, R denotes an integral domain with quotient field K, and $c(R)$ denotes the integral closure of R inside K. If I is a proper ideal of R, then $Rad(I)$ denotes the radical ideal of R. Recall from [11], that Houston and the author defined an ideal I of R to be *powerful* if, whenever $xy \in I$ for elements $x, y \in K$, we have $x \in R$ or $y \in R$. Also, recall that in [13, Theorem 4.5] Bastida and Gilmer proved that a domain R shares an ideal with a valuation domain iff each overring of R which is different from the quotient field K of R has a nonzero conductor to R. Domains with this property, called *conducive domains,* were explicity defined and studied by Dobbs and Fedder [15], and further studied by Barucci, Dobbs, and Fontana [12] and [16]. In [11, Theorem 4.1], Houston and the author proved the following result.

PROPOSITION 3.1. *([11, Theorem 4.1]) An integral domain R is a conducive domain if and only if R admits a powerful ideal.* □

The following proposition is needed in the proof of Theorem 3.2.

PROPOSITION 3.2. *([11, Theorem 1.5 and Lemma 1.1]). Suppose that I is a proper powerful ideal of R. Then $I^2 \subset (s)$ for every $s \in R \setminus Rad(I)$, and $x^{-1}I^2 \subset R$ for every $x \in K \setminus R$.* □

Now, we state the main result of this section.

THEOREM 3.3. *Suppose that R admits a nonzero proper powerful ideal I, that is, R is a conducive domain. Then exactly one of the following two statements must hold:*

1. *$\cap_{n=1}^{\infty} I^n \neq 0$ and exactly one of the following two statements must hold:*
   - *(a) R does not admit a minimal regular prime ideal and $c(R) = K$ is a valuation domain.*
   - *(b) R admits a minimal regular prime ideal Q and $c(R) = (Q : Q)$ is a valuation domain.*
2. *$\cap_{n=1}^{\infty} I^n = 0$ and $c(R) = \{x \in K : x^{-n} \notin Rad(I)$ for every $n \geq 1\}$ is a valuation domain.*

Proof: (1). Suppose that $P = \cap_{n=1}^{\infty} I^n \neq 0$. Then P is a nonzero strongly prime ideal of R by [11, Proposition 1.8]. Hence, the claim is now clear by Theorem 2.7.

(2) Suppose that $P = \cap_{n=1}^{\infty} I^n = 0$. Let $S = \{x \in K : x^{-n} \notin Rad(I)$ for every $n \geq 1\}$, and let $x \in c(R)$. We will show that $x \in S$. Since $P = 0$ and $x \in c(R)$, $x^{-n} \notin I$ for every $n \geq 1$. Hence, $x \in S$. Thus, $c(R) \subset S$. Now, let $s \in S$. We will show that $s \in c(R)$. Let d be a nonzero element of I^2. Hence, for every $n \geq 1$ we have either $s^{-n} \in K \setminus R$ or $s^{-n} \in R \setminus Rad(I)$. Thus, $ds^n \in R$ for every $n \geq 1$ by Proposition 3.2. Hence, $s \in c(R)$. Thus, $S \subset c(R)$. Therefore, $S = c(R)$. Now, we show that $c(R) = S$ is a valuation domain. Let $x \in K \setminus S$. Then $x^{-n} \in Rad(I)$ for some $n \geq 1$. Hence, $x^n \notin Rad(I)$ for every $n \geq 1$. Thus, $x^{-1} \in S$. Therefore, $c(R) = S$ is a valuation domain. □

References

[1] D. F. Anderson, *Comparability of ideals and valuation overrings*, Houston J. Math. 5(1979), 451-463.

[2] D. F. Anderson, *When the dual of an ideal is a ring*, Houston J. Math. 9(1983), 325-332.

[3] D. F. Anderson, A. Badawi, and D. E. Dobbs, *Pseudo-valuation rings, II*, Boll. Un. Mat. Ital. B(8)3(2000), 535-545.

[4] A. Badawi, D. F. Anderson, and D. E. Dobbs, *Pseudo-valuation rings*, Lecture Notes Pure Appl. Math., Vol. 185(1997), 57-67, Marcel Dekker, New York/Basel.

[5] A. Badawi, *Pseudo-valuation domains: a survey*, to appear in World Scientific, New York/London.

[6] A. Badawi, *On domains which have prime ideals that are linearly ordered*, Comm. Algebra 23(1995), 4365-4373.

[7] A. Badawi, *On divided commutative rings*, Comm. Algebra 27(1999), 1465-1474.

[8] A. Badawi, *On ϕ-pseudo-valuation rings*, Lecture Notes Pure Appl. Math., Vol. 205(1999), 101-110, Marcel Dekker, New York/Basel.

[9] A. Badawi, *On ϕ-pseudo-valuation rings, II*, Houston J. Math. 26(2000), 473-480

[10] A. Badawi, *On ϕ-chained rings and ϕ-pseudo-valuation rings*, Houston J. Math. 27(2001), 725-736

[11] A. Badawi and E. Houston, *Powerful ideals, strongly primary ideals, almost pseudo-valuation domains, and conducive domains*, to appear in Comm. Algebra.

[12] V. Barucci, D. E. Dobbs, and M. Fontana, *Conducive integral domains as pullbacks*, Manuscripta Math. 54(1986), 261-277.

[13] E. Bastida and R. Gilmer, *Overrings and divisorial ideals in rings of the form $D + M$*, Michigan Math. J. 20(1973), 79-95.

[14] D. E. Dobbs, *Divided rings and going-down*, Pacific J. Math. 67(1976), 353-363.

[15] D. E. Dobbs and R. Fedder, *Conducive integral domains*, J. Algebra 86(1984), 494-510.

[16] D. E. Dobbs, V. Barucci, and M. Fontana, *Gorenstein conducive domains*, Comm. Algebra 18(1990), 3889-3903.

[17] J. R. Hedstrom and E. G. Houston, *Pseudo-valuation domains*, Pacific J. Math. 4(1978), 551-567.

[18] J. Huckaba, *Commutative Rings with Zero Divisors*, Marcel Dekker, New York/Basel, 1988.

Frobenius Number of a Linear Diophantine Equation

ABDALLAH BADRA, Université Blaise Pascal, Laboratoire de mathématiques pures, Les Cézeaux, F 63177 Aubière.
e-mail: abdallah.badra@math.univ-bpclermont.fr

ABSTRACT. We denote by $\mathbb{N}_0$ the set of nonnegative integers. Let $d \geq 1$ and $A = \{a_1, \dots, a_d\}$ a set of positive integers. For every $n \in \mathbb{N}_0$, we write $s(n)$ for the number of solutions $(x_1, \dots, x_d) \in \mathbb{N}_0^d$ of the equation $a_1x_1 + \cdots + a_dx_d = n$. We set $g(A) = \sup\{n \mid s(n) = 0\} \cup \{-1\}$ the Frobenius number of A. Let $S(A)$ be the subsemigroup of $(\mathbb{N}_0, +)$ generated by A. We set $S'(A) = \mathbb{N}_0 \backslash S(A)$, $N'(A) = \text{Card} S'(A)$ and $N(A) = \text{Card}$ $S(A) \cap \{0, 1, .., g(A)\}$. Let p be a multiple of lcm(A) and $F_p(t) = \prod_{i=1}^d \sum_{j=0}^{\frac{p}{a_i}-1} t^{ja_i}$. We give an upper bound for $g(A)$ and reduction formulas for $g(A)$, $N'(A)$ and $N(A)$. Characterizations of these invariants as well as numerical symmetric and pseudo-symmetric semigroups in terms of $F_p(t)$, are also obtained.

1 INTRODUCTION

We denote by $\mathbb{N}_0$ (resp. $\mathbb{N}$) the set of nonnegative (resp. positive) integers. Let $d \in \mathbb{N}$ and $A = \{a_1, \dots, a_d\} \subset \mathbb{N}$. We set $\rho = \gcd(A)$ and $l = \text{lcm}(A)$. For every $n \in \mathbb{N}_0$, we write $s(n)$ for the number of solutions $(x_1, \dots, x_d) \in \mathbb{N}_0^d$ of the equation $a_1x_1 + \cdots + a_dx_d = n$. We set $g(A) = \sup\{n \mid s(n) = 0\} \cup \{-1\}$ the Frobenius number of A. Let $S(A)$ be the subsemigroup of $(\mathbb{N}_0, +)$ generated by A, $S'(A) = \mathbb{N}_0 \setminus S(A)$, $N'(A) = \text{Card } S'(A)$ and $N(A) = \text{Card } S(A) \cap \{0, 1, .., g(A)\}$. We say that $S(A)$ is symmetric (resp. pseudo-symmetric) if $\gcd(A) = 1$ and $N'(A) = N(A)$ (resp. $N'(A) = N(A) + 1$). The generating function of the $s(n)$ is

$$\Phi(t) = \frac{1}{\prod_{i=1}^d (1 - t^{a_i})}.$$

Indeed, we have

$$\frac{1}{\prod_{i=1}^d (1 - t^{a_i})} = \prod_{i=1}^d \sum_{j \geq 0} t^{ja_i} = \sum_{n \in S(A)} s(n)t^n.$$

For $p \in \mathbb{N}$, we define the Frobenius polynomial

$$F_p(t) = \prod_{i=1}^{d} \sum_{j=0}^{\frac{p}{a_i}-1} t^{ja_i} = \frac{(1-t^p)^d}{\prod_{i=1}^{d}(1-t^{a_i})}$$

and we write

$$\Phi(t) = \frac{F_p(t)}{(1-t^p)^d}. \qquad (1)$$

In theorem 3.1 we give formulas for $g(A), N'(A)$ and $N(A)$ in terms of $F_p(t)$. As a consequence we obtain an upper bound for the Frobenius number (corollary 3.2) which improves the upper bound given by Chrzastowski-Wachtel and mentioned in [9]. A characterization of numerical symmetric and pseudo-symmetric semigroups (corollary 3.4) is also obtained. In theorem 3.7 we prove reduction formulas for $g(A)$, $N'(A)$ and $N(A)$. The first one generalizes a Raczunas and Chrzastowski-Wachtel theorem [9]. As a consequence (corollary 3.10) we obtain a generalization of a Rödseth formula [10]. It is known that the Hilbert function of a graded module over a polynomial graded ring as well as $s(n)$ are numerical quasi-polynomial functions. In examples 4.9 and 4.10 we give a description of these functions in terms of the Frobenius polynomial.

2 PRELIMINARIES

Given $Q(t) = \sum_j q_j t^j \in \mathbb{Q}[t,t^{-1}]$ and an integer $p \geq 1$, there exists a unique sequence $Q_0, \dots, Q_{p-1} \in \mathbb{Q}[t,t^{-1}]$ such that $Q(t) = \sum_{r=0}^{p-1} t^r Q_r(t^p)$. Namely, $Q_r(t) = \sum_k q_{r+pk} t^k$. The Q_r are called the *p-components* of Q. We denote by $\omega(Q) = \inf\{j \mid q_j \neq 0\}$ the valuation of Q and $\deg(Q) = \sup\{j \mid q_j \neq 0\}$ the degree of Q, with $\omega(0) = +\infty$ and $\deg(0) = -\infty$. The following invariants will be associated with Q

$$\begin{aligned}
\omega_p(Q) &= \sup\{\omega(t^r Q_r(t^p)) \mid 0 \leq r \leq p-1\} \text{ the } p\text{-valuation of } Q. \\
\delta_p(Q) &= \inf\{\deg(t^r Q_r(t^p)) \mid 0 \leq r \leq p-1\} \text{ the } p\text{-degree of } Q. \\
\Omega_p(Q) &= \sum_{r=0}^{p-1} \omega(Q_r). \\
\Delta_p(Q) &= \sum_{r=0}^{p-1} \deg(Q_r).
\end{aligned}$$

Thus we have
$\omega_p(Q) = +\infty = \Omega_p(Q)$ and $\delta_p(Q) = -\infty = \Delta_p(Q)$ if $Q_r = 0$ for some r.

We fix an integer $n \in \mathbb{Z}$ and we set

$$\widehat{Q}(t) = t^n Q(t^{-1}).$$

So we have $\widehat{\widehat{Q}} = Q$ and

$$\deg(Q) + \omega(\widehat{Q}) = n = \deg(\widehat{Q}) + \omega(Q) \text{ if } Q \neq 0. \qquad (2)$$

The p-components $\widehat{Q}_r$ of $\widehat{Q}$ can be deduced from the p-components of Q. Namely, we write $n = p\lambda + \gamma$ with $0 \le \gamma < p$, so we get

$$\widehat{Q}(t) = \sum_{r=0}^{p-1} t^{p\lambda+\gamma-r} Q_r(t^{-p}) = \sum_{r=0}^{\gamma} t^{\gamma-r}(t^p)^\lambda Q_r(t^{-p}) + \sum_{r=\gamma+1}^{p-1} t^{p+\gamma-r}(t^p)^{\lambda-1} Q_r(t^{-p}).$$

It follows from the uniqueness of the p-components that

$$\widehat{Q}_r(t) = t^\lambda Q_{\gamma-r}(t^{-1}) \text{ for } 0 \le r \le \gamma \tag{3}$$

and

$$\widehat{Q}_r(t) = t^{\lambda-1} Q_{p+\gamma-r}(t^{-1}) \text{ for } r > \gamma. \tag{4}$$

So we obtain

$$\widehat{Q}_r = 0 \Leftrightarrow Q_{\gamma-r} = 0 \text{ for } 0 \le r \le \gamma \tag{5}$$

and

$$\widehat{Q}_r = 0 \Leftrightarrow Q_{p+\gamma-r} = 0 \text{ for } r > \gamma. \tag{6}$$

If $\widehat{Q}_r \neq 0$, we also deduce from (2)-(4) that

$$\lambda = \deg(\widehat{Q}_r) + \omega(Q_{\gamma-r}) \text{ when } 0 \le r \le \gamma \tag{7}$$

and

$$\lambda - 1 = \deg(\widehat{Q}_r) + \omega(Q_{p+\gamma-r}) \text{ when } r > \gamma. \tag{8}$$

Moreover, writing $n = p\lambda + r + (\gamma - r) = p(\lambda - 1) + r + (p + \gamma - r)$ we get

$$n = \deg(t^r \widehat{Q}_r(t^p)) + \omega(t^{\gamma-r} Q_{\gamma-r}(t^p)) \text{ for } 0 \le r \le \gamma$$

and

$$n = \deg(t^r \widehat{Q}_r(t^p)) + \omega(t^{p+\gamma-r} Q_{p+\gamma-r}(t^p)) \text{ for } r > \gamma.$$

Hence

$$n = \delta_p(\widehat{Q}) + \omega_p(Q) = \delta_p(Q) + \omega_p(\widehat{Q}). \tag{9}$$

Furthermore, using (7) and (8) we get

$$\sum_{r=0}^{\gamma} \left(\deg(\widehat{Q}_r) + \omega(Q_{\gamma-r})\right) + \sum_{r=\gamma+1}^{p-1} \left(\deg(\widehat{Q}_r) + \omega(Q_{p+\gamma-r})\right)$$

$$= (\gamma+1)\lambda + (p - \gamma - 1)(\lambda - 1) = n - p + 1.$$

It follows that

$$\Delta_p(\widehat{Q}) + \Omega_p(Q) = n - p + 1 = \Delta_p(Q) + \Omega_p(\widehat{Q}). \tag{10}$$

Given $m,\ j \in \mathbb{Z}$, we consider the following polynomials

$$N_{m,j}(t) = \frac{1}{(m-1)!}\prod_{i=1}^{m-1}(t-j+i) \text{ if } m>1,\ N_{m,j}(t)=0 \text{ if } m\le 0 \text{ and } N_{1,j}(t)=1.$$

For $Q(t) = \sum_j q_j t^j \in \mathbb{Q}[t,t^{-1}]$ such that $Q(1) \neq 0$, we define

$$V_m(Q,t) = \sum_j q_j N_{m,j}(t).$$

Furthermore, let $Q_0, \ldots, Q_{p-1} \in \mathbb{Q}[t,t^{-1}]$ be the p-components of Q. We consider the polynomials $U_0, \ldots, U_{p-1} \in \mathbb{Q}[t,t^{-1}]$ defined as follows $U_r = 0$ if $Q_r = 0$ and $Q_r(t) = (1-t)^{i_r}U_r(t)$ with $U_r(1) \neq 0$ otherwise. For all $0 \le r \le p-1$, we put $m_r = m - i_r$ and we define the function

$$H_m(Q,.) : \mathbb{Z} \to \mathbb{Q} \text{ by } H_m(Q, r+pk) = V_{m_r}(U_r, k).$$

In order to illustrate these definitions we give the following examples.

EXAMPLE 2.1 Let $Q(t) = F_{12} = \frac{(1-t^{12})^2}{(1-t^2)(1-t^3)} = 1+t^2+t^3+t^4+t^5+2t^6+t^7+2t^8+2t^9+2t^{10}+2t^{11}+t^{12}+2t^{13}+t^{14}+t^{15}+t^{16}+t^{17}+t^{19}$.
We take $p=12, n=19$ and $m=2$.
We write $Q(t) = (1+t^{12})+t(2t^{12})+t^2(1+t^{12})+t^3(1+t^{12})+t^4(1+t^{12})+t^5(1+t^{12})+2t^6+t^7(1+t^{12})+2t^8+2t^9+2t^{10}+2t^{11}$.
We see that the 12-components of $Q(t)$ are $Q_0(t) = Q_2(t) = Q_3(t) = Q_4(t) = Q_5(t) = Q_7(t) = (1+t), Q_1(t) = 2t$ and $Q_6(t) = Q_8(t) = Q_9(t) = Q_{10}(t) = Q_{11}(t) = 2$.
We also have
$\widehat{Q}(t) = t^{19}Q(t^{-1}) = Q(t)$.
$\omega_{12}(Q) = 13,\ \delta_{12}(Q) = 6,\ \Omega_{12}(Q) = 1,\ \Delta_{12}(Q) = 7$.
$N_{2,0}(t) = t+1, N_{2,1}(t) = t$.
$U_r = Q_r$ for all r.
$V_2(U_r,t) = 2t+1$ for $r \in \{0,2,3,4,5,7\}$, $V_2(U_1,t) = 2t$ and $V_2(U_r,t) = 2(t+1)$ for $r \in \{6,8,9,10,11\}$.
We obtain $H_2(Q,12k+r) = 2k+1$ for $r \in \{0,2,3,4,5,7\}$, $H_2(Q,12k+1) = 2k$ and $H_2(Q,12k+r) = 2(k+1)$ for $r \in \{6,8,9,10,11\}$.

EXAMPLE 2.2 Let $Q(t) = F_6(t) = 1+t^2+t^3+t^4+t^5+t^7 = \frac{(1-t^6)^2}{(1-t^2)(1-t^3)}$.
We take $p=6, n=7$ and $m=2$.
We obtain
$\omega_6(Q) = 7,\ \delta_6(Q) = 0,\ \Omega_6(Q) = 1,\ \Delta_6(Q) = 1$.
$U_r = Q_r$ for all r.
$N_{2,0}(t) = t+1, N_{2,1}(t) = t$.
$V_2(U_r,t) = t+1$ for $r \in \{0,2,3,4,5\}$ and $V_2(U_1,t) = t$.
$H_2(Q,6k+r) = k+1$ for $r \in \{0,2,3,4,5\}$ and $H_2(Q,6k+1) = k$.
We observe that $H_2(F_6,.) = H_2(F_{12},.)$.

Given $\Phi(t) \in \mathbb{Q}[[t, t^{-1}]]$, we write $\Phi(t) = \sum_n \varphi(n)t^n$ and we introduce

$$\begin{aligned} g(\Phi) &= \sup\{n \mid \phi(n) = 0\}. \\ S'(\Phi) &= \{n \geq 0 \mid \varphi(n) = 0\}. \\ S(\Phi) &= \{0 \leq n \leq g(\Phi) \mid \varphi(n) \neq 0\}. \\ N'(\Phi) &= \text{Card } S'(\Phi). \\ N(\Phi) &= \text{Card } S(\Phi). \end{aligned}$$

LEMMA 2.3 *Given $m \in \mathbb{Z}$ and $Q(t) = \sum_j q_j t^j \in \mathbb{Q}[t, t^{-1}]$ such that $Q(1) \neq 0$, we consider $\Phi(t) = \sum_n \varphi(n)t^n$ the expansion of $(1-t)^{-m}Q(t)$ as a formal power series. Then, the following conditions hold*

1. $\varphi(n) = V_m(Q, n)$ *for all* $n > \deg(Q) - m$.
2. *We suppose that $m > 0$ and $Q(t)$ has nonnegative coefficients. Then,*
   (a) $\varphi(n) = 0 \Leftrightarrow n < \omega(Q)$.
   (b) $g(\Phi) = \omega(Q) - 1$.
   (c) $N'(\Phi) = \max\{\omega(Q), 0\}$. In particular, $N'(\Phi) = \omega(Q)$ if $Q(t) \in \mathbb{Q}[t]$.

PROOF. 1. Suppose $m > 0$. We have $\Phi(t) = (1-t)^{-m}Q(t) = (\sum_j q_j t^j) \sum_{j \geq 0} \binom{j+m-1}{m-1} t^j$. So $\varphi(n) = \sum_{j=\omega(Q)}^{n} q_j \binom{n-j+m-1}{m-1}$. Moreover, we have

$$\binom{n-j+m-1}{m-1} = \frac{1}{(m-1)!} \prod_{i=1}^{m-1} (n-j+i) \text{ if } n \geq j.$$

Hence $\varphi(n) = V_m(Q, n)$ if $n \geq \deg(Q)$, in particular, the statement is true for $m = 1$. Now, suppose $m > 1$ and $\deg(Q) - m < n < \deg(Q)$ then $-m < n - \deg(Q) \leq n - j < 0$ for all j such that $n < j \leq \deg(Q)$. It follows that there exists $1 \leq i \leq m-1$ such that $n - j + i = 0$ thus $N_{m,j}(n) = 0$. So we can write

$$V_m(Q, n) = \sum_{j=\omega(Q)}^{n} q_j N_{m,j}(n) = \sum_{j=\omega(Q)}^{n} q_j \binom{n-j+m-1}{m-1} = \varphi(n).$$

Furthermore, if $m \leq 0$ then $\varphi(n) = 0$ for $n > \deg(Q) - m$ because $\Phi(t) \in \mathbb{Q}[t, t^{-1}]$ and $\deg(Q) - m = \deg \Phi(t)$.
2. Follows from the fact that $\varphi(n) = \sum_{j=\omega(Q)}^{n} q_j \binom{n-j+m-1}{m-1} > 0$ if $n \geq \omega(Q)$ and $\varphi(n) = 0$ if $n < \omega(Q)$ □

THEOREM 2.4 *Let $m \in \mathbb{Z}$ and $p \in \mathbb{N}$.* Given $Q(t) = \sum_j q_j t^j \in \mathbb{Q}[t, t^{-1}]$ *such that $Q(1) \neq 0$, we consider $\Phi(t) = \sum_n \varphi(n)t^n$ the expansion of $(1-t^p)^{-m}Q(t)$ as a formal power series. Then the following conditions hold*

1. $\varphi(n) = H_m(Q, n)$ *for all* $n > \deg(Q) - mp$.
2. *We suppose that $m > 0$ and $Q(t)$ has nonnegative coefficients. Then,*
   (a) $\varphi(pk + r) = 0 \Leftrightarrow k < \omega(Q_r)$.

(b) $g(\Phi) = \omega_p(Q) - p = \deg(Q) - p - \delta_p(\hat{Q})$ *where* $\hat{Q}(t) = t^{\deg(Q)}Q(t^{-1})$.

(c) $N'(\Phi) = \sum_{r=0}^{p-1} \max\{\omega(Q_r), 0\}$.
In particular, $N'(\Phi) = \Omega_p(Q)$ if $Q(t) \in \mathbb{Q}[t]$.

PROOF. We write $\Phi(t) = \sum_{r=0}^{p-1} t^r(1-t^p)^{-m}Q_r(t^p) = \sum_{r=0}^{p-1} t^r\,(1-t^p)^{-m_r}U_r(t^p) = \sum_{r=0}^{p-1} t^r\Phi_r(t^p)$ where $\Phi_r(t) = (1-t^p)^{-m_r}U_r(t^p) = \sum_k \varphi_r(k)t^k$. It follows from lemma 2.3.1, that $\varphi(pk+r) = \varphi_r(k) = V_{m_r}(U_r, k)$ for all $k > \deg(U_r) - m_r$. Therefore, $\varphi(n) = H_m(Q, n)$ for $n > \deg(Q) - pm$ because $n = pk + r > \deg(Q) - pm \geq p(\deg(Q_r) - m) + r \Rightarrow k > \deg(Q_r) - m = \deg(U_r) - m_r$.
2 (a) follows from lemma 2.3.2 (a).
b) We have $g(\Phi) = \max\{pg(\Phi_r) + r \mid 0 \leq r \leq p-1\} = \max\{p(\omega(Q_r) - 1) + r \mid 0 \leq r \leq p-1\} = \omega_p(Q) - p$. Moreover, if $Q_r \neq 0$ for all r we have $\omega_p(Q) - p = \deg(Q) - p - \delta_p(\hat{Q})$ by (9). Since $\omega_p(Q) = +\infty = -\delta_p(\hat{Q})$ if $Q_r = 0$ for some r, the equality is still true in this case.
c) Follows from lemma 2.3.2 (c) □

LEMMA 2.5 *Let* $\xi = e^{\frac{2i\pi}{p}}$ *be a primitive p-th root of unity and* $Q(t) = \sum_{r=0}^{p-1} t^r Q_r(t^p) \in \mathbb{Q}[t, t^{-1}]$. *Then, the following conditions are equivalent*

1. $Q(\xi^j) = 0$ *for* $0 < j < p$.
2. $Q(1) = pQ_r(1)$ *for* $0 \leq r \leq p-1$.

PROOF. By successive substitutions of $1, \xi, \ldots, \xi^{p-1}$ for t in $Q(t) = \sum_{r=0}^{p-1} t^r Q_r(t^p)$ we obtain a Vandermonde linear system $\sum_{r=0}^{p-1} \xi^{rj}Q_r(1) = Q(\xi^j)$ for $j = 0, \ldots, p-1$. If $Q(\xi) = \cdots = Q(\xi^{p-1}) = 0$, the unique solution is $Q_r(1) = \frac{1}{p}Q(1)$ for every $0 \leq r \leq p-1$. Conversely, if $\frac{Q(1)}{p}$ is the common value of the $Q_r(1)$ then $\frac{Q(1)}{p}\sum_{r=0}^{p-1} \xi^{rj} = 0 = Q(\xi^j)$ for $j = 1, \ldots, p-1$ □

LEMMA 2.6 *Let* p, q, u *be positive integers and* $Q(t), K(t) \in \mathbb{Q}[t, t^{-1}]$ *such that* $p = qu$ *and* $K(t^u) = Q(t)$. *We denote by* Q_r *(resp.* K_s*) the p-components of* Q *(resp. the q-components of* K*). Then,*

1. $Q_{su} = K_s$ *and* $Q_r = 0$ *for all* $r \notin u\mathbb{Z}$.
2. *We set* $\xi = e^{\frac{2i\pi}{p}}$, *then the following conditions are equivalent*
   (a) $Q(\xi^j) = 0$ *for* $0 < j < q$.
   (b) $Q(\xi^q) = qQ_r(1) = K(1)$ *for all* $r \in u\mathbb{Z}$.

PROOF. We can write $Q(t) = K(t^u) = \sum_{s=0}^{q-1} t^{us}K_s(t^p)$. It follows from the uniqueness of the Q_r that $Q_{su} = K_s$ for $0 \leq s < q$. Now, $Q(\xi^q) = K(1)$ and $Q(\xi^j) = K(\alpha^j)$ with $\alpha = e^{\frac{2i\pi}{q}} = \xi^u$. We apply lemma 2.5 □

For every $p \in I\!N$, we set $F_p(t) = \prod_{i=1}^{d}\sum_{j=0}^{\frac{p}{a_i}-1} t^{ja_i}$ the *Frobenius polynomial* of A. We write $F_{p,r}$ for the p-components of F_p. It is easy to see that for $n = \deg(F_p) = pd - \sum_{i=1}^{d} a_i$, we have $\widehat{F}_p(t) = t^n F_p(t^{-1}) = F_p(t)$. Let us write $p = q\rho$ and $a_i = b_i\rho$

for all $1 \le i \le d$, where $\rho = \gcd(A)$. So we can write $F_p(t) = K(t^\rho)$ with

$$K(t) = \frac{(1-t^q)^d}{\prod_{i=1}^d (1-t^{b_i})}.$$

Moreover, for $0 < j < q$ the number $\xi^j = e^{\frac{2ij\pi}{q}}$ is a root of $\prod_{i=1}^d (1-t^{b_i})$ of multiplicity $< d$ because $\gcd(b_1,..,b_d) = 1$ whereas ξ^j is a root of $(1-t^q)^d$ of multiplicity $= d$, then $K(\xi^j) = 0$. It follows from lemma 2.6 that $F_{p,r} = K_{\frac{r}{\rho}}$ if $r \in \rho\mathbb{Z}$ and $F_{p,r} = 0$ otherwise. We also deduce that $F_{p,r}(1) = \frac{1}{q}K(1) = \frac{\rho p^{d-1}}{\prod_{i=1}^d a_i}$ if $r \in \rho\mathbb{Z}$ □

3 FROBENIUS NUMBER AND NUMERICAL SEMIGROUPS

In the case of the Frobenius polynomial F_p we set $\omega_p(F_p) = \omega_p(A)$, $\delta_p(F_p) = \delta_p(A)$, $\Omega_p(F_p) = \Omega_p(A)$, $\Delta_p(F) = \Delta_p(A)$.

THEOREM 3.1 *For every $p \in I\!N$, we have*

1. $g(A) = \omega_p(A) - p = p(d-1) - \sum_{i=1}^d a_i - \delta_p(A) = l(d-1) - \sum_{i=1}^d a_i - \delta_l(A)$.
2. $N'(A) = \Omega_p(A) = \Omega_l(A)$.
3. $N(A) = \Delta_p(A) - \delta_p(A) = \Delta_l(A) - \delta_l(A)$.

PROOF. We see that for every $p \in I\!N$, the function $\Phi(t) = (1-t^p)^{-d}F_p(t) = \sum_n s(n)t^n$ is the generating function of the $s(n)$ so $g(A) = g(\Phi)$.
1. follows from theorem 2.4.2 (b).
2. follows from theorem 2.4.2 (c).
3. is a consequence of (10)□

COROLLARY 3.2

1. *For every $p \in I\!N$, we have*

$$g(A) = p(d-1) - \sum_{i=1}^d a_i \text{ if and only if } \delta_p(A) = 0.$$

2. $g(A) = +\infty$ *if and only if* $\rho > 1$.
3. *If $\rho = 1$, we have the following upper bound for the Frobenius number*

$$g(A) \le l(d-1) - \sum_{i=1}^d a_i.$$

4. *If there exists h such that $1 \le h \le d$ and* $\gcd(a_1,\dots,a_h) = 1$ *then* $g(A) \le \mathrm{lcm}(a_1,\dots,a_h)(h-1) - \sum_{i=1}^h a_i$.

REMARK 3.3 The upper bound we give in 3) improves the following inequality

$$g(A) \le l(d-1)$$

proved by Chrzastowski-Wachtel and mentioned in [9].

COROLLARY 3.4 Suppose $\gcd(A) = 1$. Then the following conditions hold

1. $S(A)$ *is symmetric* $\Leftrightarrow \Delta_p(A) = \Omega_p(A) + \delta_p(A)$ for some $p \in I\!N \Leftrightarrow \Delta_p(A) = \Omega_p(A) + \delta_p(A)$ for all $p \in I\!N$.

2. $S(A)$ *is peudo-symmetric* $\Leftrightarrow \Delta_p(A) + 1 = \Omega_p(A) + \delta_p(A)$ for some $p \in I\!N \Leftrightarrow \Delta_p(A) + 1 = \Omega_p(A) + \delta_p(A)$ for all $p \in I\!N$.

We suppose $\gcd(A) = 1$. Let $q_1, .., q_d$ be positive integers such that for all $1 \leq i \leq d$, q_i is a divisor of $\gcd(a_1, .., a_{i-1}, a_{i+1}, .., a_d)$. So $\gcd(q_i, q_j) = 1$ for $i \neq j$ because $\gcd(A) = 1$. We set $\hat{q} = \prod_{j=1}^{d} q_j, \hat{q}_i = \prod_{j \neq i} q_j, a_i = b_i \hat{q}_i$ and $B = \{b_1, .., b_d\}$. We have $\gcd(B) = 1$ and $l = \mathrm{lcm}(A) = \hat{q}\,\mathrm{lcm}(B)$. For $p \in I\!N$, we write $p = \hat{q}u$ with $u \in \mathrm{lcm}(B)\mathbb{N}$.

THEOREM 3.5 *The following formulas hold*

1. $\delta_p(A) = \hat{q}\delta_u(B)$.

2. $\omega_p(A) = \hat{q}\omega_u(B) + \sum_{i=1}^{d}(q_i - 1)a_i$.

3. $\Omega_p(A) = \hat{q}\Omega_u(B) + \frac{1}{2}\left(\sum_{i=1}^{d}(q_i - 1)a_i - \hat{q} + 1\right)$.

4. $\Delta_p(A) = \hat{q}\Delta_u(B) + \frac{1}{2}\left(\sum_{i=1}^{d}(q_i - 1)a_i - \hat{q} + 1\right)$.

In order to prove this theorem we need a lemma.

LEMMA 3.6 *Let* q *and* c *be two positive integers,* $B = \{b_1, .., b_{d-1}, c\}$, *and* $A = \{a_1, .., a_{d-1}, c\}$ *where* $a_1 = qb_1, .., a_{d-1} = qb_{d-1}$. *Suppose* $\gcd(A) = 1$ *and choose* $p \in \mathrm{lcm}(B)\mathbb{N}$ *so* $\gcd(q, c) = 1$ *and* $qp \in \mathrm{lcm}(A)\mathbb{N}$. *Then, the following formulas hold*

1. $\delta_{qp}(A) = q\delta_p(B)$.
2. $\omega_{qp}(A) = q\omega_p(B) + (q - 1)c$.
3. $\Omega_{qp}(A) = q\Omega_p(B) + \frac{1}{2}(q - 1)(c - 1)$.
4. $\Delta_{qp}(A) = q\Delta_p(B) + \frac{1}{2}(q - 1)(c - 1)$.

PROOF. We denote by

$$F(t) = F_p(t) = \frac{(1 - t^p)^d}{(1 - t^c)\prod_{i=1}^{d-1}(1 - t^{b_i})} = \sum_{r=0}^{p-1} t^r F_r(t^p)$$

the Frobenius polynomial associated with B and

$$G(t) = G_{qp}(t) = \frac{(1 - t^{qp})^d}{(1 - t^c)\prod_{i=1}^{d-1}(1 - t^{a_i})} = \sum_{s=0}^{qp-1} t^s G_s(t^{qp})$$

the Frobenius polynomial associated with A. We see that

$$G(t) = (1 + t^c + .. + t^{(q-1)c})F(t^q) = (1 + t^c + .. + t^{(q-1)c})\sum_{r=0}^{p-1} t^{qr} F_r(t^{qp}).$$

So we obtain

$$G(t) = \sum_{\substack{k=ic+jq \\ 0\le i\le q-1}} t^k F_j(t^{qp}) = \sum_{\substack{0\le k=ic+jq\le qp-1 \\ 0\le i\le q-1}} t^k F_j(t^{qp}) + \sum_{\substack{k>qp-1 \\ 0\le i\le q-1}} t^{k-qp} t^{qp} F_j(t^{qp})$$

By identification we deduce that $G_s(t^{qp}) = F_j(t^{qp})$ when $s = ic + jq$ and $G_s(t^{qp}) = t^{qp}F_j(t^{qp})$ when $s = ic + jq - qp = ic - (p - j)q$. In particular, we have $\deg(G_s) = \deg(F_j)$ and $\omega(G_s) = \omega(F_j)$ when $s = ic + jq$ and $\deg(G_s) = 1 + \deg(F_j)$ and $\omega(G_s) = 1 + \omega(F_j)$ when $s = ic + jq - qp$. Therefore, for all s which can be written in the form $s = ic + jq$ we get $\deg(t^s G_s(t^{qp})) = ic + jq + qp\deg(F_j)$ and $\omega(t^s G_s(t^{qp})) = ic + jq + qp\omega(F_j)$. For all s which can be written in the form $s = ic + jq - qp$, we get $\deg(t^s G_s(t^{qp})) = ic + jq - qp + qp(1 + \deg(F_j)) = ic + jq + qp\deg(F_j)$ and $\omega(t^s G_s(t^{qp})) = ic + jq - qp + qp(1 + \omega(F_j)$. It follows that $\delta_{qp}(G) = \min\{ic + jq + qp\deg(F_j)\} = q\min\{j + p\deg(F_j)\} = q\delta_p(F)$ and $\omega_{qp}(G) = \max\{ic+jq+qp\omega(F_j)\} = (q-1)c+q\max\{j+p\omega(F_j)\} = q\omega_p(F)+(q-1)c$. We also have

$$\Omega_{qp}(G) = \sum_{s=ic+jq} \omega(G_s) + \sum_{s=ic+jq-qp} \omega(G_s) = \sum_{s=ic+jq} \omega(F_j) + \sum_{s=ic-jq} (\omega(F_j) + 1)$$

$= q\Omega_p(F) + N'(c,q) = q\Omega_p(F) + \frac{1}{2}(q-1)(c-1)$. It follows that
$\Delta_{qp}(G) = \Omega_{qp}(G) + \delta_{qp}(G) = q(\Omega_p(F) + \delta_p(F)) + \frac{1}{2}(q-1)(c-1)$□

PROOF OF THEOREM 3.5. By induction on the number $h = d - k + 1$ such $q_1 = q_2 = .. = q_{k-1} = 1$. If $h = 1$ the result is given by lemma 3.6. Suppose that the result is true when $q_1 = q_2 = .. = q_{k-1} = 1$. We choose $p \in \text{lcm}(A)\mathbb{N}$ and we set $v = \frac{p}{q_k}$, $t_i = q_i$ for $i \neq k$ and $t_k = 1$. Then, we get $\hat{t}_i = \frac{\hat{q}_i}{q_k}$ for all $i \neq k$, $\hat{t}_k = \hat{q}_k$ and $\hat{t} = \frac{\hat{q}}{q_k}$. We also have $\frac{a_i}{q_k} = \frac{b_i\hat{q}_i}{q_k} = b_i\hat{t}_i$ for all $i \neq k$ and $a_k = b_k\hat{t}_k$. We put $c_i = b_i\hat{t}_i$ for all i and $C = \{c_1, .., c_d\}$, thus $a_i = q_k c_i$ for all $i \neq k$ and $a_k = c_k$. It follows from lemma 3.6 and the induction hypothesis that .

1) $\delta_p(A) = q_k\delta_v(C) = q_k\hat{t}\delta_u(B) = \hat{q}\delta_u(B)$.

2) $\omega_p(A) = q_k\omega_v(C) + (q_k - 1)c_k = q_k\{\hat{t}\omega_u(B) + \sum_{i=1}^d (t_i - 1)c_i\} + (q_k - 1)c_k = \hat{q}\omega_u(B) + \sum_{i=1}^d (q_i - 1)a_i$.

3) $\Omega_p(A) = q_k\Omega_v(C) + \frac{1}{2}(q_k - 1)(a_k - 1) = q_k\{\hat{t}\Omega_u(B) + \frac{1}{2}(\sum_{i=1}^d (t_i - 1)c_i - \hat{t} + 1)\} + \frac{1}{2}(q_k - 1)(a_k - 1) = \hat{q}\Omega_u(B) + \frac{1}{2}\left(\sum_{i=1}^d (q_i - 1)a_i - \hat{q} + 1\right)$.

4) $\Delta_p(A) = \Omega_p(A) + \delta_p(A) = \hat{q}\Delta_u(B) + \frac{1}{2}\left(\sum_{i=1}^d (q_i - 1)a_i - \hat{q} + 1\right)$ □

THEOREM 3.7 *The following formulas hold*

1. $g(A) = \hat{q}g(B) + \sum_{i=1}^d (q_i - 1)a_i$.

2. $N'(A) = \hat{q}N'(B) + \frac{1}{2}\left(\sum_{i=1}^d (q_i - 1)a_i - \hat{q} + 1\right)$.

3. $N(A) = \hat{q}N(B) + \frac{1}{2}\left(\sum_{i=1}^{d}(q_i - 1)a_i - \hat{q} + 1\right)$.

REMARK 3.8 In formula 1) if we take $q_1 = .. = q_{d-1} = 1$ then we obtain a Brauer and Shockley formula [5] and if we take $q_i = \gcd(A\backslash\{a_i\})$ for all i, we obtain a Raczunas and Chrzastowski-Wachtel formula [9]. Moreover formula 2) is a generalization of a Rödseth formula [10] which is obtained for $q_1 = .. = q_{d-1} = 1$.

THEOREM 3.9 *The following conditions hold*

1. $S(A)$ *is symmetric if and only if* $S(B)$ *is symmetric.*
2. *If* $\hat{q} > 1$ *then* $S(A)$ *is not pseudo-symmetric.*

COROLLARY 3.10 *Suppose there exists* i *such that* $b_i = 1$ (i.e. $a_i = \hat{q}_i$). *Then,* $S(A)$ *is symmetric* and we have

1. (a) $g(A) = \sum_{i=1}^{d}(q_i - 1)a_i - \hat{q}$.
   (b) $N(A) = N'(A) = \frac{1}{2}(\sum_{i=1}^{d}(q_i - 1)a_i - \hat{q} + 1)$.
2. *Suppose, in addition, that* $b_i = 1$ (i.e. $a_i = \hat{q}_i$) *for all* i. *Then, we have*
   (a) $g(A) = l(d-1) - \sum_{i=1}^{d} a_i$.
   (b) $N(A) = N'(A) = \frac{1}{2}(l(d-1) - \sum_{i=1}^{d} a_i + 1)$.

PROOF. Since $1 \in B$, we have $S(B) = \mathbb{N}_0$ then $g(B) = -1$ and $N(B) = N'(B) = 0$. So 1. follows from theorem 3.7. To prove 2., we observe that $q_i a_i = \hat{q} = l = \text{lcm}(A)$ if $a_i = \hat{q}_i$ for all i□

COROLLARY 3.11 *Let* b, d, h, v *be positive integers such that* $b \geq d \geq 2$ *and* $\gcd(b, v) = 1$. *Let* $B = \{b, hb + v, .., hb + (i-1)v, .., hb + (d-1)v\}$, *(*$(b_1, .., b_d)$ *is called an "almost" arithmetic sequence). Then,*
$S(A)$ *is symmetric* $\Leftrightarrow$ $S(B)$ *is symmetric* $\Leftrightarrow$ $d = 2$ or $b \equiv 2 \text{ mod}(d-1)$.

PROOF. We write $b - 1 = \beta(d-1) + \alpha$ with $0 \leq \alpha < d - 1$, and we use the following known formulas $g(B) = \left(h\left\lfloor\frac{b-2}{d-1}\right\rfloor + h - 1\right)b + bv - v$ [8] and $N'(B) = \frac{1}{2}\{(b-1)(h\beta + v + h - 1) + h\alpha(\beta + 1)\}$ [11] □

EXAMPLE 3.12 Let $A = \{150, 462, 840, 1365\} = \{5(2 \times 3 \times 5), 11(2 \times 3 \times 7), 12(2 \times 5 \times 7), 13(3 \times 5 \times 7)\}$. We set $q_1 = 7, q_2 = 5, q_3 = 3, q_4 = 2$ and $B = \{5, 11, 12, 13\}$. This is an almost arithmetic sequence with $b = 5, v = 1, h = 2, d = 4$. We see that $b \equiv 2 \text{ mod}(d-1)$ hence $S(B)$ is symmetric and we have $g(B) = 19, N'(B) = N(B) = 10$. Moreover, it follows from theorem 3.9 that $S(A)$ is symmetric. Using theorem 3.7 we obtain $g(A) = 210 \times 19 + 6 \times 150 + 4 \times 462 + 2 \times 840 + 1365 = 9783$. $N'(A) = N(A) = 210 \times 10 + \frac{1}{2}(6 \times 150 + 4 \times 462 + 2 \times 840 + 1365 - 210 + 1) = 4892$.

4 QUASI-POLYNOMIALS

DEFINITION 4.1 *A quasi-polynomial* P *of period* p *and degree* d *is a sequence* $P = (P_0, \ldots, P_{p-1})$ *with* $P_r \in \mathbb{Q}[t]$ *such that* $d = \sup\{\deg(P_r) \mid 0 \leq r \leq p-1\}$.
A quasi-polynomial P is said to be *uniform* if all the P_r have the same degree d

and the same leading coefficient $c(P)$. Given a function $h : \mathbb{Z} \to \mathbb{Q}$ and $r \in \mathbb{Z}$, we define $h_r : \mathbb{Z} \to \mathbb{Q}$, $k \mapsto h(pk + r)$. We say that h is a quasi-polynomial function if there exists a quasi-polynomial $P = (P_0, \dots, P_{p-1})$ such that $h_r(k) = P_r(k)$ for all $k \gg 0$ and $0 \leq r \leq p$. We also say that h is P-quasi-polynomial. It is easily seen that a quasi-polynomial function h has a minimal period and every period of h is a multiple of this minimal period. Furthermore, for a fixed period p, h is a P-quasi-polynomial for a unique sequence $P = (P_0, \dots, P_{p-1})$. A P-quasi-polynomial h is said to be uniform if P is uniform. We write $\deg(h) = \deg(P)$ and $c(h) = c(P)$. We denote by $F(\mathbb{Z})$ the set of all functions $h : \mathbb{Z} \to \mathbb{Q}$. For every integer $i \geq 0$ we consider the operators E^i and Δ_i, which act as follows: $(E^i h)(n) = h(n+i)$, $(\Delta_i h)(n) = h(n+i) - h(n)$. We set $E^0 = I$, $E^1 = E$ and $\Delta_1 = \Delta$ so $\Delta = E - I$, $\Delta_0 = 0$ and $\Delta_i = E^i - I$. For $a \geq 0$ and $n \geq 1$, we have $(I + E^a + \dots + E^{(n-1)a}) \circ (E^a - I) = E^{na} - I = \Delta_{na}$.

LEMMA 4.2 *Given $h \in F(\mathbb{Z})$, then the following identities hold*

1. $(E^{pi}h)_r = E^i h_r$ *for* $i \geq 0$.
2. $(\Delta_p^m h)_r = \Delta^m h_r$ *for* $m \geq 0$.

PROOF. 1. We write $(E^{pi}h)_r(k) = (E^{pi}h)(pk + r) = h(p(k+i) + r) = h_r(k+i) = (E^i h_r)(k)$.
2. We have $\Delta_p^m = (E^p - I)^m = \sum_{i=0}^m (-1)^{m-i}\binom{m}{i} E^{pi}$. Therefore, $(\Delta_p^m h)_r = \sum_{i=0}^m (-1)^{m-i}\binom{m}{i}(E^{pi}h)_r = \sum_{i=0}^m (-1)^{m-i}\binom{m}{i} E^i h_r = (E - I)^m h_r = \Delta^m h_r$□

PROPOSITION 4.3 *A function $h \in F(\mathbb{Z})$ is quasi-polynomial of period p and degree d if and only if there exists $(c_0, \dots, c_{p-1}) \neq (0, \dots, 0)$ such that $(\Delta_p^d h)_r(k) = c_r$ for all $k \gg 0$ and $0 \leq r \leq p - 1$.*

PROOF. Follows from lemma 4.2 and [6, lemma 4.1.2] □

COROLLARY 4.4 *For $h \in F(\mathbb{Z})$, if $\prod_{i=1}^d (E^{a_i} - I)(h)(n) = 0$ for $n \gg 0$, then h is quasi-polynomial of period $p \in \mathbb{N}$ and degree $< d$.*

PROOF. Follows from $\Delta_p^d = (E^p - I)^d = (\prod_{i=1}^d (\sum_{j=0}^{\frac{p}{a_i}-1} E^{ja_i})) \circ (\prod_{i=1}^d (E^{a_i} - I)$ □

EXAMPLE 4.5 Given $m \in \mathbb{Z}$ and $Q(t) \in \mathbb{Q}[t, t^{-1}]$ such that $Q(1) \neq 0$. The function $H_m(Q, .)$ associated with Q is a P−quasi-polynomial of period p, where $P = (P_0, .., P_{p-1})$ is given by $P_r = V_{m_r}(U_r, .)$.

REMARK 4.6 Suppose $m > 0$. Then, we have

1. $\deg(H_m(Q, .)) = m - 1$.
2. $m_r > 0 \Rightarrow \deg(P_r) = m_r - 1$ and $c(P_r) = \frac{U_r(1)}{(m_r - 1)!}$.
3. If $Q(1) = pQ_r(1) \neq 0$ for all $0 \leq r \leq p-1$, then $H_m(Q, .)$ is uniform of degree $m - 1$ and its leading coefficient is $c(H_m(Q, .)) = \frac{Q_r(1)}{(d-1)!} = \frac{Q(1)}{p(d-1)!}$.

4. Suppose $p = qu$ and there exists $K(t) \in \mathbb{Q}[t, t^{-1}]$ such that $K(t^u) = Q(t)$, we set $\xi = e^{\frac{2i\pi}{p}}$. If $Q(\xi^j) = 0$ for $0 < j < q$ and $Q(\xi^q) \neq 0$. Then, the following conditions hold

   (a) $P_r = 0$ if $r \notin u\mathbb{Z}$.

   (b) $\deg(P_r) = m - 1$ and $c(P_r) = \frac{Q_r(1)}{(m-1)!} = \frac{Q(\xi^q)}{p(m-1)!}$ if $r \in u\mathbb{Z}$.

PROPOSITION 4.7 *Let $m > 0$ be an integer and $h \in F(\mathbb{Z})$ be a function satisfying $h(n) = 0$ for $n \ll 0$. We consider $\Phi(t) = \sum_n h(n)t^n$. Then, the following conditions are equivalent*

1. *h is quasi-polynomial of period p and of degree $m - 1$.*
2. *$(1-t^p)^m\Phi(t) = Q(t) \in \mathbb{Z}[t, t^{-1}]$ and there exists a p-component Q_r of Q such that $Q_r(1) \neq 0$.*
3. *There exists a unique $Q(t) \in \mathbb{Z}[t, t^{-1}]$ such that $\deg H_m(Q, .) = m - 1$ and $H_m(Q, n) = h(n)$ for $n > \deg(Q) - pm$.*

In particular, h is a uniform quasi-polynomial function of period p and degree $m-1$ if and only if there exists a unique $Q(t) \in \mathbb{Z}[t, t^{-1}]$ such that $Q(1) \neq 0, Q(e^{j\frac{2i\pi}{p}}) = 0$ for all $0 < j < p$ and $H_m(Q, n) = h(n)$ for $n > \deg(Q) - pm$. In this case, the leading coefficient is $c(h) = \frac{Q(1)}{p(m-1)!}$.

PROOF. Assume 1. and set $\Phi_r(t) = \sum_n h_r(n)t^n$ for all $0 \leq r \leq p - 1$. It follows from [6, 4.1.7] that $(1 - t)^m\Phi_r(t) = Q_r(t) \in Z[t, t^{-1}]$. Since $\deg(h) = m - 1 \geq 0$, there exists $0 \leq r \leq p - 1$ such that $Q_r(1) \neq 0$. Setting $Q(t) = \sum_{r=0}^{p-1} t^r Q_r(t^p)$ we deduce 2.
2. $\Rightarrow$ 3. follows from theorem 2.4.
3. $\Rightarrow$ 1. follows from the definition of H_m.
The particular case follows from lemma 2.6 and remark 4.6 □

COROLLARY 4.8 Let $N(t)$ be an element of $\mathbb{Z}[t, t^{-1}]$ such that $N(1) \neq 0$ and $p \in I\!N$. We set $\Phi(t) = \sum_n h(n)t^n$ the expansion of

$$\frac{N(t)}{\prod_{i=1}^d (1 - t^{a_i})} = (1 - t^p)^{-d} N(t) F_p(t)$$

as a formal Laurent series. Then, $h(n) = H_d(NF_p, n)$ for $n > \deg(N) - \sum_{i=1}^d a_i$. Moreover, if in addition $\gcd(A) = 1$, then $h = H_d(NF_p, .)$ is uniform of degree $d - 1$ and its leading coefficient is $c(h) = \frac{N(1)p^{d-1}}{(d-1)! \prod_{i=1}^d a_i}$.

EXAMPLE 4.9 We write $s(n)$ for the number of solutions of the equation $a_1x_1 + .. + a_dx_d = n$ in nonnegative integers we get $s(n) = H_d(F_p, n)$ for all $n \geq 0$ where $p \in I\!N$. In particular, if $\gcd(A) = 1$ then $n \mapsto s(n)$ is a uniform quasi-polynomial function of degree $d - 1$ and of leading coefficient $c(s) = \frac{p^{d-1}}{(d-1)! \prod_{i=1}^d a_i}$.
For instance, the number of solutions of the equation $2x_1 + 3x_2 = n$ is $s(n) = H_2(F_6, n)$ (see example 2.2).

EXAMPLE 4.10 Let R_0 be a commutative ring and $R = R_0[t_1, \dots, t_d]$. Suppose that R is $\mathbb{Z}$-graded in such a way that every element of R_0 is homogeneous of degree zero and each t_i is homogeneous of degree a_i. Let $M = \oplus_{n\in\mathbb{Z}} M_n$ be a finitely generated graded R-module such that the length $l_{R_0}(M_n)$ of each M_n as an R_0- module is finite. The numerical function $H^0(M,.) : \mathbb{Z} \to \mathbb{Z}, n \mapsto l_{R_0}(M_n)$ is called the Hilbert function of M. The iterated cumulative Hilbert functions are defined by $H^{j+1}(M,n) = \sum_{i=0}^{n} H^j(M,i)$. The Poincaré series of M is denoted by $P_M(t) = \sum_n H^0(M,n)t^n$. By the Hilbert-Samuel theorem [3, 4.2 Theorem 1] there exists $Q_M(t) \in \mathbb{Z}[t,t^{-1}]$ such that $Q_M(1) \neq 0$ and $P_M(t) = \frac{Q_M(t)}{\prod_{i=1}^{d}(1-t^{a_i})}$. Moreover, it is known that $H^0(M,.)$ is quasi-polynomial [2]. Given $p \in I\!N$ and $j \geq 1$ we set $a_{d+1} = \dots = a_{d+j} = 1$. So the generating function of the $H^j(M,.)$ is

$$\sum_n H^j(M,n)t^n = \frac{P_M(t)}{(1-t)^j} = \frac{Q_M(t)}{\prod_{i=1}^{d+j}(1-t^{a_i})}.$$

It follows from corollary 4.8 that $H^j(M,n) = H_{d+j}(Q_M F_p, n)$ for all $j \geq 0$ and $n > \deg(Q_M) - \sum a_i$. Moreover, if $j > 0$ or $j = 0$ and $\gcd(A) = 1$ then $H^j(M,.)$ is a uniform quasi-polynomial function of degree $d+j-1$ and of leading coefficient $c(H^j(M,n)) = \frac{Q_M(1)p^{d+j-1}}{(d+j-1)!\prod_{i=1}^{d} a_i}$.

REFERENCES

1. V. Barucci, D. E. Dobbs and M. Fontana, Maximality Properties in Numerical Semigroups and Applications to One-Dimentional Analytically Irreducible Local Domains. Memoirs of the Amer. Math. Soc. vol 125, no 598: 1-78, 1997.

2. V.V. Bavula, Identification of the Hilbert function and Poincaré series, and the dimension of modules over filtred rings. Russian Acad. Sci. Izv. Math. 44, n° 2: 225-246, 1995.

3. N. Bourbaki, Algèbre commutative. Chapitre 8, Masson, 1983.

4. A. Brauer, On a problem of partitions, Amer. J. Math. 64: 299-312, 1942.

5. A. Brauer and J. E. Shockley, On a problem of Frobenius. J. Reine Angew. Math. 211: 215-220, 1962.

6. W. Bruns and J. Herzog, Cohen-Macaulay Rings. Cambridge University Press, Cambridge, U.K., 1993.

7. R. Fröberg, The Frobenius number of some semigroups. Comm. in Algebra 22, 14: 6021-6024, 1994.

8. M. Lewin, On a problem of Frobenius for an almost consecutive set of integers. J. Reine Angew. Math. 273: 134-137, 1975.

9. M. Raczunas and P. Chrzastowski-Wachtel, A Diophantine problem of Frobenius in terms of the least common multiple. Discrete Mathematics, 150: 347-357, 1996.

10. O.J. Rödseth, On a linear Diophantine problem of Frobenius. J. Reine Angew. Math. 301: 171-178, 1978.

11. E.S. Selmer, A linear Diophantine problem of Frobenius. J. Reine Angew. Math. 293/294: 1-17, 1977.

On Plane Algebroid Curves

V. BARUCCI, Università di Roma 1, *email*: barucci@mat.uniroma1.it

M. D'ANNA, Università di Catania, *email*: mdanna@dmi.unict.it

R. FRÖBERG, Stockholms Universitet, *email*: ralff@matematik.su.se

Abstract

Two plane analytic branches are topologically equivalent if and only if they have the same multiplicity sequence. We show that having same semigroup is equivalent to having same multiplicity sequence, we calculate the semigroup from a parametrization, and we characterize semigroups for plane branches. These results are known, but the proofs are new. Furthermore we characterize multiplicity sequences of plane branches, and we prove that the associated graded ring, with respect to the values, of a plane branch is a complete intersection.

1 INTRODUCTION

Let C and C' be two analytic plane irreducible curves (branches) defined in a neighbourhood of the origin and having singularities there. The branches are said to be topologically equivalent if there are neighbourhoods U and U' of the origin such that C is defined in U, C' in U', and there is a homeomorphism $T: U \to U'$ such that $T(C \cap U) = C' \cap U'$.

If $F(X,Y) \in \mathbb{C}[[X,Y]]$ is an irreducible formal power series, the local ring $\mathcal{O} = \mathbb{C}[[X,Y]]/(F)$ is called a (plane) algebroid branch. Two algebroid branches are formally equivalent if they have the same multiplicity sequence (see below for the definition of multiplicity sequence). Every algebroid (analytic resp.) branch is formally (topologically, resp.) equivalent to an algebraic branch, i.e. a branch defined by a polynomial [1], and if two analytic branches are formally equivalent, they are topologically equivalent. We will in the sequel consider algebroid branches.

Zariski has shown ([2]) that two branches are formally equivalent if and only if they have the same semigroup of values (see below for the definition of the value semigroup of a branch).

The crucial result of Section 2 is Proposition 2.3, which gives the relation between the value semigroups of an algebroid plane branch $\mathcal{O}$ and its blowup $\mathcal{O}'$. It is a result contained in [3]. Apéry proved that, in order to show that the value semigroup $v(\mathcal{O})$ of an algebroid plane branch $\mathcal{O}$ is symmetric. Subsequently Kunz proved that, for any analytically irreducible ring $\mathcal{O}$, $\mathcal{O}$ is Gorenstein if and only if $v(\mathcal{O})$ is symmetric. So now it is more common to say that the value semigroup of an algebroid plane branch is symmetric because the ring is Gorenstein (it is in fact a complete intersection). At any rate we are interested in Apéry's result for different reasons. By its use we give an easy proof of the fact that two plane algebroid branches are formally equivalent if and only if they have the same semigroup of values. We get also a well known formula of Hironaka and apply it again in Sections 3 and 4. The material in Section 3 is classical too and essentially contained in Enriqes-Chisini's work, but what is new, is the use of Apéry's Lemma in this context. After characterizing all possible multiplicity sequences for plane branches, we give a criterion to check if a semigroup is the value semigroup of a plane branch. In Section 4, we determine the semigroup of a plane branch from its parametrization, here also using results from [3]. This result is well known, but the proof is new as far as we know. Finally in Section 5 we show that the semigroup ring of the semigroup of a plane curve is a complete intersection.

2 PLANE BRANCHES

Starting from Apéry's article [3], we will proceed to explicate and expand various elements that are presented in the original arguments in a summary or not totally developed manner.

Let $\mathcal{O} = \mathbb{C}[[X,Y]]/(F) = \mathbb{C}[[x,y]]$, where F is irreducible in $\mathbb{C}[[X,Y]]$ be an algebroid plane branch. Since $F(X,Y)$ is irreducible, then $F(X,Y)$ must contain some term X^i and some term Y^j (otherwise F is not irreducible since we could factor out X or Y). Denote the minimal such powers by n and m respectively. Then, by the Weierstrass Preparation Theorem, the same ideal (F) can be generated by an element $X^n + \phi(X,Y)$, where $\phi(X,Y)$ is a polynomial of degree $n-1$ in X with coefficients which are power series in Y (or vice versa by an element $Y^m + \psi(X,Y)$, where $\psi(X,Y)$ is a polynomial of degree $m-1$ in Y with coefficients which are power series in X). This gives that $\mathcal{O}$ is generated by $1, x, ..., x^{n-1}$ as $\mathbb{C}[[y]]$-module (or generated by $1, y, ..., y^{m-1}$ as $\mathbb{C}[[x]]$-module).

The Puiseux Theorem gives that the branch has a parametric representation $x = t^m, y = \sum a_i t^i$ (or $x = \sum_{i \geq m} b_i t_1^i, y = t_1^n$, where $\mathbb{C}[[t]] = \mathbb{C}[[t_1]]$). Thus $\mathcal{O} = \mathbb{C}[[x,y]] \subseteq \mathbb{C}[[t]] = \bar{\mathcal{O}}$, which is a discrete valuation ring. Denote by v the valuation of such ring that consists in associating to any formal power series in $\mathbb{C}[[t]]$ its order. In particular $v(x) = m$ and $v(y) = n$. Since the fraction field of $\mathcal{O}$ equals the fraction field of $\bar{\mathcal{O}}$, there exist $f_1(t), f_2(t) \in \mathcal{O}$, such that $f_1(t)/f_2(t) = t$, so $f_1(t) = tf_2(t)$ and $v(f_1) = v(f_2)+1$. Since $\gcd(v(f_1), v(f_2)) = 1$, all sufficiently large integers belong to $v(\mathcal{O}) = \{v(z); z \in \mathcal{O} \setminus \{0\}\}$. Thus $v(\mathcal{O})$ is a numerical semigroup, i.e., a subsemigroup of $\mathbb{N}$ with finite complement to $\mathbb{N}$.

In the sequel we use the following terminology. If S is a subsemigroup of $\mathbb{N}$ and T is a subset of $\mathbb{Z}$, we call T an S-module if $s \in S, t \in T$ implies $s+t \in T$. We call T a free S-module if $T = \cup_{i=1}^k T_i$ with $T_i \cap T_j = \emptyset$ if $i \neq j$ and $T_i = n_i + S$ for some $n_i \in \mathbb{Z}$. We call $n_1, \ldots, n_k$ a basis of T.

With the hypotheses and notation above, we will construct a new basis $y_0, \ldots, y_{m-1}$ for $\mathcal{O}$ as a $\mathbb{C}[[x]]$-module, such that, for each i, $y_0, \ldots, y_i$ is a basis for $\mathcal{O}_i = \mathbb{C}[[x]] + y\mathbb{C}[[x]] + \cdots + y^i\mathbb{C}[[x]]$, and furthermore such that $v(\mathcal{O}_i) = \{v(z); z \in \mathcal{O}_i \setminus \{0\}\}$ is a free module over $v(\mathbb{C}[[x]]) = m\mathbb{N}$ with basis $\omega_0, \ldots, \omega_i$, where each $\omega_j = v(y_j)$, $j = 0, \ldots, i$ is the smallest value in $v(\mathcal{O})$ in its congruence class $\pmod m$. Let $y_0 = 1$, thus $\omega_0 = v(y_0) = 0$ and $v(\mathcal{O}_0) = v(\mathbb{C}[[x]]) = m\mathbb{N}$. Suppose that $y_0, \ldots, y_{k-1}, k < m$ have been defined such that $v(\mathcal{O}_{k-1})$ is a free $m\mathbb{N}$-module with basis $\omega_0, \ldots, \omega_{k-1}$. We claim that there exists a $\phi(x, y) \in \mathcal{O}_{k-1}$ such that $y_k = y^k + \phi(x, y)$ has a value which does not belong to $v(\mathcal{O}_{k-1})$. If $v(y^k) \notin v(\mathcal{O}_{k-1})$, we are ready. Otherwise $v(y^k) = v(z_1)$ for some $z_1 \in \mathcal{O}_{k-1}$. Then $v(y^k - c_1 z_1) > v(y^k)$ for some $c_1 \in \mathbb{C}$. If $v(y^k - c_1 z_1) \notin v(\mathcal{O}_{k-1})$, we are ready. Otherwise take $z_2 \in \mathcal{O}_{k-1}$ with $v(z_2) = v(y^k - c_1 z_1)$. Then $v(y^k - c_1 z_1 - c_2 z_2) > v(y^k - c_1 z_1)$ for some $c_2 \in \mathbb{C}$ a.s.o. Thus we see that the expansion of y^k as a power series in t must contain a term $a_i t^i$ with $a_i \neq 0$ and $i \notin v(\mathcal{O}_{k-1})$, since otherwise $y^k \in \mathcal{O}_{k-1}$.

Notice that $y_1 y_{k-1} = (y + \phi_1(x))(y^{k-1} + \phi_{k-1}(x, y)) = y^k + \psi(x, y)$, $\psi(x, y) \in \mathcal{O}_{k-1}$, so $y_k = y_1 y_{k-1} + \phi(x, y) - \psi(x, y)$ and we could equally well have defined y_k as an element of the form $y_1 y_{k-1} + \phi(x, y)$ (where $\phi(x, y) \in \mathcal{O}_{k-1}$) with a value which does not belong to $v(\mathcal{O}_{k-1})$. In such expression of y_k, $v(\phi(x, y)) \geq v(y_1 y_{k-1})$ since otherwise $v(y_k) = v(\phi(x, y)) \in v(\mathcal{O}_{k-1})$. Thus $\omega_k = v(y_k) \geq v(y_1 y_{k-1}) = v(y_1) + v(y_{k-1}) = \omega_1 + \omega_{k-1}$. In particular the sequence $\omega_0, \omega_1, \ldots$ is strictly increasing. Since $v(\mathcal{O}_{k-1})$ is free over $m\mathbb{Z}$, this shows that $\omega_k \not\equiv \omega_j$ if $j < k$. Any element $z \in \mathcal{O}_k$ can be written $z = a_0(x) y_0 + \cdots + a_k(x) y_k$. All terms in this sum have values in different congruence classes $\pmod m$. Thus $v(z) = \min v(a_i(x) y_i)$. This shows that $v(\mathcal{O}_k)$ is free with basis $\omega_0, \ldots, \omega_k$. After m steps, we get that $\mathcal{O}_{m-1} = \mathcal{O}$ is a $\mathbb{C}[[x]]$-module generated by $y_0, \ldots, y_{m-1}$ with the requested properties.

If S is a numerical semigroup and $a \in S \setminus \{0\}$, then the elements $n_0, n_1, \ldots, n_{a-1}$, where n_i is the smallest element in S congruent to $i \pmod a$, is called the *Apery set* of S with respect to a. If we order the elements in the Apery set, and then denote them $\omega_0, \ldots, \omega_{a-1}$, we have the *ordered Apery set*. We call the elements $y_0, \ldots, y_{m-1} \in \mathcal{O}$ constructed as above an *Apery basis* of $\mathcal{O}$ with respect to x. By the construction, $\omega_0 = v(y_0), \ldots, \omega_{m-1} = v(y_{m-1})$ is the ordered Apery set of $v(\mathcal{O})$.

In a similar way an Apery basis of $\mathcal{O}$ with respect to y is defined.

Example If in $\mathcal{O} = \mathbb{C}[[x, y]]$ we have $\gcd(m, n) = 1$, where $v(x) = m$ and $v(y) = n$, then $y_k = y^k$, $k = 0, \ldots, m-1$ is an Apery basis of $\mathcal{O}$, and thus $\omega_k = kn$, $k = 0, \ldots, m-1$ is the ordered Apery set of $v(\mathcal{O})$ with respect to m.

Example If in $\mathcal{O} = \mathbb{C}[[x, y]]$ we have $x = t^8$, $y = t^{12} + t^{14} + t^{15}$, then $y_0 = 1$, $y_1 = y$, $y_2 = y^2 - x^3 = 2t^{26} + \cdots$, $y_3 = y^3 - x^3 y = 2t^{38}$, $y_4 = y^4 - 2x^3y^2 - 4x^5y + 3x^6 = 8t^{53} + \cdots$, $y_5 = y^5 - 2x^3y^3 + x^6y - 4x^8 = 8t^{65} + \cdots$, $y_6 = y^6 - 3x^3y^4 - 4x^5y^3 + 3x^6y^2 + 4x^8y - x^9 = 16t^{79} + \cdots$, and $y_7 = y^7 - 3x^3y^5 + 3x^6y^3 - 4x^8y^2 - x^9y + 4x^{11} = 16t^{91} + \cdots$ is an Apery basis for $\mathcal{O}$, so the ordered Apery set of $v(\mathcal{O})$ with respect to 8 is $\{0, 12, 26, 38, 53, 65, 79, 91\}$. Thus $v(\mathcal{O})$ is minimally generated by $8, 12, 26, 53$.

If S is a numerical semigroup, we denote the *Frobenius number* of S, i.e.

$\max\{x \in \mathbb{Z}; x \notin S\}$, by $\gamma(S)$. The *conductor* of S is $c(S) = \gamma(S) + 1 = \min\{x; [x, \infty) \subseteq S\}$.

The following lemma is well known, and its easy proof is left to the reader.

LEMMA 2.1 *Let S be a numerical semigroup with Frobenius number γ and $a \in S$. If $\omega_0, \ldots, \omega_{a-1}$ is the ordered Apery set of S with respect to a, then $\gamma = \omega_{a-1} - a$.*

Now we are ready for the crucial lemma from [3]. If $\mathcal{O} = \mathbb{C}[[x, y]]$ with $v(x) < v(y)$, we denote the quadratic transform (or blowup) $\mathbb{C}[[x, y/x]]$ by $\mathcal{O}'$.

LEMMA 2.2 *If an Apery basis of $\mathcal{O}'$ with respect to x is $y'_0, \ldots, y'_{m-1}$, then $y_i = y'_i x^i$, for $i = 0, \ldots, m-1$ is an Apery basis of $\mathcal{O}$ with respect to x.*

Proof. Let $F_i(x, y/x)$ be the polynomial of degree i in y/x which defines y'_i, i.e. let $F_i(x, y/x) = (y/x)^i + \phi'_i(x, y/x)$, where $\deg(\phi'_i) < i$ in y/x. Then $y_i = x^i F(x, y/x) = y^i + \phi_i(x, y), \phi_i(x, y) \in \mathcal{O}_{i-1}$ is of the requested form and, if $v(y'_i) = \omega'_i$, then $\omega_i = v(y_i) = \omega'_i + im$, thus $\omega_i \equiv \omega'_i \pmod{m}$. We have to show that $\omega_i \notin v(\mathcal{O}_{i-1})$. This is because ω'_i is not congruent to any ω'_j, if $j < i$, and so also ω_i is not congruent to any ω_j, if $j < i$.

As a consequence we get

PROPOSITION 2.3 *[3, Lemme 2] If the ordered Apery set of $v(\mathcal{O}')$ with respect to $m = v(x)$ is $0 = \omega'_0 < \omega'_1 < \cdots < \omega'_{m-1}$, then the ordered Apery set of $v(\mathcal{O})$ with respect to m is $\omega_0 = \omega'_0 < \omega_1 = \omega'_1 + m < \omega_2 = \omega'_2 + 2m < \ldots < \omega_{m-1} = \omega'_{m-1} + (m-1)m$.*

Recall that the multiplicity of the ring $\mathcal{O} = \mathbb{C}[[x, y]]$, where $x = a_m t^m + a_{m+1} t^{m+1} + \cdots$, $a_m \neq 0$ and $y = b_n t^n + b_{n+1} t^{n+1} + \cdots$, $b_n \neq 0$, is given by $\min(m, n)$ i.e. the multiplicity of $\mathcal{O}$ is the smallest positive value in $v(\mathcal{O})$.

Set $\mathcal{O} = \mathcal{O}^{(0)}$, denote by $\mathcal{O}^{(i+1)}$ the blowup of $\mathcal{O}^{(i)}$ and by e_i the multiplicity of $\mathcal{O}^{(i)}$. The *multiplicity sequence* of $\mathcal{O}$ is by definition the sequence of natural numbers $e_0, e_1, e_2, \cdots$. Let k be the minimal index such that $e_k = 1$, i.e. such that $v(\mathcal{O}^{(k)}) = \mathbb{N}$. Two algebroid branches are *formally equivalent* if they have the same multiplicity sequence.

As a consequence of Proposition 2.3, we get easily a well known formula:

COROLLARY 2.4 *[4, Theorem 1] We have $l_{\mathcal{O}}(\bar{\mathcal{O}}/(\mathcal{O} : \bar{\mathcal{O}})) = \sum_{i=0}^{k} e_i(e_i - 1)$ and $l_{\mathcal{O}}(\mathcal{O}/(\mathcal{O} : \bar{\mathcal{O}})) = l_{\mathcal{O}}(\bar{\mathcal{O}}/\mathcal{O}) = \frac{1}{2}\sum_{i=0}^{k} e_i(e_i - 1)$.*

Proof. Let $\omega_i^{(j)}$ ($\omega_i^{(j+1)}$, resp.) be the i'th element in the ordered Apery set of $v(\mathcal{O}^{(j)})$, ($v(\mathcal{O}^{(j+1)})$, resp.), with respect to e_j and let $\mathcal{O}^{(j)} : \bar{\mathcal{O}} = t^{c_j}\mathbb{C}[[t]]$ ($\mathcal{O}^{(j+1)} : \bar{\mathcal{O}} = t^{c_{j+1}}\mathbb{C}[[t]]$, resp.). By Lemma 2.1 $c_j = \omega_{e_j-1}^{(j)} - e_j + 1$ and $c_{j+1} = \omega_{e_j-1}^{(j+1)} - e_j + 1$. Proposition 2.3 gives $\omega_{e_j-1}^{(j)} = \omega_{e_j-1}^{(j+1)} + e_j(e_j - 1)$ and so $c_j = c_{j+1} + e_j(e_j - 1)$. It follows that $c_0 = l_{\mathcal{O}}(\bar{\mathcal{O}}/(\mathcal{O} : \bar{\mathcal{O}})) = c_1 + e_0(e_0 - 1) =$

$\cdots = c_k + e_{k-1}(e_{k-1} - 1) + \cdots + e_0(e_0 - 1) = \sum_{i=0}^{k} e_i(e_i - 1)$. Since the ring $\mathcal{O}$ is Gorenstein, we get $l_{\mathcal{O}}(\mathcal{O}/(\mathcal{O} : \bar{\mathcal{O}})) = l_{\mathcal{O}}(\bar{\mathcal{O}}/\mathcal{O}) = \frac{1}{2} l_{\mathcal{O}}(\bar{\mathcal{O}}/(\mathcal{O} : \bar{\mathcal{O}}))$.

Example Not every symmetric semigroup is the value semigroup of an algebroid plane branch. The semigroup generated by 4,5,6 is symmetric and has Apery set 0,5,6,11 with respect to 4. If this were the value semigroup of a plane branch, then the Apery set of its blowup would be 0, $1 = 5-4$, $-2 = 6-8$, $-1 = 11-12$ which obviously is impossible.

THEOREM 2.5 *[2] Two algebroid plane branches are formally equivalent if and only if they have the same semigroup.*

Proof. Let $\mathcal{O} = \mathcal{O}^{(0)}, \mathcal{O}^{(1)}, \ldots$ be the sequence of blowups of $\mathcal{O}$, and let $e_0, \ldots, e_k = 1$ be the corresponding multiplicity sequence. Then $v(\mathcal{O}^{(k)}) = \mathbb{N}$ has ordered Apery set $\{0, 1, \ldots, e_{k-1} - 1\}$ with respect to e_{k-1}. Proposition 2.3 gives the ordered Apery set, hence the semigroup, of $\mathcal{O}^{(k-1)}$ with respect to e_{k-1} a.s.o. Thus the multiplicity sequence determines the semigroup of $\mathcal{O}$. On the other hand, the semigroup of $\mathcal{O}$ gives the multiplicity e_0 of $\mathcal{O}$. Proposition 2.3 gives the Apery set of $v(\mathcal{O}^{(1)})$, hence $v(\mathcal{O}^{(1)})$ and so on. Thus the semigroup $v(\mathcal{O})$ gives the multiplicity sequence.

Let c_i denote the *conductor degree* of $\mathcal{O}^{(i)}$, i.e. $\mathcal{O}^{(i)} : \bar{\mathcal{O}} = t^{c_i}\mathbb{C}[[t]]$, and call $(c_0, c_1, \ldots)$ the *conductor degree sequence* of $\mathcal{O}$. Let $f_i = l_{\mathcal{O}^{(i)}}(\bar{\mathcal{O}}/\mathcal{O}^{(i)})$, and call $(f_0, f_1, \ldots)$ the *sequence of singularity degrees* of $\mathcal{O}$.

COROLLARY 2.6 *Two algebroid plane branches are formally equivalent if and only if they have the same conductor degree sequence, and if and only if they have the same sequence of singularity degrees.*

Proof. If $\omega_0 < \omega_1 < \cdots < \omega_{e_i - 1}$ is the Apery set of $v(\mathcal{O}^{(i)})$ with respect to e_i, then by Lemma 2.1 $c_i = \omega_{e_i - 1} - e_i + 1$. Thus the multiplicity sequence of $\mathcal{O}$ determines, and is determined by, the conductor degree sequence. Since each ring $\mathcal{O}^{(i)}$ is Gorenstein, $f_i = c_i/2$ and the same is true for the sequence of singularity degrees.

Example The conductor degree of $\mathcal{O}$ does not suffice to give formal equivalence. The branches $\mathbb{C}[[t^4, t^5]]$ and $\mathbb{C}[[t^3, t^7]]$ both have conductor $t^{12}\mathbb{C}[[t]]$, but they are not formally equivalent.

3 THE MULTIPLICITY SEQUENCE FOR A PLANE BRANCH

A sequence of numbers $e_0 \geq e_1 \geq e_2 \geq \cdots$ is a multiplicity sequence of a (not necessarily plane) branch if and only if $0, e_0, e_0 + e_1, e_0 + e_1 + e_2, \ldots$ constitute a semigroup [5]. We will now determine which multiplicity sequences occur for plane branches. We will also use this result together with Proposition 2.3 and

Theorem 2.5 to get an algorithm to determine if a symmetric semigroup is the semigroup of a plane branch.

Let $\mathcal{O} = \mathbb{C}[[t^{\delta_0} + \sum_{i \geq N} a_i t^i, \sum_{i>\delta_0} b_i t^i]]$ be a branch. Let, for $i \geq 1$, $\delta_i = \min\{j;\ b_j \neq 0, \gcd(\delta_0, \dots, \delta_{i-1}, j) < \gcd(\delta_0, \dots, \delta_{i-1})\}$. Let $d_0 = \delta_0$ and $\gcd(\delta_0, \dots, \delta_i) = d_i$ for $i \geq 1$. Set also $k = \min\{i;\ d_i = 1\}$. (There exists such a k since the integral closure of $\mathcal{O}$ is $\mathbb{C}[[t]]$.) We call the parametrization *standard* if $N > \delta_k$. The numbers $\delta_0, \delta_1, \dots$ are called the *characteristic exponents* of $\mathcal{O}$. It follows from the proof of Lemma 3.1 below, that we always can get a standard parametrization from a given one.

LEMMA 3.1 *Let* $\mathcal{O} = \mathbb{C}[[t^{\delta_0} + \sum_{i \geq N} a_i t^i, \sum_{i>\delta_0} b_i t^i]]$ *be a branch with standard parametrization and with characteristic exponents* $(\delta_0, \dots, \delta_k)$. *Then the characteristic exponents of* $\mathcal{O}'$ *are:*
a) $(\delta_0, \delta_1 - \delta_0, \dots, \delta_k - \delta_0)$, *if* $\delta_0 < \delta_1 - \delta_0$.
b) $(\delta_1 - \delta_0, \delta_0, \delta_0 + \delta_2 - \delta_1, \dots, \delta_0 + \delta_k - \delta_1)$, *if* $\delta_0 > \delta_1 - \delta_0$ *and* δ_0 *is not a multiple of* $\delta_1 - \delta_0$
c) $(\delta_1 - \delta_0, \delta_0 + \delta_2 - \delta_1, \dots, \delta_0 + \delta_k - \delta_1)$, *if* δ_0 *is a multiple of* $\delta_1 - \delta_0$.

Proof We can suppose that $v(\sum_{i>\delta_0} b_i t^i) = \delta_1$. Then the blowup $\mathcal{O}'$ of $\mathcal{O}$ is $(t^{\delta_0} + \cdots, t^{\delta_1 - \delta_0} + \cdots)$. One of the following three cases will occur:
a) $\delta_0 < \delta_1 - \delta_0$
b) $\delta_0 > \delta_1 - \delta_0$ and δ_0 is not a multiple of $\delta_1 - \delta_0$
c) δ_0 is a multiple of $\delta_1 - \delta_0$.
We will in each case write $\mathcal{O}'$ in standard form and derive its characteristic exponents. In case a) $\mathcal{O}'$ is of standard form. We keep the meaning of δ_i and d_i from above and denote the corresponding entities for $\mathcal{O}'$ with δ_i' and d_i'. It follows that $d_i' = d_i$ for all i and that $\mathcal{O}'$ has characteristic exponents $(\delta_0', \delta_1', \dots, \delta_k') = (\delta_0, \delta_1 - \delta_0, \dots, \delta_k - \delta_0)$. In case b) we first make the coordinate change $X = y, Y = x$ to get $(t^{\delta_1 - \delta_0}(1 + \sum_{i \geq 1} c_i t^i), t^{\delta_0} + \cdots)$. Let $i_0 = \min\{i;\ c_i \neq 0\}$. Then we choose a new parameter t_1, by $t = t_1(1 - \frac{c_{i_0}}{\delta_1 - \delta_0} t_1^{c_{i_0}})$ to get the parametrization $(t_1^{\delta_1 - \delta_0}(1 + \sum_{i \geq 1} c_i' t_1^i), t_1^{\delta_0} + \cdots)$. Now $v(\sum_{i \geq 1} c_i' t_1^i) > v(\sum_{i \geq 1} c_i t^i)$. We continue to change parameter in this way. After a finite number of steps we get a parametrization of the branch of the type $(t^{\delta_1 - \delta_0} + \sum_{i > \delta_k} k_i t^i, t^{\delta_0} + \cdots)$ with $d_i' = d_i$ for all i, and with characteristic exponents $(\delta_1 - \delta_0, \delta_0, \delta_0 + \delta_2 - \delta_1, \dots, \delta_0 + \delta_k - \delta_1)$. In case c) finally, we use a similar reparametrization and get $d_i' = d_{i+1}$, and a branch with characteristic exponents $(\delta_1 - \delta_0, \delta_0 + \delta_2 - \delta_1, \dots, \delta_0 + \delta_k - \delta_1)$.

If $m_0, m_1, \cdots$ and $h_0, h_1, \cdots$ are natural numbers, denote by $m_0^{(h_0)}, m_1^{(h_1)}, \cdots$ the sequence of natural numbers given by m_0 repeated h_0 times, m_1 repeated h_1 times and so on. Suppose that for a couple m, n of natural numbers, the Euclidean algorithm gives

$$m = nq_1 + r_1$$

$$n = r_1 q_2 + r_2$$

$$r_{i-1} = r_i q_{i+1} + r_{i+1}$$
$$r_i = r_{i+1} q_{i+2} + 0$$

Denote by $M(m,n)$ the sequence of natural numbers $n^{(q_1)}, r_1^{(q_2)}, \dots, r_{i+1}^{(q_{i+2})}$. Of course such a sequence ends with $r_{i+1} = \gcd(m,n)$ (if $m < n$, and so $q_1 = 0$, n appears 0 times, i.e. it does not appear, hence $M(m,n) = M(n,m)$). With this notation:

THEOREM 3.2 *A sequence of natural numbers is the multiplicity sequence of an algebroid plane branch if and only if it is of the following form:*

$$M(m_0, m_1), M(m_2, m_3), \dots, M(m_{2k}, m_{2k+1}), 1, 1, \dots$$

where, for $i \geq 0$, $\gcd(m_{2i}, m_{2i+1}) = m_{2i+2}$ and m_{2i+3} is such that $m_{2i+4} < m_{2i+2}$, and finally $\gcd(m_{2k}, m_{2k+1}) = 1$.

Proof. Let $\mathcal{O}$ be an algebroid plane branch with standard parametrization. Then, by Lemma 3.1, its multiplicity sequence is

$$M(\delta_0, \delta_1), M(d_1, \delta_2 - \delta_1), M(d_2, \delta_3 - \delta_2), \dots, M(d_{k-1}, \delta_k - \delta_{k-1}), 1, 1, \dots$$

and is a sequence of the requested form. Conversely, given a sequence of natural numbers as in the statement, we can get characteristic exponents $(\delta_0, \delta_1, \dots \delta_k)$ and so an $\mathcal{O}$.

We give two concrete examples.

Example $6, 4, 2, 2, 1, 1, \dots = M(10, 6), 1, 1, \dots$ is an admissible multiplicity sequence (i.e. the multiplicity sequence of an algebroid plane branch), but $6, 4, 2, 1, 1, \dots$ is not.

Example Let $\mathcal{O} = \mathbb{C}[[x, y]]$ with

$$x = t^{2\cdot 2^n},\ y = t^{3\cdot 2^n} + t^{3\cdot 2^n + 2^{n-1}} + \dots + t^{3\cdot 2^n + 2^{n-1} + \dots + 2 + 1}.$$

The multiplicity sequence is

$$2^{n+1}, 2^n, 2^n, 2^{n-1}, 2^{n-1}, \dots, 4, 4, 2, 2, 1, \dots =$$

$$M(3 \cdot 2^n, 2^{n+1}), M(2^n, 2^{n-1}), M(2^{n-1}, 2^{n-2}), \dots, M(2, 1), \dots$$

Now we are ready to give an algorithm to determine if a symmetric semigroup is the semigroup of values of a plane curve.

LEMMA 3.3 *Let S be a symmetric semigroup, $m = \min(S \setminus \{0\})$ and let*

$$0 = \omega_0 < \omega_1 < \dots < \omega_{m-1}$$

be its ordered Apery set with respect to m. Suppose that $\omega_0 < \omega_1 - m < \dots < \omega_{m-1} - (m-1)m$ is the ordered Apery set of a semigroup S'. Then S' is symmetric.

Proof. This follows from [3].

Given a symmetric semigroup S satisfying the hypotheses of Lemma 3.3, one could repeat the process for the ordered Apery set of S' with respect to its minimal non zero element, and so on until we find either a semigroup which does not satify these hypotheses or we find $\mathbb{N}$. But even if, after a finite number of steps, we get $\mathbb{N}$, it is not true that S is a value semigroup of a plane branch, as the following example shows.

Example Let $S = \langle 6, 10, 29\rangle$; its ordered Apery set with respect to 6 is $\{0, 10, 20, 29, 39, 49\}$. The set obtained applying Lemma 3.3 is $\{0, 4 = 10 - 6, 8 = 20-12, 11 = 29-18, 15 = 39-24, 19 = 49-30\}$, hence it is the ordered Apery set of S' with respect to 6. Hence $S' = \{0, 4, 6, 8, 10, 11, 12, 14, \rightarrow \dots\} = \langle 4, 6, 11\rangle$. The ordered Apery set of S' with respect to 4 is $0, 6, 11, 17$. Hence we get the new set $\{0, 2 = 6 - 4, 3 = 11 - 8, 5 = 17 - 12\}$ which is still ordered and determines the semigroup $S'' = \langle 2, 3\rangle$. Its ordered Apery set with respect to 2 is $0, 3$. Thus we get the set $\{0, 1 = 3 - 2\}$, which is the ordered Apery set with respect to 2 of $\mathbb{N}$.

On the other hand the semigroup S is not the value semigroup of a plane branch $\mathcal{O}$ since the multiplicity sequence of $\mathcal{O}$ should be $6, 4, 2, 1, 1, \dots$ which is not admissible, since the subsequence $6, 4$ can be obtained only by $M(10, 6)$ but $M(10, 6) = 6, 4, 2, 2$.

Let S be the value semigroup of a plane branch $\mathcal{O}$. By Proposition 2.3 we get that S' (defined as in Lemma 3.3) is again a symmetric semigroup and $S' = v(\mathcal{O}')$. Repeating the process, if $S^{(0)} = S$ and $S^{(j+1)} = (S^{(j)})'$, and denoting by m_j the minimal non zero element of $S^{(j)}$ and by $\omega_0^{(j)}, \omega_1^{(j)}, \dots, \omega_{m_j-1}^{(j)}$ its ordered Apery set with respect to m_j, we get that $\omega_0^{(j)}, \omega_1^{(j)} - m, \dots, \omega_{m_j-1}^{(j)} - (m_j - 1)m_j$ is the ordered Apery set of a symmetric semigroup $S^{(j+1)}$, and $S^{(j+1)} = v(\mathcal{O}^{(j+1)})$. Since there exists an $n \geq 1$ such that $\mathcal{O}^{(n)} = \mathbb{C}[[t]]$, then $S^{(n)} = \mathbb{N}$. Moreover the sequence $m_0, \dots, m_{n-1}, 1, \dots$ is the multiplicity sequence of $\mathcal{O}$, hence is an admissible multiplicity sequence.

Conversely if $S = S_0$ is a symmetric semigroup, let $S^{(j)}, m_j, \omega_i^{(j)}$ be defined as above. If the sets $0 = \omega_0^{(j)}, \omega_1^{(j)} - m_j, \dots, \omega_{m_j-1}^{(j)} - (m_j - 1)m_j$ are ordered Apery sets for every $j = 0, \dots n-1$ and the sequence $m_0, m_1, \dots, m_{n-1}, 1, 1, \dots$ is an admissible multiplicity sequence, then S is the value semigroup of a plane branch. In fact, since the sequence $m_0, m_1, \dots, m_{n-1}, 1, 1, \dots$ is an admissible multiplicity sequence, then there exists a plane branch $\mathcal{O}$ having this sequence as multiplicity sequence. Now, by Theorem 2.5, the multiplicity sequence determines the value semigroups $v(\mathcal{O}^{(k)})$, $k = 0 \dots, n-1$, and these semigroups, by Proposition 2.3 and Lemma 3.3 have the same ordered Apery sets of the semigroups $S^{(k)}$; hence they are the same semigroups.

This discussion gives a criterion to check if S is the value semigroup of a plane branch, since we can apply repeatedly the process described in Lemma 3.3 until we find either a semigroup which does not satisfy the hypotheses in

Lemma 3.3 or we find $\mathbb{N}$. If the last case occurs, then it is enough to check if the sequence $m_0, \dots, m_{n-1}, 1, 1, \dots$ is admissible.

The condition that at each step the sequence $0 = \omega_0^{(j)}, \omega_1^{(j)} - m_j, \dots, \omega_{m_j-1}^{(j)} - (m_j - 1)m_j$ is an ordered Apery set (and not only an Apery set) is necessary as the following example shows.

Example Let $S = \{0, 4, 8, 9, 10, 12, 13, 14, 16, \rightarrow \dots\}$ be the semigroup with ordered Apery set $\{0, 9, 10, 19\}$ with respect to 4. The sequence $0, 5 = 9 - 4, 2 = 10 - 8, 7 = 19 - 12$ is not increasing. If we consider the semigroup S' with ordered Apery set $\{0, 2, 5, 7\}$ with respect to 4 it is the symmetric semigroup $\{0, 2, 4, \rightarrow \dots\}$ and then in two more steps we get $\mathbb{N}$.

Notice that the sequence $m_0, m_1, \dots$ is in this case $4, 2, 2, 1, 1, \dots$; it is admissible as multiplicity sequence since it is $M(6, 4), M(2, 1), 1, 1, \dots$. However, applying Theorem 2.5, we get the semigroup $\{0, 4, 6, 8, 10, 12, 13, 14, \rightarrow \dots\}$ with ordered Apery set $\{0, 6, 13, 19\}$ and applying Theorem 3.2 we get the parametrization $\mathcal{O} = \mathbb{C}[[t^4, t^6 + t^7]]$.

4 THE SEMIGROUP OF VALUES FOR A PLANE BRANCH

The following theorem is proved in different ways in e.g. [2], [6], [7], [8], [9].

THEOREM 4.1 *Let* $\mathcal{O} = \mathbb{C}[[t^{\delta_0} + \sum_{i \geq N} a_i t^i, \sum_{i > \delta_0} b_i t^i]]$ *be a branch with standard parametrization. Denote the minimal generators of* $v(\mathcal{O})$ *by* $\bar{\delta}_0 < \dots < \bar{\delta}_s$. *Then* $s = k$, $\bar{\delta}_0 = \delta_0, \bar{\delta}_1 = \delta_1$ *and* $\bar{\delta}_i = \bar{\delta}_{i-1}\frac{d_{i-2}}{d_{i-1}} + \delta_i - \delta_{i-1}$ *if* $i = 2, \dots, k$.

We will divide the proof into several steps. From now on we will, for a plane branch with characteristic exponents $(\delta_0, \delta_1, \dots)$, let $\bar{\delta}_i$ denote the numbers defined in Theorem 4.1. It is clear that $d_i = \gcd(\bar{\delta}_0, \dots, \bar{\delta}_i) = \gcd(\delta_0, \dots, \delta_i)$. We keep also this notation in the sequel.

LEMMA 4.2 *The conductor of* $S = \langle \bar{\delta}_0, \dots, \bar{\delta}_k \rangle$ *is*

$$(\frac{d_0}{d_1} - 1)(\bar{\delta}_1 - d_1) + (\frac{d_1}{d_2} - 1)(\bar{\delta}_2 - d_2) + \dots + (\frac{d_{k-1}}{d_k} - 1)(\bar{\delta}_k - d_k),$$

and S *is symmetric.*

Proof. Since $\gcd(\bar{\delta}_0, \bar{\delta}_1) = d_1$, we have that $i\bar{\delta}_1, 0 \leq i \leq \frac{d_0}{d_1} - 1$, are all different (mod $\bar{\delta}_0$). They are also all smaller than $\bar{\delta}_2$, since $\bar{\delta}_2 > \frac{d_0}{d_1}\bar{\delta}_1$. In the same way all $i\bar{\delta}_1 + j\bar{\delta}_2, 0 \leq i \leq \frac{d_0}{d_1} - 1, 0 \leq j \leq \frac{d_1}{d_2} - 1$ are all different (mod $\bar{\delta}_0$), and they are all smaller than $\bar{\delta}_3$, since $\bar{\delta}_3 > \frac{d_1}{d_2}\bar{\delta}_2 > (\frac{d_1}{d_2} - 1)\bar{\delta}_2 + (\frac{d_0}{d_1} - 1)\bar{\delta}_1$ a.s.o. In this way we see that the Apery set of S with respect to $\bar{\delta}_0$ is $\{j_1\bar{\delta}_1 + j_2\bar{\delta}_2 + \dots + j_k\bar{\delta}_k;\ 0 \leq j_i < \frac{d_{i-1}}{d_i}, i = 1, \dots, k\}$ and $i_1\bar{\delta}_1 + i_2\bar{\delta}_2 + \dots + i_k\bar{\delta}_k > j_1\bar{\delta}_1 + j_2\bar{\delta}_2 + \dots + j_k\bar{\delta}_k$ if and only if $i_k = j_k, \dots, i_s = j_s, i_{s-1} > j_{s-1}$ for some s, i.e., if the last nonzero coordinate of $(i_1 - j_1, \ldots, i_k - j_k)$ is positive. (We have found $\frac{d_0}{d_1}\frac{d_1}{d_2}\cdots\frac{d_{k-1}}{d_k} = \frac{d_0}{d_k} = d_0 = \bar{\delta}_0$ elements which are smallest in their congruence classes $\pmod{\bar{\delta}_0}$.) Hence, the largest number in the Apery set is $\omega_{\bar{\delta}_0-1} = (\frac{d_0}{d_1} - 1)\bar{\delta}_1 + (\frac{d_1}{d_2} - 1)\bar{\delta}_2 + \cdots + (\frac{d_{k-1}}{d_k} - 1)\bar{\delta}_k$. Since the conductor equals $\omega_{\bar{\delta}_0-1} - (\bar{\delta}_0 - 1)$ (cf. Lemma 2.1), we get the first statement after a small calculation. If $\omega_i = i_1\bar{\delta}_1 + \cdots + i_k\bar{\delta}_k$, it is easy to see that $\omega_{\bar{\delta}_0-1-i} = (\frac{d_0}{d_1} - 1 - i_1)\bar{\delta}_1 + \cdots + (\frac{d_{k-1}}{d_k} - 1 - i_k)\bar{\delta}_k$. Thus $\omega_i + \omega_{\bar{\delta}_0-1-i} = \omega_{\bar{\delta}_0-1}$, which gives that S is symmetric (cf. [3]).

For a semigroup S and an integer $d > 0$, we define the d-conductor of S to be $c_d(S) = \min\{nd;\ md \in S \text{ if } m \geq n\}$. Thus $c_1(S)$ is the usual conductor of S.

COROLLARY 4.3 *Let $S = \langle\bar{\delta}_0, \ldots, \bar{\delta}_k\rangle$ and let $d_i = \gcd(\bar{\delta}_0, \ldots, \bar{\delta}_i)$. Then*

$$c_{d_i}(S) = (\frac{d_0}{d_1} - 1)(\bar{\delta}_1 - d_1) + (\frac{d_1}{d_2} - 1)(\bar{\delta}_2 - d_2) + \cdots + (\frac{d_{i-1}}{d_i} - 1)(\bar{\delta}_i - d_i)$$

for every $i \leq k$.

Proof. By the proof of Lemma 4.2, the semigroup $S_i = \langle\frac{\bar{\delta}_0}{d_i}, \ldots, \frac{\bar{\delta}_i}{d_i}\rangle$ has conductor $c(S_i) =$

$$(\frac{d_0/d_i}{d_1/d_i} - 1)(\frac{\bar{\delta}_1}{d_i} - \frac{d_1}{d_i}) + (\frac{d_1/d_i}{d_2/d_i} - 1)(\frac{\bar{\delta}_2}{d_i} - \frac{d_2}{d_i}) + \cdots + (\frac{d_{i-1}/d_i}{d_i/d_i} - 1)(\frac{\bar{\delta}_i}{d_i} - \frac{d_i}{d_i}).$$

Then $c_{d_i}(\langle\bar{\delta}_0, \ldots, \bar{\delta}_k\rangle) = d_i c(S_i)$. A calculation gives that $\bar{\delta}_{i+1} > c_{d_i}(\langle\bar{\delta}_0, \ldots, \bar{\delta}_k\rangle)$, hence $\bar{\delta}_j > c_{d_i}(\langle\bar{\delta}_0, \ldots, \bar{\delta}_k\rangle)$ if $j > i$. Thus $c_{d_i}(S) = c_{d_i}(\langle\bar{\delta}_0, \ldots, \bar{\delta}_k\rangle)$.

LEMMA 4.4 *For $i = 2, \ldots, k$ we have $\bar{\delta}_i = \frac{1}{d_{i-1}}\sum_{j=1}^{i-1}(d_{j-1} - d_j)\delta_j + \delta_i$. Thus the conductor of $S = \langle\bar{\delta}_0, \ldots, \bar{\delta}_k\rangle$ is $\sum_{i=1}^{k}(d_{i-1} - d_i)\delta_i + (1 - d_0)$. Furthermore $c_{d_i}(S) = \frac{1}{d_i}\sum_{j=1}^{i}\delta_j(d_{j-1} - d_j) + d_i - d_0$.*

Proof. By a calculation, replacing in Lemma 4.2 and in Corollary 4.3 $\bar{\delta}_i$ with $\frac{1}{d_{i-1}}\sum_{j=1}^{i-1}(d_{j-1} - d_j)\delta_j + \delta_i$, we get the claim.

For the next proposition, we need a technical lemma. Let $g(t) = \sum_{i\geq 0} a_i t^i$, $a_0 \neq 0$ be a power series such that $\gcd(\{i;\ a_i \neq 0\}) = 1$. Let, for $i = 1, \ldots, k-1$, $\mathbf{d}_i = (d_i, \ldots, d_{k-1})$, and let $\mathbf{d}_i(g(t)) = (\epsilon_i(g), \ldots, \epsilon_{k-1}(g))$, where $\epsilon_s(g) = \min\{j;\ a_j \neq 0,\ d_s \text{ does not divide } j\}$. The easy proof of the next lemma is left to the reader.

LEMMA 4.5 *Let $g(t) = \sum_{i>0} a_i t^i, a_0 \neq 0$, $h(t) = \sum_{i>0} b_i t^i, b_0 \neq 0$, be power series such that $\gcd(\{i;\ a_i \neq 0\} = \gcd(\{i;\ b_i \neq 0\} = 1$. Then*
(a) $\mathbf{d}_i(gh) \geq \min(\mathbf{d}_i(g), \mathbf{d}_i(h))$ (coefficientwise).
(b) If $g = h$ there is equality in (a).
(c) If $\mathbf{d}_i(g(t)) = (\epsilon_1, \ldots, \epsilon_{k-1})$, then $\mathbf{d}_{i+1}((\sum_{i\geq\epsilon_1} a_i t^i)/t^{\epsilon_1}) = (\epsilon_2(g) - \epsilon_1(g), \ldots, \epsilon_s(g) - \epsilon_1(g))$.

We will call a power series *monic* if its least nonzero coefficient is 1.

PROPOSITION 4.6 *Let* $\mathcal{O} = \mathbb{C}[[t^{\delta_0} + \sum_{i \geq N} a_i t^i, \sum_{i > \delta_0} b_i t^i]]$ *be a branch of standard parametrization and with characteristic exponents* $(\delta_0, \ldots, \delta_k)$. *Let* $\bar{\delta}_i$ *be defined as in Theorem 4.1. Then we have* $\langle \bar{\delta}_0, \ldots, \bar{\delta}_k \rangle \subseteq v(\mathcal{O})$, *i.e.* $\bar{\delta}_i \in v(\mathcal{O})$ *for* $i = 0, \ldots, k$.

Proof. Let, for $i = 1, \ldots, k-1$, $\mathbf{d}_i = (d_i, \ldots, d_{k-1})$, where $d_i = \gcd(\delta_0, \ldots, \delta_i)$ as above. We will, by induction, construct monic elements $f_i \in \mathcal{O}$ such that $v(f_i) = \bar{\delta}_i$ and such that $\mathbf{d}_i(f_i/t^{\bar{\delta}_i}) = (\delta_{i+1} - \delta_i, \delta_{i+2} - \delta_i, \ldots, \delta_k - \delta_i)$ if $1 \leq i < k$. We let $f_0 = t^{\delta_0} + \sum_{i \geq N} a_i t^i$. If $v(\sum_{i > \delta_0} b_i t^i)$ is not a multiple of δ_0, then $o(\sum_{i > \delta_0} b_i t^i) = \delta_1$ and we let $f_1 = b_{\delta_1}^{-1} \sum_{i > \delta_0} b_i t^i$. If $v(\sum_{i > \delta_0} b_i t^i) = m_0 \delta_0$, let $f_1' = \sum_{i > \delta_0} b_i t^i - c f_0^{m_0}$, where $c \neq 0$ is chosen so that $v(f_1') > o(\sum_{i > \delta_0} b_i t^i)$. Repeat this until $v(f_1^{(n)}) = \delta_1$, and let $f_1 = c' f_1^{(n)}$, where c' is chosen so that f_1 is monic. It is clear that $\mathbf{d}_1(f_1/t^{\bar{\delta}_1}) = (\delta_2 - \delta_1, \delta_3 - \delta_1, \ldots, \delta_k - \delta_1)$. Suppose we have constructed $f_0, f_1, \ldots, f_i \in \mathcal{O}$ so that the conditions in the proposition are fulfilled. Then $f_i^{d_{i-1}/d_i}$ has value $\gamma_i = \bar{\delta}_i \frac{d_{i-1}}{d_i}$, which is a multiple of d_{i-1}. A simple calculation, using Lemma 4.4, shows that $\gamma_i - c_{d_i}(\langle \bar{\delta}_0, \ldots, \bar{\delta}_i \rangle) = \delta_i - d_i + d_0 > 0$. Thus of course $\gamma_i > c_{d_{i-1}}(\langle \bar{\delta}_0, \ldots, \bar{\delta}_{i-1} \rangle)$. This last means that $\gamma_i = \sum_{j=0}^{i-1} n_j \bar{\delta}_j$ for some $n_j \geq 0$. We choose $f_{i+1}' = f_i^{d_{i-1}/d_i} - f_0^{n_0} \cdots f_{i-1}^{n_{i-1}}$. From Lemma 4.5(b) it follows that $\mathbf{d}_i(f_i^{d_{i-1}/d_i} / t^{\bar{\delta}_i^{d_{i-1}/d_i}}) = \mathbf{d}_i(f_i/t^{\bar{\delta}_i})$. Since, for $j < i$, $\mathbf{d}_j(f_j/t^{\bar{\delta}_j}) = (\delta_{j+1} - \delta_j, \ldots, \delta_k - \delta_j)$, we have $\mathbf{d}_j(f_j/t^{\bar{\delta}_j}) = (\delta_{i+1} - \delta_j, \ldots, \delta_k - \delta_j) > (\delta_{i+1} - \delta_i, \ldots, \delta_k - \delta_i)$ (coefficientwise). Lemma 4.5(a) and (b) shows that $\mathbf{d}_i(f_0^{n_0} \cdots f_{i-1}^{n_{i-1}} / t^{\bar{\delta}_i}) > (\delta_{i+1} - \delta_i, \ldots, \delta_k - \delta_i)$. Thus the smallest power in f_{i+1}' which is not a multiple of d_i and has nonzero coefficient is $\bar{\delta}_{i+1}$. If $v(f_{i+1}')$ is not a multiple of d_i, we choose $f_{i+1} = c f_{i+1}'$ (c chosen so that f_{i+1} is monic). If $v(f_{i+1}')$ is a multiple of d_i, then $\gamma_i > c_{d_i}(\bar{\delta}_0, \ldots, \bar{\delta}_i \rangle)$ shows that $v(f_{i+1}' - f_0^{m_0} \cdots f_i^{m_i}) = v(f_{i+1}'') > v(f_{i+1}')$ for some $m_0, \ldots, m_i \geq 0$. We repeat until $v(f_{i+1}^{(n)}) = \bar{\delta}_{i+1}$, and let $f_{i+1} = c' f_{i+1}^{(n)}$, where c' is chosen so that f_{i+1} is monic. It follows from Lemma 4.5(c) that $\mathbf{d}_{i+1}(f_{i+1}/t^{\bar{\delta}_{i+1}}) = (\delta_{i+2} - \delta_{i+1}, \ldots, \delta_k - \delta_{i+1})$.

LEMMA 4.7 *Let* $\mathcal{O}$ *be a branch with characteristic exponents* $(\delta_0, \ldots, \delta_k)$. *Then the semigroup* $v(\mathcal{O})$ *has conductor* $\sum_{i=1}^{k} (d_{i-1} - d_i)\delta_i + (1 - d_0)$.

Proof. We make induction over the number l of blowups we need to get a regular branch. If $l = 1$, then $\mathcal{O} = \mathbb{C}[[t^{\delta_0}, t^{\delta_0+1} + \cdots]]$. It follows from Proposition 2.3 that $v(\mathcal{O}) = \langle \delta_0, \delta_0 + 1 \rangle$, which has conductor $(\delta_0 - 1)\delta_0 = (\delta_0 - 1)(\delta_0 + 1) + 1 - \delta_0 = (d_0 - d_1)\delta_1 + 1 - d_0$. Suppose the claim is proved for $l - 1$. Let c and c' denote the conductors of $v(\mathcal{O})$ and $v(\mathcal{O}')$, respectively. In case a) of Lemma 3.1, a calculation using Lemma 4.4 gives $c - c' = \sum_{i=1}^{k} (d_{i-1} - d_i)\delta_{i-1} = \delta_0^2 - \delta_0$. By induction the statement is true for $v(\mathcal{O}')$. Proposition 2.3 shows it is true for $v(\mathcal{O})$. A similar calculation in case b) of Lemma 3.1 shows that $c - c' = \delta_0^2 - \delta_0$ also in this case. In case c) of Lemma 3.1 finally, we get, by using $\delta_1 - \delta_0 = d_1, \delta_0 = k d_1, \delta_1 = (k+1) d_1$ for some k, that $c - c' = k^2 d_1^2 + k d_1 = \delta_0^2 - \delta_0$ also in case c).

Proof of Theorem 4.1. We know that $\langle\bar{\delta}_0, \dots, \bar{\delta}_k\rangle \subseteq v(\mathcal{O})$ and that by Lemmas 4.4 and 4.7 these two semigroups have the same conductor. Since $\langle\bar{\delta}_0, \dots, \bar{\delta}_k\rangle$ is symmetric, all strictly larger semigroups have smaller conductor. This gives that the two semigroups are in fact the same.

We get an easy criterion for a semigroup $\langle a_0, \dots, a_k\rangle$ to be a semigroup for a plane branch. The following seems to be a simpler characterization of the semigroup of a plane branch, with respect to equivalent characterizations found in [2] or [10].

PROPOSITION 4.8 *Let S be a semigroup which is minimally generated by $a_0 < a_1 < \cdots < a_k$ and let $d_i = \gcd(a_0, \dots, a_i), i = 0, \dots, k$. Then S is the semigroup of a plane branch if and only if the following conditions are satisfied.*
(a) $d_0 > d_1 > \cdots > d_k = 1$.
(b) $a_i > \operatorname{lcm}(d_{i-2}, a_{i-1})$ for $i = 2, \dots, k$.

Proof. The necessity follows from Theorem 4.1, the sufficiency from the branch $\mathbb{C}[[t^{a_0}, t^{a_1} + t^{a_1+a_2-\operatorname{lcm}(d_0,a_1)} + \cdots + t^{a_1+\cdots+a_k-(\operatorname{lcm}(d_0,a_1)+\cdots+\operatorname{lcm}(d_{k-2},a_{k-1}))}]]$.

We give two concrete examples.

Example Let $S = \langle 30, 42, 280, 855\rangle$. Then S satisfies the conditions in Proposition 4.8, so $S = v(\mathcal{O})$ for some $\mathcal{O}$. We can choose e.g. $\mathcal{O} = [[t^{30}, t^{42} + t^{112} + t^{127}]]$. The conductor equals $t^{1554}\mathbb{C}[[t]]$. With the notation of the previous section, the multiplicity sequence is $M(30, 42), M(6, 70), M(2, 15), \dots$, which is $30, 12^{(2)}, 6^{(13)}, 4, 2^{(9)}, 1^{(2)}, \dots$.

Example Let $\mathcal{O} = \mathbb{C}[[x, y]]$ with

$$x = t^{2\cdot 2^n},\ y = t^{3\cdot 2^n} + t^{3\cdot 2^n + 2^{n-1}} + \cdots + t^{3\cdot 2^n + 2^{n-1} + \cdots + 2 + 1}.$$

The generators of $v(\mathcal{O})$ are $\bar{\delta}_0 = 2^{n+1}, \bar{\delta}_i = 2^{n-i+1}(3 \cdot 2^{2i-2} + (4^{i-1} - 1)/3)$ for $i = 1, \dots, n+1$.

5 COMPLETE INTERSECTION RINGS ARISING FROM THE SEMIGROUP OF A PLANE BRANCH

Let $S = \langle\bar{\delta}_0, \dots, \bar{\delta}_k\rangle = v(\mathcal{O})$ be the semigroup of a plane branch, where $\bar{\delta}_0 < \bar{\delta}_1 < \dots < \bar{\delta}_k$ is a minimal set of generators of S, and let $\mathbb{C}[S] = \mathbb{C}[t^{\bar{\delta}_0}, \dots, t^{\bar{\delta}_k}] = \mathbb{C}[Y_0, \dots, Y_k]/I = T$. We will show that T has an associated graded ring (in the $(Y_0, \dots, Y_k)$-filtration), which is a complete intersection. In particular this implies that T is a complete intersection [11]. We will use [12, Theorem 1] which states that if all elements in $\operatorname{Ap}(S, \bar{\delta}_0)$, the Apery set of S with respect to $\bar{\delta}_0$, have unique expressions as linear combinations of the generators of S, then the relations are determined by the minimal elements above the Apery set. In the following results, we suppose $S = v(\mathcal{O})$, where $\mathcal{O}$ is a plane branch. We also keep the notation of the previous sections.

LEMMA 5.1 *All elements in* $\mathrm{Ap}(S, \bar{\delta}_0)$ *have unique expressions.*

Proof. The elements in $\mathrm{Ap}(S, \bar{\delta}_0)$ are of the form $i_1\bar{\delta}_1 + \cdots + i_k\bar{\delta}_k$, with $0 \leq i_j < d_{j-1}/d_j$ (cf. proof of Lemma 4.2). Suppose $i_1\bar{\delta}_1 + \cdots + i_k\bar{\delta}_k = j_0\bar{\delta}_0 + \cdots + j_k\bar{\delta}_k$. Then $i_k\bar{\delta}_k \equiv j_k\bar{\delta}_k \pmod{d_{k-1}}$. Since $i_1\bar{\delta}_1 + \cdots + i_{k-1}\bar{\delta}_{k-1} < \bar{\delta}_k$, this implies that $i_k = j_k$. If $k > 1$ we get $i_{k-1}\bar{\delta}_{k-1} = j_{k-1}\bar{\delta}_{k-1} \pmod{d_{k-2}}$, which gives $i_{k-1} = j_{k-1}$ a.s.o. Finally $0 = j_0\bar{\delta}_0$, so $j_0 = 0$.

Next we determine the "minimals" (cf. [12]), i.e. the minimal elements $(n_1, \cdots, n_k) \in \mathbb{N}^k$ such that $n_1\bar{\delta}_1 + \cdots + n_k\bar{\delta}_k \notin \mathrm{Ap}(S, \bar{\delta}_0)$ (the order in $\mathbb{N}^k$ is the usual one). Some n_j must be at least d_{j-1}/d_j, otherwise the element belongs to $\mathrm{Ap}(S, \bar{\delta}_0)$. On the other hand at most one $n_j \geq d_{j-1}/d_j$ and there must be equality, if the element is minimal outside $\mathrm{Ap}(S, \bar{\delta}_0)$. Thus the minimals are

$$\{(d_0/d_1, 0, \cdots, O), (0, d_1/d_2, 0, \cdots, 0), \cdots, (0, \cdots, 0, d_{k-1}/d_k)\}.$$

Thus the following theorem follows from [12, Theorem 1].

THEOREM 5.2 *A minimal presentation for* $\mathbb{C}[S]$ *is*

$$\mathbb{C}[S] = \mathbb{C}[Y_0, \ldots, Y_k]/(Y_1^{d_0/d_1} - m_1, \ldots, Y_k^{d_{k-1}/d_k} - m_k)$$

where m_j *is a monomial in* $Y_0, \ldots, Y_j$ *for* $j = 1, \ldots, k$. *Thus* $\mathbb{C}[S]$ *is a complete intersection.*

COROLLARY 5.3 *The associated graded ring of* $\mathbb{C}[S]$ *with respect to the filtration given by powers of* $(Y_0, \ldots, Y_k)$ *is* $\mathbb{C}[Y_0, \ldots, Y_k]/(Y_1^{d_0/d_1}, \ldots, Y_k^{d_{k-1}/d_k})$. *Thus it is a complete intersection.*

Proof. Since $m_j = Y_0^{n_0} \cdots Y_{j-1}^{n_{j-1}}$ and $n_0\bar{\delta}_0 + \cdots + n_{j-1}\bar{\delta}_{j-1} = (d_{j-1}/d_j)\bar{\delta}_j$, it is clear that $n_0 + \cdots + n_{j-1} > (d_{j-1}/d_j)$, so $\mathrm{in}(Y_j^{d_{j-1}/d_j} - m_j) = Y_j^{d_{j-1}/d_j}$. Since $Y_1^{d_0/d_1}, \ldots, Y_k^{d_{k-1}/d_k}$ is a regular sequence, we get the result, cf. [11].

Remark. Notice that not only for semigroups of plane branches the two results above hold. For example, if $S = \langle 4, 6, 7\rangle$, then S is not the semigroup of a plane branch, but $\mathbb{C}[S] = \mathbb{C}[X, Y, Z]/(Y^2 - X^3, Z^2 - X^2Y)$ is a complete intersection and also its associated graded ring is a complete intersection.

COROLLARY 5.4 *The generating function for* S, *i.e.* $\sum_{i \in S} t^i$, *equals*

$$(1 - t^{(d_0/d_1)\bar{\delta}_1}) \cdots (1 - t^{(d_{k-1}/d_k)\bar{\delta}_k})/((1 - t^{\bar{\delta}_0}) \cdots (1 - t^{\bar{\delta}_k})).$$

Proof. As graded algebra $\mathbb{C}[S]$ is generated by $k + 1$ elements of degrees $\bar{\delta}_i$, $i = 0, \ldots, k$ and has k minimal relations of degrees $(d_{i-1}/d_i)\bar{\delta}_i$, $i = 1, \ldots, k$, which constitute a regular sequence.

Examples. If $\mathcal{O} = \mathbb{C}[[t^8, t^{12} + t^{14} + t^{15}]]$, then $v(\mathcal{O}) = \langle 8, 12, 26, 53\rangle$ so the generating function is $(1-t^{24})(1-t^{52})(1-t^{106})/((1-t^8)(1-t^{12})(1-t^{26})(1-t^{53}))$.

If $\mathcal{O} = \mathbb{C}[[t^{30}, t^{42}+t^{112}+t^{127}]]$, then $v(\mathcal{O}) = \langle 30, 42, 280, 855\rangle$ so the generating function is $(1-t^{210})(1-t^{840})(1-t^{1710})/((1-t^{30})(1-t^{42})(1-t^{280})(1-t^{855}))$.

References

[1] Samuel, P. Algébricité de certains points singuliers algébroïd. J. Math. Pures Appl. **1956**, *35*, 1–6.

[2] Zariski, O. Le problème des modules pour les branches planes. Hermann, Paris, 1986.

[3] Apéry, R. Sur les branches superlinéaires des courbes algébriques. C. R. Acad. Sci. Paris **1946**, *222*, 1198–2000.

[4] Hironaka, H. On the arithmetic and effective genera of algebraic curves. Mem. Coll. Sci. Univ. Kyoto **1957**, *30*, 177–195.

[5] du Val, P. The Jacobian algorithm and the multiplicity sequence of an algebraic branch. Revue Fac. Sci. Univ. Istanbul, Ser. A, **1942**, *tome VII, fasc. 3-4*, 107–112.

[6] Campillo, A. Algebroid curves in positive characteristic. Lect. Notes in Math. **1980**, *813*, Springer.

[7] Abhyankar, S. Desingularization of plane curves. Proc. Sympos. Pure Math. **1983**, *40*, 1–45.

[8] Bertin, J.; Carbonne, P. Semi-groupes d'entiers et application aux branches. J. Algebra **1977**, *49*, 81–95.

[9] Merle, M. Invariants polaires des courbes planes. Invent. Math. **1977**, *41*, 103–111.

[10] Bresinsky, H. Semigroups corresponding to algebroid branches in the plane. Proc. Am. Math. Soc. **1972**, *32*, 381–384.

[11] Valabrega, P.; Valla, G. Form rings and regular sequences. Nagoya Math. J. **1978**, *72*, 93–101.

[12] Rosales, J.-C. Numerical semigroups with Apéry sets of unique expression. J. Algebra **2000**, *226*, 479–487.

On Radical Operations

ALI BENHISSI: Depart. Math., Faculty of Sciences, 5000 Monastir, Tunisia

INTRODUCTION. In the first half of the paper, we recall some facts about radical operations, especially the acc property for $*-$ideals and its transfer to the polynomials and power series rings. Many examples are given illustrating that radical operations behave in different fashions. Now, it is a standard theorem of ordinary commutative ring theory that any radical ideal is an intersection of prime ideals. We extend this fact to $*-$ideals in a ring equiped with specific radical operation $*$. Even though finite decomposition is guaranteed by acc for $*-$ideals, we don't know if this condition characterize the finitude of the decomposition. After we look at the following situation: Let $A \subset B$ be an extension of rings and $*$ a radical operation on B. Let P a prime ideal in A and I a $*-$ideal in B which contracts to P. We ask two questions: (1) Can I be enlarged to $*-$prime ideal which also contracts to P ? (2) Is I the intersection of $*-$prime ideals contracting to P ? To answer these questions affirmatively requires additional hypothesis.

1. GENERALITIES. Recalling from [3] the following definition.

DEFINITION 1.1: A radical operation on a ring A assigns to each ideal I of A an ideal I^* of A, called the $*-$radical of I, subject of the conditions (i) $I \subseteq I^*$; $I^{**} = I^*$ and (ii) $(I \cap J)^* = I^* \cap J^* = (IJ)^*$. If $I^* = I$, then I is said to be a $*-$ideal.

EXAMPLE 1: For any ideal I of a ring A, let $\mathcal{J}(I)$ be the $\mathcal{J}-$radical of I; i.e., the intersection of all maximal ideals containing I. It is easy to see that $I \longrightarrow \mathcal{J}(I)$ is a radical operation. If $\mathcal{J}(I) = I$, we say that I is a $\mathcal{J}-$radical ideal.

The following lemma is proved in [3].

LEMMA 1.2: The following properties hold for each pair of ideals I and J of A

(iii) $I \subseteq J \Longrightarrow I^* \subseteq J^*$ (iv) $(I+J)^* = (I+J^*)^* = (I^*+J^*)^*$.

(v) $(IJ)^* = (IJ^*)^* = (I^*J^*)^*$ (vi) $I^* = \sqrt{I^*}$.

PROPOSITION 1.3: If the prime spectrum of a ring A is totally ordered by inclusion, then for any radical operation $*$ on A, each proper $*$−ideal of A is prime.

Proof: By (vi), if I is a proper $*$−ideal, then I is radical, so $I = \bigcap P$, where P ranges over all the primes containing I. Since the family (P) is totally ordered, then $\bigcap P$ is prime.

2. CHAIN CONDITION FOR $*$−IDEALS. In this section, we illustrate some results proved in [3] by examples. The first one shows that the acc for $*$−ideals can hold in a non noetherian ring and the second shows that in the same ring, the acc can hold for some radical operation and not for another.

EXAMPLE 2: Let A be a non noetherian ring with a totally ordered finite prime spectrum. Then for any radical operation $*$ on A, the acc for $*$−ideals is satisfied.

EXAMPLE 3: Since every $\mathcal{J}$−ideal is a radical ideal, then the acc for radical ideals implies the acc for $\mathcal{J}$−ideals. The converse is false and any ring with prime spectrum an infinite strictly increasing sequence of type $P_1 \subset P_2 \subset \ldots \subset M$ is a counterexample.

Some radical operations satisfy a further axiom: (vii) for each ideal I of A, $I^* = \cup J^*$, where J ranges over all finitely generated sub-ideals $J \subseteq I$.

An ideal I of A is called $*$−finitely generated if there exists a finite subset S of A such that $I = (S)^*$.

EXAMPLE 4: Let A be a ring and P a cone of A; i.e., a subset of A stable by addition and multiplication and contains the squares of A. For any ideal I of A, put $\sqrt[P]{I} = \{a \in A;\ \exists m \in \mathbb{N},\ \exists p \in P,\ a^{2m} + p \in I\}$. In [3], we showed that $\sqrt[P]{.}$ is a radical operation. If $\sqrt[P]{I} = I$, we say that I is a P−ideal. Any P−finitely generated ideal of A is P−generated by one element. Indeed,

$\sqrt[P]{(a_1,\ldots,a_n)} = \sqrt[P]{(a_1^2+\ldots+a_n^2)}$.

The following lemma is proved in [3].

LEMMA 2.1: If (vii) is satisfied, then A has the acc for $*-$ideals if and only if each $*-$ideal of A is $*-$finitely generated.

As noted in [3], the implication $\Longrightarrow$ is true without (vii), but the reverse implication needs this hypothesis as will be shown by the following example.

EXAMPLE 5: Let A be the ring of continuous real-valued functions on the unit interval. By [7, Example 2.10], A does not have the acc for the $\mathcal{J}$-radical ideals, but every $\mathcal{J}$-radical ideal of A is $\mathcal{J}$-finitely generated. So (vii) is not satisfied.

3. PASSAGE TO POLYNOMIAL AND POWER SERIES RINGS.

Some radical operations satisfy a further axiom: (viii) for any ideal I of A, $(IA[X])^* = I^*A[X]$. This means the radical operations in $A[X]$ and in A should fit together. In [3], we prove the following theorem.

THEOREM 3.1: If (vii) and (viii) are additionally satisfied, then A has the acc for $*-$ideals if and only if the same is true for $A[X]$.

EXAMPLE 6: Let A be a valuation domain with an infinite ascending chain of primes $(0) = P_1 \subset P_2 \subset \ldots$, which cannot be refined. (Take for example a valuation having value group the weak direct sum of countably many copies of the integers ordered lexicographically.) Since the maximal spectrum of A reduces to one point, A has the acc for $\mathcal{J}-$ideals and each $\mathcal{J}-$ideal is $\mathcal{J}-$finitely generated. By [7, Example 2.9], $A[X]$ does not have the acc for $\mathcal{J}-$ideals and it contains $\mathcal{J}-$ideal which is not $\mathcal{J}-$finitely generated.

EXAMPLE 7: It is well known [3] that if A is a ring with the acc for radical ideals, then the same is tue for $A[X]$. By lemma 2.1, if K is any commutative field and S an infinite subset of $K[X_1,\ldots,X_n]$, then there is a finite subset of S with the same zeros. Naturally, this can be seen more directly from the classical Hilbert's basis theorem than from our theorem.

Now, we turn our attention to the transfer of the acc property to formal power series rings and we start by some examples.

EXAMPLE 8: For any non SFT-ring A, $A[[X]]$ does not satisfy the acc for prime ideals. For example, if A is a non discrete valuation domain of rank 1, then *spec*(A) is noetherian, but *spec*$(A[[X]])$ has not this property. This is also the case for the ring $A = \mathbb{Q}[X_i, i \in \mathbb{N}]/(X_i^n, i \in \mathbb{N})$, with $n \geq 2$, whose prime spectrum is reduced to the singleton $(\bar{X}_i, i \in \mathbb{N})$, but *spec*$(A[[X]])$ is not noetherian.

EXAMPLE 9: For any ring A, $P = \Sigma A^2$ is a positive cone. An ideal I of A is $P-$real if and only if it is real; i.e., if $a_1, \ldots, a_n \in A$ are such that $a_1^2 + \ldots + a_n^2 \in I$, then $a_1, \ldots, a_n \in I$. In [8, §5], Ribenboim showed that if A is a valuation domain of some valuation v of value group $\mathbb{R}$ and residue field $\simeq \mathbb{R}$, then A satisfies the acc for real ideals but $A[[X]]$ does not have the acc for real ideals.

EXAMPLE 10: If A satisfies the acc for the $\mathcal{J}-$ideals, then $A[[X]]$ does also. Indeed, let $Max(A) = \{M_\lambda;\ \lambda \in \Lambda\}$, then $Max(A[[X]]) = \{M_\lambda + XA[[X]];\ \lambda \in \Lambda\}$. Let $(I_i)_{i \in \mathbb{N}}$ an increasing sequence of $\mathcal{J}-$ideals of $A[[X]]$. For each i, there is a subset $\Lambda_i \subseteq \Lambda$ such that $I_i = \bigcap_{\lambda \in \Lambda_i} (M_\lambda + XA[[X]])$, so $I_i \cap A = \bigcap_{\lambda \in \Lambda_i} M_\lambda$ is $\mathcal{J}-$ideal of A as an intersection of maximal ideals. There exists $n \in \mathbb{N}$, such that $I_i \cap A = I_n \cap A$, for each $i \geq n$. For $i \geq n$, $\bigcap_{\lambda \in \Lambda_i} M_\lambda = \bigcap_{\lambda \in \Lambda_n} M_\lambda$, so $\bigcap_{\lambda \in \Lambda_i} (M_\lambda + XA[[X]]) = \bigcap_{\lambda \in \Lambda_n} (M_\lambda + XA[[X]])$, then $I_i = I_n$.

However there are some other positive results in the formal power series rings based on the notion of Zariski topology defined on the prime spectrums of rings.

1) Let $A \subset B$ domains not fields. By [4, Theorem 2.3], if *spec*(A) = *spec*(B), then the contraction map *spec*$(B[[X]]) \longrightarrow$ *spec*$(A[[X]])$ is a homeomorphism. So *spec*$(B[[X]])$ is noetherian if and only if *spec*$(A[[X]])$ is noetherian. By [1, Proposition 3.3], A and B are local so are $A[[X]]$ and $B[[X]]$ and the acc for $\mathcal{J}-$ideals is trivial. In particular, if A is a PVD with associated valuation domain V, then *spec*(A) = *spec*(V), so *spec*$(A[[X]])$ is noetherian if and only if *spec*$(V[[X]])$ is noetherian. By [1, Proposition 3.5], if $A \subseteq B$ is an extension of rings such that *spec*(A) = *spec*(B), the Zariski topologies on *spec*(A) and *spec*(B) coincide. The hypothesis *spec*(A) = *spec*(B) can't be weakned in the remark; since there exists an extension of domains $A \subset B$ such that the contraction map *spec*$(B) \longrightarrow$ *spec*(A) is a homeomorphism, $A[[X]]$ is noetherian, but *spec*$(B[[X]])$

is not noetherian. Resume Example 2.2 of [4], let $A \subset B$ an extension of valuation domains such that B dominates A, the value group of A is $\mathbb{Z}$ and the value group of B is any noncyclic subgroup of $\mathbb{R}$ containing $\mathbb{Z}$, for example $\mathbb{R}$, itself; A is a discrete valuation domain, so $A[[X]]$ is noetherian, but B is not an SFT-ring, so $spec(B[[X]])$ is not noetherian. Note that each of the maximal spectrums of $A[[X]]$ and $B[[X]]$ reduces to one point, so these rings satisfy the acc for the $\mathcal{J}-$ideals.

EXAMPLE 11: The ring $A = \mathbb{Q} + T\mathbb{C}[T]_{(T)}$ is a non-noetherian PVD of dimension 1 associated to the discrete valuation domain $\mathbb{C}[T]_{(T)}$, so $spec(A[[X]])$ is noetherian.

2) Recalling that a globalized pseudo-valuation domain (GPVD) is a subring A of a Prüfer domain B such that the extension $A \subseteq B$ be unibranched (i.e., the contraction map $spec(B) \longrightarrow spec(A)$ is a homeomorphism) and there exists a nonzero radical ideal I common to A and B such that any prime ideal of B (resp. A) containing I is maximal in A (resp. B). By [6, Theorem 2.4], if A is a GPVD with the SFT-property, the contraction map $\psi : spec(B[[X]]) \longrightarrow spec(A[[X]])$ is a homeomorphism, so $spec(B[[X]])$ is noetherian if and only if $spec(A[[X]])$ is noetherian. Since ψ is an order-theoretic isomorphism betwen the two sets, then ψ induces a homeomorphism $Max(B[[X]]) \longrightarrow Max(A[[X]])$. On the other hand, by [7], a ring satisfies the acc for the $\mathcal{J}-$ideals if and only if its maximal spectrum is noetherian; i.e., satisfies the dcc for closed sets. So in the preceding situation $A[[X]]$ satisfies the acc for the $\mathcal{J}-$ideals if and only if $B[[X]]$ satisfies the acc for the $\mathcal{J}-$ideals.

EXAMPLE 12: Resume Example 3.3 of [5]. Let $m \geq 2$ be a positive integer. Consider a field k with the following two properties

(1) There exist m pairwise incomparable one-discrete valuation domains $V_i = k + M_i$ (with maximal ideal M_i) and a common quotient field.

(2) There exist m distinct proper subfields k_i of k such that for at least one i, $[k : k_i] = \infty$, $1 \leq i \leq m$.

Then $R = \bigcap_{i:1}^{m}(k_i + M_i)$ is a GPVD, with the associated Prüfer domain $T = \bigcap_{i:1}^{m} V_i$. Since T is a Dedekind domain (so noetherian), then T and R have the SFT-property, so their spectrums are noetherian. Note that R is not PVD, non-noetherian not Prüfer. Since $T[[X]]$ is noetherian, then $spec(R[[X]])$ is noetherian.

3) Let T be a local domain, with residue field K, $\phi : T \longrightarrow K$ the natural surjection, k a subfield of K and $R = \phi^{-1}(k)$. By [2, Theorem 3.4], if the contraction map $spec(K[[X_1, \ldots, X_n]]) \longrightarrow spec(k[[X_1, \ldots, X_n]])$ is a homeomorphism, for each $n \geq 1$, then the same is true for the contraction map $spec(T[[X_1, \ldots, X_n]]) \longrightarrow spec(R[[X_1, \ldots, X_n]])$. In this case, $spec(T[[X_1, \ldots, X_n]])$ is noetherian if and only if $spec(R[[X_1, \ldots, X_n]])$ is noetherian. The hypothesis in this assertion holds if for example, the extension $k \subset K$ is purely inseparable of finite exponent.

EXAMPLE 13: Resume Example 3.6 of [2]. Let $k \subset K$ be fields of characteristic $p \neq 0$ such that $[K : k] = \infty$ and $K^p \subseteq k$. Let $T = K[[Y_1, \ldots, Y_m]]$ and $R = k + (Y_1, \ldots, Y_m)K[[Y_1, \ldots, Y_m]]$, $m \geq 1$. Then R is non-noetherian, so $R[[X_1, \ldots, X_n]]$ is non-noetherian, but its prime spectrum is noetherian.

Note that the conditions $charact\ k = p \neq 0$ and $K^p \subseteq k$, in order that the contraction map $spec(K[[X_1, \ldots, X_n]]) \longrightarrow spec(k[[X_1, \ldots, X_n]])$ be a homeomorphism are not very restrictive, since in [2, Theorem 4.6], one shows that if $k \subset K$ are fields such that the contraction map $spec(K[[X_1, \ldots, X_n]]) \longrightarrow spec(k[[X_1, \ldots, X_n]])$ is injective, for some positive integer $n \geq 2$, then K/k is purely inseparable.

4. DECOMPOSITION OF $*$−IDEALS AS INTERSECTION OF $*$−PRIME IDEALS.

We recall the following theorem from [3].

THEOREM 4.1: If A satisfies the acc for $*$−ideals, then any $*$−ideal is the intersection of a finite number of $*$−prime ideals.

The uniqueness of the decomposition in the theorem is a matter of commutative algebra. Call a representation of I as an intersection of prime ideals irredundant if none of the ideals can be omitted.

PROPOSITION 4.2: Let I be an ideal of a commutative ring. Suppose that I can be expressed in two ways as an irredundant intersection of primes $I = P_1 \cap \ldots \cap P_r = Q_1 \cap \ldots \cap Q_s$. Then $r = s$ and, after perhaps renumbering, $P_i = Q_i$.

Proof: We have $P_1 \cap \ldots \cap P_r \subseteq Q_1$, hence one of the P_i's is contained in Q_1. We can suppose $P_1 \subseteq Q_1$. Similarly one of the Q_i's is contained in P_1. By the irredundancy, this must Q_1, and we have $Q_1 = P_1$. Similarly each Q_i gets equated to a unique P_i and vice versa.

The first part of the following corollary is proved in [3] and the second derives from the theorem and the proposition.

COROLLARY 4.3: If A satisfies the acc for $*-$ideals, then any $*-$ideal I of A has a finite number of minimal prime ideals, each of them is $*-$ideal and their intersection is I.

In connection with the corollary but without the acc, we start by some concrete examples and a general result will be proved later.

EXAMPLE 14: Let A be a ring and I a $\mathcal{J}-$ideal of A having a finite number $Q_1, \ldots, Q_s$ of minimal prime ideals. Suppose that Q_1 is not a $\mathcal{J}-$ideal and let $a \in \mathcal{J}(Q_1) \setminus Q_1$. If $s \geq 2$, take $b_i \in Q_i \setminus Q_1$, for $i = 2, \ldots, s$. We will show that $ab_2 \ldots b_s \in \mathcal{J}(I) = I \subseteq Q_1$, which is impossible. Let M be any maximal ideal of A containing I, one of the Q_i' s is contained in M. If $Q_1 \subseteq M$, then $a \in \mathcal{J}(Q_1) \subseteq M$. If $Q_i \subseteq M$, for some $i \geq 2$, then $b_i \in M$. So $ab_1 \ldots b_s \in \bigcap M = \mathcal{J}(I)$.

EXAMPLE 15: Let P be a cone of a ring A, I a $P-$ideal of A and $Q_1, \ldots, Q_s$ primes, with $Q_1 \cap \ldots \cap Q_s = I$. Suppose for example that Q_1 is not $P-$ideal, we can find $a \in A \setminus Q_1$, $m \in \mathbb{N}$ and $p \in P$ such that $a^{2m} + p \in Q_1$. Let $b_i \in Q_i \setminus Q_1$, $i = 2, \ldots, s$ and $b = b_2 \ldots b_s$, then $(ab)^{2m} + pb^{2m} \in Q_1 \cap \ldots \cap Q_s = I$, so $ab \in \sqrt[P]{I} = I \subseteq Q_1$, which is impossible.

PROPOSITION 4.4: Let I be a $*-$ideal of A such that $I = P_1 \cap \ldots \cap P_n$, where the P_i's are prime ideals and $P_i \not\subseteq P_j$, for $i \neq j$. Then the P_i's are $*-$ideals and they are the only minimal elements in the set of $*-$prime ideals of A containing I.

Proof: By (ii), we have $I = I^* = P_1^* \cap \ldots \cap P_n^*$. Fix i, $1 \leq i \leq n$, there is some j such that $P_j \subseteq P_j^* \subseteq P_i$. Since the intersection in irredundent, $i = j$ and $P_i = P_j^*$ is $*-$ideal. For any $*-$prime Q containing $I = P_1 \cap \ldots \cap P_n$, there exits i such that $P_i \subseteq Q$.

As a consequence, if A satifies the acc for $*-$ideals, then any $*-$ideal of A has a finite number of minimal prime ideals, each of them is a $*-$ideal.

COROLLARY 4.5: Suppose that each $*-$ideal of A has a finite number of minimal primes. Then for each ideal I of A, the set of $*-$prime ideals containing I has a finite number of minimal elements.

Proof: The sets of $*-$prime ideals containing I and I^* respectively are the same. By hypothesis, I^* has a finite number of minimal primes say $P_1, \ldots, P_n$. Since I^* is radical by (vi), then $I^* = P_1 \cap \ldots \cap P_n$ and the proposition is applied.

LEMMA 4.6: Let $*$ be a radical operation on a ring A and T a multiplicative set of A.

a) If Q is a $*-$ideal of A maximal in the set of $*-$ideals with respect to the exclusion of T, then Q is prime.

b) If (vii) is satisfied and if the set $\mathcal{F}$ of all the $*$-ideals of A disjoint with T is not empty, then $\mathcal{F}$ is inductive hence it admits a maximal element.

Proof: a) Suppose the contrary that $ab \in Q$, $a \notin Q$, $b \notin Q$. Then $(Q + aA)^*$ and $(Q+bA)^*$ are $*$-ideals properly large than Q, hence they contain elements of T, say t_1 and t_2, we have $t_1t_2 \in (Q+aA)^* \cap (Q+bA)^* = (Q^2+aQ+bQ+abA)^* \subseteq Q^* = Q$, by (ii), a contradiction.

b) Let $(I_\lambda)_{\lambda \in \Lambda}$ be a totally ordered family of elements of $\mathcal{F}$ and $I = \bigcup_{\lambda \in \Lambda} I_\lambda$. We have $I^* = \cup J^*$, where J ranges over all finitely generated sub-ideals $J \subseteq I$. Since (I_λ) is totally ordered, for each J, there is a $\lambda \in \Lambda$ such that $J \subseteq I_\lambda$, so $I^* = I$.

THEOREM 4.7: Let $*$ be a radical operation on a ring A such that (vii) is satisfied, then any proper $*$-ideal I of A is an intersection of $*$-prime ideals.

Proof: Given an element x not in I, we have to produce a $*$-prime ideal containing I but not containing x. Take T to be the powers of x. Since I is radical by (vi), then $T \cap I = \emptyset$. By the lemma, there is a $*$-ideal Q containing I and maximal with respect to the exclusion of T and Q is prime.

THEOREM 4.8: Let $A \subset B$ be an extension of rings and $*$ a radical operation on B satisfying (vii). Let I be a $*-$ideal in B such that $P = I \cap A$ is a prime ideal in A. Then I can be enlarged to a $*-$prime ideal in B which also contracts to P.

Proof: T is taken to be the complement of P in A and the lemma is applied.

THEOREM 4.9: Let $A \subset B$ be an extension of rings and and $*$ a radical operation on B satisfying (vii). Let I a $*-$ideal of B such that $ab \in I$, $a \in A$, $b \in B$ implies that a or b is in I. Then I can be expressed as an intersection of $*-$primes in B each of which also contracts to $P = I \cap A$.

Proof: Let x be an element in B but not in I. We must construct a $*-$prime ideal in B which contains I, contracts to P, and fails to contain x. Take T to be the set of all elements ax^n where a is in A but not in P. Then T is multiplicatively closed, and it follows from our hypothesis that it is disjoint from I. The lemma then provides us with a $*-$prime ideal Q which contains I and is disjoint from T. The element x is not in Q, since it is in T. Finally to see that $Q \cap A = P$, let $a \in Q \cap A$. Then $ax \in Q$, and this is a contradiction unless $a \in P$.

REFERENCES

[1] D.F. Anderson, D.E. Dobbs; Pairs of rings with the same prime ideals; Canad. J. Math. 32 (2) (1980), 362-384.

[2] D.F. Anderson, D.E. Dobbs, M. Fontana, M. Khalis; Catenarity of formal power series rings over a pullback; J. Pure Appl. Algebra 78 (1992), 109-122.

[3] A. Benhissi; Hilbert's basis theorem for $*-$radical ideals; J. Pure Appl. Algebra 161 (2001), 245-253.

[4] D.E. Dobbs; Rings of formal power series with homeomorphic prime spectra, Rend. Circ. Mat. Palermo, Serie II, T. XLI (1992), 55-61.

[5] F. Girolami; Power series rings over globalized pseudo-valuation domains; J. Pure Appl. Algebra 50 (1988), 259-269.

[6] F. Girolami; The catenarian property of power series rings over a globalized pseudo-valuation domain; Rend. Circ. Mat. Palermo, Serie II, T. XXXVIII (1989), 5-12.

[7] J. Ohm, R.L. Pendleton; Rings with noetherian spectrum; Duke Math. J. 35 (1968), 631-639.

[8] P. Ribenboim; The ascending chain condition for real ideals; Manuscripta Math. 57 (1986), 109-124.

A Splitting Property Characterizing Artinian Principal Ideal Rings

ABDELMALEK BOUANANE and RAJA EDDAHABI
Département de Mathématiques et Informatique
Université Abdelmalek Essaâdi
Faculté des Sciences de Tétouan, B.P. 2121
Tétouan, Morocco

Throughout, all rings considered are commutative with unity. Recall that a module is uniserial if its lattice of submodules forms a chain. Let R be a ring and A an R-module. We shall use $Soc(A)$ to denote the socle of A and $U(A)$ to denote the sum of all uniserial submodules of A. If G is an abelian group, $U(G)$ coincides with the torsion part of G.

In his paper [3], Dickson has introduced the notion of splitting ring. Similarly, in the present context we call a ring R a splitting ring if for each R-module A, $U(A)$ is a direct summand of A.

The purpose of this note is to prove that a ring R is a splitting ring if and only if R is an Artinian principal ideal ring.

For the terminology and unproved statements in this paper the reader is referred to [1] or [5].

Before proving the main theorem we establish several preliminary results.

LEMMA 1 Let R be a splitting ring. Then there are no ideals properly between M and M^2 for any maximal ideal M of R.

Proof. First note that any local R-module K with non-zero socle is uniserial, as $Soc(K) \subseteq U(K)$ and $U(K)$ is a direct summand of K. If M is a maximal ideal of R and if we take for K the R-module R/M^2, we get that M/M^2 is trivial or simple.

Let R be a splitting ring .Then, in view of Lemma 1, we have the following well-known results, which will be used repeatedly.

(1) For any maximal ideal M and any positive integer n, the only ideals between M and M^n are powers of M.

(2) An R-module $A \neq 0$ is uniserial if and only if $A \cong R/M^n$ for some maximal ideal M and some integer $n \geq 1$.

The next result investigate the local case.

LEMMA 2 A local ring R is a splitting ring if and only if R is a special primary ring.

Proof. Let M be the maximal ideal of R and suppose that R is a special primary ring. If A is an R-module, then for any non-zero $x \in A$, $Ann(x)$ is a power of M, hence $x \in U(A)$. Thus $A = U(A)$. Therefore R is a splitting ring.

Conversely, suppose that R is a splitting ring. First of all note that for any cyclic R-module A, $U(A)$ is uniserial. Next we show that R is a valuation ring, i.e., the set of ideals of R is totally ordered under inclusion. It suffices to show that any two principal ideals are comparable. Let $x, y \in R$ and let $A = R/(Mx + My)$. Since the submodule $N = (Rx + Ry)/(Mx + My)$ of A is annihilated by M, then it is semi-simple and hence $N \subseteq U(A)$. Since $U(A)$ is uniserial, then N is trivial or simple. If $\overline{x}$ and $\overline{y}$ are the images of x and y in N, respectively, then $R\overline{x}$ and $R\overline{y}$ are comparable. We may assume $R\overline{x} \subseteq R\overline{y}$. Then there are elements $a \in R$ and $b, c \in M$ such that $x - ay = bx + cy$. Since $1 - b$ is a unit, we get $x \in Ry$, i.e., $Rx \subseteq Ry$. Next we distinguish two cases:

Case 1: $M = M^2$. Here any uniserial R-module is simple so that for any R-module A we have $U(A) = Soc(A)$. Let $x \in M$ and consider the module R/Mx. We have $Rx/Mx \subseteq Soc(R/Mx)$, hence Rx/Mx is a direct summand of $Soc(R/Mx)$. Therefore Rx/Mx is a direct summand of R/Mx since $Soc(R/Mx)$ is itself a direct summand of R/Mx. But any cyclic module over a local ring is indecomposable, so Rx/Mx is trivial since $Rx/Mx \neq R/Mx$. Thus $Rx = Mx$ and this clearly implies $x = 0$. Since x is arbitrary in M, then $M = 0$. Thus R is a field.

Case 2: $M \neq M^2$. Let $a \in M - M^2$. Since R is a valuation ring and there are no ideals properly between M and M^2, then $M = (a)$, the ideal generated by a. Here we consider the module $P = \prod_{n=1}^{\infty} R/M^n$. First note that $\bigcap_{k=1}^{\infty} M^k P = 0$. Take $x = (x_n)_n \in P$ defined by $x_n = 0$ if n is odd and $x_n = a^{n/2} + M^n$ if n is even. For any $k \geq 1$ the element $y_k = (0, ...0, x_{2k}, x_{2k+1}, ...)$ is in $M^k P$, the element $z_k = (x_1, ..., x_{2k-1}, 0, ...)$ is in $U(P)$ since it is annihilated by M^{k-1}, and we have $x = y_k + z_k$. Therefore, if $\overline{x}$ is the image of x in $P/U(P)$, then $\overline{x} \in M^k(P/U(P))$ for any $k \geq 1$. Since $U(P)$ is a direct summand of P, then $P/U(P)$ is isomorphic to a submodule of P and hence $\bigcap_{k=1}^{\infty} M^k(P/U(P))$ is trivial. Thus $\overline{x} = 0$ and this means $x \in U(P)$. Consequently, there exists $r \geq 1$ such that $a^r x = 0$; in particular $a^r x_{2r+2} = 0$, so $a^{2r+1} \in (a^{2r+2})$, which clearly gives $a^{2r+1} = 0$. Thus M is a nilpotent principal ideal. Therefore any ideal of R is a power of M, and hence R is a special primary ring. The lemma is completely proved.

Our next goal is to show that the splitting property is inherited by localizations at maximal ideals. We need the following simple lemma.

LEMMA 3 Let R be a ring, let S be a multiplicatively closed subset of R and let A be an R_S-module. Let A_R be the R-module obtained from A by restriction of scalars, relatively to the canonical homomorphism $R \rightarrow R_S$. Then any direct summand of A_R is an R_S-submodule of A, and hence a direct summand of A.

Proof. Let K be a direct summand of A_R, say $A_R = K \oplus L$ for some submodule L of A_R. Let $x \in K$ and let $a/s \in R_S$, where $a \in R$ and $s \in S$. Then $(a/s)x = y + z$ for some $y \in K$, $z \in L$. If we multiply by $s/1$, we obtain $ax = sy + sz$, hence $ax = sy$, since $A_R = K \oplus L$. This in turn gives $(a/s)x = y \in K$. This proves that K is an R_S-submodule of A, since it is already a subgroup of the additive group A. By symmetry, L is also an R_S-submodule of A and hence K is a direct summand of A.

COROLLARY 4 Let R be a splitting ring. Then R_M is a special primary ring for any maximal ideal M of R.

Proof. Let us first recall a well-known fact. Let A be an R -module such that $M^k A = 0$ for some maximal ideal M of R and positive integer k. It is well- known (without any assumption on R) that for any multiplicative subset S of R with $S \cap M = \emptyset$, the localization $A_S = A$, and A is a uniserial R-module if and only if A is a uniserial R_S-module. Using this and Lemma 3 it follows that if R is a splitting ring, then so is R_M for every maximal ideal M of R. Then we apply Lemma 2.

Let R be a ring. Recall from [4] that R is said to be FGS if all cyclic R-modules have a finitely generated socle. R is called TC if whenever A is a submodule of a cyclic module and has a simple essential socle, then A is finitely generated. R is TC if and only if every finitely embedded cyclic R-module is Noetherian [4, Lemma 2.6].

We are now ready to prove our main result.

THEOREM 5 A ring R is a splitting ring if and only if R is an Artinian principal ideal ring.

Proof. If R is an Artinian principal ideal ring, then each proper ideal is a product of maximal ideals. Therefore any cyclic R-module is a direct sum of uniserial submodules, and hence $A = U(A)$ for any R-module A. In fact, it is well- known that any module over an Artinian principal ideal ring is a direct sum of cyclic submodules, so it is a direct sum of uniserial submodules. Conversely, suppose that R is a splitting ring. Let A be a cyclic R -module. Since $U(A)$ is a direct summand of A, then $U(A)$ is cyclic. Therefore, $U(A)$ is a finite sum of uniserial submodules and is hence of finite length. Since $Soc(A) \subseteq U(A)$, then $Soc(A)$ is finitely generated. If, in addition, $Soc(A)$ is essential in A, then $A = U(A)$ and hence A is of finite length. Thus R is FGS and TC and so R is Noetherian by [4, Theorem 2.4]. Let P be any prime ideal of R and let M be a maximal ideal containing P. Since, by Corollary 4, R_M is a special primary ring, then MR_M is the unique prime ideal of R_M. Therefore $PR_M = MR_M$ and so $P = M$. Thus any prime ideal of R is maximal. By [6, p. 203], R is then Artinian. Now the zero ideal is a product of maximal ideals, hence R is an Artinian principal ideal ring, since for each maximal ideal M of R there are no ideals properly between M and M^2.

ACKNOWLEDGMENT The authors would like to thank the referee for his valuable suggestions and comments.

References

[1] F.W. Anderson and K.R. Fuller, Rings and categories of modules, Springer Verlag, New York, 1974.

[2] K. Benabdallah, A. Bouanane and Surjeet Singh, On sums of uniserial modules, Rocky Mountain J. Math. 20 (1) (1990) 15 - 29.

[3] S.E. Dickson, Noetherian splitting rings are Artinian, J. London Math. Soc. 42 (1967) 732 - 736.

[4] R.P. Kurshan, Rings whose cyclic modules have finitely generated socle, J. Algebra 15 (1970) 376 -386.

[5] M.D. Larsen and P.J. McCarthy, Multiplicative theory of ideals, Academic Press, New York and London, 1971.

[6] O. Zariski and P. Samuel, Commutative algebra, Vol. 1, D. Van Nostrand Company, Inc., Princeton, New Jersey, 1958.

Rings of Integer-Valued Polynomials and the bcs-Property

JAMES BREWER and LEE KLINGLER

Department of Mathematical Sciences
Florida Atlantic University
Boca Raton, Florida 33431
E-mail address: brewer@fau.edu

1. INTRODUCTION

For a commutative ring R, recall that a submodule B of a projective module P is called *basic* if $B_{\mathfrak{M}}$ contains a nontrivial summand of $P_{\mathfrak{M}}$, (or, equivalently, if the image of B in $P/\mathfrak{M}P$ is nonzero), for every maximal ideal $\mathfrak{M}$ of R. A commutative ring R is called a *bcs-ring* if and only if for every positive integer n, every finitely generated basic submodule of R^n contains a rank one projective summand of R^n. (This notion was introduced by M. Hautus and E. Sontag in [14] and called "*property* (†)." In [1], J. Brewer, D. Katz, and W. Ullery referred it to as the "*UCS-property.*" It was called the "*bcs-property*" and considered systematically by C. Weibel and W. Vasconcelos in [15].) For over a decade, the present authors have been interested in knowing whether all Prüfer domains are bcs-rings. This question is analogous to the question, "Is every Bézout domain an elementary divisor ring?" One place one might look for a counterexample is in "rings of integer-valued polynomials."

If D is an integral domain with field of fractions K and if E is a subset of K, then

$$\mathrm{Int}(E, D) = \{f \in K[X] : f(E) \subseteq D\}$$

is called the *ring of integer-valued polynomials on the subset E*. The case when $E = D$ will be denoted $\mathrm{Int}(D)$. Under certain conditions, rings of integer-valued polynomials are Prüfer domains. Are such rings bcs-rings?

The present paper is our second concerning this question. In the first paper [3], we showed that, if D is a semi-local principal ideal domain with each residue field finite (a technical condition which implies that $\mathrm{Int}(D)$ is a two-dimensional Prüfer domain), then $\mathrm{Int}(D)$ is a bcs-ring. In this paper, we study two additional situations where the question has an affirmative answer. Specifically, we consider $\mathrm{Int}(E, V)$, where E is a subset of the quotient field of a valuation domain V. Rings of this type were studied in [5] by P.-J. Cahen, J.-L. Chabert, and A. Loper and were shown to be higher dimensional Prüfer domains having many nice properties. We prove that they are bcs-rings. We also study $\mathrm{Int}(E, D)$ when D is a Bézout domain and E is a finite set. Using results of S. Chapman, A. Loper, and W. Smith [6], we show that these rings also are bcs-rings. In both cases, we do this by showing that the rings are almost local-global rings, that is, that each proper homomorphic image is a local-global ring. (See definition below.) Finally, we return to the most fundamental example of all, $\mathrm{Int}(\mathbb{Z})$. We give an example which shows that $\mathrm{Int}(\mathbb{Z})$ has proper homomorphic images which are not local-global and hence, that if one is to show that $\mathrm{Int}(\mathbb{Z})$ is a bcs-ring, a new technique will be required.

2. PRELIMINARY RESULTS

It is often convenient to have a matrix-theoretic interpretation of basic modules and the bcs-property. By the *content* of a vector $v \in R^n$ we mean the ideal of R generated by the entries of v, which we denote by $c(v)$. The content of a set of vectors is then the ideal sum of the contents of the individual vectors. Thus, a submodule B of the free module R^n is basic if and only if the content of the vectors in B is the unit ideal R. Clearly if B is a basic submodule of R^n, then only finitely many vectors from B are required to generate the unit ideal R, and hence there is a finitely generated submodule $B_0 \subseteq B$ such that B_0 is basic. Therefore, to establish the bcs-property for a ring R, one need only consider finitely generated basic R-modules.

If B is a finitely generated submodule of R^n, by a slight abuse of notation, we also denote by B an $n \times m$ matrix whose columns span the module B. Thus, the content of the module B is the same as the content of the matrix B, and the module B is basic if and only if the matrix B has *unit content.* Set in this context, the bcs-property says that, if B is a matrix with unit content, then there exists a matrix A such that the product BA has the following two properties: BA has unit content; and all 2×2 minors of BA are zero. This follows from a lemma of R. Gilmer and R. Heitmann [12].

We shall also need a strong form of the bcs-property. A commutative ring R is said to be a *bcu-ring* if and only if given an $n \times m$ matrix B of unit content, there exists a *vector* $u \in R^m$ such that Bu is unimodular, that is, such that Bu has unit content.

A commutative ring R is said to be a *local-global ring* if each polynomial over R (in several variables) admitting unit values locally, admits unit values. (See [8] for more detail.) Following the terminology of [9], the commutative ring R is said to be *almost local-global* if every proper homomorphic image of R is local-global. One of the principal results of [2] is *Theorem 3: If R is an almost local-global ring, then R is a bcs-ring* . We present next a slight sharpening of this theorem. Although we do not need its full strength in this paper, the proof is instructive, and so we include it.

THEOREM 1 *Let R be a ring such that*

1. *For every (nonzero) finitely generated ideal I of R, the residue class ring R/I is a bcu-ring.*
2. *For every (nonzero) finitely generated ideal I of R, if $M = \langle x, y\rangle$, is a free (R/I)-module of rank one, then there exists an*

element $r \in R$ *such that* $M = \langle x + \bar{r}y \rangle$.

Then R *is a bcs-ring. In particular, if* R *is an almost local-global ring, then* R *is a bcs-ring.*

Proof. Let B be a basic submodule of R^n, with x a non-zero element of B. If x is unimodular, then $\langle x \rangle$ is already a rank one free summand of B, and we are finished.

So, suppose that x is not unimodular, and consider the ring $\tilde{R} = R/c(x)$. By hypothesis (1), the image $\tilde{B}$ in $\tilde{R}^n$ contains a unimodular vector, say $\tilde{y}$. If y is any preimage of $\tilde{y}$ in B, then we have that $c(x) + c(y) = R$. Next, consider the (basic) submodule $\langle x, y \rangle$ of B, and let I be the ideal generated by the 2×2 minors of the matrix $[x\ y]$. If $I = (0)$, then $\langle x, y \rangle$ is a rank one projective summand of B, and we are finished, while if $I = R$, then x would have be a unimodular vector, contrary to assumption.

Thus, suppose that $(0) \neq I \neq R$, and consider the residue class ring $\bar{R} = R/I$. Let $\bar{B}$ be the image of B in $\bar{R}^n$, and note that $\langle \bar{x}, \bar{y} \rangle \subseteq \bar{B}$ is a rank one projective summand, because $c(\bar{x}) + c(\bar{y}) = R/I$, and the 2×2 minors of $[\bar{x}\ \bar{y}]$ vanish. Since rank one projective modules over a bcu-ring are free, we have that $\langle \bar{x}, \bar{y} \rangle$ is a free module. By hypothesis (2), there exists an element $r \in R$ such that $\langle \bar{x}, \bar{y} \rangle = \langle \bar{x} + \bar{r}\bar{y} \rangle$, and hence $c(x + ry) + I = R$. We claim that $c(x + ry) = R$. Suppose, by way of contradiction, that there exists a maximal ideal $\mathfrak{M}$ of R such that $c(x + ry) \subseteq \mathfrak{M}$. Then over the residue field $R/\mathfrak{M}$, the vectors $\bar{x}$ and $\bar{y}$ are linearly dependent, from which it follows that the 2×2 minors of $[x\ y]$ belong to $\mathfrak{M}$. Thus, $I \subseteq \mathfrak{M}$, which is impossible since $c(x + ry) + I = R$. It follows that $c(x + ry) = R$ as claimed, and therefore $x + ry$ is a unimodular vector belonging to B.

To prove the last statement of the theorem, we need to show that a local-global ring satisfies both hypotheses (1) and (2). First, we note that a local-global ring must be a bcu-ring. For let A be an $n \times m$ matrix having unit content, let $X_1, X_2, \ldots, X_n$ and $Y_1, Y_2, \ldots, Y_m$

be indeterminates, and set

$$g(X_1, X_2, \ldots, X_n, Y_1, Y_2, \ldots, Y_m) = [X_1 \cdots X_n] \cdot A \cdot \begin{bmatrix} Y_1 \\ \vdots \\ Y_m \end{bmatrix}$$

Now, A has a unimodular vector in its image if and only g takes on a unit value. But A has unit content, so g takes on unit values locally. Thus, over a local-global ring, g must take on a unit value, so that A would have to have a unimodular vector in its image.

Similarly, if $M = \langle x, y \rangle \subseteq R^n$, is a free module of rank one, with $x \neq 0$, then let $X_1, X_2, \ldots, X_n$, and Y be indeterminates, and set

$$g(X_1, X_2, \ldots, X_n, Y) = [X_1 \cdots X_n] \cdot [x \; y] \begin{bmatrix} 1 \\ Y \end{bmatrix}$$

Then the vector $x + ry$ is unimodular (for some scalar $r \in R$) if and only if g takes on a unit value. But g does take on unit values locally (because locally, at least one of x or y is unimodular). Therefore over a local-global ring, g must take on a unit value, and hence the vector $x + ry$ is unimodular for some scalar $r \in R$. ∎

Our next result is a generalization of the lemma in [3]. It is essentially nothing more than the Chinese Remainder Theorem, which itself is at the core of why semi-quasi-local rings are local-global. (For simplicity, and by a slight abuse of notation, we write $Max(R/A)$ for the set of maximal ideals $\mathfrak{M}$ of a ring R that contain the ideal A.)

LEMMA 2 *Let R be a ring, and suppose that R has non-zero ideals $J_1, J_2, \ldots, J_n$ such that $Max(R) = Max(R/J_1) \cup Max(R/J_2) \cup \ldots \cup Max(R/J_n)$, where, for $i \neq k$, $Max(R/J_i) \cap Max(R/J_k) = \emptyset$. If R/J_i is a local-global ring for $1 \leq i \leq n$, then R is a local-global ring.*

Proof. Let $f \in R[X_1, \ldots, X_m]$ be a polynomial that represents a unit locally at each maximal ideal of R. We must find an m-tuple $(a_1, a_2, \ldots, a_m) \in R^m$ such that $f(a_1, a_2, \ldots, a_m) \notin \mathfrak{M}$ for all maximal ideals $\mathfrak{M}$ of R. Since R/J_1 is local-global, there exists an m-tuple $(b_1, b_2, \ldots, b_m)$ such that $f(b_1, b_2, \ldots, b_m) \notin \mathfrak{M}$ for all

$\mathfrak{M} \in Max(R/J_1)$. Suppose, inductively, that $f(b_1, b_2, \ldots, b_m) \notin \mathfrak{M}$ for all $\mathfrak{M} \in Max(R/J_1) \cup \ldots \cup Max(R/J_k)$. If $f(b_1, b_2, \ldots, b_m) \notin \mathfrak{M}$ for all $\mathfrak{M} \in Max(R/J_{k+1})$, we can continue the induction. On the other hand, if $f(b_1, b_2, \ldots, b_m) \in \mathfrak{M}$ for some $\mathfrak{M} \in Max(R/J_{k+1})$, we must modify $(b_1, b_2, \ldots, b_m)$. Since $J_1, J_2, \ldots, J_k, J_{k+1}$ are pairwise comaximal,we can find an element $a \in R$ such that $a \equiv 0 \pmod{J_i}$ for $1 \leq i \leq k$ and $a \equiv 1 \pmod{J_{k+1}}$. Note then that $a \equiv 0 \pmod{\mathfrak{M}}$, for all $\mathfrak{M} \in Max(R/J_i)$ for $1 \leq i \leq k$, and $a \equiv 1 \pmod{\mathfrak{M}}$, for all $\mathfrak{M} \in Max(R/J_{k+1})$. Also, since R/J_{k+1} is local-global, there exists an m-tuple $(c_1, c_2, \ldots, c_m) \in R^m$ such that $f(c_1, c_2, \ldots, c_m) \notin \mathfrak{M}$ for all $\mathfrak{M} \in Max(R/J_{k+1})$. Consider the m-tuple

$$v = (b_1 + a(c_1 - b_1), b_2 + a(c_2 - b_2), \ldots, b_m + a(c_m - b_m))$$

Working modulo $\mathfrak{M}$, one easily checks that $f(v) \notin \mathfrak{M}$ for all $\mathfrak{M} \in Max(R/J_1) \cup Max(R/J_2) \cup \ldots Max(R/J_k) \cup Max(R/J_{k+1})$, which completes the induction and the proof. ■

The following well known result is often useful when trying to verify that a ring is a local-global ring.

LEMMA 3 *Let R be a ring, $\mathfrak{M}$ a maximal ideal of R, and f a polynomial in finitely many variables with coefficients from R. Then f takes on a unit value locally at $\mathfrak{M}$ if and only if f takes on a unit value residually in $R/\mathfrak{M}$.*

Proof. The polynomial f represents a unit modulo $\mathfrak{M}$ if and only if there exist elements $a_1, a_2, \ldots, a_n \in R$ such that $f(a_1 + \mathfrak{M}, a_2 + \mathfrak{M}, \ldots, a_n + \mathfrak{M}) \neq 0 \pmod{\mathfrak{M}}$. The latter condition holds if and only if $f(a_1, a_2, \ldots, a_n) \notin \mathfrak{M}$. Said otherwise, f represents a unit in the ring $R/\mathfrak{M}$ if and only if there exists an n-tuple $\bar{\alpha}$ in $(R/\mathfrak{M})^n \cong (R_\mathfrak{M}/\mathfrak{M}R_\mathfrak{M})^n$ such that $f(\bar{\alpha}) \neq 0$. On the other hand, f represents a unit in the quasi-local ring $R_\mathfrak{M}$ if and only if there exists an n-tuple $\bar{\alpha} \in (R_\mathfrak{M}/\mathfrak{M}R_\mathfrak{M})^n$ such that $f(\bar{\alpha}) \neq 0$ if and only if $f(\alpha) \notin \mathfrak{M}R_\mathfrak{M}$ if and only if $f(\alpha)$ is a unit in $R_\mathfrak{M}$. ■

3. MAIN RESULTS

We begin this section by recalling for the reader the essential facts about "higher dimensional integer-valued polynomial Prüfer domains," culled from [5].

NOTATION 4 *Let V be a valuation domain, with maximal ideal $\mathfrak{M}$, and denote by K the quotient field of V. In order to use Cauchy sequences to define the completion $\hat{K}$ of K, with respect to the topology induced by the valuation associated with V, one must assume a little something about V to make the topology metrizable. (One could assume, for example, that V has finite or countable rank, or that V contains a height-one prime.) Then denote by $\hat{V}$ the completion of V under this topology. We assume also that K contains a subset E whose completion $\hat{E}$ is compact. (If in fact E is infinite, then this assumption is enough to force the topology to be metrizable.) It turns out that E is fractional (that is, there is a nonzero element $d \in V$ such that $dE \subseteq V$), and hence we may assume that E is contained in V. We define*

$$R = Int(E, V) = \{f \in K[X] : f(E) \subset V\}$$

the ring of integer-valued polynomials on the subset E of V. This ring has many lovely properties, including:

1. R is a Prüfer domain with the $1\frac{1}{2}$-generator property.

2. R has dimension one more than the dimension of V. Hence, one can construct such rings of arbitrary dimension. (In some sense, that was the point of [5].)

3. The structure of Spec(R) is as follows. Let $\mathfrak{P}$ be a non-zero prime ideal of R. If $\mathfrak{P} \cap V = 0$, then $\mathfrak{P} = (p)^c = pK[X] \cap R$ for some irreducible polynomial p of $K[X]$ (just as for Int($\mathbb{Z}$)). If $\mathfrak{P} \cap V = \mathfrak{p}$, for some non-zero prime ideal $\mathfrak{p}$ of V, then for some element α of the completion $\hat{E}$ of E, the prime ideal $\mathfrak{P} = \mathfrak{P}_{\mathfrak{p},\alpha} = \{f \in R : f(\alpha) \in \hat{\mathfrak{p}}\}$.

4. Therefore, if $\mathfrak{P}$ is a maximal ideal of R, then either $\mathfrak{P} = \mathfrak{P}_{\mathfrak{M},\alpha}$ for some element $\alpha \in \hat{E}$, or $\mathfrak{P} = (p)^c$ for some irreducible polynomial p of $K[X]$. Furthermore, we note the obvious but useful fact that $\mathfrak{M}R \subseteq \mathfrak{P}_{\mathfrak{M},\alpha}$ for any element $\alpha \in \hat{E}$. Moreover, $R/\mathfrak{M}R$ is a zero-dimensional ring.

We are now able to prove that such rings of integer-valued polynomials are bcs-rings.

THEOREM 5 *If R is a ring of integer-valued polynomials as in Notation 4, then R is an almost local-global ring. In particular, R is a bcs-ring.*

Proof. Let I be a non-zero proper ideal of R. The maximal ideals of R are of the form $\mathfrak{P}_{\mathfrak{M},\alpha}$ for elements $\alpha \in \hat{E}$, or of the form $(p)^c$ for irreducible polynomials p of $K[X]$. For each such $\alpha \in \hat{E}$, we have that $\mathfrak{M} \subseteq \mathfrak{P}_{\mathfrak{M},\alpha}$, and there exist only finitely many maximal ideals $\mathfrak{M}_1, \ldots, \mathfrak{M}_n$ of the type $(p)^c$ that contain I. Set $J_1 = (\mathfrak{M}_1 \cap \ldots \cap \mathfrak{M}_n)/I$ and $J_2 = (\mathfrak{M}R + I)/I$. Then $(R/I)/J_1 \cong R/(\mathfrak{M}_1 \cap \ldots \cap \mathfrak{M}_n)$ is semi-quasi-local and hence local-global. Likewise, $(R/I)/J_2 \cong R/(\mathfrak{M}R + I)$ is a homomorphic image of the zero-dimensional ring $R/\mathfrak{M}R$; as zero-dimensional rings are local-global, it follows that R/J_2 is also local-global. Applying Lemma 2 completes the proof of the first assertion. The second assertion follows from Theorem 1 above (or Theorem 3 of [2]). ∎

We can also prove a partial non-noetherian generalization of the main result of [3], which stated that if D is a semi-local principal ideal domain with each residue field finite, then Int(D) is a bcs-ring. We can drop not only the noetherian assumption, but also the one-dimensional assumption, if we change from Int(D) to Int(E, D) for some finite subset $E \subseteq D$.

PROPOSITION 6 *Let D be a semi-quasi-local Bézout domain with maximal ideals $\mathfrak{M}_1, \mathfrak{M}_2, \ldots, \mathfrak{M}_n$ and let $E = \{e_1, e_2, \ldots, e_k\}$ be a finite subset of D. Suppose that Int$(D) \neq D[X]$. Then $R =$ Int(E, D) is both a Prüfer domain and an almost local global ring, and therefore also a bcs-ring.*

Proof. By Theorem 4 of [6], R has the strong two-generator property and hence by Proposition 1 of [6] is a Prüfer domain. If I is a non-zero ideal of R, we show that R/I is a semi-quasi-local ring and hence a local-global ring. A maximal ideal $\mathfrak{P}$ of R has the property that $\mathfrak{P} \cap D = P$ for some non-zero prime ideal P of D, or $\mathfrak{P} = (p)^c$ for some irreducible polynomial p of $K[X]$, where K denotes the quotient field of D. By a result of McQuillan (see [4, page 114, Exercise 2]), the prime ideals of R having non-zero contraction to D are of the form $\mathfrak{P}_{P,e_j} = \{f \in R \mid f(e_j) \in P\}$, where P is a non-zero prime ideal of D. It follows that the maximal ideals of R having non-zero contraction to D are the ideals $\mathfrak{P}_{\mathfrak{M}_i,e_j}$ and hence are finite in number. Moreover, there are only finitely many maximal ideals of the type $(p)^c$ that contain I. It follows that R/I is a semi-quasi-local ring and thus is a local-global ring. Therefore, R is almost local-global. That R is a bcs-ring follows from Theorem 1 above (or Theorem 3 of [2]).

■

Perhaps the most interesting open question in this area is the question, "Is Int($\mathbb{Z}$) a bcs-ring?" We next give an example which demonstrates that, if Int($\mathbb{Z}$) is a bcs-ring, then a new method of proof will be required.

EXAMPLE 7 *Let $f(X) = X^2 + 14$. Then Int($\mathbb{Z}$)/($f(X)$Int($\mathbb{Z}$)) is not a bcu-ring, and, in particular, Int($\mathbb{Z}$)/($f(X)$Int($\mathbb{Z}$)) is not a local-global ring. Thus, Int($\mathbb{Z}$) is not almost local-global.*

Proof. We draw on results and ideas from [1] and [13]. Consider the prime ideal $\mathfrak{P} = f(X)\mathbb{Q}(X) \cap \text{Int}(\mathbb{Z})$ of Int($\mathbb{Z}$). Then

$$\begin{array}{c} \mathbb{Q}(X)/(f(X)\mathbb{Q}(X)) = \mathbb{Q}(\sqrt{-14}) \\ \cup \\ \text{Int}(\mathbb{Z})/\mathfrak{P} \\ \cup \\ \mathbb{Z}[X]/(f(X)\mathbb{Z}[X]) = \mathbb{Z}[\sqrt{-14}] \end{array}$$

and so $D = \text{Int}(\mathbb{Z})/\mathfrak{P}$ is a ring between the ring of algebraic integers $\mathbb{Z}[\sqrt{-14}]$ and its quotient field $\mathbb{Q}(\sqrt{-14})$. Since Int($\mathbb{Z}$) is a Prüfer domain, so is D, and hence D contains the ring Z^* of all algebraic integers in $\mathbb{Q}(\sqrt{-14})$ and must therefore be a Dedekind domain. The mapping $\bar{F} \to \bar{F}\bar{D}$ is a surjection from the class group of Z^* to the class group of D [10, Theorem 40.4]. The kernel of the map is generated by $\{\bar{P} : P \cap \mathbb{Z}$ ramifies with respect to $\mathbb{Z}^*\}$ and is an elementary abelian 2-group. (See [13] and [7].) By [7, pages 261-274], the class group of $\mathbb{Z}^*$ is not an elementary abelian 2-group, and so the kernel of the map is a proper subgroup of the class group of $\mathbb{Z}^*$. Thus, D

is not a principal ideal domain, and hence D is not a bcu-ring [1, Proposition 4].

It follows that D is not a local-global ring, for, as shown in the proof of Theorem 1, a local-global ring is a bcu-ring. This already shows that $\text{Int}(\mathbb{Z})$ has a homomorphic image $D = \text{Int}(\mathbb{Z})/\mathfrak{P}$ which is not a local-global ring, nor even a bcu-ring. However, this ideal $\mathfrak{P}$ need not be finitely generated. On the other hand, the bcu-property is clearly preserved under homomorphic images, so it follows that that $\text{Int}(\mathbb{Z})/(f(X)\text{Int}(\mathbb{Z}))$ (which maps onto D) also cannot be a bcu-ring, nor a local-global ring. ■

REMARK 8 *In the other direction, the usual method of demonstrating that a ring is* not *a bcs-ring is to apply Theorem 2.3 of [15]:* If R is a bcs-ring, then the natural map $Pic(R) \to Pic(R/I)$ must be onto for every ideal I of R. *(For example, $Pic(\mathbb{Z}[X])$ is trivial, but $\mathbb{Z}[X]$ admits proper homomorphic images which have nontrivial Picard groups, so that $\mathbb{Z}[X]$ is not a bcs-ring.) Trying to use this approach to show that Int($\mathbb{Z}$) is not a bcs-ring will be difficult, because of a theorem of R. Gilmer, W. Heinzer, D. Lantz, and W. Smith [11]: Pic(Int($\mathbb{Z}$))* is a free abelian group on a countably infinite basis.

References

1. J. Brewer, D. Katz, and W. Ullery, *Pole assignability in polynomial rings, power series rings, and Prüfer domains*, J. Algebra **106**(1987), 265–286.
2. J. Brewer and L. Klingler, *Pole assignability and the invariant factor theorem in Prüfer domains and Dedekind domains*, J. Algebra **111**(1987), 536–545.
3. J. Brewer and L. Klingler, *The ring of integer-valued polynomials of a semi-local principal ideal domain*, Linear Algebra and its Applications **157**(1991), 141–145.
4. P.-J. Cahen, J.-L. Chabert, *Integer-Valued Polynomials*, Mathematical Surveys and Monographs, Vol. 48, American Mathematical Society, U.S.A., 1997.
5. P.-J. Cahen, J.-L. Chabert, and A. Loper, *High dimension*

Prüfer domains of integer-valued polynomials, J. Korean Math. Soc. **38**(2001), 915–935.

6. S. Chapman, A. Loper, and W. Smith, *The strong two-generator property in rings of integer-valued polynomials determined by finite sets*, Archiv. Math., to appear.
7. H. Cohn, *A Classical Invitation to Algebraic Numbers and Class Fields*, Springer-Verlag, New York, 1978.
8. D. Estes and R. Guralnick, *Module equivalences: local to global when primitive polynomials represent units*, J. Algebra **7**(1982), 138–157.
9. L. Fuchs and L. Salce, *Modules over Non-Noetherian Domains,* Mathematical Surveys and Monographs, Vol. 84, American Mathematical Society, U.S.A., 2001.
10. R. Gilmer, *Multiplicative Ideal Theory*, Marcel Dekker, New York, 1972.
11. R. Gilmer, W. Heinzer, D. Lantz, and W. Smith, *The ring of integer-valued polynomials of a Dedekind domain,* Proc. Amer. Math. Soc. **108**(1990), 673-681.
12. R. Gilmer and R. Heitmann, *On* $Pic(R[X])$ *for* R *seminormal*, J. Pure and Applied Algebra, **16**(1980), 251–257.
13. R. Gilmer and W. Smith, *Integer-valued polynomials and the strong two-generator property*, Houston J. of Mathematics **11**(1985), 65–74.
14. M. L. J. Hautus and E. Sontag, *New results on pole-shifting for parameterized families of systems*, J. Pure and Applied Algebra **40**(1986), 229–244.
15. W. Vasconcelos and C. Weibel, *BCS-rings*, J. Pure and Applied Algebra **52**(1988), 173–185.

Factorial Groups and Pólya Groups in Galoisian Extension of Q

JEAN-LUC CHABERT

DEPARTMENT OF MATHEMATICS, UNIVERSITÉ DE PICARDIE, 80039 AMIENS, FRANCE, LAMFA CNRS-FRE 2270

E-mail address: jlchaber@worldnet.fr

ABSTRACT. The classical factorials in $\mathbb{Z}$ may be generalized by factorial ideals in every integral domain D. When D is completely integrally closed, these factorial ideals generate a subgroup of the divisorial group, the factorial group $\mathcal{F}act(D)$, and, if D is a Krull domain, $\mathcal{F}act(D)$ is a free abelian group. The classes of the factorial ideals generate the Pólya group $\mathcal{P}o(D)$ and we give some properties of this group $\mathcal{P}o(D)$ especially when D is the ring of integers of a finite Galoisian extension of $\mathbb{Q}$.

1. DEFINITIONS

We first recall the definition of the factorial ideals of a domain (see [5]). Let D be an *integral domain* with quotient field K. Consider the ring of integer-valued polynomials on D, that is,

$$\mathrm{Int}(D) = \{f \in K[X] \mid f(D) \subseteq D\}$$

and, for each $n \in \mathbb{N}$, let $\mathfrak{I}_n(D)$ be the fractional ideal of D formed by the leading coefficients of the polynomials in $\mathrm{Int}(D)$ with degree $\leq n$.

Definition 1.1. The *factorial ideal* of index n of the domain D is the entire ideal:

$$(n!)_D = \mathfrak{I}_n^{-1}(D) = \{a \in D \mid a\mathfrak{I}_n(D) \subseteq D\}.$$

Obviously, $(0!)_D = D$ and the sequence $\{(n!)_D\}_{n \in \mathbb{N}}$ is a decreasing sequence of divisorial ideals (that is, of ideals which are intersection of principal fractional ideals).

For instance, $(n!)_{\mathbb{Z}} = n!\mathbb{Z}$. Such factorial ideals were introduced and studied in [1] and [11]. We are going to strengthen step by step the hypotheses on the domain D and finally D will be the ring of integers of a Galoisian extension K of $\mathbb{Q}$.

Assume now that the domain D is *completely integrally closed*. One knows that, if we define the product $I * J$ of two divisorial ideals of D as the smallest divisorial ideal containing the ideal $I \cdot J$, then the set $\mathcal{D}(D)$ of divisorial fractional ideals of D is a group [6, Theorem 34.3], the *divisorial group*.

Recall that the smallest divisorial ideal containing an ideal $\mathfrak{I}$ is the ideal $(\mathfrak{I}^{-1})^{-1}$ where

$$\mathfrak{I}^{-1} = \{x \in K \mid x\mathfrak{I} \subseteq D\}.$$

Definition 1.2. The *factorial group* of a completely integrally closed domain D is the subgroup $\mathcal{F}act(D)$ of the divisorial group $\mathcal{D}(D)$ generated by the factorial ideals of D.

Examples. 1) $\mathcal{F}act(\mathbb{Z}) = \mathcal{D}(\mathbb{Z}) \simeq \mathbb{Q}/\{\pm 1\}$.
2) Let V be a rank-one valuation domain. Then, $\mathcal{F}act(V) \simeq \mathbb{Z}$ if the valuation is discrete with finite residue field and $\mathcal{F}act(V) \simeq \{1\}$ otherwise.

The obvious containment

$$\mathfrak{I}_n(D) \cdot \mathfrak{I}_m(D) \subseteq \mathfrak{I}_{n+m}(D) \quad \text{for } n, m \in \mathbb{N}$$

leads to the following:

$$(n!)_D^{-1} * (m!)_D^{-1} \subseteq ((n+m)!)_D^{-1},$$

and hence,

$$((n+m)!)_D \subseteq (n!)_D * (m!)_D.$$

Note that, in the case where D is a Dedekind domain, the previous containment means that the ideal product $(n!)_D\,(m!)_D$ divides the ideal $((n+m)!)_D$.

Denote by $\mathcal{P}(D)$ the subgroup of $\mathcal{D}(D)$ formed by the nonzero principal fractional ideals of D and consider the class group $\mathcal{D}(D)/\mathcal{P}(D)$.

Definition 1.3. The *Pólya-Ostrowski group* (or shortly, the *Pólya group*) of the domain D is the image $\mathcal{P}o(D)$ of the factorial group $\mathcal{F}act(D)$ in the class group $\mathcal{D}(D)/\mathcal{P}(D)$:

$$\mathcal{P}o(D) = \mathcal{F}act(D)/(\mathcal{F}act(D) \cap \mathcal{P}(D)).$$

This definition generalizes that given in [4, II. §3] for Dedekind domains.

2. The factorial group of a Krull domain

Hypothesis and notation. Assume now that D is a *Krull domain* and denote by $spec^1(D)$ the set formed by the height-one prime ideals of D.

One knows that the divisorial group $\mathcal{D}(D)$ of D is a free abelian group with a basis formed by the elements of $spec^1(D)$. Recall that, in a Krull domain D, for each $n \in \mathbb{N}$ and each $\mathfrak{p} \in Spec(D)$, one has [5, Corollary 5.2]:

$$\mathfrak{I}_n(D)_\mathfrak{p} = \mathfrak{I}_n(D_\mathfrak{p}) \quad \text{and} \quad ((n!)_D)_\mathfrak{p} = (n!)_{D_\mathfrak{p}}$$

and that, for each $\mathfrak{p} \in spec^1(D)$ [4, II.2.9]:

$$\mathfrak{I}_n(D_\mathfrak{p}) = \mathfrak{p}^{-w_{N(\mathfrak{p})}(n)} D_\mathfrak{p}$$

where

$$w_q(n) = \sum_{k>0} \left[\frac{n}{q^k}\right] \quad \text{and} \quad N(\mathfrak{p}) = |D/\mathfrak{p}|.$$

Note moreover that $w_q(n) = 0$ for $n < q$.

Proposition 2.1. *If D is a Krull domain, then the ideals $\mathfrak{J}_n(D)$ are divisorial ideals, more precisely:*

$$\mathfrak{J}_n(D) = \prod_{\mathfrak{p}\in spec^1(D),\, N(\mathfrak{p})\leq n} \mathfrak{p}^{-w_{N(\mathfrak{p})}(n)}.$$

Proof. Let $\mathfrak{m}$ be a maximal ideal of D. If $\mathfrak{m} \notin spec^1(D)$, then $\text{Int}(D_\mathfrak{m}) = D_\mathfrak{m}[X]$ [4, I.3.5] and $\mathfrak{J}_n(D_\mathfrak{m}) = D_\mathfrak{m}$. If $\mathfrak{m} \in spec^1(D)$, then $\mathfrak{J}_n(D)_\mathfrak{m} = \mathfrak{m}^{-w_{N(\mathfrak{m})}(n)} D_\mathfrak{m}$ when $N(\mathfrak{m}) \leq n$ and $\mathfrak{J}_n(D)_\mathfrak{m} = D_\mathfrak{m}$ when $N(\mathfrak{m}) > n$. Consequently, for each $\mathfrak{m} \in \max(D)$,

$$\mathfrak{J}_n(D)_\mathfrak{m} = \mathfrak{J}_n(D_\mathfrak{m}) = (n!)_{D_\mathfrak{m}}^{-1} = ((n!)_D)_\mathfrak{m}^{-1}.$$

Since in a Krull domain, the localization of the inverse of an ideal $\mathfrak{J}$ is equal to the inverse of the localization of $\mathfrak{J}$, we see that $\mathfrak{J}_n(D)$ is the inverse of $(n!)_D$, and hence, that $\mathfrak{J}_n(D)$ is a divisorial ideal. Moreover,

$$\mathfrak{J}_n(D) = \prod_{\mathfrak{p}\in spec^1(D),\, N(\mathfrak{p})\leq n} \mathfrak{p}^{-w_{N(\mathfrak{p})}(n)}.$$

□

It follows from the previous proposition that the factorial group $\mathcal{F}act(D)$ of a Krull domain D is also the subgroup of the divisorial group $\mathcal{D}(D)$ generated by the fractional ideals $\mathfrak{J}_n(D)$. In fact, the abelian group $\mathcal{F}act(D)$ is free and we are going to describe a basis.

Notation. For each integer $q \geq 2$, let $\Pi_q(D)$, or shortly Π_q, be the product of all height-one prime ideals of D with norm q, that is,

$$\Pi_q(D) = \prod_{\mathfrak{p}\in spec^1(D),\, N(\mathfrak{p})=q} \mathfrak{p}.$$

If there is no $\mathfrak{p} \in spec^1(D)$ with norm q (in particular, if q is not a prime power), then $\Pi_q(D) = D$. Note that, in a Krull domain, there are at most finitely many prime ideals with the same finite norm q.

Proposition 2.2. *If D is a Krull domain, then the factorial group $\mathcal{F}act(D)$ of D is the free abelian subgroup of the divisorial group $\mathcal{D}(D)$ with basis $\{\Pi_q \mid q \geq 2,\, \Pi_q \neq D\}$.*

Proof.

$$(n!)_D = \prod_{\mathfrak{p}\in spec^1(D),\, N(\mathfrak{p})\leq n} \mathfrak{p}^{w_{N(\mathfrak{p})}(n)}$$

$$= \prod_{q\geq 2} \left(\prod_{\mathfrak{p}\in spec^1(D),\, N(\mathfrak{p})=q} \mathfrak{p} \right)^{w_q(n)}$$

Finally,

$$(n!)_D = \prod_{q\geq 2} \Pi_q^{w_q(n)}.$$

Thus, $\mathcal{F}act(D)$ is contained in the subgroup of $\mathcal{D}(D)$ generated by the Π_q's. Conversely, $w_n(n) = 1$ for each $n \in \mathbb{N}$ implies that $(1!)_D = D$, $(2!)_D = \Pi_2, \ldots$ and $(n!)_D = \Pi_n \times \prod_{2 \leq q < n} \Pi_q^{w_q(n)}$. We then may prove by induction on n that all the Π_q's belong to $\mathcal{F}act(D)$. Finally, $\mathcal{F}act(D)$ is exactly the subgroup generated by the Π_q's.
Moreover, there are no relations between the ideals Π_q's (if we omit the Π_q's such that $\Pi_q = D$) because the height-one prime ideals involved in distinct Π_q's are also distinct. □

Recall also the following definition given by Pólya [9]:

Definition 2.3. We say that the ring $\mathrm{Int}(D)$ of integer-valued polynomials on D has a *regular basis* if the D-module $\mathrm{Int}(D)$ admits a basis $(f_n)_{n \in \mathbb{N}}$ where $\deg(f_n) = n$.

Proposition 2.4 (Pólya). [4, II.1.4] *Without any hypothesis on the domain D, $\mathrm{Int}(D)$ has a regular basis if and only if all the fractional ideals $\mathfrak{I}_n(D)$ are principal.*

Corollary 2.5. *If D is a Krull domain, then the following conditions are equivalent:*

1. $\mathrm{Int}(D)$ *has a regular basis,*
2. *for each $n \in \mathbb{N}$, the fractional ideal $\mathfrak{I}_n(D)$ is principal,*
3. *for each $n \in \mathbb{N}$, the entire ideal $(n!)_D$ is principal,*
4. *for each $q \geq 2$, the ideal Π_q is principal,*
5. $\mathcal{F}act(D) \subseteq \mathcal{P}(D)$,
6. $\mathcal{P}o(D) \simeq \{1\}$.

Examples. 1) If D is a unique factorization domain, then $\mathrm{Int}(D)$ has a regular basis [4, Exercise II.23], and hence, $\mathcal{P}o(D) \simeq \{1\}$.
2) There exist Krull domains with height-one prime ideals of finite norm whose dimension is strictly greater than 1 [3, Example III.4.2]. For instance, let p be a prime number and $\alpha \in \mathbb{Q}_p$ be transcendental over $\mathbb{Q}$. Denote by v the valuation of $\mathbb{Q}(X)$ defined by $v(\varphi) = v_p(\varphi(\alpha))$ for each $\varphi \in \mathbb{Q}(X)$ where v_p is the p-adic valuation of $\mathbb{Q}_p$. Let V be the corresponding valuation domain and let $D = \mathbb{Z}\left[\frac{1}{p}\right][X] \cap V$. Then, D is a two-dimensional Noetherian integrally closed domain and $D/pD \simeq \mathbb{F}_p$. In fact, D is a unique factorization domain, and hence $\mathcal{P}o(D) \simeq \{1\}$.

Remark 2.6. In the case where D is a Krull domain, we have other interpretations of the ideals $\mathfrak{I}_n(D)$ and $(n!)_D$.
1)Let $\mathfrak{I}_n(D)$ be the fractional ideal of D generated by all the coefficients of all the polynomials of $\mathrm{Int}(D)$ with degree $\leq n$. Then, $\mathfrak{I}_n(D) = \mathfrak{I}_n(D)$ because $\mathfrak{I}_n(D) \subseteq \mathfrak{I}_n(D) \subseteq (n!)_D^{-1}$ (see [5, Remark 5.10]) and $\mathfrak{I}_n(D)$ is divisorial (Proposition 2.1).

2) For each polynomial $f \in K[X]$, let $d(f, D)$ be the divisorial ideal of D generated by the values of f on D. Then [5, Proposition 5.9]:

$$(n!)_D = \cap\{d(f, D) \mid f \in D[X],\ f \ monic,\ \deg(f) \leq n\}.$$

3. Pólya groups in Galoisian extensions of $\mathbb{Q}$

If D is a Dedekind domain, then $\mathcal{D}(D)$ is also the group $\mathcal{I}(D)$ formed by the nonzero fractional ideals of D and $\mathcal{D}(D)/\mathcal{P}(D)$ is the class group $\mathcal{C}l(D) = \mathcal{I}(D)/\mathcal{P}(D)$ of D. The factorial group $\mathcal{F}act(D)$ of D is then the free abelian subgroup of $\mathcal{I}(D)$ generated by the ideals Π_q ($\Pi_q \neq D$, $q \geq 2$) where Π_q denotes the product of all prime ideals of D of norm q. The Pólya group $\mathcal{P}o(D)$ of D is the subgroup of the class group $\mathcal{C}l(D)$ generated by the classes of the Π_q's. In this section, we denote by $\mathbb{P}$ the set of prime numbers.

Example [4, Exercise II.31]. The Pólya group of $\mathbb{Z}[\sqrt{-29}]$ is $\{1, \overline{\mathfrak{p}}\}$ where $\overline{\mathfrak{p}}$ denotes the class of the prime ideal lying over 2, it is strictly contained in $\mathcal{C}l(\mathbb{Z}[\sqrt{-29}])$.

If K is a number field or a finite separable extension of $\mathbb{F}_q(T)$, we may consider for D the ring of integers $\mathcal{O}_K$, that is, the integral closure in K of $\mathbb{Z}$ or $\mathbb{F}_q[T]$. Instead of $\mathcal{P}(\mathcal{O}_K)$, $\mathfrak{I}(\mathcal{O}_K)$, $\mathcal{F}act(\mathcal{O}_K)$, $\mathcal{P}o(\mathcal{O}_K)$, $\Pi_q(\mathcal{O}_K), \ldots$ we write shortly $\mathcal{P}(K)$, $\mathfrak{I}(K)$, $\mathcal{F}act(K)$, $\mathcal{P}o(K)$, $\Pi_q(K), \ldots$ and, following Zantema [11], we say that K is a *Pólya-field* if $\mathrm{Int}(\mathcal{O}_K)$ has a regular basis, that is, $\mathcal{P}o(K) \simeq \{1\}$. In fact, we are going to only consider for K number fields that are Galoisian extensions of $\mathbb{Q}$.

One knows that, if K is a finite Galoisian extension of $\mathbb{Q}$, for each prime p, the g_p maximal ideals $\mathfrak{m}$ of $\mathcal{O}_K$ lying over p are conjugate with respect to the Galois group $Gal(K/\mathbb{Q})$, they have the same ramification index $e_p = e_p(K/\mathbb{Q})$ and the same residuel degree $f_p = f_p(K/\mathbb{Q})$, and hence,

$$e_p f_p g_p = [K : \mathbb{Q}].$$

Moreover,

$$p\mathcal{O}_K = \prod_{\mathfrak{m}|p} \mathfrak{m}^{e_p} = (\Pi_q(K))^{e_p} \quad \text{where} \quad q = p^{f_p}.$$

Consequently, following Ostrowski [8]:

Proposition 3.1. *Assume $K/\mathbb{Q}$ is a finite Galoisian extension.*

1. *If $q = p^f$ where the prime number p is not ramified in the extension $K/\mathbb{Q}$, then $\Pi_q(K)$ is principal.*
2. *The Pólya group $\mathcal{P}o(K)$ is generated by the classes of the $\Pi_q(K)$'s where $q = p^f$ and p is ramified in the extension $K/\mathbb{Q}$.*

Corollary 3.2. *Let $K/\mathbb{Q}$ be a finite Galoisian extension. For each $p \in \mathbb{P}$, let $e_p = e_p(K/\mathbb{Q})$. The natural morphism from $\mathcal{F}act(K)$ onto $\mathcal{P}o(K)$ factors through $\oplus_{p\in\mathbb{P}}\mathbb{Z}/e_p\mathbb{Z}$:*

$$\mathcal{F}act(K) \xrightarrow{\psi} \oplus_{p\in\mathbb{P}}\mathbb{Z}/e_p\mathbb{Z} \xrightarrow{\varphi} \mathcal{P}o(K).$$

Moreover, the following sequence of abelian groups is exact:

$$1 \to \mathbb{Q}^*/\{\pm 1\} \to \mathcal{F}act(K) \xrightarrow{\psi} \oplus_{p\in\mathbb{P}} \mathbb{Z}/e_p\mathbb{Z} \to 0.$$

Proof. As $\mathcal{F}act(K)$ is a free group generated by the ideals $\Pi_q(K)$'s where $q = p^{f_p}$, for each $p \in \mathbb{P}$, it is isomorphic to the direct sum $\oplus_{p\in\mathbb{P}} \mathbb{Z}$ of copies of $\mathbb{Z}$: an ideal $\mathfrak{J} = \prod_p (\Pi_q(K))^{k_p}$ of $\mathcal{F}act(K)$ corresponds to the element with k_p in the component corresponding to p. As the ideal $(\Pi_q(K))^{e_p}$ is principal, the natural morphism from $\mathcal{F}act(K)$ onto $\mathcal{P}o(K)$ (hence also from $\oplus_{p\in\mathbb{P}} \mathbb{Z}$ onto $\mathcal{P}o(K)$) factors through $\oplus_{p\in\mathbb{P}} \mathbb{Z}/e_p\mathbb{Z}$.

Obviously, the corresponding morphism $\psi : \mathcal{F}act(K) \to \oplus_{p\in\mathbb{P}} \mathbb{Z}/e_p\mathbb{Z}$ is onto. On the other hand, an ideal $\mathfrak{J} = \prod_p (\Pi_q(K))^{k_p}$ of $\mathcal{F}act(K)$ is in $Ker(\psi)$ if and only if, for each prime p, $k_p = e_p m_p$ with $m_p \in \mathbb{Z}$, that is, $\mathfrak{J} = \left(\prod_p p^{m_p}\right) \mathcal{O}_K$. In other words, $\mathfrak{J}$ is generated by a rational number and $Ker(\psi)$ corresponds to the ideal group of $\mathbb{Z}$. □

In particular, the order of $\mathcal{P}o(K)$ is a divisor of $\prod_p e_p$. Of course, φ is onto. We will describe $Ker(\varphi)$ later.

Remark 3.3. When $K/\mathbb{Q}$ is not Galoisian, $Po(K)$ is not necessarily generated by the Π_{p^f}'s where p is ramified (see [4, Exercise II.32]).

Notation. Let L/K be a finite extension. On the one hand, the injective morphism

$$j_K^L = \mathfrak{J} \in \mathcal{I}(K) \mapsto \mathfrak{J}\mathcal{O}_L \in \mathcal{I}(L)$$

induces a (non necessarily injective) morphism

$$\varepsilon_K^L : \overline{\mathfrak{J}} \in \mathcal{C}l(K) \mapsto \overline{\mathfrak{J}\mathcal{O}_L} \in \mathcal{C}l(L).$$

On the other hand, the morphism norm (see [10, I, §5])

$$N_L^K : \mathcal{I}(L) \to \mathcal{I}(K)$$

which is determined by its values on the maximal ideals $\mathfrak{n}$ of $\mathcal{O}_L$:

$$N_L^K(\mathfrak{n}) = \mathfrak{m}^{f_{\mathfrak{n}}(L/K)}$$

where $\mathfrak{m} = \mathfrak{n} \cap \mathcal{O}_K$ and $f_{\mathfrak{n}}(L/K) = [\mathcal{O}_L/\mathfrak{n} : \mathcal{O}_K/\mathfrak{m}]$, induces a morphism:

$$\nu_L^K : \overline{\mathfrak{J}} \in \mathcal{C}l(L) \mapsto \overline{N_L^K(\mathfrak{J})} \in \mathcal{C}l(K).$$

Recall that, if the extension L/K is separable, then for each $\mathfrak{J} \in \mathcal{I}(K)$:

$$N_L^K \circ j_K^L(\mathfrak{J}) = \mathfrak{J}^{[L:K]}.$$

The factorial subgroups and the Pólya subgroups behave well with respect to these morphisms as soon as K and L are Galoisian extensions of $\mathbb{Q}$:

Proposition 3.4. *If $K \subseteq L$ are two Galoisian extensions of $\mathbb{Q}$, then*

1.

$$j_K^L(\mathcal{F}act(K)) \subseteq \mathcal{F}act(L) \quad \text{and} \quad \varepsilon_K^L(\mathcal{P}o(K)) \subseteq \mathcal{P}o(L)$$

2.
$$N_L^K(\mathcal{F}act(L)) \subseteq \mathcal{F}act(K) \quad and \quad \nu_l^K(\mathcal{P}o(L)) \subseteq \mathcal{P}o(K).$$

Proof. If $q = p^{f_p(K/\mathbb{Q})}$, then
$$j_K^L(\Pi_q(K)) = (\Pi_{q'}(L))^{e_p(L/K)} \quad \text{where } q' = q^{f_p(L/K)}$$
and if $q' = p^{f_p(L/\mathbb{Q})}$, then
$$N_L^K(\Pi_{q'}(L)) = (\Pi_q(K))^{[L:K]/e_p(L/K)} \quad \text{where } q = p^{f_p(K/\mathbb{Q})}.$$
□

From now on we use the techniques of the proofs given by Zantema [11] to characterize the Pólya-fields, that is, the number fields K such that $\mathcal{P}o(K) \simeq \{1\}$ in order to obtain some description of the Pólya group $\mathcal{P}o(K)$ itself.

Proposition 3.5. *Let $K_1/\mathbb{Q}$ and $K_2/\mathbb{Q}$ be two finite Galoisian extensions and let $L = K_1K_2$. If $[K_1 : \mathbb{Q}]$ and $[K_2 : \mathbb{Q}]$ are relatively prime, then:*
$$j_{K_1}^L(\mathcal{F}act(K_1)) \cdot j_{K_2}^L(\mathcal{F}act(K_2)) = \mathcal{F}act(L)$$
and
$$\nu_{L/K_i}(\mathcal{P}o(L)) = \mathcal{P}o(K_i).$$

Proof. Let $n_1 = [K_1 : \mathbb{Q}]$ and $n_2 = [K_2 : \mathbb{Q}]$. For a fixed prime number p, let $e_i = e_p(K_i/\mathbb{Q})$ and $f_i = f_p(K_i/\mathbb{Q})$ $(i = 1, 2)$. Then $e_p(L/\mathbb{Q}) = e_1e_2$ and $f_p(L/\mathbb{Q}) = f_1f_2$. Let $\Pi_i = \Pi_{p^{f_i}}(K_i)$ $(i = 1, 2)$ and $\Pi = \Pi_{p^{f_1f_2}}(L)$. Then
$$p\mathcal{O}_{K_i} = \Pi_i^{e_i}, \; p\mathcal{O}_L = \Pi^{e_1e_2}, \; \Pi_i\mathcal{O}_L = \Pi^{e_{3-i}} \; (i = 1, 2).$$
Writing $n_i = e_id_i$ $(i = 1, 2)$, on the one hand one has:
$$\Pi_1^{u_2d_2}\Pi_2^{u_1d_1}\mathcal{O}_L = \Pi^{e_2u_2d_2+e_1u_1d_1} = \Pi.$$
this is the first assertion. On the other hand,
$$N_L^{K_1}(\Pi)^{u_2e_2} = N_L^{K_1}(\Pi^{e_2})^{u_2} = N_L^{K_1}(\Pi_1\mathcal{O}_L)^{u_2} = \Pi_1^{n_2u_2} =$$
$$\Pi_1^{1-n_1u_1} = \Pi_1 \times (\Pi_1^{e_1})^{-d_1u_1} = \Pi_1 \times (p\mathcal{O}_{K_1})^{-d_1u_1}.$$
This is the last assertion. □

Proposition 3.6. *Let $K_1/\mathbb{Q}$ and $K_2/\mathbb{Q}$ be two finite Galoisian extensions and let $L = K_1K_2$. If $[K_1 : \mathbb{Q}]$ and $[K_2 : \mathbb{Q}]$ are relatively prime, then:*

1. *The morphisms $\varepsilon_{K_1}^L$ and $\varepsilon_{K_2}^L$ are injective.*
2. *The Pólya group $\mathcal{P}o(L)$ is the direct product of its subgroups $\varepsilon_{K_i}^L(\mathcal{P}o(K_i))$.*
3. *Consequently, one has the isomorphism*
$$\mathcal{P}o(L) \simeq \mathcal{P}o(K_1) \times \mathcal{P}o(K_2).$$

Proof. Let $n_1 = [K_1 : \mathbb{Q}]$ and $n_2 = [K_2 : \mathbb{Q}]$.
1. Since the order of $\mathcal{P}o(K_1)$ is a divisor of the product of the ramification indices in the extension $K_1/\mathbb{Q}$, the order of any of its elements is a divisor of a power of n_1, and hence is prime to n_2. Consequently, the morphism $\nu_L^{K_1} \circ \varepsilon_{K_1}^L : \overline{\mathfrak{I}} \in \mathcal{P}o(K_1) \mapsto \overline{\mathfrak{I}}^{n_2}$ is injective, and $\varepsilon_{K_1}^L$ also.
2. By considering the orders of the elements we see that

$$\varepsilon_{K_1}^L(\mathcal{P}o(K_1)) \cap \varepsilon_{K_2}^L(\mathcal{P}o(K_2)) = \{1\}.$$

On the other hand, assertion 1 of Proposition 3.5 obviously implies

$$\varepsilon_{K_1}^L(\mathcal{P}o(K_1)) \cdot \varepsilon_{K_2}^L(\mathcal{P}o(K_2)) = \mathcal{P}o(L).$$

3. This is an easy consequence of 1 and 2. □

Corollary 3.7. *Assume that the extension $K/\mathbb{Q}$ is cyclic of degree n. Write $n = \prod_{p|n} p^{v_p(n)}$ and, for each prime p dividing n, let K_p be the unique subextension of K such that $[K_p : \mathbb{Q}] = p^{v_p(n)}$. Then*

$$\mathcal{P}o(K) \simeq \prod_{p|n} \mathcal{P}o(K_p).$$

Of course, as a corollary we obtain Zantema's result [11, Theorem 3.4]:

Corollary 3.8. *Let $K_1/\mathbb{Q}$ and $K_2/\mathbb{Q}$ be two Galoisian extensions whose degrees are relatively prime. Then K_1K_2 is a Pólya field if and only if K_1 and K_2 are Pólya fields.*

We still assume that $K/\mathbb{Q}$ is a Galoisian extension and, for the sake of completeness, we recall now a cohomological description of the kernel of the surjective morphism

$$\varphi : \oplus_{p\in\mathbb{P}} \mathbb{Z}/e_p\mathbb{Z} \to \mathcal{P}o(K).$$

The Galois group $G = Gal(K/\mathbb{Q})$ acts on the groups K, K^*, $\mathcal{O}_K$, $\mathcal{O}_K^\times$, $\mathcal{P}(K)$, $\mathcal{I}(K)$ (where $\mathcal{O}_K^\times$ denotes the unit group of $\mathcal{O}_K$). Obviously, the ideals $\Pi_q(K)$ are invariant, and hence, in this case, the factorial group $\mathcal{F}act(K)$ is nothing else than $\mathcal{I}(K)^G$, the subgroup of $\mathcal{I}(K)$ formed by the ambiguous ideals of K (that is, the ideals of K invariant by G). Consequently,

$$\mathcal{F}act(K) \cap \mathcal{P}(K) = \mathcal{I}(K)^G \cap \mathcal{P}(K) = \mathcal{P}(K)^G,$$

and

$$\mathcal{P}o(K) \simeq \mathcal{F}act(K)/\mathcal{P}(K)^G = \mathcal{I}(K)^G/\mathcal{P}(K)^G.$$

Thus, for Galoisian extensions $K/\mathbb{Q}$, the Pólya group $\mathcal{P}o(K)$ is the subgroup of $\mathcal{C}l(K)$ formed by the classes of the ambiguous ideals of K.

From the short exact sequence

$$1 \to \mathcal{O}_K^\times \to K^* \to \mathcal{P}(K) \to 1,$$

the left exactness of the functor $U \mapsto U^G$ on the abelian category of G-modules leads, with Hilbert 90, to the following exact cohomological sequence:

Lemma 3.9 (Brumer and Rosen). [2, Lemma 2.1] *The following sequence of abelian groups is exact:*

$$1 \to \mathbb{Q}^*/\{\pm 1\} \to \mathcal{P}(K)^G \xrightarrow{\delta} H^1(G, \mathcal{O}_K^\times) \to 1.$$

With the exact sequence given in Corollary 3.2, we obtain the following commutative diagram

$$\begin{array}{ccccccccc}
1 & \longrightarrow & \mathbb{Q}^*/\{\pm 1\} & \longrightarrow & \mathcal{P}(K)^G & \xrightarrow{\delta} & H^1(G, \mathcal{O}_K^\times) & \longrightarrow & 1 \\
\downarrow & & \downarrow & & \downarrow & & \downarrow{\scriptstyle\theta} & & \downarrow \\
1 & \longrightarrow & \mathbb{Q}^*/\{\pm 1\} & \longrightarrow & \mathcal{F}act(K) & \xrightarrow{\psi} & \oplus_{p \in \mathbb{P}} \mathbb{Z}/e_p\mathbb{Z} & \longrightarrow & 0
\end{array}$$

Let us recall the construction of θ. Let $f \in Z^1(G, \mathcal{O}_K^\times)$ and let $\overline{f}$ be its image in $H^1(G, \mathcal{O}_K^\times)$. There is $x \in K^*$ such that $x\mathcal{O}_K \in \mathcal{P}(K)^G$ and $\delta(x\mathcal{O}_K) = \overline{f}$. Since f and $\sigma \in G \mapsto \sigma(x)/x \in \mathcal{O}_K^\times$ are congruent modulo $B^1(G, \mathcal{O}_K^\times)$, we may define θ by $\theta(\overline{f}) = \psi(x\mathcal{O}_K)$.

By the snake's lemma, we have:

$$Ker(\theta) = \{1\} \text{ and } Coker(\theta) = \frac{\mathcal{F}act(K)}{\mathcal{P}(K)^G} = \mathcal{P}o(K).$$

Consequently,

Proposition 3.10. [11, p. 163] *If $K/\mathbb{Q}$ is Galoisian with Galois group G, the following sequence of abelian groups is exact:*

$$1 \to H^1(G, \mathcal{O}_K^\times) \xrightarrow{\theta} \oplus_{p \in \mathbb{P}} \mathbb{Z}/e_p\mathbb{Z} \xrightarrow{\varphi} \mathcal{P}o(K) \to 1.$$

In particular,

$$|\mathcal{P}o(K)| \times |H^1(G, \mathcal{O}_K^\times)| = \prod_{p \in \mathbb{P}} e_p,$$

and hence, if $K/\mathbb{Q}$ is a finite Galoisian extension, K is a Pólya-field if and only if $|H^1(G, \mathcal{O}_K^\times)| = \prod_{p \in \mathbb{P}} e_p$.

If G is cyclic generated by σ, one knows that [7, IV.3.7]:

$$H^1(G, \mathcal{O}_K^\times) \simeq H^{-1}(G, \mathcal{O}_K^\times) = \frac{\{a \in \mathcal{O}_K^\times \mid N_{K/\mathbb{Q}}(a) = 1\}}{\{\sigma(a)/a \mid a \in \mathcal{O}_K^\times\}}.$$

The cardinality of this last group is known, this is $2[K:\mathbb{Q}]$ if K is real and $N(\mathcal{O}_K^\times) = \{1\}$, this is $[K:\mathbb{Q}]$ otherwise. Thus, we have:

Corollary 3.11. *Assume that the extension $K/\mathbb{Q}$ is cyclic of degree n.*

1. *If K is real and $N(\mathcal{O}_K^\times) = \{1\}$, then*

$$|\mathcal{P}o(K)| = \frac{1}{2n} \times \prod_p e_p$$

2. *Else*

$$|\mathcal{P}o(K)| = \frac{1}{n} \times \prod_p e_p.$$

Acknowledgments. The author thanks F. Halter-Koch for interesting references.

REFERENCES

[1] M. Bhargava, Generalized Factorials and Fixed Divisors over Subsets of a Dedekind Domain, *J. Number Theory* **72** (1998), 67–75.

[2] A. Brumer and M. Rosen, Class number and ramification in number fields, *Nagoya Math. J.* **23** (1963), 97–101.

[3] P.-J. Cahen, Polynômes à valeurs entières, Thèse, Université Paris 11, 1973.

[4] P.-J. Cahen and J.-L. Chabert, Integer-Valued Polynomials, *Amer. Math. Soc. Surveys and Monographs*, **48**, Povidence, R.I., 1997.

[5] J.-L. Chabert, Generalized factorial ideals, *Arabian J. Sc. and Eng.* **26** (2001), 51–68.

[6] R. Gilmer, *Multiplicative Ideal Theory*, Marcel Dekker, New York, 1972.

[7] J. Neukirch, *Algebraic Number Theory*, Springer, Berlin, 1999.

[8] A. Ostrowski, Über ganzwertige Polynome in algebraischen Zahlkörpern, *J. reine angew. Math.* **149** (1919), 117–124.

[9] G. Pólya, Über ganzwertige Polynome in algebraischen Zahlkörpern, *J. reine angew. Math.* **149** (1919), 97–116.

[10] J.-P. Serre, *Corps locaux*, Hermann, Paris, 1962.

[11] H. Zantema, Integer valued polynomials over a number field, *Manuscr. Math.* **40** (1982), 155–203.

Monomial Ideals and the Computation of Multiplicities

D. DELFINO, A. TAYLOR,* W. V. VASCONCELOS,† N. WEININGER‡
Department of Mathematics
Rutgers University
110 Frelinghuysen Rd
Piscataway, New Jersey 08854-8019
e-mail: {delfino,ataylor,vasconce,nweining}@math.rutgers.edu

R. H. VILLARREAL§

Departamento de Matemáticas
Centro de Investigación y de Estudios Avanzados del IPN
Apartado Postal 14–740
07000 México City, D.F.
e-mail: vila@math.cinvestav.mx

Abstract

The theory of the integral closure of ideals has resisted direct approaches to some of its basic questions (membership and completeness tests, and construction). We mainly treat the membership problem in the monomial case by exploiting the connection with multiplicities and its linkage to the computation of volumes of polyhedra. We discuss several existent software packages and introduce our own contribution, a *Monte Carlo* based approach to the computation of volumes. Finally, we make comparisons of multiplicities of general ideals and of their initial ideals.

Introduction

Let R denote a Noetherian ring and I one of its ideals. The *integral closure* of I is the ideal $\overline{I}$ of all elements of R that satisfy an equation of the form

$$z^n + a_1 z^{n-1} + \cdots + a_{n-1} z + a_n = 0, \qquad a_i \in I^i.$$

There are several issues associated with this notion, from which we single out the following. Let $R = k[x_1, \ldots, x_n]$ be a ring of polynomials over the field k, let

*Partially supported by NSF-VIGRE grant.
†Partially supported by the NSF.
‡Partially supported by NSF-VIGRE grant.
§Partially supported by CONACyT and SNI, México.

$I = (f_1, \ldots, f_m) \subset R$, and let $f \in R$. Our main concern is how to carry out the following tests/construction:

- *Membership Test*: $f \in \overline{I}$?
- *Completeness Test*: $I = \overline{I}$?
- *Construction Task*: $I \leadsto \overline{I}$?
- *Complexity Cost*: $\mathrm{cx}(I \leadsto \overline{I})$?

In the literature one does not find effective methods to generally deal with these problems. The difficulty arises, partly, from the specialized nature of the equations the elements need to satisfy. The exception, when we understand the problem fully, is the case of monomial ideals. In this case, $\overline{I}$ is the monomial ideal defined by the integral convex hull of the exponent vectors of I (see [8, p. 140]). Through the techniques of integer programming, all four problems can, theoretically and often in practice, be solved. For non-monomial ideals, only specialized cases of some of these questions have been dealt with ([7] treats the completeness test for generic complete intersections).

Our interest in these questions is reinforced by its connections to another issue, which has not been adequately dealt with either, the computation of multiplicities in local rings. If $(R, \mathfrak{m})$ is a Noetherian local ring of Krull dimension d, and I is an $\mathfrak{m}$–primary ideal then $e(I)$, the *multiplicity* of I, is the integer

$$\lim_{n\to\infty} \frac{\lambda(R/I^n)}{n^d} d!,$$

where $\lambda(\cdot)$ is the length function. The Hilbert function of the ideal is $\lambda(R/I^n)$, which is given by a polynomial of degree d for $n \gg 0$ (see [2], [8]). When the ideal is monomial, the limit can be interpreted as a Riemann sum of volumes (normalized by the factor $d!$) and we exploit this connection.

These numbers are not easily captured, if at all, by Gröbner bases computations. In part this is because a large number of indeterminates are required to frame the calculation. A simplified version occurs when I is the maximal ideal and a conversion to a monomial ideal is possible through a theorem of Macaulay (Theorem 4.1). The connection between the two sets of issues, integral closure and multiplicity, rests primarily on a well-known theorem of Rees ([15], and its generalizations): For $I \subset L$, $e(I) = e(L)$ if and only if $L \subset \overline{I}$.

We briefly describe our results. The first section is an elementary recasting of the description of $\overline{I}$ for a monomial ideal I. It is mainly used to recast the interpretation of multiplicity as a volume. It also exhibits the fact that the degrees of the generators of $\overline{I}$ do not exceed the top degree of a generating set of I by more than $d-1$. In some sense this solves the complexity count of the determination of $\overline{I}$ by placing a bounding box around I, according to Corollary 1.3

Equipped with the understanding of multiplicities of monomial ideals of finite co-length, in section 2 we introduce a *Monte Carlo* method for the computation of volumes of polytopes and report on our experience with it. It is simple to set up and we found it comparable (in deriving estimates) to the more technical approaches aimed at exact computation. One of our goals is to explore the existing library of software to deal with these questions. We are particularly interested in problems in large numbers of indeterminates, obviously beyond the horizon of symbolic computation engines based on Gröbner basis techniques.

In section 3, we use standard linear programming techniques to deal with the four tests above. Ideally, one would like to answer the first two tests through an oracle matrix. For instance, in the membership test: Given a monomial ideal I, there is a matrix $\mathbb{A}$ and a vector $\mathbf{b}$ such that a monomial $\mathbf{x}^{\mathbf{v}} \in \overline{I}$ if and only if

$$\mathbb{A} \cdot \mathbf{v} \geq \mathbf{b}.$$

We show how to do this with off-the-shelf software, and rather efficiently for ideals of finite co-length. For this class of ideals, we also show how any membership oracle can be used as a completeness test and as a path to the construction task using exclusively monomial arithmetic.

The last section is an exploration of the relationships between the multiplicities of an ideal I of finite co-length and of its initial ideal $in_>(I)$, for some term ordering. It always holds that $e(I) \leq e(in_>(I))$, with equality meaning that for each integer n, $in_>(I^n)$ is integral over $(in_>(I))^n$ (Theorem 4.3). Note that in this case, the initial algebra of the Rees algebra $R[It]$ is Noetherian (a very infrequent occurrence).

Regrettably, the methods developed to compute multiplicities and treat integral closure issues do not extend to general ideals of rings of polynomials, or to affine algebras. In these cases, one can still appeal to Gröbner bases methods for small-scale examples.

1 Integral closure of monomial ideals

To set up the framework, we recall some general facts about the integral closure of monomial ideals that are required for our treatment of multiplicity. Let $R = k[x_1, \ldots, x_n]$ be a ring of polynomials over the field k, and let I be the ideal generated by the set of monomials $\mathbf{x}^{v_1}, \ldots, \mathbf{x}^{v_m}$. First we recall two descriptions of the integral closure of I.

One standard way to describe the integral closure of a monomial ideal is ([8, Exercise 4.23]):

Proposition 1.1 *Suppose $R = k[x_1, \ldots, x_n]$, and I is generated by a set of monomials $\mathbf{x}^{v_1}, \ldots, \mathbf{x}^{v_m}$. Let Γ be the set of exponents of monomials in I,*

$$\Gamma = \bigcup_{i=1}^{m} v_i + \mathbb{N}^n.$$

Regarding Γ as a subset of $\mathbb{R}^n_+$, let Λ be the convex hull of $\mathbb{R}^n_+ + \Gamma$, and let Γ^* be the set of integral points in Λ. Then $\overline{I}$ is the ideal generated by $\mathbf{x}^v$, $v \in \Gamma^*$.

We will use a second description (see [20, Section 6.6], [21, Section 7.3]) of the generators of the integral closure. If $\mathbf{x}^v \in \overline{I}$, it will satisfy an equation

$$(\mathbf{x}^v)^\ell \in I^\ell,$$

and therefore we have the following equation for the exponent vectors,

$$\ell \cdot v = u + \sum_{i=1}^{m} r_i \cdot v_i, \quad r_i \geq 0, \quad \sum_{i=1}^{m} r_i = \ell.$$

This means that $v = \frac{u}{\ell} + \alpha$, where α belongs to the convex hull $\mathrm{Conv}(v_1, \ldots, v_m)$ of $v_1, \ldots, v_m$. The vector v can be written as (set $w = \frac{u}{\ell}$)

$$v = \lfloor w \rfloor + (w - \lfloor w \rfloor) + \alpha,$$

and it is clear that the integral vector

$$v_0 = (w - \lfloor w \rfloor) + \alpha$$

also has the property that $\mathbf{x}^{v_0} \in \overline{I}$.

Proposition 1.2 *Let I be an ideal generated by the monomials $\mathbf{x}^{v_1}, \ldots, \mathbf{x}^{v_m}$. Let C be the rational convex hull of $V = \{v_1, \ldots, v_m\}$ and*

$$B = [0,1) \times \cdots \times [0,1) = [0,1)^n.$$

Then $\overline{I}$ is generated by $\mathbf{x}^v$, where $v \in (C+B) \bigcap \mathbb{N}^n$.

For simplicity we set $B(V) = (C+B) \bigcap \mathbb{N}^n$.

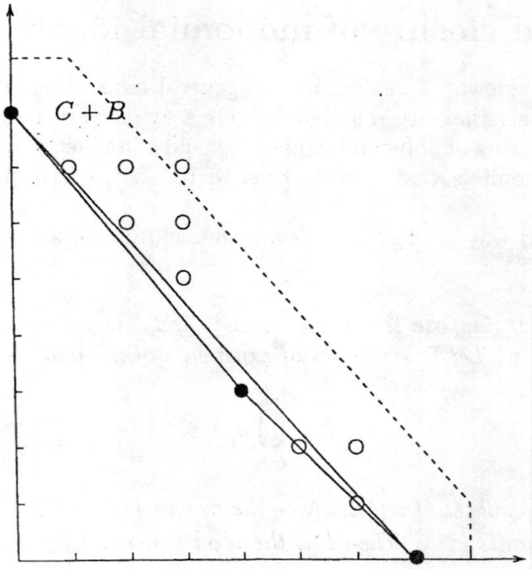

Figure 1: $B(V)$: The dotted lines indicate the boundary of $C+B$. The open circles are those lattice points which give elements in the integral closure of I. The lattice point that is not in $(C+B)\bigcap \mathbb{N}^n$, is in the ideal generated by the lattice points in $(C+B)\bigcap \mathbb{N}^n$.

The following help illustrate some of the issues with computing the integral closure of an ideal. First, a degree bound for the generators of the integral closure arises directly from Proposition 1.2. A sharper bound might depend on the codimension of the ideal.

Corollary 1.3 *Let I be a monomial ideal of $k[x_1,\ldots,x_n]$, generated by monomials of degree at most d. Then $\overline{I}$ is generated by monomials of degree at most $d+n-1$.*

The following example shows that the integral closure of a monomial ideal I, although by Proposition 1.1 defined by the integral convex hull of all the exponent vectors of I, may not be generated by the monomials defined by the integral convex hull of the exponent vectors of a minimal set of generators of I. The vector $\langle 3,5\rangle$ in Figure 1 also illustrates this possibility. This, of course, makes the determination of $\overline{I}$ a great deal harder. We will revisit this example when we give two membership tests.

Example 1.4 Let I be the ideal of the ring of polynomials $R = k[x_1,\ldots,x_8]$ defined by the monomials given through exponent vectors $v_1,\ldots,v_8$:

$$\begin{matrix} 1 & 1 & 1 & 1 & 1 & 0 & 0 & 0 \\ 1 & 1 & 1 & 1 & 0 & 1 & 0 & 0 \\ 1 & 1 & 1 & 1 & 0 & 0 & 1 & 0 \\ 1 & 1 & 1 & 1 & 0 & 0 & 0 & 1 \\ 1 & 0 & 0 & 0 & 1 & 1 & 1 & 1 \\ 0 & 1 & 0 & 0 & 1 & 1 & 1 & 1 \\ 0 & 0 & 1 & 0 & 1 & 1 & 1 & 1 \\ 0 & 0 & 0 & 1 & 1 & 1 & 1 & 1 \end{matrix}$$

Let $L = I^3$, and consider the vector $v = (2,\ldots,2)$. In view of the equality

$$8v = (1,\ldots,1) + 3(v_1 + \cdots + v_8),$$

one has

$$v = (\frac{1}{8}(3v_1) + \cdots + \frac{1}{8}(3v_8)) + (\frac{1}{8},\ldots,\frac{1}{8}),$$

which shows that $\mathbf{x}^v$ lies in the integral closure of L; note that this monomial has degree 16 while L is generated by monomials of degree 15. Since the vectors v_i are linearly independent, one can easily check with *Maple* that decrementing v, in any coordinate, by 1 produces elements that do not lie in the convex hull of $\{3v_1,\ldots,3v_8\}$. It follows that $\overline{L}$ requires minimal generators of degree at least 16.

We next recall the connection between integral closure, multiplicity and the computation of volumes of polyhedra. Let $f_1 = \mathbf{x}^{v_1},\ldots,f_r = \mathbf{x}^{v_r}$ be a set of monomials generating the ideal I. The convex hull $C(V)$ of the v_i's partitions the positive quadrant into 3 regions: an unbounded connected region, $C(V)$ itself and the complement $\mathcal{P}$ of the other two. The bounded region $\mathcal{P}$ is the region most pertinent to our calculation (see also [18, p. 235], [19]).

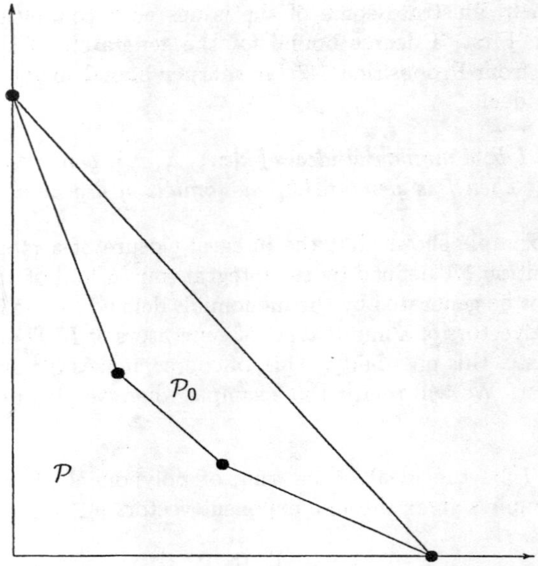

Figure 2: The polytope boundary is $C(V)$ and $\mathcal{P}$ is as marked. We will also refer to the polytope marked by $\mathcal{P}_0$ and the simplex that bounds $\mathcal{P}$ and $\mathcal{P}_0$ will be referred to as Δ.

The integral closure $\overline{I}$ is generated by the monomials whose exponents have the form

$$\sum_{i=1}^{r} r_i v_i + \epsilon \subset \mathbb{N}^d$$

such that $r_i \geq 0$ and $\sum r_i = 1$ and ϵ is a positive vector with entries in $[0, 1)$. Suppose that I is of finite co-length, then, using the notation of Proposition 1.1, $\lambda(R/\overline{I})$ is the number set of lattice points not in $C(\mathbb{R}^n_+ + \Gamma)$.

Consider the integral closure of I^n. According to the valuative criterion ([22, p. 350]), $\overline{I^n}$ is equal to the integral closure of the ideal generated by the nth powers of the f_i's. This means that the generators of $\overline{I^n}$ are defined by the exponent vectors of the form

$$\sum r_i n v_i + \epsilon,$$

with r_i and ϵ as above. We rewrite

$$n(\sum r_i v_i + \frac{\epsilon}{n})$$

so the vectors enclosed must have denominators dividing n. To deal with $\overline{I^n}$ we are going to use the set of vectors v_i, but change the scale by $1/n$. This means that each I^n determines the same $\mathcal{P}$. The length ℓ_n of $R/\overline{I^n}$ is the number of scaled lattice points in $\mathcal{P}$. Placing the lower left corner of a hypercube of side $1/n$ at each lattice point we see that the sum of the volumes of the hypercubes is equal to the number of lattice points times $(1/n)^d$ which in turn is $(1/n)^d \ell_n = (1/n)^d \lambda(R/\overline{I^n})$. However, this sum is also a Riemann sum approximating the volume of $\mathcal{P}$ and thus the limit of this quantity as $n \to \infty$ is just the exact volume of $\mathcal{P}$ (see Figure 3). This number, multiplied by $d!$, is the multiplicity of the ideal.

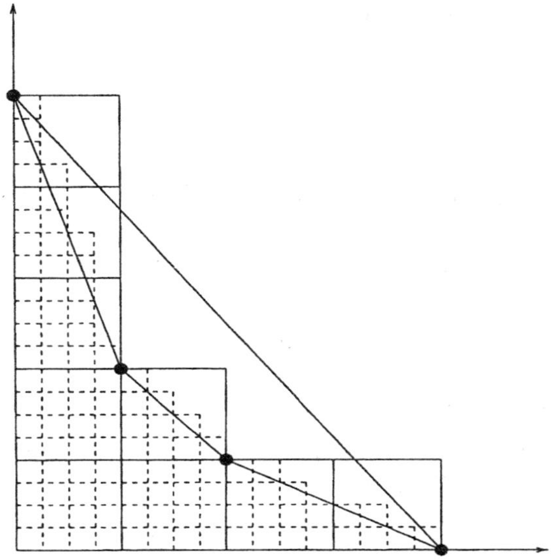

Figure 3: The cubes of side one and side $\frac{1}{4}$ are shown.

Let us sum up some of these relationships between multiplicities and volumes of polyhedra (see [19, p. 131]).

Proposition 1.5 *Let I be a monomial ideal of $R = k[x_1, \ldots, x_d]$ generated by $\mathbf{x}^{v_1}, \ldots, \mathbf{x}^{v_m}$. Suppose that $\lambda(R/I) < \infty$. If $\mathcal{P}$ is the region of $\mathbb{N}^d$ defined by I then*

$$e(I) = d! \cdot \text{Vol}(\mathcal{P}). \tag{1}$$

Example 1.6 Our first example is an ideal of $k[x, y, z]$. Suppose

$$I = (x^a, y^b, z^c, x^\alpha y^\beta z^\gamma), \quad \frac{\alpha}{a} + \frac{\beta}{b} + \frac{\gamma}{c} < 1.$$

The inequality ensures that the fourth monomial does not lie in the integral closure of the other three. A direct calculation shows that the multiplicity is indeed the volume of the region $\mathcal{P}$ times $d!$, which in this case is given by a nice formula

$$e(I) = ab\gamma + bc\alpha + ac\beta.$$

We observe that $\mathcal{P}$ is not a polytope, but can be expressed as the difference between two polytopes directly determined by the set of exponents vectors defining I, $V = \{v_1, \ldots, v_m\}$, $v_i \neq 0$. Since I has finite co-length, suppose the first d exponent vectors correspond to the generators of $I \cap k[x_i]$, $i = 1, \ldots, d$. Let Δ be the polyhedron defined by these vectors, $\Delta = C(0, v_1, \ldots, v_d)$, and denote by $\mathcal{P}_0$ the convex hull of V (see Figure 2). We note

$$\mathcal{P} = \Delta \setminus \mathcal{P}_0,$$

and therefore

$$\mathrm{Vol}(\mathcal{P}) = \mathrm{Vol}(\Delta) - \mathrm{Vol}(\mathcal{P}_0) = \frac{|v_1|\cdots|v_d|}{d!} - \mathrm{Vol}(\mathcal{P}_0).$$

We use this relationship to compute multiplicities. If we set

$$p = \frac{\mathrm{Vol}(\mathcal{P}_0)}{\mathrm{Vol}(\Delta)}$$

then the proposition follows.

Proposition 1.7 *Let I be a monomial ideal of finite colength generated by the monomials $\mathbf{x}^{v_1}, \ldots, \mathbf{x}^{v_m}$. With the notation above, we have*

$$e(I) = (1-p)|v_1|\cdots|v_d|.$$

2 A probabilistic approach to volumes and multiplicities

There is an extensive literature on the computation of volumes of polyhedra. We benefited from the discussion of volume computation in [4]. The associated costs of the various methods depend on how the convex sets are represented. They often require conversion from one representation to another. We propose a manner in which to approach the calculation of p, i.e. the calculation of $\mathrm{Vol}(\mathcal{P}_0)$ as a fraction of $\mathrm{Vol}(\Delta)$. First, note that Δ is defined by the equations

$$\Delta : \frac{x_1}{|v_1|} + \cdots + \frac{x_d}{|v_d|} \leq 1, \quad x_i \geq 0. \tag{2}$$

According to [6, pp. 284–285], since $\mathcal{P}_0$ is the convex hull of the vectors v_i, $i = 1, \ldots, m$, there are standard linear programming techniques to convert the convex hull description of $\mathcal{P}_0$ into an intersection of halfspaces

$$\mathcal{P}_0 : \mathbb{A} \cdot \mathbf{x} \leq \mathbf{b}. \tag{3}$$

Equally important, these linear programming techniques have been converted into very efficient routines in several programming environments. We will focus on those routines found in the collection [5].

Our statistical approach is based on classical *Monte Carlo* quadrature methods ([17]). Sampling a very large number of points in Δ, and checking when those points lie in $\mathcal{P}_0$ are both computationally straight forward because of the ease of the descriptions given in equations (2) and (3).

Our proposal consists of making a series of N independent trials, keeping track of the number of hits H, and using the frequency $\frac{H}{N}$ as an approximation for p. According to basic probability theory, these approximations come with an attached probability in the sense that for small $\epsilon > 0$

$$\text{Probability}\left\{\left|\frac{H}{N} - p\right| < \epsilon\right\}$$

is high. This estimation is based on Chebyshev's inequality ([10, p. 233]). We briefly review this inequality. If X is a random variable with finite second moment $\mathrm{E}(X^2)$, then for any $t > 0$

$$P\{|X| \geq t\} \leq t^{-2}E(X^2).$$

In particular for a variable X of mean $E(X) = \mu$ and finite variance $\mathrm{Var}(X)$, for any $t > 0$

$$P\{|X - \mu| \geq t\} \leq t^{-2}\mathrm{Var}(X). \tag{4}$$

For a set of N independent trials $x_1, \ldots, x_N$ of probability p, the random variable we are interested in is the average number of hits

$$X = \frac{x_1 + \cdots + x_N}{N} = \frac{H}{N}.$$

We have $E(X) = p$ and $\mathrm{Var}(X) = \sqrt{\frac{p(1-p)}{N}}$. If we set $\epsilon = t^{-2}\mathrm{Var}(X)$, and substitute into (4), we obtain

$$P\left\{\left|\frac{H}{N} - p\right| < \sqrt{\frac{p(1-p)}{\epsilon N}}\right\} > 1 - \epsilon.$$

Since $p(1-p) \leq \frac{1}{4}$, it becomes easy to estimate the required number of trials to achieve a high degree of confidence. Thus, for instance, a crude application shows that in order to obtain a degree of confidence of 0.95, and $\epsilon = 0.02$, the required number of trials should be $N \geq 12,500$ (Actually, a refined analysis, using the law of large numbers, cuts this estimate by $\frac{4}{5}$).

We have implemented this probabilistic approach to multiplicity. Our implementation uses off-the-shelf software. We illustrate our implementation though the discussion of some examples.

Example 2.1 Computing multiplicity using probability requires a conversion that uses PORTA (see [5]), a collection of transformation techniques in linear programming.

We will illustrate an application of the probabilistic method for the calculation of multiplicity in the setting of Proposition 1.5. Let $I = (x^3, y^4, z^5, w^6, xyzw)$. Proposition 1.5 gives $e(I) = 342$. To apply the probabilistic method, the exponents are written into a matrix and PORTA is used to obtain the inequalities defining the convex hull. The PORTA input and output are recorded below.

```
The points defining the convex hull must be written in a file with
the extension .poi [say mult1.poi] and the routine ''traf'' is called

traf mult1.poi

-----------------------------------
The content of mult1.poi is:
-----------------------------------
DIM = 4

CONV_SECTION
3 0 0 0
0 4 0 0
0 0 5 0
0 0 0 6
1 1 1 1

END
-------------------------------------
The output file is the desired set of linear inequalities
and it is put in the file mult1.poi.ieq:
-------------------------------------
DIM = 4

VALID
1 1 1 1

INEQUALITIES_SECTION
(  1) -23x1-15x2-12x3-10x4 <= -60
(  2) -20x1-15x2-12x3-13x4 <= -60
(  3) -10x1- 9x2- 6x3- 5x4 <= -30
(  4) - 4x1- 3x2- 3x3- 2x4 <= -12
(  5) +20x1+15x2+12x3+10x4 <=  60

END
```

A C^{++} program is then used to calculate the probability. Testing with 10,000 points gives a probability of .04989 and a multiplicity of 342.04.

Now we present more examples utilizing our implementation of our proposed probability based algorithm for computing the multiplicity of monomial ideals. We include an analysis of their run times and probable accuracy of results. In presenting these examples, the dimension-independence of the method is clear. However, the differences between theoretical results and implemented results are also clear. All results listed were obtained on a Pentium III processor that runs at 900 MHZ, has 256 MB RAM and is operating under Red Hat Linux.

We illustrate the results of our algorithm using three examples. For the purposes of the examples, we will refer to our algorithm as *POLYPROB.* We revisit Example 2.1 and give two other examples for comparing MACAULAY2 [12], VINCI [4],

NORMALIZ [3], and POLYPROB. VINCI is an alternate program for computing the volume of a polytope, while our computations in MACAULAY2 are classical, meaning we compute the leading coefficient of the Hilbert Polynomial of the associated graded ring.

Example 2.2 Our second example is again in a four dimensional ring

$$I = (x^4, y^5, z^6, w^7, xz^2w, y^2zw^2, xyzw).$$

This example is more complicated, but we can still use MACAULAY2, VINCI and POLYPROB to compute the multiplicity.

Example 2.3 Last we present an example where MACAULAY2 fails, and the issues of accuracy and speed in POLYPROB and VINCI are also illustrated. This example is sixteen dimensional

$$(x_1^2, x_2^3, x_3^4, x_4^5, x_5^6, x_6^7, x_7^8, x_8^9, x_9^{10}, x_{10}^{11}, x_{11}^{12}, x_{12}^{13}, x_{13}^{14}, x_{14}^{15}, x_{15}^{16}, x_{16}^{17}, x_3x_5x_8x_{10}x_{12}x_{14}x_{16},$$
$$x_2x_5x_7x_{13}, x_3x_7x_9x_{10}x_{13}^2x_{15}^2x_{16}, x_2x_4x_6x_{11}^2x_{14}^2, x_4x_6x_8x_{11}, x_1x_9, x_1x_{15}).$$

For this ideal, while we have

$$P\left\{\left|\frac{H}{N} - p\right| < \sqrt{\frac{p(1-p)}{\epsilon N}}\right\} > 1 - \epsilon,$$

when we multiply the probability by 17! to get the multiplicity, we also multiply the error by this same number. In Example 2.3 for $\epsilon = .02$ and $N = 20000$ we get $H = 16618$ and $H/N = .8309$ in one trial. The formula states that the probability that $|.8309 - p| < \frac{1}{4(.01)(20000)} = .000625$ is greater than .98. However, we can only say that $|(1 - .8309)17! - e(I)| < (.00125)(17!) = 2.22305(10^{11})$. Even with everything else the same and $N = 1{,}000{,}000$, $|.8309 - p| < .000025$, but $|(1 - .8309)17! - e(I)| < (.0000125)(17!) = 4.44609(10^9)$. We would need $N = 10^{14}$ to get the error on the multiplicity, using POLYPROB, to around 10. Unfortunately, the numbers we are dealing with mean that using standard floating point arithmetic there will be large computer precision error involved. The program VINCI also has this problem for large computations. We have been able to implement POLYPROB using GMP [11] aribitrary precision arithmetic and these are the numbers we include here. Unfortunately, it would take days to run 2.3 in POLYPROB with $N = 10^{14}$.

For each example, NORMALIZ computed the multiplicity (342, 546, and 60012790921296 respectively) in a negligable amount of time (less than .01 in each of the first two examples) so we don't list this in the chart to save space. This table lists the exact (up to computer precision error) multiplicity as computed by MACAULAY2 or VINCI (VI) and the POLYPROB (PP) results for different values of N. The entries in the "PP result" column are an average of 10 trials. Averaging trials appears to give slightly more accurate results. Last we include the CPU run times for each of the calculations.

Ex.	M2 re-sult	VI result	PP result	N	# ineq.	PP time	M2 time	VI time
2.1	342	342	342.202	5000	5	.05	.25	.07
			342.04	10000		.07		
			342.043	20000		.10		
			342.093	50000		.18		
2.2	546	546	547.688	10000	14	.07	1.37	.07
			547.642	20000		.10		
			545.946	50000		.18		
			545.553	100000		.31		
2.3	-	$(6.001279)*(10^{13})$	$(5.95065)(10^{13})$	10000	494	.73	-	.13
			$(5.98462)(10^{13})$	20000		1.42		
			$(6.0257)(10^{13})$	50000		3.47		
			$(6.01912)(10^{13})$	100000		6.98		
			$(6.01127)(10^{13})$	500000		34.93		
			$(6.00291)(10^{13})$	1000000		1:10.57		

At this point NORMALIZ seems to outperform all of the programs on these examples. POLYPROB is clearly much better than M2 on even medium problems. In terms of time VINCI appears to be the best of those programs in the chart. However, we note that POLYPROB will work as accurately if we give it an ideal of the form $(x_1^{\alpha_1}, \ldots, x_n^{\alpha_n}, f_1, \ldots f_n)$ where f_i for at least one i is in the integral closure of the ideal $(x_1^{\alpha_1}, \ldots, x_n^{\alpha_n})$, but VINCI will fail to give the correct multiplicity in this case and as noted before, VINCI is only written using standard floating point arithmetic.

POLYPROB Implementation

The fundamental operation of our POLYPROB algorithm is a random trial: that is, generating a random vector within the simplex containing the polytope, and testing whether the vector is in the polytope. Thus, POLYPROB requires an efficient way to get random vectors uniformly distributed over a simplex. To see how to do this, first consider the general problem of generating a vector $(x_1, \ldots, x_n)$ uniformly distributed over an n–dimensional polytope $\mathcal{P}$. Given a description of the polytope, say as the convex hull of a set of vertices, we can calculate the minimum and maximum values for each coordinate of a vector in the polytope. That is, we can determine that the polytope lies within the hypercube $\prod_{i=1}^{n}[a_i, b_i]$. Our first task, then, is to pick $x_1 \in [a_1, b_1]$ according to an appropriate probability distribution.

Thus, for any $c \in [a_1, b_1]$ we can calculate $f(c) = \Pr(x_1 \in [a_1, c])$ by calculating the volume of $\mathcal{P} \cap \{(x_1, \ldots, x_n) : a_1 \leq x_1 \leq c\}$ as a percentage of the volume of $\mathcal{P}$. This gives us a monotone increasing distribution function $f : [a_1, b_1] \to [0, 1]$. It is from this distribution function that we want to sample x_1. If we can pick a random real number X uniformly distributed over $[0, 1]$, then we can just take $x_1 = f^{-1}[X]$. Once x_1 has been sampled, its value determines an $(n-1)$-dimensional cross-section of $\mathcal{P}$ so we have now reduced the problem to picking a smaller random vector $(x_2, \ldots, x_n)$ uniformly distributed over that cross-section. Thus we can iteratively pick $x_2, \ldots, x_n$ by the same algorithm used to pick x_1.

For general polytopes, there are very large practical problems with this algorithm. However, for simplices all of these problems disappear. Since Δ is a simplex we only need to perform this algorithm for simplices. Consider a simplex with one vertex at the origin and vertices $v_1, \dots, v_n$ where $v_i = (0, \dots, a_i, 0, \dots, 0)$. Then

$$\Pr(x_1 \in [c, a_1]) = \left(\frac{a_1 - c}{a_1}\right)^n$$

so the inverse of the distribution function is just $f^{-1}(X) = 1 - X^{1/n}$ times the scaling factor a_i. And if we sample x_1, the cross-section of the simplex at x_1 is just the $(n-1)$-dimensional simplex with vertices at $(x_1, 0, 0, \dots, 0)$ and $v_2 \dots v_n$ where $v_i = (x_1, 0, \dots, (1 - \frac{x_1}{a_1})a_i, \dots)$.

The source code for our implementation of POLYPROB illustrates our application of this method; it is available at

`http://www.math.rutgers.edu/~nweining/polyprob.tar.gz`

3 Membership test for integral closure of monomial ideals

In this section we provide a linear programming solution to the membership test '$f \in \overline{I}$?'

Monomial ideals of finite co-length

We will provide now membership & completeness tests and a construction of the integral closure of monomial ideals of finite co-length. Our treatment is a by-product of the half-spaces description of the convex hull given in Eq. (3). We point out how the following oracle gives a solution to the membership and completeness tests and the construction task in case of an ideal of finite co-length.

Proposition 3.1 *Let I be a monomial ideal of finite co-length as above, and let f be a monomial. Denote by $\mathbf{e} = (e_1, \dots, e_n)$ the exponent vector of f, by $\mathbf{v} = (v_1, \dots, v_n)$, A and $\mathbf{b}$ the vectors and matrices associated to I as discussed above. Then f is integral over I if one of the two conditions holds:*

$$A \cdot \mathbf{e} \leq \mathbf{b},$$

$$\sum_{i=1}^{n} \frac{e_i}{v_i} \geq 1.$$

Proof. These conditions simply express the fact that either $\mathbf{e}$ lies in the convex hull of the vectors $\mathbf{v}_1, \dots, \mathbf{v}_n$ (in which case f would lie in the integral closure of $(\mathbf{x}^{v_1}, \dots, \mathbf{x}^{v_n})$), or that adding f to I does not affect the volume of $\mathcal{P}$. In the second case, $e(I) = e(I, f)$, f is integral over I by Rees' theorem. □

Definition 3.2 A *membership oracle* for the integral closure of an ideal I is a boolean function $\mathcal{A}$ such that $f \in \overline{I}$ if and only if $\mathcal{A}(f) = \text{true}$.

Proposition 3.1 above shows that monomial ideals of finite co-length admit such oracles. We show now how given any membership oracle $\mathcal{A}$ for a monomial ideal I of finite co-length leads also to a completeness test. We begin with a general observation that shows some of the opportunities and difficulties in developing such tests.

Proposition 3.3 *Let $(R, \mathfrak{m})$ be a Noetherian local ring and let I be an ideal of finite co-length. Denote by $L = I : \mathfrak{m}$ the socle ideal of I. Then I is complete if and only if no element of $L \setminus I$ is integral over I.*

Proof. If $f \in \overline{I} \setminus I$, then for some power of $\mathfrak{m}$, $\mathfrak{m}^r f$ will contain non-trivial elements in the socle of I. The converse is clear. □

Proposition 3.4 *Let I be a monomial ideal in $k[x_1, \ldots, x_d]$ of finite co-length and let $\mathcal{A}$ be a membership oracle for the integral closure of I. Let $\{f_1, \ldots, f_s\}$ be the monomials in $I : (x_1, \ldots, x_d) \setminus I$. Then*

$$I = \overline{I} \Longleftrightarrow \mathcal{A}(f_i) = \text{false},\ i = 1, \ldots, s.$$

Proof. First, we consider the reverse direction. Let $L = I : (x_1, \ldots, x_d)$ be the socle ideal of I. L is generated by the f_i and monomials in I. Since $\overline{I}$ is a monomial ideal, if f is a monomial $\in \overline{I} \setminus I$, by multiplying by another monomial g, we obtain gf generating a nonzero element in the vector space L/I. This means that gf must be one of the f_i. Since gf is also integral over I, the assertion follows. The other assertion is obvious. □

The construction of $\overline{I}$ follows in a straightforward manner:

If $I \neq \overline{I}$, define

$$I_1 = (I, \mathcal{A}(f_i) = \text{true},\ i = 1, \ldots, s).$$

Since $\overline{I_1} = \overline{I}$, $\mathcal{A}$ is still a membership oracle for the integral closure of I_1 and we can repeat until $I_n = \overline{I}$. The program terminates by Proposition 3.4 and is has been implemented in MACAULAY2.

General monomial ideals

A more comprehensive membership test for the question "$f \in \overline{I}$", valid for any monomial ideal, is the following. However, this test lacks the effectivity of the method in the previous section.

Proposition 3.5 *Let $v_1, \ldots, v_m$ be a set of vectors in $\mathbb{N}^n$ and let A be the $n \times m$ matrix whose columns are the vectors $v_1, \ldots, v_m$. If $I = (\mathbf{x}^{v_1}, \ldots, \mathbf{x}^{v_m})$, then a monomial $\mathbf{x}^b$ lies in the integral closure of I if and only if the linear program:*

$$\text{Maximize } x_1 + \cdots + x_m \qquad (*)$$

$$\text{Subject to } Ax \leq b \text{ and } x \geq 0$$

has an optimal value greater or equal than 1, which is attained at a vertex of the rational polytope $P = \{x \in \mathbb{R}^m \mid Ax \leq b \text{ and } x \geq 0\}$.

Proof. $\Rightarrow$) Let $\mathbf{x}^b \in \overline{I}$, that is, $\mathbf{x}^{pb} \in I^p$ for some positive integer p. There are non-negative integers r_i satisfying

$$\mathbf{x}^{pb} = \mathbf{x}^{\delta} (\mathbf{x}^{v_1})^{r_1} \cdots (\mathbf{x}^{v_m})^{r_m} \text{ and } r_1 + \cdots + r_m = p.$$

Hence the column vector c with entries $c_i = r_i/p$ satisfies

$$Ac \leq b \text{ and } c_1 + \cdots + c_m = 1.$$

This means that the linear program has an optimal value greater or equal than 1.

$\Leftarrow$) Observe that the vertices of P have rational entries (see [6, Theorem 18.1]) and that the maximum of $x_1 + \cdots + x_m$ is attained at a vertex of the polytope$P$, thus there are non-negative rational numbers $c_1, \ldots, c_m$ such that

$$c_1 + \cdots + c_m \geq 1 \text{ and } c_1 v_1 + \cdots + c_m v_m \leq b.$$

By induction on m it follows rapidly that there are rational numbers $\epsilon_1, \ldots, \epsilon_m$ such that

$$0 \leq \epsilon_i \leq c_i \ \forall i \text{ and } \sum_{i=1}^{m} \epsilon_i = 1.$$

Therefore there is a vector $\delta \in \mathbb{Q}^n$ with non-negative entries satisfying

$$b = \delta + \epsilon_1 v_1 + \cdots + \epsilon_m v_m.$$

Thus there is an integer $p > 0$ such that

$$pb = \underbrace{p\delta}_{\in \mathbb{N}^n} + \underbrace{p\epsilon_1}_{\in \mathbb{N}} v_1 + \cdots + \underbrace{p\epsilon_m}_{\in \mathbb{N}} v_m,$$

and consequently $\mathbf{x}^b \in \overline{I}$. □

Remark 3.6 According to [6, Theorem 5.1] if the primal problem $(*)$ has an optimal solution x, then the dual problem

Minimize $b_1 y_1 + \cdots + b_n y_n$

Subject to $yA \geq \mathbf{1}$ and $y \geq 0$

has an optimal solution y such that the optimal values of the two problems coincide. Thus one can also use the dual problem to test whether x^b is in $\overline{I}$. Here $\mathbf{1}$ denotes the vector with all its entries equal to 1. The advantage of considering the dual is that one has a fixed polyhedron

$$Q = \{y \in \mathbb{R}^n \mid yA \geq \mathbf{1} \text{ and } y \geq 0\}$$

that can be used to test membership of any monomial $\mathbf{x}^b$, while in the primal problem the polytope P depends on b. Using *PORTA* one can readily obtain the vertices of the polyhedral set Q. The matrix M whose rows are the vertices of Q is a "membership test matrix" in the sense that a monomial $\mathbf{x}^b$ lies in $\overline{I}$ iff $Mb \geq \mathbf{1}$.

Let us illustrate the criterion with a previous example.

Example 3.7 Consider the ideal I of Example 1.4. To verify that $\mathbf{x}^b = x_1^2 \cdots x_8^2$ is in $\overline{I^3}$ one uses the following procedure in *Mathematica*

```
ieq:={
3x1 + 3x2 + 3x3 + 3x4 + 3x5<=2,
3x1 + 3x2 + 3x3 + 3x4 + 3x6<=2,
3x1 + 3x2 + 3x3 + 3x4 + 3x7<=2,
3x1 + 3x2 + 3x3 + 3x4 + 3x8<=2,
3x1 + 3x5 + 3x6 + 3x7 + 3x8<=2,
```

```
3x2 + 3x5 + 3x6 + 3x7 + 3x8<=2,
3x3 + 3x5 + 3x6 + 3x7 + 3x8<=2,
3x4 + 3x5 + 3x6 + 3x7 + 3x8<=2}

vars:={x1,x2,x3,x4,x5,x6,x7,x8}

f:=x1+x2+x3+x4+x5+x6+x7+x8

ConstrainedMax[f,ieq,vars]
```

The answer is:

```
{16/15,
{x1 -> 2/15, x2 -> 2/15, x3 -> 2/15, x4 -> 2/15, x5 -> 2/15,
   x6 -> 2/15, x7 -> 2/15, x8 -> 2/15}}
```

where the first entry is the optimal value and the other entries correspond to a vertex of the polytopes P. Using the criterion and the procedure above one rapidly verifies that $\mathbf{x}^b$ is a minimal generator of $\overline{I^3}$.

4 Computation of general multiplicities

We will make general observations about the computation of the multiplicity of arbitrary primary ideals. The input data is usually the following. Let $A = k[x_1, \ldots, x_r]/L$ be an affine algebra and let I be a primary ideal for some maximal ideal $\mathfrak{M}$ of A. The Hilbert–Samuel polynomial is the function, $n \gg 0$

$$n \mapsto \lambda(A/I^n) = \frac{e(I)}{d!} n^d + \text{lower terms}, \quad \dim A_{\mathfrak{M}} = d.$$

In other words, $e(I)$ is the ordinary multiplicity of the standard graded algebra

$$\mathrm{gr}_I(A) = \sum_{n \geq 0} I^n / I^{n+1}.$$

For the actual computation, ordinarily one needs a presentation of this algebra

$$\mathrm{gr}_I(A) = k[T_1, \ldots, T_m]/H,$$

where the right side is not always a standard graded algebra. In the special case of $I = (x_1, \ldots, x_r)A$ and L is a homogeneous ideal, one has that

$$\mathrm{gr}_M(A) \simeq A,$$

and therefore it can be computed in almost all computer algebra systems by making use of:

Theorem 4.1 (Macaulay Theorem) *Given an ideal I and a term ordering $>$, the mapping*

$$\text{NormalForm}\colon R/I \longrightarrow R/in_>(I) \tag{5}$$

is an isomorphism of k-vector spaces. If I is a homogeneous ideal and $>$ is a degree term ordering, then NormalForm *is an isomorphism of graded k–vector spaces, in particular the two rings have the same Hilbert function.*

For our case, this implies that

$$e(I) = \deg(A) = \deg(k[x_1, \ldots, x_r]/\mathrm{in}_>(L)),$$

where $>$ is any degree term ordering of the ring of polynomials $k[x_1, \ldots, x_r]$. We can turn the general problem into this case by the following observation (which hides the difficulties of the conversion). Let $(R, \mathfrak{m})$ be a Noetherian local ring and let I be an $\mathfrak{m}$–primary ideal. To calculate the multiplicity $e(I)$ we need some form of access to a presentation of the associated graded ring $\mathrm{gr}_I(R)$,

$$\mathrm{gr}_I(R) = k[T_1, \ldots, T_s]/(f_1, \ldots, f_m),$$

in order to avail ourselves of the programs that determine Hilbert functions. A proposed solution, that uses heavily, Gröbner basis theory, is given in [14].

Alternatively, one can turn to indirect means. For instance, suppose $R = k[x_1, \ldots, x_d]$ is a ring of polynomials and I is an $(x_1, \ldots, x_d)$–primary ideal. Let J be a minimal reduction of I, then

$$e(I) = \lambda(R/J).$$

(A similar approach works whenever R is a Cohen–Macaulay ring.) If $>$ is a term order of R, then

$$\lambda(R/I) = \lambda(R/\mathrm{in}_>(J)).$$

The difficulty is to obtain J. It usually arises by taking a set of d generic linear combination of a generating system of I. In addition, even when I is homogeneous, J will not be homogeneous (often it is forbidden to be). One positive observation that can be made is:

Proposition 4.2 *Let I be an $(x_1, \ldots, x_d)$–primary ideal. For any term order $>$ of R,*

$$e(I) \leq e(in_>(I)) \leq d! \cdot e(I). \qquad (6)$$

Proof. Denote $L = \mathrm{in}_>(I)$. The multiplicities are read from the leading coefficients of the Hilbert polynomials $\lambda(R/I^n)$ and $\lambda(R/L^n)$, $n \gg 0$. We note however that while $\lambda(R/I) = \lambda(R/L)$, for large n we can only guarantee

$$\lambda(R/I^n) = \lambda(R/\mathrm{in}_>(I^n)) \leq \lambda(R/L^n),$$

since the inclusion

$$(\mathrm{in}_>(I))^n \subset \mathrm{in}_>(I^n)$$

may be proper.

The other inequality will follow from Lech's formula ([13]) applied to the ideal L:

$$e(L) \leq d!\lambda(R/L)e(R) \leq d!e(I),$$

since $e(R) = 1$ and $\lambda(R/L) = \lambda(R/I) \leq e(I)$. □

As an illustration, let $I = (xy, x^2 + y^2) \subset k[x, y]$. Picking the deglex ordering with $x > y$, gives $L = \mathrm{in}_>(I) = (xy, x^2, y^3)$. We thus have

$$4 = e(I) < e(L) = 5.$$

We are now going to explain the equality $e(I) = e(L)$. Set $L_n = \text{in}_>(I^n)$. Note that $B = \sum_{n\geq 0} L_n t^n$ is the Rees algebra of the filtration defined by L_n's. Actually, B is the initial algebra $\text{in}_>(R[It])$ of the Rees algebra $R[It]$ for the extended term order of $R[t]$:

$$ft^r > gt^s \Leftrightarrow r > s \quad \text{or} \quad r = s \quad \text{and} \quad f > g.$$

In general, B is not Noetherian (which is the case in the simple example above, according to [9]).

Theorem 4.3 *Let I be an $(x_1, \ldots, x_d)$-primary ideal of the polynomial ring $k[x_1, \ldots, x_d]$, and let $>$ be a term ordering. The following conditions are equivalent:*

(a) $e(I) = e(in_>(I))$.

(b) *B is integral over $R[Lt]$, in particular B is Noetherian.*

Proof. (a) $\Rightarrow$ (b): To prove that B is contained in the integral closure of $R[Lt]$ it will be enough to show that for each s, the algebra $R[L_s t]$ is integral over $R[L^s t]$, in other words, to prove the assertion (b) for corresponding Veronese subalgebras.

Since, by hypothesis, the functions $\lambda(R/L^n)$ and $\lambda(R/I^n) = \lambda(R/L_n)$, for $n \gg 0$, are polynomials of degree d with the same leading coefficients, and we have

$$\lambda(R/(L^s)^n) \geq \lambda(R/L_s^n) \geq \lambda(R/L_{sn}) = \lambda(R/I^{sn}) = \lambda(R/(I^s)^n),$$

and

$$e(L^s) = s^d e(L) = s^d e(I) = e(I^s),$$

it follows that L^s and L_s have the same multiplicities. By Rees theorem ([15]), L_s is integral over L^s.

(b) $\Rightarrow$ (a): It is immediate. □

Some of these facts can be extended to more general affine algebras. Suppose I is a monomial ideal of finite co-length and $L \subset I$ is a monomial subideal. The multiplicity of I/L arises from the function

$$n \mapsto \lambda(R/(I^n + L)).$$

We will argue that there is a 'volume formula', similar to Proposition 1.5 that holds in this case. It is an application of the associativity formula for multiplicities: If $\mathfrak{p}_1, \ldots, \mathfrak{p}_r$ are the minimal prime ideals of L of dimension $s = \text{height } L$, then

$$e(I/L) = \sum_{i=1}^{r} \lambda((R/L)_{\mathfrak{p}_i}) \cdot e((I + \mathfrak{p}_i)/\mathfrak{p}_i).$$

Once the $\mathfrak{p}_i$ have been found, we may apply Proposition 1.5 to each monomial ideal $I_i = (I + \mathfrak{p}_i)/\mathfrak{p}_i$. The other terms are co-lengths of monomial ideals. Indeed, the length l_i of the localization $(R/L)_{\mathfrak{p}_i}$ is obtained by setting to 1 in R and in L all the variables which do not belong to $\mathfrak{p}_i$. On the other hand, the ideal $I/\mathfrak{p}_i$ is obtained by setting to 0 the variables that lie in $\mathfrak{p}_i$.

Proposition 4.4 *The multiplicity of the 'monomial' ideal I/L is given by*

$$e(I/L) = \sum_{i=1}^{r} l_i \cdot e(I_i).$$

We can also make comparisons between multiplicities of ideals in general affine rings and the monomial case. Consider an ideal

$$I/L \subset A = R/L = k[x_1, \ldots, x_n]/L$$

of codimension d. For some term order, let L' and I' be the corresponding initial ideals. Denoting by $(\cdot)'$ the initial ideal operation, we have

$$\frac{I'^n + L'}{L'} \subset \frac{(I^n + L)'}{L'}, \quad n \geq 0.$$

As in the case when $L = (0)$, we have

$$\lambda(R/(I^n + L)) = \lambda(R/(I^n + L)') \leq \lambda(R/(I'^n + L')), \ n \geq 0,$$

and consequently,

$$e(I/L) \leq e(I'/L').$$

On the other hand, by Lech's formula ([13]),

$$e(I'/L') \leq d! \cdot \lambda(R/I') \cdot e(R/L') = d! \cdot \lambda(R/I) \cdot e(R/L),$$

the substitution $e(R/L') = e(R/L)$ by Macaulay's theorem.

References

[1] D. Bayer and M. Stillman, *Macaulay*: A system for computation in algebraic geometry and commutative algebra, 1992. Available at `http://www.math.harvard.edu`.

[2] W. Bruns and J. Herzog, *Cohen–Macaulay Rings*, Cambridge University Press, 1993.

[3] W. Bruns and R. Koch, *Normaliz*: A program to compute normalizations of semigroups. Available by anonymous `ftp from ftp.mathematik.Uni-Osnabrueck.DE/pub/osm/kommalg/software/`

[4] B. Büeler, A. Enge and K. Fukuda, Exact volume computations for convex polytopes: a practical study. In G. Kalai and G. Ziegler, editors, *Polytopes-Combinatorics and Computation*, DMV-Seminar **29**, Birkhäuser Verlag, 2000.

[5] T. Christof, revised by A. Löbel and M. Stoer, *PORTA*, 1997. Available at `http://www.zib.de/Optimization/Software/Porta`.

[6] V. Chvátal, *Linear Programming*, W. H. Freeman and Company, New York, 2000.

[7] A. Corso, C. Huneke and W. V. Vasconcelos, On the integral closure of ideals, *Manuscripta Math.* **95** (1998), 331–347.

[8] D. Eisenbud, *Commutative Algebra with a view toward Algebraic Geometry*, Springer, Berlin Heidelberg New York, 1995.

[9] S. Eliahou, A problem about polynomial ideals, *Contemporary Math.* **58** (1986), 107–120.

[10] W. Feller, *An Introduction to Probability Theory and its Applications*, 3rd Edition, Wiley, New York, 1967.

[11] *GMP Libraries*, 1997. Available at `http://www.swox.com/gmp/`.

[12] D. Grayson and M. Stillman, Macaulay 2, a software system for research in algebraic geometry. Available at `http://www.math.uiuc.edu/Macaulay2`.

[13] C. Lech, Note on multiplicities of ideals, *Arkiv för Matematik* 4 (1960), 63–86.

[14] T. Mora and M. E. Rossi, An algorithm for the Hilbert function of a primary ideal, *Comm. Algebra* **23** (1995), 1899–1911.

[15] D. Rees, α–transforms of local rings and a theorem on multiplicities of ideals, *Math. Proc. Camb. Phil. Soc.* **57** (1961), 8–17.

[16] L. Robbiano and M. Sweedler, Subalgebra bases, *in Commutative Algebra*, Proceedings, Salvador 1988 (W. Bruns and A. Simis, Eds.), Lecture Notes in Mathematics **1430**, Springer, Berlin Heidelberg New York, 1990, 61–87.

[17] Yu. A. Shreider, *The Monte Carlo Method*, Pergamon Press, Oxford, 1966.

[18] R. Stanley, *Enumerative Combinatorics*, Vol. I, Wadsworth & Brooks/Cole, Monterey, California, 1986.

[19] B. Teissier, Monômes, volumes et multiplicités, in *Introduction à la theorie des Singularitiés*, Vol. II (Lê Dung Trang, Ed.), Travaux en Cours **37**, Hermann, Paris, 1988.

[20] W. V. Vasconcelos, *Computational Methods in Commutative Algebra and Algebraic Geometry*, Springer, Heidelberg, 1998.

[21] R. Villarreal, *Monomial Algebras*, Monographs and Textbooks in Pure and Applied Mathematics **238**, Marcel Dekker, New York, 2001.

[22] O. Zariski and P. Samuel, *Commutative Algebra*, Vol. II, Van Nostrand, Princeton, 1960.

Analytic Spread of a Pregraduation

YOUSSOUF M. DIAGANA Université d'Abobo-Adjamé, 03 BP 1644 Abidjan 03, Côte d'Ivoire. Email address : y_diagana@yahoo.com

ABSTRACT We define a pregraduation of a commutative ring A by a family $g = (G_n)_{n\in\mathbb{Z}\cup\{+\infty\}}$ of ideals of A such that $G_0 = A$, $G_\infty = (0)$ and $G_pG_q \subseteq G_{p+q}$, for all $p, q \in \mathbb{Z}$. The notion of $J-$independence of order k with respect to a pregraduation of a ring A is defined as in [1]. We will show that r elements of G_1 are $J-$independent of order k with respect to a pregraduation g if and only if there exist isomorphisms from the polynomial ring with r indeterminates over $\dfrac{A}{J+G_k}$ to some $\dfrac{A}{J+G_k}-$algebras. A weak notion of $J-$independence called the regular $J-$independence will allow to define the analytic spread of a pregraduation on a ring.

INTRODUCTION

The purpose of this paper is to define and study the analytic independence of a family $(G_n)_{n\in\mathbb{Z}\cup\{+\infty\}}$ of ideals of a ring A such that $G_0 = A$ and $G_pG_q \subseteq G_{p+q}$ for all $p, q \in \mathbb{Z}$, called a pregraduation of A, with the convention that $G_{+\infty} = (0)$ and to give extensions of the analytic spread for a pregraduation.

Theorem 2.4 gives criteria of $J-$independence of order k with respect to a pregraduation of a ring A, where $k \in \overline{\mathbb{N}^*} = \mathbb{N}^* \cup \{+\infty\}$ and J is an ideal of A. The maximum number of elements of J which are $J-$independent of order k with respect to $g = (G_n)$ will be denoted by $\ell_J(g, k)$. We show that this number is different from the analytic spread of a filtration as defined by J.S. Okon in [4].

The notion of regular $J-$independence of order k with respect to a pregraduation g will allow to define the regular $J-$analytic spread of g, denoted by $\ell^a_J(g, k)$, and the regular analytic spread $\ell^a(g, k)$ of g as $\ell^a(g, k) =$

$\sup\{\ell^a_{\mathcal{M}}(g,k) : \mathcal{M} \in \text{Max } A\}$.

Corollary 3.4 gives a characterization of $\ell^a_J(f,k)$ by some integers $\ell_J(I_p,k)$ when $f=(I_n)$ is a nœtherian filtration. When $(A,\mathcal{M})$ is a nœtherian local ring with infinite residue field and f a nœtherian filtration on A, Proposition (3.7) shows that the integer $\ell^a_{\mathcal{M}}(f,k)$ coincides with the various extensions of the analytic spread obtained in [1] except for $\ell_{\mathcal{M}}(f,k)$.

Throughout this paper all rings are commutative and filtrations are decreasing pregraduations. The I-adic filtration is the family (I^n), denoted f_I, where I is an ideal of A and $I^n = A$ for $n \le 0$. A filtration $f=(I_n)$ is said to be nœtherian if its generalized Rees ring $\Re(A,f) = \sum_{n\in\mathbb{Z}} I_n X^n$ is nœtherian.

1. GENERALIZED ANALYTIC INDEPENDENCE

DEFINITIONS 1.1. **(1.1.1)** Let A be a ring and $(G_n)_{n\in\mathbb{Z}\cup\{+\infty\}}$ a family of ideals of A. We say that (G_n) is a pregraduation of A if $G_0 = A$, $G_\infty = (0)$ and $G_pG_q \subseteq G_{p+q}$ for all $p,q\in\mathbb{Z}$.

(1.1.2) Let $g=(G_n)_{n\in\mathbb{Z}\cup\{+\infty\}}$ be a pregraduation of a ring A, J an ideal of A and $k\in\overline{\mathbb{N}^*} = \mathbb{N}^*\cup\{+\infty\}$ such that $J+G_k \ne A$. Elements $a_1,\dots,a_r$ of G_1 are said to be J-independent of order k with respect to g if for each homogeneous polynomial $F(X_1,\dots,X_r)$ of degree s with coefficients in A, the relation $F(a_1,\dots,a_r)\in JG_s+G_{s+k}$ implies that F has its coefficients in $J+G_k$.

PROPOSITION 1.2. *Let A be a ring, $g=(G_n)_{n\in\mathbb{Z}\cup\{+\infty\}}$ a pregraduation of A and $k\in\overline{\mathbb{N}^*}$. Let J be an ideal of A such that $J+G_k\ne A$, $a_1,\dots,a_r$ elements which are J-independent of order k with respect to g and I the ideal $(a_1,\dots,a_r)$. We have :*

(i) *If J contains G_k, then the elements $a_1,\dots,a_r$ are J-independent with respect to g and to $f_I=(I^n)$.*

(ii) *If there exists an integer i such that $a_i\in J+G_k$, then*
$$I^p \subseteq G_p \subseteq J+G_k \quad \text{for } p\ge 1.$$

(iii) *If $r\ge 2$, then $I^p\subseteq G_p\subseteq J+G_k$ for $p\ge 1$.*

Proof. *(i)* Let $x=F(a_1,\dots,a_r)$ where F is an homogeneous polynomial of degree s with coefficients in A. Suppose that J contains G_k.
$[x\equiv 0 \pmod{JG_s}$ or $x\equiv 0 \pmod{JI^s}]$ implies that $x\in JG_s+G_{s+k}$.
Hence $F\in (J+G_k)[X_1,\dots,X_r]$. Furthermore, $J+G_k = J+I^k = J$.
Therefore the elements $a_1,\dots,a_r$ are J-independent (of order $+\infty$) with respect to g and to f_I.

(ii) If $a_i\in J+G_k$ then for $p\ge 1$ and $y\in G_p$, we have :
$$ya_i^p \in G_p(J+G_k)\subseteq (JG_p+G_{p+k}).$$

$a_1, \ldots, a_r$ being $J-$independent of order k with respect to g, we have : $y \in J + G_k$. Therefore $G_p \subseteq J + G_k$ and we have $I^p \subseteq G_1^p \subseteq G_p \subseteq J + G_k$.

(iii) If $r \geq 2$, for $i \neq j \in [1, \ldots, r]$, let $F(X_1, \ldots, X_r) = a_i X_j - a_j X_i$. Then F is homogeneous of degree 1 and $F(a_1, \ldots, a_r) = 0$. So, $a_i \in J + G_k$. The result follows by (ii).

PROPOSITION 1.3. *Let A be a ring, $g = (G_n)$ a pregraduation of A and $k \in \overline{\mathbb{N}^*}$. Let J be an ideal of A such that $J + G_k \neq A$. Assume that $(G_{n+k})_{n \geq 0}$ is decreasing or that $G_{n+k} \subseteq G_n$ for $n \geq 1$. Let $a_1, \ldots, a_r$ be $J-$independent elements of order k with respect to g. Then for $p \geq 1$ we have :*

(i) *If $G_k \subseteq J + G_{pk}$, then $a_1^p, \ldots, a_r^p$ are $J-$independent of order k with respect to the pregraduation $g^{(p)} = (G_{pn})$.*

(ii) *If $G_k \subseteq J + G_p^k$, then $a_1^p, \ldots, a_r^p$ are $J-$independent of order k with respect to the pregraduation $f_{Gp} = (G_p^n)$.*

(iii) *If $G_k \subseteq J + G_k^p$, then $a_1^p, \ldots, a_r^p$ are $J-$independent of order k with respect to the pregraduation $g^p = (G_n^p)$.*

(iv) *In particular, if $J \supseteq G_k$, $a_1^p, \ldots, a_r^p$ are $J-$independent of order k with respect to the pregraduations $g^{(p)}$, g^p and f_{G_p}.*

Proof. The Proof follows from definitions of (1.1) and from the fact that, under the hypotheses, for all $n \geq 0$ we have :

$$\begin{cases} JG_p^n + G_p^{n+k} \subseteq JG_{pn} + G_{p(n+k)} \subseteq JG_{pn} + G_{pn+k} \\ JG_n^p + G_{n+k}^p \subseteq JG_{pn} + G_{p(n+k)} \subseteq JG_{pn} + G_{pn+k}. \end{cases}$$

NOTATIONS 1.4. Let A be a ring, $g = (G_n)_{n \in \mathbb{Z} \cup \{+\infty\}}$ a pregraduation of A and $k \in \overline{\mathbb{N}^*}$. Let J be an ideal of A such that $J + G_k \neq A$. We put : $\ell_J(g, k) = \sup\{r \in \mathbb{N} \;/\; \exists\, a_1, \ldots, a_r \in J,\ J-$independent of order k with respect to $g\}$, $\ell(g, k) = \sup\{\ell_{\mathcal{M}}(g, k) : \mathcal{M}$ maximal over $G_k\}$, and $\sup(J) = \sup\{r \in \mathbb{N} \;/ \exists\, a_1, \ldots, a_r \in J,\ J - \text{independent}\}$.

REMARK 1.5. (i) If $\mathcal{M}$ is a maximal ideal of A and $\mathcal{M} \not\supseteq G_k$, then $\ell_{\mathcal{M}}(g, k) = 0$. Consequently, $\ell(g, k) = \sup\{\ell_{\mathcal{M}}(g, k) : \mathcal{M} \in \text{Max } A\}$.

(ii) Assume that $(G_{n+k})_{n \in \mathbb{N}}$ is decreasing or that $G_{n+k} \subseteq G_n$ for all $n \geq 1$. If J contains G_k, then for all p and $n \in \mathbb{N}^*$ we have :

a) $\ell_J(g, k) \leq \ell_J(g^{(p)}, k) \leq \ell_J(G_p, k) \leq \ell_J(G_p^n, k)$,

b) $\ell_J(g, k) \leq \ell_J(g^{(p)}, k) \leq \ell_J(G_{pn}, k)$. (See (1.3)-(iv))

PROPOSITION 1.6. *Under hypotheses and notations of (1.4), if P is a prime ideal over $J+G_k$, we have :*

(1.6.1) $\ell_J(g,k) \leq \sup(J+G_k) \leq ht(J+G_k) \leq \dim A_P = ht\ P.$

(1.6.2) *If J contains G_k, then*

(i) $\ell_J(g,k) \leq \ell_J(g) \leq ht\ J \leq \dim A.$

(ii) *Moreover, if A is a nœtherian ring and k is such that $(G_{n+k})_{n\in\mathbb{N}}$ is decreasing or such that $G_{n+k} \subseteq G_n$ for all $n \geq 1$, then the sequence $n \mapsto \ell_J(G_p^{n!}, k)$ is an increasing and eventually stationary sequence.*

2. CRITERIA OF J–INDEPENDENCE

2.1.– Let $k \in \overline{\mathbb{N}^*}$, $g = (G_n)$ and $f = (I_n)$ be two pregraduations of A such that $I_n \subseteq G_n$ for all n. Let J be an ideal of A.

For each n, we put $K_n = I_n \cap (JG_n + G_{n+k})$; then we have : $K_n I_m \subseteq K_{n+m}$ for all n, $m \in \mathbb{Z}$. So, $\sum\limits_n \frac{I_n}{K_n}$ is a graded ring where $(a+K_n)(b+K_m) = (ab+K_{n+m})$ for all $a \in I_n$ and $b \in I_m$. Let $F = \sum\limits_n I_n X^n$ and $G = \sum\limits_n G_n X^n$. Then $L = \sum\limits_n K_n X^n = F \cap (u^k, J)G$ is a graded ideal of F where $u = \frac{1}{X}$ with the convention that $u^\infty = 0$ and we have : $\frac{F}{L} \approx \sum\limits_n \frac{I_n}{K_n}$.

In particular, if $f = g$ we have :

(2.1.1) $$\frac{G}{G \cap (u^k, J)G} \approx \sum_n \frac{G_n}{G_n \cap (JG_n + G_{n+k})}$$

(2.1.2) If $G_{n+k} \subseteq G_n$ for all $n \geq 1$, then $\frac{G}{(u^k, J)G} \approx \sum\limits_{n\in\mathbb{Z}} \frac{G_n}{JG_n + G_{n+k}}$

(2.1.3) $\frac{G^+}{G^+ \cap (u^k, J)G} \approx \sum\limits_{n\geq 0} \frac{G_n}{G_n \cap (JG_n + G_{n+k})}$ where $G^+ = \sum\limits_{n\geq 0} G_n X^n$

(2.1.4) $$\frac{G^+}{JG^+} \approx \sum_n \frac{G_n}{JG_n}.$$

2.2.– Let $R(A,I) = \sum\limits_{n\geq 0} I^n X^n$ be the Rees ring of an ideal $I \subseteq G_1$. We have :

(2.2.1) $$\frac{R(A,I)}{R(A,I) \cap (u^k, J)G} \approx \sum_{n\geq 0} \frac{I^n}{I^n \cap (JG_n + G_{n+k})}$$

(2.2.2) $$\frac{R(A,I)}{R(A,I) \cap JG^+} \approx \sum_{n\geq 0} \frac{I^n}{I^n \cap JG_n}.$$

(2.2.3) If $I \subseteq J$, then $\dfrac{\Re(A,I)}{(u,J)\Re(A,I)} \approx \sum_{n\geq 0} \dfrac{I^n}{JI^n} \approx \dfrac{R(A,I)}{JR(A,I)}$,

where $\Re(A,I) = \sum_{n\in\mathbb{Z}} I^n X^n$ is the generalized Rees ring of the ideal I.

2.3.– Let J be an ideal of A such that $J + G_k \neq A$, $a_1, \ldots, a_r \in G_1$ and $I = (a_1, \ldots, a_r)$. Put $K_n = I^n \cap (JG_n + G_{n+k})$ for $n \geq 0$ and $Q_J(g,k)$ the graded ring $\sum_n \dfrac{I^n}{K_n}$. Let $t_i = a_i + K_1$ for $i = 1, \ldots, r$. We have :

$$Q_J(g,k) = \frac{A}{K_0}[t_1, \ldots, t_r] = \left\{ \sum_{i_1,\ldots,i_r} (\alpha_{i_1,\ldots,i_r} + K_0)\, t_1^{i_1} \cdots t_r^{i_r} : \alpha_{i_1,\ldots,i_r} \in A \right\}.$$

Let $\varphi_J(g,k)$ be the graded morphism from $\dfrac{A}{J+G_k}[X_1, \ldots, X_r]$ to $Q_J(g,k)$ such that $\varphi_J(g,k)(X_i) = t_i$ for each i and $\varphi_J(g,k)(\alpha) = \alpha$ for $\alpha \in \dfrac{A}{J+G_k}$. There exists an isomorphism ψ_k from $Q_J(g,k)$ to $\dfrac{R(A,I)}{R(A,I) \cap (u^k, J)G}$ such that $\psi_k(t_i) = a_i X + R(A,I) \cap (u^k, J)G$ and $\psi_k(\alpha) = \alpha$ for $\alpha \in \dfrac{A}{J+G_k}$.

Let $u_i = \psi_k(t_i)$ for all i. Then $\dfrac{R(A,I)}{R(A,I) \cap (u^k, J)G} = \dfrac{A}{J+G_k}[u_1, \ldots, u_r]$.

Put $\varphi_k = \varphi_J(g,k)$ and $\theta_k = \psi_k \circ \varphi_k$. We have the following theorem :

THEOREM 2.4. *The following statements are equivalent :*

(i) $a_1, \ldots, a_r$ *are* $J-$*independent of order* k *with respect to* g;

(ii) θ_k *is an isomorphism of* $\dfrac{A}{J+G_k}[X_1, \ldots, X_r]$ *over* $\dfrac{R(A,I)}{R(A,I) \cap (u^k, J)G}$;

(iii) *The elements* $t_1, \ldots, t_r$ *are algebraically independent over* $\dfrac{A}{J+G_k}$;

(iv) *The elements* $u_1, \ldots, u_r$ *are algebraically independent over* $\dfrac{A}{J+G_k}$.

Proof. For all $F = \sum_{i_1+\cdots+i_r=s} \lambda_{i_1,\ldots,i_r} X_1^{i_1} \cdots X_r^{i_r}$ where $\lambda_{i_1,\ldots,i_r} \in A$, we put

$$\overline{F}(X_1, \ldots, X_r) = \sum_{i_1+\cdots+i_r=s} (\lambda_{i_1,\ldots,i_r} + K_0) X_1^{i_1} \cdots X_r^{i_r}.$$

(i) $\Longleftrightarrow$ *(ii)* The elements $a_1, \ldots, a_r$ are $J-$independent of order k with respect to g if and only if for all $F \in A[X_1, \ldots, X_r]$ homogeneous of degree s,

$F(a_1,\dots,a_r) \equiv 0 \pmod{JG_s + G_{s+k}}$ implies that F has all of its coefficients in $J + G_k$, i.e., $F(a_1,\dots,a_r) \in I^s \cap (JG_s + G_{s+k}) = K_s$ implies that $F \in K_0[X_1,\dots,X_r]$ this means that for all $F \in A[X_1,\dots,X_r]$, $\varphi_k(\overline{F}) = 0$ implies that $\overline{F} = 0$, i.e., φ_k is an isomorphism and θ_k is an isomorphism.

(ii) $\Longleftrightarrow$ *(iv)* $\theta_k(X_i) = u_i$ and, for $G = \sum_{i_1+\cdots+i_r=s} \alpha_{i_1,\dots,i_r} X_1^{i_1}\cdots X_r^{i_r}$,

$$\theta_k(\overline{G}) = \psi_k(\overline{G}(t_1,\dots,t_r)) = \sum_{i_1+\cdots+i_r=s} (\alpha_{i_1,\dots,i_r} + K_0) u_1^{i_1}\cdots u_r^{i_r} = \overline{G}(u_1,\dots,u_r).$$

θ_k being surjective, (ii) holds if and only if θ_k is injective if and only if for all $G \in A[X_1,\dots,X_r]$, $\theta_k(\overline{G}) = 0$ implies that $\overline{G} = 0$ this means that the elements $u_1,\dots,u_r$ are algebraically independent over $\dfrac{A}{J+G_k}$.

(ii) $\Longleftrightarrow$ *(iii)* (ii) holds if and only if φ_k is an isomorphism. Replacing θ_k by φ_k in the proof of ((ii) $\Leftrightarrow$ (iv)) and u_i by t_i one obtains : φ_k is an isomorphism if and only if (iii) holds.

3. REGULAR INDEPENDENCE AND ANALYTIC SPREAD

DEFINITIONS 3.1. **(3.1.1)** Let A be a ring, $k \in \overline{\mathbb{N}^*}$, $g = (G_n)$ a pregraduation and J an ideal of A. We say that $a_1,\dots,a_r$ are regular $J-$independent of order k with respect to g if there exists $p \in \mathbb{N}^*$ with $J + G_{pk} \neq A$ such that $a_1,\dots,a_r$ are $J-$independent of order k with respect to $g^{(p)} = (G_{pn})$.

(3.1.2) The regular $J-$analytic spread of order k of g is the integer $\ell_J^a(g,k) = \sup\{r \in \mathbb{N}$ such that there exist $a_1,\dots,a_r \in J$, regular $J-$independent of order k with respect to $g\}$.

(3.1.3) The regular analytic spread of order k of g is the integer $\ell^a(g,k) = \sup\{\ell^a_{\mathcal{M}}(g,k) : \mathcal{M} \in \operatorname{Max} A \text{ and } \mathcal{M} \supseteq G_k\}$.

REMARK 3.2. Let A be a ring and $g = (G_n)$ a pregraduation of A. Then we have :

(i) If the elements $a_1,\dots,a_r$ are $J-$independent of order k with respect to g, then they are regular $J-$independent of order k with respect to g. Consequently $\ell_J(g,k) \leq \ell_J^a(g,k)$.

(ii) Assume that J contains G_{ik} for all $i \geq 1$. If $a_1,\dots,a_r$ are regular $J-$independent of order k with respect to g then they are $J-$independent of order nk for $n \geq 1$ and of order $+\infty$; consequently,

a) if P is a prime ideal over J, then we have :

$$\ell_J(g,k) \leq \ell_J^a(g,k) \leq \sup(J) \leq ht\ J) \leq \dim A_p$$

b) if A is a nœtherian ring, then $\ell_J^a(g,k) < \infty$.

(iii) $\ell^a(g,k) = \sup\{\ell_{\mathcal{M}}^a(g,k) : \mathcal{M} \in \text{Max } A\}$.

(iv) If A is a nœtherian semi-local ring, then $\ell^a(g,k)$ is finite and there exists a maximal ideal $\mathcal{M}$ over G_k such that $\ell^a(g,k) = \ell_{\mathcal{M}}^a(g,k)$.

THEOREM 3.3. *Let A be a ring, $k \in \overline{\mathbb{N}^*}$ and $g = (G_n)$ a pregraduation of A. We assume that the sequence $(G_{n+k})_{n \geq 0}$ is decreasing or that $G_{n+k} \subseteq G_n$ for each $n \geq 1$. Let J be an ideal of A which contains G_k.*

(i) *We have $\ell_J(g^{(n)},k) \leq \ell_J(g^{(pn)},k) \leq \ell_J^a(g,k)$ for $n, p \in \mathbb{N}^*$.*

If A is nœtherian, then we have :

(ii) *There exists $p \in \mathbb{N}^*$ such that $\ell_J^a(g,k) = \ell_J(g^{(pn)},k)$ for $n \in \mathbb{N}^*$*

(iii) $\ell_J^a(g,k) = \sup\{\ell_J(g^{(n)},k) : n \in \mathbb{N}^*\}$

(iv) $\ell_J^a(g,k) = \ell_J^a(g^{(p)},k)$ *for $p \in \mathbb{N}^*$*

(v) *The sequence $n \mapsto \ell_J(g^{(n!)},k)$ is an increasing and eventually stationary sequence of limit $\ell_J^a(g,k)$.*

Proof. *(i)* For $n \geq 1$, if elements $a_1, \dots, a_r$ are in J and are $J-$independent of order k with respect to $g^{(n)}$, then they are regular $J-$independent of order k with respect to g. Hence $\ell_J^a(g,k) \geq r$; so, $\ell_J\left(g^{(n)},k\right) \leq \ell_J^a(g,k)$.
Furthermore, from (1.5)-(ii) we have :

$$\ell_J\left(g^{(n)},k\right) \leq \ell_J\left(\left[g^{(n)}\right]^{(p)},k\right) = \ell_J\left(g^{(pn)},k\right) \leq \ell_J^a(g,k), \text{ for all } p \geq 1.$$

Hence we have (i).

Suppose that A is nœtherian. Then we have :

(ii) If $a_1, \dots, a_r \in J$ are $J-$independent of order k with respect to $g^{(p)}$ with $p \geq 1$, we have from (1.2)-(ii) : J contains all G_n, $n \geq 1$ and from (3.2)-(ii), $\ell_J^a(g,k)$ is finite.

Let $r = \ell_J^a(g,k)$. There exists $p \geq 1$ and elements $b_1, \dots, b_r \in J$ which are $J-$independent of order k with respect to $g^{(p)}$. Therefore $\ell_J^a(g,k) \leq \ell_J(g^{(p)},k)$. Using (i) we have : $\ell_J^a(g,k) = \ell_J\left(g^{(pn)},k\right)$ for $n \geq 1$.

(iii) is a consequence of (i) and (ii).

(iv) (i) and (iii) $\Rightarrow$

$$\begin{aligned} \ell_J^a(g,k) &= \sup\{\ell_J(g^{(n)},k) : n \in \mathbb{N}^*\} \\ &\leq \sup\{\ell_J(g^{(pn)},k) : n \in \mathbb{N}^*\} \leq \ell_J^a(g,k). \end{aligned}$$

$$\text{So, } \ell_J^a(g,k) = \sup\left\{\ell_J(g^{(pn)},k) : n \in \mathbb{N}^*\right\} = \ell_J^a(g^{(p)},k).$$

(v) is a consequence of (i) and (ii).

Applying Theorem 3.3 to nœtherian filtrations we obtain the following result which gives a characterization of the regular $J-$analytic spread of a nœtherian filtration $f = (I_n)$ in terms of $\ell_J(I_n)$.

COROLLARY 3.4. *Let A be a nœtherian ring, $f = (I_n)$ a nœtherian filtration on A and J an ideal of A over I_1. Let k be an integer in $\overline{\mathbb{N}^*}$. We assume that m is a positive integer such that $I_{mn} = I_m^n$ for all $n \in \mathbb{N}$. Then*

(i) $\ell_J^a(f,k) = \ell_J^a(I_{mn},k)$ *for* $n \in \mathbb{N}^*$

(ii) *There exists* $p \in \mathbb{N}^*$, *which is a multiple of* m *such that*
$$\ell_J^a(f,k) = \ell_J(I_{pn},k) \quad \text{for } n \in \mathbb{N}^*$$

(iii) $\ell_J^a(f,k) = \sup\{\ell_J(I_{mn},k) : n \in \mathbb{N}^*\}$

(iv) *the sequence* $n \mapsto \ell_J(I_{mn!},k)$ *is an increasing and eventually stationary sequence of limit* $\ell_J^a(f,k)$

(v) *the sequence* $n \mapsto \ell_J(I_{n!},k)$ *is eventually stationary of limit* $\ell_J^a(f,k)$.

COROLLARY 3.5. *Let A be a nœtherian ring, $f = (I_n)$ a nœtherian filtration on A and $m \geq 1$ an integer such that $I_{mn} = I_m^n$ for $n \geq 0$. Let $k \in \overline{\mathbb{N}^*}$.*

(i) *If $\mathcal{M}$ is a maximal ideal over $\sqrt{f}$, we have* $\ell_{\mathcal{M}}^a(f,k) = \ell_{\mathcal{M}}^a(f)$

(ii) $\ell^a(f,k) = \ell^a(f)$.

Proof. *(i)* From the definition, $\ell_{\mathcal{M}}^a(f,k) = \sup\{\ell_{\mathcal{M}}(I_m^n,k) : n \in \mathbb{N}^*\}$.
If $J \supseteq I_p$, $JI_p^s + I_p^{s+k} = JI_p^s$ and $J + I_p^k = J$. Therefore the elements of I_p are $J-$independent of order k with respect to f_{I_p} if and only if they are $J-$independent with respect to f_{I_p}. Thus we have : $\ell_{\mathcal{M}}(I_m^n,k) = \ell_{\mathcal{M}}(I_m^n)$ for $n \geq 1$ and $\ell_{\mathcal{M}}^a(f,k) = \sup\{\ell_{\mathcal{M}}(I_m^n) : n \in \mathbb{N}^*\} = \ell_{\mathcal{M}}^a(f)$.

(ii) If $k \in \mathbb{N}^*$ and $\mathcal{M}$ a maximal ideal over I_k then $\sqrt{f} \subseteq \mathcal{M}$.
Taking the supremum we have (ii).

EXAMPLE 3.6. Here we prove that $\ell(g,k)$ can be different from $\ell^a(g,k)$ even for a nœtherian filtration of a local ring with infinite residue field.

Let $(A,\mathcal{M})$ be a nœtherian local ring and I a proper ideal of A the analytic spread of which is given by $\gamma(I) = \dim \sum_{n \geq 0}(I^n/\mathcal{M}I^n) = 2$.

(For instance it is the case if A is equal to the localized of $K[X_1,X_2,X_3]$ at (X_1,X_2,X_3), I the localized of (X_1,X_2) and K a field).

If $A/\mathcal{M}$ is infinite, then by D. G. Northcott and D. Rees [3, §3, Theorem 3], $\ell(I)$ is equal to the analytic spread of I, i.e., $\ell(I) = \gamma(I) = 2$. (See [1, §0] for various interpretations of analytic spread).
Thus, there exist $a_1, a_2 \in I$ which are $\mathcal{M}$-independent with respect to f_I.

a) Let $g = (G_n)$ be such that $G_n = A$ for $n \leq 0$, $G_{4n} = I^{2n}$ for $n \geq 1$, $G_{4n+1} = G_{4n+2} = I^{2n+2}$ and $G_{4n+3} = I^{2n+3}$ for $n \geq 0$. One verifies that g is a pregraduation of A and is not a filtration. For all $a \in G_1$ we have :

$$1\, a^{4n+2} = aa^{4n+1} \in G_1 G_{4n+2} \subseteq \mathcal{M}G_{4n+2} + G_{4n+2+k}, \text{ for } n \geq 1.$$

But $1 \notin \mathcal{M} = \mathcal{M} + G_k$ therefore $\ell(g,k) = 0$.
From Theorem 3.3, Corollary 3.4 and [1, (2.4)] we have :

$\ell^a_{\mathcal{M}}(g,k) = \ell^a_{\mathcal{M}}(g^{(4)},k) = \ell^a_{\mathcal{M}}(I^2,k) = \ell^a(I,k) = \sup\{\ell(I^n) : n \in \mathbb{N}^*\} = \sup\{\gamma(I^n) : n \in \mathbb{N}^*\} = \gamma(I)$. Thus $\ell^a(g,k) = 2 \neq \ell(g,k) = 0$.

b) Let $f = (I_n)$ be a filtration such that $I_n = A$ for $n \leq 0$, $I_{2n} = I^n$ for $n \geq 1$ and $I_{2n+1} = I^{n+1}$ for $n \geq 0$. One verifies that f is a nœtherian filtration.
For $k \in \overline{\mathbb{N}^*}$, $a \in G_1$ and $n \geq 1$, we have : $a^{2n+2} = a\, a^{2n+1} \in I_1 I_{2n+2}$.
Thus $a^{2n+2} \in \mathcal{M}I_{2n+2} + I_{2n+2+k}$. As $1 \notin \mathcal{M}$, we have : $\ell(f,k) = 0$.
Furthermore, $\ell^a(f,k) = \ell^a(I,k) = \sup\{\ell(I^n) : n \in \mathbb{N}^*\} = \gamma(I) = 2 \neq \ell(f,k)$.

Okon defined in [4, (2.1.7)] the analytic spread of a filtration f by the integer :

$$\gamma(f,1) = \sup\left\{\dim \frac{\Re(A,f)}{(u,\mathcal{M})\Re(A,f)} : \mathcal{M} \in \operatorname{Max} A\right\} \text{ where } \Re(A,f) = \sum_{n\in\mathbb{Z}} I_n X^n$$

for $f = (I_n)$ and $u = X^{-1}$. By (2.1.2), this means that

$$\gamma(f,1) = \sup\left\{\dim \sum_{n\geq 0} \frac{I_n}{\mathcal{M}I_n + I_{n+1}} : \mathcal{M} \in \operatorname{Max} A\right\}.$$

PROPOSITION 3.7. *Let $(A,\mathcal{M})$ be a nœtherian local ring, $f = (I_n)$ a nœtherian filtration on A and I an ideal of A. If $A/\mathcal{M}$ is infinite then :*

(i) $\ell(f,k) \leq \ell(f)$ *and* $\ell(f,k) \leq \ell^a(f,k) = \ell^a(f) = \gamma(f,1) = \gamma_{\mathcal{M}}(f) = \tau_{\mathcal{M}}(f)$

(ii) $\ell(I) = \ell^a(I) = \gamma(I,1) = \gamma_{\mathcal{M}}(I) = \tau_{\mathcal{M}}(I)$

$$\textit{with } \gamma_J(f) = \dim \frac{R(A,f)}{JR(A,f)} \textit{ and } \tau_J(f) = \operatorname{tr}\deg_{\frac{A}{J+G_k}} \frac{R(A,f)}{JR(A,f)}$$

where dim *designates the Krull dimension and* tr deg *designates the transcendence degree (See* [2, §1]*).*

Proof. *(i)* Let $p \geq 1$ be such that $I_{pn} = I_p^n$ for $n \geq 0$. Then

$$\ell^a(f) = \ell^a(I_p) = \sup\{\ell(I_p^n) : n \in \mathbb{N}^*\}.$$

By Northcott and Rees [3, §3, Theorem 3], as $A/\mathcal{M}$ is infinite, we have $\ell(I_p^n) = \gamma(I_p^n)$.

From Propositions (2.4) and (2.12) of [1], there exists $q \geq 1$ such that
$$\gamma(I_p^n) = \gamma(I_p) = \gamma(I_{pq}) = \gamma_{\mathcal{M}}(f) = \gamma(f,1) = \tau_{\mathcal{M}}(f).$$
With Corollary 3.5, Remark 3.2-(i) and Proposition 1.2-(i) we have :
$$\ell^a(f,k) = \ell^a(f) \geq \ell(f,k) \text{ and } \ell(f,k) \leq \ell(f).$$

(ii) If $A/\mathcal{M}$ is infinite, then $\ell(I) = \gamma(I)$ and the conclusion follows from (i).

References

[1] Y. Diagana, H. Dichi and D. Sangaré, Filtrations, generalized analytic independence, analytic spread, Afrika Matematika, Série 3, 4 (1994) 101-114.

[2] E. Hamann, Krull dimension and transcendence degree, Comm. Algebra 18 (4) (1990) 1189-1198.

[3] D. G. Northcott and D. Rees, Reduction of ideals in local rings, Proc. Cambridge Philos. Soc. 50 (1954) 145-158.

[4] J. S. Okon, Prime divisors, analytic spread and filtrations, Pacific J. Math. 113 (1984) 451-462.

Extension of the Hilbert–Samuel Theorem

HENRI DICHI, *Laboratoire de Mathématiques pures, Université Blaise Pascal, Les Cézeaux, 63177 Aubière Cedex, France.*
e-mail: Henri.Dichi@math.univ-bpclermont.fr

INTRODUCTION

Let $A = \bigoplus_{n\in\mathbb{N}} A_n$ be a noetherian graded ring of finite Krull dimension where A_0 is an artinian ring and $M = \bigoplus_{n\in\mathbb{N}} M_n$ be a graded A-module of finite type with Krull dimension d. The Hilbert function $H(M,-)$ of M is defined by $H(M,n) = \ell_{A_0}(M_n)$. The Hilbert-Theorem asserts that $H(M,-)$ is a polynomial function of degree $d-1$ when A is a homogeneous graded ring. Here we are concerned with the case of a not necessarily homogeneous graded ring A. We prove that the Hilbert function $H(M,-)$ and the cumulative Hilbert function $H^*(M,-)$ are quasi-polynomial functions and in addition that $H^*(M,-)$ is a uniform quasi-polynomial function. Then it is possible to define the multiplicity of a graded module of finite type by the asymptotic formula $e_A(M) = \lim_{n\to\infty} \frac{d!}{n^d} H^*(M,n)$ as in the homogeneous case. Another point of view is to consider the Hilbert series $SH_M(T) = \sum_{n\in\mathbb{N}} H(M,n)T^n$ of M. In the second section some well-known results concerning $SH_M(T)$ and some arithmetic and geometric examples are given. In the last part, we give an extension of the Hilbert-Samuel Theorem to good filtrations. In particular we prove that if A is a noetherian semi local ring, M an A-module of finite type with Krull dimension d and $f = (I_n)$ a noetherian filtration on A such that $\sqrt{I_1} = r(A)$ the Jacobson ideal of A, then the function $n \mapsto \ell_A(M/I_nM)$ is a uniform quasi-polynomial function of degree d.

1 HILBERT FUNCTIONS

Let $A = \bigoplus_{n\in\mathbb{N}} A_n$ be a noetherian graded ring of finite Krull dimension where A_0 is an artinian ring. Let $M = \bigoplus_{n\in\mathbb{N}} M_n$ be a graded A-module of finite type with Krull dimension d.

Then we can define the following Hilbert functions

$$H(M,-) : \mathbb{N} \to \mathbb{N},\ H^*(M,-) : \mathbb{N} \to \mathbb{N}$$

by

$$H(M,n) = \ell_{A_0}(M_n)$$

$$H^*(M,n) = \sum_{k=0}^{n} H(M,k)$$

We know that if A is a homogeneous graded ring, that is if $A = A_0[x_1, x_2, ..., x_r]$ and the degree of x_i is 1 for all i, the Hilbert functions are polynomial with degree respectively $d-1$ and d. This is false if degree of $x_i > 1$ for some i.

DEFINITION 1 *A quasi-polynomial P of degree $d \geq -1$ and period $p \geq 1$ is a sequence $(P_0, P_1, ..., P_{p-1})$, $P_i \in \mathbb{Q}[X]$ such that* $\max(d^\circ P_i) = d$. *Then P is called a uniform quasi-polynomial if all the P_i have the same degree and the same leading coefficient. We write $P(n) = P_i(n)$ if $n \in \mathbb{Z}$ and $n \equiv i\ [p]$ for some $i = 0, 1, ..., p-1$*

DEFINITION 2 *A function $f : \mathbb{N} \to \mathbb{N}$ is said to be a quasi-polynomial function of degree d and period p if there exists a quasi-polynomial $P = (P_0, P_1, ..., P_{p-1})$ of degree $d \geq -1$ and period $p \geq 1$ such that*

$$f(n) = P(n) \ for\ n >> 0$$

Then, we write : $f \sim P$. If in addition P is a uniform quasi-polynomial, f is called a uniform quasi-polynomial function.

We have the following result which is a generalization of the classical Hilbert Theorem. A proof is given in [6]. See also [1] and [2].

THEOREM 1 *Let A be a noetherian graded ring of finite Krull dimension. We assume that $A = A_0[x_1, x_2, ..., x_r]$, where each x_i is homogeneous of degree $d_i \geq 1$ and that A_0 is an artinian ring. Let $M = \oplus_{n\in\mathbb{N}} M_n$ be a graded A-module of finite type with Krull dimension d. Let us put $p = lcm(d_1, d_2, ..., d_r)$. Then the functions $H(M,-)$ and $H^*(M,-)$ are quasi-polynomial functions of period p and of degree $d-1$ and d respectively. Moreover*
(1) if $H(M,-) \sim P$, P_0 have degree $d-1$.

(2) $H^*(M,-)$ is a uniform quasi-polynomial function.

Under the conditions of Theorem 1, we define the multiplicity of M by the following asymptotic formula

$$e_A(M) = \lim_{n\to\infty} \frac{d!}{n^d} H^*(M,n)$$

In the particular case where $0 \to M' \to M \to M'' \to 0$ is an exact sequence of graded A-modules with the same Krull dimension, we have

$$e_A(M) = e_A(M') + e_A(M'')$$

The following example was suggested to me by L. Avramov during the Fez meeting. It proves that the quasi-polynomial function $H(M,-)$ is not necessarily a uniform function, and that the P_i's do not have alternatively degree $d-1$ or -1 as we could expect.

Here if $M = \oplus_{n\in\mathbb{N}} M_n$ is a graded ring and d is an integer, $M(d)$ is the graded ring the graduation of which is defined by $M(d)_n = M_{n+d}$.

EXAMPLE 1 Let A be the graded ring $k[X,Y]$ where k is a field and where X and Y have the same degree 2. We consider the graded A-module $M = M_1 \oplus M_2$ where $M_1 = A(1) = \bigoplus_{n\in\mathbb{N}} M_{1,n}$ and $M_2 = A/(Y) = \bigoplus_{n\in\mathbb{N}} M_{2,n}$. Then the Krull dimension of M is 2 and we have

$$\dim_k M_{1,n} = \begin{cases} 0 & \text{if } n \equiv 0\ [2], \\ \dfrac{n+3}{2} & \text{if } n \equiv 1\ [2]. \end{cases}$$

$$\dim_k M_{2,n} = \begin{cases} 0 & \text{if } n \equiv 1\ [2], \\ 1 & \text{if } n \equiv 0\ [2]. \end{cases}$$

So the numerical function $H(M,-) : n \mapsto \dim_k M_n$ is a quasi-polynomial function of period 2 and degree 1, and if $H(M,-) \sim P = (P_0, P_1)$ we have

$$P_0(T) = 1,\ P_1(T) = \frac{T+3}{2}$$

The degree and the leading coefficient of P_0 and P_1 are different. The multiplicity of M is 4^{-1} □

Here we give a geometric example. When $\mathcal{P}$ is a rational polytope, we are interested in the numerical function which associates at each integer n the number of integral points belonging to the polytope $n\mathcal{P}$. This function has been intensively studied by E. Ehrhart [7].

EXAMPLE 2 Let $\mathcal{P}$ be a rational polytope of dimension d in $\mathbb{R}^s$, that is a polytope whose vertices have rational components . We consider for any integer $m \in \mathbb{N}$, the cardinal $\mathcal{P}_m$ of the set $m\mathcal{P} \cap \mathbb{Z}^s$ and we want to study the function $\varphi : n \mapsto \mathcal{P}_n$. We define

$$\mathcal{C} = \mathbb{R}_+\{(p, 1) : p \in \mathcal{P}\} \text{ and } \mathbb{M} = \mathcal{C} \cap \mathbb{Z}^{s+1}$$

$\mathbb{M}$ is a monoid of finite type [3, Proposition 6.1.2 (b)]. So the ring $A = \mathbb{R}[\mathbb{M}]$ is a normal $\mathbb{R}$-algebra of finite type generated by $\{X^{m_i} : i = 1, 2, ..., j\}$ if $m_1, m_2, ..., m_j$ generate $\mathbb{M}$ and its Krull dimension is $d + 1$ [3, Chapter 6]. We graduate A by setting $d^\circ m = x_{s+1}$ for each $m = (x_1, x_2, ..., x_{s+1}) \in \mathbb{M}$. Then A is a noetherian graded ring and we have

$$\dim_{\mathbb{R}} A_m = \mathcal{P}_m \text{ as a vector space}$$

Hence the function $n \mapsto \mathcal{P}_n$ is a quasi-polynomial function with degree the Krull dimension $\mathrm{Kdim}\mathbb{R}[\mathbb{M}]$ of $\mathbb{R}[\mathbb{M}]$ minus 1 and period $p = \mathrm{lcm}(d^\circ m_1, d^\circ m_2, ..., d^\circ m_j)$. Furthermore as $\mathbb{M}$ is a normal affine monoid, we know from a result proved by Hochster [8, Theorem 1] that A is a Cohen-Macaulay ring. So if we assume that A has a regular homogeneous sequence $(x_1, x_2, ..., x_{d+1})$ such that the degrees of the x_i's are relatively prime integers, then we have

$$e_A(A) = d!\ (\mathrm{relvol}(\mathcal{P}))$$

where $\mathrm{relvol}(\mathcal{P})$ is the relative volume of $\mathcal{P}$. Indeed, under the above hypotheses, we know that φ is a uniform quasi-polynomial function (See the Corollary of Proposition 4). So if D is a multiple of the period of φ such that $D\mathcal{P}$ is an integral polytope and if $P = (P_0, P_1, ..., P_{D-1})$ is the quasi-polynomial associated with φ, then the polynomial associated with the polynomial function $n \mapsto \mathcal{P}_{nD}$ is $Q(X) = P_0(DX)$. Hence the conclusion follows from [13, 4.6.30] and from the definition of multiplicity □

APPLICATIONS It is well known that if $(A, \mathfrak{M})$ is a local noetherian ring, I an ideal of A and M a A-module of finite type, the Bass functions

$$n \mapsto \nu_A^i(I^n M) \text{ and } n \mapsto \nu_A^i(I^n M / I^{n+1} M)$$

are polynomial functions.

Recall that for a module M of finite type on a local ring A with residual field k, the i-th Bass number of M is defined to be

$$\nu_A^i(M) = \dim_k Ext_A^i(k, M)$$

the dimension of the k-vector space $Ext_A^i(k, M)$.

Instead of an adic filtration, we consider here a noetherian filtration $f = (I_n)$ on A. Its Rees ring is the graded noetherian ring $R(f) = \sum_{n \in \mathbb{N}} I_n X^n$. Let us put

$R(f) = A[x_1, x_2, ..., x_r]$ where $x_1, x_2, ..., x_r$ are homogeneous elements of $R(f)$. The degree of each x_i is $d_i \geq 1$ for $i = 1, 2, ..., r$. Then we have

PROPOSITION 1 *The functions* $n \mapsto \nu_A^i(I_n M)$ *and* $n \mapsto \nu_A^i(I_n M / I_{n+1} M)$ *are quasi-polynomial functions of period* $p = lcm(d_1, d_2, ..., d_r)$.

PROOF. Let $\mathbf{A}$ be the Rees ring of f and $\mathbf{M} = \oplus_{n \in \mathbb{N}} I_n M$. It follows from the properties of noetherian filtrations (see [12]) that $\mathbf{A}$ is a noetherian graded ring and that $\mathbf{M}$ is a finite graded $\mathbf{A}$-module. Let us take a free finite resolution (L_n) of the A-module $A/\mathfrak{M}$:

$$(*) \qquad \cdots \to L_n \to L_{n-1} \to \cdots \to L_1 \to L_0 \to A/\mathfrak{M} \to 0$$

This yields the complex of A-modules:

$$0 \to Hom_A(A/\mathfrak{M}, \mathbf{M}) \to Hom_A(L_0, \mathbf{M}) \to Hom_A(L_1, \mathbf{M}) \to Hom_A(L_2, \mathbf{M}) \to$$

$$\cdots \to Hom_A(L_{n-1}, \mathbf{M}) \to Hom_A(L_n, \mathbf{M}) \to \cdots$$

By hypothesis each L_n is a finite free A-module, hence $Hom_A(L_n, \mathbf{M}) \simeq \mathbf{M}^{d_n}$ where d_n is the rank of L_n and $Ext_A^n(A/\mathfrak{M}, \mathbf{M})$ is a $\mathbf{A}$-module of finite type. So it follows from Theorem 1 that the function $n \mapsto \nu_i^A(I_n M)$ is a quasi-polynomial function of period p. In the same way, to prove that the function $n \mapsto \nu_i^A(I_n M / I_{n+1} M)$ is a quasi-polynomial function, we consider the graded noetherian ring $\mathbf{A} = G_f(A) = \bigoplus_{n \in \mathbb{N}} I_n / I_{n+1}$ and the graded $\mathbf{A}$-module $\mathbf{M} = G_f(M) = \bigoplus_{n \in \mathbb{N}} I_n M / I_{n+1} M$ □

PROPOSITION 2 *Let* $(A, \mathfrak{M})$ *be a noetherian local ring,* M *an* A*-module of finite type and* $f = (I_n)$ *a noetherian filtration on* A. *Then the functions* $n \mapsto depth(I_n M)$ *and* $n \mapsto depth(I_n M / I_{n+1} M)$ *are periodic functions of period* $p = lcm(d_1, d_2, ..., d_r)$ *for large* n.

PROOF. We know that for each A-module M of finite type the depth of M is $\inf\{i : \nu_i^A(M) \neq 0\}$. We put $M_n = I_n M$ or $I_n M / I_{n+1} M$ and $d' =$ Kdim$A$. Then it follows from Proposition 1 that the function $n \mapsto \nu_i^A(M_n)$ is a quasi-polynomial function. So we consider the quasi-polynomial $Q_i = (Q_{i,0}, Q_{i,1}, ..., Q_{i,p-1})$ associated with the function $n \longmapsto \nu_A^i(M_n)$ and for each integer j we put $K_j = \{i : i \leq d' \text{ and } Q_{i,j} \neq 0\}$ and i_j for the least integer of K_j. Since depth$(M) \leq d'$ for each nonzero A-module of finite type, we have for $n \gg 0$, depth$(M_{j+np}) = +\infty$ if $K_j = \emptyset$ and depth$(M_{j+np}) = i_j$ otherwise □

2 HILBERT SERIES

Let $A = \bigoplus A_n$ be a graded noetherian ring of finite Krull dimension with A_0 an artinian ring and $M = \bigoplus M_n$ a nonzero graded module of finite type and

Krull dimension d. We know that $H(M,-) \sim P$ where P is a quasi-polynomial of period p and degree $d-1$.

The Hilbert series of M is the formal series $SH_M(T)$ defined by

$$SH_M(T) = \sum_{n\in\mathbb{N}} H(M,n)T^n \in \mathbb{Z}[[T]]$$

We recall that the valuation of a polynomial $Q(T) = \sum_0^n a_i T^i$ or of a series $S(T) = \sum_0^\infty a_i T^i$ is the least integer i such that $a_i \neq 0$. The valuation of the nul polynomial is $+\infty$.

We have the following results :

PROPOSITION 3 *With the notations as above, we have*

(1) $SH_M(T)$ is a rational function given by $\dfrac{Q(T)}{\prod_{i=1}^d (1-T^{\alpha_i})}$ with $\alpha_i \geq 1$ for $i = 1,2,...,d$, and the period p of the quasi-polynomial associated with $SH_M(T)$ belongs to $\{\alpha_1, \alpha_2, ..., \alpha_d\}$, $Q(T) \in \mathbb{Z}[T]$ and $Q(1) \neq 0$.

(2) $v(Q(T)) = v(SH_M(T)) = \min\{n \in \mathbb{N} : M_n \neq 0\}$ where v is the valuation of the polynomial $Q(T)$ (resp. of the formal series $SH_M(T)$).

(3) the degree of the rational function $SH_M(T)$ is $\max\{n \in \mathbb{N} : P(n) \neq H(M,n)\}$ with the convention $\max(\emptyset) = -1$.

SKETCH OF PROOF. (1) holds by induction on d, by using the fact that a numerical function is a quasi-polynomial function of degree $d \geqslant 1$ and period p if and only if the numerical function $\Delta : n \to f(n+p) - f(n)$ is a quasi-polynomial function of degree $d-1$ and period p. See another proof in [3, Proposition 4.3.3].
(2) Let $Q = \sum_{n\geqslant v(Q)} q_n T^n$, then we have

$$\prod_{i=1}^{d}(1-T^{\alpha_i})SH_M(T) = (1+\sum_{n\geqslant 1} b_n T^n)(\sum_{n\geqslant v(SH_M)} a_n T^n) = \sum_{n\geqslant v(Q)} q_n T^n$$

so that $a_{v(SH_M)} = q_{v(Q)}$.
(3) is proved by induction on d and from the following remarks:
(i) If $f : \mathbb{N} \to \mathbb{N}$ is a quasi-polynomial function of period p and if $f \sim P$, then $\Delta f \sim \Delta P = P(X+p) - P(X)$ and $\max\{n : f(n) \neq P(n)\} = \max\{n : \Delta f(n) \neq \Delta P(n)\}$
(ii) If

$$\sum_{n\in\mathbb{N}} a_n T^n = \frac{Q(T)}{\prod_1^{d-1}(1-T^{\alpha_i})}\left(\frac{1}{1-T^\alpha}\right) = \sum_{n\in\mathbb{N}} b_n T^n \frac{1}{1-T^\alpha}$$

then

– $n \mapsto b_n$ is a quasi-polynomial function. We set $b_n \sim B$

– $a_n = b_n + b_{n-\alpha} + ... + b_{n-m\alpha}$ if $n = m\alpha + r$ with $0 \leqslant r < \alpha$

– $n \mapsto a_n$ is a quasi-polynomial function. If we put $a_n \sim A$ we have

$$\Delta A(X) = B(X+\alpha) + B(X+2\alpha) + ...B(X+p\alpha)$$

$$\Delta A(d_1 - \alpha) \neq \Delta a_{d_1-\alpha} \text{ if } d_1 = \max\{n : b_n \neq B(n)\}$$

$$\Delta A(n) = \Delta a_n, \text{ for any } n > d_1 - \alpha \, \square$$

In the next propositions, we determine a sequence (α_i) and a polynomial $Q(T)$ when M is a graded Cohen-Macaulay ring or when we know a minimal free resolution of M.

PROPOSITION 4 *Let A be a noetherian graded ring where A_0 is a local artinian ring. Let M be a Cohen-Macaulay graded A-module of finite type and Krull dimension d. Then there exist an M-regular sequence $\underline{x} = (x_1, x_2, ..., x_d)$ where each x_i is homogeneous of degree $\alpha_i \geq 1$ and a polynomial $Q(T) \in \mathbb{Z}[T]$ such that*

$$SH_M(T) = \frac{Q(T)}{\prod_{i=1}^{d}(1 - T^{\alpha_i})}$$

Furthermore, $Q(T) = SH_{M/\underline{x}M}(T)$ and the coefficients of Q are nonnegative integers.

PROOF. The case $d = 0$ is clear. The proof follows from an induction on d and from the fact that if M is a Cohen-Macaulay module of Krull dimension d and if x is a homogeneous regular element of M, then M/xM is a Cohen-Macaulay module of Krull dimension $d-1$ $\square$

COROLLARY *Let A be a noetherian graded ring where A_0 is a local artinian ring and $M \neq (0)$ a Cohen-Macaulay graded A-module of finite type and Krull dimension d. We assume that there exist an M-regular sequence $\underline{x} = (x_1, x_2, ..., x_d)$ where each x_i is homogeneous of degree $\alpha_i \geq 1$ and that the integers α_i are relatively prime. Then the Hilbert function of M is a uniform quasi-polynomial function of period $p = \prod_{i=1}^{d} \alpha_i$. Moreover,*
(i) $Q_1(T) = (1 - T^p)^d SH_M(T) \in \mathbb{Z}[T]$, $Q_1(1) \neq 0$ and $Q_1(e^{2ki\pi/p}) = 0$ for $0 < k < p$.
(ii) The multiplicity of M is given by

$$e_A(M) = \frac{Q_1(1)}{p^d}$$

PROOF. (i) We know from Proposition 4 that $Q_1(T) = \dfrac{Q(T)(1-T^p)^d}{\prod_{i=1}^{d}(1-T^{\alpha_i})}$ where $Q(T)$ is a polynomial with positive coefficients. In particular $Q_1(1) \neq 0$ since 1 is a root of multiplicity d of the polynomials $(1-T^p)^d$ and $\prod_{i=1}^{d}(1-T^{\alpha_i})$. Furthermore each complex number $\xi_k = e^{2ik\pi/p}$ with $0 < k < p$ is a root of $(1-T^p)^d$ of multiplicity d and a root of the polynomial $\prod_{i=1}^{d}(1-T^{\alpha_i})$ of multiplicity $< d$ since the integers α_i are relatively prime. So (i) is proved and it follows from [1, 3.2, Prop 6] that $H(M,-)$ is a uniform quasi-polynomial. The leading coefficient of the quasi-polynomial associated with $H(M,-)$ being $\frac{Q_1(1)}{p^d(d-1)!}$, we obtain (ii) □

PROPOSITION 5 Let $A = k[x_1, x_2, ..., x_r]$ be a noetherian graded ring over a field k where the graduation is given by : $d^\circ x_i = \alpha_i \geq 1$, for $i = 1, 2, ..., r$ and let M be a graded A-module of finite type. Let

$$0 \to L_s \to L_{s-1} \to \cdots \to L_0 \to M \to 0$$

be a minimal free graded resolution of M. Put $L_i = \bigoplus_{j\in\mathbb{N}}(A(-j))^{\beta_A^{i,j}}$ where $\beta_A^{i,j} \geq 0$ for each $j \in \mathbb{N}$, $i = 1, 2, ..., r$. Then we have

$$SH_M(T) = \sum_{\substack{j\in\mathbb{N}\\ i=1,2,..,n}} \frac{(-1)^i \beta_A^{i,j} T^j}{\prod_{i=1}^{r}(1-T^{\alpha_i})}$$

PROOF. It is an easy consequence of the additive property of the length function □

The following example shows that the Hilbert function of a graded ring A need not be a uniform quasi-polynomial function even if A is a Cohen-Macaulay ring.

EXAMPLE 3 Let $A = k[X_1, X_2, ..., X_d]$ be the ring of polynomials graded by $d^\circ X_i = \alpha \geq 1$ for some integer α and $i = 1, 2, ..., d$. Then

$$SH_A(T) = \frac{1}{(1-T^\alpha)^d}$$

Furthermore, if $H(A,-) \sim P$, we have

(i) $H(A,n) = P(n)$ for $n \in \mathbb{N}$

(ii) $P_0(T) = Q(\frac{T}{\alpha})$ and $P_i(T) = 0$ if $i \not\equiv 0\ [\alpha]$ where $Q(T) = \binom{T+d-1}{d}$

(iii) $e_A(A) = \dfrac{1}{\alpha^d} \in \mathbb{Q}_\star^+$.

In example 4 we compute the number of solutions of a diophantine equation. We give a geometric interpretation of the leading coefficient of the uniform quasi-polynomial associated with the numerical function $n \longmapsto \sharp\{(x_1, x_2, ..., x_d) \in \mathbb{N}^d : \alpha_1 x_1 + \alpha_2 x_2 + ... + \alpha_d x_d \leq n\}$ where α_i is a positive integer for $i = 1, 2, ..., d$. Here $\sharp B$ is the cardinality of a finite set B.

EXAMPLE 4 Let $(\alpha_1, \alpha_2, ..., \alpha_d)$ be a sequence of positive integers. Consider the diophantine equation

$$(S): \qquad \alpha_1 x_1 + \alpha_2 x_2 + ... + \alpha_d x_d = n$$

with $n \in \mathbb{N}$. Let

$$S_n = \{(x_1, x_2, ..., x_d) \in \mathbb{N}^d \text{ which satisfies } (S)\}$$

We are interested in the behaviour of the function $\varphi : n \mapsto \sharp S_n$ for large n. So, we consider a field k and the graded ring $A = k[X_1, X_2, ..., X_d]$ the graduation of which is given by $d^\circ X_i = \alpha_i$ for each $i = 1, 2, ..., d$. Then $H(A, n) = \dim_k A_n = \sharp S_n$. So φ is a quasi-polynomial function of degree $d - 1$ and of period $p = \operatorname{lcm}(\alpha_1, \alpha_2, ..., \alpha_d)$. The Hilbert series associated with A (the generating function of (S)) is

$$SH_A(T) = \frac{1}{\prod_{i=1}^{d}(1 - T^{\alpha_i})}$$

Indeed,

$$SH(T) = \sum_{n \geq 0} (\sharp S_n) T^n = \prod_{i=1}^{d} \sum_{j \geq 0} T^{j\alpha_i} = \frac{1}{\prod_{i=1}^{d}(1 - T^{\alpha_i})}$$

Moreover if $H(A, -) \sim P$, we have $\sharp S_n = P(n)$ for all $n \in \mathbb{N}$. Similarly, the number of solutions in $\mathbb{N}^d$ of

$$(S_{\leq}): \qquad \alpha_1 x_1 + \alpha_2 x_2 + ... + \alpha_d x_d \leq n$$

is a uniform quasi-polynomial function of degree d and of period p given by $H^*(A, n)$. Let P_i be the rational points defined by $P_0 = O$ and $P_i = (x_{i,j})$ with $x_{i,j} = \delta_{i,j}/\alpha_i$ for all $i, j = 1, 2, ..., d$ where $\delta_{i,j}$ is the Kronecker symbol. It follows from example 2 that $H^*(A, n) = \mathcal{P}_n$ where $\mathcal{P}$ is the d-dimensional rational polytope in $\mathbb{R}^d$ whose vertices are the points $P_0, P_1, ..., P_d$. Then, as the volume $v(\mathcal{P})$ of $\mathcal{P}$ is given by $\dfrac{1}{d! \prod_{i=1}^{d} \alpha_i} = \lim_{n \to \infty} \frac{H^*(M,n)}{n^d}$, $v(\mathcal{P})$ is the leading coefficient of the uniform quasi-polynomial associated with the function $n \longmapsto \sharp\{(x_1, x_2, ..., x_d) \in \mathbb{N}^d : \alpha_1 x_1 + \alpha_2 x_2 + ... + \alpha_d x_d \leq n\}$.

If in addition the integers $\alpha_1, \alpha_2, ..., \alpha_d$ are relatively prime, it follows from [1, 3.2, Proposition 6] that the function $n \mapsto \sharp S_n$ is a uniform quasi-polynomial function. So its leading coefficient is $d.v(\mathcal{P}) = \dfrac{1}{(d-1)! \prod_{i=1}^{d} \alpha_i}$.

For instance, by using a maple algorithm, we obtain with $d = 2$, $\alpha_1 = 2$, $\alpha_2 = 3$:

$$SH_A(T) = 1+T^2+T^3+T^4+T^5+2T^6+T^7+2T^8+2T^9+2T^{10}+2T^{11}+...$$

and the quasi-polynomial P associated with the function $n \mapsto \sharp S_n$ has degree 1 and period 6. P is given by

$$P_0(T) = (1/6)T + 1$$
$$P_1(T) = (1/6)T - 1/6$$
$$P_2(T) = (1/6)T + 2/3$$
$$P_3(T) = (1/6)T + 1/2$$
$$P_4(T) = (1/6)T + 1/3$$
$$P_5(T) = (1/6)T + 1/6$$

REMARK Let $f = (I_n)$ be a noetherian filtration on a local noetherian ring $(A, \mathfrak{M})$ with residual field k. Let $R(f) = A[x_1, x_2, ..., x_r]$ be its Rees ring with $d^\circ x_i = d_i$. For each $n \in \mathbb{N}$, let $\mu(I_n)$ be the cardinal of a minimal set of generators of the ideal I_n. Then

$$\varphi : n \longmapsto \mu(I_n) = \dim_k I_n/\mathfrak{M}I_n$$

is a quasi-polynomial function of period $p = \text{lcm}(d_1, d_2, ..., d_r)$. Indeed $\mu(I_n) = H(G_f(A), n)$ where $G_f(A)$ is the graded ring $\bigoplus_{n\in\mathbb{N}} I_n/\mathfrak{M}I_n \simeq R(f)/\mathfrak{M}R(f)$. The degree d of φ is the Krull dimension of $G_f(A)$ minus 1. The integer $d+1$ is *the analytic spread of* f and is one of the possible extension to filtrations of the notion of analytic spread of an ideal introduced by Northcott and Rees [9]. For another generalizations, see [5],[10].

PROPOSITION 6 *Let* $A = k[X_1, X_2, ..., X_d]$ *be the polynomial ring over a field* k *graded by* $d^\circ X_i = \alpha_i \geq 1$ *for all* $i = 1, 2, ..., d$. *Then the multiplicity of* A *is given by*

$$e_A(A) = \frac{1}{\prod_{i=1}^d \alpha_i}$$

Moreover, if the integers $\alpha_1, \alpha_2, ..., \alpha_d$ *are relatively prime, the Hilbert function of* A *is a uniform quasi-polynomial function of degree* $d-1$ *and period* $\prod_{i=1}^d \alpha_i$. *Its leading coefficient is* $\dfrac{1}{(d-1)!\prod_{i=1}^d \alpha_i}$.

PROOF. Use example 4 □

3 EXTENSION OF THE HILBERT-SAMUEL THEOREM

DEFINITION 3 *Let $f = (I_n)$ be a filtration on the ring A and $\Phi = (M_n)$ a filtration on the A-module M. We say that (f, Φ) is a good pair of filtrations if the following conditions hold:*
(i) $I_n M_p \subset M_{n+p}$, for all $n, p \geq 0$
(ii) $G_f(A) = \oplus_{n \geq 0} I_n/I_{n+1}$ is a noetherian graded ring of finite Krull dimension
(iii) $G_\Phi(M) = \oplus_{n \geq 0} M_n/M_{n+1}$ is a finite $G_f(A)$-module.

When A is a noetherian ring, an I-adic filtration $\Phi = (M_n)$ of M gives a pair (f_I, Φ) of good filtrations where f_I is the I-adic filtration on A. More generally, if f is a noetherian filtration on A, the notion of f-good filtration Φ on M introduced by Ratliff also gives a pair (f, Φ) of good filtrations. See [11].

THEOREM 2 *Let $f = (I_n)$ be a filtration on the noetherian ring A and $\Phi = (M_n)$ a filtration on the A-module M. Assume that (f, Φ) is a good pair of filtrations and that $\ell_A(M/I_1 M)$ is finite. Then the numerical functions $\varphi : n \longmapsto \ell_A(M_n/M_{n+1})$ and $\psi : n \longmapsto \ell_A(M/M_n)$ are quasi-polynomial functions respectively of degree $\mathrm{Kdim} G_\Phi(M) - 1$ and $\mathrm{Kdim} G_\Phi(M)$. Furthermore, (i) if $G_f(A) = A/I_1[z_1, z_2, ..., z_r]$ where each z_i is homogeneous of degree $d_i \geq 1$, $p = \mathrm{lcm}(d_1, d_2, ..., d_r)$ is a period for φ and ψ. (ii) ψ is a uniform quasi-polynomial function.*

PROOF. Since $I_n M = I_n M_0 \subset M_n$ and $\mathrm{Supp}(M/I_n M) = \mathrm{Supp}(M/I_1 M) \subset \mathrm{Max}(A)$, we have $\ell_A(M/I_n M) < \infty$ for $n \in \mathbb{N}$ and as M_n/M_{n+1} is a submodule of M/M_{n+1}, $\ell_A(M_n/M_{n+1})$ is finite. We put $K = \mathrm{Ann} M$ and $J_n = (I_n + K)/K$ for all n. Then $g = (J_n)$ is a filtration on B. We consider the graded ring $G_g(B) = \bigoplus_{n \geq 0} J_n/J_{n+1}$. Since $I_1 M = J_1 M$ and $\ell_B(M/J_1 M) < \infty$, we have $\ell_B(B/J_1) < \infty$ hence B/J_1 is an artinian ring. Moreover as (Φ, f) is a pair of good filtrations, (Φ, g) is also a pair of good filtrations, so $G_\Phi(M) = \bigoplus_{n \geq 0} M_n/M_{n+1}$ is a graded $G_g(B)$-module of finite type. Then it follows from the Hilbert Theorem that the functions $\varphi = H(G_g(M), -)$ and $\psi = H^*(G_g(M), -)$ are quasi-polynomial functions satisfying (i) and (ii) □

THEOREM 3 *Let A be a noetherian semi-local ring, M an A-module of finite type with Krull dimension d and $f = (I_n)$ a noetherian filtration on A such that $\sqrt{I_1} = r(A)$ the Jacobson ideal of A. Then the function $\varphi : n \mapsto \ell_A(M/I_n M)$ is a uniform quasi-polynomial function of degree d and of period $p = \mathrm{lcm}(d_1, d_2, ..., d_r)$ if the Rees ring of f is generated over A by homogeneous elements $x_1, x_2, ..., x_r$ with $d^\circ x_i = d_i$ for $i = 1, 2, ..., r$.*

PROOF. (f, Φ) being a pair of good filtrations when $f = (I_n)$ is a noetherian filtration on A with $\Phi = (I_n M)$, the proof is a consequence of Theorem 2 since $\mathrm{Kdim} M = \mathrm{Kdim} G_\Phi(M)$ □

In particular, we can define the f-multiplicity of M of respect a noetherian filtration $f = (I_n)$ by

$$e_f(M) = \lim_{n\to\infty} \frac{d!}{n^d}\ell_A(M/I_nM)$$

EXAMPLE 5 Let A be the ring of formal series $k[[X_1, X_2, ..., X_d]]$ over the field k, $I = (X_1, X_2, ..., X_d)$ and $f = (I_n)$ the filtration on A given by its Rees ring $R(f) = A[IX, IX^2, IX^3, ..., IX^s]$ for some integer s. Then the function

$$\varphi : n \mapsto \dim_k A/I_n$$

is a quasi-polynomial function of degree d and period s. If $\varphi \sim P = (P_0, P_1, ..., P_{s-1})$, we have

$$P_0(n) = \sum_{i=0}^{\frac{n}{s}-1} \binom{d+i-1}{i} \text{ if } n \equiv 0\ [s]$$

$$P_i(X) = P_0(X+i), \text{ for } i = 0, 1, ..., s-1$$

$$e_f(A) = \frac{1}{s^d} \in \mathbb{Q}_*^+$$

REFERENCES

1. A. Badra, *Frobenius number, linear diophantine equations and Hilbert function*, (to appear)
2. V.V. Bavula, *Identification of the Hilbert function and Poincaré series, the dimension of modules over filtered rings*, Russian Acad. Sci. Izv. Math. 44 (2) (1995) 225-246
3. W. Bruns, J. Herzog, *Cohen-Macaulay rings*, Cambridge stud. Adv. Math., 39, Cambridge University Press (1993)
4. M. Brodmann, *The asymptotic nature of the analytic spread*, Math. Proc. Camb. Philos. Soc., 86 (1979) 35-39
5. Y. Diagana, H. Dichi, D. Sangaré, *Filtrations, Generalized analytic independence, analytic spread*, Afrika Matematika, serie 3, vol.4 (1994)101-114
6. H. Dichi, D. Sangaré, *Hilbert functions, Hilbert-Samuel quasi-polynomials with respect to f-good filtrations, multiplicities* , J. Pure Appl. Algebra. 138 (1999) 205-213

7. E. Ehrhart, *Démontration de la loi de réciprocité du polyèdre rationnel*, C.R. Acad. Sci. Paris 265A (1967) 91-94

8. M. Hochster, *Rings of invariants of tori, Cohen-Macaulay rings generated by monomials, and polytopes*, Annals of Math. 96 (1972) 318-337

9. D.G. Northcott and D. Rees, *Reductions of ideals in local rings*, Proc. Cambridge Philos. Soc. 50 (1954) 145-158

10. J.S. Okon, *Prime divisors, analytic spread and filtrations*, Pacific J. Math., 113, 2 (1984) 451-462

11. L.J. Ratliff,Jr., D.E. Rush, *Note on I-good filtrations and noetherian Rees rings*, Comm. Algebra 16 (1988) 955-975

12. L.J. Ratliff,Jr., *Notes on essentially powers filtrations*, Michigan Math. J. 26 (1979) 313-324

13. R.P. Stanley, *Enumerative combinatorics*, Vol I., Cambridge stud. Adv. Math., 49, Cambridge University Press (1997)

On the Integral Closure of Going-Down Rings

DAVID E. DOBBS Department of Mathematics, University of Tennessee, Knoxville, Tennessee 37996-1300, U. S. A., dobbs@math.utk.edu

ABSTRACT

If R is an n-dimensional going-down ring (resp., an n-dimensional locally divided ring), with $0 < n < \infty$, in which each zero-divisor is nilpotent and if P is a prime ideal of R of height $n-1$ such that $P \subseteq J(R)$, then the integral closure of R in R_P is a going-down ring (resp., a locally divided ring). Consequently, the question of whether the integral closure of a two-dimensional going-down domain R is a going-down domain is reduced to the subcase in which R is a divided domain that is integrally closed in R_P, where P is the unique height 1 prime ideal of R. The question of whether the integral closure of a going-down domain is a going-down domain is shown to be equivalent to the question of whether the integral closure of a going-down ring (resp., locally divided ring) in which each zero-divisor is nilpotent is a going-down ring (resp., locally divided ring).

1 INTRODUCTION

All rings considered are commutative with identity, typically nonzero. As in [5] and [13], a (commutative) integral domain R is said to be a *going-down domain* in case $R \subseteq T$ satisfies the going-down property (dubbed GD, as in [16, p. 28]) for all integral domains T containing R. The most natural examples of going-down domains are the integral domains of (Krull) dimension 1 and the Prüfer domains. As both of these classes are stable under the formation of integral closure (in the quotient field), it seems natural to ask if the same holds for the class of going-down domains. Such "ascent" questions have been of continuing interest (cf. [6, Corollary 2.5, Proposition 3.1 and Theorem 3.3], [8, Theorem 3.2 and Corollary 3.4], [18, Propositions 1 and 2]).

Our interest here is in finding equivalents of the question, (Q), of whether the integral closure of a going-down domain R must be a going-down domain. (Note that (Q) remains open even for two-dimensional domains: cf. [6, Corollary 3.5].) Using ideas that go back to [7], Corollary 2.2 reduces question (Q) to the subcase in which R is a seminormal divided domain. (Recall that an integral domain D is called a *divided domain* if $PD_P = P$ for each prime ideal P of D; and that D is called a *locally divided domain* if D_N is a divided domain for each prime ideal N of D. Any locally divided domain is a going-down domain [7, Remark 2.7 (b)]; while the converse holds for seminormal domains (cf. [7, Corollary 2.6]), it fails in general ([7, Example 2.9], [4, Remark 2.3]). One of our main results, Theorem 2.6, includes an "ascent" result for finite-dimensional "going-down domain" (resp., "locally divided domain"). Its local case states that if R is an $n-$dimensional quasilocal going-down domain (resp., an $n-$dimensional divided domain), with $0 < n < \infty$, and if P is the unique prime ideal of R of height $n-1$, then the integral closure of R in R_P is a going-down domain (resp., a locally divided domain). (Such an ideal P is uniquely determined since any going-down domain must be a treed domain [5, Theorem 2.2].) As a consequence of Theorem 2.6, Corollary 2.7 reduces the question (Q) for integral domains R with $\dim(R) \leq n$, where $n < \infty$, to the subcase in which R is a divided domain (hence quasilocal) that is integrally closed in a certain localization.

Theorem 2.6 is to be contrasted with [11, Example 3.1], which showed that a finite-type (hence integral) overring of a two-dimensional divided (hence quasilocal going-down domain) domain need not be a going-down domain. Although the latter result showed that one cannot naively attack the question (Q) by using the fact that the class of going-down domains is stable under direct limit [12, Corollary 2.7], Theorem 2.6 gives reason to hope for additional positive "ascent" results relative to naturally occurring subalgebras of the integral closure of a given going-down domain.

For any ring A, let $Z(A)$ (resp., $\text{nil}(A)$) denote the set of zero-divisors (resp., nilpotent elements) of A; and let $\text{tq}(A)$ denote the total quotient ring of A. Recall from [10] that a ring A is said to be a *going-down ring* if A/P is a going-down domain for each prime ideal P of A. Any zero- or one-dimensional ring is a going-down ring, as is any chained ring [10, Proposition 2.1 (c), (d)]. An integral domain is a going-down ring if and only if it is a going-down domain. Generalizing the case of domains, [10, Corollary 2.6] established that if a ring A satisfies $Z(A) = \text{nil}(A)$, then A is a going-down ring if and only if $A \subseteq B$ satisfies GD for all overrings B of A (that is, for all $A-$subalgebras B of $\text{tq}(A)$). Following [10], the study of going-down rings (and of "locally divided rings") A satisfying $Z(A) = \text{nil}(A)$ was pursued in [2] and [3]. In fact, the present work is also couched in this generality, as the formulations of Theorem 2.6 and Corollary 2.7 involve, not just going-down domains and locally divided domains, but going-down rings and locally divided rings A such that $Z(A) = \text{nil}(A)$. (Background on divided rings and locally divided rings from [1] and [2] is deferred to Section 2 in order to limit this Introduction to a manageable size.) Our other main result, Theorem 2.4, establishes that the question

(Q) is equivalent to the (ostensibly more general) question of whether the class of going-down rings (resp., locally divided rings) A satisfying $Z(A) = \text{nil}(A)$ is stable under the formation of integral closure (in the ambient total quotient ring).

The proofs of Theorems 2.4 and 2.6 use a result characterizing when certain pullbacks are going-down rings (resp., locally divided rings) [10, Proposition 2.2] (resp., [2, Theorem 2.7]). For readers interested only in the domain-theoretic cases, note that the special cases of these pullback results involving going-down domains and locally divided domains appear in [9]. Appropriate background for what we use on pullbacks can be found in [14]. Proposition 2.1 and Corollary 2.2 refer to the seminormalization R^+ of an integral domain R. Appropriate background on seminormalization and seminormality can be found in [19].

In addition to the notation introduced above, if A is a ring, we let $\text{Spec}(A)$ denote the set of prime ideals of A; $\text{Max}(A)$ the set of maximal ideals of A; $J(A)$ the Jacobson radical of A; $\dim(A)$ the Krull dimension of A; $\text{ht}(P) = \text{ht}_A(P)$ the height (in A) of a prime ideal P of A; and A' the integral closure of A (in $\text{tq}(A)$). Any unexplained material is standard, as in [15], [16].

2 RESULTS

We begin with a result on integral domains that could have been published 25 years ago. In fact, it was recently obtained as a corollary of work on arbitrary reduced rings (possibly with nontrivial zero-divisors) [2, Theorem 3.4], but it seems appropriate to record the streamlined proof given below for the domain-theoretic case.

PROPOSITION 2.1. *[2, Corollary 3.6] Let R be an integral domain. Then R is a going-down domain if and only if R^+ is a locally divided domain.*

Proof. Since R^+ is a seminormal integral domain, it follows from the comments concerning [7, Corollary 2.6] in the Introduction that R^+ is a locally divided domain if and only if it is a going-down domain. Therefore, it suffices to show that R is a going-down domain if and only if R^+ is a going-down domain. This, in turn, follows from [7, Lemma 2.3] since $R \subseteq R^+$ is an integral extension such that the canonical map $\text{Spec}(R^+) \to \text{Spec}(R)$ is an injection (cf. [19, Lemma 2.2]). □

It has been observed that sharper "ascent" results for "going-down domain" relative to integral closure are often available for integral domains that are locally finite-dimensional (cf. [5, Proposition 2.7], [18, Proposition 3]). In this spirit, we record in Corollary 2.2 (a) that the question (Q) can be considered for integral domains of a fixed (Krull) dimension, a theme that we return to in Corollaries 2.7 and 2.8 in the context of bounded (Krull) dimension. Corollary 2.2 (b) reduces (Q) to the case of a seminormal divided (hence, quasilocal) domain.

COROLLARY 2.2. *(a) Let $0 \leq n \leq \infty$. Then the following conditions are equivalent:*

(1) If R is a going-down domain such that $dim(R) = n$, then R' is a going-down domain;

(2) If T is a seminormal locally divided domain such that $dim(T) = n$, then T' is a going-down domain.

(b) The following conditions are equivalent:

(1) If R is a going-down domain, then R' is a going-down domain;

(2) If T is a seminormal divided domain, then T' is a going-down domain.

Proof. (a) Since any locally divided domain is a going-down domain, (1) $\Rightarrow$ (2) trivially. Conversely, suppose (2), and let R be as in (1). Put $T := R^+$. As T is integral over R, $\dim(T) = \dim(R) = n$ (cf. [16, Theorem 48]). Also, by Proposition 2.1, T is a locally divided domain. Since T is seminormal, (2) ensures that T' is a going-down domain. But $T' = R'$, and so (1) follows.

(b) As above, (1) $\Rightarrow$ (2) trivially. Conversely, suppose (2) and let R be as in (1), once again putting $T := R^+$. As $T' = R'$, we may suppose that $R = T$, that is, that R is a seminormal locally divided domain, by Proposition 2.1. Now, for each $M \in \mathrm{Max}(R)$, R_M inherits seminormality from R (cf. [19, Corollary 2.10]) and is a divided domain. Hence, by (2), $(R_M)'$ is a going-down domain for each $M \in \mathrm{Max}(R)$. But $(R_M)' = R'_{R\setminus M}$ (cf. [15, Proposition 10.2]). It follows easily that R' is a going-down domain. □

In view of the extensive literature on "ascent" of the "going-down domain" property, it seems natural to ask for "ascent" results on "going-down ring" (with respect to integral extensions). In this regard, we show in Theorem 2.4 that for going-down rings A such that $Z(A) = \mathrm{nil}(A)$, the question that is analogous to (Q) is actually equivalent to (Q). In order to present similar results for the "locally divided ring" concept at the same time, we pause first to review some background on divided rings and locally divided rings from [1] and [2].

Let A be a ring. A prime ideal P of A is said to be (a) *divided (prime ideal) in* A if P is comparable under inclusion with each ideal of A; if A is an integral domain, this condition is equivalent to $PA_P = P$. We say that A is a *divided ring* if each prime ideal of A is divided in A; and that A is a *locally divided ring* if A_P is a divided ring for each $P \in \mathrm{Spec}(A)$. An integral domain is a divided (resp., locally divided) ring if and only if it is a divided (resp., locally divided) domain. Divided rings are the same as the quasilocal locally divided rings. Any locally divided ring is a treed going-down ring; and any zero-dimensional ring is a locally divided ring. The class of divided rings (resp., locally divided rings) is stable under the formation of rings of fractions and factor rings.

We next isolate some facts that will be used in the proofs of Theorems 2.4 and 2.6.

LEMMA 2.3. *(a) Let A be a ring. Then the following conditions are equivalent:*

(1) A is a going-down ring (resp., locally divided ring);

(2) A_P is a going-down ring (resp., locally divided ring) for each $P \in Spec(A)$;

(3) A_M is a going-down ring (resp., locally divided ring) for each $M \in Max(A)$;

(b) Let $A \subseteq B$ be an integral extension of rings. If $B_{A\setminus M}$ is a going-down ring (resp., locally divided ring) for each $M \in Max(A)$, then B is a going-down ring (resp., locally divided ring).

Proof. (a) The assertion for going-down rings (resp. locally divided rings) was proved in [10, Proposition 2.1 (b)] (resp., [2, Proposition 2.1 (a), (c)]).

(b) By (a), it is enough to show that B_N is a going-down ring (resp., divided ring) for each $N \in \text{Max}(B)$. Let $M := N \cap A$. Then $M \in \text{Max}(A)$ by integrality (more precisely, by the going-up property, as in [15, Corollary 11.6], [16, Theorem 44]). By hypothesis, $B_{A\setminus M}$ is a going-down ring (resp., locally divided ring). As the class of going-down rings (resp., locally divided rings) is stable under localization [10, Proposition 2.1 (b)] (resp., [2, Proposition 2.1 (a)]), it follows that $B_N \cong (B_{A\setminus M})_{NB_{A\setminus M}}$ is, indeed, a going-down ring (resp., locally divided ring). □

THEOREM 2.4. *The following conditions are equivalent:*

(1) If R is a going-down domain, then R' is a going-down domain;

(2) If R is a going-down domain, then R' is a locally divided domain;

(3) If R is a locally divided domain, then R' is a locally divided domain;

(4) If A is a quasilocal going-down ring such that $Z(A) = nil(A)$, then A' is a going-down ring;

(5) If A is a quasilocal going-down ring such that $Z(A) = nil(A)$, then A' is a locally divided ring;

(6) If A is a divided ring such that $Z(A) = nil(A)$, then A' is a locally divided ring;

(7) If B is a going-down ring such that $Z(B) = nil(B)$, then B' is a going-down ring;

(8) If B is a going-down ring such that $Z(B) = nil(B)$, then B' is a locally divided ring;

(9) If B is a locally divided ring such that $Z(B) = nil(B)$, then B' is a locally divided ring.

Proof. Note, as a special case of Proposition 2.1 (cf. also [17, Corollary 11], [7, Corollary 2.8]) that an integrally closed integral domain is a going-down domain if and only if it is a locally divided domain. Consequently, (1) ⇔ (2). Moreover, the implications divided domain ⇒ locally divided domain ⇒ going-down domain and Corollary 2.2 (b) yield that (2) ⇒ (3) ⇒ (1). Similarly, (6) ⇒ (1).

Next, since any divided ring is quasilocal, the implications divided ring ⇒ locally divided ring ⇒ going-down ring yield that (5) ⇒ (6), (4); and, for the same reason, (8) ⇒ (9), (7). Moreover, it is trivial that (7) ⇒ (4), (1); that (8) ⇒ (5), (2); and that (9) ⇒ (6), (3). We turn next to the more substantial implications.

(2) ⇒ (5): Assume (2), and let A be as in (5). Since $Z(A) = \text{nil}(A)$, A has a unique minimal prime ideal, say P_0, and $\text{tq}(A) = A_{P_0}$ [10, Proposition 2.3 (b)]. Put

$B := A'$ and $N := P_0 A_{P_0}$. Now, the hypotheses on A ensure that $A + N$ is integral over A [3, Theorem 2.1], whence $(A+N)' = A' = B$. It is easy to see that $N \in \operatorname{Spec}(B)$ and $B_N = A_{P_0}$. As $\dim(B_N) = \dim(A_{P_0}) = \operatorname{ht}_A(P_0) = 0$, [2, Corollary 2.2] gives that B_N is a locally divided ring. Moreover, since the class of going-down rings is stable under the formation of factor rings, the integral domain $R := A/P_0$ is also a going-down ring, and hence a going-down domain. Therefore, by (2), R' is a locally divided domain, and hence a locally divided ring. Notice that $R' = B/N$ by, for instance, applying [14, Corollary 1.5 (5)] to the pullback description

$$A + N = ((A+N)/N) \times_{A_{P_0}/N} A_{P_0} \cong R \times_{tq(R)} A_{P_0}.$$

In particular, B/N is a locally divided ring. Accordingly, to show that B is a locally divided ring (and thus complete the proof of (5)), it suffices, by [2, Theorem 2.7], to observe that $NB_N = NA_{P_0} = N \subseteq B$.

(4) $\Rightarrow$ (7); (5) $\Rightarrow$ (8): Assume (4) (resp., (5)), and let B be as in (7) (resp., (8)). Consider $T := \operatorname{tq}(B) = B_{Q_0}$, where Q_0 is the unique minimal prime ideal of B [10, Proposition 2.3 (b)]. By Lemma 2.3 (b), applied to the ring extension $B \subseteq B'$, it is enough to show that if $M \in \operatorname{Max}(B)$, then $E := (B')_{B \setminus M}$ is a going-down ring (resp., locally divided ring). Now, by [15, Proposition 10.2], E is the integral closure of B_M in $T_{B \setminus M}$. Note that B_M is a going-down ring [10, Proposition 2.1 (b)] and an overring of B [2, Proposition 2.5 (a)], whence $Z(B_M) = \operatorname{nil}(B_M)$ by [2, Lemma 2.6]. In particular, $\operatorname{tq}(B_M) = \operatorname{tq}(B) = T$. Also, $B \setminus M \subseteq B \setminus Q_0 = B \setminus Z(B)$, whence each element of $B \setminus M$ is a unit of T, and so $T_{B \setminus M} = T$. Thus, E is the integral closure of the quasilocal going-down ring B_M in its total quotient ring, and we have seen that $Z(B_M) = \operatorname{nil}(B_M)$. Accordingly, by (4) (resp., (5)), applied to $A := B_M$, we conclude that E is a going-down ring (resp., locally divided ring), as desired. □

REMARK 2.5. (a) For each $n, 1 \leq n \leq \infty$, a construction is given in [2, Example 3.10] of an n–dimensional quasilocal integrally closed going-down ring A which is not a locally divided ring. The existence of this ring A does not answer question (Q) in the negative (indeed, this question remains open), for A does not violate condition (5) in the statement of Theorem 2.4. The point is that $Z(A) \neq \operatorname{nil}(A)$. Unfortunately, it was erroneously claimed in [2, Remark 3.11] that $Z(A) = \operatorname{nil}(A)$, the error occurring because of a miscalculation of $Z(A)$. The authors of [2] wish to express their regret for this error in [2, Remark 3.11].

(b) In view of (a), it is natural to ask what *can* be concluded about integrally closed rings satisfying condition (5) in the statement of Theorem 2.4. One possible answer to this question is given by the following result. If A is a quasilocal integrally closed going-down ring such that $P_0 := Z(A) = \operatorname{nil}(A)$, then P_0 is a divided prime ideal in A; that is, $P_0 A_{P_0} = P_0$.

For a proof, let $B := A + P_0 A_{P_0}$. Thanks to [3, Theorem 2.1], the hypotheses ensure that B is integral over A. As $A \subseteq B \subseteq A_{P_0} = \operatorname{tq}(A)$ [10, Proposition 2.3 (b)] and A is integrally closed, $B = A$, whence $P_0 A_{P_0} = P_0 A_{P_0} \cap A = P_0$. By [2, Proposition

2.5 (c)], this means that P_0 is a divided prime ideal in A, to complete the proof.

Unlike most of our other results (2.2, 2.4, 2.7, 2.8) which find equivalents to the "ascent" question (Q) or its variants, Theorem 2.6 is itself an "ascent" result. Its proof adapts techniques that have been used in the proofs of Lemma 2.3 and Theorem 2.4.

THEOREM 2.6. *Let n be a positive integer. Let A be a going-down ring (resp., a locally divided ring) such that* $dim(A) = n$ *and* $Z(A) = nil(A)$. *Let* $P \in Spec(A)$ *be a prime ideal of A of height* $n-1$ *and suppose that* $P \subseteq J(A)$. *Then the integral closure of A in* A_P *is a going-down ring (resp., a locally divided ring).*

Proof. The proof will repeatedly use the fact that the class of going-down rings (resp., locally divided rings) is stable under the formation of rings of fractions [10, Proposition 2.1 (b)] (resp., [2, Proposition 2.1 (a)]). Let us begin with the case in which A is quasilocal. For this case, we may adapt the proof that (2) $\Rightarrow$ (5) in the proof of Theorem 2.4, as in the following sketch. Let B denote the integral closure of A in A_P, and put $N := PA_P$. As before, $A+N$ is integral over A (since locally divided ring $\Rightarrow$ going-down ring). Then $N \in \mathrm{Spec}(B)$ and $B_N = A_P$ is a going-down ring (resp., a locally divided ring). As $R := A/P$ is one-dimensional, so is the integral domain $R' = B/N$, whence B/N is also a going-down ring (resp., a locally divided ring). Since $NB_N = NA_{P_0} = N \subseteq B$, we have that N is comparable under inclusion with each ideal of B. Hence, B is a going-down ring (resp., a locally divided ring), by [10, Proposition 2.2] (resp., [2, Theorem 2.7]). This completes the proof in case A is quasilocal.

We turn now to the general case. Its proof has some of the flavor of the proofs that (4) $\Rightarrow$ (7) and (5) $\Rightarrow$ (8) in Theorem 2.4. Once again, let B denote the integral closure of A in A_P. By Lemma 2.3 (b), it is enough to show that if $M \in \mathrm{Max}(A)$, then $B_{A\setminus M}$ is a going-down ring (resp., locally divided ring). Observe that A_M is an overring of A and that $Z(A_M) = \mathrm{nil}(A_M)$ [2, Proposition 2.5 (a) and Lemma 2.6]. Moreover, A_M is a quasilocal going-down ring (resp., a divided ring). Also, since $P \subseteq J(A) \subseteq M$, we have that $\mathrm{ht}_{A_M}(PA_M) = \mathrm{ht}_A(P) = n-1$. Next, observe via [15, Proposition 10.2] that $B_{A\setminus M}$ is the integral closure of A_M in $(A_P)_{A\setminus M}$. As $A\setminus M \subseteq A\setminus P \subseteq A_P/PA_P$, we see that $(A_P)_{A\setminus M} = A_P = (A_M)_{PA_M}$. Accordingly, it follows from the above local case (with A_M playing the earlier role of A) that $B_{A\setminus M}$ is a going-down ring (resp., locally divided ring), as desired. □

In Corollary 2.7, we return to the type of theme that was addressed prior to the statement of Corollary 2.2. In the spirit of Theorems 2.4 and 2.6, it is no harder to move beyond integral domains and work in the context of rings A such that $Z(A) = \mathrm{nil}(A)$. The impact of Corollary 2.7 is to reduce question (Q) in the finite-dimensional case to the study of quasilocal going-down domains R for which the conductor $(R : R')$ is contained in the prime ideal of R that is adjacent to the maximal ideal of R.

COROLLARY 2.7. *Let n be a positive integer. Then the following conditions are equivalent:*

(1) If R is a going-down domain such that $dim(R) \leq n$*, then* R' *is a going-down domain;*

(2) If T is a divided domain such that $1 \leq dim(T) = m \leq n$ *and if T is integrally closed in* T_Q*, where Q is the unique prime ideal of T of height* $m-1$*, then* T' *is a going-down domain;*

(3) If A is a quasilocal going-down ring such that $1 \leq dim(A) = m \leq n$*,* $Z(A) = nil(A)$*, and A is integrally closed in* A_P*, where P is any prime ideal of A of height* $m-1$*, then* A' *is a locally divided ring;*

(4) If B is a quasilocal going-down ring such that $1 \leq dim(B) = m \leq n$ *and* $Z(B) = nil(B)$*, then* B' *is a locally divided ring.*

Proof. (1) $\Rightarrow$ (4): As noted earlier, an integrally closed integral domain is a going-down domain if and only if it is a locally divided domain. Thus, (1) is equivalent to the following condition: "If R is a going-down domain such that $\dim(R) \leq n$, then R' is a locally divided domain." Therefore, to prove that (1) $\Rightarrow$ (4), one need only rework the proof that (2) $\Rightarrow$ (5) in Theorem 2.4. Indeed, that proof carries over (with appropriate changes in notation) after observing (in the notation of Theorem 2.4) that $\dim(A/P_0) = \dim(A) \leq n$.

Next, note that the prime Q in condition (2) is uniquely determined, since each divided domain is a quasilocal going-down domain and each going-down domain is a treed domain. These facts also explain why the implications (4) $\Rightarrow$ (3) $\Rightarrow$ (2) are trivial. It now suffices to prove that (2) $\Rightarrow$ (1).

Assume (2), and let R be as in (1). To show that R' is a going-down domain (equivalently, a locally divided domain), it is enough (cf. Lemma 2.3 (b)) to draw the same conclusion about $(R')_{R\setminus M} = (R_M)'$ for each $M \in \text{Max}(R)$. Since R_M inherits the "going-down domain" property from R and $\dim(R_M) \leq \dim(R) \leq n$, there is no harm in replacing R with R_M. In other words, without loss of generality (R, M) is quasilocal. Since [7, Theorem 2.5] ensures that R has a divided unibranched integral overring (necessarily of the same dimension and with the same integral closure), there is now no harm in supposing that R is a divided domain. Put $m := \dim(R)$. Without loss of generality, $m > 0$ (since the integral closure of a field is the given field, and fields are trivially going-down domains). Let P denote the (unique) prime ideal of R having height $m-1$.

Let S denote the integral closure of R in R_P. By Theorem 2.6, S is a locally divided domain, with $\dim(S) = \dim(R) = m \leq n$. Since $S' = R'$, it suffices to show that S' is a going-down domain. Hence, by Lemma 2.3 (b), we need only show that if $N \in \text{Max}(S)$, then $T := S_N$ is such that T' $(= (S')_{S\setminus N})$ is a going-down domain. We shall do this by showing that T is the type of ring considered in condition (2).

Of course, T is a divided domain, being a localization of a locally divided domain; and $d := \dim(T) \leq \dim(S) = m \leq n$. Moreover, $d \neq 0$ since S is not a field (since $m \neq 0$). Accordingly, by (2), it suffices to show that if Q is the (unique) prime ideal of S of height $d-1$ such that $Q \subset N$, then T is integrally closed in $T_{QT} = S_Q$.

By integrality (actually, the going-up property), $N \cap R = M$. Therefore, since $R \subseteq S$ satisfies GD (because R is a going-down domain) and the prime ideals of S that are contained in N are linearly ordered by inclusion (since S, being a locally divided domain, is necessarily a treed domain), it follows from the incomparability property of integral extensions [16, Theorem 44] that $Q \cap R = P$. We claim that $Q = P$.

Indeed, since R is a divided domain, $P = PR_P \in \text{Spec}(R_P)$. Applying the canonical map $\text{Spec}(R_P) \to \text{Spec}(S)$, we have that $P = P \cap S \in \text{Spec}(S)$. As $R \subseteq S$ satisfies incomparability, it follows that P is the only prime ideal I of S such that $I \cap R = P$. In particular, $Q = P$, as claimed above. In addition, it now follows, by combining [15, Corollary 5.2] with the fact that $R \subseteq S$ satisfies the going-up property, that $S_{R\setminus P} = S_{S\setminus P} =: S_P$. However, the containments $R \subseteq S \subseteq R_P$ easily lead to $S_{R\setminus P} = R_P$, whence $R_P = S_P$.

We pause to note an alternate proof that $R_P = S_P$. Once again using the fact that $P = PR_P$, we see that S is the pullback of the inclusion map $S/P \to R_P/P$ and the canonical projection map $\pi : R_P \to R_P/P$. Applying [14, Proposition 1.9] to the multiplicatively closed subset $S \setminus P$ of S, we see that S_P is the pullback of the inclusion map $\text{tq}(S/P) \to R_P/P$ and π. Of course, $\text{tq}(S/P) = R_P/P$ since S/P is an overring of R/P, and so the second proof of the equality $R_P = S_P$ comes from the categorical triviality that the pullback of an isomorphism is an isomorphism.

We now complete the proof that T is integrally closed in S_Q. Since the definition of S ensures that S is the integral closure of S in R_P, [15, Proposition 10.2] yields that $T = S_N$ is the integral closure of T in $(R_P)_{S\setminus N}$. It therefore suffices to show that $(R_P)_{S\setminus N} = S_Q$. In fact, $S \setminus N \subseteq R_P \setminus P$, whence $(R_P)_{S\setminus N} = R_P = S_P = S_Q$, as desired. □

In the spirit of Theorem 2.4, the interested reader may add to the list of equivalent conditions in Corollary 2.7.

Note that, apart from the Jaffard case that was resolved in [6, Corollary 3.5], question (Q) remains open for two-dimensional going-down domains. For this reason, we next record the upshot of Corollary 2.7 in the two-dimensional case.

COROLLARY 2.8. *The following conditions are equivalent:*

(1) If R is a going-down domain such that $\dim(R) = 2$, then R' is a going-down domain;

(2) If T is a divided domain such that $\dim(T) = 2$ and if T is integrally closed in T_Q, where Q is the (unique) height 1 prime ideal of T, then T' is a going-down domain.

Proof. If an integral domain T satisfies $\dim(T) = 1$, then T' is a one-dimensional integral domain [16, Theorem 48], and hence a going-down domain. Accordingly, the proof is immediate from the equivalence $(1) \Leftrightarrow (2)$ in Corollary 2.7, with $n := 2$. □

In closing, we speculate on some directions for further work on going-down rings. Surely, one could obtain a result like Corollary 2.2 in the context of reduced rings, rather than just for integral domains. It would be more interesting to study ascent (or descent) for going-down rings A that do not satisfy $Z(A) = \mathrm{nil}(A)$. Partial work on other questions for such arbitrary going-down rings (and arbitrary locally divided rings) appeared in [3], with particular success in the treed case. Keeping in mind the recent work of McAdam [18] on going-down domains, we believe that it seems reasonable to expect an "ascent/descent" result for "going-down ring", analogous to that in [18, Proposition 3], when working within a suitable class of locally finite-dimensional treed rings.

References

[1] A. Badawi, *On divided commutative rings*, Comm. Algebra, **27** (1999), 1465–1474.

[2] A. Badawi and D. E. Dobbs, *On locally divided rings and going-down rings*, Comm. Algebra, **29** (2001), 2805–2825.

[3] A. Badawi and D. E. Dobbs, *Some examples of locally divided rings*, in Ideal Theoretic Methods in Commutative Algebra, Lecture Notes Pure Appl. Math., **220**, Dekker, New York (2001), pp. 73–83.

[4] P.-J. Cahen, J.-L. Chabert, D. E. Dobbs and F. Tartarone, *On locally divided domains of the form Int(D)*, Arch. Math., **74** (2000), 183–191.

[5] D. E. Dobbs, *On going-down for simple overrings, II*, Comm. Algebra, **1** (1974), 439–458.

[6] D. E. Dobbs, *Ascent and descent of going-down rings for integral extensions*, Bull. Austral. Math. Soc., **15** (1976), 253–264.

[7] D. E. Dobbs, *Divided rings and going-down*, Pac. J. Math., **67** (1976), 353–363.

[8] D. E. Dobbs, *Coherence, ascent of going-down and pseudo-valuation domains*, Houston J. Math., **4** (1978), 551–567.

[9] D. E. Dobbs, *On Henselian pullbacks*, in Factorization in Integral Domains, Lecture Notes Pure Appl. Math., **189**, Dekker, New York (1997), pp. 317–326.

[10] D. E. Dobbs, *Going-down rings with zero-divisors*, Houston J. Math., **23** (1997), 1–12.

[11] D. E. Dobbs and M. Fontana, *Integral overrings of two-dimensional going-down domains*, Proc. Amer. Math. Soc., **115** (1992), 655–662.

[12] D. E. Dobbs, M. Fontana and I. J. Papick, *Direct limits and going-down*, Comm. Math. Univ. St. Pauli, **31** (1982), 129–135.

[13] D. E. Dobbs and I. J. Papick, *On going-down for simple overrings, III*, Proc. Amer. Math. Soc., **54** (1976), 35–38.

[14] M. Fontana, *Topologically defined classes of commutative rings*, Ann. Mat. Pura Appl., **123** (1980), 331–355.

[15] R. Gilmer, Multiplicative Ideal Theory, Dekker, New York, 1972.

[16] I. Kaplansky, Commutative Rings, rev. ed., Univ. Chicago Press, Chicago, 1974.

[17] S. McAdam, *Simple going down*, J. London Math. Soc., **13** (1976), 167–173.

[18] S. McAdam, *Going down: ascent/descent*, Comm. Algebra, to appear.

[19] R. G. Swan, *On seminormality*, J. Algebra, **67** (1980), 210–229.

Generalized Going-Down Homomorphisms of Commutative Rings

DAVID E. DOBBS Department of Mathematics, University of Tennessee, Knoxville, Tennessee 37996-1300, U. S. A., dobbs@math.utk.edu

MARCO FONTANA Dipartimento di Matematica, Università degli Studi Roma Tre, 00146 Roma, ITALY, fontana@mat.uniroma3.it

GABRIEL PICAVET Laboratoire de Mathématiques Pures, Université Blaise Pascal (Clermont II), 63177 Aubière Cedex, FRANCE, Gabriel.Picavet@math.univ-bpclermont.fr

ABSTRACT

Sufficient conditions are given for a (unital) homomorphism $f\colon A \to B$ of (commutative) rings to be a chain morphism, in the sense that ${}^a f\colon \mathrm{Spec}(B) \to \mathrm{Spec}(A)$ permits the covering of chains of arbitrary cardinality. One such sufficient condition is that f satisfies lying-over, ${}^a f$ be open in the flat (resp., Zariski) topology, and that each reduced fiber of ${}^a f$ be quasilocal (resp., an integral domain). Sufficient conditions are given for f to have the generalized going-down property GGD (that is, "going-down" predicated for chains of arbitrary cardinality). Typical of such sufficient conditions are the following: f is a chain morphism and B is quasilocal treed; f satisfies going-down and either the reduced fibers of ${}^a f$ are integral domains or A is a going-down ring. "Universally going-down" is equivalent to "universally GGD"; in particular, if f is flat, then f satisfies GGD. The universally subtrusive homomorphisms are the same as the universally chain morphisms, and these descend the GGD property.

1 INTRODUCTION

All rings considered below are commutative with identity, and all ring homomorphisms are unital. Adapting the notation in [10, p. 28], we let GU, GD, LO, and

INC denote the going-up, going-down, lying-over, and incomparable properties, respectively, for ring homomorphisms. As in [3], our interests here include the following strengthening of the LO property. A ring homomorphism $f\colon A \to B$ is called a *chain morphism* if the associated map ${}^a f\colon \mathrm{Spec}(B) \to \mathrm{Spec}(A)$, $Q \mapsto f^{-1}(Q)$, permits each chain (of arbitrary cardinality) of prime ideals of A to be "covered" by some chain of prime ideals of B. Theorem 2.3 gives our main sufficient conditions for a ring homomorphism f to be a chain morphism, namely, that f satisfy GU (resp., GD) and LO, with reduced fibers that are quasilocal (resp., integral domains). Theorem 2.8 is essentially a corollary giving sufficient conditions that are couched topologically. (For background on the flat spectral topology, see [8], [4]; for background on the patch, or constructible, topology, see [8], [7].) Proposition 2.2 collects the sufficient conditions for chain morphisms that were established in [3], most notably, that f be injective and integral. Theorem 3.26 presents a significant generalization: any universally subtrusive f (in the sense of [14], for instance, any pure or faithfully flat f) is universally a chain morphism. This result depends on the heart of the paper, Section 3, which develops the theory of the GGD (*generalized going-down*) concept, the property that "going-down" behavior hold for chains of arbitrary cardinality.

Proposition 3.1 states the sufficient condition for a ring homomorphism $f\colon A \to B$ to satisfy GGD that was obtained in [3]. Numerous other sufficient conditions are given, including that f satisfy GD with reduced fibers that are integral domains (Corollary 3.6); and that B be a quasilocal treed ring with f satisfying either GD or both LO and GU (Corollary 3.4). Theorem 3.9 identifies a context for which GGD and GD are equivalent, namely, where A is a going-down ring (in the sense of [2]) in which each maximal ideal of A contains a unique minimal prime ideal of A. A noteworthy upshot appears in Corollary 3.14: a weak Baer ring A is a going-down ring if and only if $A \hookrightarrow B$ satisfies GD for each overring B of A. Despite the nomenclature, such an assertion fails if A is not a weak Baer ring [2, Examples 1 and 2, pp. 9-12]. Accordingly, since our present focus is on properties of homomorphisms, we intend to devote a subsequent paper to weak Baer going-down rings and related themes.

As a companion for the characterization of universally chain morphisms in Theorem 3.26, we also characterize the universally GGD ring homomorphisms (in Theorem 3.16): they are precisely the universally going-down maps. In particular, each flat ring homomorphism is (universally) GGD. Among other sufficient conditions for a ring homomorphism f to satisfy GGD is that f satisfy GD and ${}^a f$ be injective (Corollary 3.21), in which case ${}^a f$ is a topological immersion (relative to the Zariski topology). Finally, we note a consequence of Theorem 3.26: each universally subtrusive ring homomorphism descends the universally going-down property.

We next describe notational conventions. Unless otherwise specified, maps of the form ${}^a f$ are considered relative to the Zariski topology. As usual, a typical closed set in that topology on Spec(A) is $V(I) = \{P \in \text{Spec}(A) : P \supseteq I\}$, where I is an ideal of A. We denote the closure of a set X in the Zariski topology by $\overline{X}$, with X^c denoting the closure of X in the patch topology. By a *patch*, we mean a set that is closed in the patch topology. A ring A is *treed* if no maximal ideal of A contains incomparable prime ideals of A. If A is a ring, then $U(A)$ denotes the set of units of A and $tq(A)$ denotes the total quotient ring of A. By an *overring of* a ring A, we mean any A-subalgebra of $tq(A)$; or, more intuitively, any ring B such that $A \subseteq B \subseteq tq(A)$. Finally, $\subset$ denotes proper containment, and $|I|$ denotes the cardinality of the set I.

Background is recalled as needed. Any unexplained material is standard, as in [10], [7], [6].

Dobbs was supported in part by a University of Tennessee Faculty Development Award and a Visiting Professorship funded by the Istituto Nazionale di Alta Matematica. He thanks the Università degli Studi di Roma Tre for the warm hospitality accorded during his visit in October-November, 2000.

As this paper went to press (May, 2002), Kang-Oh have announced a preprint, "Lifting Up a Tree of Prime Ideals to a Going-Up Extension," whose methods, we have determined, can be extended to show that GGD and GD are equivalent for ring homomorphisms $A \to B$ such that both A and B are locally irreducible. In particular, this equivalence holds if A and B are each integral domains.

2 CHAIN MORPHISMS

Let A be a ring and X a subset of Spec(A). Following [15], we define $\mathcal{U}(X) := \cup\{P : P \in X\}$ and $\mathcal{R}(X) := \cap\{P : P \in X\}$. Observe that if X is a chain, then $\mathcal{U}(X), \mathcal{R}(X) \in \text{Spec}(A)$ [10, Theorem 9]. A chain X is called a *local chain* if X has a (necessarily unique) maximal element. If X is a chain, then $X \cup \{\mathcal{U}(X)\}$ is a local chain; in fact, a chain X is a local chain if and only if $\mathcal{U}(X) \in X$. By reworking the proof of [15, Proposition 2.2], we see that each local chain is quasi-compact in the Zariski topology.

We next introduce the key concepts of this section. Suppose that $f \colon A \to B$ is a ring homomorphism. Consider $X = \{P_i : i \in I\}$, a subset of Spec(A). (The notation is generally taken so that $P_i \neq P_j$ whenever $i \neq j$; as a result, $|X| = |I|$.) A subset $Y = \{Q_i : i \in I\}$ of Spec(B) is said to *cover* (or to *dominate*) X if $f^{-1}(Q_i) = P_i$ for each $i \in I$. (By the notational convention, $Q_i \neq Q_j$ if $i \neq j$, and so $|Y| = |I|$.) We say that f is a *chain morphism* if, for each chain X in Spec(A), there exists a chain Y in Spec(B) such that Y covers X. As partial motivation for this definition, we make two observations: each chain Y of Spec(B) has a subchain that covers the chain $\{P : P = f^{-1}(Q) \text{ for some } Q \in Y\}$ of Spec(A); and, by focusing on singleton chains, we see that any chain morphism satisfies LO.

Proposition 2.1 collects some easy but useful facts, and Proposition 2.2 gives examples of chain morphisms that are essentially already known.

PROPOSITION 2.1. *Let $f: A \to B$ be a ring homomorphism. Then:*

(a) If a local chain Y in Spec(B) *covers a subset X of* Spec(A)*, then X is a local chain.*

(b) If a chain Y in Spec(B) *covers a local chain X in* Spec(A)*, then Y is a local chain and $f^{-1}(\mathcal{U}(Y)) = \mathcal{U}(X)$.*

(c) If f is a chain morphism and X is a local chain in Spec(A)*, then X is covered by some local chain Y in* Spec(B) *and $f^{-1}(\mathcal{U}(Y)) = \mathcal{U}(X)$.*

Proof. (a) By the above observation, X is a chain. If $P \in X$, there exists $Q \in Y$ such that $f^{-1}(Q) = P$, whence $P \subseteq f^{-1}(\mathcal{U}(Y)) \in X$. It follows that $\mathcal{U}(X) = f^{-1}(\mathcal{U}(Y)) \in X$, and so X is a local chain.

(b) Choose $Q \in Y$ such that $f^{-1}(Q) = \mathcal{U}(X)$. If $Q \subset Q_1 \in Y$, then $f^{-1}(Q) \subset f^{-1}(Q_1)$, contradicting the fact that $f^{-1}(Q) = \mathcal{U}(X) \supseteq f^{-1}(Q_1)$. Thus, $\mathcal{U}(Y) = Q \in Y$, and so Y is a local chain. Then $f^{-1}(\mathcal{U}(Y)) = \mathcal{U}(X)$ by the proof of (a).

(c) Apply (b). □

PROPOSITION 2.2. *(Dobbs [3]) Let $f: A \to B$ be a ring homomorphism. Then f is a chain morphism in each of the following four cases:*

(i) f is injective and integral;

(ii) f satisfies LO and GU, and each chain in Spec(A) *is well-ordered via inclusion;*

(iii) f satisfies LO and GD, and each chain in Spec(A) *is well-ordered via reverse inclusion;*

(iv) A is Noetherian, and f satisfies LO and either GU or GD.

Proof. (i) was proved in [3, Remark(d)]; (ii) and (iii) follow from what was proved in [3, Theorem] and [3, Remark(a)], respectively, as those proofs, although given for injective f, carry over to the general case; and (iv) follows from (ii) and (iii), since A Noetherian ensures that each chain in Spec(A) is finite (hence, well-ordered with respect to both inclusion and reverse inclusion). □

As noted above, each chain morphism satisfies LO. Partial converses were given in Proposition 2.2 (ii)-(iv). Before deriving additional partial converses (in Theorems 2.3 and 2.8), we interpret topologically some conditions appearing in those results. Let $f: A \to B$ be a ring homomorphism and let $P \in$ Spec(A). It is well known that ${}^a f^{-1}(P)$, the so-called *topological fiber of P (with respect to f)*, is homeomorphic to Spec$((A_P/PA_P) \otimes_A B)$ in both the Zariski topology and the flat topology. One calls $(A_P/PA_P) \otimes_A B \cong B_P/PB_P$ *the fiber of f at P*; its associated reduced ring, $B_P/\sqrt{PB_P}$, is called the *reduced fiber (of f at P)*. It is easy to show, via Zorn's Lemma and [10, Theorem 9], that each element of ${}^a f^{-1}(P)$ is contained in some maximal element of ${}^a f^{-1}(P)$ and contains some minimal element of ${}^a f^{-1}(P)$. It follows that ${}^a f^{-1}(P)$ has a unique maximal (resp., unique minimal) element if and only if the reduced fiber of f at P is a quasilocal ring (resp., an integral domain);

that is (cf. [15, Lemme 2.5]), if and only if ${}^a f^{-1}(P)$ is irreducible in the flat (resp., Zariski) topology.

THEOREM 2.3. *Let $f: A \to B$ be a ring homomorphism that satisfies at least one of the following two conditions:*

(i) f satisfies GU and each reduced fiber of f is quasilocal;

(ii) f satisfies GD and each reduced fiber of f is an integral domain.

Then:

(a) For each chain $X \subseteq \mathrm{Im}({}^a f)$, there exists a chain Y in Spec(B) *such that Y covers X.*

(b) If, in addition, f satisfies LO, then f is a chain morphism.

Proof. It suffices to establish (a). Assume (i) (resp., (ii)). Consider any chain $X = \{P_i : i \in I\}$. By the above comments, we can choose Q_i to be the unique maximal (resp., unique minimal) element of ${}^a f^{-1}(P_i)$. Evidently, $Y := \{Q_i : i \in I\}$ covers X. It remains only to verify that Y is a chain. In fact, if $P_i \subset P_j$, then if follows from GU (resp., GD) and the maximality of Q_j (resp., minimality of Q_i) that $Q_i \subset Q_j$. □

We pause to note additional topological interpretations for some conditions in the statement of Theorem 2.3. Let $f: A \to B$ be a ring homomorphism. Then ${}^a f$ is closed in the Zariski (resp., flat) topology if and only if f satisfies GU (resp., GD) [4, Proposition 2.7]. It now seems natural to ask for "open" analogues of the "closed" assertions in Theorem 2.3. We provide such analogues in Theorem 2.8, which is really just a corollary of Theorem 2.3. In order to give an alternate approach to Theorem 2.8, we first develop some topological results. We also take advantage of this opportunity to introduce some deeper results on chains that will be useful in Section 3.

PROPOSITION 2.4. *Let A be a ring and let X be a subset of* Spec(A)*. Then:*

(a) If X is a chain, then its patch closure X^c is also a chain.

(b) X is a chain if and only if there exist a ring homomorphism $A \to V$ and a chain Y in Spec(V) *such that V is a valuation domain and Y covers X.*

(c) X is a local chain if and only if there exists a ring homomorphism $f: A \to V$ and a local chain Y in Spec(V) *such that V is a valuation domain, Y covers X, and $f^{-1}(\mathcal{U}(Y)) = \mathcal{U}(X)$.*

Proof. (b) The "if" assertion is clear. Conversely, suppose that X is a chain. As in the proof of [3, Remark(d)], there is no harm in replacing A with $A/\mathcal{R}(X)$, and so we may suppose that A is an integral domain. The lifting result of Kang-Oh [9, Theorem] provides a valuation domain V containing A and a chain Y in Spec(V) that covers X.

(a) Let X, V and A be as in the proof of (b). Let f be the composite $A \twoheadrightarrow A/\mathcal{R}(X) \hookrightarrow V$; put $Z := \mathrm{Im}({}^a f) \subseteq \mathrm{Spec}(A)$. By definition of the patch (constructible) topology, Z is patch-closed; that is, $Z^c = Z$. Moreover, Z is a chain since V is quasilocal treed. As $X \subseteq Z$, we have $X^c \subseteq Z^c = Z$. Then X^c, being a subset of a chain, is itself a chain.

(c) The "if" assertion follows from Proposition 2.1 (a). The "only if" assertion follows by combining (b) with Proposition 2.1 (b). □

COROLLARY 2.5. *Let A be a ring, X a chain in* Spec(A), *and* $P \in$ Spec(A) *such that* $\mathcal{U}(X) \subseteq P$. *Then there exist a ring homomorphism* $f\colon A \to V$, *a chain* Y *in* Spec(V), *and* $Q \in$ Spec(V) *such that* V *is a valuation domain,* Y *covers* X, $f^{-1}(Q) = P$, *and* $\mathcal{U}(Y) \subseteq Q$.

Proof. Apply Proposition 2.4 (c) to the local chain $X \cup \{P\}$, to obtain a suitable local chain Z in Spec(V). It suffices to take $Q := \mathcal{U}(Z)$; and $Y := Z$ (resp., $Z\backslash\{Q\}$) if $P \in X$ (resp., $P \notin X$). □

PROPOSITION 2.6. *Let A be a ring and X a chain in* Spec(A). *Then:*

(a) If X is a maximal chain in Spec(A), *then X is stable under unions and intersections and, moreover, X is a patch and a local chain.*

(b) There exists a maximal chain X' in Spec(A) *such that* $X \subseteq X'$. *For any such* X', *there exist* $P \in$ Spec(A) *and a minimal valuation overring W of A/P such that* $P \subseteq \mathcal{R}(X)$ *and* Im(Spec$(W) \to$ Spec$(A)) = X'$.

(c) Let $f\colon A \to V$ be a ring homomorphism such that V is a valuation domain and $X \subseteq$ Im$({}^a f)$. *Then there exist* $P \in$ Spec(A) *and a minimal valuation overring W of A/P such that* $X \subseteq$ Im(Spec$(W) \to$ Spec$(A))$.

Proof. (a), (b): By the reasoning in [3, pp. 3888-3889], if X is any chain, then $X \cup \{\mathcal{R}(Z) : \phi \neq Z \subseteq X\}$ is a chain. It follows that any maximal chain is stable under intersections. By reasoning similarly with $X \cup \{\mathcal{U}(Z) : \phi \neq Z \subseteq X\}$, we see that any maximal chain is stable under unions. Of course, considering $X \cup \{\mathcal{U}(X)\}$ shows that any maximal chain is a local chain.

It follows easily via Zorn's Lemma that each chain X is contained in a maximal chain. Consider any maximal chain $X' \supseteq X$. By the proof of Proposition 2.4 (a), there exists a valuation overring V of $D := A/\mathcal{R}(X')$ so that the composite ring homomorphism $g\colon A \twoheadrightarrow D \hookrightarrow V$ satisfies $X' \subseteq$ Im$({}^a g)$. By Zorn's Lemma (cf. [6, p. 231]), V contains a minimal valuation overring W of D. If f denotes the composite $A \twoheadrightarrow D \hookrightarrow W$, then Im$({}^a g) \subseteq$ Im$({}^a f)$ since Spec$(V) \subseteq$ Spec(W) (cf. [6, Theorem 26.1]). However, Im$({}^a f)$ is a chain (since W is quasilocal treed), and so $X' =$ Im$({}^a f)$ by the maximality of X'. Then X' is a patch, by the definition of the patch (constructible) topology. Finally, note that $P := \mathcal{R}(X') \subseteq \mathcal{R}(X)$.

(c) Observe that $P :=$ ker$(f) \in$ Spec(A). Let k (resp., K) denote the quotient field of A/P (resp., of V). Then the canonical ring inclusion $A/P \hookrightarrow V$ extends to an inclusion of fields, $k \hookrightarrow K$. Since $V \cap k$ is a valuation overring of A/P, another application of Zorn's Lemma produces a minimal valuation overring W of A/P such that $W \subseteq V \cap k$. By hypothesis, $X \subseteq$ Im(Spec$(V \cap k) \to$ Spec$(A))$. It remains only to note that Spec$(V \cap k) \subseteq$ Spec(W). □

LEMMA 2.7. *Let $f: A \to B$ be a ring homomorphism such that ${}^af$ is open in the flat topology and each reduced fiber of f is quasilocal. Then a subset X of* $\text{Im}({}^af)$ *is irreducible in the flat topology on* Spec(*A*) *if and only if* ${}^af^{-1}(X)$ *is irreducible in the flat topology on* Spec(*B*).

Proof. By the above comments, ${}^af^{-1}(P)$ is irreducible in the flat topology, for all $P \in \text{Spec}(A)$. Hence, by [7, Proposition 2.1.14, p. 54], we need only verify that ${}^af$ induces a map ${}^af^{-1}(X) \to X$ that is continuous, surjective and open in the subspace topology induced by the flat topology. Both "continuous" and "surjective" are clear. As for "open", consider any (flat-)open set U in $\text{Spec}(B)$, and observe that

$$ {}^af(U \cap {}^af^{-1}(X)) = {}^af(U) \cap X. $$

Since the hypothesis ensures that ${}^af(U)$ is (flat-) open in $\text{Spec}(A)$, the assertion follows. □

THEOREM 2.8. *Let f be a ring homomorphism that satisfies at least one of the following two conditions:*

(i) ${}^af$ is open in the flat topology and each reduced fiber of f is quasilocal;

(ii) ${}^af$ is open in the Zariski topology and each reduced fiber of f is an integral domain.

Then:

(a) For every chain $X \subseteq$ $\text{Im}({}^af)$*, there exists a chain Y in* $\text{Spec}(B)$ *such that Y covers X.*

(b) If, in addition, f satisfies LO, then f is a chain morphism.

Proof. (b) is an immediate consequence of (a). As for (a), if (ii) holds, the assertion may be proved exactly as in Theorem 2.3 (ii), since Zariski-open ${}^af$ entails going-down f [7]. A parallel proof is also available if (i) holds, since flat-open ${}^af$ entails going-up f (that is, Zariski-closed f) [13, Remarque, p. 2252].

Alternate, more topological proofs are available for Theorem(s 2.3 and) 2.8. We illustrate such methods with another proof for case (i). As in the earlier proof, it suffices to show that if P_i and P_j are distinct elements of a chain in $\text{Im}({}^af)$ and if Q_i (resp., Q_j) is the maximal element in ${}^af^{-1}(P_i)$ (resp., in ${}^af^{-1}(P_j)$), then Q_i and Q_j are comparable under inclusion. As P_i and P_j are comparable and flat-closed sets are stable under generization [4, Lemma 2.1], it follows that $Z := \{P_i, P_j\}$ is irreducible in the flat topology (cf. also [15, Proposition 2.4]). Hence, by Lemma 2.7, $Y := {}^af^{-1}(Z)$ is also irreducible in the flat topology. As Y is a patch (being the spectral image of $B_{P_i}/P_iB_{P_i} \times B_{P_j}/P_jB_{P_j}$), [15, Lemme 2.5] ensures that Y is directed via inclusion. Thus, Q_i and Q_j are each contained in some prime $Q \in Y$ such that ${}^af(Q) \in \{P_i, P_j\}$. Without loss of generality, ${}^af(Q) = P_i$, whence $Q \subseteq Q_i$ by choice of Q_i. Then $Q_i = Q \supseteq Q_j$. □

There are useful algebraic sufficient conditions for the "open" properties in the statement of Theorem 2.8. For instance, if a ring homomorphism f is integral (resp., flat) and of finite presentation, then ${}^a f$ is open in the flat (resp., Zariski) topology, by [13, Proposition 6] (resp., [7, Corollaire 3.9.4(i), p. 254]). We close the section by using this fact to give an application of Theorem 2.8.

COROLLARY 2.9. *Let $P_1, \ldots, P_m \in \mathbb{Z}[X_1, \ldots, X_n]$ be such that $(P_1, \ldots, P_m)$ is a prime ideal in $K[X_1, \ldots, X_n]$ for any field K. Suppose that A is a ring and $f \colon A \to B := A[X_1, \ldots, X_n]/(P_1, \ldots, P_m)$ is such that ${}^a f$ is open in the Zariski topology (for instance, take f to be flat). Then each chain in $\mathrm{Im}({}^a f)$ can be covered by a chain in* Spec(B).

Proof. The hypothesis ensures that each (reduced) fiber of f is an integral domain. Apply Theorem 2.8 (a), using condition (ii). □

A concrete illustration of Corollary 2.9 is provided by $n = 2, m = 1, P_1 = X_1^2 - X_2^3$.

3 GENERALIZED GOING-DOWN

We begin with the key definition of this paper. A ring homomorphism $f \colon A \to B$ is said to satisfy the *generalized going-down* property (GGD) if the following holds: for each local chain X in Spec(A) and each $Q \in$ Spec(B) such that $f^{-1}(Q) = \mathcal{U}(X)$, there exists a local chain Y in Spec(B) such that $\mathcal{U}(Y) = Q$ and Y covers X. Evidently, GGD $\Rightarrow$ GD. We next record the only instance of GGD that has appeared in the literature.

PROPOSITION 3.1. *(Dobbs [3, proof of Remark (a)]). Let A be a ring such that each chain in* Spec(A) *is well-ordered via reverse inclusion. Then a ring homomorphism $f \colon A \to B$ satisfies GGD if (and only if) f satisfies GD.*

In comparing Propositions 3.1 and 2.2 (iii), one suspects that the notions of GGD and chain morphism are closely related. Proposition 3.2 states some evident connections, with less evident connections in the subsequent results. Of course, the two concepts are logically independent: if f is an injective integral ring homomorphism that does not satisfy GD, then f is a chain morphism that does not satisfy GGD; and if S is a multiplicatively closed subset of a ring A such that S contains a nonunit of A, then the canonical map $A \to A_S$ satisfies GGD but (as it fails to have LO) is not a chain morphism.

PROPOSITION 3.2. *Let $f \colon A \to B$ be a ring homomorphism. Then:*

(a) If f satisfies LO and GGD, then f is a chain morphism.

(b) If ${}^a f$ is injective and f is a chain morphism, then f satisfies GGD.

PROPOSITION 3.3. *If B is a quasilocal treed ring and $f \colon A \to B$ is a chain morphism, then f satisfies GGD.*

Proof. Consider a local chain $X = \{P_i : i \in I\}$ in Spec(A) and $Q \in$ Spec(B) such that $f^{-1}(Q) = \mathcal{U}(X)$. Since f is a chain morphism, Proposition 2.1 (c) provides a local chain $Y = \{Q_i : i \in I\}$ in Spec(B) that covers X, with $f^{-1}(\mathcal{U}(Y)) = \mathcal{U}(X)$. Choose (the unique) $j \in I$ such that $P_j = \mathcal{U}(X)$. Then $Q_j = \mathcal{U}(Y)$. If $Q_j = Q$, then Y is the desired local chain Z in Spec(B) such that $\mathcal{U}(Z) = Q$ and Z covers X. If $Q_j \subset Q$, then $Z := (Y \backslash \{Q_j\}) \cup \{Q\}$ suffices. Since B is quasilocal treed, there is only one remaining case, namely, $Q \subset Q_j$. For this case, it suffices to take $Z := \{Q_i \cap Q : i \in I\}$. □

COROLLARY 3.4. *Let B be a quasilocal treed ring. Let $f\colon A \to B$ be a ring homomorphism that satisfies either GD or both LO and GU. Then f satisfies GGD.*

Proof. If $P \in \text{Im}(^a f)$, then B quasilocal treed implies that $^a f^{-1}(P)$ has a unique maximal element and a unique minimal element; that is, each reduced fiber of f is a quasilocal integral domain. The conclusion therefore follows by combining Theorem 2.3 and the proof of Proposition 3.3. □

By reworking the proof of Proposition 3.3, we next find two companion results. Just as Corollary 3.4 issued from combining Proposition 3.3 with Theorem 2.3, one can produce additional applications by combining part (a) or part (b) of Corollary 3.5 with Theorem 2.3. We leave such formulations to the reader.

COROLLARY 3.5. *Let $f\colon A \to B$ be a chain morphism that satisfies at least one of the following two conditions:*

(i) B is treed and each reduced fiber of f is quasilocal;

(ii) Each (Zariski-) irreducible component of Spec(B) *is a chain (via inclusion) and each reduced fiber of f is an integral domain.*

Then f satisfies GGD.

Proof. We proceed to rework the proof of Proposition 3.3. It suffices to verify that Q_j and Q are comparable via inclusion. In case (i), this follows since B is treed and Q_j, Q are each contained in (any maximal ideal of B that contains) the unique maximal element of $^a f^{-1}(P_j)$. An essentially "dual" proof is available if (ii) holds. Indeed, Q_j, Q each contain the unique minimal element I of $^a f^{-1}(P_j)$. Using Zorn's Lemma, choose a minimal prime ideal N of B such that $N \subseteq I$ [10, Theorem 10]. Then Q_j, Q are each in the (Zariski-) irreducible set $V(N)$, which is a chain by hypothesis, whence Q_j and Q are comparable. □

COROLLARY 3.6. *Let $f\colon A \to B$ be a ring homomorphism such that each reduced fiber of f is an integral domain and f satisfies GD. Then f satisfies GGD.*

Proof. Once again, we rework the proof of Proposition 3.3. Even if f is not a chain morphism, the requisite chain $Y = \{Q_i\}$ is provided by Theorem 2.3 (with emphasis on its condition (ii)). By taking Q_i to be the unique minimal element of $^a f^{-1}(P_i)$, we are assured that $Q_j \subseteq Q$, and so $Z := (Y \backslash \{Q_j\}) \cup \{Q\}$ suffices. □

REMARK 3.7. (a) Recall from [4, pp. 567-568] that there is a "weak going down" concept that can be used to characterize the flat topology. In a different vein, we can also use an ostensibly "weaker" property to characterize GGD. Indeed, it is not difficult to show that a ring homomorphism $f\colon A \to B$ satisfies GGD if and only if the following holds: for each chain X in $\mathrm{Spec}(A)$, each $P \in \mathrm{Spec}(A)$ such that $\mathcal{U}(X) \subseteq P$, and each $Q \in {}^af^{-1}(P)$, there exists a chain Y in $\mathrm{Spec}(B)$ such that $\mathcal{U}(Y) \subseteq Q$ and Y covers X.

(b) On the other hand, a related property that is ostensibly "stronger" than GGD may actually be stronger. For instance, consider the following property, say (*), that a ring homomorphism $f\colon A \to B$ can satisfy: for each chain X in $\mathrm{Spec}(A)$, with $P := \mathcal{U}(X)$, and each $Q \in {}^af^{-1}(P)$, there exists a chain Y in $\mathrm{Spec}(B)$ such that $\mathcal{U}(Y) = Q$ and Y covers X. It is straightforward to verify that property (*) implies GGD. However, unlike the situation in (a), the converse is false. In other words, GGD fails to imply property (*). To see this, take f to be an inclusion map $A \to B$, where B is a valuation domain with prime spectrum $0 = Q_0 \subset Q_1 \subset \dots \subset Q_n \subset \dots \subset Q' \subset Q$, such that B/Q' is a $K-$algebra for some field K, and define A to be the pullback $B \times_{B/Q'} K$. Put $P := Q \cap A (= Q' \cap A)$. Then, by a standard gluing argument, with $X := \mathrm{Spec}(A) \setminus \{P\}$, one checks that f fails to satisfy property (*), for the only chain Y in $\mathrm{Spec}(B)$ that covers X is $Y = \mathrm{Spec}(B) \setminus \{Q', Q\}$, with $\mathcal{U}(Y) = Q' \neq Q$.

Next, we collect some elementary but useful facts indicating that *GGD* behaves rather similarly to known behavior of GD.

PROPOSITION 3.8. *(a) Let $f\colon A \to B$ and $g\colon B \to C$ be ring homomorphisms. If f and g each satisfies GGD, so does $g \circ f$. If g satisfies LO and $g \circ f$ satisfies GGD, then f satisfies GGD.*

(b) If f is a ring homomorphism, then the following seven conditions are equivalent:

(1) f satisfies GGD;

(2) $f_S\colon A_S \to B_S := B \otimes_A A_S$ satisfies GGD for each multiplicatively closed subset S of A;

(3) $f_P\colon A_P \to B_P := B \otimes_A A_P$ satisfies GGD for each $P \in \mathrm{Spec}(A)$;

(4) $A_P \to B_Q$ satisfies GGD for each $Q \in \mathrm{Spec}(B)$ and $P := f^{-1}(Q)$;

(5) $A/I \to B/IB$ satisfies GGD for each ideal I of A;

(6) $A/P \to B/PB$ satisfies GGD for each minimal prime ideal P of A;

(7) f_{red} satisfies GGD.

(c) Let $f_i\colon A_i \to B_i\ (i = 1, \dots, n)$ be finitely many ring homomorphisms. Then the induced map $A_1 \times \dots \times A_n \to B_1 \times \dots \times B_n$ satisfies GGD if and only if f_i satisfies GGD for each i. If $A_1 = \dots = A_n =: A$, then the induced map $A \to B_1 \times \dots \times B_n$ satisfies GGD if and only if f_i satisfies GGD for each i.

A principal theme of this section is that the classic sources of going-down homomorphisms (namely, going-down domains and flat maps) give rise to GGD behavior. We pursue this point somewhat more generally in Theorems 3.9 and 3.16 after giving some background material and applications.

Recall from [1] and [5] that an integral domain A is called a *going-down domain* if $A \subseteq B$ satisfies GD for each overring B of A. The most natural examples of going-down domains are arbitrary valuation domains and the integral domains of (Krull) dimension at most 1. As in [2], a ring A is called a *going-down ring* if A/P is a going-down domain for each (equivalently, each minimal) prime ideal P of A. Any integral domain is a going-down ring if and only if it is a going-down domain [2, Remark (a), p. 4]; any ring of dimension at most 1 is a going-down ring [2, Proposition 2.1 (c)]; a finite ring product $A_1 \times \cdots \times A_n$ is a going-down ring if and only if each A_i is a going-down ring [2, Proposition 2.1 (b)]; but there exists a going-down ring A and an overring B of A such that $A \subseteq B$ does not satisfy GD [2, Example 1, p. 9]. Adapting terminology from [12], we say that a ring homomorphism $f\colon A \to B$ is a *min morphism* if $f^{-1}(Q)$ is a minimal prime ideal of A for each minimal prime ideal Q of B. It is evident that if a ring homomorphism f satisfies GD, then f is a min morphism. In the theory of Krull domains, an example of min morphisms is proved by the classical condition of *pas d'éclatement*, PDE (also known as *no blowing up*, NBU). Finally, recall that a ring A is said to be *locally irreducible* if each maximal ideal of A contains a unique minimal prime ideal of A.

THEOREM 3.9. *Let A be a locally irreducible ring and a going-down ring and let $f\colon A \to B$ be a ring homomorphism. Then the following conditions are equivalent:*

(1) f is a min morphism;

(2) f satisfies GD;

(3) f satisfies GGD.

Proof. By the above comments, (3) $\Rightarrow$ (2) $\Rightarrow$ (1). It remains to show that if f is a min morphism, X a local chain in $\mathrm{Spec}(A)$ and $Q \in \mathrm{Spec}(B)$ such that $f^{-1}(Q) = \mathcal{U}(X)$, then there exists a local chain Y in $\mathrm{Spec}(B)$ such that $\mathcal{U}(Y) = Q$ and Y covers X. By [10, Theorem 10], Q contains some minimal prime ideal J of B. Since f is a min morphism, $I := J \cap A$ is a minimal prime ideal of A. Of course, $I \subseteq Q \cap A = \mathcal{U}(X)$ and so, since A is locally irreducible, I is the only minimal prime ideal of A that is contained in $\mathcal{U}(X)$. As each $P \in X$ contains a minimal prime ideal of A, it follows that $I \subseteq P$, whence $I \subseteq \mathcal{R}(X)$. There is no harm in replacing f with $A/I \hookrightarrow B/J$. Hence, without loss of generality, $A \subseteq B$ are integral domains and A is a going-down domain (cf. [2, Proposition 2.1 (b) and Remark (a), p. 4]). Choose a valuation overring (V,N) of B such that $N \cap B = Q$ (cf. [10, Theorem 56]). Of course, V is quasilocal and treed. Moreover, $A \subseteq V$ satisfies GD since A is a going-down domain. Hence, by Corollary 3.4, there exists a local chain $Z = \{Q_i\}$ in $\mathrm{Spec}(V)$ such that Z covers X. Then, by Proposition 2.1 (b), $Y := \{Q_i \cap B\}$ has the desired properties. $\square$

Recall that $A \subseteq B$ need not satisfy GD when B is an overring of a going-down ring A [2, Example 1, p. 9]. By avoiding one feature of that example, we have the following pleasant consequence.

COROLLARY 3.10. *If $f: A \to B$ is an injective ring homomorphism such that A is a going-down ring and B has a unique minimal prime ideal, then f satisfies GGD.*

Proof. If P is a minimal prime ideal of A, then [10, Exercise 1, p. 41] ensures that $f^{-1}(Q) = P$ for some prime ideal Q of B. By [10, Theorem 10], we can take Q to be the unique minimal prime ideal Q_0 of B. Hence, P is uniquely determined as $f^{-1}(Q_0)$; that is, A has a unique minimal prime ideal and f is a min morphism. In particular, A is locally irreducible. An application of Theorem 3.9 completes the proof. □

COROLLARY 3.11. *Let $f: A \to B$ be a ring homomorphism such that A is a going-down ring. Then f satisfies GGD if and only if f satisfies GD.*

Proof. The "only if" assertion is valid even without the hypothesis on A. Conversely, suppose that f satisfies GD. It follows that if $P \in \mathrm{Spec}(A)$, then the induced map $g: A/P \to B/PB$ is a min morphism (by the proof of [10, Exercise 37, p. 44]). As A/P is a going-down ring [2, Proposition 2.1 (b)], Theorem 3.9 yields that g satisfies GGD. By Proposition 3.8 (b), so does f. □

Corollary 3.12 isolates the most important instance of Corollaries 3.10 and 3.11. This result was actually established in the proof of Theorem 3.9.

COROLLARY 3.12. *If $A \subseteq B$ are integral domains and A is a going-down domain, then $A \hookrightarrow B$ satisfies GGD.*

Corollary 3.13 will present a more concrete application of Theorem 3.9 in the context of rings with nontrivial zero-divisors. Recall that a ring A is called a *weak Baer ring* if, for each $a \in A$, the annihilator of A is generated by an idempotent; that is, $\{b \in A : ba = 0\} = Ae$ for some $e = e^2 \in A$. Among many known characterizations is the following: A is a weak Baer ring if and only if A is a (necessarily reduced) locally irreducible ring such that $tq(A)$ is von Neumann regular. An example of a weak Baer ring that is a going-down ring but not an integral domain is provided by any finite product $A_1 \times \cdots \times A_n$ where each A_i is a weak Baer ring and a going-down ring (for instance, a going-down domain) and $n \geq 2$; to see this, recall that the class of weak Baer rings (resp., going-down rings) is stable under arbitrary (resp., finite) products [12, p. 28] (resp., [2, Proposition 2.1 (b)]).

COROLLARY 3.13. *Let $f: A \to B$ be a ring homomorphism such that A is a weak Baer ring and a going-down ring. Then f satisfies GGD if and only if f is a min morphism.*

Proof. Apply Theorem 3.9. □

COROLLARY 3.14. *(a) If A is a ring and B is an overring of A, then $A \hookrightarrow B$ is a min morphism.*
(b) Let A be a weak Baer ring. Then the following conditions are equivalent:
(1) $A \hookrightarrow B$ satisfies GD for each overring B of A;
(2) $A \hookrightarrow B$ satisfies GGD for each overring B of A;
(3) A is a going-down ring.

Proof. (a) Let B be an overring of A; that is, $A \subseteq B \subseteq T := tq(A)$. Let P be a minimal prime ideal of B. Then there exists a minimal prime ideal Q of T such that $Q \cap B = P$ (by [10, Exercise 1, p. 41 and Theorem 10]). As T is a ring of fractions of A, it follows that T is A-flat, so that $A \hookrightarrow T$ satisfies GD (cf. [10, Exercise 37, p. 44]), whence $P \cap A = Q \cap A$ is a minimal prime ideal of A, as desired.

(b) Since weak Baer rings are locally irreducible, (a) combines with Theorem 3.9 to yield that (3) $\Rightarrow$ (2). As (2) $\Rightarrow$ (1) trivially, it remains only to prove that (1) $\Rightarrow$ (3). Suppose (1). By [2, Proposition 2.1 (b)], it suffices to establish that if $P \in \text{Spec}(R)$, then A_P is a going-down ring. The hypothesis on A ensures that A_P is an integral domain, since A is reduced and locally irreducible. Therefore, by a characterization of going-down domains (cf. [1], [5]), it is enough to show that $A_P \hookrightarrow E$ satisfies GD for each overring E of A_P. Now, since T is von Neumann regular, [16, Proposition 1.4(2)] gives an identification $tq(A_P) \cong T_P$, whence $E = B_P$ for some suitable overring B of A. Then $A_P \hookrightarrow E$ inherits GD from $A \hookrightarrow B$, to complete the proof. □

Recall from [1, Proposition 3.2] and [5, Theorem 1] that in order to determine whether a given integral domain A is a going-down domain, it suffices to verify that GD is satisfied by all inclusions $A \hookrightarrow V$ for which V is a valuation domain. In this spirit, we next provide characterizations of the "universally chain morphism" and "universally GGD" properties. Theorems 3.26 and 3.16 establish that these properties are equivalent to "universally subtrusive" and "universally going-down," respectively.

PROPOSITION 3.15. *Let $f \colon A \to B$ be a ring homomorphism. Then the following conditions are equivalent:*

(1) f is universally GGD (resp., is a universally chain morphism), in the sense that the induced map $D \to D \otimes_A B$ satisfies GGD (resp., is a chain morphism) for all ring homomorphisms $A \to D$;

(2) The induced map $V \to V \otimes_A B$ satisfies GGD (resp., is a chain morphism) for all ring homomorphisms $A \to V$ for which V is a valuation domain.

Proof. We treat the assertion about "universally GGD" first. Of course, (1) $\Rightarrow$ (2) trivially. Assume (2). It suffices to show that f satisfies GGD. (Indeed, given ring homomorphisms $A \to D \to V$, observe the canonical isomorphism $V \otimes_D (D \otimes_A B) \cong V \otimes_A B$.) Consider a local chain X in $\text{Spec}(A)$ and $Q \in \text{Spec}(B)$ such that $f^{-1}(Q) = \mathcal{U}(X) =: P$. Our task is to produce a local chain Y in $\text{Spec}(B)$ such that $\mathcal{U}(Y) = Q$ and Y covers X. By Proposition 2.4 (c), there exists a ring homomorphism $g: A \to V$ and a local chain W in $\text{Spec}(V)$ such that V is a valuation domain, W covers X, and $g^{-1}(\mathcal{U}(W)) = P$. Put $E := V \otimes_A B$. In the category of affine schemes, we have $\text{Spec}(E) \cong \text{Spec}(V) \times_{\text{Spec}(A)} \text{Spec}(B)$. Therefore, by a property of pullbacks of schemes [7, Corollaire 3.2.7.1(i), p. 235], there exists $J \in \text{Spec}(E)$ such that J lies over $\mathcal{U}(W)$ (in $\text{Spec}(V)$) and J lies over Q (in $\text{Spec}(B)$). Moreover, by hypothesis, the induced map $V \to E$ satisfies GGD. Therefore, there exists a local chain Z in $\text{Spec}(E)$ such that $\mathcal{U}(Z) = J$ and Z covers W. By applying ${}^a(B \to V \otimes_A B)$ to the elements of Z, we obtain the elements of a chain Y with the desired properties. To prove the assertion about a "universally chain morphism," adapt the above proof, replacing the appeal to Proposition 2.4 (c) with a citation of Proposition 2.4 (b). □

THEOREM 3.16. *A ring homomorphism $f: A \to B$ is universally GGD if and only if f is universally going-down.*

Proof. As GGD ⇒ GD, the "only if" assertion is immediate. For the converse, suppose that f is universally going-down. Consider any ring homomorphism $A \to V$ for which V is a valuation domain. By the hypothesis on f, the induced map $h: V \to V \otimes_A B$ satisfies GD. Since V is a going-down ring, it follows from Corollary 3.11 (also from Theorem 3.9) that h satisfies GGD. An application of Proposition 3.14 yields that f is universally GGD, as desired. □

As noted prior to Proposition 3.2, the structure map of any ring of fractions $A \to A_S$ satisfies GGD. We next obtain a substantial generalization of this fact.

COROLLARY 3.17. *Each flat ring homomorphism satisfies (universally) GGD.*

Proof. Each flat ring homomorphism is universally going-down (cf. [10, Exercise 37, p. 44]). Apply Theorem 3.16. □

Proposition 3.18 will present another class of ring homomorphisms satisfying GGD. First, we adapt some terminology introduced in [11, p. 123]. A ring homomorphism $f: A \to B$ is called a *prime morphism* if the following condition is satisfied: if $f(a)b \in PB$ where $a \in A, b \in B$ and $P \in \text{Spec}(A)$, then either $a \in P$ or $b \in PB$; equivalently, if B/PB is a torsion-free A/P-module for each $P \in \text{Spec}(A)$. In general,

$$f \text{ is a flat ring-homomorphism} \Rightarrow f \text{ is a prime morphism} \Rightarrow f \text{ satisfies GD}$$

(cf. [11, Proposition 2], [10, Exercise 37, p. 44]). An interesting example is provided by $g: A \to A[T]/(pT) =: B$, where p is a prime integer, $A := \mathbb{Z}/p^2\mathbb{Z}$ and T is an indeterminate. Indeed, g is not flat (since $(p+p^2\mathbb{Z}) \otimes (T+pT)$ is a nonzero element of the kernel of $pA \otimes_A B \to B$), g is a prime morphism, and g satisfies GGD by Theorem 3.9. Finally, recall from [7, p. 145] that a *normal ring* is, by definition, a ring A such that A_P is an integrally closed integral domain for each $P \in \text{Spec}(A)$.

PROPOSITION 3.18. *If A is a normal ring and a prime morphism $f: A \to B$ is integral, then f satisfies GGD.*

Proof. By Proposition 3.8 (b), it suffices to show that if P is any minimal prime ideal of A, then the induced map $g: A/P \to B/PB$ satisfies GGD. Observe that A/P is an integrally closed integral domain since A is a normal ring [7, p. 145]; B/PB is a torsion-free A/P-module since f is a prime morphism; and g is integral. Accordingly, by Seydi's generalization of the classical Going-down Theorem [19], g is universally (Zariski-) open. It follows that g is universally going-down [7, Corollaire 3.9.4 (i), p. 254] and, hence, by Theorem 3.16, that g is (universally) GGD. □

We say that a ring homomorphism $f: A \to B$ is *prime-producing* if, for each $P \in \mathrm{Spec}(A)$, either $PB \in \mathrm{Spec}(B)$ or $PB = B$. Examples of prime-producing maps f include the structure maps of arbitrary rings of fractions $A \to A_S$ and the weak content maps of Rush [17]. It is evident that if a prime-producing map f satisfies LO, then f is a prime morphism and, hence, satisfies GD. A generalization of this fact will be given in Proposition 3.19. First, it is convenient to say that a ring homomorphism $f: A \to B$ satisfies the CNI property (so dubbed because it is a sort of "dual" of the INC property) if the following condition is satisfied: whenever $P \subseteq Q$ are prime ideals of A such that $PB = QB \neq B$, then $P = Q$. It is clear that if f satisfies LO, then f satisfies CNI (for then $f^{-1}(\mathfrak{p}B) = \mathfrak{p}$ for each $\mathfrak{p} \in \mathrm{Spec}(A)$).

PROPOSITION 3.19. *If a ring homomorphism $f: A \to B$ is prime-producing and satisfies CNI, then f satisfies GGD.*

Proof. Consider a local chain $X = \{P_i : i \in I\}$ in $\mathrm{Spec}(A)$ and $Q \in \mathrm{Spec}(B)$ such that $f^{-1}(Q) = \mathcal{U}(X) = P_j$. We seek a local chain Y in $\mathrm{Spec}(B)$ such that $\mathcal{U}(Y) = Q$ and Y covers X. Now, for each $i \in I$, we have $P_iB \subseteq P_jB \subseteq Q \subset B$. Hence, $P_iB \in \mathrm{Spec}(B)$, since f is prime-producing. Moreover, the CNI property ensures that P_i coincides with $Q_i := f^{-1}(P_iB)$, since $P_i \subseteq Q_i$ and $P_iB = Q_iB$. It therefore suffices to take $Y := \{P_iB : i \in I, i \neq j\} \cup \{Q\}$. □

Proposition 3.2 (b) illustrated that GGD-theoretic consequences can ensue in the presence of a ring homomorphism f for which ${}^af$ is injective. We next pursue this theme by enhancing the set-theoretic restriction with a topological one. Specifically, we say that a continuous function $f: X \to Y$ of topological spaces is a *topological immersion* if the induced map $X \to f(X)$ is a homeomorphism (that is, injective and either open or closed). It is straightforward to verify that a continuous map $f: X \to Y$ is a topological immersion if and only if f is injective and $f^{-1}(\overline{f(Z)}) = \overline{Z}$ for each subset Z of X. Our main interest here concerns ring homomorphisms $f: A \to B$ for which ${}^af: \mathrm{Spec}(B) \to \mathrm{Spec}(A)$ is a topological immersion (relative to the Zariski topology); in such a case, we also call f a *topological immersion*. There are many ring-theoretic characterizations of such f. A particularly useful characterization is given next.

PROPOSITION 3.20. *Let* $f\colon A \to B$ *be a ring homomorphism. Then:*
(a) The following two conditions are equivalent:

(1) If Q_1 *and* Q_2 *are prime ideals of* B *such that* $f^{-1}(Q_1) \subseteq f^{-1}(Q_2)$, *then* $Q_1 \subseteq Q_2$;

(2) f *is a topological immersion.*

(b) Suppose that the equivalent conditions in (a) hold and that a subset Y *of* $\mathrm{Spec}(B)$ *covers a subset* X *of* $\mathrm{Spec}(A)$. *Then* Y *is a chain (resp., local chain) if and only if* X *is a chain (resp., local chain).*

Proof. (a) (2) $\Rightarrow$ (1): Consider $Q_1, Q_2 \in \mathrm{Spec}(B)$, with $f^{-1}(Q_1) \subseteq f^{-1}(Q_2)$. Then, by the definition of the Zariski topology and the above characterization of topological immersions, we have

$$Q_2 \in {}^a f^{-1}({}^a f(Q_2)) \subseteq {}^a f^{-1}(\overline{{}^a f(Q_1)}) = \overline{Q_1};$$

that is, $Q_1 \subseteq Q_2$.

(1) $\Rightarrow$ (2): Assume (1). If ${}^a f(Q_1) = {}^a f(Q_2)$, then (1) yields that $Q_1 \subseteq Q_2$ and $Q_2 \subseteq Q_1$. Therefore, ${}^a f$ is injective. It remains to prove that if F is a (Zariski-) closed subset of $\mathrm{Spec}(B)$, then $G := {}^a f(F)$ is (Zariski-) closed in $\mathrm{Im}({}^a f)$. We shall show, in fact, that $G = \overline{G} \cap \mathrm{Im}({}^a f)$. One conclusion is obvious. For the reverse inclusion, consider $P \in \overline{G} \cap \mathrm{Im}({}^a f)$; pick $Q \in \mathrm{Spec}(B)$ such that $f^{-1}(Q) = P$. Now, observe that F is a patch (since $\mathrm{Im}(\mathrm{Spec}(B/J) \to \mathrm{Spec}(B)) = V(J)$ for each ideal J of B), and hence so is its spectral image, G. Thus, $\overline{G}$ is the union of the specializations of the points of G [7, Corollaire 7.3.2, p. 339]. In particular, $\mathfrak{p} \subseteq P$ for some $\mathfrak{p} \in G$. Pick $\mathfrak{q} \in F$ such that $f^{-1}(\mathfrak{q}) = \mathfrak{p}$. Using (1), we infer that $\mathfrak{q} \subseteq Q$, whence $Q \in F$, since Zariski-closed sets are stable under specialization. Therefore, $P = {}^a f(Q) \in {}^a f(F) = G$, as desired.

(b) In view of Proposition 2.1 (a),(b), it remains only to show that if $X =: \{P_i\}$ is a chain, then so is $Y =: \{Q_i\}$. As $f^{-1}(Q_i) = P_i$ for each i, the conclusion follows from condition (1) in (a). □

We next mention two families of examples of ring homomorphisms that induce/are topological immersions; the verifications follow most readily by checking condition (1) in Proposition 3.20. The first family consists of the flat epimorphisms (that is, the flat maps $A \to B$ such that the induced multiplication map $B \otimes_A B \to B$ is an isomorphism). In particular, the structure map of any ring of fractions $A \to A_S$ is a topological immersion. The second family consists of the ring homomorphisms $f\colon A \to B$ with the following property: for each $b \in B$, there exists $a \in A$ and $u \in \mathcal{U}(B)$ such that $b = f(a)u$. Besides rings of fractions, this second family includes all surjective ring homomorphisms and, for each field k and analytic indeterminate T, the inclusion map $k[T] \hookrightarrow k[[T]]$.

COROLLARY 3.21. *Let* $f\colon A \to B$ *be a ring homomorphism. Then the following conditions are equivalent:*

(1) ${}^a f$ *is injective and* f *satisfies GD;*

(2) f is a topological immersion and satisfies GGD.

Proof. (2) $\Rightarrow$ (1) trivially. Conversely, assume (1). One then readily verifies condition (1) in Proposition 3.20, and so f is a topological immersion. Next, to verify that f satisfies GGD, consider a local chain $X = \{P_i\}$ in Spec(A) and $Q \in$ Spec(B) such that $f^{-1}(Q) = \mathcal{U}(X)$. For each i, take Q_i to be the unique element of ${}^af^{-1}(P_i)$. It follows from (1) that $P_i \subseteq P_j$ entails $Q_i \subseteq Q_j$. Accordingly, $Y := \{Q_i\}$ is a local chain in Spec(B) such that $\mathcal{U}(Y) = Q$ and Y covers X, as desired. □

COROLLARY 3.22. *Let $f: A \to B$ be a ring homomorphism such that ${}^af$ is a topological immersion with closed image. Then the induced inclusion of rings $A/\ker(f) \hookrightarrow B$ satisfies GGD.*

Proof. Put $I := \ker(f)$. We begin with a fact that depends only on f being a ring homomorphism, namely, that $\overline{\mathrm{Im}({}^af)} = V(I)$. (To fashion a proof, recall that minimal prime ideals of a base ring are lain over from any ring extension [10, Exercise 1, p. 41] and Zariski-closed sets are stable under specialization.) Under the given assumptions, it follows that $\mathrm{Im}({}^af) = V(I)$.

Our task is to show that if X is a local chain in Spec(A/I) and $Q \in$ Spec(B) lies over $\mathcal{U}(X)$, then there exists a local chain Y in Spec(B) such that $\mathcal{U}(Y) = Q$ and Y covers X. Of course, X induces a local chain Z in Spec(A) such that $Z \subseteq V(I)$ and Q lies over $\mathcal{U}(Z)$. We shall show that $Y := {}^af^{-1}(Z)$ has the asserted properties. Indeed, since ${}^af$ is a topological immersion, it follows via condition (1) in Proposition 3.20 that Y is a chain. Moreover, Y is a local chain, with $\mathcal{U}(Y) = Q$. Now, ${}^af(Y) = Z \cap \mathrm{Im}({}^af) = Z \cap V(I) = Z$. Finally, Y covers X since Spec$(B) \to$ Spec(A/I) is an injection. □

COROLLARY 3.23. *Let f be a ring homomorphism. Then:*

(a) If f is an injection and ${}^af$ is a topological immersion with closed image, then f satisfies GGD.

(b) If f is an injection satisfying GU and ${}^af$ is an injection, then f satisfies GGD.

(c) Suppose that for all $Q \in$ Spec(B) and $P := f^{-1}(Q)$, the induced map $A_P \to B_Q$ is an injection whose corresponding map Spec$(B_Q) \to$ Spec(A_P) is a topological immersion with closed image. Then f satisfies GGD.

Proof. (a) is immediate from Corollary 3.22; (b) admits a simple direct proof but can also be obtained as a corollary of (a); to prove (c), combine (a) and Proposition 3.8 (b). □

For applications of the next result, it is useful to have examples of ring homomorphisms $g: A \to D$ that are universally topological immersions. Among these, we mention flat epimorphic g, surjective g, and g such that ${}^ag$ is a universal homeomorphism.

COROLLARY 3.24. *Let $f: A \to B$ be a ring homomorphism such that ${}^a f$ is injective and f satisfies GD. Let $g: A \to D$ be a ring homomorphism that is universally a topological immersion. Then the induced ring homomorphism $h: D \to D \otimes_A B$ satisfies GGD.*

Proof. Put $E := D \otimes_A B$. Our task is to show that if X is a local chain in $\mathrm{Spec}(D)$ and $Q \in \mathrm{Spec}(E)$ satisfies $h^{-1}(Q) = \mathcal{U}(X)$, then there exists a local chain Y in $\mathrm{Spec}(E)$ such that $\mathcal{U}(Y) = Q$ and Y covers X. As ${}^a g$ is injective, it follows from Proposition 2.1 (a),(b) that $W := {}^a g(X)$ is a local chain in $\mathrm{Spec}(A)$ such that $g^{-1}(\mathcal{U}(X)) = \mathcal{U}(W)$. Now, since Corollary 3.21 ensures that f satisfies GGD, there exists a local chain Z in $\mathrm{Spec}(B)$ such that $f^{-1}(\mathcal{U}(Z)) = \mathcal{U}(W)$ and Z covers W. Next, since X and Z have the same index set, we can use a result on pullbacks of schemes [7, Corollaire 3.2.7.1(i), p. 235] to produce the individual elements of a subset Y of $\mathrm{Spec}(E)$ such that Y covers X (relative to h) and Y covers Z (relative to the canonical ring homomorphism $j: B \to E$). As the hypothesis on g ensures that j is a topological immersion, Proposition 3.20 (b) yields that Y is a local chain. Finally, we shall show that $\mathcal{U}(Y) = Q$. By Proposition 2.1 (b), $j^{-1}(\mathcal{U}(Y)) = \mathcal{U}(Z)$. Therefore,

$$
\begin{aligned}
&{}^a(j \circ f)(\mathcal{U}(Y)) = f^{-1}(j^{-1}(\mathcal{U}(Y))) = f^{-1}(\mathcal{U}(Z)) = \\
&\mathcal{U}(W) = g^{-1}(\mathcal{U}(X)) = g^{-1}(h^{-1}(Q)) = {}^a(h \circ g)(Q) = {}^a(j \circ f)(Q).
\end{aligned}
$$

Since ${}^a(j \circ f) = {}^a f \circ {}^a j$ is a composite of injections, $\mathcal{U}(Y) = Q$. □

By analogy with the earlier definition of "chain morphism", we say that a ring homomorphism $f: A \to B$ is a *2-chain morphism* (or, as in [14, p. 528], *subtrusive*) if the following condition is satisfied: for all prime ideals $P_1 \subseteq P_2$ of A, there exist prime ideals $Q_1 \subseteq Q_2$ of B such that $f^{-1}(Q_i) = P_i$ for $i = 1, 2$. It is easy to see that any ring homomorphism f that satisfies LO and either GU or GD must be a 2-chain morphism. As noted in [14, p. 538], examples of universally 2-chain morphisms include the ring homomorphisms f that are pure; the f that satisfy LO and are universally going-down; and the f that satisfy LO and are integral. For us, the most important examples of universally 2-chain morphisms are special cases of the last two classes just mentioned, namely, the faithfully flat ring homomorphisms and (thanks to a result on pullbacks of schemes [7, Corollaire 3.2.7.1(i), p. 235] and the Lying-over Theorem [10, Theorem 44]) the injective integral ring homomorphisms.

Before stating a useful characterization of universally 2-chain morphisms, we recall the following definitions. If $f: A \to B$ is a ring homomorphism, the *torsion ideal of f* is $T(f) := \{b \in B :$ there exists a non-zero-divisor $c \in A$ such that $cb = 0\}$; and f is called *torsion-free* if $T(f) = 0$.

PROPOSITION 3.25. *(Picavet [14, Théorème 37(a), p. 556 and Proposition 16, p. 543]) Let $f: A \to B$ be a ring homomorphism. Then the following conditions are equivalent:*

(1) If $A \to V$ is a ring homomorphism for which V is a valuation domain and the induced map $V \to V \otimes_A B =: E$ has torsion ideal T, then the induced ring homomorphism $V \to E/T$ is faithfully flat;

(2) f is a universally 2-chain morphism.

Observe that LO is a universal property (as can be seen via [7, Corollaire 3.2.7.1(i), p. 235]); and, of course, so is "integral". Accordingly, the proof of our motivating result Proposition 2.2 (i) in [3, Remark (d)] actually establishes that any integral ring homomorphism that satisfies LO (for instance, any injective integral map) must be a universally chain morphism. We next present a substantial generalization of this fact.

THEOREM 3.26. *A ring homomorphism $f: A \to B$ is a universally chain morphism if and only if f is a universally 2-chain morphism.*

Proof. Any chain morphism is a 2-chain morphism, and so the "only if" assertion is trivial. For the converse, it suffices to show that if f is a universally 2-chain morphism, then f is a chain morphism. Our task is to show that if X is a chain in $\mathrm{Spec}(A)$, then there exists a chain Y in $\mathrm{Spec}(B)$ such that Y covers X. By Proposition 2.4 (b), we find a valuation domain V and a ring homomorphism $g: A \to V$ such that some chain W in $\mathrm{Spec}(V)$ covers X. Put $E := V \otimes_A B$. By Proposition 3.25, the induced ring homomorphism $h: V \to E/T$ is faithfully flat, where T denotes the torsion ideal of the canonical map $V \to E$. Accordingly, by Corollary 3.17, h satisfies GGD; and, being faithfully flat, h also satisfies LO. Therefore, by Proposition 3.2 (a), h is a chain morphism. In particular, some chain Z in $\mathrm{Spec}(E/T)$ covers W. If j denotes the composite $B \to E \to E/T$, it follows from the fact that $h \circ g = j \circ f$ and the functoriality of Spec that $Y := {}^a j(Z)$ covers X, as desired. □

COROLLARY 3.27. *Universally (2-) chain morphisms descend both GGD and GD. More precisely: if $f: A \to B$ is a ring homomorphism and $g: A \to D$ is a universally (2-) chain morphism such that the induced map $h: D \to D \otimes_A B =: E$ satisfies GGD (resp., GD), then f satisfies GGD (resp., GD).*

Proof. We give a proof for the "GGD" assertion, as it carries over for the "GD" assertion. Consider a local chain X in $\mathrm{Spec}(A)$ and $Q \in \mathrm{Spec}(B)$ such that $f^{-1}(Q) = \mathcal{U}(X)$. Since g is a chain morphism, there exists a chain Z in $\mathrm{Spec}(D)$ such that Z covers X. By Proposition 2.1 (b), Z is a local chain and $g^{-1}(\mathcal{U}(Z)) = \mathcal{U}(X)$. As $\mathcal{U}(Z)$ and Q each lie over $\mathcal{U}(X)$, the oft-used fact about pullbacks of schemes [7, Corollaire 3.2.7.1(i), p. 235] supplies $J \in \mathrm{Spec}(E)$ such that J lies over $\mathcal{U}(Z)$ in $\mathrm{Spec}(D)$ and J lies over Q in $\mathrm{Spec}(B)$. Since h satisfies GGD, there exists a local chain W in $\mathrm{Spec}(E)$ such that $\mathcal{U}(W) = J$ and W covers Z. If j denotes the canonical ring homomorphism $B \to E$, then the chain $Y := {}^a j(W)$ covers X. Moreover, by Proposition 2.1 (b), Y is a local chain satisfying $Q = j^{-1}(J) = j^{-1}(\mathcal{U}(W)) = \mathcal{U}(Y)$. Therefore, f satisfies GGD. □

COROLLARY 3.28. *Universally (2-) chain morphisms descend universally going-down (universally GGD).*

Proof. It follows from Corollary 3.27 via standard tensor product identities that any universally (2-) chain morphism descends universally GGD. An application of Theorem 3.16 permits the "universally going-down" formulation. □

COROLLARY 3.29. *Let* $f: A \to B$ *be a ring homomorphism, and let* $a_1, \ldots a_n$ *be finitely many elements of* A *such that* $(a_1, \ldots, a_n) = A$. *Then* f *satisfies GGD if and only if the induced ring homomorphism* $f_i : A_{a_i} \to B_{a_i}$ *satisfies GGD for all* $i = 1, \ldots, n$.

Proof. The "only if" assertion is immediate from Proposition 3.8 (b). For the converse, assume that each f_i satisfies GGD. By Proposition 3.8 (c), so does the induced map $\prod A_{a_i} \to \prod B_{a_i}$. Of course, $\prod B_{a_i} \cong (\prod A_{a_i}) \otimes_A B$; and $A \to \prod A_{a_i}$ is faithfully flat, hence a universally 2-chain morphism. Hence, by Corollary 3.27, f satisfies GGD. □

REMARK 3.30. In view of the diversity of contexts identified above which give sufficient conditions for GGD, one might well ask if any traditional construction can produce a ring A supporting a ring homomorphism $f: A \to B$ that satisfies GD but not GGD. In this regard, one could consider $A = C(X)$, the ring of continuous real-valued functions defined on a topological space X. However, such A cannot support f with the above properties. Indeed, any ring of the form $C(X)$ is a real closed ring, in the sense of Schwartz [18]. By [18, Propositions 1.4 and 1.5], it follows that any real closed ring is a locally irreducible ring and a going-down ring. Thus, if A is a real closed ring (for instance, a ring of the form $C(X)$) and a ring-homomorphism $f: A \to B$ satisfies GD, then each of Theorem 3.9, Corollary 3.10, Corollary 3.11 implies that f satisfies GGD.

References

[1] D. E. Dobbs, *On going-down for simple overrings, II*, Comm. Algebra **1** (1974), 439–458.

[2] D. E. Dobbs, *Going-down rings with zero-divisors*, Houston J.Math **23** (1997), 1–12.

[3] D. E. Dobbs, *A going-up theorem for arbitrary chains of primeideals*, Comm. Algebra **27** (1999), 3887–3894. Corrigendum **28** (2000),1653–1654.

[4] D. E. Dobbs, M. Fontana and I. J. Papick, *On the flat spectral topology*, Rend. Mat. **1**(4) (1981), 559–578.

[5] D. E. Dobbs and I. J. Papick, *On going-down for simple overrings, III*, Proc. Amer. Math. Soc. **54** (1976), 35–38.

[6] R. Gilmer, Mutiplicative Ideal Theory, Dekker, New York, 1972.

[7] A. Grothendieck and J. Dieudonné, Eléments de Géométrie Algébrique, Springer-Verlag, Berlin, 1971.

[8] M. Hochster, *Prime ideal structure in commutative rings*, Trans. Amer. Math. Soc. **142** (1969), 43–60.

[9] B. G. Kang and D. Y. Oh, *Lifting up an infinite chain of prime ideals to a valuation ring*, Proc. Amer. Math. Soc. **126** (1998), 645–646.

[10] I. Kaplansky, Commutative Rings, rev. ed., Univ. Chicago Press, Chicago, 1974.

[11] D. L. McQuillan, *On prime ideals in ring extensions*, Arch. Math. **33** (1979) 121–126.

[12] G. Picavet, *Ultrafiltres sur un espace spectral. Anneaux de Baer. Anneaux à spectre minimal compact*, Math. Scand **46** (1980), 23–53.

[13] G. Picavet, *Propriétés et applications de la notion de contenu*, Comm. Algebra **13** (1985), 2231–2265.

[14] G. Picavet, *Submersion et descente*, J. Algebra **103** (1986), 527–591.

[15] G. Picavet, *Parties sondables d'un spectre et profondeur*, Boll. U.M.I. (7) **8-B** (1994), 677–730.

[16] G. Picavet, *Seminormal or t-closed schemes and Rees rings*, Algebras and representation theory **1** (1998), 255–309.

[17] D. E. Rush, *Content algebras*, Canad. Math. Bull. **21** (1978), 329–334.

[18] N. Schwartz, *Gabriel filters in real closed rings*, Comment. Math. Helv. **72** (1997), 434–465.

[19] H. Seydi, *Sur deux théorèmes d'algèbre commutative*, C. R. Acad. Paris, Sér. A **271** (1970), 1105–1108.

The Class Group of the Composite Ring of a Pair of Krull Domains and Applications

SAID EL BAGHDADI Department of Mathematics, Faculté des Sciences et Techniques, P. O. Box 523, Beni Mellal, Morocco

1. INTRODUCTION

Let $A \subset B$ be an extension of integral domains. Then $R = A + XB[X]$ is called the composite ring of the pair (A, B). This construction has been extensively studied by several authors for its pliability in providing examples and counterexamples. The class group of the composite ring $A + XB[X]$ has been investigated in [4] and [5].

Let R be an integral domain. We recall from [7] that the class group of R, $Cl(R)$, is the group of t-invertible fractional t-ideals of R, under the t-product $I * J = (IJ)_t$, modulo its subgroup of principal fractional ideals. For a Krull domain R, $Cl(R)$ is the usual divisor class group of R (cf. [9]); while for R a Prüfer domain, $Cl(R) = Pic(R)$, the ideal class group of R. The interest of this concept resides in the fact that divisibility properties of a domain R are often reflected in group-theoretic properties of $Cl(R)$. For instance, if R is a Krull domain, then $Cl(R) = 0$ if and only if R is factorial. If R is a Prüfer domain, then $Cl(R) = 0$ if and only if R is a Bézout domain. For more details on class groups, see [3].

This paper aims to give a full description of the class group of the composite ring of a pair of Krull domains. A first key result is Proposition 2.3, where we show that there is an exact sequence of canonical homomorphisms

$$0 \to Cl(A) \to Cl(R) \to \prod_{m \in \Lambda} Cl(R_m),$$

where $\Lambda = \{m \in t - Max(A) \mid (mB)_t \neq B\}$. As an immediate corollary of this result, we extend [4, Theorem 4.4] to the case where B is a t-flat overring of A. The main result of this paper, Theorem 2.8, shows that if (A, B) is a pair of Krull domains satisfying (PDE), then there is an exact sequence of canonical homomorphisms

$$0 \to Cl(A) \to Cl(R) \to \oplus_{m\in\Lambda} Z^n/e_m Z \to 0.$$

As an application, we compute the class group of some well-known constructions such as $Cl(Z + XO_K[X])$, where O_K is the ring of algebraic integers of a number field K, and $Cl(Z+XZ[\sqrt{d}][X])$, where d is an integer with $d \equiv 2,3 \pmod 4$ square free.

2. MAIN RESULTS

Let D be an integral domain with quotient field k. By an ideal of D we mean an integral ideal of D. Let I be a nonzero fractional ideal of D; we denote by I^{-1} the inverse of I and by $I_v = (I^{-1})^{-1}$ the v-closure of I. We say that I is a divisorial or v-ideal if $I_v = I$ and v-finite if $I = J_v$ for some finitely generated fractional ideal J of D. We define the t-closure of I by $I_t = \cup\{J_v \mid J \subset I \text{ finitely generated}\}$. We say that I is a t-ideal if $I_t = I$ and that I is t-invertible if $(II^{-1})_t = D$. For more details about these notions, see [12, Sections 32 and 34]. The set $\mathcal{T}(D)$ of t-invertible fractional t-ideals of D is a group under the t-product $I * J = (IJ)_t$, and the set $\mathcal{P}(D)$ of nonzero principal fractional ideals of D is a subgroup of $\mathcal{T}(D)$. The class group of D is defined as the group $Cl(D) = \mathcal{T}(D)/\mathcal{P}(D)$. As usual, $t - Max(D)$ denotes the set of all t-maximal ideals of D.

To avoid unnecessary repetition, let us fix notation for the rest of the paper. Data will consist of an extension of integral domains $A \subset B$, K the quotient field of B, and $R = A + XB[X]$.

We say that $A \subset B$ is a t-linked extension [2] if for any finitely generated ideal F of A such that $F^{-1} = A$, $(B : FB) = B$. Any flat extension is t-linked. If B is a generalized ring of fractions of A, then $A \subset B$ is t-linked (cf. [14, Lemma 2.26]).

LEMMA 2.1 $A \subset R$ is t-linked if and only if $A \subset B$ is t-linked.

Proof. Let F be a finitely generated ideal of A such that $F^{-1} = A$. By [4, Lemma 2.1], $(FR)^{-1} = F^{-1} + X(FB)^{-1}[X]$. Hence $(FR)^{-1} = R$ if and only if $(FB)^{-1} = B$. Therefore, $A \subset R$ is t-linked if and only if $A \subset B$ is t-linked. ◊

Assume that $A \subset B$ is a t-linked extension. By Lemma 2.1 and [2, Theorem 2.2], the canonical homomorphism $\varphi : Cl(A) \to Cl(R)$, $[H] \mapsto [(HR)_t]$ is well-defined. Now let $m \in t - Max(A)$; we denote by B_m the quotient ring $B_{A\setminus m}$. Then $R_m = A_m + XB_m[X]$. For $m \in t - Max(A)$, there is a canonical group homomorphism $\psi_m : Cl(R) \to Cl(R_m)$, $[I] \mapsto [IR_m]$ (cf. [6, Proposition 2.2]).

Let $\Lambda = \{m \in t - Max(A) \mid (mB)_t \neq B\}$ and $\Lambda' = \{m \in t - Max(A) \mid (mB)_t = B\}$. Then $t - Max(A) = \Lambda \cup \Lambda'$.

LEMMA 2.2 Assume B is integrally closed and let $m \in \Lambda'$. Then the homomorphism ψ_m is the zero map.

Proof. Let $m \in \Lambda'$ and set $M = (mR)_t$. By [5, Lemma 2.8], $M = m + XB[X]$ and M is a t-maximal ideal of R. Now let $I = H + XJ[X]$ be a t-invertible t-ideal of R, where J is a t-invertible t-ideal of B and $H \subset J$ is a nonzero ideal of A (cf. [4, Corollary 2.4]). Then $II^{-1} \not\subseteq M$, and hence $(II^{-1}) \cap A \not\subseteq m$. So $(II^{-1})R_m = R_m$. Thus IR_m is an invertible ideal of R_m. On the other hand, since B is integrally closed, then $Pic(R_m) = Pic(A_m) = 0$ (cf. [1, page 113]). Hence IR_m is a principal ideal of R_m. Therefore, $\psi_m = 0$. ◇

The set of homomorphisms $(\psi_m)_{m\in\Lambda}$ induces a homomorphism $\psi : Cl(R) \to \prod_{m\in\Lambda} Cl(R_m)$. Inspired by work on integer-valued polynomials (cf. [8, Proposition 3.4]), we state the following key result.

PROPOSITION 2.3 Assume B is integrally closed and t-linked over A. Then the following sequence is exact

$$0 \to Cl(A) \overset{\varphi}{\to} Cl(R) \overset{\psi}{\to} \prod_{m\in\Lambda} Cl(R_m).$$

Proof. The injectivity of φ follows from [4, Lemma 3.8(2)]. Now let H be a t-invertible t-ideal of A. Then HA_m is principal for each $m \in \Lambda$, and hence $(HR_m)_t$ is principal for each $m \in \Lambda$. Thus $Im\varphi \subset Ker\psi$. For the reverse inclusion, we adapt the proof of [8, Proposition 3.4]. Let $[I] \in Ker\psi$. By [4, Corollary 2.4], we may assume that $I = H + XJ[X]$, where J is a t-invertible t-ideal of B and $H \subset J$ is a nonzero ideal of A. Next we show that $I = (HR)_t$. For this it suffices to show that $IR_M = (HR)_tR_M$ for each t-maximal ideal M of R. If $M \cap A = 0$, then $IR_M = (HR)_tR_M = R_M$. Suppose that $M \cap A \neq 0$. Since $A \subset R$ is t-linked, then $(M \cap A)_t \neq A$ (cf. [2, Proposition 2.1]). Let m be a maximal t-ideal of A such that $M \cap A \subset m$. By using Lemma 2.2 and the fact that $[I] \in Ker\psi$, we deduce that $IR_m = \alpha_m R_m$ for some $\alpha_m \in H$. Hence $IR_m = (HR)_tR_m$. Thus $IR_M = (HR)_tR_M$, and hence $I = (HR)_t$. Similarly, $I^{-1} = (H^{-1}R)_t$. Now, we have $(HH^{-1}R)_t = (II^{-1})_t = R$. Hence by [4, Lemma 3.8(2)], $(HH^{-1})_t = A$. Therefore, H is a t-invertible ideal of A. Thus $[I] = \varphi([H_t])$ (cf. [2, Theorem 2.1(6)]). Hence $Ker\psi \subset Im\varphi$. ◇

Assume B is an overring of A. Recall that the extension $A \subset B$ is t-flat [15] if and only if for each t-maximal ideal M of B, $B_M = A_{M\cap A}$. Every t-flat overring B of A is t-linked. Flatness implies t-flatness, but the converse does not hold in general (cf. [15, Remark 2.12]). The class of t-flat overrings of a domain A constitutes a large class of generalized rings of fractions of A. In particular, if A is Noetherian or a Mori domain, then any generalized ring of fractions of A is a t-flat overring of A. This follows from [11, Propositions 1.5 and 1.8] and [15, Proposition 2.5(v)].

We have the following corollary which recovers [4, Theorem 4.4].

COROLLARY 2.4 Assume that B is integrally closed and a t-flat overring of A. Then $Cl(R) \cong Cl(A)$.

Proof. Let $m \in \Lambda$. Since $A \subset B$ is t-flat, then $A_m = B_m = B_M$ for some t-maximal ideal M of B such that $M \cap A = m$. Hence $R_m = B_M[X]$. Now let I be a t-invertible t-ideal of R. By [4, Corollary 2.4], we may assume that $I = H + XJ[X]$, where J is a t-invertible t-ideal of B and $H \subset J$ is a nonzero ideal of A. Since B_M is integrally closed and $IR_m = IB_M[X]$ is a t-invertible t-ideal of $B_M[X]$, then, by [16, Lemme 2, section 3], $IR_m = JB_M[X]$. Hence IR_m is a principal ideal (since J is a t-invertible t-ideal of B). Thus $\psi_m([I]) = 0$, and hence $\psi = 0$. Therefore, φ is an isomorphism. $\diamondsuit$

Let $A \subset B$ be an extension of integral domains. Assume that A is a Mori domain. Since in a Mori domain every nonzero ideal is contained in only finitely many maximal t-ideals, then the image of the homomorphism ψ is contained in the direct sum of the $Cl(R_m)$'s, where m ranges over Λ. In this case, we can obtain interesting results on the surjectivity of ψ. Before this we need the following lemma:

LEMMA 2.5 Assume that A and B are Mori domains. If I is a t-ideal of R of the form $H + XJ[X]$, where J is an ideal of B and $H \subset J$ is a nonzero ideal of A, then I is a v-finite v-ideal.

Proof. Since A and B are Mori domains, then $H_t = F_v$ for some finitely generated ideal $F \subset H$ of A, and $J_t = G_v$ for some finitely generated ideal $G \subset J$ of B. We may assume that $F \subset G$. Then $I_v = (F + XG[X])_v$ by [5, Lemma 2.2(ii)]. On the other hand, as in the proof of [5, Proposition 2.6], there exists $I' \subset F + XG[X]$, a finitely generated ideal of R, such that $(F + XG[X])_v = I'_v$. Hence $I = I_v = I'_v$ is a v-finite v-ideal. $\diamondsuit$

A domain A is said to have t-dimension 1 (written t-dim$A = 1$) if each prime t-ideal of A has height one. Krull domains have t-dimension 1.

THEOREM 2.6 Let $A \subset B$ be a t-linked extension of Mori domains with t-dim$A = 1$ and B integrally closed. Then the following sequence is exact

$$0 \to Cl(A) \xrightarrow{\varphi} Cl(R) \xrightarrow{\psi} \oplus_{m \in \Lambda} Cl(R_m) \to 0.$$

Proof. By Proposition 2.3, it suffices to show that ψ is surjective. Let $m \in \Lambda$ and let I_m be a proper t-invertible t-ideal of R_m. By [4, Corollary 2.4], we may assume that $I_m = H_m + XJ_m[X]$, where J_m is a t-invertible t-ideal of B_m and $H_m \subset J_m$ is a nonzero ideal of A_m. Set $I = I_m \cap R$. Then I is a t-ideal of R and $I = H + XJ[X]$, where $H = H_m \cap A$ and $J = J_m \cap B$. By [5, Lemma 2.1] and Lemma 2.5, I, $(II^{-1})_t$ and $(II^{-1})^{-1}$ are v-finite v-ideals of R. Hence $(II^{-1})_t R_m = (II^{-1}R_m)_t = (I_m I_m^{-1})_t = R_m$. We claim that $(II^{-1})_t = R$. Suppose, on the contrary, that $(II^{-1})_t \subset M$ for some t-maximal ideal M of R. Then $H \subset M \cap A$. Since $\sqrt{H_m} = mA_m$ (dim$A_m = 1$), one can easily show that $\sqrt{H} = m$. Hence $m \subset M \cap A$. On the other hand, since $A \subset R$ is t-linked, then $(M \cap A)_t \neq A$;

so $m = M \cap A$ (since t-dim$A = 1$). Hence $(II^{-1})_t \cap A \subset m$, a contradiction since $(II^{-1})_t R_m = R_m$. Hence I is a t-invertible t-ideal of R. Now let $([I_m])_m$ in the direct sum of $Cl(R_m)$'s, $m \in \Lambda$. We may assume that each I_m has the form $H_m + XJ_m[X]$, where J_m is a t-invertible t-ideal of B_m and $H_m \subset J_m$ is a nonzero ideal of A_m. Set $F = (\prod_m (I_m \cap R))_t$, where m runs over the t-maximal ideals in Λ such that I_m is not principal. Then, it is easy to see that F is a t-invertible t-ideal of R and $\psi([F]) = ([I_m])_m$. Hence ψ is surjective. $\diamondsuit$

For an integral domain A, $X^{(1)}(A)$ will denote the set of height-one prime ideals of A. Recall that an extension $A \subset B$ of Krull domains is said to satisfy (PDE) if for each $Q \in X^{(1)}(B)$, we have $ht(Q \cap A) \leq 1$ (cf. [9, Chapter 2]).

We have the following corollary of Theorem 2.6:

COROLLARY 2.7 Let $A \subset B$ be an extension of Krull domains satisfying (PDE). Then the sequence $0 \to Cl(A) \xrightarrow{\varphi} Cl(R) \xrightarrow{\psi} \oplus_{m \in \Lambda} Cl(R_m) \to 0$ is exact.

Proof. By [2, Proposition 2.1(3)], the extension of Krull domains $A \subset B$ is t-linked if and only if it satisfies (PDE). The corollary now follows from Theorem 2.6. $\diamondsuit$

We next give our main result of this paper.

THEOREM 2.8 Let $A \subset B$ be an extension of Krull domains satisfying (PDE). For each $m \in \Lambda$, let $(mB)_t = (m_1^{\alpha_1} m_2^{\alpha_2} \dots m_n^{\alpha_n})_t$, where $n \geq 1$ (which depends on m), each $\alpha_i \geq 1$, and $m_1, m_2, \dots, m_n$ are distinct elements of $X^{(1)}(B)$. Set $e_m = (\alpha_1, \alpha_2, \dots, \alpha_n)$. Then we have an exact sequence:

$$0 \to Cl(A) \to Cl(R) \to \oplus_{m \in \Lambda} Z^n / e_m Z \to 0.$$

Proof. By Corollary 2.7, it suffices to prove the theorem for A a rank-one discrete valuation ring (DVR). Thus let $m = pA$ (p prime element) be the maximal ideal of A with $(mB)_t = (m_1^{\alpha_1} m_2^{\alpha_2} \dots m_n^{\alpha_n})_t$, and set $e_m = (\alpha_1, \alpha_2, \dots, \alpha_n)$. We must show that $Cl(R) \cong Z^n / e_m Z$. Since A is a DVR, then $qf(A) \subset B$ or $qf(A) \cap B = A$. If $qf(A) \subset B$, then $Cl(R) = Cl(A) = 0$ (cf. [4, Theorem 4.7]) and we are done. Thus we may assume $qf(A) \cap B = A$.

For $i = 1, \dots, n$, set $I_i = pA + Xm_i[X]$. By [5, Theorem 4.8], the I_i's are t-invertible t-ideals of R. Note that the I_i's are t-primes since $I_i = m_i[X] \cap R$. We next show that $Cl(R)$ is generated by the classes $[I_i]$ for $i = 1, \dots, n$. Let I be a t-invertible t-ideal of R. Then, by [5, Theorem 1.1 and Theorem 4.8], $I = u(J \cap A + XJ[X])$ for some $u \in qf(R)$ and some t-invertible t-ideal J of B such that $J \cap A \neq 0$. Since B is a Krull domain and $J \cap A \neq 0$, then $J = (m_1^{\gamma_1} m_2^{\gamma_2} \dots m_n^{\gamma_n})_t$ for some integers $\gamma_1, \gamma_2, \dots, \gamma_n$. Thus $I = u(I_1^{\gamma_1} I_2^{\gamma_2} \dots I_n^{\gamma_n})_t$ (cf. [5, Lemmas 2.1 and 4.6]). Hence $[I] = \sum \gamma_i [I_i]$.

Now let $\{e_1, \ldots, e_n\}$ be the canonical free basis of Z^n and consider the homomorphism $\mu : Z^n \to Cl(R)$ determined by $\mu(e_i) = [I_i]$. By the previous paragraph, μ is surjective. Next we show that $Ker(\mu) = e_m Z$. Clearly, $e_m Z \subset Ker(\mu)$. For the reverse inclusion, assume that $\sum \gamma_i[I_i] = 0$ in $Cl(R)$ for some integers $\gamma_1, \gamma_2, \ldots, \gamma_n$. Then $(m_1^{\gamma_1} m_2^{\gamma_2} \ldots m_n^{\gamma_n})_t = cB$ for some $c \in qf(A)$. Hence there exists an integer r such that $\gamma_i = r\alpha_i$ for each $i = 1, \ldots, n$. Thus $Ker(\mu) \subset e_m Z$. Hence $Cl(R) \cong Z^n/e_m Z$. ◇

We have the following immediate corollary of Theorem 2.8:

COROLLARY 2.9 If A is factorial, then $Cl(R) \cong \oplus_{m \in \Lambda} Z^n/e_m Z$.

Let A be a Dedekind domain with quotient field k and let K be a finite-dimensional extension of k. Then the integral closure B of A in K is a Dedekind domain. One of the most important examples is when $A = Z$ is the ring of integers and K is a number field; then $B = O_K$ is the ring of algebraic integers of K. We denote by Δ the set of positive primes of Z which ramify in B. Then Δ is finite (cf. [13, Theorem 7.3, chap.I]).

COROLLARY 2.10 Let K be a number field and O_K the ring of algebraic integers of K. Then $Cl(Z + XO_K[X]) \cong (\oplus_{p\in\Delta} Z^n/e_p Z) \oplus (\oplus Z)$.

Proof. Let p be a positive prime element of Z such that $p \notin \Delta$. Then $pO_K = m_1 m_2 \ldots m_n$ for some pairwise distinct prime ideals $m_1, m_2, \ldots, m_n \in X^{(1)}(O_K)$. Hence $e_p = (1, 1, \ldots, 1)$ and $Z^n/e_p Z = Z^{n-1}$. The conclusion now follows from Corollary 2.9. ◇

EXAMPLE 2.11 Let d be an integer with $d \equiv 2, 3 \ (mod\, 4)$ square free, and let Δ be the set of positive prime divisors of $4d$. Then $Cl(Z + XZ[\sqrt{d}][X]) \cong (\oplus_\Delta Z/2Z) \oplus (\oplus Z)$. Indeed, for the quadratic field $K = Q(\sqrt{d})$, the discriminant $D = 4d$. Then, by [13, Theorem 7.3, chap.I], a prime number $p \in Z$ is ramified in $O_K = Z[\sqrt{d}]$ if and only if p divides $4d$. Now let $p \in \Delta$; then $pZ[\sqrt{d}] = P^2$ for some $P \in X^{(1)}(Z[\sqrt{d}])$ (cf. [13, Exercise 3, p. 28]). Hence $n = 1$ and $e_p = 2$.

The case $d = -1$ of Example 2.11 was obtained in [4].

EXAMPLE 2.12 Let k be a perfect field and K a field containing k. Let Y be an indeterminate over K. Then the extension $k[Y] \subset K[Y]$ satisfies the hypotheses of Corollary 2.9, and we have $Cl(k[Y] + XK[Y][X]) \cong \oplus Z$. Indeed, let f be an irreducible polynomial of $k[Y]$. Then $f = uf_1^{\alpha_1} \ldots f_n^{\alpha_n}$ for some $u \in K$ and some pairwise distinct monic irreducible polynomials $f_1, \ldots, f_n \in K[Y]$. Since k is perfect, then $\alpha_1 = \cdots = \alpha_n = 1$. Hence $e_f = (1, 1, \ldots, 1)$ and $Z^n/e_f Z = Z^{n-1}$. Thus, by Corollary 2.9, $Cl(k[Y] + XK[Y][X]) \cong \oplus Z$.

REMARK 2.13 1) By [5, Corollary 4.15], we have the following exact sequence:

$$0 \to K^*/Q^*U(O_K) \to Cl(Z + XO_K[X]) \to Cl(O_K) \to 0.$$

As an application of the above results, we can use this exact sequence to deduce information on the size of the class number of the number field K. In particular, let $d > 0$, $d \equiv 1, 2 \ (mod\, 4)$ square free, i.e., $Z[\sqrt{-d}]$ is Dedekind. Then $Z[\sqrt{-d}]$ is factorial if and only if $d = 1, 2$. Indeed, if $d > 2$, one can easily show that the subgroup of elements of order 2 in $Q(\sqrt{-d})^*/Q^*$ (note that in this case $U(O_K) = \{\pm 1\}$) is cyclic of order 2, while it is of order ≥ 4 in $Cl(Z + XZ[\sqrt{-d}][X]) = (\oplus_\triangle Z/2Z) \oplus (\oplus Z)$ (since $\#\triangle \geq 2$). Hence $Cl(O_K) \neq 0$. The factoriality in the two cases $d = 1, 2$ is well known.

2) Recall that if B is an integrally closed domain, then $Cl(B[X]) = Cl(B)$ (cf. [10, Theorem 3.6], see also Corollary 2.4). The computation of $Cl(B[X])$ has not been investigated for B not integrally closed and we don't know any example. From Example 2.12 we get the following example: Let $B = k + XK[X]$ (notation of Example 2.12) with k not algebraically closed in K (and hence B is not integrally closed). In this case, $Cl(B[Y]) = \oplus Z$ (a free abelian group).

REFERENCES

[1] D.D. Anderson, D.F. Anderson, and M. Zafrullah, Rings between D[X] and K[X], Houston J. Math. 17 (1991) 109-129.

[2] D.D. Anderson, E.G. Houston, and M. Zafrullah, t-linked extensions, the t-class group, and Nagata's Theorem, J. Pure Appl. Algebra 86 (1993) 109-124.

[3] D.F. Anderson, The class group and local class group of an integral domain, Mathematics and its Applications, Kluwer Academic Publishers, 520 (2000) 33-55.

[4] D.F. Anderson, S. EL Baghdadi, and S. Kabbaj, On the class group of A+XB[X] domains, Lecture notes in Pure Appl. Math., Marcel Dekker, 205 (1999) 73-85.

[5] D.F. Anderson, S. EL Baghdadi, and S. Kabbaj, The homogeneous class group of A+XB[X] domains, International Journal of Commutative Rings, to appear.

[6] D.F. Anderson and A. Ryckaert, The class group of $D + M$, J. Pure Appl. Algebra 52 (1988) 199-212.

[7] A. Bouvier, Le groupe des classe d'un anneau intègre, 107ème congrès des sociétés savantes, Brest, vol. 4 (1982) 85-92.

[8] P.-J. Cahen, S. Gabelli, and E. G. Houston, Mori domains of integer-valued polynomials, J. Pure Appl. Algebra 153 (2000) 1-15.

[9] R.M. Fossum, The divisor class group of a Krull domain, Springer, New York, 1973.

[10] S. Gabelli, On divisorial ideals in polynomial rings over Mori domains, Comm. Algebra 15 (1987) 2349-2370.

[11] S. Gabelli, On Nagata's theorem for the class group, II, Lecture notes in Pure Appl. Math., Marcel Dekker, 206 (1999) 117-142.

[12] R. Gilmer, Multiplicative ideal theory, Marcel Dekker, New York, 1972.

[13] G. J. Janusz, Algebraic number fields, Academic Press, New York, 1973.

[14] B. G. Kang, $*$-operations on integral domains, Ph.D. Thesis, The University of Iowa, 1987.

[15] D. J. Kwak and Y. S. Park, On t-flat overrings, Ch. J. Math. 23(1) (1995) 17-24.

[16] J. Querré, Idéaux divisoriels d'un anneau de polynômes, J. Algebra 64 (1980) 270-284.

Complete Integral Closure and Noetherian Property for Integer-Valued Polynomial Rings

S. Gabelli Dipartimento di Matematica, Università degli Studi Roma Tre, Roma, ITALY

F. Tartarone Dipartimento di Matematica, Università degli Studi Roma Tre, Roma, ITALY

Abstract

Denote by A' the integral closure of the ring A and by A^* its complete integral closure. Let D be a domain with quotient field K and $E \subseteq K$ be a subset. We study the integer-valued polynomial rings $\mathrm{Int}(E, D)'$, $\mathrm{Int}(E, D)^*$, $\mathrm{Int}(E, D')$, $\mathrm{Int}(E, D^*)$ and the relations of their properties with the noetherianity of $\mathrm{Int}(E, D)$.

INTRODUCTION

Throughout, we denote by A' the integral closure of the ring A and by A^* its complete integral closure. If D is a domain with quotient field K, the *integer-valued polynomial ring* over D is the domain $\mathrm{Int}(D) := \{f \in K[X]; f(D) \subseteq D\}$. Clearly $D[X] \subseteq \mathrm{Int}(D) \subseteq K[X]$.

Since D is a homomorphic image of $\mathrm{Int}(D)$, if $\mathrm{Int}(D)$ is Noetherian, then also D is Noetherian. On the contrary, when $\mathrm{Int}(D) \neq D[X]$, the noetherianity of D is scarcely a sufficient condition for $\mathrm{Int}(D)$ to be Noetherian. For example, it is well known that $\mathrm{Int}(\mathbb{Z})$ is not Noetherian, being a two-dimensional Prüfer domain [5, ch. VI]. As a matter of fact, if D is Noetherian and it is also one-dimensional or integrally closed, then $\mathrm{Int}(D)$ is Noetherian if and only if it is trivial, that is it coincides with $D[X]$ [5, Corollary VI.2.6]. More generally, if $\mathrm{Int}(D)$ is Noetherian, then $\mathrm{Int}(D') = D'[X]$ [5, Proposition VI.2.4] and this means that each height-one prime of D' has infinite residue field [5, Corollary IV.4.10]. The converse holds if the conductor of D' into D is not zero.

The aim of this paper is to study how the noetherianity of the ring $\mathrm{Int}(E, D) := \{f \in K[X]; f(E) \subseteq D\}$ of the *integer-valued polynomials over a subset* E of K is related to the properties of $\mathrm{Int}(E, D')$, in particular in the case when E is a nonzero fractional ideal of D.

To this extent, it is useful to consider the ring $D[X/E] := \bigcap_{a \in E \setminus (0)} D[a^{-1}X]$. This graded ring plays, with respect to $\mathrm{Int}(E, D)$, the role of $D[X]$ with respect to $\mathrm{Int}(D)$. In fact, for example, when $\mathrm{Int}(D) = D[X]$, then $\mathrm{Int}(E, D) = D[X/E]$

[8, Lemma 4.5]. Another similarity of behaviour happens when the conductor of D' into D is not zero. In this case, if $\text{Int}(D)$ is Noetherian, then $\text{Int}(D)$ is a pullback of $D'[X]$; in the same way, if $\text{Int}(E, D)$ is Noetherian, then $\text{Int}(E, D)$ is a pullback of $D'[X/E]$.

In Section 3, we use these facts to prove that if E is a fractional ideal of D and either $\text{Int}(D)$ or $\text{Int}(E, D)$ is Noetherian, then $\text{Int}(E, D') = D'[X/E]$ (Proposition 3.9); thus extending the result for $\text{Int}(D)$ in [5, Proposition VI.2.4] mentioned above. We also prove that if $\text{Int}(D)$ is Noetherian and the conductor of D' into D is not zero, then $\text{Int}(E, D)$ is Noetherian if and only if $D[X/E]$ is Noetherian (Proposition 3.11). The hypothesis that $\text{Int}(D)$ is Noetherian is not too much restrictive. In fact, if D' has nonzero conductor into D and $\text{Int}(E, D)$ is Noetherian, then $\text{Int}(D)$ is necessarily Noetherian (Proposition 3.8).

Establishing whether in general the graded ring $D[X/E]$ is Noetherian is an open problem. We give an example, which can be easily generalized, of a domain D such that $D[X/E]$ is Noetherian while $\text{Int}(E, D)$ need not be Noetherian. However we show that if D is a Krull domain, then $D[X/E]$ is a Krull domain (Proposition 3.12). It follows that if either $\text{Int}(D)$ or $\text{Int}(E, D)$ is Noetherian, then $D'[X/E] = \text{Int}(E, D')$ is always a Krull domain (Corollary 3.13).

In the first two sections of the paper, we study the complete integral closure of $\text{Int}(E, D)$. We recall that integer-valued polynomial rings have been used to put in evidence the anomalous behavior of the complete integral closure. For instance, if V is a complete rank one discrete valuation domain with finite residue field, then $\text{Int}(V)$ is a completely integrally closed domain which is not the intersection of rank one valuation domains and which has localizations which are not completely integrally closed (see [5] pp. 130-131).

We show that the integral closure of $\text{Int}(E, D)$ and its complete integral closure do not have similar properties. Among other results, we prove for example that when D is Noetherian or D^* has nonzero conductor, then $\text{Int}(E, D^*) = \text{Int}(E, D)^*$ (Corollary 1.5 and Proposition 2.1), while it is known that an analogous result is not true for the integral closure, even when both the hypotheses are satisfied. A nice consequence of this equality is that if D is Noetherian, then $\text{Int}(D)' = \text{Int}(D)^*$ (Corollary 1.6). But we give an example showing that, even when D is Noetherian and has nonzero conductor, it can happen that $\text{Int}(E, D)' \subsetneq \text{Int}(E, D)^*$.

We thank the referee for his very careful reading of the manuscript and his comments.

1 GENERAL FACTS, REMARKS AND EXAMPLES

A nonzero element $x \in K$ is said to be *almost integral* over D if there is a nonzero element $d \in D$ such that $dx^i \in D$, for $i \geq 0$. This is equivalent to saying that all the powers of x belong to a finite D-module or that they generate a fractional ideal of D. The *complete integral closure* of D in K, which we denote by D^*, is the set of the elements of K that are almost integral over D. It is not difficult to verify that $D^* = \bigcup(I : I)$, where I ranges among the fractional ideals of D. D is *completely integrally closed* if $D = D^*$.

If an element is integral over D, then it is almost integral, and if D is Noetherian the converse holds. Thus, if D' denotes the integral closure of D in K, one has $D \subseteq D' \subseteq D^*$, and if D is Noetherian then $D' = D^*$. It is well known that D^* is always integrally closed, but many examples have been given to show that it might not be completely integrally closed. However, D^* is completely integrally closed if the conductor $(D : D^*)$ is a nonzero ideal, that is if D^* is contained in a finite D-submodule of K (see for example [1]).

We recall that a nonempty subset E of K is a *D-fractional subset* if there exists a nonzero element $d \in D$ such that $dE \subseteq D$. The *fractional ideals* of D are exactly the D-fractional D-submodules of K.

If E is D-fractional, then $\mathrm{Int}(E, D)$ contains the nonconstant polynomial dX and so it is not trivial. Conversely, if $\mathrm{Int}(E, D) \neq D$, then it is known that E must be a D-fractional subset of the integral closure D' of D [5, Proposition I.1.9]. It follows that $\mathrm{Int}(E, D') \neq D'$ if and only if E is D'-fractional.

We are mainly interested in the study of the complete integral closure of $\mathrm{Int}(E, D)$ when E is a D-fractional subset of K, namely a fractional ideal of D. In this case we might as well assume that E is contained in D. In fact, for any $a \in K \setminus (0)$, $f(X) \in \mathrm{Int}(aE, D)$ if and only if $f(aX) \in \mathrm{Int}(E, D)$. Hence the isomorphism of $K[X]$ onto itself which maps $f(X)$ to $f(aX)$ induces an isomorphism between $\mathrm{Int}(aE, D)$ and $\mathrm{Int}(E, D)$.

LEMMA 1.1. *Let E be an infinite additive subgroup of K. If $f(X) \in K[X] \setminus (0)$ and $f(a) \in D$ for almost all $a \in E$ (that is $f(a) \notin D$ for at most finitely many $a \in E$) then $f \in \mathrm{Int}(E, D)$.*

Proof. Apply the same argument given in [5, Proposition I.1.5] for $E = D$. □

PROPOSITION 1.2. *Let E be an infinite additive subgroup of K, then*

$$\mathrm{Int}(E, D)^* \subseteq \mathrm{Int}(E, D^*).$$

In particular

$$\mathrm{Int}(D)^* \subseteq \mathrm{Int}(D, D^*).$$

Proof. Since $K[X]$ is completely integrally closed, then $\mathrm{Int}(E, D)^* \subseteq K[X]$. Let $f(X) \in K[X]$ be almost integral over $\mathrm{Int}(E, D)$. Then, for some $g(X) \in \mathrm{Int}(E, D) \setminus \{0\}$ and all $n \geq 0$, $g(X)f(X)^n \in \mathrm{Int}(E, D)$. It follows that, for all $a \in E$ such that $g(a) \neq 0$, $g(a)f(a)^n \in D$. Hence $f(a) \in D^*$ for almost all $a \in D$ and, by applying Lemma 1.1, $f(X) \in \mathrm{Int}(E, D^*)$. □

COROLLARY 1.3. *Let E be an infinite additive subgroup of K. Then $\mathrm{Int}(E, D)$ is completely integrally closed if and only if D is completely integrally closed.*

In particular, $\mathrm{Int}(D)$ is completely integrally closed if and only if D is completely integrally closed.

Proof. By Proposition 1.2, if D is completely integrally closed, then $\mathrm{Int}(E, D)$ is completely integrally closed. Conversely, if $\mathrm{Int}(E, D)$ is completely integrally closed, then $D = \mathrm{Int}(E, D) \cap K$ is completely integrally closed. □

We recall that, for any D-fractional subset E of K, a similar result holds for the integral closure; namely, we have that

$$\text{Int}(E,D)' \subseteq \text{Int}(E,D')$$

and in particular $\text{Int}(E,D)$ is integrally closed if and only if D is integrally closed [5, Proposition IV.4.1].

Since for Noetherian domains the integral closure and the complete integral closure coincide, then, if D is Noetherian, we have that $D' = D^*$ and $D[X]' = D'[X] = D^*[X] = D[X]^*$. We now show that, in this case, we also have that $\text{Int}(D)' = \text{Int}(D)^*$, even though $\text{Int}(D)$ need not be Noetherian.

Recall that, if D is Noetherian, then, by [5, Theorem IV.4.7],

$$\text{Int}(D)' = \text{Int}(D,D').$$

PROPOSITION 1.4. [5, Exercise IV.24] *Let D be a Noetherian domain and E a D-fractional subset of K. Then*

$$\text{Int}(E,D') \subseteq \text{Int}(E,D)^*.$$

Proof. By [5, Lemma IV.4.2], for each $f \in \text{Int}(E,D')$, there exists a D-algebra R, finitely generated as a D-module, such that $D \subseteq R \subseteq D'$ and $f(E) \subseteq R$.

Since R is a finitely generated D-module, the conductor $I := (D\colon R)$ is a nonzero common ideal of D and R. Then $\text{Int}(E,D)$ and $\text{Int}(E,R)$ share the ideal $J := \text{Int}(E,I) := \{g(X) \in K[X]; g(E) \subseteq I\}$. It follows that

$$f \in \text{Int}(E,R) \subseteq (J\colon J) \subseteq \text{Int}(E,D)^*$$

and so $\text{Int}(E,D') \subseteq \text{Int}(E,D)^*$. □

COROLLARY 1.5. *Let D be a Noetherian domain and E a D-fractional infinite additive subgroup of K. Then*

$$\text{Int}(E,D') = \text{Int}(E,D)^*.$$

Proof. By Proposition 1.4, we have $\text{Int}(E,D') \subseteq \text{Int}(E,D)^*$. The opposite inclusion holds by Proposition 1.2, because $D' = D^*$. □

COROLLARY 1.6. *If D is a Noetherian domain, then*

$$\text{Int}(D)' = \text{Int}(D,D') = \text{Int}(D)^*.$$

Proof. By [5, Theorem IV.4.7] we have $\text{Int}(D)' = \text{Int}(D,D')$ and, by Corollary 1.5, $\text{Int}(D,D') = \text{Int}(D)^*$. □

If D is Noetherian and E is a proper infinite additive subgroup of K, the containement $\text{Int}(E,D)' \subseteq \text{Int}(E,D)^*$ may be proper (see Remark 1.8 (c) below). However we now show that if $E = aD$ is a nonzero principal fractional ideal of D, then the equality holds.

PROPOSITION 1.7. *Let D be a Noetherian domain and let $a \in K\setminus(0)$. Then:*

$$\operatorname{Int}(aD, D') = \operatorname{Int}(aD, D)^* = \operatorname{Int}(aD, D)'.$$

Proof. The automorphism of $K[X]$ which maps $f(X)$ to $f(aX)$ induces an isomorphism between $\operatorname{Int}(aD, D')$ and $\operatorname{Int}(D, D')$ and between $\operatorname{Int}(aD, D)$ and $\operatorname{Int}(D)$.

We conclude because, by Corollary 1.6, $\operatorname{Int}(D, D') = \operatorname{Int}(D)^* = \operatorname{Int}(D)'$. □

Remark 1.8.

(a) If D is not Noetherian, it may happen that $\operatorname{Int}(D)' \subsetneq \operatorname{Int}(D, D')$. This is shown by the following example, given in [5, Exercise IV.27].

Let $\mathbb{F}_q$ be the finite field with q elements and $\overline{\mathbb{F}}_q$ its algebraic closure. Consider the rank-one valuation domain $V := \overline{\mathbb{F}}_q[[t]]$ and the domain $D := \mathbb{F}_q + tV$; by [9] D is local, one-dimensional and non-Noetherian. Moreover, the integral closure of D is V. If $f := \frac{X^q - X}{t}$, then $f \in \operatorname{Int}(D, V)$ but $f \notin \operatorname{Int}(D)'$.

(b) Since $D \subseteq D^*$, it is obvious that $\operatorname{Int}(D^*) \subseteq \operatorname{Int}(D, D^*)$, but in general this inclusion is strict. In fact, [5, Exercise IV.29] gives the following example of a one-dimensional Noetherian domain such that $\operatorname{Int}(D) \not\subseteq \operatorname{Int}(D') = \operatorname{Int}(D^*)$. Since $\operatorname{Int}(D) \subseteq \operatorname{Int}(D, D^*)$, it follows that $\operatorname{Int}(D^*) \neq \operatorname{Int}(D, D^*)$.

Let k be a finite field, t an indeterminate over k and $D := k[t^2, t^3]$. It is easy to see that D is a one-dimensional Noetherian domain with integral closure $k[t]$. In addition, if $f = \prod_{a \in k}(X - a)$ and $g = f^3/t^4$, than $g \in \operatorname{Int}(D)$ but $g \notin \operatorname{Int}(D')$.

(c) We recall that, if D is a one-dimensional Noetherian domain with finite residue field, then $\operatorname{Int}(E, D)' = \operatorname{Int}(E, D')$ [5, Proposition IV.4.3]. But in general, if the residue field of D is not finite, it may happen that $\operatorname{Int}(E, D)' \subsetneq \operatorname{Int}(E, D')$. For instance, if $D := \mathbb{R} + t\mathbb{C}[[t]]$, then $D' = \mathbb{C}[[t]]$ and, setting $\mathfrak{m} := tD'$, we have that $\operatorname{Int}(\mathfrak{m}, D)' \subsetneq \operatorname{Int}(\mathfrak{m}, D')$ [5, Exercise IV.26]. We will illustrate in detail this fact in the next section, using technics different from the ones used in [5].

The same example also shows that Corollary 1.6 is no longer true if $\operatorname{Int}(D)$ is replaced by $\operatorname{Int}(E, D)$. In fact $\operatorname{Int}(\mathfrak{m}, D') = \operatorname{Int}(\mathfrak{m}, D)^*$ by Proposition 1.5. Thus

$$\operatorname{Int}(\mathfrak{m}, D)' \subsetneq \operatorname{Int}(\mathfrak{m}, D') = \operatorname{Int}(\mathfrak{m}, D)^*.$$

2 THE NONZERO CONDUCTOR CASE

We now turn to the case when D^* has nonzero conductor $I := (D\colon D^*) := \{x \in K; xD^* \subseteq D\}$, that is when D^* is a fractional ideal of D. In this hypothesis, we prove results similar to those obtained in Section 1 when D is Noetherian.

PROPOSITION 2.1. *If $I := (D\colon D^*) \neq (0)$ and E is an infinite additive subgroup of K, then*

$$\operatorname{Int}(E, D)^* = \operatorname{Int}(E, D^*).$$

In particular

$$\operatorname{Int}(D)^* = \operatorname{Int}(D, D^*).$$

Proof. Since $\mathrm{Int}(E, D)$ and $\mathrm{Int}(E, D^*)$ share the ideal $J := \mathrm{Int}(E, I)$, we have that

$$\mathrm{Int}(E, D^*) \subseteq (J : J) \subseteq \mathrm{Int}(E, D)^*.$$

The opposite inclusion follows from Proposition 1.2. □

We recall that, if $I := (D\colon D^*) \neq (0)$, then D^* is completely integrally closed [1] and so $\mathrm{Int}(E, D)^* = \mathrm{Int}(E, D^*)$ is completely integrally closed by Corollary 1.3.

The inclusion $\mathrm{Int}(D) \subseteq \mathrm{Int}(D, D^*)$ is always verified, since $D \subseteq D^*$. But, in general, it is not true that $\mathrm{Int}(D) \subseteq \mathrm{Int}(D^*)$, even when D is Noetherian and $I \neq (0)$, as we have seen in Remark 1.8(b).

COROLLARY 2.2. *If* $I := (D : D^*) \neq (0)$, *then*

$$\mathrm{Int}(D) \subseteq \mathrm{Int}(D^*) \iff \mathrm{Int}(D^*) = \mathrm{Int}(D)^*.$$

Proof. Assume that $\mathrm{Int}(D) \subseteq \mathrm{Int}(D^*)$. By Proposition 2.1, we have

$$\mathrm{Int}(D) \subseteq \mathrm{Int}(D^*) \subseteq \mathrm{Int}(D, D^*) = \mathrm{Int}(D)^*.$$

Since $\mathrm{Int}(D^*)$ is completely integrally closed, we conclude that $\mathrm{Int}(D^*) = \mathrm{Int}(D)^*$. The converse is clear. □

COROLLARY 2.3. *Assume that* $I := (D\colon D^*) \neq (0)$ *and consider the following conditions:*

(i) $\mathrm{Int}(D) \subseteq D^*[X]$;

(ii) $\mathrm{Int}(D, D^*) = D^*[X]$;

(iii) $\mathrm{Int}(D^*) = D^*[X]$.

Then (i) $\iff$ (ii) $\Longrightarrow$(iii).

Proof. We have $D^*[X] \subseteq \mathrm{Int}(D^*) \subseteq \mathrm{Int}(D, D^*)$ and $\mathrm{Int}(D) \subseteq \mathrm{Int}(D, D^*)$. Since D^* is completely integrally closed, $D^*[X]$ is completely integrally closed. Besides, $\mathrm{Int}(D)^* = \mathrm{Int}(D, D^*)$ by Proposition 2.1. Hence (i) $\Longrightarrow$(ii) $\Longrightarrow$(iii). That (ii) $\Longrightarrow$(i) is clear. □

Remark 2.4.

(a) Substituting the complete integral closure with the integral closure, Proposition 2.1 is no longer true, even when D is Noetherian (see Remarks 1.8(a) and (c)).

(b) If D is Noetherian, Proposition 2.1 and Corollary 2.2 remain true relaxing the hypothesis that $I \neq (0)$.

In fact, in this case, we have $\mathrm{Int}(E, D)^* = \mathrm{Int}(E, D') = \mathrm{Int}(E, D^*)$ by Proposition 1.5. In addition, $\mathrm{Int}(D)^* = \mathrm{Int}(D)' = \mathrm{Int}(D, D')$ by Corollary 1.6 and so we have

$$\mathrm{Int}(D) \subseteq \mathrm{Int}(D') \iff \mathrm{Int}(D') = \mathrm{Int}(D, D'),$$

which is (i) $\iff$ (ii) of [5, Theorem IV.4.9].

(c) The three conditions of Corollary 2.3 are all equivalent in the Noetherian case, with no assumption on the conductor.

In fact, if D is Noetherian, (i) and (iii) are equivalent by [5, Corollary IV.4.10]. Hence, if (i) holds, $\mathrm{Int}(D) \subseteq \mathrm{Int}(D') \subseteq D'[X]$ and so, by [5, Corollary IV.4.9], $\mathrm{Int}(D, D') = \mathrm{Int}(D') = D'[X]$, which is condition (ii). We also note that the implication (ii) $\Longrightarrow$(iii) is always true.

However, in general it is possible to have $I := (D : D^*) \neq (0)$ and:

$$\mathrm{Int}(D)^* = \mathrm{Int}(D, D^*) \supsetneq \mathrm{Int}(D^*) = D^*[X].$$

Hence, in Corollary 2.3, condition (iii) does not imply conditions (i) and (ii); in this case, by Corollary 2.2, also $\mathrm{Int}(D) \not\subseteq \mathrm{Int}(D^*)$.

For example, take $D := \mathbb{F}_q + t\overline{\mathbb{F}}_q[[t]]$, where $\overline{\mathbb{F}}_q$ is the algebraic closure of $\mathbb{F}_q$ (see Remark 1.8(a)). In this case, $D^* = D' = \overline{\mathbb{F}}_q[[t]] =: V$ and $I = tV \neq (0)$. Hence $\mathrm{Int}(D)^* = \mathrm{Int}(D, V)$ by Proposition 2.1. Moreover, denoting by $d_0 = 0, d_1, \ldots, d_{q-1}$ the elements of $\mathbb{F}_q$ and setting $\varphi = \frac{\prod_{i=0}^{q-1}(X-d_i)}{t}$, by [7, Lemma 2.2], we have $\mathrm{Int}(D, V) = V[X][\varphi]$. But, since V is a one-dimensional, local, Noetherian domain with infinite residue field, by [5, Corollary I.3.15] we also have $\mathrm{Int}(V) = V[X]$. Hence

$$\mathrm{Int}(D^*) = \mathrm{Int}(V) = V[X] \subsetneq V[X][\varphi] = \mathrm{Int}(D, V) = \mathrm{Int}(D, D^*).$$

If E is a subset of K, we set

$$D[X/E] := \bigcap_{a \in E\setminus(0)} D[a^{-1}X].$$

Since $D[X] \subseteq \mathrm{Int}(D)$, applying the automorphism of $K[X]$ defined by $f(X) \mapsto f(aX)$, we have that $D[a^{-1}X] \subseteq \mathrm{Int}(aD, D)$, for each nonzero element $a \in K$. Thus, if E is a D-module,

$$\mathrm{Int}(E, D) = \bigcap_{a \in E\setminus(0)} \mathrm{Int}(aD, D) \supseteq \bigcap_{a \in E\setminus(0)} D[a^{-1}X] = D[X/E]$$

and similarly, denoting by ED^* (resp. ED') the D^*-module (resp. D'-module) which is the extension of E to D^* (resp. D'),

$$\mathrm{Int}(E, D') \supseteq \mathrm{Int}(ED', D') \supseteq D'[X/E]$$

and

$$\mathrm{Int}(E, D^*) \supseteq \mathrm{Int}(ED^*, D^*) \supseteq D^*[X/E].$$

If, in addition, $\mathrm{Int}(D) = D[X]$, then $\mathrm{Int}(E, D) = D[X/E]$ [8, Lemma 4.5].

The ring $D[X/E]$ behaves with respect to $\mathrm{Int}(E, D)$ mostly like $D[X]$ with respect to $\mathrm{Int}(D)$. An example of this fact is given by the following result, which is the analogue of Corollary 2.3.

PROPOSITION 2.5. *Assume that either D is Noetherian or $I := (D\colon D^*) \neq (0)$ and let $E \subseteq K$ be a nonzero D-module. Consider the following conditions:*

(i) $\mathrm{Int}(E, D) \subseteq D^*[X/E]$;

(ii) $\mathrm{Int}(E, D^\star) = D^\star[X/E]$;

(iii) $\mathrm{Int}(ED^\star, D^\star) = D^\star[X/E]$.

Then (i) $\iff$ (ii) $\Longrightarrow$(iii).

Proof. Since $\mathrm{Int}(E, D) \subseteq \mathrm{Int}(E, D^\star)$, obviously (ii) $\Longrightarrow$(i).

(i) $\Longrightarrow$(ii). If D is Noetherian or $I := (D\colon D^\star) \neq (0)$, then $\mathrm{Int}(E, D^\star) = \mathrm{Int}(E, D)^\star$ by Corollary 1.5 or Proposition 2.1 respectively. From (i) it follows that $\mathrm{Int}(E, D) \subseteq D^\star[a^{-1}X]$, for each nonzero element $a \in E$. Since $D^\star[a^{-1}X]$ is completely integrally closed, then $\mathrm{Int}(E, D)^\star \subseteq D^\star[a^{-1}X]$ and $\mathrm{Int}(E, D^\star) = \mathrm{Int}(E, D)^\star \subseteq D^\star[X/E]$. The opposite containment $\mathrm{Int}(E, D^\star) \supseteq D^\star[X/E]$ always holds.

(ii) $\Longrightarrow$(iii). We have the inclusions: $D^\star[X/E] \subseteq \mathrm{Int}(ED^\star, D^\star) \subseteq \mathrm{Int}(E, D^\star)$. If $\mathrm{Int}(E, D^\star) \subseteq D^\star[X/E]$, then $\mathrm{Int}(ED^\star, D^\star) = D^\star[X/E]$. □

Remark 2.6. In [14] it is shown that, if D is a domain and E is a nonzero D-module, then $D[X/E]$ is a graded ring of the form $\bigoplus_{n\geq 0} E_nX^n$, where $E_0 = D$ and $E_n = \bigcap_{u\in E\setminus(0)} u^{-n}D$ for $n \geq 1$. In particular, if $E = aD$, for some nonzero $a \in K$, then $E_n = a^{-n}D$ for all $n \geq 0$. We observe that $\{E_n\}_{n\geq 0}$ is a sequence of D-modules such that $E_iE_j \subseteq E_{i+j}$.

In P.L. Kiihne's Ph.D Thesis [12], there are considered graded rings of the type:

$$\mathcal{M}[X] := \bigoplus_{n\geq 0} M_nX^n,$$

where M_0 is a domain and $\mathcal{M} := \{M_n\}_{n\geq 0}$ is a sequence of M_0-modules such that $M_iM_j \subseteq M_{i+j}$. The author studies conditions on the sequence $\mathcal{M}$ for which the polynomial ring $\mathcal{M}[X]$ is integrally closed, completely integrally closed, etc...

In particular she shows that if there exists an integer n such that $M_k = M_n$ for each $k \geq n$ (i. e. the sequence is stationary), then:

- *$\mathcal{M}[X]$ is integrally closed if and only if $M_n = M_1$, for each $n \geq 1$, M_0 is integrally closed in M_1 and M_1 is integrally closed.*

- *$\mathcal{M}[X]$ is completely integrally closed if and only if $M_n = M_0$, for each $n \geq 0$ and M_0 is completely integrally closed.*

In our case, if we set $\mathcal{M} := \{E_n\}_{n\geq 0}$, we get that $D[X/E]$ is exactly the graded ring $\mathcal{M}[X]$. If $\mathrm{Int}(D) = D[X]$, we also have that $\mathrm{Int}(E, D) = D[X/E]$ and we know that $\mathrm{Int}(E, D)$ is integrally closed (respectively completely integrally closed) if and only if D is integrally closed (respectively completely integrally closed) by [5, Proposition IV.4.1] (respectively Corollary 1.3). Thus, $D[X/E]$ gives a natural example of the fact that the two results above no longer hold if the sequence $\mathcal{M}$ is not assumed to be stationary.

We now describe in detail the example mentioned in Remark 1.8(c).

Let $D := \mathbb{R}+t\mathbb{C}[[t]]$. Then $D' = \mathbb{C}[[t]]$ and the conductor of D' into D is $\mathfrak{m} := tD'$. Since D and D' are one-dimensional local Noetherian domains with infinite residue field, then $\mathrm{Int}(D) = D[X]$ and similarly $\mathrm{Int}(D') = D'[X]$ [5, Corollary I.3.15]. Hence, via the isomorphism defined by $f(X) \mapsto f(tX)$, we have that $\mathrm{Int}(tD, D) =$

$D[X/t]$ and $\mathrm{Int}(\mathfrak{m}, D') = \mathrm{Int}(tD', D') = D'[X/t]$. We also have that $\mathrm{Int}(\mathfrak{m}, D) = D[X/\mathfrak{m}]$ [8, Lemma 4.5].

On the other hand we now show that $\mathrm{Int}(\mathfrak{m}, D) = D + \mathfrak{m}D[X/t]$. It is easy to check that $D+\mathfrak{m}D[X/t] \subseteq \mathrm{Int}(\mathfrak{m}, D)$. For the opposite inclusion, since $tD \subseteq \mathfrak{m}$, then $\mathrm{Int}(\mathfrak{m}, D) \subseteq \mathrm{Int}(tD, D) = D[X/t]$. Thus, $\mathrm{Int}(\mathfrak{m}, D) = D[X/\mathfrak{m}]$ is a graded ring such that $D + \mathfrak{m}D[X/t] \subseteq \mathrm{Int}(\mathfrak{m}, D) \subseteq D[X/t]$. Hence $\mathrm{Int}(\mathfrak{m}, D) = \bigoplus_{n\geq 0} M_n X^n$, where $t^{-n}\mathfrak{m} \subseteq M_n \subseteq t^{-n}D$. This means that $M_n = t^{-n}\mathfrak{m}$ or $M_n = t^{-n}D$. Suppose that $M_n = t^{-n}D$, for a certain $n > 0$ and choose $\alpha \in D\backslash\mathfrak{m}$. Then $t^{-n}\alpha X^n \in \mathrm{Int}(\mathfrak{m}, D)$. Now, there exists $z \in D'$ such that $\alpha z^n \notin D$, because $\alpha \notin \mathfrak{m} = (D\colon D')$. But, since $tz \in tD' = \mathfrak{m}$, then $t^{-n}\alpha(tz)^n = \alpha z^n \in D$, which is a contradiction. Thus, for each $n \geq 1$, $M_n = t^{-n}\mathfrak{m}$, $\mathrm{Int}(\mathfrak{m}, D) = D + \mathfrak{m}D[X/t]$ and $\mathrm{Int}(\mathfrak{m}, D)' = D' + \mathfrak{m}D'[X/t]$.

It follows that

$$\mathrm{Int}(\mathfrak{m}, D)' = D' + \mathfrak{m}D'[X/t] \subsetneq D'[X/t] = \mathrm{Int}(\mathfrak{m}, D').$$

This example also shows that in general $D[X/E]$ does not behave exactly like the polynomial ring $D[X]$. For instance, while $D[X]' = D'[X]$, the containment $D[X/E]' \subseteq D'[X/E]$ may be proper. In fact we have seen that:

$$D[X/\mathfrak{m}]' = \mathrm{Int}(\mathfrak{m}, D)' \subsetneq \mathrm{Int}(\mathfrak{m}, D') = D'[X/t] = D'[X/\mathfrak{m}].$$

3 STUDYING NOETHERIANITY THROUGH PULLBACKS

The relationship between the noetherianity of $\mathrm{Int}(D)$ and the properties of the integral closure of D is given by the following proposition, which is obtained by putting together [5, Corollary IV.4.10] and [5, Proposition VI.2.4].

PROPOSITION 3.1. *Let D be a Noetherian domain and consider the following conditions:*

(i) $\mathrm{Int}(D)$ *is Noetherian;*

(ii) $\mathrm{Int}(D) \subseteq D'[X]$;

(iii) $\mathrm{Int}(D') = D'[X]$;

(iv) *each height-one prime ideal of D' has infinite residue field.*

Then (i) $\Longrightarrow$(ii) $\Longleftrightarrow$ (iii) $\Longleftrightarrow$ (iv). □

In general, conditions (i) and (ii) are not equivalent, as it is shown in [10].

PROPOSITION 3.2. [5, Remarks VI.2.5] *Assume that D is Noetherian and $I := (D\colon D') \neq (0)$. Then* $\mathrm{Int}(D)$ *is Noetherian if and only if* $\mathrm{Int}(D) \subseteq D'[X]$.

Proof. Assume that, $\mathrm{Int}(D) \subseteq D'[X]$. Since D' is a finite-type D-module, then $D'[X]$ is Noetherian and finite over $D[X]$. Hence $D'[X]$ is finite over $\mathrm{Int}(D)$ and, by Eakin-Nagata's Theorem, $\mathrm{Int}(D)$ is Noetherian. The converse follows from Proposition 3.1. □

Now we want to show that, in case $I := (D\colon D') \neq (0)$, Propositions 3.1 and 3.2 give a way to construct a Noetherian domain D with the property that $\mathrm{Int}(D)$ is either Noetherian and not trivial or not Noetherian.

We start by using pullback diagrams to prove that when D is Noetherian and $I := (D\colon D') \neq (0)$ then $\mathrm{Int}(D)$ is Noetherian if and only if $\mathrm{Int}(D') = D'[X]$.

We recall that if A is a domain sharing an ideal I with its overring B, then A is a *pullback* of B in the sense that the following diagram is commutative:

$$\begin{array}{ccc} A & \longrightarrow & A/I \\ \downarrow & & \downarrow \\ B & \longrightarrow & B/I \end{array}$$

where the vertical arrows are the natural inclusions and the horizontal arrows are the natural projections.

The following well-known result establishes necessary and sufficient conditions in order to have that A is a Noetherian domain:

PROPOSITION 3.3. [9, Proposition 1.8] *With the notation above, A is Noetherian if and only if B and A/I are Noetherian and B is a finite type A-algebra.* □

In our case, if $I := (D\colon D') \neq (0)$, then I is a nonzero common ideal of D and D' and we have a pullback diagram:

$$\begin{array}{ccc} D & \longrightarrow & D/I \\ \downarrow & & \downarrow \\ D' & \longrightarrow & D'/I \end{array}$$

Assume that $\mathrm{Int}(D)$ is Noetherian, then $\mathrm{Int}(D) \subseteq D'[X] \subseteq \mathrm{Int}(D')$ [5, Proposition VI.2.4]. Since $J := \mathrm{Int}(D, I) = \{f \in K[X]; f(D) \subseteq I\}$ is a nonzero common ideal of $\mathrm{Int}(D)$, $D'[X]$ and $\mathrm{Int}(D')$, we also have pullback diagrams as follows:

$$\begin{array}{ccc} \mathrm{Int}(D) & \longrightarrow & \mathrm{Int}(D)/J \\ \downarrow & & \downarrow \\ D'[X] & \longrightarrow & D'[X]/J \\ \downarrow & & \downarrow \\ \mathrm{Int}(D') & \longrightarrow & \mathrm{Int}(D')/J \end{array}$$

Thus, $\mathrm{Int}(D')$ is Noetherian and, since D' is noetherian and integrally closed, then $\mathrm{Int}(D') = D'[X]$ [5, Proposition VI.2.6].

On the contrary, if $\mathrm{Int}(D') = D'[X]$, then $\mathrm{Int}(D, D') = D'[X]$ by Corollary 2.3. Since $I[X]$ is a common ideal of $D[X]$ and $D'[X]$, we have the pullback diagrams:

$$\begin{array}{ccc} D[X] & \longrightarrow & D[X]/I[X] \\ \downarrow & & \downarrow \\ \mathrm{Int}(D) & \longrightarrow & \mathrm{Int}(D)/I[X] \\ \downarrow & & \downarrow \\ \mathrm{Int}(D,D') = D'[X] & \longrightarrow & D'[X]/I[X]. \end{array}$$

where $D[X]$ is Noetherian. Whence $\mathrm{Int}(D)$ is Noetherian.

We also observe that, in the situation above, if D/I is finite, then $\mathrm{Int}(D) \neq D[X]$ [5, Remark I.3.13 (iii)]. In fact, let $\{d_0, \dots, d_{q-1}\}$ be a set of representatives of D modulo I and let $x \in D'$. Since I is contained in the ideal $(D :_D x)$, then the polynomial $f(X) = x\prod_{0\leq k<q-1}(X - d_k)$ is in $\mathrm{Int}(D)$. But $f(X) \notin D[X]$ if $x \in D' \setminus D$.

The following result is a direct consequence of Proposition 3.1, Proposition 3.2 and the last observation.

COROLLARY 3.4. *Let D be a Noetherian domain such that $I := (D\colon D') \neq (0)$ and D/I is finite. If each height-one prime ideal of D' has infinite residue field, then* $\mathrm{Int}(D)$ *is Noetherian and* $\mathrm{Int}(D) \neq D[X]$. *Otherwise* $\mathrm{Int}(D)$ *is not Noetherian.*

COROLLARY 3.5. *Let D be a Noetherian domain such that $I := (D\colon D') \neq (0)$, D/I is finite and $(D'\colon I) \neq D'$. Then* $\mathrm{Int}(D)$ *is not Noetherian.*

Proof. Since D' is a finitely generated D-module, then D'/I is also finite. Since D' is a Krull domain and $(D'\colon I) \neq D'$, then the divisorial closure of I is different from D' and so I is contained in a height-one prime of D'. For such a prime P, D'/P also is finite. Hence we conclude by applying Corollary 3.4. □

Remark 3.6. In the previous corollary, the hypothesis that $I := (D'\colon I) \neq D'$ is necessary. In fact the following example of a Noetherian domain D such that $I := (D\colon D') \neq (0)$, D/I is finite, $(D'\colon I) = D'$, and $\mathrm{Int}(D)$ is Noetherian can be found on page 8 of [10] (see also [5, Exercise VI.15]).

Let k be a finite field, u, v two indeterminates over k and $B = k[[u,v]]$. B is a two-dimensional, Noetherian, integrally closed, local domain with maximal ideal $M = (u,v)$. Moreover, each height-one prime ideal of B has infinite residue field. Consider the following pullback diagram:

$$\begin{array}{ccc} D := k + M^2 & \longrightarrow & k \\ \downarrow & & \downarrow \\ B & \longrightarrow & B/M^2 \end{array}$$

We have that $D' = B$, $I := (D\colon D') = M^2$ and $D/M^2 = k$ is finite. Hence $\mathrm{Int}(D)$ is Noetherian and not trivial by Corollary 3.4. However, since M has height two, $(D'\colon I) = D'$.

In the rest of this section, we want to investigate the noetherianity of $\mathrm{Int}(E, D)$ when E is a nonzero fractional ideal of D. In this case $\mathrm{Int}(E, D)$ is not equal to D as we have observed in Section 1. We note that if $\mathrm{Int}(E, D)$ is Noetherian, then D is Noetherian, being an homomorphic image of $\mathrm{Int}(E, D)$.

LEMMA 3.7. [5, Exercise VI.13] *Let E be a nonzero fractional ideal of D. If $\mathrm{Int}(E, D)$ is Noetherian, then each height one prime ideal of D' has infinite residue field.*

Proof. If $\mathrm{Int}(E, D)$ is Noetherian, then D is Noetherian. Replacing $\mathrm{Int}(D)$ by $\mathrm{Int}(E, D)$ in the proof of [5, Proposition VI.2.4] and using [5, Excercise V.8] instead of [5, Corollary V.2.4] we get that each height one prime ideal of D' has infinite residue field. □

By using the last lemma, Propositions 3.1 and 3.2, we immediately have the following proposition.

PROPOSITION 3.8. *Assume that $I := (D : D') \neq (0)$ and let E be a nonzero fractional ideal of D. If $\mathrm{Int}(E, D)$ is Noetherian, then $\mathrm{Int}(D)$ is Noetherian.*

The following result generalizes Proposition 3.1.

PROPOSITION 3.9. *Let E be a nonzero fractional ideal of D and assume that either $\mathrm{Int}(D)$ or $\mathrm{Int}(E, D)$ is Noetherian. Then the two following equivalent conditions hold:*

(i) $\mathrm{Int}(E, D) \subseteq D'[X/E]$;

(ii) $\mathrm{Int}(E, D') = D'[X/E] = \mathrm{Int}(ED', D')$.

Proof. If either $\mathrm{Int}(D)$ or $\mathrm{Int}(E, D)$ is Noetherian, then each height one prime ideal of D' has infinite residue field by Proposition 3.1 or Lemma 3.7 respectively. Hence $\mathrm{Int}(D') = D'[X]$, again by Proposition 3.1, and $\mathrm{Int}(ED', D') = D'[X/E]$ by [8, Lemma 4.5]. To finish, conditions (i) and (ii) are equivalent by Proposition 2.5. □

If I is an ideal of D and E is a nonzero D-module, we set

$$I[X/E] := \bigcap_{a \in E \setminus (0)} I[a^{-1}X].$$

It is easy to check that $I[X/E]$ is an ideal of $D[X/E]$.

LEMMA 3.10. *Suppose that $I := (D \colon D') \neq (0)$ and let E be a nonzero D-module. Then*

$$I[X/E] \subseteq (D[X/E] \colon D'[X/E]).$$

Proof. If $\alpha \in I \setminus (0)$, then $\alpha D' \subseteq D$. Thus, $\alpha D'[a^{-1}X] \subseteq D[a^{-1}X]$, for each $a \in E$. Hence $\alpha D'[X/E] \subseteq D[X/E]$. □

PROPOSITION 3.11. *Assume that $I := (D \colon D') \neq (0)$, let E be a nonzero fractional ideal of D and consider the following conditions:*

(i) $\mathrm{Int}(E, D)$ *is Noetherian;*

(ii) $D[X/E]$ *is Noetherian;*

(iii) $D'[X/E]$ *is Noetherian.*

Then (i) $\Longrightarrow$(ii) $\Longleftrightarrow$ (iii). *If in addition* $\mathrm{Int}(D)$ *is Noetherian, then* (ii) $\Longrightarrow$(i) *and all the three conditions are equivalent.*

Proof. By Lemma 3.10, $D[X/E]$ and $D'[X/E]$ share the nonzero ideal $I[X/E]$. Hence we have the following pullback diagram:

$$\begin{array}{ccc} D[X/E] & \longrightarrow & D[X/E]/I[X/E] \\ \downarrow & & \downarrow \\ D'[X/E] & \longrightarrow & D'[X/E]/I[X/E] \end{array}$$

Since D' is a finitely generated D-module, then $D'[X/E]$ is a finitely generated $D[X/E]$-module. Whence (ii) $\Longleftrightarrow$ (iii) (Proposition 3.3).

Now assume that $\mathrm{Int}(E, D)$ is Noetherian. By Proposition 3.9, we have that $D[X/E] \subseteq \mathrm{Int}(E, D) \subseteq D'[X/E]$. Hence $\mathrm{Int}(E, D)$ and $D'[X/E]$ also share the nonzero ideal $I[X/E]$ and we have the pullback diagrams:

$$\begin{array}{ccc} D[X/E] & \longrightarrow & D[X/E]/I[X/E] \\ \downarrow & & \downarrow \\ \mathrm{Int}(E, D) & \longrightarrow & \mathrm{Int}(E, D)/I[X/E] \\ \downarrow & & \downarrow \\ D'[X/E] & \longrightarrow & D'[X/E]/I[X/E] \end{array}$$

It follows that, if $\mathrm{Int}(E, D)$ is Noetherian, then also $D'[X/E]$ is Noetherian, that is (i) $\Longrightarrow$(iii).

On the other hand, if $\mathrm{Int}(D)$ is Noetherian, again by Proposition 3.9, we have the same pullbak diagrams as before. Hence, if $D[X/E]$ is Noetherian, $\mathrm{Int}(E, D)$ is Noetherian. That is, in this case, (ii) $\Longrightarrow$(i). □

By the previous proposition, if E is a nonzero fractional ideal of D, $I := (D \colon D') \neq (0)$ and $\mathrm{Int}(D)$ is Noetherian, then $\mathrm{Int}(E, D)$ is Noetherian if and only if $D[X/E]$ is Noetherian. On the other hand, by using Proposition 3.8 we get that when $D[X/E]$ is Noetherian, then $\mathrm{Int}(E, D)$ is Noetherian if and only if $\mathrm{Int}(D)$ is Noetherian. In Remark 3.14 (c) below, we give an example where $D[X/E]$ is Noetherian but $\mathrm{Int}(E, D)$ is not Noetherian.

We now prove that if D is a Krull domain, then $D[X/E]$ is a Krull domain, for every fractional ideal E of D. We recall that $\mathrm{Int}(D)$ is Krull if and only if D is Krull and $\mathrm{Int}(D) = D[X]$ [7, Corollary 2.7]. In this case $\mathrm{Int}(E, D) = D[X/E]$.

PROPOSITION 3.12. *Let E be a fractional ideal of D. If D is a Krull domain, then $D[X/E]$ is a Krull domain.*

Proof. Without loss of generality we can assume that $E \subseteq D$. We can represent D as the intersection $\bigcap_p D_p$, where p runs among the height-one prime ideals of D, each domain D_p is a DVR and this intersection has finite character. Moreover, $D[X/E] = \bigcap_p D_p[X/E]$.

Now, E is contained at most in finitely many height-one primes, say $p_1, \cdots, p_n$. Then, if $p \neq p_i$ for $i = 1, \cdots, n$, we have $ED_p = D_p$ and $D_p[X/E] = D_p[X]$. Instead, for $i = 1, \cdots, n$, since D_{p_i} is a DVR, we have $ED_{p_i} = x_i D_{p_i}$, for some $x_i \in E$, and $D_{p_i}[X/E] = D_{p_i}[X/x_i] \cong D_{p_i}[X]$.

The domain $R := \bigcap_{p \neq p_1, \cdots, p_n} D_p[X/E] = \bigcap_{p \neq p_1, \cdots, p_n} D_p[X]$, is a generalized ring of quotients of $D[X]$. Since $D[X]$ is Krull, then R is Krull. For the same reason, $D_{p_i}[X/E] \cong D_{p_i}[X]$ is Krull.

It follows that $D[X/E] = R \cap (\bigcap_i D_{p_i}[X/x_i])$ is a finite intersection of Krull domains and then it is Krull. □

Recalling that the integral closure of a Noetherian domain is Krull and by using Proposition 3.1 and Lemma 3.7, we immediately get the following corollary.

COROLLARY 3.13. *If either* $\text{Int}(D)$ *or* $\text{Int}(E, D)$ *is Noetherian, then* $\text{Int}(E, D') = D'[X/E]$ *is Krull, for each fractional ideal* E *of* D. □

Remark 3.14.

(a) To prove that $\text{Int}(E, D) \subseteq D'[X/E]$ when $\text{Int}(D)$ is Noetherian (Proposition 3.9), we can also proceed in the following way.

If $\text{Int}(D)$ is Noetherian, then $\text{Int}(D) \subseteq D'[X]$ (Proposition 3.1). Hence, by the isomorphism $f(X) \mapsto f(a^{-1}X)$, we have that

$$\text{Int}(aD, D) \subseteq D'[a^{-1}X],$$

for all $a \in K \backslash (0)$. It follows that

$$\text{Int}(E, D) = \bigcap_{a \in E \backslash (0)} \text{Int}(aD, D) \subseteq \bigcap_{a \in E \backslash (0)} D'[a^{-1}X] = D'[X/E].$$

(b) In the hypotheses of Proposition 3.11, the study of the noetherianity of $\text{Int}(E, D)$ is reduced to investigating the noetherianity of the graded ring $D[X/E]$. Unfortunatly this fact does not simplify the problem. In fact, following the notation of Remark 2.6, the characterization of the rings $\mathcal{M}[X]$ which are Noetherian is still an open question.

Y. Haouat studied in his Ph.D Thesis [11], the noetherianity for graded rings of the form:

$$\mathcal{A}[X] := \bigoplus_{n \geq 0} A_n X^n,$$

where $\mathcal{A} := \{A_n\}_{n \geq 0}$ is an increasing sequence of rings. In this case, he showed that $\mathcal{A}[X]$ is Noetherian if and only if A_0 is Noetherian, the sequence $\{A_n\}_{n \geq 0}$ is stationary and every ring A_n is a finite-type A_0-module.

Later, P.L. Kiihne studied in [12] the Noetherian property for the rings $\mathcal{M}[X]$, when the sequence of modules $\mathcal{M} = \{M_n\}_{n \geq 0}$ is stationary.

We immediately see that the stationarity of the sequence $\mathcal{M}$ is not a necessary condition for the noetherianity of $\mathcal{M}[X]$. In fact, the domain $D[a^{-1}X]$ is Noetherian if D is Noetherian, but it is generated by the strictly increasing sequence of D-modules $\{a^{-n}D\}_{n\geq 0}$.

(c) In general we know that $D[X/E] = \bigoplus_{n\geq 0} E_n X^n$ is Noetherian if and only if $E_0 = D$ is Noetherian and $\bigoplus_{n\geq 1} E_n X^n$ is a finite type D-algebra.

This last condition is satisfied for example when E is a fractional ideal of D such that the modules E_n become principal (generated by a power of the same element), for all $n >> 0$ and in addition $E_k E_h = E_{h+k}$, for $h, k >> 0$. In fact, assume that there exists $N > 0$ such that $E_n = b^{i_n} D$ and $E_h E_k = E_{h+k}$, for each $n, h, k > N$, and let $\alpha_{h1}, \cdots, \alpha_{hj_h}$ be the generators of E_h, for $h = 1, \cdots, N$. Then:

$$D[X/E] = D[\alpha_{11}X, \cdots, \alpha_{1j_1}X, \alpha_{21}X^2, \cdots, \alpha_{2j_2}X^2, \cdots,$$
$$\alpha_{N1}X^N, \cdots, \alpha_{Nj_N}X^N, b^{i_{N+1}}X^{N+1}, \cdots, b^{i_m}X^m],$$

for a certain $m > N$.

For instance, let k be a field and consider the domain $D := k[[X^3, X^7, X^{11}]]$. Then D is a one-dimensional, local, Noetherian, domain with maximal ideal $M := (X^3, X^7, X^{11})$. The integral closure of D is $V := k[[X]]$ and V has nonzero conductor $I = M^9$ into D. Hence D is a particular example of a pullback domain of the following type (see, for example, [3], [4]):

$$\begin{array}{ccc} D & \longrightarrow & D/I \\ \downarrow & & \downarrow \\ V & \longrightarrow & V/I \end{array}$$

where V is a rank-one discrete valuation domain with maximal ideal $\mathfrak{m}$, $I := \mathfrak{m}^r$, for some $r > 0$ and V and D have the same residue field.

Let us consider the ideal $E := (X^3, X^7)D$. Setting $E(n) := \{u^n; u \in E \setminus (0)\}D$, we have that $E_n = \bigcap_{u\in E\setminus(0)} u^{-n}D = (D\colon E(n))$; thus E_n is principal if and only if $E(n)$ is principal.

Now, if v is the value function of V, we have that $v(D) = \{0, 3, 6, 7, 9 \rightarrow\}$ (all integers equal or bigger than 9 are in $v(D)$) and $v(E) = \{0, 3, 6, 7, 9, 10, 12 \rightarrow\}$. We observe that $v(E(n)) = nv(E) + v(D)$ and, making some computations, it is easy to check that, for each $n \geq 3$, $v(E(n)) = 3n + v(D)$. This means that $v(E(n)) = v(X^{3n}D)$. By [13, Proposition 1], for all $n, m \geq 3$, we have that $E(n) = X^{3n}D$ and $E(n)E(m) = E(n+m)$.

More generally, it can be proved that if D is a pullback as above, then, for every fractional ideal E, the modules E_n are principal of type $z^n D$, for some $z \in E$ and $n >> 0$. This fact will be illustrated in details in a forthcoming paper.

We note that if D is as before, the noetherianity of $D[X/E]$ follows more easily from Proposition 3.11. In fact, observing that $EV = aV$ is principal in V, we get that $V[X/E] = \text{Int}(aV, V) = V[X/a]$ is Noetherian. Thus we conclude that $D[X/E]$ also is Noetherian.

However, since D is a one-dimensional, local, Noetherian domain, then $\text{Int}(D)$ is Noetherian if and only if $\text{Int}(D) = D[X]$ [5, Corollary VI.2.6] if and only if

$k = D/M$ is infinite [5, Corollary I.3.15]. Hence, if $k = D/M$ is finite, $\text{Int}(E, D)$ is not Noetherian by Proposition 3.8. On the other hand, if $k = D/M$ is infinite, then $\text{Int}(E, D) = D[X/E]$ is Noetherian.

(d) Another proof of Corollary 3.13 can be obtained as follows.

Since $\text{Int}(D)$ or $\text{Int}(E, D')$ is Noetherian, D is Noetherian and $\text{Int}(E, D') = D'[X/E] = \text{Int}(ED', D')$ (Proposition 3.9). Assume that $E = a_1D + \cdots + a_nD$. Then:

$$\text{Int}(ED', D') = \text{Int}(\bigcup_{i=1,\cdots,n} a_iD', D') = \bigcap_{i=1,\cdots,n} \text{Int}(a_iD', D')$$

[6, Corollary 3.12], whence $\text{Int}(E, D') = \bigcap_{i=1,\cdots,n} \text{Int}(a_iD', D')$.

Since $\text{Int}(a_iD', D') \cong \text{Int}(D') = D'[X]$ is a Krull domain (Proposition 3.1 or Lemma 3.7), then $\text{Int}(E, D')$ is a finite intersection of Krull domain and so it is Krull.

(e) We recall that a *Mori domain* is a domain satisfying the ascending chain condition on divisorial ideals; in particular Noetherian and Krull domains are Mori domains. A recent general reference for Mori domains is [2].

If D is Mori, then $D = \bigcap_{p \in t_m(D)} D_p$, where $t_m(D)$ is the set of maximal divisorial ideals of D, and this intersection has finite character [2, Theorem 3.3]. If $D[X]$ is Mori (for example if D is Noetherian or integrally closed [2, Theorem 6.1]) and E is a fractional ideal of D such that ED_p is principal for each $p \in t_m(D)$ (for example if E is divisorial and v-invertible), arguing exactly as in the proof of Proposition 3.12, we get that $D[X/E]$ is Mori. In fact a generalized ring of quotients of a Mori domain is Mori [2, Theorem 2.5] and a finite intersection of Mori domains is Mori [2, Theorem 2.4].

Slightly modifying tha same proof, we also get that if $\text{Int}(D)$ is Mori and ED_p is principal for each $p \in t_m(D)$, then $\text{Int}(E, D)$ is a Mori domain. (We recall that if $\text{Int}(D)$ is a Mori domain, then D is Mori [7, Proposition 2.6].) To see this, it is enough to observe that $\text{Int}(E, D) = \bigcap_{p \in t_m(D)} \text{Int}(ED_p, D_p)$ [6, Lemma 3.4]. Now, assuming $E \subseteq D$, if $E \nsubseteq p$, we have $ED_p = D_p$ and $\text{Int}(ED_p, D_p) = \text{Int}(D_p) = \text{Int}(D)_p$ (the last equality follows from [7, Proposition 2.1]). Instead, if $E \subseteq p$, we have $ED_p = xD_p$, for some $x \in E$, and $\text{Int}(ED_p, D_p) = \text{Int}(xD_p, D_p) \cong \text{Int}(D_p) = \text{Int}(D)_p$.

The domain $R := \bigcap_{E \nsubseteq p} \text{Int}(ED_p, D_p) = \bigcap_{E \nsubseteq p} \text{Int}(D)_p$, is Mori as a generalized ring of quotients of a Mori domain and, for the same reason, if $E \subseteq p$, we have that $\text{Int}(ED_p, D_p) \cong \text{Int}(D)_p$, is Mori. It follows that $\text{Int}(E, D) = R \cap (\bigcap_{E \nsubseteq p} \text{Int}(ED_p, D_p)$ is a finite intersection of Mori domains and then it is Mori.

References

[1] V.Barucci, Strongly divisorial ideals and complete integral closure of an integral domain, *J. Algebra* **99** (1986), 132–142.

[2] V.Barucci, Mori domains, *Non-Noetherian Commutative Ring Theory,* Chapter 3, Kluwer Academic Publishers, 2000.

[3] V. Barucci, D.E. Dobbs and M. Fontana, Conducive integral domains as pull-backs, *Manuscripta Math.*, **54**, (1986), 261–277.

[4] V. Barucci, D.E. Dobbs and M. Fontana, Gorenstein conducive domains, *Comm. Algebra*, **18**, (1990), 3889–3903.

[5] P.-J. Cahen and J.-P. Chabert, *Integer valued polynomials*, Amer. Math. Soc. Surveys and Monographs, **48**, Providence, 1997.

[6] P.-J. Cahen, Polynomial closure, *J. Number Theory*, **61**, n. 2 (1996), 226–247.

[7] P.-J. Cahen, S. Gabelli and E. Houston, Mori domains of integer valued polynomials, *J. Pure Appl. Algebra*, **153** (2000), 1-15.

[8] M. Fontana, L. Izelgue, S. Kabbaj and F. Tartarone, On the Krull dimension of domains of integer-valued polynomials, *Expositiones Math.*, **15**, (1997), 433–465.

[9] M. Fontana, Topologically defined classes of commutative rings, *Ann. Mat. Pura Appl. IV*, **123**, (1980), 331–355.

[10] R. G. Gilmer, W. Heinzer and D. Lantz, The noetherian property in rings of integer valued polynomials, *Trans. Amer. Math. Soc.*, **338** (1993), 187–199.

[11] Y. Haouat, *Anneaux de Polynômes*, Ph.D Thesis, Faculté des Sciences de Tunis, 1988.

[12] P.L. Kiihne, *Factorization and Ring-Theoretic properties of subrings of polynomials over a field and power series over a field*, Ph.D Thesis, University of Tennessee, Knoxville, 1999.

[13] T. Matsuoka, On the degree of singularity of one-dimensional analytically irreducible Noetherian local rings, *J. Math. Kyoto Univ.*, **11**, (1971), 485–494.

[14] F. Tartarone, Polynomial closure in pseudo-valuation domains, *preprint*.

Controlling the Zero Divisors of a Commutative Ring

SARAH GLAZ

Department of Mathematics, University of Connecticut, Storrs, CT 06269
glaz@uconnvm.uconn.edu

Abstract. This article explores nine zero divisor controlling conditions, their impact on the domain-like behavior of rings with zero divisors, their interrelations, and their applications to the extension of domain properties to rings with zero divisors.

1. INTRODUCTION

Commutative algebraists often consider the following question: given a property P that holds for a domain, how best can one extend P to rings with zero divisors? The answer may sometimes depend completely on the particular property. But more often, especially when P involves all the entities of a ring R, where by entities we mean elements or ideals or finitely generated ideals, etc., one may take one of the following three approaches:

a. Require that all entities of R satisfy P, and place no other restrictions on R.

b. Require that all regular entities of R satisfy P.

c. Require that all entities of R satisfy P, <u>and</u> place some conditions on R to control the behavior of its zero divisors, that is, to make R share certain characteristics of a domain.

Each of these approaches have been used for various properties P with certain degrees of success. Which approach one uses depends on personal taste, and on the desired properties of the generalization. For example, if one wishes to have some control and knowledge of the behavior of non regular entities of the ring, one will try to avoid approach (b), which puts no restrictions on the non regular entities of the ring. Approach (a) is not always feasible, as certain conditions cannot be imposed on non regular entities without additional hypotheses.

The aim of this article is to consider a large number of zero divisor controlling conditions found in the literature. Some of these conditions were used in generalizing a domain property to rings with zero divisors using approach (c); other of these conditions have interesting and deep implications on the nature of the ring itself and were investigated in their own right. In each case, we will explore briefly the origin of the condition and its use for generalizations of domain properties to a ring R with zero divisors. We will point out the impact of each condition on sets of zero divisors, localizations by prime ideals, total ring of quotients, or sets of minimal prime ideals of the ring, all of which determine the closeness of the ring behavior to that of a domain. To the extent that it is possible, we will point out the relations between the various conditions, and provide examples and counterexamples.

Throughout the paper, all rings are commutative with identity. We will use the following notation and basic notions regarding a ring R:

$Z(R) = \{a \in R \mid ax = 0 \text{ for some } 0 \neq x \in R\}$ denotes **the set of zero divisors of R**.

$Nil(R) = \{a \in R \mid a^n = 0 \text{ for some positive integer } n\}$ denotes **the set of nilpotent elements of R**.

$Min(R) = \{P \in \text{spec } R \mid P \text{ is a minimal prime ideal of } R\}$ denotes **the set of minimal primes of R**.

Q(R) denotes **the total ring of quotients of R**, that is, the localization of R by the set of all its non zero divisors.

A non zero divisor of R will be called **a regular element,** and an ideal of R which contains a regular element will be called **a regular ideal**.

The oldest and most extensively studied zero divisor controlling condition is that which asks R to be a reduced ring, that is, a ring with no nonzero nilpotent elements. The literature pertaining to the nature of reduced rings is too vast to mention. For that reason, we will consider this condition only when it appears in the context of the other zero divisor controlling conditions considered in this paper. We will explore the following zero divisor controlling conditions on a ring R. Several other, less used conditions will show up throughout the exposition.

1. R is locally a domain.
2. Q(R) is a Von Neumann regular ring.
3. Min(R) is a compact subspace of Spec(R) in the Zariski topology.
4. R is a PP ring (sometimes called a weak Baer ring).
5. R is a ring with few zero divisors.
6. R is an additively regular ring.
7. R is a Marot ring.
8. R is a ZD ring.
9. Every zero divisor of R is nilpotent.

The interrelations between these nine properties makes it hard to separate the exposition into independent sections. Nevertheless, for the sake of readability, we divided the conditions into two groups. Section 2 considers conditions 1 - 4, and Section 3 considers conditions 5 - 9. Thus, Sections 2 and 3 explore the conditions themselves and the relations between them. Section 4 is devoted to a representative sample of applications of the zero divisor controlling conditions to the extension of domain properties to rings with zero divisors. We cover briefly the extended notions of valuation, Prufer and Krull rings; finite conductor, quasi-coherent and G-GCD rings; and Going Down and related rings. Interested readers are provided with a bibliography for further reading on each topic.

2. MIN(R) AND LOCALIZATIONS OF R

We first consider conditions under which a commutative ring R is locally a domain.

THEOREM 2.1 [G1], [M1], [M2] *Let R be a ring. The following conditions are equivalent:*

1. *R_P is a domain for every prime ideal P of R.*
2. *R_m is a domain for every maximal ideal m of R.*
3. *Every principal ideal of R is flat.*
4. *R is reduced and every prime ideal of R contains a unique minimal prime ideal.*
5. *R is reduced and every maximal ideal m of R contains a unique minimal prime ideal P.*

In this case, $P = \{r \in R \mid$ there is a $u \in R - m$ such that $ur = 0\}$ and $R_P = Q(R_m)$, the quotient field of R_m.

This theorem appears with somewhat different proofs in Matlis [M2], and in Glaz [G1]. (4) does not appear in either of these sources, but appears in [M1], along with a few other of the conditions of this theorem, proved under the restriction that Q(R) is Von Neumann regular. As this assumption is unnecessary, we included (4) here. Rings satisfying that every principal ideal is flat are sometimes called **PF (or PIF) rings**.

EXAMPLE 2.2 The following examples are taken from several sources: [E], [G1], [M1], [M2], [Q1], [Q2], and [V]:

1. Coherent local rings satisfying that every principal ideal has finite projective dimension are locally domains.

2. Coherent **regular rings**, that is, coherent rings satisfying that every finitely generated ideal has finite projective dimension, are locally domains. In particular, this class of rings includes all coherent rings R of finite weak global dimension. It is worthwhile mentioning that the class of coherent rings of finite weak global dimension includes the classical non Noetherian rings such as Von Neumann regular, semihereditary, and hereditary rings. **Von Neumann regular rings** are rings R satisfying that for every $a \in R$, there is a $b \in R$ such that $ab^2 = a$. Equivalently, R is Von Neumann regular iff w.gl.dim R = 0. Such rings are automatically coherent. **Semihereditary rings** are rings in which every finitely generated ideal is projective. Equivalently, R is semihereditary iff R is coherent and w.gl.dim R $\leq$ 1. **Hereditary rings** are rings with gl.dim R = 1. They are always coherent.

3. Rings of global dimension 2 are locally domains.

4. A ring R has w.gl.dim R $\leq$ 1 iff R_P is a valuation domain for every prime ideal P of R. Therefore rings R with w.gl.dim R $\leq$ 1 are always locally domains, regardless of their coherence status.

We next consider the condition: Min(R) is compact in the induced Zariski topology from Spec(R). If R is Noetherian or a domain Min(R) is finite and therefore compact. In general Min(R) does not have to inherit the compactness of Spec(R). For a reduced ring one has a way of testing the minimality of a prime ideal P.

PROPOSITION 2.3 *Let R be a reduced ring, and let P be a prime ideal of R. Then P is a minimal prime ideal iff for all* $x \in P$, $(0 : x) \not\subset P$.

As a consequence of this proposition for a reduced ring R, Min(R) is a Hausdorff space in the induced Zariski topology, but it still might not inherit the compactness of Spec(R).

We exhibit a number of conditions under which Min(R) is indeed compact when R is a reduced ring.

THEOREM 2.4 [G1], [M2] *Let R be a reduced ring, and let S be a ring extension of R. Then:*

1. *If every prime ideal of S contracts to a minimal prime ideal of R, then Min(R) is compact.*
2. *If S is Von Neumann regular and flat over R, then every prime ideal of S contracts to a minimal prime ideal of R. Hence, in this case, Min(R) is compact.*

COROLLARY 2.5 [G1], [M2], [Q1] *Let R be a reduced ring with Q(R), the total ring of quotients of R, Von Neumann regular. Then Min(R) is compact.*

EXAMPLE 2.6 Quentel [Q1] provides an example of a reduced ring with compact minimal spectrum, but not Von Neumann regular total ring of quotients. The version presented below is from [G1]:

Let K be a countable algebraically closed field, and let I be an infinite set. Denote by K^I the set of all set maps from I to K. For a map $f \in K^I$, let $supp\, f = \{a \in I \mid f(a) \neq 0\}$, and let $cosupp\, f = \{a \in I \mid f(a) = 0\}$. A K subalgebra R of K^I is called a T algebra if it satisfies the following two conditions:
(i) R is countable and contains all the constant maps.
(ii) Every $f \in R$ which is not constant satisfies $cosupp\, f \neq \phi$.

The construction described in detail in [G1], shows the existence of a T algebra W, $W \subset K^{IxN^N}$, and for every $g \in W$ there exist g_1 and $g_2 \in W$ such that $cosupp\, g = supp\, g_1 \cup supp g_2$. Such an algebra W is a reduced ring equal to its own total ring of quotients, it has compact *Min(W)*, but it is not Von Neumann regular.

Before presenting the exact connection between the compactness of Min(R), the Von Neumann regularity of Q(R), and the locally domain property of R, we will present several other conditions under which Min(R) is compact or Q(R) is Von Neumann regular when R is a reduced ring.

THEOREM 2.7 [G1], [M2], [O], [Q1] *Let R be a reduced ring. The following conditions are equivalent:*

1. *Min(R) is compact.*
2. *For every element $b \in R$, there exists a finitely generated ideal $J \subset (0{:}b)$ such that $(0{:}bR + J) = 0$.*
3. *ΠR_P, where P runs over Min(R), is a flat R module.*

4. E(R), the injective envelope of R, is a flat R module.
5. M(R), the maximal flat epimorphic extension of R, is Von Neumann regular.

Theorem 2.7 collects a number of conditions, scattered throughout the mentioned sources, under which Min(R) is compact for a reduced ring R. To clarify the statements of this theorem, we remind the reader of the definitions of E(R) and M(R) mentioned in (4) and (5).

Let R be a ring and let M be an R module. An R module E is called **an essential extension** of M, if $M \subset E$, and for any nonzero submodule E' of E we have $E' \cap M \neq 0$. Every R module admits an essential injective extension E(M), which is unique up to isomorphism. This extension is called **the injective envelope** of M. If E(M) is the injective envelope of M, there is no injective proper submodule between M and E(M). Matlis [M2] approaches both the compactness of Min(R) and the Von Neumann regularity of Q(R) via the exploration of E(R).

Let R be a ring. Denote by M(R), **the maximal flat epimorphic extension of R**, the unique (up to isomorphism) ring satisfying:
1. $R \subset M(R)$, and M(R) is a flat epimorphism of R.
2. If $R \subset S$ and S is a flat epimorphism of R, then $S \subset M(R)$.
The approach to the compactness of Min(R) and to the Von Neumann regularity of Q(R) via the investigation of M(R) is due to Quentel [Q1] and Olivier [O], [O1].

EXAMPLES 2.8 The above theorem guarantees that if R is a reduced coherent ring, then Min(R) is compact.

The next theorem collects the conditions under which the total ring of quotients of a reduced ring is Von Neumann regular.

THEOREM 2.9 [G1], [H], [M2], [O], [Q1] *Let R be a reduced ring. The following conditions are equivalent:*
1. Q(R) is a Von Neumann regular ring.
2. If I is an ideal of R contained in the union of the minimal prime ideals of R, then I is contained in one of them.
3. If J is a finitely generated ideal of R, then there exists a $b \in J$ and an $a \in (0{:}J)$ such that $a + b$ is a regular element of R.
4. If $b \in R$, then there exists $a \in (0{:}b)$ such that $(0{:}\ aR+bR) = 0$.
5. $Q(R) = M(R)$.
6. Min(R) is compact and if a finitely generated ideal is contained in the union of the minimal prime ideals of R, then it is contained in one of them.
7. Min(R) is compact and each finitely generated ideal consisting entirely of zero

divisors has a nonzero annihilator.

EXAMPLE 2.10 The above theorem guarantees that if R is a coherent regular ring, then Q(R) is Von Neumann regular.

At this stage we are ready to present the main result connecting the zero divisor controlling conditions presented in this section.

A ring R is called **a PP ring** (or **a weak Baer ring**) if every principal ideal of R is projective.

THEOREM 2.11 [G1] *Let R be a ring. The following conditions are equivalent:*
1. Min(R) is compact and every principal ideal of R is flat.
2. R is a PP ring.
3. Q(R) is Von Neumann regular and every principal ideal of R is flat.

The ideas for the proof of this theorem also appear in [Q1] and [V], excellent sources, where some errors crept into the approach to this particular result. A correct version appears in [G1].

EXAMPLES 2.12 Two examples [G1] show that the two conditions in (1), and the two conditions in (3) of Theorem 2.11, are independent of each other.

1. [V] Let M be a countable direct sum of copies of Z/2Z with addition and multiplication defined componentwise. Let R = Z ⊕ M and define multiplication as follows: for m, n ∈ M and a, b ∈ Z
(a,m)(b,n) = (ab, an + bm + mn). Every principal ideal of R is flat, but Min(R) is not compact.

2. [M2] Let R be a Noetherian, local, reduced ring which is not a domain. Then Min(R) is compact (actually finite), and Q(R) is a finite direct sum of the domains R_P, where P runs over Min(R). Therefore Q(R) is Von Neumann regular, but R is not a domain, so not every principal ideal of R is flat.

PP rings appear in the literature in expected and unexpected places. From Theorem 2.11, we know that such rings are locally domains and possess Von Neumann regular total ring of quotients. Another way of looking at such a ring is observing that a principal ideal aR is projective iff (0:a) is generated by an idempotent. Hence PP rings have plenty of idempotents. In fact, every element a ∈ R can be expressed as a = a′e, where a′ is a non zero divisor in R, and e is an idempotent in R. Another useful property of PP rings is:

THEOREM 2.13 [G3] *Let R be a PP ring, and let I be a finitely generated flat ideal of R. Then I is projective.*

There are many PP rings, for example, all coherent regular rings are such, and in the application section, we will exhibit a class of not necessarily coherent PP rings, namely G-GCD rings.

3. MAROT RINGS AND RELATED CONDITIONS

The notion of a ring with few zero divisors was introduced by Davis [D]. **A maximal ideal of zero** is an ideal (necessarily prime) maximal with respect to not containing regular elements. The set of zero divisors of R, Z(R) = $\cup$ P, as P runs over all maximal ideals of zero of R. A ring R is said to have **few zero divisors** if it has only finitely many maximal ideals of zero, equivalently, Z(R) is a union of finitely many prime ideals.

Because of the one-to-one correspondence between the prime ideals of a ring R which contain no regular elements and the prime ideals of Q(R), it is evident that R has few zero divisors iff Q(R) is semilocal. It follows that if R has few zero divisors, then any **overring of R,** that is, any ring between R and Q(R), has few zero divisors. In particular, any overring of a Noetherian ring has few zero divisors, providing a large family of examples of rings of this kind.

A ring R with total ring of quotients Q(R) is said to be **additively regular** if for each z $\in$ Q(R), there exists a u $\in$ R such that z + u is a regular element in Q(R). This condition appears first in [M1], and is named in [GH].

PROPOSITION 3.1 [GH] *Let R be a ring. The following conditions are equivalent:*

1. *R is additively regular.*
2. *For each $z \in Q(R)$ and each regular element b of R, there exists a $u \in R$ such that $z + bu$ is a regular element of Q(R).*
3. *For each $a \in R$ and each regular element b of R, there exists a $u \in R$ such that $a + ub$ is regular in R.*

Additively regular rings have the following useful property:

THEOREM 3.2 [M3] *Let R be an additively regular ring. Let $I_1, \ldots, I_n$ and I be regular ideals of R. Denote by Reg(I) the set of regular elements of I. Then $Reg(I) \subset \cup \{I_i \mid 1 \leq i \leq n\}$ iff $I \subset \cup \{I_i \mid 1 \leq i \leq n\}$.*

THEOREM 3.3 [H1], [M1] *Let R be a ring with total ring of quotients Q(R). If Q(R)/Nil(Q(R)) is Von Neumann regular, then R is additively regular. In particular, if Q(R) is Von Neumann regular, then R is additively regular.*

EXAMPLES 3.4

1. [PS] It follows from Theorem 3.3 that any ring whose total ring of quotients has Krull dimension zero is additively regular.

2. [M1] For any ring R, the polynomial ring in any number of variables is an additively regular ring.

3. [GH] Let $R = \Pi R_\alpha$ for an arbitrary set $\{\alpha\}$. Then R is an additively regular ring iff R_α is an additively regular ring for every α.

A ring R is called **a Marot ring** if every regular ideal can be generated by a set of regular elements. This property was defined by Marot [M1].

THEOREM 3.5 [M1] *Let R be a ring with total ring of quotients Q(R). The following conditions are equivalent:*
1. R is a Marot ring.
2. Any two-generated ideal (a, b) with b regular can be generated by a finite set of regular elements.
3. Every regular fractional ideal of R, that is, every regular R module contained in Q(R), can be generated by a set of regular elements.

The next result connects the zero divisor controlling conditions exhibited, so far, in this section.

THEOREM 3.6 [D], [H], [M1] *Let R be a ring. Consider the following conditions:*
1. R has few zero divisors.
2. R is an additively regular ring.
3. R is a Marot ring.
Then (1) $\Rightarrow$ *(2)* $\Rightarrow$ *(3).*

None of the implications of Theorem 3.6 is reversible, as the two examples below show.

EXAMPLE 3.7 [H] There are additively regular rings which do not have few zero divisors.

Let $\{R_\alpha\}$ be an infinite family of rings with few zero divisors. The product ring $R = \Pi R_\alpha$ is an additively regular ring which is not a ring with few zero divisors.

EXAMPLE 3.8 [H], [M3] There are Marot rings which are not additively regular.

The following example was constructed by Matsuda [M3]. Let k be a finite field of characteristic $p > 0$. Let A be the subring of the polynomial ring k[x], $A = k[x^p, x^{p+1}, x^{p+2}, \ldots]$. Let $\{F_0, F_1, \ldots, F_n, G_1, G_2, \ldots\}$ be a set of irreducible polynomials in k[x] such that:

1. $F_0 = x$, and $F_1 = 1 + x$.
2. $\deg F_i < 2p$ for all i.
3. $\deg G_j \geq 2p$ for all j.
4. No two elements of the set are associated.
5. Each irreducible element of k[x] is associated with an element of the set.

Let $K_j = k[x]/(G_j)$. Then K_j is naturally an A module. Let M be the direct sum of the modules K_j, and let $R = A \alpha M$ be **the trivial ring extension of A by M** (sometimes called **the idealization of M in A** [H]), that is, R is the set $A \oplus M$ with addition defined componentwise and multiplication defined by $(a, m)(a', m') = (aa', am' + a'm)$ for all $a, a' \in A$ and $m, m' \in M$. Then R is a Marot ring for any p. For $p = 2$, for example, R is not an additively regular ring.

It is interesting to note that Matsuda's exploration of the additively regular and Marot properties of trivial ring extensions also yielded an example of a ring satisfying the condition of Theorem 3.2, but which is not additively regular.

EXAMPLE 3.9 [H], [M1], [SP] In addition to Noetherian rings, domains, and all additively regular rings, Marot rings can be generated in several other ways:

1. Any overring of a Marot ring is a Marot ring.

2. $R = R_1 \oplus \ldots \oplus R_n$ is a Marot ring iff R_i is a Marot ring for every i.

3. If every regular finitely generated ideal of a ring R is principal, then R is a Marot ring.

We now present a few other properties of Marot rings which makes this condition particularly useful when generalizing domain properties to rings with zero divisors.

An ideal P of a ring R is **prime** (respectively, **primary) for its regular elements** if whenever a and b are regular elements of R such that $ab \in P$, then $a \in P$ or $b \in P$ (respectively, then $a \in P$ or $b^n \in P$ for some positive integer n).

THEOREM 3.10 [H], [SP] *Let R be a Marot ring. Then a regular ideal P of R is prime (respectively, primary) iff P is prime (respectively, primary) for its regular elements.*

If a ring contains zero divisors, one may define invertibility of nonzero (fractional) ideals I of R, that is of R submodules of Q(R), in a way that resembles the definition for the case R is a domain. Let $I^{-1} = \{x \in Q(R) \mid xI \subset R\}$ denote the inverse of I. Then I is **invertible** if $II^{-1} = R$. The relation between invertibility, projectivity, and the property of being locally principal of an ideal can be summarized as follows:

PROPOSITION 3.11 *Let R be a ring and let I be an ideal of R. Then:*

1. *If I is invertible, then I is projective.*
2. *If I is projective, then I is locally principal.*
3. *If I is a finitely generated regular ideal, then I is invertible iff I is projective, iff I is locally principal. (In particular, if R is a domain, 3 holds for every nonzero finitely generated ideal I of R.)*

A (fractional) ideal I of R is **divisorial** if $(I^{-1})^{-1} = I$.

Let I be a regular (fractional) ideal of R satisfying that $dI \subset R$ for some regular element $d \in R$. Then $I \subseteq (I^{-1})^{-1} \subseteq \cap \{Ra \mid I \subseteq Ra\}$. Note that since I is regular, the ideals Ra are necessarily regular. Hence, if $I = \cap \{Ra / I \subseteq Ra\}$, then I is divisorial. If I is generated by regular elements, say $I = \Sigma Ra_\alpha$, then $I^{-1} = \cap Ra_\alpha^{-1}$. It follows that:

PROPOSITION 3.12 [AM1] *Let R be a ring. Then $(I^{-1})^{-1} = \cap \{Ra \mid I \subseteq Ra\}$ for every regular (fractional) ideal I of R satisfying $dI \subset R$ for some regular element $d \in R$ iff for every regular (fractional) ideal I of R satisfying $dI \subset R$ for some regular element $d \in R$, there exists a (fractional) ideal J generated by regular elements and satisfying $dJ \subset R$ for some regular element $d \in R$, such that $(I^{-1})^{-1} = (J^{-1})^{-1}$.*

COROLLARY 3.13 [AM1] *Let R be a Marot ring, and let I be a regular (fractional) ideal of R satisfying $dI \subset R$ for some regular element $d \in R$. Then I satisfies $(I^{-1})^{-1} = \cap \{Ra \mid I \subseteq Ra\}$.*

D. D. Anderson and Markanda provide an example in [AM1] that shows that if R is not a Marot ring, then the conclusion of Corollary 3.13 does not need to hold.

A related zero divisor controlling condition was defined by Evans [E1]. A ring R is called **a ZD ring**, if R/I is a ring with few zero divisors for every ideal I of R. Examples of ZD rings abound. Below is a sample found in the literature.

EXAMPLES 3.14

1.Noetherian rings are ZD rings.

2. [E1] A ring is called **Laskerian** if every ideal is the intersection of a finite number of primary ideals. Laskerian rings are ZD rings.

3. [E1] Localizations of ZD rings are ZD rings.

4. [HO] A ring R is Noetherian iff the polynomial ring R[x] is a ZD ring.

It is interesting to note that if the power series ring R[[x]] is a ZD ring, R may not be Noetherian, but has Noetherian prime spectrum [GH].

We conclude this section with a different zero divisor controlling condition defined by Dobbs [D1]. Let R be a ring. In general, $\mathrm{Nil}(R) \subseteq Z(R)$. On one end of the spectrum $\mathrm{Nil}(R) = 0$, that is, R is reduced. The other extreme is to ask that every zero divisor of R be nilpotent, that is $\mathrm{Nil}(R) = Z(R)$. This condition is equivalent to 0 being a primary ideal of R.

Let R be a ring in which 0 is a primary ideal. Then R has a unique minimal prime ideal P, and $R_P = Q(R)$, the total ring of quotients of R. On the other hand, if R is a ring with a unique prime ideal, then 0 is a primary ideal of R. This observation [D1] lead to a number of examples of rings in which 0 is a primary ideal.

EXAMPLES 3.15 [D1] Examples of rings R satisfying $\mathrm{Nil}(R) = Z(R)$:

1. Artinian local rings.

2. Let A be a domain, and let a be a nonzero prime element of A. Then the ring $R = A/(a^2)$ has 0 primary ideal.

4. APPLICATIONS

In this section, we will exhibit applications of the zero divisor controlling conditions discussed in Sections 2 and 3 to extension of domain properties to rings with zero divisors. The literature in this direction is vast. The main scope of this article is the exploration of the zero divisor controlling conditions themselves. We therefore restrict ourselves to a representative sample of applications which we describe without going into too many details.

4.1 Valuation Rings, Prufer Rings, and Krull Rings

Valuation rings with zero divisors were defined by Manis [M]. **A valuation** is a map ν from a ring K onto a totally ordered group G and a symbol ∞, such that for all x and y in K:

1. $\nu(xy) = \nu(x) + \nu(y)$.
2. $\nu(x + y) \geq \min\{\nu(x), \nu(y)\}$.
3. $\nu(1) = 0$ and $\nu(0) = \infty$.

The ring $A = A_\nu = \{x \in K \mid \nu(x) \geq 0\}$, together with the ideal $P = P_\nu = \{x \in K \mid \nu(x) > 0\}$, denoted (A, P), is called **a valuation pair (of K)**. A is called **a valuation ring (of K).** G is called **the value group** of A.

In the presence of the Marot property, valuation rings share some properties of valuation domains:

PROPOSITION 4.1 [H], [PS] *Let A be a Marot ring. Assume that $A \neq Q(A)$. Then the following conditions are equivalent:*

1. *A is a valuation ring.*
2. *For each regular element $x \in Q(A)$, either $x \in A$ or $x^{-1} \in A$.*
3. *A has only one regular maximal ideal and each of its finitely generated regular ideals is principal.*

Let R be a ring with total ring of quotients Q(R). The ideal $C(R) = \{x \in R \mid xQ(R) \subset R\}$ is called **the core of R**. Note that C(R) can be obtained as the intersection of all regular ideals of R, or as the intersection of all regular principal ideals of R. A valuation ring (R, P) is said to be **discrete** if each primary ideal Q of R such that $C(R) \subseteq Q \subseteq P$, is a power of its radical. A valuation ring (R, P) has **rank n** if the rank of the value group G is n.

A few other pertinent results concerning valuation rings will appear as we discuss generalizations of Prufer domains and Krull domains. For further results regarding valuation theory in rings with zero divisors, see, for example, [D], [H],

[G5], and [PS].

Next, we briefly examine a few possible generalization of the property "R is a Prufer Domain" to rings with zero divisors. There are at least fourteen different characterizations of Prufer domains (see [G] and [FHP]) which may be generalized to rings with zero divisors; some of them may be generalized in several ways. There is an extensive literature exploring all generalizations available to date. We will bring up here three of the most popular generalizations. These happen to coincide with the three types of approaches mentioned in the introduction (to clarify this interpretation consider "all entities" to be "all finitely generated ideals", and see Proposition 3.11). It is instructive to see the three approaches "in action", and see how some of the zero divisor controlling conditions described in the previous sections bridge between the three generalizations. In all the three cases, we will generalize the following characterization of Prufer domains:

A domain D is a **Prufer domain** if every nonzero finitely generated ideal of D is invertible.

Generalizing Prufer domains using approach (a): A ring R is called **an arithmetical ring** if every finitely generated ideal of R is locally principal.

This kind of generalization satisfies some, but not all, of the equivalent conditions defining a Prufer domain.

PROPOSITION 4.2 [G5], [J] *Let R be a ring. The following conditions are equivalent:*

1. *R is an arithmetical ring.*
2. *The ideals of R_m are totally ordered by inclusion for each maximal ideal m.*
3. *The ideals of R form a distributive lattice, that is, for all ideals I, J and L of R, we have* $I + J \cap L = (I + J) \cap (I + L)$.
4. *For all ideals I and J, and any finitely generated ideal L of R, we have* $((I + J) : L) = (I : L) + (J : L)$.

Other interesting Prufer-like properties of arithmetical rings may be found in [J].

Generalizing Prufer domains using approach (b): A ring is called a **Prufer ring** if every finitely generated regular ideal is invertible.

This definition is due to Griffin [G5], and he prefers this generalization of a Prufer domain to other generalizations, as it seems to be the one whose relation with its total ring of quotients is similar to that of a Prufer domain to its field of quotients. In [G5], Griffin exhibits 15 equivalent conditions to the property of being a Prufer ring, among them the conditions in Proposition 4.2 restricted (at least partially) to regular ideals. I find this generalization of a Prufer domain to be somewhat unsatisfactory as the similarity with a Prufer domain breaks down on an important point. A valuation ring need not be a Prufer ring [BL], [H].

A Prufer valuation pair is closer to what one would expect a valuation pair to be, namely:

PROPOSITION 4.3 [H] *Let R be a ring, and let P be a prime ideal of R. The following conditions are equivalent:*
1. (R, P) is a Prufer valuation pair.
2. R is a Prufer ring and P is the unique regular maximal ideal of R.
3. R is a valuation ring and P is the unique regular maximal ideal of R.

If we add the Marot condition to the definition of a Prufer ring, we eliminate most difficulties.

THEOREM 4.4 [G5], [H] *Let R be a Marot ring. Then R is a valuation ring iff R is a Prufer valuation ring.*

Griffin [G5] showed that an arithmetical ring can be obtained from a Prufer ring R by imposing some zero divisor restricting conditions on Q(R).

PROPOSITION 4.5 [G5] *A ring R is arithmetical iff R is a Prufer ring and Q(R) satisfies that ideals of $Q(R)_P$ are totally ordered by inclusion for all maximal prime ideals of zero P in Q(R).*

Additional information about Prufer rings can be found in, for example, [H], [G5], [BL].

Generalizing Prufer domains using approach (c): R is a **semihereditary ring**, that is, every finitely generated ideal of R is projective.

Given that this condition implies, in particular, that principal ideals are projective, the zero divisor controlling condition imposed with this generalization is the PP condition. We will actually see that the zero divisor controlling condition of this generalization can be viewed in a different way.

THEOREM 4.6 [G1], [G5], [M1] *Let R be a ring. The following conditions are equivalent:*

1. *R is semihereditary.*
2. *R is coherent and w.gl.dim R ≤ 1.*
3. *Q(R) is Von Neumann regular and R_m is a valuation domain for every maximal ideal m of R.*
4. *R is a Prufer ring and Q(R) is Von Neumann regular.*

A semihereditary ring R shares the following property with a Prufer domain: R_P is a valuation domain for all prime ideals P of R.

Marot's investigation [M1] into the zero divisor controlling condition "Q(R) is Von Neumann regular" not only yielded the rich characterization of semihereditary rings of Theorem 4.6 (3), but also added an equally useful and interesting characterization of hereditary rings.

Theorem 4.7 [G1], [M1], [V] *Let R be a ring. Then R is hereditary iff Q(R) is hereditary and any ideal of R that is not contained in any minimal prime ideal of R is projective.*

Additional results about semihereditary rings and related homological conditions can be found in [G1] and [V].

We next examine some aspects of the generalization of the concept of being a Krull domain to rings with zero divisors. Let R be a ring and let Q(R) be the total ring of quotients of R. Assume R $\neq$ Q(R). (This is a technical condition making the statements of some theorems cleaner and the statements of other theorems messier. Some authors prefer to make this distinction, other prefer not to.) R is called **a Krull ring** if there exists a family $\{(V_\alpha, P_\alpha) \mid \alpha \in I\}$ of discrete rank one valuation pairs of Q(R) with associated valuations $\{v_\alpha \mid \alpha \in I\}$ such that:
1. $R = \cap \{V_\alpha \mid \alpha \in I\}$.
2. $v_\alpha(a) = 0$ for almost all α for each regular element $a \in Q(R)$, and each P_α is a regular ideal of V_α.

There are three definitions of a Krull ring with zero divisors in the literature. Our definition is by Kennedy [K4], and was also adopted by Kang [K], [K1], [K2]. Another definition is by Huckaba [H]. Huckaba's definition is identical with a Marot Krull ring in the Kennedy sense. The third definition is by Portelli and Spangler [PS]. Portelli and Spangler's definition can be shown to be equivalent to Huckaba's definition.

Perhaps here is the place to say that adding the Marot condition to the definition of a Krull ring brings this generalization closer to sharing many of a

Krull domain properties. For example, even the definition itself becomes smoother as the notion of a rank 1 discrete valuation ring is more manageable: A Marot valuation ring (V, P) with associated valuation v and value group G is a discrete rank one valuation ring (respectively, v is a discrete rank one valuation) if G is isomorphic to the group of integers. In this case, P is the unique regular prime ideal of V and there exists a regular element x of P such that P = (x). Marot Krull rings share many properties of Krull domains. [H] provides a particularly clear and detailed exposition of these results. [K], [AM], and [AM1] take up the notion of a UFD and extend it (in a number of ways) to rings with zero divisors. It is the Marot property that allows for the conclusion of Corollary 3.13. As a consequence, only under the additional assumption that a ring is both Marot and Krull do the authors get a relation between factoriality properties and Krull ring behavior reminiscent of the domain case. A good survey article on the extension of the UFD notion to rings with zero divisors is [A].

It is possible to define Krull rings without resorting to valuation rings.

Recall that a ring R is **completely integrally closed** if, for $0 \neq a$ and u in Q(R), $au^n \in R$ for all n implies $u \in R$.

Let I be a nonzero fractional ideal of R. I_t is defined to be $\Sigma\,(I_0^{-1})^{-1}$, where I_0 runs over the nonzero finitely generated R submodules contained in I. We say that I is **t invertible** if $(II^{-1})_t = R$.

THEOREM 4.8 [H], [K2], [M4] *The following conditions are equivalent for a ring R:*

1. *R is a Krull ring.*
2. *[M4] R is completely integrally closed and satisfies the ascending chain condition on divisorial ideals.*
3. *[K2] Every regular ideal of R is t-invertible.*
4. *[K2] Every regular prime ideal of R is t invertible.*
5. *[K2] Every regular prime ideal of R contains a t invertible regular prime ideal.*

All references mentioned above contain additional information on Krull rings.

A related topic is the investigation into the behavior of the integral closure of a ring with zero divisors. Tom Lucas and others have done fundamental work in this direction. Interested readers are referred to [L] for a good survey and additional references on this topic.

4.2 Finite Conductor Rings and G-GCD Rings

The next application of the zero divisor controlling conditions described in the previous sections involves the use of the PP condition in a recent generalization by Glaz [G3], [G4] of finite conductor properties to rings with zero divisors.

Let R be a ring. R is **a finite conductor ring** if (a) ∩ (b) and (0:c) are finitely generated ideals of R for all elements a, b, and c of R. R is a **quasi-coherent ring** if $(a_1) \cap \ldots \cap (a_n)$ and (0:c) are finitely generated for all elements $a_1, \ldots, a_n$ and c of R.

This definition [G3] extends the notion of finite conductor and quasi-coherence of domains by adding the zero divisor controlling condition: "(0:c) is a finitely generated ideal for all c". Though this is a much weaker condition than the PP condition, it does affect the domain-like behavior of some rings. For example:

PROPOSITION 4.9 [G3] *Let R be a ring with weak global dimension one. The following conditions are equivalent:*
1. R is a semihereditary ring.
2. R is a coherent ring.
3. (0:c) is a finitely generated ideal of R for every element c of R.

With these definitions finite conductor and quasi-coherent rings accept several equivalent domain-like characterizations [G3], [G4].

A particular case of a finite conductor ring is the recently defined G-GCD ring. **A G-GCD domain** is defined by the condition that intersections of two invertible ideals is an invertible ideal [AA]. Glaz [G3], [G4] generalized this condition to rings with zero divisors as follows:

A ring R is called **a G-GCD ring** if the following two conditions hold:
1. R is a PP ring.
2. The intersection of any two finitely generated flat ideals of R is a finitely generated flat ideal of R.

At first glance, it seems that one may replace condition 2 by other, similar conditions and obtain different generalizations of the G-GCD domain notion. But in fact the PP condition is powerful enough to make all these generalizations coincide:

THEOREM 4.10 [G3], [G4] *Let R be a ring. The following conditions are equivalent:*

1. *R is a G-GCD ring.*
2. *R is a PP ring and the intersection of any two principal (fractional) ideals of R is a finitely generated flat (fractional) ideal of R.*
3. *R is a PP ring and the intersection of any two finitely generated projective ideals of R is a finitely generated projective ideal of R.*
4. *R is a PP ring and the intersection of two invertible ideals of R is an invertible ideal of R.*

G-GCD rings are reduced rings which are locally GCD domains, they are integrally closed in their total ring of quotients, and they possess compact Min(R) and Von Neumann regular total ring of quotients. Coherent regular rings are G-GCD rings, but not all G-GCD rings are coherent [G3]. On the other hand, it was through this definition that coherent-like and regularity-like properties of polynomial rings over coherent rings were discovered. Namely:

THEOREM 4.11 [G3], [G4]

1. *Let R be an integrally closed coherent domain. Then the polynomial ring R[x] is a quasi-coherent domain.*
2. *Let R be a coherent regular ring. Then the polynomial ring R[x] is a G-GCD ring.*

For additional properties of finite conductor, quasi-coherent, and G-GCD rings see [G3], [G4]. A different definition of a G-GCD ring, and a study of this class of G-GCD rings, appear in Ali and Smith [AS].

4.3 Going Down and Related Rings

Another application of the zero divisor controlling conditions described in the previous sections is the use of the condition Nil(R) = Z(R) for the extension of several related domain conditions to rings with zero divisors.

A ring extension $R \subseteq T$ satisfies **Going Down (GD)** if given prime ideals $P \subset P_1$ in R, and Q_1 in T satisfying $Q_1 \cap R = P_1$, there is a prime ideal Q in T such that $Q \subset Q_1$ and $Q \cap R = P$. A domain R is called **a Going Down domain** in case the extension $R \subseteq T$ satisfies GD for each overring T of R. Dobbs [D1] extended the Going Down notion to rings with zero divisors as follows: A ring R is a Going Down ring if R/P is a Going Down domain for every prime ideal P of R.

Under the zero divisor controlling condition Nil(R) = Z(R), this notion becomes a natural generalization of the Going Down property for domains.

PROPOSITION 4.12 [D1] *Let R be a ring in which 0 is a primary ideal. Then R is a Going Down ring iff R $\subseteq$ T satisfies GD for each overring T of R.*

Examples are provided in [D1] that show that the assumption Nil(R) = Z(R) is necessary.

Two related notions in which the condition Nil(R) = Z(R) played a role are the notions of divided and locally divided rings introduced in [B] and [BD]. A ring R is **a divided ring** (respectively, **a locally divided ring**) if each prime ideal is comparable under inclusion with each ideal of R (respectively, if R_P is a divided ring for every prime ideal P of R). Divided rings are Going Down rings, though the converse is false even for domains [D1]. David Anderson, Badawi and Dobbs [BAD], [ABD] extended another domain notion to rings with zero divisors, namely the PVD notion. **PVDs** were first defined by Hedstrom and Houston in [HH] as those domains for which every prime ideal is strongly prime. In a domain D a prime ideal P is **strongly prime** if xy $\in$ P, for x and y in the field of quotients of D, implies x $\in$ P or y $\in$ P. This notion was extended to rings with zero divisors as follows: Let R be a ring. A prime ideal P of R is said to be **strongly prime** if aP and (b) are comparable for all a, b $\in$ R. A ring is called a **pseudo valuation ring (PVR)** if each prime ideal is strongly prime. It is interesting to note that if (R, m) is a local PVR, then Z(R) can be Nil(R), m, or any prime ideal properly in between [ABD]. A detailed analysis of these rings can be found in [B], [BD], [BAD], and [ABD].

REFERENCES

[A] D. D. Anderson, *Extension Of Unique Factorization: A Survey*, Marcel Dekker Lecture Notes in Pure Appl. Math. **205** (1999), 31 - 53

[AA] D. D. Anderson and D. F. Anderson, *Generalized GCD Domains*, Comm. Math. Univ. St. Pauli **XXVIII** (1979), 215 - 221

[ABD] D. F. Anderson, A. Badawi and D. E. Dobbs, *Pseudo Valuation Rings, II*, Bollettino U. M. I. **(8) 3-B** (2000), 535 - 545

[AM] D. D. Anderson and R. Markanda, *Unique Factorization Rings With Zero Divisors*, Houston J. Math. **11** (1985), 15 - 30

[AM1] D. D. Anderson and R. Markanda, *Unique Factorization Rings With Zero Divisors: Corrigendum*, Houston J. Math. **11** (1985), 423 - 426

[AS] M. M. Ali and D. J. Smith, *Generalized GCD Rings*, Beiträge zur

Algebra und Geometrie **42** (2001), 219 - 233

[B] A. Badawi, *On Divided Commutative Rings*, Comm. Algebra **27** (1999), 1465 - 1474

[BAD] A. Badawi, D. F. Anderson and D. E. Dobbs, *Pseudo Valuation Rings*, Marcel Dekker Lecture Notes in Pure Appl. Math. **185** (1997), 57 - 67

[BD] A. Badawi and D. E. Dobbs, *On Locally Divided Rings And Going Down Rings*, Comm. Alg., to appear

[BL] M. B. Boisen and M. D. Larsen, *Prufer And Valuation Rings With Zero Divisors*, Pacific J. Math. 40 (1972), 7 - 12

[D] E. D. Davis, *Overrings Of Commutative Rings II*, Transactions Amer. Math. Soc. **110** (1964), 196 -212

[D1] D. E. Dobbs, *Going Down Rings With Zero Divisors*, Houston J. Math. 23 (1997), 1 - 11

[E] S. Endo, *On Semihereditary Rings*, J. Math. Soc. Japan **13** (1961), 109 - 119

[E1] E. G. Evans, *Zero Divisors In Noetherian Like Rings*, Trans. American Math. Soc. **155** (1971), 505 - 512

[FHP] M. Fontana, J. A. Huckaba and I. J. Papick, *Prufer Domains*, Marcel Dekker (1997)

[G] R. Gilmer, *Multiplicative Ideal Theory*, Queen's Papers on Pure Appl. Math. **12** (1968)

[GH] R. Gilmer and J. A. Huckaba, *Δ Rings*, J. Algebra **28** (1974), 414 - 432

[GH1] R. Gilmer and W. Heinzer, *The Laskerian Property, Power Series Rings And Noetherian Spectra*, Proc. American Math. Soc. **79** (1980), 13 - 16

[G1] S. Glaz, *Commutative Coherent Rings*, Springer Verlag Lecture Notes in Math. # **1371,** (1989)

[G2] S. Glaz, *Commutative Coherent Rings: Historical Perspective And Current Developments*, Nieuw Arch. Wiskunde **10** (1992), 37 - 56

[G3] S. Glaz, *Finite Conductor Rings*, Proc. American Math. Soc. **129** (2001), 2833 - 2843

[G4] S. Glaz, *Finite Conductor Rings With Zero Divisors*, Kluwer Academic Publisher Math and Appl. # **520** (2000), 251 - 271

[G5] M. Griffin, *Prufer Rings With Zero Divisors*, J. Reine Angew. Math. **239/240** (1970), 55 - 67

[H] J. A. Huckaba, *Commutative Rings With Zero Divisors*, Marcel Dekker, Inc. (1988)

[H1] J. A. Huckaba, *On Valuation Rings That Contain Zero Divisors*, Proc. Amer. Math. Soc. **40** (1973), 9 - 15

[HH] J. R. Hedstrom and E. G. Houston, *Pseudo-Valuation Domains*, Pacific

J. Math. **75** (1978), 137 - 147

[HO] W. Heinzer and J. Ohm, *On The Noetherian Like Rings Of E. G. Evans*, Proc. American Math. Soc. **34** (1972), 73 - 74

[J] C. U. Jensen, *Arithmetical Rings*, Acta Math. Acad, Sci. Hungaricae **17** (1966), 115 - 123

[K] B. G. Kang, *A Characterization Of Krull Rings With Zero Divisors*, J. Pure Appl. Algebra **72** (1991), 33 - 38

[K1] B. G. Kang, *Integral Closure Of Rings With Zero Divisors*, J. Algebra **162** (1993), 317 - 323

[K2] B. G. Kang, *Characterizations Of Krull Rings With Zero Divisors*, J. Pure Appl. Algebra **146** (2000), 283 - 290

[K3] I. Kaplansky, *Commutative Rings*, Allyn and Bacon, Inc. (1970)

[K4] R. E. Kennedy, *Krull Rings*, Pacific J. Math. 89 (1980), 131 - 136

[L] T. G. Lucas, *Integrality Properties In Rings With Zero Divisors*, Marcel Dekker Lecture Notes Pure Appl. Math. **220** (2001), 265 - 280

[M] M. E. Manis, *Extension Of Valuation Theory*, Bull. Amer. Math. Soc. **73** (1967), 735 - 736

[M1] J. Marot, *Ph.D Thesis: Sur Les Anneaux Universeillement Japonais*, Univ. Paris-Sud (1977)

[M2] E. Matlis, *The Minimal Prime Spectrum Of A Reduced Ring*, Illinois J. Math. **27** (1983), 353 - 391

[M3] R. Matsuda, *On Marot Rings*, Proc. Japan Acad. **60** (1984), 134 - 137

[M4] R. Matsuda, *On Kennedy's Problem*, Comm. Math. Univ. Sancti Pauli **31** (1981), 13 - 24

[O] J. P. Olivier, *Anneaux Absolument Plats Universels Et Épimorphismes D'Anneaux*, C. R. Acad. Sci. Paris Ser. A **266** (1968), 317 - 318

[O1] J. P. Olivier, *Anneaux Absolument Plats Universels Et Épimorphismes Á Buts Reduits*, Sém. Samuel, Paris (1967 - 1968), 6.1 - 6.12

[PS] D. Portelli and W. Spangler, *Krull Rings With Zero Divisors*, Comm. Algebra **11** (1983), 1817 - 1851

[Q1] Y. Quentel, *Sur La Compacité Du Spectre Minimal D'un Anneau*, Bull. Soc. Math. France **99** (1971), 265 - 272

[Q2] Y. Quentel, *Erratum, Sur La Compacité Du Spectre Minimal D'un Anneau*, Bull. Soc. Math. France **100** (1972), 461

[V] W. V. Vasconcelos, *The Rings Of Dimension Two*, Marcel Dekker Lecture Notes in Pure Appl. Math # **22** (1976)

Weak Module Systems and Applications: A Multiplicative Theory of Integral Elements and the Marot Property

FRANZ HALTER-KOCH Institut für Mathematik,
Karl-Franzens-Universität, Heinrichstrasse 36, A-8010 Graz, Austria,
franz.halterkoch@kfunigraz.ac.at

Abstract

We introduce the concept of weak module systems for commutative monoids. This concept is a common generalization of the notion of a weak ideal system as presented in the author's book "Ideal Systems" (M. Dekker, 1998) and the notion of a module system as presented in the author's article "Localizing Systems, Module Systems, and Semistar Operations" (J. Algebra **238**, 2001). With the aid of this concept, we first develop a purely multiplicative theory of integral elements, valid for commutative monoids and rings without any cancellation assumptions. Next, we generalize the Marot property (used in the theory of rings with zero divisors) to monoids and shed new light on the theory of Dedekind and Prüfer monoids without cancellation.

1 INTRODUCTION AND NOTATIONS

We present the concept of a weak module system on a commutative monoid. This concept is a final step in a series of generalizations of the concepts of star and semistar operations in the theory of integral domains and that of Lorenzen's r-systems and Aubert's x-systems in the theory of commutative monoids. The basis of our considerations are the theory of ideal systems as presented in [3] and the theory of module systems as presented in [4]. We will, however, introduce all concepts from the very beginning, and only in proofs or for motivations and examples we will refer the reader to the above-mentioned sources. In section 2 and section 3 we present the formalism. In section 4 we apply our concept to give a purely multiplicative theory of integral elements, valid for (commutative) monoids and rings. It turns out that the fundamental properties of integrality can be derived using an appropriate weak module system. In section 5 we introduce the Marot property for weak module systems to obtain a close connection between the ideal theory of a monoid and that of the submonoid of its cancellative elements. Indeed, the Marot property has its origin in the theory of rings with zero divisors, where it serves for a similar purpose (see [5]). In the case of ideal systems, we use this connection to obtain a characterization of Dedekind and Prüfer monoids.

Throughout this paper, a *monoid* D means a multiplicative commutative semigroup containing a unit element $1 \in D$ (such that $1a = a$ for all $a \in D$) and a zero element $0 \in D$ (such that $0a = 0$ for all $a \in D$). For any subsets $X, Y \subset D$ and $n \in \mathbb{N}$, we set

$$XY = \{xy \mid x \in X,\, y \in Y\}, \quad (X : Y) = (X :_D Y) = \{z \in D \mid zY \subset X\}$$

and

$$X^n = X \cdot \ldots \cdot X = \{x_1 \cdot \ldots \cdot x_n \mid x_\nu \in X\}.$$

For a subset $X \subset D$ and $y \in D$, we set $yX = \{y\}X$ and $(X : y) = (X : \{y\})$. A subset $S \subset D$ is called *multiplicatively closed* if $1 \in S$ and $SS = S$. By a *submonoid* of D we mean a multiplicatively closed subset $H \subset D$ such that $0 \in H$. For any subset $X \subset D$, we denote by $[X]$ the smallest submonoid of D containing X. An element $a \in D$ is called *cancellative* or *regular* if, for all $b, c \in D$, $ab = ac$ implies $b = c$. We denote by $D^\bullet$ the (multiplicatively closed) subset of all cancellative elements of D and by $D^\times$ the group of all invertible elements of D. A monoid D is called *cancellative* if $D^\bullet = D \setminus \{0\}$, and D is called a *groupoid* if $D^\times = D \setminus \{0\}$.

Every ring D is (disregarding the additive structure) a monoid. Our main reference for rings is [2].

For a set X, we denote by $\mathbb{P}(X)$ the power set of X and by $\mathbb{P}_f(X)$ the set of all finite subsets of X.

Throughout this paper, let D be a monoid.

2 WEAK MODULE SYSTEMS: DEFINITION AND ELEMENTARY PROPERTIES

DEFINITION 2.1. A *weak module system* on D is a map

$$r : \begin{cases} \mathbb{P}(D) & \to \mathbb{P}(D) \\ X & \mapsto X_r \end{cases}$$

such that the following properties are fulfilled for all $X, Y \in \mathbb{P}(D)$ and $c \in D$,

(M1) $X \cup \{0\} \subset X_r$.
(M2) $X \subset Y_r$ implies $X_r \subset Y_r$.
(M3) $cX_r \subset (cX)_r$.

A weak module system r on D is called a *weak ideal system on* D if

(IS) $\{1\}_r = D$.

Let r be a weak module system on D. An element $c \in D$ is called *regular for* r, if

$$(cX)_r = cX_r \quad \text{for all subsets} \quad X \subset D.$$

We denote by $\mathrm{Reg}(r)$ the set of all regular elements for r. A weak module system r is called a *module system* if $\mathrm{Reg}(r) = D$, and it is called an *ideal system on* D if it is a module system and a weak ideal system.

Let r be a weak module system on D. By an *r-module* we mean a subset $J \subset D$ satisfying $J_r = J$, and we denote by $\mathcal{M}_r = \mathcal{M}_r(D)$ the set of all r-modules. An r-module J is called *r-finitely generated* if $J = E_r$ for some $E \in \mathbb{P}_{\mathrm{f}}(D)$. We denote by $\mathcal{M}_{r,\mathrm{f}} = \mathcal{M}_{r,\mathrm{f}}(D)$ the set of all r-finitely generated r-modules. By an *r-monoid* we mean an r-module which is a submonoid of D.

Observe that our concept of a module system coincides with that presented in [4] for cancellative monoids, and that our concept of a weak ideal system coincides with that presented in [3]. We refer the reader to the examples discussed there. Before we present further examples, we gather the most important elementary properties of weak module systems and develop some related concepts. The properties listed in the following proposition will be used freely in the sequel.

PROPOSITION 2.2. *Let r be a weak module system on D, and let X, Y be subsets of D.*

1. $\emptyset_r = \{0\}_r = \{0\}$.

2. $X \subset Y$ *implies* $X_r \subset Y_r$.

3. $(X_r)_r = X_r = (X \setminus \{0\})_r = (X \cup \{0\})_r$. *In particular, X_r is an r-module.*

4. $(XY)_r = (XY_r)_r = (X_rY_r)_r$.

5. *For any family $(X_\alpha)_{\alpha\in A}$ of subsets of D, we have*

$$\bigcup_{\alpha\in A}(X_\alpha)_r \subset \Big(\bigcup_{\alpha\in A} X_\alpha\Big)_r = \Big(\bigcup_{\alpha\in A}(X_\alpha)_r\Big)_r.$$

6. *The intersection of any family of r-modules is again an r-module.*

7. *We have*

$$X_r = \bigcap_{\substack{J\in\mathcal{M}_r \\ J\supset X}} J,$$

and X_r is the smallest r-module containing X.

8. $(X : Y) \subset (X_r : Y) = (X_r : Y_r)$, *and if J is an r-module, then $(J : Y)$ is also an r-module.*

9. $D^\times \cup \{0\} \subset \mathrm{Reg}(r)$, *and for all $a, b \in D$, we have $ab \in \mathrm{Reg}(r)$. If $a \in D^\bullet$, $b \in D$ and $ab \in \mathrm{Reg}(r)$, then $b \in \mathrm{Reg}(r)$.*

Proof. The proofs of 1. to 8. are literally the same as those for the corresponding properties of weak ideal systems (see [3], Propositions 2.1, 2.3 and 2.4).

9. We clearly have $\{0,1\} \subset \mathrm{Reg}(r)$, and $\mathrm{Reg}(r)$ is multiplicatively closed. Therefore it is sufficent to prove that, for all $a \in D^\bullet$ and $b \in D$, $ab \in \mathrm{Reg}(r)$ implies $b \in \mathrm{Reg}(r)$. If $ab \in \mathrm{Reg}(r)$ and $X \subset D$, then $(abX)_r \supset a(bX)_r \supset abX_r = (abX)_r$. Therefore equality holds. Since $a \in D^\bullet$, we obtain $(bX)_r = bX_r$, and consequently $b \in \mathrm{Reg}(r)$. □

COROLLARY 2.3. *Every weak module system on a groupoid is a module system.*

Proof. Obvious by Proposition 2.2.9. □

DEFINITION 2.4. Let r be a weak module system on D. For $I, J \in \mathcal{M}_r$, we call

$$I \cdot_r J = (IJ)_r$$

the *r-product* of I and J. The composition $\cdot_r$ is called *r-multiplication* .

COROLLARY 2.5. *Let r be a weak module system on D.*

1. *For any family* $(X_\alpha)_{\alpha\in A}$ *of subsets of* D *and* $Y \subset D$, *we have*

$$\bigcup_{\alpha\in A} \big((X_\alpha)_r \cdot_r Y_r\big) = \Big(\bigcup_{\alpha\in A} X_\alpha\Big)_r \cdot_r Y_r\,.$$

2. *If* $H \subset D$ *is a submonoid, then* H_r *is an* r*-monoid. In particular,* $\{1\}_r = \{0,1\}_r$ *is the smallest* r*-monoid in* D.

3. $(\mathcal{M}_r, \cdot_r)$ *is a monoid with unit element* $\{1\}_r$ *and zero element* $\{0\}$. $\mathcal{M}_{r,\mathrm{f}}$ *is a submonoid of* $\mathcal{M}_r$.

Proof. 1. is proved in precisely the same way as [3], Proposition 2.3, and 2. and 3. are proved as [4], Corollary 1.4. □

DEFINITION 2.6. Let r be a weak module system on D and $H \subset D$ a submonoid. For $X \subset D$, we define

$$X_{r[H]} = (XH)_r\,,$$

and we call $r[H] : \mathbb{P}(D) \to \mathbb{P}(D)$ the *extension* of r with H.

PROPOSITION 2.7. *Let* r *be a weak module system on* D *and* $H \subset D$ *a submonoid.*

1. $r[H]$ *is a weak module sytem on* D, $r[H] = r[H_r]$ *and* $\{1\}_{r[H]} = H_r$.

2. $\mathcal{M}_{r[H]} = \{J \in \mathcal{M}_r \mid JH_r = J\}$.

3. $r[H] = r$ *holds if and only if* $H \subset \{1\}_r$, *and in this case* $H_r = \{1\}_r$.

4. *If* r *is a module system, then so is* $r[H]$.

5. *If* H *is an* r*-monoid, then* r_H, *the restriction of* $r[H]$ *to* $\mathbb{P}(H)$, *is a weak ideal system on* H. *In particular, if* r *is a module system, the* r_H *is an ideal system.*

6. *If* $H_1, H_2 \subset D$ *are submonoids, then* $r[H_1H_1] = r[H_1][H_2]$.

Proof. The proofs of 1., 2. and 3. are literally the same as those of [4], Proposition 1.6, and 4., 5. and 6. follow immediately from the definitions. □

EXAMPLES 2.8 (Remarks and Conventions). 1. Let r be a weak module system on D and $H \subset D$ an r-monoid. We shall usually write $r[H]$ or simply r instead of r_H to denote the weak ideal system induced by r on H. For weak ideal systems, we use the terminology introduced in [3]. In particular, we denote by

$$\mathcal{I}_r(H) = \mathcal{M}_{r[H]} \cap \mathbb{P}(H) = \{I \in \mathcal{M}_r \mid I \subset H \text{ and } IH = I\}$$

the set of all r-ideals of H, by $\mathcal{I}_{r,\mathrm{f}}(H) = \mathcal{M}_{r[H],\mathrm{f}} \cap \mathbb{P}(H)$ the set of all r-finitely generated r-ideals of H, by r-spec(H) the set of all prime r-ideals and by r-max(H) the set of all r-maximal r-ideals of H.

In particular, if r is a weak ideal system on D, then D is the only r-submonoid of D, and $r[D] = r$.

2. The *trivial module system* s on D is defined by

$$X_s = X \cup \{0\} \quad \text{for} \quad X \subset D\,.$$

It is easily checked that s is indeed a module system. The s-monoids in D are just the submonoids of D. If $H \subset D$ is a submonoid, then the ideal system $s[H]$ coincides with the ideal system $s(H)$ of ordinary semigroup ideals introduced in [3], 2.2.

3. Let D be a ring. For $X \subset D$, we denote by $X_d \subset D$ the additive group generated by X. Then it is easily checked that $d : \mathbb{P}(D) \to \mathbb{P}(D)$ is a module system on D, and we call it the *additive system* on D. The d-modules in D are just the additive subgroups of D, and the d-monoids are the subrings of D. In particular, $\{1\}_d$ is the prime ring of D. For any subring $H \subset D$, the ideal system $d[H]$ coincides with the ideal system $d(H)$ of ordinary ring ideals as considered in [3], 2.2. More generally, $\mathcal{M}_{r[H]}$ consists of all H-submodules of D.

4. Let D be a topological monoid. For $X \subset D$, we set $X_\tau = \overline{[X]}$ (the smallest closed submonoid of D containing X). Then $\tau : \mathbb{P}(D) \to \mathbb{P}(D)$ is a weak module system which (in general) is not a module system.

DEFINITION 2.9. Let q and r be weak module systems on D. We say that q is *finer* than r, or r is *coarser* than q, and we write $q \le r$, if $\mathcal{M}_r \subset \mathcal{M}_q$.

PROPOSITION 2.10. *Let q and r be weak module systems on D, and let $H \subset D$ be a submonoid.*

1. *$s \le r \le r[H]$ (where s is the trivial module system on D).*

2. *If $q \le r$, then $q[H] \le r[H]$.*

3. Then the following assertions are equivalent:

(a) $q \leq r$.

(b) For all $X \subset D$, *we have* $X_q \subset X_r$.

(c) For all $X \subset D$, *we have* $X_r = (X_q)_r$.

Proof. 1. and 2. follow immediately from the definitions, and the proof of 3. is literally the same as that for the corresponding statement concerning weak ideal sytems (see [3], Proposition 5.1). □

DEFINITION 2.11. A weak module system r on D is called *finitary* if, for all subsets $X \subset D$,

$$X_r = \bigcup_{E \in \mathbb{P}_{\mathrm{f}}(X)} E_r .$$

EXAMPLES 2.12. 1. The trivial module system s on D is finitary.

2. If D is a ring, then the additive system d on D is finitary.

3. Let r be a weak module system on D, and define $r_{\mathrm{f}} : \mathbb{P}(D) \to \mathbb{P}(D)$ by

$$X_{r_{\mathrm{f}}} = \bigcup_{E \in \mathbb{P}_{\mathrm{f}}(X)} E_r .$$

Then r_{f} is a finitary weak module system on D, and if r is a module system (a weak ideal system), then so is r_{f}. Moreover, $\mathcal{M}_{r,\mathrm{f}} = \mathcal{M}_{r_{\mathrm{f}},\mathrm{f}}$ and for all $E \in \mathbb{P}_{\mathrm{f}}$ we have $E_r = E_{r_{\mathrm{f}}}$. These assertions are proved as for finitary ideal systems (see [3], Proposition 3.1).

PROPOSITION 2.13. *Let* r *be a weak module system on* D.

1. The following assertions are equivalent:

(a) r *is finitary.*

(b) For every subset $X \subset D$, *we have*

$$X_r \subset \bigcup_{E \in \mathbb{P}_{\mathrm{f}}(X)} E_r .$$

(c) For every directed family $(X_\alpha)_{\alpha \in A}$ *of subsets of* D, *we have*

$$\Big(\bigcup_{\alpha \in A} X_\alpha\Big)_r = \bigcup_{\alpha \in A} (X_\alpha)_r$$

(d) *The union of every directed family of* r*-modules is an* r*-module.*

(e) *For all* $X \subset D$ *and* $J \in \mathcal{M}_{r,\mathrm{f}}$ *such that* $J \subset X_r$*, there exists some* $E \in \mathbb{P}_\mathrm{f}(X)$ *such that* $J \subset E_r$*.*

2. *Let* r *be finitary,* $X \subset D$ *and* $X_r \in \mathcal{M}_{r,\mathrm{f}}$*. Then there exists some* $E \in \mathbb{P}_\mathrm{f}(X)$ *such that* $E_r = X_r$*.*

3. *If* r *is finitary and* $H \subset D$ *is a submonoid, then* $r[H]$ *is also finitary.*

4. *For any finitary weak module system* q *on* D*, the following assertions are equivalent:*

   (a) $q \leq r$.

   (b) *For all finite subsets* $E \subset D$*, we have* $E_q \subset E_r$*.*

   (c) $\mathcal{M}_{r,\mathrm{f}} \subset \mathcal{M}_q$*.*

Proof. 1., 2. and 3. are proved as for module systems or weak ideal systems (see [3], Proposition 3.1 and [4], Proposition 2.3).

4. is proved as the corresponding statement for weak ideal systems (see [3], Proposition 5.1). □

3 QUOTIENT MONOIDS AND THEIR MODULE SYSTEMS

For a multiplicatively closed subset $S \subset D^\times$ and $X \subset D$, we set

$$S^{-1}X = \{s^{-1}x \mid s \in S\,, x \in X\}.$$

Observe that $1 \in S$ implies $X \subset S^{-1}X$, and if $X \subset D$ is a submonoid, then so is $S^{-1}X$.

Let $H \subset D$ be a submonoid and $S \subset D^\times \cap H$ a multiplicatively closed subset. Then $S^{-1}H$ is called the *quotient monoid of* H *with respect to* S. The set

$$\overline{S} = \{z \in H \mid zu \in S \text{ for some } u \in H\}$$

is called the *saturation of* S *in* H. It is easily seen that $\overline{S} \subset D^\times \cap H$ is again a multiplicatively closed subset, $S^{-1}H = \overline{S}^{-1}H$, $(S^{-1}H)^\times = \overline{S}^{-1}\overline{S}$, $\overline{S} = (S^{-1}H)^\times \cap S$, and if $S_1 \subset D^\times \cap H$ is another multiplicatively closed subset, then $S_1^{-1}H = S^{-1}H$ if and only if $\overline{S} = \overline{S_1}$. D is called a *total quotient monoid of* H if $H^\bullet \subset D^\times$ and $D = H^{\bullet -1}H$. Every monoid possesses a total quotient monoid, which is uniquely determined up to (canonical) isomorphisms. If D is cancellative, then its total quotient monoid is a groupoid and will be called the *quotient groupoid of* D. The total quotient monoid of a ring is its total quotient ring, disregarding the additive structure.

THEOREM 3.1 **(Construction of weak module systems)**. *Let $H \subset D$ be a submonoid and $S \subset D^\times \cap H$ a multiplicatively closed subset such that $D = S^{-1}H$. Let $r : \mathbb{P}_f(H) \to \mathbb{P}(D)$ be a map such that the conditions* **(M1)**, **(M2)** *and* **(M3)** *of Definition 2.1 are satisfied for all $X, Y \in \mathbb{P}_f(H)$ and all $c \in H$. Assume moreover that $(cX)_r = cX_r$ for all $X \in \mathbb{P}_f(H)$ and all $c \in S$. Then there exists a unique finitary weak module system $\overline{r}$ on D such that $\overline{r} \mid \mathbb{P}_f(H) = r$. If $E \in \mathbb{P}_f(D)$ and $c \in S$ is such that $cE \subset H$, then*

$$E_{\overline{r}} = c^{-1}(cE)_r\,, \tag{$*$}$$

and for an arbitrary subset $X \subset D$, we have

$$X_{\overline{r}} = \bigcup_{E \in \mathbb{P}_f(X)} E_{\overline{r}}\,. \tag{$**$}$$

Proof. If $\overline{r}$ is a finitary module system such that $\overline{r} \mid \mathbb{P}_f(H) = r$, then $(*)$ and $(**)$ hold, since $S \subset \operatorname{Reg}(\overline{r})$, and thus $\overline{r}$ is uniquely determined by r.

To prove existence, we first define $\widetilde{r} : \mathbb{P}_f(D) \to \mathbb{P}(D)$ as follows: If $E \in \mathbb{P}_f(D)$ and $c \in S$ is such that $cE \subset H$, we set $E_{\widetilde{r}} = c^{-1}(cE)_r$. Since $(cX)_r = cX_r$ for all $c \in S$ and $X \in \mathbb{P}_f(H)$, this definition is independent of the choice of c. Moreover, **(M1)**, **(M2)** and **(M3)** are satisfied for $\widetilde{r}$ and all $X, Y \in \mathbb{P}_f(D)$ and $c \in D$. Now we define $\overline{r} : \mathbb{P}(D) \to \mathbb{P}(D)$ by

$$X_{\overline{r}} = \bigcup_{E \in \mathbb{P}_f(X)} E_{\widetilde{r}}\,.$$

Then we have $\overline{r} \mid \mathbb{P}_f(D) = \widetilde{r}$ and hence $\overline{r} \mid \mathbb{P}_f(H) = r$. Therefore it remains to prove that $\overline{r}$ is a weak module system. This is done in essentially the same way as in the proof of [3], Proposition 3.3. □

COROLLARY 3.2. *Let $H \subset D$ be a submonoid and $S \subset D^\times \cap H$ a multiplicatively closed subset such that $D = S^{-1}H$.*

1. *Let q and r be finitary weak module systems on D. Then $q \le r$ if and only if $E_q \subset E_r$ for all $E \in \mathbb{P}_f(H)$.*

2. *Let $\mathfrak{X}$ be the set of all finitary weak module systems r on D such that $\{1\}_r = H$, and let $\mathfrak{Y}$ be the set of all weak ideal systems r_0 on H such that $S \subset \operatorname{Reg}(r_0)$. Then*

$$\begin{cases} \mathfrak{X} & \to \quad \mathfrak{Y} \\ r & \mapsto \quad r \mid H \end{cases}$$

is a bijective map.

Proof. Obvious by the Theorem. □

CONVENTION 3.3. Let $H \subset D$ be a submonoid and $S \subset D^\times \cap H$ a multiplicatively closed subset such that $D = S^{-1}H$. According to Corollary 3.2 we will henceforth not distinguish between finitary weak module systems r on D such that $\{1\}_r = H$ and weak ideal systems r_0 on H such that $S \subset \mathrm{Reg}(r_0)$.

DEFINITION 3.4. Let r be a weak module system on D, $H = \{1\}_r$ and $S \subset D^\times \cap H$ a multiplicatively closed subset. Then the weak module system

$$S^{-1}r = r[S^{-1}H]$$

is called the *quotient system of r with respect to S.*

If $P \in s\text{-spec}(H)$, then $H \setminus S$ is a multiplicatively closed subset of H. If, moreover, $H \setminus S \subset D^\times$, then we set

$$r_P = (H \setminus P)^{-1}r \quad \text{and} \quad X_P = (H \setminus P)^{-1}X \quad \text{for every subset} \quad X \subset D.$$

We call r_P the *localization of r with respect to P*, and for any subset $X \subset D$, we call X_P the *localization of X with respect to P.*

PROPOSITION 3.5. *Let r be a weak module system on D, $H = \{1\}_r$ and $S \subset D^\times \cap H$ a multiplicatively closed subset.*

1. *For every r-monoid $M \subset D$, we have*

$$S^{-1}\big(r[M]\big) = (S^{-1}r)[M] = (S^{-1}r)[S^{-1}M] = r[S^{-1}M].$$

2. *For every subset $X \subset D$, we have*

$$(S^{-1}X)_{S^{-1}r} = (S^{-1}X)_r \supset S^{-1}X_r,$$

and if r is finitary, then even

$$(S^{-1}X)_r = S^{-1}X_r.$$

3. *If r is finitary, then $J \in \mathcal{M}_r$ implies $S^{-1}J \in \mathcal{M}_{S^{-1}r}$. In particular, $S^{-1}H$ is an $S^{-1}r$-monoid and hence an r-monoid.*

Proof. 1. This follows from Proposition 2.7.6, since $S^{-1}D = (S^{-1}H)D$.

2. By definition,

$$(S^{-1}X)_{S^{-1}r} = \big((S^{-1}X)(S^{-1}H)\big)_r = \big(S^{-1}X\{1\}_r\big)_r = (S^{-1}X)_r\,.$$

If $z \in S^{-1}X_r$, then there exists some $s \in S$ such that $sz \in X_r \subset (S^{-1}X)_r$, and hence $z \in s^{-1}(S^{-1}X)_r = (s^{-1}S^{-1}X)_r = (S^{-1}X)_r$.

Let now r be finitary and $E \subset X$ a finite subset. Then there exists some $s \in S$ such that $sE \subset SX \subset \{1\}_r X \subset X_r$, hence $sE_r \subset X_r$ and $E_r \subset s^{-1}X_r \subset S^{-1}X_r$. Thus we obtain

$$(S^{-1}X)_r = \bigcup_{E \in \mathbb{P}_f(S^{-1}X)} E_r \subset S^{-1}X_r\,.$$

3. This follows immediately from 2. above. □

COROLLARY 3.6. *Let r be a finitary weak module system on D, $H = \{1\}_r$ and $S \subset D^\times \cap H$ a multiplicatively closed subset. Then the map*

$$\iota : \begin{cases} \mathcal{M}_r & \to \ \mathcal{M}_{S^{-1}r} \\ J & \mapsto \ S^{-1}J \end{cases}$$

is a surjective monoid homomorphism (with respect to $\cdot_r$ and $\cdot_{S^{-1}r}$), $\iota \mid \mathcal{M}_{S^{-1}r} = \mathrm{id}$, and $\iota(\mathcal{M}_{r,\mathrm{f}}) = \mathcal{M}_{S^{-1}r,\mathrm{f}}$.

Proof. We use Proposition 3.5. If $J \in \mathcal{M}_r$, then $S^{-1}J \in \mathcal{M}_{S^{-1}r}$, and if $J \in \mathcal{M}_{r,\mathrm{f}}$, say $J = E_r$ for some $E \in \mathbb{P}_f(D)$, then $S^{-1}J = E_{S^{-1}r} \in \mathcal{M}_{S^{-1}r,\mathrm{f}}$. If $I \in \mathcal{M}_{S^{-1}r} \subset \mathcal{M}_r$, then $I = I_{S^{-1}r} = S^{-1}I_r = \iota(I)$. If $I, J \in \mathcal{M}_r$, then $S^{-1}(I \cdot_r J) = S^{-1}(IJ)_r = (S^{-1}IJ)_r = \big((S^{-1}I)(S^{-1}J)\big)_{S^{-1}r} = S^{-1}I \cdot_{S^{-1}r} S^{-1}J$. □

THEOREM 3.7. *Let r be a finitary weak module system on D, $H = \{1\}_r$, $H^\bullet \subset D^\times$ and $X \subset D$. Then we have*

$$X_r = \bigcap_{P \in r\text{-}\max(H)} X_{r_P}\,.$$

Proof. By definition, we have $X_r \subset (X_r)_P = X_{r_P}$ for all $P \in r\text{-}\max(H)$. Suppose now that $z \in X_{r_P}$ for all $P \in r\text{-}\max(H)$. Then $J = (X_r : z) \cap H \in \mathcal{I}_r(H)$, and it is sufficient to prove that $1 \in J$. If $1 \notin J$, then there exists some $P \in r\text{-}\max(H)$ such that $J \subset P$ (see [3], Theorem 6.4). Since $z \in (X_r)_P$, there exists some $s \in H \setminus P$ such that $sz \in X_r$ and hence $s \in J$, a contradiction. □

4 CANCELLATION PROPERTIES AND INTEGRAL ELEMENTS

PROPOSITION 4.1. *Let r be a weak module system on D.*

1. *For $J \in \mathcal{M}_{r,\mathrm{f}}$, the following assertions are equivalent:*

   (a) $J \in \mathcal{M}_{r,\mathrm{f}}^{\bullet}$.

   (b) For all $I, I' \in \mathcal{M}_{r,\mathrm{f}}$, $J \cdot_r I' \subset J \cdot_r I$ implies $I' \subset I$.

   (c) For all $I \in \mathcal{M}_{r,\mathrm{f}}$ and all $c \in D$, $cJ \subset J \cdot_r I$ implies $c \in I$.

   (d) For all $I \in \mathcal{M}_{r,\mathrm{f}}$, we have $(J \cdot_r I : J) \subset I$.

2. *If $J \in \mathcal{M}_{r,\mathrm{f}}^{\bullet}$, then $(J : J) \subset \{1\}_r$.*

Proof. 1. (a) $\Longrightarrow$ (b). If $J \cdot_r I' \subset J \cdot_r I$, then $J \cdot_r I = \big((J \cdot_r I) \cup (J \cdot_r I')\big)_r = J \cdot_r (I \cup I')_r$ and therefore $I = (I \cup I')_r \supset I'$.

(b) $\Longrightarrow$ (c). If $cJ \subset J \cdot_r I$, then $\{c\}_r \cdot_r J = (cJ)_r \subset J \cdot_r I$, and therefore $c \in \{c\}_r \subset I$.

(c) $\Longrightarrow$ (d). Obvious.

(d) $\Longrightarrow$ (a). Let $I, I' \in \mathcal{M}_{r,\mathrm{f}}$ be such that $J \cdot_r I = J \cdot_r I'$ and $x \in I'$. Then $Jx \in J \cdot_r I' = J \cdot_r I$ implies $x \in (J \cdot_r I : J) \subset I$. Thus we obtain $I' \subset I$, and hence equality holds by symmetry.

2. Apply 1. (d) with $I = \{1\}_r$. □

DEFINITION 4.2. A weak module system r on D is called *finitely cancellative* if, for every $J \in \mathcal{M}_{r,\mathrm{f}}$, $J \cap D^{\bullet} \neq \emptyset$ implies $J \in \mathcal{M}_{r,\mathrm{f}}^{\bullet}$.

COROLLARY 4.3. *Let r be a weak module system on D. Then r is finitely cancellative if and only if $\big((EF)_r : E\big) \subset F_r$ for all $E, F \in \mathbb{P}_{\mathrm{f}}(D)$ such that $E \cap D^{\bullet} \neq \emptyset$.*

Proof. Obvious by condition (d) of Proposition 4.1.1. □

DEFINITION 4.4. Let r be a weak module system on D. Then we define a map $\overline{r} : \mathbb{P}(D) \to \mathbb{P}(D)$ by

$$X_{\overline{r}} = \bigcup_{\substack{B \in \mathbb{P}_{\mathrm{f}}(D) \\ B \cap D^{\bullet} \neq \emptyset}} \big((XB)_r : B\big),$$

and we call $\overline{r}$ the *cancellative extension of r*.

REMARK 4.5. Let r be a weak module system on D and r_{f} the finitary weak module system defined in Example 2.12. Then we have $\overline{r} = \overline{r_{\mathrm{f}}}$, and therefore it suffices to study $\overline{r}$ for weak module systems.

THEOREM 4.6. *Let r be a finitary weak module system on D.*

1. *$\overline{r}$ is a finitary weak module system on D, and $r \le \overline{r}$. If r is a weak ideal system, then so is $\overline{r}$.*

2. *$\overline{r}$ is cancellative. If r' is another cancellative finitary weak module system on D such that $r \le r'$, then $\overline{r} \le r'$. In particular, r is cancellative if and only if $\overline{r} = r$, and $\overline{\overline{r}} = \overline{r}$.*

3. *If $H \subset D$ is a submonoid, then $\overline{r}[H] = \overline{r[H]}$. In particular, if $S \subset \{1\}_r \cap D^\times$ is a multiplicatively closed subset, then $S^{-1}\overline{r} = \overline{S^{-1}r}$.*

Proof. 1. We check the properties **(M1)**, **(M2)**, **(M3)** for all $X, Y \subset D$ and all $c \in D$.

(M1) is obvious.

(M2) Suppose that $X \subset Y_{\overline{r}}$ and $x \in X_{\overline{r}}$. Then there exists some $B \in \mathbb{P}_{\mathrm{f}}(D)$ such that $B \cap D^\bullet \neq \emptyset$ and $xB \subset (XB)_r$. Since $(XB)_r \subset (Y_r B)_r = (YB)_r$, we obtain $x \in \big((YB)_r : B\big) \subset Y_{\overline{r}}$.

(M3) We have

$$cX_{\overline{r}} = \bigcup_{\substack{B\in\mathbb{P}_{\mathrm{f}}(D)\\ B\cap D^\bullet\neq\emptyset}} c\big((XB)_r : B\big) \subset \bigcup_{\substack{B\in\mathbb{P}_{\mathrm{f}}(D)\\ B\cap D^\bullet\neq\emptyset}} \big(c(XB)_r : B\big)$$
$$\subset \bigcup_{\substack{B\in\mathbb{P}_{\mathrm{f}}(D)\\ B\cap D^\bullet\neq\emptyset}} \big((cXB)_r : B\big) = (cX)_{\overline{r}}\,.$$

For any subsets $X, B \subset D$ we have $X_r \subset \big((XB)_r : B\big)$, and therefore $X_r \subset X_{\overline{r}}$. In particular, $\{1\}_r \subset \{1\}_{\overline{r}}$. Hence $r \le \overline{r}$, and if r is a weak ideal system, then so is $\overline{r}$.

2. We use Corollary 4.3. If $E, F \in \mathbb{P}_{\mathrm{f}}(D)$ and $E \cap D^\bullet \neq \emptyset$, then

$$\big((EF)_{\overline{r}} : E\big) = \Big(\bigcup_{\substack{B\in\mathbb{P}_{\mathrm{f}}(D)\\ B\cap D^\bullet\neq\emptyset}} \big((EFB)_r : B\big) : E\Big)$$
$$= \bigcup_{\substack{B\in\mathbb{P}_{\mathrm{f}}(D)\\ B\cap D^\bullet\neq\emptyset}} \big((EFB)_r : EB\big) \subset F_{\overline{r}}\,,$$

since $B \cap D^\bullet \neq \emptyset$ implies $BE \cap D^\bullet \neq \emptyset$.

Let now r' be another cancellative finitary weak module system on D such that $r \le r'$. For $X \in \mathbb{P}_{\mathrm{f}}(D)$ we obtain

$$X_{\overline{r}} = \bigcup_{\substack{B \in \mathbb{P}_{\mathrm{f}}(D) \\ B \cap D^\bullet \neq \emptyset}} ((XB)_r : B) \subset \bigcup_{\substack{B \in \mathbb{P}_{\mathrm{f}}(D) \\ B \cap D^\bullet \neq \emptyset}} ((XB)_{r'} : B) \subset X_{r'},$$

again by Corollary 4.3, and hence $\overline{r} \le r'$ by Proposition 2.13.

3. For $X \subset D$, we obtain

$$X_{\overline{r}[H]} = (XH)_{\overline{r}} = \bigcup_{\substack{B \in \mathbb{P}_{\mathrm{f}}(D) \\ B \cap D^\bullet \neq \emptyset}} ((XHB)_r : B) = \bigcup_{\substack{B \in \mathbb{P}_{\mathrm{f}}(D) \\ B \cap D^\bullet \neq \emptyset}} ((XB)_{r[H]} : B) = X_{\overline{r[H]}},$$

and therefore $\overline{r}[H] = \overline{r[H]}$. □

DEFINITION 4.7. Let r be a finitary weak module system on D, and let $H \subset M \subset D$ be submonoids. An element $x \in D$ is called *r-integral over H* if $x \in H_{\overline{r}}$. A subset $X \subset D$ is called *r-integral over H* if $X \subset H_{\overline{r}}$. Moreover,

$$\mathrm{cl}_r^M(H) = H_{\overline{r}} \cap M$$

is called the *r-(integral) closure of H in M*. H is called *r-(integrally) closed in M*, if $\mathrm{cl}_r^M(H) = H$.

REMARKS 4.8. 1. Let r be a finitary weak module system on D, $H \subset D$ a submonoid and $x \in D$. Then x is r-integral over D if and only if

$$xE \subset (EH)_r = E_{r[H]} \quad \text{for some} \quad E \in \mathbb{P}_{\mathrm{f}}(D).$$

In particular, if D is a groupoid, r is a module system and H is an r-monoid, then our notion of r-integralitiy coincides with that introduced in [4], Definition 8.1.

2. Suppose that D is a groupoid and $H \subset D$ is a submonoid such that D is a quotient groupoid of H. Let r be an ideal system on H, viewed as a finitary module system on D such that $\{1\}_r = H$. Then an element $x \in D$ is r-integral over H if and only if there exists some $J \in \mathcal{I}_{r,\mathrm{f}}(H)$ such that $x \in (J : J)$. In this case, notice that the notion of r-integrality coincides with that introduced in [3], Ch. 14.

Before we prove the main results on r-integrality (Theorem 4.11) in order to demonstrate the power of the module system approach, we investigate the meaning of this concept in the case of the trivial module system s and the additive system d on a ring.

PROPOSITION 4.9. *Let s be the trivial module system on D, $H \subset D$ a submonoid and $x \in D$. Then x is s-integral over H if and only if there exists some $k \in \mathbb{N}_0$ and $l \in \mathbb{N}$ such that $x^{k+l} \in x^k H$.*

Proof. Suppose first that x is s-integral over H, and let $E \in \mathbb{P}_{\mathrm{f}}(D)$ be such that $E \cap D^\bullet \neq \emptyset$ and $xE \subset EH$, say $E = \{b_1, \ldots, b_m\}$, where $m \in \mathbb{N}$ and $b_1 \in D^\bullet$. Then there exists a map $\sigma : \{1, \ldots, m\} \to \{1, \ldots, m\}$ and there exist elements $x_1, \ldots, x_m \in H$ such that

$$xb_i = b_{\sigma(i)} x_i \quad \text{for all} \quad i \in \{1, \ldots, m\},$$

and consequently

$$x^n b_i = b_{\sigma^n(i)} \prod_{\nu=0}^{n-1} x_{\sigma^\nu(i)} \quad \text{for all} \quad n \in \mathbb{N}.$$

Let $k \in \mathbb{N}_0$ and $l \in \mathbb{N}$ be such that $\sigma^{k+l}(1) = \sigma^k(1)$ Setting $j = \sigma^k(1)$, we then obtain

$$x^{k+l} b_1 = b_j \prod_{\nu=0}^{k+l-1} x_{\sigma^\nu(1)} = x^l b_1 \prod_{\nu=l}^{k+l-1} x_{\sigma^\nu(1)} \in x^l b_1 H.$$

Since $b_1 \in D^\bullet$, we get $x^{k+l} \in x^k H$.

Let now $k \in \mathbb{N}_0$ and $l \in \mathbb{N}$ be such that $x^{k+l} \in x^k H$. Setting $E = \{1, x, x^2, \ldots, x^{k+l-1}\}$ completes the proof. □

PROPOSITION 4.10. *Let D be a ring and d the additive system on D. Let $H \subset D$ be a subring and $x \in D$. Then x is d-integral over H if and only if x is integral over D in the usual sense.*

Proof. We use [2], Theorem (9.3) and (9.2).

If x is integral over H in the usual sense, then the ring $H[x]$ is a finitely generated H-module, and therefore there exists some $n \in \mathbb{N}$ such that $H[x] = H + Hx + \ldots + Hx^n = E_{d[H]}$, where $E = \{1, x, \ldots, x^n\} \subset D^\bullet$. Since $xE \subset xH[x] \subset H[x] = E_{d[H]}$, x is d-integral over H.

If x is d-integral over H, then there exists some $E \in \mathbb{P}_{\mathrm{f}}(D)$ such that $E \cap D^\bullet \neq \emptyset$ and $xE \subset E_{d[H]}$. Hence $E_{d[H]}$ is a finitely generated faithful $H[x]$-module, and therefore x is integral over H in the usual sense. □

THEOREM 4.11. *Let r be a finitary weak module system on D, and let $H \subset M \subset D$ be submonoids.*

1. *If M is an r-monoid, then $H' = \mathrm{cl}_r^M(H)$ is an r-monoid which is r-closed in M.*

2. *If M is r-integral over H and x is r-integral over M, then x is r-integral over H.*

3. *If $T \subset \{1\}_r \cap D^\times$ is a multiplicatively closed subset, then*

$$\mathrm{cl}_{T^{-1}r}^{T^{-1}M}(T^{-1}H) = T^{-1}\mathrm{cl}_r^M(H).$$

4. *If M is an r-monoid and $H_0 = \{1\}_r$, then*

$$\mathrm{cl}_r^M(H) = \bigcap_{P \in r\text{-max}(H_0)} \mathrm{cl}_{r_P}^{M_P}(H_P).$$

Proof. 1. Since $r \leq \overline{r}$, $H_{\overline{r}}$ is an r-monoid, and therefore $H' = H_{\overline{r}} \cap M$ is also an r-monoid. Since $H' \subset \mathrm{cl}_r^M(H') = H'_{\overline{r}} \cap M \subset H_{\overline{r}} \cap M = H'$, equality follows.

2. If $M \subset H_{\overline{r}}$ and $x \in M_{\overline{r}}$, then $x \in (H_{\overline{r}})_{\overline{r}}$.

3. By Theorem 4.6.3 and Proposition 3.5.2, we obtain

$$\begin{aligned}\mathrm{cl}_{T^{-1}r}^{T^{-1}M} &= T^{-1}M \cap (T^{-1}H)_{\overline{T^{-1}r}} = T^{-1}M \cap (T^{-1}H)_{T^{-1}\overline{r}} \\ &= T^{-1}M \cap T^{-1}H_{\overline{r}} = T^{-1}(M \cap H_{\overline{r}}) = \mathrm{cl}_r^M(H).\end{aligned}$$

4. Since $\mathrm{cl}_r^M = M \cap H_{\overline{r}}$ is an r-monoid, Theorem 3.7 together with Theorem 4.6.3 and Proposition 3.5.2 implies

$$\begin{aligned}\mathrm{cl}_r^M(H) &= \bigcap_{P \in r\text{-max}(H_0)} (M \cap H_{\overline{r}})_P = \bigcap_{P \in r\text{-max}(H_0)} (M_P \cap H_{\overline{r}_P}) \\ &= \bigcap_{P \in r\text{-max}(H_0)} (M_P \cap H_{\overline{r_P}}) = \bigcap_{P \in r\text{-max}(H_0)} \mathrm{cl}_{r_P}^{M_P}(H_P).\end{aligned}$$

□

5 THE MAROT PROPERTY

DEFINITION 5.1. The submonoid $D^* = D^\bullet \cup \{0\} \subset D$ is called the *cancellative kernel* of D. For a weak module system r on D, we define $r^* : \mathbb{P}(D^*) \to \mathbb{P}(D^*)$ by

$$X_{r^*} = X_r \cap D^* \quad \text{for all} \quad X \subset D^*,$$

and we call r^* the *cancellative kernel* of r. We further set

$$\mathcal{M}_r^* = \{J \in \mathcal{M}_r \mid J \cap D^\bullet \neq \emptyset\} \cup \{\{0\}\},$$

and we call the non-zero elements in $\mathcal{M}_r^*$ the *regular r-modules.*

PROPOSITION 5.2. *Let r be a finitary weak module system on D. Then*

1. *r^* is a weak module system on D^*. If r is a module system or a weak ideal system, then so is r^*. If r is finitary, then so is r^*.*

2. *$\mathcal{M}_r^* \subset \mathcal{M}_r$ is a submonoid, and*

$$\rho_r : \begin{cases} \mathcal{M}_r^* & \to \quad \mathcal{M}_{r^*} \\ J & \mapsto \quad J \cap D^* \end{cases}$$

is a surjective map which is bijective if and only if $(J \cap D^*)_r = J$ for all $J \in \mathcal{M}_r^*$.

3. *If ρ_r is bijective, then ρ_r is a monoid isomorphism. If moreover r is finitary, then $\rho_r(\mathcal{M}_r^* \cap \mathcal{M}_{r,\mathrm{f}}) = \mathcal{M}_{r^*,\mathrm{f}}$.*

4. *If $S \subset D^\times \cap \{1\}_r$ is a multiplicatively closed subset, then $(T^{-1}r)^* = T^{-1}r^*$.*

5. *$\overline{r^*} = \overline{r}^*$.*

Proof. 1. We must first check the properties **(M1)**, **(M2)**, **(M3)** for all $X, Y \subset D^*$ and all $c \in D^*$.

(M1) is obvious.

(M2) If $X \subset Y_{r^*} = Y_r \cap D^*$, then $X_r \subset Y_r$ and therefore $X_{r^*} = X_r \cap D^* \subset Y_r \cap D^* = Y_{r^*}$.

(M3) $cX_{r^*} \subset cX_r \cap D^* \supset (cX)_r \cap D^* = (cX)_{r^*}$.

If r is a module system, $X \subset D^*$, $0 \neq c \in D^*$ and $z \in (cX)_{r^*} = cX_r \cap D^*$, then $z = cy$ for some $y \in X_r$, and $z \in D^*$ implies $y \in D^*$. Therefore it follows that $z \in c(X_r \cap D^*) = cX_{r^*}$.

If r is an ideal system, then $\{1\}_{r^*} = \{1\}_r \cap D^* = D \cap D^* = D^*$.

If r is finitary and $X \subset D^*$, then

$$X_{r^*} = X_r \cap D^* = \bigcup_{E \in \mathbb{P}_\mathrm{f}(X)} E_r \cap D^* = \bigcup_{E \in \mathbb{P}_\mathrm{f}(X)} E_{r^*}\,.$$

2. If $J \in \mathcal{M}_r^*$, then $(J \cap D^*)_{r^*} = (J \cap D^*)_r \cap D^* \subset J_r \cap D^* = J \cap D^*$ implies $J \cap D^* \in \mathcal{M}_{r^*}$. Hence ρ_r is a map, as asserted, and it is injective if and only if $(J \cap D^*)_r = J$ for all $J \in \mathcal{M}_r^*$.

If $I \in \mathcal{M}_{r^*}$, then $I = I_{r^*} = I_r \cap D^* = \rho_r(I_r)$, and $I_r \in \mathcal{M}_r^*$. Hence ρ_r is surjective.

3. Let ρ_r be bijective. For $I, J \in \mathcal{M}_r$ we obtain, using 2., $\rho_r(I) \cdot_{r^*} \rho_r(J) = \big((I \cap D^*)(J \cap D^*)\big)_r \cap D^* = (I \cap D^*)_r \cdot_r (J \cap D^*)_r \cap D^* = (I \cdot_r J) \cap D^* = \rho_r(I \cdot_r J)$.

Moreover, suppose that r is finitary and $J \in \mathcal{M}_r^* \cap \mathcal{M}_{r,\mathrm{f}}$. Since $J = (J \cap D^*)_r$, by Proposition 2.13.2, there exists a finite subset $E \subset J \cap D^*$ such that $J = E_r$. Hence $J \cap D^* = E_r \cap D^* = E_{r^*}$.

4. Observe that $D^\times \cap \{1\}_r \subset \{1\}_{r^*}$. For $X \subset D^*$, we obtain $X_{(S^{-1}r)^*} = X_{S^{-1}r} \cap D^* = (S^{-1}X)_r \cap D^* = (S^{-1}X)_{r^*} = X_{S^{-1}r^*}$.

5. Observe first that, for all subsets $U, V \subset D^*$, $(U_{r^*} :_{D^*} V) = (U_r : V) \cap D^*$. Indeed, if $x \in D^*$, then $xV \subset D^*$, and therefore $xV \subset U_r$ if and only if $xV \subset U_{r^*}$. Now we obtain, for $X \subset D^*$,

$$X_{\overline{r^*}} = \bigcup_{\substack{B \in \mathbb{P}_{\mathrm{f}}(D^*) \\ B \cap D^\bullet \neq \{0\}}} ((XB)_{r^*} :_{D^*} B) = \bigcup_{\substack{B \in \mathbb{P}_{\mathrm{f}}(D) \\ B \cap D^\bullet \neq \{0\}}} ((XB)_r : B) \cap D^* = X_{\overline{r}^*}.$$

□

DEFINITION 5.3. Let r be a weak module system on D. We say that r has the *Marot property* if $(J \cap D^*)_r = J$ for all $J \in \mathcal{M}_r^*$ (that means, every regular r-module is generated by its regular elements).

A ring D is called a *Marot ring* if $d[D]$ has the Marot property.

PROPOSITION 5.4. *Let q and r be weak module systems on D such that $q \le r$. If q has the Marot property, then r also has the Marot property.*

Proof. If $J \in \mathcal{M}_r \subset \mathcal{M}_q$, then $(J \cap D^*)_r = \big((J \cap D^*)_q\big)_r = (J_q)_r = J_r$. □

DEFINITION 5.5. Let Q be a total quotient monoid of D and r a finitary ideal system on D (viewed as a finitary module system on Q such that $\{1\}_r = D$).

An r-ideal $J \in \mathcal{I}_r(D)$ is called r-invertible, if there exists some $J' \in \mathcal{M}_r$ such that $J \cdot_r J' = D$.

D is called an *r-Dedekind monoid* if every regular r-ideal is r-invertible.

D is called an *r-Prüfer monoid* if every regular r-finitely generated r-ideal is r-invertible.

THEOREM 5.6. *Let Q be a total quotient monoid of D and r a finitary ideal system on D (viewed as a finitary module system on Q such that $\{1\}_r = D$) satisfying the Marot property. Then D is an r-Prüfer monoid (an r-Dedekind monoid) if and only if D^* is an r^*-Prüfer monoid (an r^*-Dedekind monoid).*

Proof. This follows from Proposition 5.2 after applying the isomorphism ρ_r. □

References

[1] M. Fontana and J. A. Huckaba, *Localizing Systems and Semistar Operations*, Commutative Rings in a Non-Noetherian Setting (S. T. Chapman and S. Glaz, eds.), Kluwer, 2000.

[2] R. Gilmer, *Multiplicative Ideal Theory*, M. Dekker 1972.

[3] F. Halter-Koch, *Ideal Systems*, M. Dekker 1998.

[4] F. Halter-Koch, *Localizing Systems, Module Systems and Semistar Operations*, J. Algebra **238** (2001), 723 - 761.

[5] J. A. Huckaba, *Commutative Rings with Zero Divisors*, M. Dekker 1988.

[6] P. Jaffard, *Les Systèmes d'Idéaux*, Dunod 1960.

[7] A. Okabe and R. Matsuda, *Semistar-operations on integral domains*, Math. J. Toyama Univ. **17** (1994), 1–21.

Examples of Integral Domains Inside Power Series Rings

WILLIAM HEINZER, CHRISTEL ROTTHAUS AND SYLVIA WIEGAND

DEPARTMENT OF MATHEMATICS, PURDUE UNIVERSITY, WEST LAFAYETTE, IN 47907-1395

DEPARTMENT OF MATHEMATICS, MICHIGAN STATE UNIVERSITY, EAST LANSING, MI 48824-1027

DEPARTMENT OF MATHEMATICS AND STATISTICS, UNIVERSITY OF NEBRASKA, LINCOLN, NE 68588-0323

ABSTRACT. We present examples of Noetherian and non-Noetherian integral domains which can be built inside power series rings. Given a power series ring R^* over a Noetherian integral domain R and given a subfield L of the total quotient ring of R^* with $R \subseteq L$, we construct subrings A and B of L such that B is a localization of a nested union of polynomial rings over R and $B \subseteq A := L \cap R^*$. We show in certain cases that flatness of a related map on polynomial rings is equivalent to the Noetherian property for B. Moreover if B is Noetherian, then $B = A$. We use this construction to obtain for each positive integer n an explicit example of a 3-dimensional quasilocal unique factorization domain B such that the maximal ideal of B is 2-generated, B has precisely n prime ideals of height two, and each prime ideal of B of height two is not finitely generated.

1. INTRODUCTION

This paper is a continuation of our study of a technique for constructing integral domains by (1) intersecting a power series ring with a field to obtain an integral domain A as in the abstract, and (2) approximating the domain A with a nested

The authors would like to thank the National Science Foundation and the National Security Agency for support for this research. In addition they are grateful for the hospitality and cooperation of Michigan State, Nebraska and Purdue, where several work sessions on this research were conducted.

union of localized polynomial rings to obtain an integral domain B as in the abstract. Classical examples such as those of Akizuki [A] and Nagata [N, pages 209-211] use the second (nested union) description of this construction. It is possible to also realize these classical examples as the intersection domains of the first description [HRW6].

In this paper we observe that, in certain applications of this technique, flatness of a map of associated polynomial rings implies the constructed domains are Noetherian and that $A = B$. We also in the present paper apply this observation to the construction of examples of both Noetherian and non-Noetherian integral domains.

We begin by describing the technique.

1.1 General Setting

Let R be a commutative Noetherian integral domain. Let a be a nonzero nonunit of R and let R^* be the (a)-adic completion of R. Then R^* is isomorphic to $R[[y]]/(y-a)$, where y is an indeterminate; thus we consider R^* as $R[[a]]$, the "power series ring" in a over R. The intersection domain (type 1 above) and the approximation domain (type 2) of the construction are inside R^*. Let $\tau_1, \ldots, \tau_n \in aR^*$ be algebraically independent over the fraction field K of R and let $\underline{\tau}$ abbreviate the list $\tau_1, \ldots, \tau_n$. By Theorem 2.2, also known as [HRW1, Theorem 1.1], $A_{\underline{\tau}} := K(\tau_1, \ldots, \tau_n) \cap R^*$ is simultaneously Noetherian and computable as a nested union $B_{\underline{\tau}}$ of certain associated localized polynomial rings over R using $\underline{\tau}$ if and only if the extension $T := R[\underline{\tau}] := R[\tau_1, \ldots, \tau_n] \overset{\psi}{\hookrightarrow} R_a^*$ is flat.

In the case where $\psi : T \hookrightarrow R_a^*$ is flat, so that the intersection domain $A_{\underline{\tau}}$ is Noetherian and computable, we construct new "insider" examples inside $A_{\underline{\tau}}$. We choose elements $f_1, \ldots, f_m$ of T, considered as polynomials in the τ_i with coefficients in R and abbreviated by $\underline{f}$. Assume that $f_1, \ldots, f_m$ are algebraically independent over K; thus $m \leq n$. If $S := R[\underline{f}] := R[f_1, \ldots, f_m] \overset{\varphi}{\hookrightarrow} T = R[\underline{\tau}]$ is flat, we observe in Section 3 that the "insider ring" $A_{\underline{f}} := K(\underline{f}) \cap R^*$ is Noetherian and computable; that is, $A_{\underline{f}}$ is equal to an approximating union $B_{\underline{f}}$ of localized polynomial rings constructed using the f_i. Moreover, we can often identify conditions on the map φ

which imply $B_{\underline{f}}$ and $A_{\underline{f}}$ are not Noetherian. Thus the "insider" examples $A_{\underline{f}}$ and $B_{\underline{f}}$ are inside intersection domains $A_{\underline{\tau}}$ known to be Noetherian; the new insider is Noetherian if the associated extension $S \to T$ of polynomial rings is flat. The insider examples are examined in more detail in Section 3.

In Section 2 we give background and notation for the construction and for flatness of polynomial extensions in greater generality: Suppose that $\underline{x} := (x_1, \ldots, x_n)$ is a tuple of indeterminates over R and that $\underline{f} := (f_1, \ldots, f_m)$ consists of elements of the polynomial ring $R[\underline{x}]$ that are algebraically independent over K. We consider flatness of the following map of polynomial rings.

$$\varphi : S := R[\underline{f}] \hookrightarrow T := R[\underline{x}]. \tag{1.2}$$

In Section 4 we continue the analysis of the flatness of (1.2) and the nonflat locus. We discuss results of [P], [W] and others.

In Section 5 we present for each positive integer n an insider example B such that:

(1) B is a 3-dimensional quasilocal unique factorization domain,

(2) B is not catenary,

(3) the maximal ideal of B is 2-generated,

(4) B has precisely n prime ideals of height two,

(5) Each prime ideal of B of height two is not finitely generated,

(6) For every non-maximal prime P of B the ring B_P is Noetherian.

2. BACKGROUND AND NOTATION

We begin this section by recalling some details for the approximation to the intersection domain A_τ of (1.1).

2.1 Notation for approximations

Assume that R, K, a, $\tau_1, \ldots, \tau_n$, $\underline{\tau}$ and $A_{\underline{\tau}}$ are as in General Setting 1.1. Then the (a)-adic completion of R is $R^* = R[[x]]/(x-a) = R[[a]]$. Write each $\tau_i :=$

$\sum_{j=1}^{\infty} b_{ij}a^j$, with the $b_{ij} \in R$. There are natural sequences $\{\tau_{ir}\}_{r=0}^{\infty}$ of elements in A, called the rth *endpieces* for the τ_i, which "approximate" the τ_i, defined by:

(2.1.1) For each $i \in \{1, \dots, n\}$ and $r \geq 0$, $\tau_{ir} := \sum_{j=r+1}^{\infty} (b_{ij}a^j)/a^r$.

Now for each r, $U_r := R[\tau_{1r}, \dots, \tau_{nr}]$ and B_r is U_r localized at the multiplicative system $1 + aU_r$. Then define $U_{\underline{\tau}} := \cup_{r=1}^{\infty} U_r$ and $B_{\underline{\tau}} := \cup_{r=1}^{\infty} B_r$. Thus $U_{\underline{\tau}}$ is a nested union of polynomial rings over R and $B_{\underline{\tau}}$ is a nested union of localized polynomial rings over R. The definition of the U_r (and hence also of B_r and $U_{\underline{\tau}}$ and $B_{\underline{\tau}}$) are independent of the representation of the τ_i as power series with coefficients in R [HRW1, Proposition 2.3].

The following theorem is the basis for our construction of examples.

2.2 Theorem. [HRW1, Theorem 1.1] Let R be a Noetherian integral domain with fraction field K. Let a be a nonzero nonunit of R. Let $\tau_1, \dots, \tau_n \in aR[[a]] = aR^*$ be algebraically independent over K, abbreviated by $\underline{\tau}$. Let $U_{\underline{\tau}}$ and $B_{\underline{\tau}}$ be as in (2.1). Then the following statements are equivalent:

(1) $A_{\underline{\tau}} := K(\underline{\tau}) \cap R^*$ is Noetherian and $A_{\underline{\tau}} = B_{\underline{\tau}}$.

(2) $U_{\underline{\tau}}$ is Noetherian.

(3) $B_{\underline{\tau}}$ is Noetherian.

(4) $R[\underline{\tau}] \to R_a^*$ is flat.

Since flatness is a local property, the following two propositions are immediate corollaries of [HRW5, Theorem 2.1]; see also [P, Théorème 3.15].

2.3 Proposition. Let T be a Noetherian ring and suppose $R \subseteq S$ are Noetherian subrings of T. Assume that $R \to T$ is flat with Cohen-Macaulay fibers and that $R \to S$ is flat with regular fibers. Then $S \to T$ is flat if and only if, for each prime ideal P of T, we have $\text{ht}(P) \geq \text{ht}(P \cap S)$.

As a special case we have:

2.4 Proposition. Let R be a Noetherian ring and let $x_1, \dots, x_n$ be indeterminates over R. Assume that $f_1, \dots, f_m \in R[x_1, \dots, x_n]$ are algebraically independent over R. Then

(1) $\varphi : S := R[f_1, \dots, f_m] \hookrightarrow T := R[x_1, \dots, x_n]$ is flat if and only if, for each prime ideal P of T, we have $\text{ht}(P) \geq \text{ht}(P \cap S)$.

(2) For $Q \in \operatorname{Spec} T$, $\varphi_Q : S \to T_Q$ is flat if and only if for each prime ideal $P \subseteq Q$ of T, we have $\operatorname{ht}(P) \geq \operatorname{ht}(P \cap S)$.

2.5 Definitions and Remarks

(1) The *Jacobian ideal* J of the extension (1.2) is the ideal generated by the $m \times m$ minors of the $m \times n$ matrix $\mathcal{J}$ given below:

$$\mathcal{J} := \left(\frac{\partial f_i}{\partial x_j} \right)_{i,j} .$$

(2) For the extension (1.2), the *nonflat locus* of φ is the set $\mathcal{F}$, where

$$\mathcal{F} := \{ Q \in \operatorname{Spec}(T) : \text{the map} \quad \varphi_Q : S \to T_Q \text{ is not flat } \}.$$

For convenience, we also define the set $\mathcal{F}_{\min}$ and the ideal F of T:

$$\mathcal{F}_{\min} := \{ \text{ minimal elements of } \mathcal{F} \} \text{ and } F := \cap\{Q : Q \in \mathcal{F}\}.$$

By [M2, Theorem 24.3], the set $\mathcal{F}$ is closed in the Zariski topology and hence is equal to $\mathcal{V}(F)$, the set of primes of T that contain the ideal F. Thus the set $\mathcal{F}_{\min}$ is a finite set and consists precisely of the minimal primes of the ideal F.

Moreover, Proposition 2.4 implies $\mathcal{F}_{\min} \subseteq \{Q \in \operatorname{Spec} T : \operatorname{ht} Q < \operatorname{ht}(Q \cap S)\}$ and for every prime ideal $P \subsetneq Q \in \mathcal{F}_{\min}$, $\operatorname{ht} P \geq \operatorname{ht}(P \cap S)$.

(3) In general for a commutative ring T and a subring R, we say that elements $f_1, \ldots, f_m \in T$ are *algebraically independent* over R if, for indeterminates $t_1, \ldots, t_m$ over R, the only polynomial $G(t_1, \ldots, t_m) \in R[t_1, \ldots, t_m]$ with $G(f_1, \ldots, f_m) = 0$ is the zero polynomial.

2.6 Example and Remarks

(1) Let k be a field, let x and y be indeterminates over k and set $f = x$, $g = (x-1)y$. Then $k[f, g] \xrightarrow{\varphi} k[x, y]$ is not flat.

Proof. For the prime ideal $P := (x - 1) \in \operatorname{Spec}(k[x, y])$, we see that $\operatorname{ht}(P) = 1$, but $\operatorname{ht}(P \cap k[f, g]) = 2$; thus the extension is not flat by Proposition 2.4.

(2) The Jacobian ideal J of f and g in (1) is given by:

$$J = (\det \begin{pmatrix} \frac{\delta f}{\delta x} & \frac{\delta f}{\delta y} \\ \frac{\delta g}{\delta x} & \frac{\delta g}{\delta y} \end{pmatrix}) = (\det \begin{pmatrix} 1 & 0 \\ y & x-1 \end{pmatrix}) = (x-1).$$

(3) In this example the nonflat locus is equal to the set of prime ideals Q of $k[x,y]$ which contain the Jacobian ideal $(x-1)k[x,y]$, thus $J = F$.

We record in Proposition 2.7 observations about flatness that follow from well-known properties of the Jacobian.

2.7 Proposition. Let R be a Noetherian ring, let $x_1, \ldots, x_n$ be indeterminates over R, and let $f_1, \ldots, f_m \in R[x_1, \ldots, x_n]$ be algebraically independent over R. Consider the embedding $\varphi : S := R[f_1, \ldots, f_m] \hookrightarrow T := R[x_1, \ldots, x_n]$. Let J denote the Jacobian ideal of φ and let $Q \in \operatorname{Spec} T$. Then

(1) Q does not contain J if and only if $\varphi_Q : S \to T_Q$ is essentially smooth.

(2) If Q does not contain J, then $\varphi_Q : S \to T_Q$ is flat. Thus $J \subseteq F$.

(3) $\mathcal{F}_{\min} \subseteq \{Q' \in \operatorname{Spec} T : J \subseteq Q' \text{ and } \operatorname{ht}(Q' \cap S) > \operatorname{ht} Q'\}$.

Proof. For item 1, we observe that our definition of the Jacobian ideal J given in (2.4) agrees with the description of the smooth locus of an extension given in [E], [S, Section 4].

To see this, let $u_1, \ldots, u_m$ be indeterminates over $R[x_1, \ldots, x_n]$ and identify

$$R[x_1, \ldots, x_n] \quad \text{with} \quad \frac{R[u_1, \ldots, u_m][x_1, \ldots, x_n]}{(\{u_i - f_i\}_{i=1,\ldots,m})}.$$

Since $u_1, \ldots, u_m$ are algebraically independent, the ideal J generated by the minors of $\mathcal{J}$ is the Jacobian ideal of the extension (1.2) by means of this identification. We make this more explicit as follows.

Let $U_1 := R[u_1, \ldots, u_m, x_1, \ldots, x_n]$ and $I = (\{f_i - u_i\}_{i=1,\ldots,m})U_1$. Consider the following commutative diagram

$$\begin{array}{ccc} S := R[f_1, \ldots, f_m] & \longrightarrow & T := R[x_1, \ldots, x_n] \\ \cong\downarrow & & \cong\downarrow \\ S_1 := R[u_1, \ldots, u_m] & \longrightarrow & T_1 := R[u_1, \ldots, u_m, x_1, \ldots, x_n]/I \end{array}$$

Define as in [E], [S, Section 4]

$$H = H_{T_1/S_1} := \text{the radical of } \Sigma\Delta(g_1, \ldots, g_s)[(g_1, \ldots, g_s) : I],$$

where the sum is taken over all s with $0 \leq s \leq m$, for all choices of s polynomials $g, \ldots, g_s$ from $I = (\{f_1 - u_1, \ldots, f_m - u_m\})U_1$, where $\Delta := \Delta(g_1, \ldots, g_s)$ is the ideal of $T \cong T_1$ generated by the $s \times s$-minors of $\left(\frac{\partial g_i}{\partial x_j}\right)$, and $\Delta = T$ if $s = 0$.

To establish (2.7.1), we show that $H = \operatorname{rad}(J)$. Since u_i is a constant with respect to x_j, we have $\left(\frac{\partial(f_i - u_i)}{\partial x_j}\right) = \left(\frac{\partial f_i}{\partial x_j}\right)$. Thus $J \subseteq H$.

For $g_1, \ldots, g_s \in I$, the $s \times s$-minors of $\left(\frac{\partial g_i}{\partial x_j}\right)$ are contained in the $s \times s$-minors of $\left(\frac{\partial f_i}{\partial x_j}\right)$. Thus it suffices to consider s polynomials $g_1, \ldots, g_s$ from the set $\{f_1 - u_1, \ldots, f_m - u_m\}$. Now $f_1 - u_1, \ldots, f_m - u_m$ is a regular sequence in $R[u_1 \ldots u_r, x_1, \ldots, x_n]$. Thus for $s < m$, $[(g_1, \ldots, g_s) : I] = (g_1, \ldots, g_s)$. Thus the $m \times m$-minors of $\left(\frac{\partial f_i}{\partial x_j}\right)$ generate H up to radical, and so $H = \operatorname{rad}(J)$.

Hence by [E] or [S, Theorem 4.1], T_Q is essentially smooth over S if and only if Q does not contain J.

Item 2 follows from item 1 because essentially smooth maps are flat. In view of Proposition 2.4 and (2.5.2), item 3 follows from item 2. □

2.8 Remarks

(1) For φ as in (1.2), it would be interesting to identify the set $\mathcal{F}_{\text{min}}$. In particular we are interested in conditions for $J = F$ and/or conditions for $J \subsetneq F$.

(2) If $\operatorname{char} R = 0$, then the zero ideal is not in $\mathcal{F}_{\text{min}}$ and so $F \neq \{0\}$.

(3) In view of (2.7.3), we can describe $\mathcal{F}_{\text{min}}$ exactly as

$\mathcal{F}_{\text{min}} = \{Q \in \operatorname{Spec} T : J \subseteq Q, \operatorname{ht}(Q \cap S) > \operatorname{ht} Q \text{ and } \forall P \subsetneq Q, \operatorname{ht}(P) \geq \operatorname{ht}(P \cap S)\}$.

(4) By (2.8.3), every prime ideal Q of $\mathcal{F}_{\text{min}}$ contains two primes $P_1 \subsetneq P_2$ of S such that Q is minimal above both P_1T and P_2T.

3. EXPLICIT CONSTRUCTIONS INSIDE SIMPLER EXTENSIONS

Using Theorem 2.2 and intersection domains inside the completion which are known to be Noetherian, we formulate a shortcut method for the construction of "insider" examples.

3.1 General Method

Let R be a Noetherian integral domain. Let a be a nonzero nonunit of R and let $R^* = R[[x]]/(x-a)$ be the (a)-adic completion of R. Let $\tau_1, \ldots, \tau_n \in aR^*$, abbreviated by $\underline{\tau}$, be algebraically independent over the fraction field K of R. Assume that the extension $T := R[\tau_1, \ldots, \tau_n] \overset{\psi}{\hookrightarrow} R_a^*$ is flat. Thus by Theorem 2.2, $D := A_{\underline{\tau}} = K(\tau_1, \ldots, \tau_n) \cap R^*$ is Noetherian and computable as a nested union of localized polynomial rings over R using the τ's.

Let $f_1, \ldots, f_m$ be elements of T, abbreviated by $\underline{f}$ and considered as polynomials in the τ_i with coefficients in R. Assume that $f_1, \ldots, f_m$ are algebraically independent over K; thus $m \leq n$. Let $S := R[\underline{f}] \overset{\varphi}{\hookrightarrow} T = R[\underline{\tau}]$; put $\alpha := \psi \circ \varphi : S \to R_a^*$. That is, we have:

$$\begin{array}{ccccc} & & & & R_a^* \\ & & \alpha:=\psi\varphi & & \\ & & & & \psi \\ R & \subseteq & S := R[\underline{f}] & \overset{\varphi}{\longrightarrow} & T := R[\underline{\tau}] \end{array}$$

Using the f's in place of the τ's, we define the ring $A := A_{\underline{f}} := K(\underline{f}) \cap R^*$ and the approximation rings $U_r, B_r, U_{\underline{f}}$ and $B = B_{\underline{f}}$, as in (2.1). Let

$$F := \cap\{P \in \operatorname{Spec}(T) \mid \varphi_P : S \to T_P \text{ is not flat }\}.$$

Thus, as in (2.5.2), the ideal F defines the nonflat locus of the map $\varphi : S \to T$. For $Q^* \in \operatorname{Spec}(R_a^*)$, we consider whether the localized map $\varphi_{Q^* \cap T}$ is flat:

(3.1.1) $$\varphi_{Q^* \cap T} : S \to T_{Q^* \cap T}$$

3.2 Theorem. With the notation of (3.1) we have

(1) For $Q^* \in \operatorname{Spec}(R_a^*)$, the map $\alpha_{Q^*} : S \to (R_a^*)_{Q^*}$ is flat if and only if the map $\varphi_{Q^* \cap T}$ in (3.1.1) is flat.

(2) The following are equivalent:

(i) A is Noetherian and $A = B$.

(ii) B is Noetherian.

(iii) The map $\varphi_{Q^* \cap T}$ in (3.1.1) is flat for every maximal $Q^* \in \operatorname{Spec}(R_a^*)$.

(iv) $FR_a^* = R_a^*$.

(3) $\varphi_a : S \to T_a$ is flat if and only if $FT_a = T_a$. Moreover, either of these conditions implies B is Noetherian and $B = A$.

Proof. For item (1), we have $\alpha_{Q^*} = \psi_{Q^*} \circ \varphi_{Q^* \cap T} : S \to T_{Q^* \cap T} \to (R_a^*)_{Q^*}$. Since the map ψ_{Q^*} is faithfully flat, the composition α_{Q^*} is flat if and only if $\varphi_{Q^* \cap T}$ is flat [M1, page 27]. For item (2), the equivalence of (i) and (ii) is part of Theorem 2.2. The equivalence of (ii) and (iii) follows from item (1) and Theorem 2.2. For the equivalence of (iii) and (iv), we use $FR^* \neq R^* \iff F \subseteq Q^* \cap T$, for some Q^* maximal in $\operatorname{Spec}(R^*)_a) \iff$ the map in (3.1.1) fails to be flat. Item (3) follows from the definition of F and the fact that the nonflat locus of $\varphi : S \to T$ is closed. $\square$

To examine the map $\alpha : S \to R_a^*$ in more detail, we use the following terminology.

3.3 Definition

For an extension of Noetherian rings $\varphi : A' \hookrightarrow B'$ and for $d \in \mathbb{N}$, we say that $\varphi : A' \hookrightarrow B'$, *satisfies* LF_d if for each $P \in \operatorname{Spec}(B')$ with $\operatorname{ht}(P) \leq d$, the composite map $A' \to B' \to B'_P$ is flat.

3.4 Corollary. With the notation of (3.1), we have $\operatorname{ht}(FR_a^*) > 1 \iff \varphi : S \to R_a^*$ satisfies $\mathrm{LF}_1 \iff B = A$.

Proof. The first equivalence follows from the definition of LF_1 and the second equivalence from [HRW4,Theorem 5.5].

3.5 A more concrete situation

Let $R := k[x, y_1, \ldots, y_s]$, where k is a field and $x, y_1, \ldots, y_s$ are indeterminates over k with the y_i abbreviated by $\underline{y}$. Let $R^* = k[\underline{y}][[x]]$, the (x)-adic completion of

R. Let $\tau_1, \ldots, \tau_n$, abbreviated by $\underline{\tau}$, be elements of $xk[[x]]$ which are algebraically independent over $k(x)$. Let $D := A_{\underline{\tau}} := k(x, \underline{y}, \underline{\tau}) \cap R^*$. Let $T = R[\underline{\tau}]$. Then $T \to R^*_x$ is flat, D is a nested union of localized polynomial rings obtained using the τ_i and D is a Noetherian regular local ring; moreover, if $\operatorname{char} k = 0$, then D is excellent [HRW3, Proposition 4.1].

We now use the procedure of (3.1) to construct examples inside D. Let $f_1, \ldots, f_m$, abbreviated by $\underline{f}$, be elements of T considered as polynomials in $\tau_1, \ldots, \tau_n$ with coefficients in R, that are algebraically independent over $k(x, \underline{y})$. We assume the constant terms in $R = k[x, \underline{y}]$ of the f_i are zero. Let $S := R[\underline{f}]$. The inclusion map $S \hookrightarrow T$ is an injective R-algebra homomorphism, and $m \leq n$.

Let $A := \mathcal{Q}(S) \cap R^*$ and let B be the nested union domain associated to the $\underline{f}$, as in (2.1). By Theorem 2.2, B is Noetherian and $B = A$ if and only if the map $\alpha : S \to R^*_x$ is flat. Furthermore, by Theorem 3.2, we can recover information about flatness of α by considering the map $\varphi : S \to T$.

The following remark describes how the f_i are chosen in several classical examples:

3.6 Remark

With the notation of (3.5).

(1) Nagata's famous example [N1], [N2, Example 7, page 209], [HRW6, Example 3.1], may be described by taking $n = s = m = 1, y_1 = y, \tau_1 = \tau$, and $f_1 = f$ and localizing. Then $R = k[x, y]_{(x,y)}$, $T = k[x, y, \tau]_{(x,y,\tau)}$, $f = (y + \tau)^2$, $S = k[x, y, f]_{(x,y,f)}$ and $A = k(x, y, (y + \tau)^2) \cap R^*$. The Noetherian property of B is implied by the flatness property of the map $S \to T_x$. Thus $B = A$. In this case, T is actually a free S-module with $< 1, y + \tau >$ as a free basis.

(2) An example of Rotthaus [R1],[HRW6, Example 3.3], may be described by taking $n = s = 2$, and $m = 1$ and localizing. Then $R = k[x, y_1, y_2]_{(x,y_1,y_2)}$, $T = R[\tau_1, \tau_2]_{(\mathbf{m},\tau_1,\tau_2)}$, $f_1 = (y_1 + \tau_1)(y_2 + \tau_2)$, $S = R[f_1]_{(\mathbf{m},f_1)}$ and $A = k(x, y_1, y_2, (y_1 + \tau_1)(y_2 + \tau_2)) \cap R^*$. Since the map from $R[f_1] \to R_x[\tau_1, \tau_2] = T_x$ is flat, the associated nested union domain B is Noetherian.

(3) The following example is given in [HRW5, Section 4]. Let $n = s = m = 2$, let $f_1 = (y_1 + \tau_1)^2$ and $f_2 = (y_1 + \tau_1)(y_2 + \tau_2)$. It is shown in [HRW5] for this example that $B \subsetneq A$ and that both A and B are non-Noetherian.

The following lemma follows from [P, Proposition 2.1] in the case of one indeterminate x, so in the case where $T = R[x]$.

3.7 Lemma. Let R be a Noetherian ring, let $x_1, \dots, x_n$ be indeterminates over R, and let $T = R[x_1, \dots x_n]$. Suppose $f \in T - R$ is such that the constant term of f is zero. Then the following are equivalent:

(1) $R[f] \to T$ is flat.

(2) $R[f] \to T$ is faithfully flat.

(3) For each maximal ideal q of R, we have $qT \cap R[f] = qR[f]$.

(4) The coefficients of f generate the unit ideal of R.

Proof. (1) $\implies$ (2): It suffices to show for $P \in \operatorname{Spec}(R[f])$ that $PT \neq T$. Let $q = P \cap R$ and let $k(q)$ denote the fraction field of R/q. Since $R[f] \to T$ is flat, tensoring with $k(q)$ gives injective maps

$$k(q) \to k(q) \otimes_R R[f] \cong k(q)[f'] \xrightarrow{\varphi} k(q) \otimes_R T \cong k(q)[x_1, \dots, x_n],$$

where f' is the image of f in $k(q)[x_1, \dots, x_n]$. The injectivity of φ implies f' has positive total degree as a polynomial in $k(q)[x_1, \dots, x_n]$.

The image p' of P in $k(q)[f']$ is either zero or a maximal ideal of $k(q)[f']$. It suffices to show $p'k(q)[x_1, \dots, x_n] \neq k(q)[x_1, \dots, x_n]$. If $p' = 0$, this is clear. Otherwise p' is generated by a nonconstant polynomial $h(f')$ and $p'k(q)[x_1, \dots, x_n]$ is generated by $h(f'(x_1, \dots, x_n))$ which has total degree equal to $\deg(h)\deg(f') > 0$. Thus (1) implies (2).

(2) $\implies$ (3): This follows from Theorem 7.5 (ii) of [M2].

(3) $\implies$ (4) : If the coefficients of f were contained in a maximal ideal q of R, then $f \in qT \cap R[f]$, but $f \notin qR[f]$.

(4) $\implies$ (1): Let v be another indeterminate and consider the commutative diagram

$$\begin{array}{ccc} R[v] & \longrightarrow & T[v] = R[x_1, \dots, x_n, v] \\ \pi \downarrow & & \pi' \downarrow \\ R[f] & \xrightarrow{\varphi} & \frac{R[x_1, \dots, x_n, v]}{(v - f(x_1, \dots, x_n))}. \end{array}$$

where π maps $v \to f$ and π' is the canonical quotient homomorphism. By [M1, Corollary 2, p. 152], φ is flat if the coefficients of $f-v$ generate the unit ideal of $R[v]$. Moreover, the coefficients of $f-v$ as a polynomial in $x_1,\dots,x_n$ with coefficients in $R[v]$ generate the unit ideal of $R[v]$ if and only if the nonconstant coefficients of f generate the unit ideal of R. □

We observe in Proposition 3.8 that one direction of (3.7) holds for more than one polynomial: see also [P, Theorem 3.8] for a related result concerning flatness.

3.8 Proposition. Assume the notation of (3.7) except that $f_1,\dots,f_m \in T$ are polynomials in $x_1,\dots,x_n$ with coefficients in R and $m \geq 1$. If the inclusion map $\varphi : S = R[f_1,\dots f_m] \to T$ is flat, then the nonconstant coefficients of each of the f_i generate the unit ideal of R.

Proof. Since $f_1,\dots,f_m$ are algebraically independent over $\mathcal{Q}(R) = K$, for every $1 \leq i \leq m$, the inclusion $R[f_i] \hookrightarrow R[f_1,\dots,f_m]$ is flat. If $S \longrightarrow T$ is flat, so is the composition $R[f_i] \longrightarrow S = R[f_1,\dots,f_m] \longrightarrow T$ and the statement follows from Proposition 3.7. □

3.9 Theorem. Assume the notation of (3.1). If $m = 1$, that is, there is only one polynomial $f_1 = f$, then

(1) The map $S \to T_a$ is flat $\iff$ the nonconstant coefficients of f generate the unit ideal in R_a,

(2) Either of the conditions in (1) implies the constructed ring A is Noetherian and $A = B$.

(3) B is Noetherian and $A = B$ $\iff$ for every prime ideal Q^* in R^* with $a \notin Q^*$, the nonconstant coefficients of f generate the unit ideal in R_q, where $q := Q^* \cap R$.

(4) If the nonconstant coefficients of $f_1 = f$ generate an ideal L of R_a of height d, then the map $S \to R^*_a$ satisfies LF_{d-1}, but not LF_d.

Proof. Item (1) follows from Lemma 3.7 for the ring R_a with $x_i = \tau_i$.

By Theorems 2.2 and 3.2, the first condition in item (1) implies item (2).

For item (3), suppose the nonconstant coefficients of f generate the unit ideal of R_q. Then by Lemma 3.7, $R_q[f] \to R_q[\tau_1,\dots,\tau_n]$ is flat. Since $R_q[\tau_1,\dots,\tau_n] \to$

$R^*_{Q^*}$ is flat, $R_q[f] \to R^*_{Q^*}$ is also flat. For the other direction, suppose there exists $Q^* \in \operatorname{Spec} R^*$ with $a \notin Q^*$ such that the nonconstant coefficients of f are in qR_q, where $q = Q^* \cap R$. If $R[f] \to R^*_{Q^*}$ were flat, then, since $qR^*_{Q^*} \neq R^*_{Q^*}$, we would have $qR^*_{Q^*} \cap R[f] = qR[f]$. This would imply $f \in qR^*_{Q^*} \cap R[f]$, but $f \notin qR[f]$, a contradiction.

For item (4), if $Q^* \in \operatorname{Spec}(R^*_a)$ the map $S \to (R^*_a)_{Q^*}$ is not flat if and only if $L \subseteq Q^*$. By hypothesis there exists such a prime ideal of height d, but no such prime ideal of height less than d.

3.10 Example

With the notation of (3.5), let $m = 1$ and assume that n and s are each greater than or equal to d. Then $f_1 = f := y_1\tau_1 + \cdots + y_d\tau_d$ gives an example where $S \to T_x$ satisfies LF_{d-1}, but fails to satisfy LF_d. For $d \geq 2$ this gives examples where $A = B$, i.e., A is "limit-intersecting", but is not Noetherian.

The following is a related even simpler example: In the notation of (3.5), let $m = 1, n = 1$, and $s = 2$; that is, $R = k[x, y_1, y_2]_{(x,y_1,y_2)}$ and $\tau \in xk[[x]]$. If $f_1 = f = y_1\tau + y_2\tau^2$, then the constructed intersection domain $A := R^* \cap k(x, y_1, y_2, f)$ is not Noetherian. Thus we have a situation where $B = A$ is not Noetherian. This gives a simpler example of such behavior than the example given in Section 4 of [HRW2].

In dimension two (the two variable case), Valabrega proved the following.

3.11 Proposition [V, Prop. 3]. For $R = k[x, y]_{(x,y)}$ with completion $\widehat{R} = k[[x, y]]$, if L is a field between the fraction field of R and the fraction field F of $k[y]\,[[x]]$, then $A = L \cap \widehat{R}$ is a two-dimensional regular local domain with completion $\widehat{R}$.

Example 3.10 shows that the dimension three analog to Valabrega's result fails. With $R = k[x, y_1, y_2]_{(x,y_1,y_2)}$ the field $L = k(x, y_1, y_2, f)$ is between $k(x, y_1, y_2)$ and the fraction field of $k[y_1, y_2]\,[[x]]$, but $L \cap \widehat{R} = L \cap R^*$ is not Noetherian.

3.12 Remark

With the notation of (3.1), it can happen that $\varphi_a : S \to T_a$ is not flat, but α :

$S \to R_a^*$ is flat. For example, using the notation of (3.5), let $R := k[x,y]$, where k is a field and x, y are indeterminates over k. Let $\sigma, \tau \in xk[[x]]$ be such that x, σ, τ are algebraically independent over k, let $T := R[\sigma, \tau]$, and let $S := R[\sigma, \sigma\tau]$. Then $\varphi_x : S \to T_x$ is not flat since σT_x is a height-one prime such that $\sigma T_x \cap S = (\sigma, \sigma\tau)S$ has height 2. To see that R_x^* is flat over S, observe that $\dim R_x^* = 1$ and if $Q^* \in \operatorname{Spec} R_x^*$, then $Q^* \cap k[x, \sigma, \sigma\tau] = (0)$. Therefore $\operatorname{ht}(Q^* \cap S) \leq 1$.

4. FLATNESS OF MAPS OF POLYNOMIAL RINGS

4.1 Proposition. Let k be a field, let $x_1, \ldots, x_n$ be indeterminates over k, and let $f_1, \ldots, f_m \in k[x_1, \ldots, x_n]$ be algebraically independent over k. Consider the embedding $\varphi : S := k[f_1, \ldots, f_m] \hookrightarrow T := k[x_1, \ldots, x_n]$ and let J denote the Jacobian ideal of φ. Then

(1) $\mathcal{F}_{\min} \subseteq \{Q \in \operatorname{Spec} T : J \subseteq Q, \operatorname{ht} Q \leq m-1 \text{ and } \operatorname{ht} Q < \operatorname{ht}(Q \cap S)\}$.

(2) φ is flat $\iff$ for every $Q \in \operatorname{Spec}(T)$ such that $\operatorname{ht}(Q) \leq m-1$ and $J \subseteq Q$ we have $\operatorname{ht}(Q \cap S) \leq \operatorname{ht}(Q)$.

(3) If $\operatorname{ht} J \geq m$, then φ is flat.

Proof. For item 1, if $\operatorname{ht}(Q) \geq m$, then $\operatorname{ht}(Q \cap S) \leq \dim(S) = m \leq \operatorname{ht}(Q)$, so by (2.3) $S \to T_Q$ is flat. Therefore $Q \notin \mathcal{F}_{\min}$. Item 1 now follows from (2.7.3).

The ($\Longrightarrow$) direction of item 2 is clear [M2, Theorem 9.5]. For ($\Longleftarrow$) of item 2 and for item 3, it suffices to show $\mathcal{F}_{\min}$ is empty and this holds by item 1. □

The following is an immediate corollary to (4.1).

4.2 Corollary. Let k be a field, let $x_1, \ldots, x_n$ be indeterminates over k and let $f, g \in k[x_1, \ldots, x_n]$ be algebraically independent over k. Consider the embedding $\varphi : S := k[f, g] \hookrightarrow T := k[x_1, \ldots, x_n]$ and let J be the associated Jacobian ideal. Then

(1) $\mathcal{F}_{\min} \subseteq \{\text{minimal primes } Q \text{ of } J \text{ with } \operatorname{ht}(Q \cap S) > \operatorname{ht} Q = 1\}$.

(2) φ is flat $\iff$ for every height-one prime ideal $Q \in \operatorname{Spec} T$ such that $J \subseteq Q$ we have $\operatorname{ht}(Q \cap S) \leq 1$.

(3) If $\operatorname{ht}(J) \geq 2$, then φ is flat.

In the case where k is algebraically closed, another argument can be used for (4.2.2): Each height-one prime ideal $Q \in \operatorname{Spec} T$ has the form $Q = hT$ for some element $h \in T$. If $\operatorname{ht}(P \cap S) = 2$, then $Q \cap S$ has the form $(f - a, g - b)S$, where $a, b \in k$. Thus $f - a = f_1 h$ and $g - b = g_1 h$ for some $f_1, g_1 \in T$. Now the Jacobian ideal of f, g is the same as the Jacobian ideal of $f - a, g - b$ and an easy computation shows this has h as a factor. Thus Q contains the Jacobian ideal, and so by assumption, $\operatorname{ht}(Q \cap S) \leq 1$, a contradiction.

4.3 Examples

Let k be a field of characteristic different from 2 and let x, y, z be indeterminates over k.

(1) With $f = x$ and $g = xy^2 - y$, consider $S := k[f, g] \xrightarrow{\varphi} T := k[x, y]$. Then $J = (2xy - 1)T$. Since $\operatorname{ht}((2xy - 1)T \cap S) = 1$, φ is flat. Hence $J \subsetneq F = T$.

(2) With $f = x$ and $g = yz$, consider $S := k[f, g] \xrightarrow{\varphi} T := k[x, y, z]$. Then $J = (y, z)T$. Since $\operatorname{ht} J \geq 2$, φ is flat. Again $J \subsetneq F = T$.

We are interested in extending Prop. 4.1 to the case of polynomial rings over a Noetherian domain. In this connection we first consider behavior with respect to prime ideals of R in a situation where the extension (1.2) is flat.

4.4 Proposition. Let R be a commutative ring, let $x_1, \ldots, x_n$ be indeterminates over R, and let $f_1, \ldots, f_m \in R[x_1, \ldots, x_n]$ be algebraically independent over R. Consider the embedding $\varphi : S := R[f_1, \ldots, f_m] \hookrightarrow T := R[x_1, \ldots, x_n]$.

(1) If $\mathbf{p} \in \operatorname{Spec} R$ and $\varphi_{\mathbf{p}T} : S \to T_{\mathbf{p}T}$ is flat, then $\mathbf{p}S = \mathbf{p}T \cap S$ and the images $\overline{f_i}$ of the f_i in $T/\mathbf{p}T \cong (R/\mathbf{p})[x_1, \ldots, x_n]$ are algebraically independent over $R/\mathbf{p}$.

(2) If φ is flat, then for each $\mathbf{p} \in \operatorname{Spec}(R)$ we have $\mathbf{p}S = \mathbf{p}T \cap S$ and the images $\overline{f_i}$ of the f_i in $T/\mathbf{p}T \cong (R/\mathbf{p})[x_1, \ldots, x_n]$ are algebraically independent over $R/\mathbf{p}$.

Proof. Item 2 follows from item 1, so it suffices to prove item 1. Assume that $T_{\mathbf{p}T}$ is flat over S. Then $\mathbf{p}T \neq T$ and it follows from [M2, Theorem 9.5] that $\mathbf{p}T \cap S = \mathbf{p}S$. If the $\overline{f_i}$ were algebraically dependent over $R/\mathbf{p}$, then there exist indeterminates $t_1, \ldots, t_m$ and a polynomial $G \in R[t_1, \ldots, t_m] - \mathbf{p}R[t_1, \ldots, t_m]$ such that $G(\overline{f_1}, \ldots, \overline{f_m}) \in \mathbf{p}T$. This implies $G(f_1, \ldots, f_m) \in \mathbf{p}T \cap S$. But $f_1, \ldots, f_m$ are algebraically independent over R and $G(t_1, \ldots, t_m) \notin \mathbf{p}R[t_1, \ldots, t_m]$ implies $G(f_1, \ldots, f_m) \notin \mathbf{p}S = \mathbf{p}T \cap S$, a contradiction. $\square$

4.5 Proposition. Let R be a Noetherian integral domain, let $x_1, \ldots, x_n$ be indeterminates over R and let $f_1, \ldots, f_m \in R[x_1, \ldots, x_n]$ be algebraically independent over R. Consider the embedding $\varphi : S := R[f_1, \ldots, f_m] \hookrightarrow T := R[x_1, \ldots, x_n]$ and let J denote the Jacobian ideal of φ. Then

(1) $\mathcal{F}_{\min} \subseteq \{Q \in \operatorname{Spec} T : J \subseteq Q, \dim(T/Q) \geq 1 \text{ and } \operatorname{ht}(Q \cap S) > \operatorname{ht} Q\}$.

(2) φ is flat $\iff$ $\operatorname{ht}(Q \cap S) \leq \operatorname{ht}(Q)$ for every nonmaximal $Q \in \operatorname{Spec}(T)$ with $J \subseteq Q$.

(3) If $\dim R = d$ and $\operatorname{ht} J \geq d + m$, then φ is flat.

Proof. For item 1, suppose $Q \in \mathcal{F}_{\min}$ is a maximal ideal of T. Then $\operatorname{ht} Q < \operatorname{ht}(Q \cap S)$ by (2.4.2). By localizing at $R - (R \cap Q)$, we may assume that R is local with maximal ideal $Q \cap R := \mathbf{m}$. Since Q is maximal, T/Q is a field finitely generated over $R/\mathbf{m}$. By the Hilbert Nullstellensatz [M2, Theorem 5.3], T/Q is algebraic over $R/\mathbf{m}$ and $\operatorname{ht}(Q) = \operatorname{ht}(\mathbf{m}) + n$. It follows that $Q \cap S = P$ is maximal in S and $\operatorname{ht}(P) = \operatorname{ht}(\mathbf{m}) + m$. But the algebraic independence hypothesis for the f_i implies $m \leq n$. This is a contradiction. Therefore item 1 follows from (2.7.3).

The ($\Longrightarrow$) direction of item 2 is clear. For ($\Longleftarrow$) of item 2 and for item 3, it suffices to show the set $\mathcal{F}_{\min}$ is empty, and this follows from item 1. $\square$

As an immediate corollary to (2.7) and (4.5), we have:

4.6 Corollary. Let R be a Noetherian integral domain, let $x_1, \ldots, x_n$ be indeterminates over R and let $f_1, \ldots, f_m \in R[x_1, \ldots, x_n]$ be algebraically independent over R. Consider the embedding $\varphi : S := R[f_1, \ldots, f_m] \hookrightarrow T := R[x_1, \ldots, x_n]$ and let J be the associated Jacobian ideal. Then φ is flat if for every nonmaximal $Q \in \operatorname{Spec}(T)$ such that $J \subseteq Q$ we have $\operatorname{ht}(Q \cap S) \leq \operatorname{ht}(Q)$.

Also as a corollary of (2.7) and (4.5) we have:

4.7 Corollary. Let R be a Noetherian ring, let $x_1, \ldots, x_n$ be indeterminates over R and let $f_1, \ldots, f_m \in R[x_1, \ldots, x_n]$ be algebraically independent over R. Consider the embedding $\varphi : S := R[f_1, \ldots, f_m] \hookrightarrow T := R[x_1, \ldots, x_n]$, let J be the Jacobian ideal of φ and let F be the (reduced) ideal which describes the nonflat locus of φ as in (2.4.2). Then $J \subseteq F$ and either $F = T$, that is, φ is flat, or $\dim(T/Q) \geq 1$, for all $Q \in \operatorname{Spec}(T)$ which are minimal over F.

4.8 Proposition. Let R be a Noetherian integral domain containing a field of characteristic zero. Let $x_1, \ldots, x_n$ be indeterminates over R and let $f_1, \ldots, f_m \in R[x_1, \ldots, x_n]$ be algebraically independent over R. Consider the embedding $\varphi : S := R[f_1, \ldots, f_m] \hookrightarrow T := R[x_1, \ldots, x_n]$ and let J be the associated Jacobian ideal. Then

(1) If $\mathbf{p} \in \operatorname{Spec} R$ and $J \subseteq \mathbf{p}T$, then $\mathbf{p}T \in \mathcal{F}$, i.e., $\varphi_{\mathbf{p}T}S \to T_{\mathbf{p}T}$ is not flat.

(2) If the embedding $\varphi : S \hookrightarrow T$ is flat, then for every $\mathbf{p} \in \operatorname{Spec}(R)$ we have $J \not\subseteq \mathbf{p}T$.

Proof. Item 2 follows from item 1, so it suffices to prove item 1. Let $\mathbf{p} \in \operatorname{Spec} R$ with $J \subseteq \mathbf{p}T$, and suppose $\varphi_{\mathbf{p}T}$ is flat. Let $\overline{f_i}$ denote the image of f_i in $T/\mathbf{p}T$. Consider

$$\overline{\varphi} : \overline{S} := (R/\mathbf{p})[\overline{f_1}, \ldots, \overline{f_m}] \to \overline{T} := (R/\mathbf{p})[x_1, \ldots, x_n].$$

By Proposition 4.4, $\overline{f_1}, \ldots, \overline{f_m}$ are algebraically independent over $\overline{R} := R/\mathbf{p}$. Since the Jacobian ideal commutes with homomorphic images, the Jacobian ideal of $\overline{\varphi}$ is zero. Thus for each $Q \in \operatorname{Spec} \overline{T}$ the map $\overline{\varphi}_Q : \overline{S} \to \overline{T_Q}$ is not smooth. But taking $Q = (0)$ gives $\overline{T_Q}$ which is a field separable over the fraction field of $\overline{S}$ and hence $\overline{\varphi_Q}$ is a smooth map. This contradiction completes the proof. □

5. EXAMPLES

5.1 Examples

For each positive integer n, we present an example of a 3-dimensional quasilocal unique factorization domain B such that

(1) B is not catenary,

(2) the maximal ideal of B is 2-generated,

(3) B has precisely n prime ideals of height two,

(4) Each prime ideal of B of height two is not finitely generated,

(5) For every non-maximal prime P of B the ring B_P is Noetherian.

The notation for this construction is a localized version of the notation of Section 3.5, with $s = 1$. Thus k is a field, $R = k[x, y]_{(x,y)}$ is a 2-dimensional regular local ring and $R^* = k[y]_{(y)}[[x]]$ is the (x)-adic completion of R. Let $\tau = \sum_{j=1}^{\infty} c_j x^j \in xk[[x]]$ be algebraically independent over $k(x)$. Let $p_i \in R - xR$ be such that $p_i R^*$ are n distinct prime ideals. For example, we could take $p_i = y - x^i$. Let $q = p_1 \cdots p_n$. We set $f := q\tau$ and consider the injective R-algebra homomorphism $S = R[f] \hookrightarrow R[\tau] = T$.

Let B be the nested union domain associated to f as in (2.1). If $\tau_r = \sum_{j=r+1}^{\infty} \frac{c_j x^j}{x^r}$ is the rth endpiece of τ, then $\rho_r := q\tau_r$ is the rth endpiece of f. For each $r \in \mathbb{N}$, let $B_r = R[\rho_r]_{(x,y,\rho_r)}$. Then each B_r is a 3-dimensional regular local ring and $B = \bigcup_{r=1}^{\infty} B_r$.

The map $\alpha : S \to R_x^*$ is not flat since $p_i R_x^*$ is a height-one prime and $p_i R_x^* \cap S = (p_i, f)S$ is of height two. By Theorem 2.2, B is not Noetherian. By [HRW4, Theorem 4.5], B is a quasilocal unique factorization domain. Moreover, by [HRW4, Theorem 4.4], for each $t \in \mathbb{N}, x^t B = x^t R^* \cap B$ and $R/x^t R = B/x^t B = R^*/x^t R^*$. It follows that the maximal ideal of B is $(x, y)B$. If $P \in \operatorname{Spec} B$ is such that $P \cap R = (0)$, then because the field of fractions $K(f)$ of B has transcendence degree one over the field of fractions K of R, $\operatorname{ht}(P) \leq 1$ and hence because B is a UFD, P is principal.

Claim 1. Let I be an ideal of B and let $t \in \mathbb{N}$. If $x^t \in IR^*$, then $x^t \in I$.

Proof. There exist elements $b_1, \ldots, b_s \in I$ such that $IR^* = (b_1, \ldots, b_s)R^*$. If $x^t \in IR^*$, there exist $\alpha_i \in R^*$ such that

$$x^t = \alpha_1 b_1 + \cdots + \alpha_s b_s.$$

We have $\alpha_i = a_i + x^{t+1}\lambda_i$, where $a_i \in B$ and $\lambda_i \in R^*$. Thus

$$x^t[1 - x(b_1\lambda_1 + \cdots + b_s\lambda_s)] = a_1 b_1 + \cdots + a_s b_s \in B.$$

Since $x^t R^* \cap B = x^t B$, $\gamma := 1 - x(b_1\lambda_1 + \cdots + b_s\lambda_s) \in B$. Moreover, γ is invertible in R^* and hence also in B. It follows that $x^t \in I$. $\square$

To examine more closely the prime ideal structure of B, it is useful to consider the inclusion map $B \hookrightarrow A := R^* \cap K(f)$ and the map $\operatorname{Spec} A \to \operatorname{Spec} B$.

5.2 Proposition. With the notation of Example 5.1 and $A = R^* \cap K(f)$, we have

(1) A is a two-dimensional regular local domain with maximal ideal $\mathbf{m}_A = (x, y)A$.

(2) $\mathbf{m}_A$ is the unique prime of A lying over $\mathbf{m}_B = (x, y)B$, the maximal ideal of B.

(3) If $P \in \operatorname{Spec} B$ is nonmaximal, then $\operatorname{ht}(PR^*) \leq 1$ and $\operatorname{ht}(PA) \leq 1$. Thus every nonmaximal prime of B is contained in a nonmaximal prime of A.

(4) If $P \in \operatorname{Spec} B$ and $xq \notin P$, then $\operatorname{ht} P \leq 1$.

(5) If $P \in \operatorname{Spec} B$, $\operatorname{ht} P = 1$ and $P \cap R \neq 0$, then $P = (P \cap R)B$.

Proof. By Proposition 3.11 (the result of Valabrega) $A := R^* \cap K(f)$ is a two-dimensional regular local domain having the same completion as R and R^*. This proves item 1. Since $B/xB = A/xA = R^*/xR^*$, $\mathbf{m}_A = (x, y)A$ is the unique prime of A lying over $\mathbf{m}_B = (x, y)B$. Thus item 2 holds and also item 3 if $x \in P$. To see (3), it remains to consider $P \in \operatorname{Spec} B$ with $x \notin P$. By Claim 1, for all $t \in \mathbb{N}$, $x^t \notin PR^*$. Thus $\operatorname{ht}(PR^*) \leq 1$. Since $A \hookrightarrow R^*$ is faithfully flat, $\operatorname{ht}(PA) \leq 1$.

For (4), we see by (3) that $\operatorname{ht}(PA) \leq 1$. Let $Q \in \operatorname{Spec} A$ be a height-one prime ideal such that $P \subseteq Q$. Since $xq \notin P$, we have $B_P = S_{P \cap S} = T_{Q \cap T} = A_Q$, where $S = R[f]$ and $T = R[\tau]$. Thus $\operatorname{ht}(P) \leq 1$. For (5), if $x \in P$, then $P = xB$ and the statement is clear. Assume $x \notin P$. Since B_x is a localization of $(B_r)_x$, we have $(P \cap R)B_r = P \cap B_r$ for all $r \in \mathbb{N}$. Thus $P = (P \cap R)B$. $\square$

We observe that the DVRs B_{xB} and A_{xA} are equal. Moreover, A is the nested union $\bigcup_{r=1}^{\infty} R[\tau_n]_{(x,y,\tau_n)}$ of 3-dimensional regular local domains. Since A is a two-dimension regular local domain each nonmaximal prime of A is principal. If pA is a height-one prime of A with $pA \notin \{p_1A, \ldots p_nA\}$, then $A_{pA} = B_{pA \cap B}$ and $\operatorname{ht}(pA \cap B) = 1$. We observe in Claim 2 that $p_iA \cap B$ has height two and is not finitely generated.

Claim 2. Let p_i be one of the prime factors of q. Then p_iB is prime in B. Moreover

(1) p_iB and $Q_i := (p_i, \rho_1, \rho_2, \dots)B = p_iA \cap B$ are the only primes of B lying over p_iR in R,

(2) Q_i is of height two and is not finitely generated.

Proof. We use that $B = \bigcup_{r=1}^{\infty} B_r$, where $B_r = R[\rho_r]_{(x,y,\rho_r)}$ is a 3-dimensional regular local ring. For each $r \in \mathbb{N}$, p_iB_r is prime in B_r. Hence p_iB is a height-one prime ideal of B, for $i = 1, \dots, n$. Since $\rho_r = q\tau_r$, $p_iA \cap B_r = (p_i, \rho_r)B_r$ is a height-two prime ideal of the 3-dimensional regular local domain B_r. Therefore $Q_i := (p_i, \rho_1, \rho_2, \dots)B = p_iA \cap B$ is a nested union of prime ideals of height two, so $\text{ht}(Q_i) \leq 2$. Since p_iB is a nonzero prime ideal properly contained in Q_i, $\text{ht}(Q_i) = 2$. Moreover $x \notin (p_i, \rho_r)B_r$ for each r, so $x \notin Q_i$. Hence for each $r \in \mathbb{N}$, $\rho_{r+1} \notin (p_i, \rho_r)B$ and Q_i is not finitely generated. □

Since $x \notin Q_i$ and $B[1/x]$ is a localization of the Noetherian domain $B_n[1/x]$, we see that B_{Q_i} is Noetherian. Since the Q_i are the only prime ideals of B of height two and B is a UFD, B_P is Noetherian for every non-maximal prime P.

This completes the presentation of Examples 5.1. With regard to the birational inclusion $B \hookrightarrow A$ and the map $\operatorname{Spec} A \to \operatorname{Spec} B$, we remark that the following holds: Each Q_i contains infinitely many height-one primes of B that are the contraction of primes of A and infinitely many that are not. Among the primes that are not contracted from A are the p_iB. In the terminology of [ZS, page 325], P is *not lost* in A if $PA \cap B = P$. Since $p_iA \cap B = Q_i$ properly contains p_iB, p_iB is lost in A. Since $(x, y)B$ is the maximal ideal of B and $(x, y)A$ is the maximal ideal of A and B is integrally closed, a version of Zariski's Main Theorem [Pe], [Ev], implies that A is not essentially finitely generated as a B-algebra.

REFERENCES

A. Y. Akizuki (1935). Einige Bemerkungen über primäre Integritatsbereiche mit Teilerketten satz, *Proc. Phys-Math. Soc. Japan, 17:* 327-336.

BR1. M. Brodmann and C. Rotthaus (1982). Local rings with bad sets of formal prime divisors, *J. Algebra, 75:* 386-94.

BR2. M. Brodmann and C. Rotthaus (1983). A peculiar unmixed domain, *Proc.*

AMS, *87:* 596-600.

E. R. Elkik (1973,). Solutions d'equations a coefficients dans un anneau Hensélian, *Ann. scient. Ec. Norm. Sup. 4th serie, t.6:* 553-604.

Ev. E. Evans (1970). A generalization of Zariski's Main Theorem, *Proc. Amer. Math. Soc.*, *26:* 45-48.

Gr. A.Grothendieck (1967). Élements de Géometrie algebrique IV_4, *Publ. Math. Inst. Hautes Etud. Sci.*, *32.*

HRW1. W. Heinzer, C. Rotthaus and S. Wiegand (1998). Building Noetherian domains inside an ideal-adic completion, *Abelian Groups, Module Theory and Topology Proceedings in Honor of Adalberto Orsatti's 60th Birthday* (Dikranjan and Salce, ed.), Dekker Inc, New York, pp. 279-287.

HRW2. ______ (1997). Noetherian rings between a semilocal domain and its completion, *J. of Algebra*, *198:* 627-655.

HRW3. ______ (1999). Noetherian domains inside a homomorphic image of a completion, *J. of Algebra*, *215:* 666-681.

HRW4. ______ (1999). Intermediate rings between a local domain and its completion, *Illinois J. Math.*, *43:* 19-46.

HRW5. ______. Intermediate rings between a local domain and its completion II, to appear, *Illinois J. Math.*.

HRW6. ______. Building Noetherian and non-Noetherian integral domains using power series, *Proceedings of Conference in Honor of James Huckaba.*

H1. R. Heitmann (1979). Examples of noncatenary rings, *Trans. Amer. Mat. Soc.*, *247:* 125-136.

M1. H. Matsumura (1980). *Commutative Algebra 2nd Edition*, Benjamin-Cummings, Reading, MA.

M2. ______ (1986). *Commutative Ring Theory*, Cambridge University Press, Cambridge.

N. M. Nagata (1962). *Local Rings*, John Wiley, New York.

O1. T. Ogoma (1980). Non-catenary pseudo-geometric normal rings, *Japan J. Math.*, *6:* 147-163.

O2. ______ (1982). Cohen-Macaulay factorial domain is not necessarily Gorenstein, *Mem. Fac. Sci. Kochi Univ.*, *3:* 65-74.

Pe. C. Peskine (1966). Une généralsation du "main theorem" de Zariski, *Bull. Sci. Math.*, *90:* 377-408.

P. G. Picavet (1991). Endomorphismes polynomiaux plats, *J. Algebra*, *138:* 195-225.

R1. C. Rotthaus (1977). Nicht ausgezeichnete, universell japanische Ringe, *Math. Z.*, *152:* 107-125.

R2. ______ (1979). Universell japanische Ringe mit nicht offenen regulärem Ort, *Nagoya Math. J.*, *74:* 123-135.

R3. ______ (1980). Komplettierung semilokaler quasiausgezeichneter Ringe, *Nagoya Math. J.*, *76:* 173-180.

S. R. Swan. Néron-Popescu desingularization, preprint.

V. P. Valabrega (1973). On two-dimensional regular local rings and a lifting problem, *Ann. Scuola Norm. Sup. Pisa*, *27:* 1-21.

W. S. Wang (1980). A Jacobian criterion for separability, *J. Algebra*, *65:* 453-494.

ZS. O. Zariski and P. Samuel (1960). *Commutative Algebra volume II*, D. Van Nostrand, New York.

Generalized Going-Up Homomorphisms of Commutative Rings

ANDREW J. HETZEL Department of Mathematics, University of Tennessee, Knoxville, Tennessee 37996-1300, U. S. A., ahetzel@math.utk.edu

ABSTRACT

Dobbs, Fontana, and Picavet have recently proved a variety of results concerning generalized going-down homomorphisms. In this paper, we provide many analogous results for generalized going-up homomorphisms.

1 INTRODUCTION

All rings considered below are commutative with identity, and all ring homomorphisms are unital. Adapting the notation in [9, p. 28], we let GU, GD, LO, and INC denote the going-up, going-down, lying-over, and incomparable properties, respectively, for ring homomorphisms. As well, suppose that $f\colon A \to B$ is a ring homomorphism. Consider $X = \{P_i : i \in I\}$, a subset of $\text{Spec}(A)$. (The notation is generally taken so that $P_i \neq P_j$ whenever $i \neq j$; as a result, $|X| = |I|$.) A subset $Y = \{Q_i : i \in I\}$ of $\text{Spec}(B)$ is said to *cover* (or to *dominate*) X if $f^{-1}(Q_i) = P_i$ for each $i \in I$. (By the notational convention, $Q_i \neq Q_j$ if $i \neq j$, and so $|Y| = |I|$.) As in [4], we say that f is a *chain morphism* if, for each chain X in $\text{Spec}(A)$, there exists a chain Y in $\text{Spec}(B)$ such that Y covers X. Furthermore, as in [4], a chain X is called a *local chain* if X has a (necessarily unique) maximal element, namely $\mathcal{U}(X)$, and a ring homomorphism $f\colon A \to B$ is said to satisfy the *generalized going-down* property (GGD) if the following holds: for each local chain X in $\text{Spec}(A)$ and each $Q \in \text{Spec}(B)$ such that $f^{-1}(Q) = \mathcal{U}(X)$, there exists a local chain Y in $\text{Spec}(B)$ such that $\mathcal{U}(Y) = Q$ and Y covers X. Finally, as in [4], let ${}^a f\colon \text{Spec}(B) \to \text{Spec}(A)$ be the associated map that takes Q to $f^{-1}(Q)$.

During the past 25 years, the notion of rings $A \subset B$ satisfying GD has attracted considerable attention. Recently, in [4], Dobbs, Fontana, and Picavet have considered the question of when a ring homomorphism $f\colon A \to B$ satisfies GGD. In this paper, we consider the question of when a ring homomorphism $f\colon A \to B$ satisfies the analogous generalized going-up property (GGU).

The work presented below is a portion of the author's dissertation work currently in progress under the direction of Dr. David E. Dobbs at the University of Tennessee, Knoxville. The author would like to thank Dr. Dobbs for his guidance and, most of all, his patience.

As this paper went to press (May, 2002), Kang-Oh have announced a preprint whose methods, we have determined, can be extended to show that GGU and GU are equivalent for arbitrary ring homomorphisms. Accordingly, in this late revision, we have deleted the statements and proofs of special cases given in our earlier draft for the implication GU $\Rightarrow$ GGU. Nevertheless, we have retained the GGU formulations of our earlier draft, to best convey the intended spirit and the role of chain morphisms.

2 GENERALIZED GOING-UP

Let A be a ring and X a subset of $\mathrm{Spec}(A)$. Following [11], we define $\mathcal{R}(X) := \cap\{P : P \in X\}$. Observe that if X is a chain, then $\mathcal{R}(X) \in \mathrm{Spec}(A)$ [9, Theorem 9]. A chain X is called a *rooted chain* if X has a (necessarily unique) minimal element. If X is a chain, then $X \cup \{\mathcal{R}(X)\}$ is a rooted chain; in fact, a chain X is a rooted chain if and only if $\mathcal{R}(X) \in X$.

We begin with a result whose statement and proof are dual to those of [4, Proposition 2.1].

PROPOSITION 2.1. *Let $f \colon A \to B$ be a ring homomorphism. Then:*

(a) If a rooted chain Y in $\mathrm{Spec}(B)$ *covers a subset X of* $\mathrm{Spec}(A)$, *then X is a rooted chain.*

(b) If a chain Y in $\mathrm{Spec}(B)$ *covers a rooted chain X in* $\mathrm{Spec}(A)$, *then Y is a rooted chain and $f^{-1}(\mathcal{R}(Y)) = \mathcal{R}(X)$.*

(c) If f is a chain morphism and X is a rooted chain in $\mathrm{Spec}(A)$, *then X is covered by some rooted chain Y in* $\mathrm{Spec}(B)$ *and $f^{-1}(\mathcal{R}(Y)) = \mathcal{R}(X)$.*

Proof. (a) By the above observation, X is a chain. If $P \in X$, there exists $Q \in Y$ such that $f^{-1}(Q) = P$, whence $P \supseteq f^{-1}(\mathcal{R}(Y))$. It follows that $\mathcal{R}(X) = f^{-1}(\mathcal{R}(Y)) \in X$, and so X is a rooted chain.

(b) Choose $Q \in Y$ such that $f^{-1}(Q) = \mathcal{R}(X)$. If $Q_1 \subset Q \in Y$, then $f^{-1}(Q_1) \subset f^{-1}(Q)$, contradicting the fact that $f^{-1}(Q) = \mathcal{R}(X) \subseteq f^{-1}(Q_1)$. Thus, $\mathcal{R}(Y) = Q \in Y$, and so Y is a rooted chain. Then $f^{-1}(\mathcal{R}(Y)) = \mathcal{R}(X)$ by the proof of (a).

(c) Apply (b). $\square$

The next proposition provides for the lifting of chains of prime ideals in a ring to a chain of prime ideals in a valuation domain. Its statement and proof are dual to those of [4, Proposition 2.4].

PROPOSITION 2.2. *Let A be a ring and let X be a subset of* $\mathrm{Spec}(A)$. *Then:*

(a) [4, Proposition 2.4 (a)] If X is a chain, then its patch closure X^c is also a chain.

(b) [4, Proposition 2.4 (b)] X is a chain if and only if there exists a ring homomorphism $A \to V$ and a chain Y in $\text{Spec}(V)$ *such that V is a valuation domain and Y covers X.*

(c) X is a rooted chain if and only if there exists a ring homomorphism $f: A \to V$ and a rooted chain Y in $\text{Spec}(V)$ *such that V is a valuation domain, Y covers X, and $f^{-1}(\mathcal{R}(Y)) = \mathcal{R}(X)$.*

Proof. (c) The "if" assertion follows from Proposition 2.1 (a). The "only if" assertion follows by combining (b) with Proposition 2.1 (b). □

COROLLARY 2.3. *Let A be a ring, X a chain in* $\text{Spec}(A)$, *and $P \in$* $\text{Spec}(A)$ *such that $\mathcal{R}(X) \supseteq P$. Then there exists a ring homomorphism $f: A \to V$, a chain Y in* $\text{Spec}(V)$, *and $Q \in$* $\text{Spec}(V)$ *such that V is a valuation domain, Y covers X, $f^{-1}(Q) = P$, and $\mathcal{R}(Y) \supseteq Q$.*

Proof. We dualize the proof of [4, Corollary 2.5]. Apply Proposition 2.2 (c) to the rooted chain $X \cup \{P\}$, to obtain a suitable rooted chain Z in $\text{Spec}(V)$. It suffices to take $Q = \mathcal{R}(Z)$; and $Y = Z$ (resp., $Z \setminus \{Q\}$) if $P \in X$ (resp., $P \notin X$). □

We refer the reader to [4] for further results on chain morphisms and coverings.

We now proceed to the key definition of this paper. A ring homomorphism $f: A \to B$ is said to satisfy the *generalized going-up* property (GGU) if the following holds: for each rooted chain X in $\text{Spec}(A)$ and each $Q \in \text{Spec}(B)$ such that $f^{-1}(Q) = \mathcal{R}(X)$, there exists a rooted chain Y in $\text{Spec}(B)$ such that $\mathcal{R}(Y) = Q$ and Y covers X. In their preprint, Kang-Oh have identified that, for extensions of commutative rings, what we are calling the GGU-property is equivalent to what they have called the SCLO-property. It is straightforward to verify that this equivalence holds for arbitrary ring homomorphisms. Evidently, GGU $\Rightarrow$ GU. For historical reasons, we now record an instance where the reverse implication was shown to be true.

PROPOSITION 2.4. *[3, Theorem] Let A be a ring such that each chain in* $\text{Spec}(A)$ *is well-ordered via inclusion. Then a ring homomorphism $f: A \to B$ satisfies GGU if (and only if) f satisfies GU.*

Of course, the two concepts of "GGU" and "chain morphism" are logically independent: if R is a ring of non-zero (Krull) dimension and P is not a minimal prime of R, then the canonical projection $R \twoheadrightarrow R/P$ is not a chain morphism but does satisfy GGU; if R is a ring of non-zero (Krull) dimension and x is an indeterminate over R, then [9, Exercise 3, p. 41] gives that the canonical injection $R \hookrightarrow R[x]$ is a chain morphism that does not satisfy GU and, hence, does not satisfy GGU. Nevertheless, we do have the following connection between "GGU" and "chain morphism" which dualizes [4, Proposition 3.2].

PROPOSITION 2.5. *Let $f\colon A \to B$ be a ring homomorphism. Then:*

(a) If f satisfies LO and GGU, then f is a chain morphism.

(b) If ${}^a f$ is injective and f is a chain morphism, then f satisfies GGU.

Let $f\colon A \to B$ be a ring homomorphism and let $P \in \mathrm{Spec}(A)$. It is well known that ${}^a f^{-1}(P)$, the so-called *topological fiber of P (with respect to f)*, is homeomorphic to $\mathrm{Spec}((A_P/PA_P)\otimes_A B)$ in both the Zariski topology and the flat topology. One calls $(A_P/PA_P)\otimes_A B \cong B_P/PB_P$ *the fiber of f at P*; its associated reduced ring, $B_P/\sqrt{PB_P}$, is called the *reduced fiber (of f at P)* as noted in [4]. It is easy to show, via Zorn's Lemma and [9, Theorem 9], that each element of ${}^a f^{-1}(P)$ is contained in some maximal element of ${}^a f^{-1}(P)$ and contains some minimal element of ${}^a f^{-1}(P)$. It follows that ${}^a f^{-1}(P)$ has a unique maximal (resp., unique minimal) element if and only if the reduced fiber of f at P is a quasilocal ring (resp., an integral domain); that is (cf. [11, Lemme 2.5]), if and only if ${}^a f^{-1}(P)$ is irreducible in the flat (resp., Zariski) topology.

The next result, which dualizes [4, Proposition 3.3], introduces a useful argument.

PROPOSITION 2.6. *If B is a quasilocal treed ring and $f\colon A \to B$ is a chain morphism, then f satisfies GGU.*

Proof. Consider a rooted chain $X = \{P_i : i \in I\}$ in $\mathrm{Spec}(A)$ and $Q \in \mathrm{Spec}(B)$ such that $f^{-1}(Q) = \mathcal{R}(X)$. Since f is a chain morphism, Proposition 2.1 (c) provides a rooted chain $Y = \{Q_i : i \in I\}$ in $\mathrm{Spec}(B)$ that covers X, with $f^{-1}(\mathcal{R}(Y)) = \mathcal{R}(X)$. Choose (the unique) $j \in I$ such that $P_j = \mathcal{R}(X)$. Then $Q_j = \mathcal{R}(Y)$. If $Q_j = Q$, then Y is the desired rooted chain Z in $\mathrm{Spec}(B)$ such that $\mathcal{R}(Z) = Q$ and Z covers X. If $Q \subset Q_j$, then $Z := (Y\backslash\{Q_j\}) \cup \{Q\}$ suffices. Since B is quasilocal treed, there is only one remaining case, namely, $Q_j \subset Q$. For this case, it also suffices to take $Z := (Y\backslash\{Q_j\}) \cup \{Q\}$. □

We next infer a dual of [4, Corollary 3.4].

COROLLARY 2.7. *Let B be a quasilocal treed ring. Let $f\colon A \to B$ be a ring homomorphism that satisfies both LO and GD. Then f satisfies GGU.*

Proof. If $P \in \mathrm{Im}({}^a f)$, then B quasilocal treed implies that ${}^a f^{-1}(P)$ has a unique maximal element and a unique minimal element; that is, each reduced fiber of f is a quasilocal integral domain. The conclusion therefore follows by combining [4, Theorem 2.3] and the proof of Proposition 2.6. □

By reworking the proof of Proposition 2.6, we next find a companion result; this is the dual of [4, Corollary 3.5].

COROLLARY 2.8. *Let $f\colon A \to B$ be a chain morphism that satisfies at least one of the following two conditions:*

(i) B is treed and each reduced fiber of f is quasilocal;

(ii) Each (Zariski-) irreducible component of $\mathrm{Spec}(B)$ *is a chain (via inclusion)*

and each reduced fiber of f is an integral domain.

Then f satisfies GGU.

Proof. We proceed to rework the proof of Proposition 2.6. It suffices to verify that Q_j and Q are comparable via inclusion. In case (i), this follows since B is treed and Q_j, Q are each contained in (any maximal ideal of B that contains) the unique maximal element of ${}^af^{-1}(P_j)$. An essentially "dual" proof is available if (ii) holds. Indeed, Q_j, Q each contain the unique minimal element I of ${}^af^{-1}(P_j)$. Using Zorn's Lemma, choose a minimal prime ideal N of B such that $N \subseteq I$ [9, Theorem 10]. Then Q_j, Q are each in the (Zariski-) irreducible set $V(N)$, which is a chain by hypothesis, whence Q_j and Q are comparable. □

In [4, Remark 3.7], Dobbs, Fontana, and Picavet showed that it is possible to characterize GGD in terms of chains that are not necessarily local chains. We next note that there is an analogous characterization of GGU in terms of chains that are not necessarily rooted chains. Indeed, it is easy to see that a ring homomorphism $f: A \to B$ satisfies GGU if and only if the following holds: for each chain X in $\mathrm{Spec}(A)$, each $P \in \mathrm{Spec}(A)$ such that $P \subseteq \mathcal{R}(X)$, and each $Q \in {}^af^{-1}(P)$, there exists a chain Y in $\mathrm{Spec}(B)$ such that $Q \subseteq \mathcal{R}(Y)$ and Y covers X.

We next collect some useful facts about the GGU property which are dual to those in [4, Proposition 3.8].

PROPOSITION 2.9. *(a) Let $f: A \to B$ and $g: B \to C$ be ring homomorphisms. If f and g each satisfies GGU, so does $g \circ f$. If g satisfies LO and $g \circ f$ satisfies GGU, then f satisfies GGU.*

(b) If f is a ring homomorphism, then the following seven conditions are equivalent:

(1) f satisfies GGU;

(2) $f_S: A_S \to B_S := B \otimes_A A_S$ satisfies GGU for each multiplicatively closed subset S of A;

(3) $f_P: A_P \to B_P := B \otimes_A A_P$ satisfies GGU for each $P \in \mathrm{Spec}(A)$;

(4) $A_P \to B_Q$ satisfies GGU for each $Q \in \mathrm{Spec}(B)$ and $P := f^{-1}(Q)$;

(5) $A/I \to B/IB$ satisfies GGU for each ideal I of A;

(6) $A/P \to B/PB$ satisfies GGU for each minimal prime ideal P of A;

(7) f_{red} satisfies GGU.

(c) Let $f_i: A_i \to B_i$ $(i = 1, \ldots, n)$ be finitely many ring homomorphisms. Then the induced map $A_1 \times \cdots \times A_n \to B_1 \times \cdots \times B_n$ satisfies GGU if and only if f_i satisfies GGU for each i. If $A_1 = \ldots = A_n =: A$, then the induced map $A \to B_1 \times \cdots \times B_n$ satisfies GGU if and only if f_i satisfies GGU for each i.

Recall from [1] and [6] that an integral domain A is called a *going-down domain* if $A \subseteq B$ satisfies GD for each overring B of A. The most natural examples of going-down domains are arbitrary valuation domains and the integral domains of (Krull) dimension at most 1. As in [2], a ring A is called a *going-down ring* if A/P is a going-down domain for each (equivalently, each minimal) prime ideal P of A.

Any integral domain is a going-down ring if and only if it is a going-down domain [2, Remark (a), p. 4]; any ring of dimension at most 1 is a going-down ring [2, Proposition 2.1 (c)]; a finite ring product $A_1 \times \cdots \times A_n$ is a going-down ring if and only if each A_i is a going-down ring [2, Proposition 2.1 (b)]; but there exists a going-down ring A and an overring B of A such that $A \subseteq B$ does not satisfy GD [2, Example 1, p. 9]. We will say that a ring homomorphism $f\colon A \to B$ is a *max morphism* if $f^{-1}(Q)$ is a maximal ideal of A for each maximal prime ideal Q of B. It is evident that if a ring homomorphism f satisfies GU, then f is a max morphism. As in [5], we call a ring a *pm-ring* if every prime ideal is contained in a unique maximal ideal (for example, any quasilocal ring is a pm-ring). We then have the following analogue of [4, Theorem 3.9].

THEOREM 2.10. *Let A be a pm-ring ring and a going-down ring and let* $f\colon A \to B$ *be a ring homomorphism. Then the following conditions are equivalent:*

(1) f is a max morphism;

(2) f satisfies GU;

(3) f satisfies GGU.

Proof. By the above comments, (3) ⇒ (2) ⇒ (1). It remains to show that if f is a max morphism, X a rooted chain in $\mathrm{Spec}(A)$ and $Q \in \mathrm{Spec}(B)$ such that $f^{-1}(Q) = \mathcal{R}(X)$, then there exists a rooted chain Y in $\mathrm{Spec}(B)$ such that $\mathcal{R}(Y) = Q$ and Y covers X. There is no harm in replacing A with $A/\mathcal{R}(X)$, B with B/Q, and f with $A/\mathcal{R}(X) \hookrightarrow B/Q$. Hence, without loss of generality, $A \subseteq B$ are integral domains and A is a quasilocal going- down domain. Now, take M to be a maximal ideal of B such that $M \cap A$ is the maximal ideal of A. Choose a valuation overring (V,N) of B such that $N \cap B = M$ (cf. [7, Theorem 19.6]). Of course, V is quasilocal and treed. Moreover, $A \subseteq V$ satisfies GD since A is a going-down domain. As well, since $A \subseteq V$ satisfies GD and every prime ideal of A survives in V, $A \subseteq V$ satisfies LO. Hence, by Corollary 2.7, $A \subseteq V$ satisfies GGU. Thus, there exists a rooted chain $Z = \{Q_i\}$ in $\mathrm{Spec}(V)$ such that Z covers X and $\mathcal{R}(Z) = \{0\}$. Then, by Proposition 2.1 (b), $Y := \{Q_i \cap B\}$ has the desired properties. □

Proposition 2.5 (b) illustrated that GGU-theoretic consequences can ensue in the presence of a ring homomorphism f for which ${}^a f$ is injective. We next pursue this theme by enhancing the set-theoretic restriction with a topological one. Specifically, as in [4], we say that a continuous function $f\colon X \to Y$ of topological spaces is a *topological immersion* if the induced map $X \to f(X)$ is a homeomorphism (that is, injective and either open or closed). It is straightforward to verify that a continuous map $f\colon X \to Y$ is a topological immersion if and only if f is injective and $f^{-1}(\overline{f(Z)}) = \overline{Z}$ for each subset Z of X. Our main interest here concerns ring homomorphisms $f\colon A \to B$ for which ${}^a f\colon \mathrm{Spec}(B) \to \mathrm{Spec}(A)$ is a topological immersion (relative to the Zariski topology); in such a case, we also call f a *topological immersion*. There are many ring-theoretic characterizations of such f. A particularly useful characterization is given next.

PROPOSITION 2.11. *Let $f: A \to B$ be a ring homomorphism. Then:*

(a) [4, Proposition 3.20] The following two conditions are equivalent:

(1) If Q_1 and Q_2 are prime ideals of B such that $f^{-1}(Q_1) \subseteq f^{-1}(Q_2)$, then $Q_1 \subseteq Q_2$;

(2) f is a topological immersion.

(b) Suppose that the equivalent conditions in (a) hold and that a subset Y of $\mathrm{Spec}(B)$ *covers a subset X of* $\mathrm{Spec}(A)$. *Then Y is a chain (resp., rooted chain) if and only if X is a chain (resp., rooted chain).*

Proof. (b) In view of Proposition 2.1 (a), (b), it remains only to show that if $X =: \{P_i\}$ is a chain in $\mathrm{Spec}(A)$, then so is $Y =: \{Q_i\}$ where Y is a subset of $\mathrm{Spec}(B)$ that covers X. As $f^{-1}(Q_i) = P_i$ for each i, the conclusion follows from condition (1) in Proposition 2.11 (a). □

We next mention a family of examples of ring homomorphisms that were noted in [4] to induce topological immersions; the verifications follow most readily by checking condition (1) in Proposition 2.11. The family consists of the flat epimorphisms (that is, the flat maps $A \to B$ such that the induced multiplication map $B \otimes_A B \to B$ is an isomorphism). In particular, the structure map of any ring of fractions $A \to A_S$ is a topological immersion.

The next result is the dual of [4, Corollary 3.21].

COROLLARY 2.12. *Let $f: A \to B$ be a ring homomorphism. Then the following conditions are equivalent:*

(1) ${}^af$ is injective and f satisfies GU;

(2) f is a topological immersion and satisfies GGU.

Proof. (2) $\Rightarrow$ (1) trivially. Conversely, assume (1). One then readily verifies condition (1) in Proposition 2.11, and so f is a topological immersion. Next, to verify that f satisfies GGU, consider a rooted chain $X = \{P_i\}$ in $\mathrm{Spec}(A)$ and $Q \in \mathrm{Spec}(B)$ such that $f^{-1}(Q) = \mathcal{R}(X)$. For each i, take Q_i to be the unique element of ${}^af^{-1}(P_i)$. It follows from (1) that $P_i \subseteq P_j$ entails $Q_i \subseteq Q_j$. Accordingly, $Y := \{Q_i\}$ is a rooted chain in $\mathrm{Spec}(B)$ such that $\mathcal{R}(Y) = Q$ and Y covers X, as desired. □

The next result is the dual of [4, Corollary 3.22].

COROLLARY 2.13. *Let $f: A \to B$ be a ring homomorphism such that ${}^af$ is a topological immersion with closed image. Then the induced inclusion of rings* $A/\ker(f) \hookrightarrow B$ *satisfies GGU.*

Proof. Put $I := \ker(f)$. We begin with a fact that depends only on f being a ring homomorphism, namely, that $\overline{\mathrm{Im}({}^af)} = V(I)$. (To fashion a proof, recall that minimal prime ideals of a base ring are lain over from any ring extension [9, Exercise 1, p. 41] and Zariski-closed sets are stable under specialization.) Under the given assumptions, it follows that $\mathrm{Im}({}^af) = V(I)$.

Our task is to show that if X is a rooted chain in $\mathrm{Spec}(A/I)$ and $Q \in \mathrm{Spec}(B)$

lies over $\mathcal{R}(X)$, then there exists a rooted chain Y in $\mathrm{Spec}(B)$ such that $\mathcal{R}(Y) = Q$ and Y covers X. Of course, X induces a rooted chain Z in $\mathrm{Spec}(A)$ such that $Z \subseteq V(I)$ and Q lies over $\mathcal{R}(Z)$. We shall show that $Y := {}^a f^{-1}(Z)$ has the asserted properties. Indeed, since ${}^a f$ is a topological immersion, it follows via condition (1) in Proposition 2.11 that Y is a chain. Moreover, Y is a rooted chain, with $\mathcal{R}(Y) = Q$. Now, ${}^a f(Y) = Z \cap \mathrm{Im}({}^a f) = Z \cap V(I) = Z$. Finally, Y covers X since $\mathrm{Spec}(B) \to \mathrm{Spec}(A/I)$ is an injection. □

Next, we have the dual of [4, Corollary 3.23].

COROLLARY 2.14. *Let f be a ring homomorphism. Then:*

(a) If f is an injection and ${}^a f$ is a topological immersion with closed image, then f satisfies GGU.

(b) Suppose that for all $Q \in \mathrm{Spec}(B)$ and $P := f^{-1}(Q)$, the induced map $A_P \to B_Q$ is an injection whose corresponding map $\mathrm{Spec}(B_Q) \to \mathrm{Spec}(A_P)$ is a topological immersion with closed image. Then f satisfies GGU.

Proof. (a) is immediate from Corollary 2.13; to prove (b), combine (a) and Proposition 2.9 (b). □

The next result is the dual of [4, Corollary 3.24]. For applications of Corollary 2.20, it is useful to have examples of ring homomorphisms $g: A \to D$ that are universally topological immersions. Among these, we mention flat epimorphic g, surjective g, and g such that ${}^a g$ is a universal homeomorphism.

COROLLARY 2.15. *Let $f: A \to B$ be a ring homomorphism such that ${}^a f$ is injective and f satisfies GU. Let $g: A \to D$ be a ring homomorphism that is universally a topological immersion. Then the induced ring homomorphism $h: D \to D \otimes_A B$ satisfies GGU.*

Proof. Put $E := D \otimes_A B$. Our task is to show that if X is a rooted chain in $\mathrm{Spec}(D)$ and $Q \in \mathrm{Spec}(E)$ satisfies $h^{-1}(Q) = \mathcal{R}(X)$, then there exists a rooted chain Y in $\mathrm{Spec}(E)$ such that $\mathcal{R}(Y) = Q$ and Y covers X. As ${}^a g$ is injective, it follows from Proposition 2.1 (a), (b) that $W := {}^a g(X)$ is a rooted chain in $\mathrm{Spec}(A)$ such that $g^{-1}(\mathcal{R}(X)) = \mathcal{R}(W)$. Now, since Corollary 2.12 ensures that f satisfies GGU, there exists a rooted chain Z in $\mathrm{Spec}(B)$ such that $f^{-1}(\mathcal{R}(Z)) = \mathcal{R}(W)$ and Z covers W. Next, since X and Z have the same index set, we can use a result on pullbacks of schemes [8, Corollaire 3.2.7.1(i), p. 235] to produce the individual elements of a subset Y of $\mathrm{Spec}(E)$ such that Y covers X (relative to h) and Y covers Z (relative to the canonical ring homomorphism $j: B \to E$). As the hypothesis on g ensures that j is a topological immersion, Proposition 2.11 (b) yields that Y is a rooted chain. Finally, we shall show that $\mathcal{R}(Y) = Q$. By Proposition 2.1 (b), $j^{-1}(\mathcal{R}(Y)) = \mathcal{R}(Z)$. Therefore,

$$
\begin{aligned}
&{}^a(j \circ f)(\mathcal{R}(Y)) = f^{-1}(j^{-1}(\mathcal{R}(Y))) = f^{-1}(\mathcal{R}(Z)) = \\
&\mathcal{R}(W) = g^{-1}(\mathcal{R}(X)) = g^{-1}(h^{-1}(Q)) = {}^a(h \circ g)(Q) = {}^a(j \circ f)(Q).
\end{aligned}
$$

Since ${}^a(j \circ f) = {}^a f \circ {}^a j$ is a composite of injections, $\mathcal{R}(Y) = Q$. □

By analogy with the definition of "chain morphism", we say, as in [4], that a ring homomorphism $f: A \to B$ is a *2-chain morphism* (or, as in [10, p. 528], *subtrusive*) if the following condition is satisfied: for all prime ideals $P_1 \subseteq P_2$ of A, there exist prime ideals $Q_1 \subseteq Q_2$ of B such that $f^{-1}(Q_i) = P_i$ for $i = 1, 2$. It is easy to see that any ring homomorphism f that satisfies LO and either GU or GD must be a 2-chain morphism. As noted in [10, p. 538], examples of universally 2-chain morphisms include the ring homomorphisms f that are pure; the f that satisfy LO and are universally going-down; and the f that satisfy LO and are integral. For us, the most important examples of universally 2-chain morphisms are special cases of the last two classes just mentioned, namely, the faithfully flat ring homomorphisms and (thanks to a result on pullbacks of schemes [8, Corollaire 3.2.7.1(i), p. 235] and the Lying-over Theorem [9, Theorem 44]) the injective integral ring homomorphisms.

Before stating a useful characterization of universally 2-chain morphisms, we recall the following definitions. If $f: A \to B$ is a ring homomorphism, the *torsion ideal of* f is $T(f) := \{b \in B :$ there exists a non-zero-divisor $c \in A$ such that $cb = 0\}$; and f is called *torsion-free* if $T(f) = 0$.

PROPOSITION 2.16. *(Picavet [10, Théorème 37(a), p. 556 and Proposition 16, p. 543]) Let $f: A \to B$ be a ring homomorphism. Then the following conditions are equivalent:*

(1) If $A \to V$ is a ring homomorphism for which V is a valuation domain and the induced map $V \to V \otimes_A B =: E$ has torsion ideal T, then the induced ring homomorphism $V \to E/T$ is faithfully flat;

(2) f is a universally 2-chain morphism.

Observe that LO is a universal property (as can be seen via [8, Corollaire 3.2.7.1(i), p. 235]); and, of course, so is "integral". Accordingly, [3, Remark (d)] actually establishes that any integral ring homomorphism that satisfies LO (for instance, any injective integral map) must be a universally chain morphism. We next record a substantial generalization of this fact.

THEOREM 2.17. *[4, Theorem 3.26] A ring homomorphism $f: A \to B$ is a universally chain morphism if and only if f is a universally 2-chain morphism.*

We next infer the duals of [4, Corollary 3.27-Corollary 3.29].

COROLLARY 2.18. *Universally (2-) chain morphisms descend both GGU and GU. More precisely: if $f: A \to B$ is a ring homomorphism and $g: A \to D$ is a universally (2-) chain morphism such that the induced map $h: D \to D \otimes_A B =: E$ satisfies GGU (resp., GU), then f satisfies GGU (resp., GU).*

Proof. We give a proof for the "GGU" assertion, as it carries over for the "GU" assertion. Consider a rooted chain X in $\operatorname{Spec}(A)$ and $Q \in \operatorname{Spec}(B)$ such that $f^{-1}(Q) = \mathcal{R}(X)$. Since g is a chain morphism, there exists a chain Z in $\operatorname{Spec}(D)$ such that Z covers X. By Proposition 2.1 (b), Z is a rooted chain and $g^{-1}(\mathcal{R}(Z)) = \mathcal{R}(X)$. As

$\mathcal{R}(Z)$ and Q each lie over $\mathcal{R}(X)$, the oft-used fact about pullbacks of schemes [8, Corollaire 3.2.7.1(i), p. 235] supplies $J \in \mathrm{Spec}(E)$ such that J lies over $\mathcal{R}(Z)$ in $\mathrm{Spec}(D)$ and J lies over Q in $\mathrm{Spec}(B)$. Since h satisfies GGU, there exists a rooted chain W in $\mathrm{Spec}(E)$ such that $\mathcal{R}(W) = J$ and W covers Z. If j denotes the canonical ring homomorphism $B \to E$, then the chain $Y := {}^a j(W)$ covers X. Moreover, by Proposition 2.1 (b), Y is a rooted chain satisfying $Q = j^{-1}(J) = j^{-1}(\mathcal{R}(W)) = \mathcal{R}(Y)$. Therefore, f satisfies GGU. □

COROLLARY 2.19. *Universally (2-) chain morphisms descend universally going-up (universally GGU).*

Proof. It follows from Corollary 2.18 via standard tensor product identities that any universally (2-) chain morphism descends universally GGU. □

COROLLARY 2.20. *Let* $f: A \to B$ *be a ring homomorphism, and let* $a_1, \ldots a_n$ *be finitely many elements of* A *such that* $(a_1, \ldots, a_n) = A$. *Then* f *satisfies GGU if and only if the induced ring homomorphism* $f_i : A_{a_i} \to B_{a_i}$ *satisfies GGU for all* $i = 1, \ldots, n$.

Proof. The "only if" assertion is immediate from Proposition 2.9 (b). For the converse, assume that each f_i satisfies GGU. By Proposition 2.9 (c), so does the induced map $\prod A_{a_i} \to \prod B_{a_i}$. Of course, $\prod B_{a_i} \cong (\prod A_{a_i}) \otimes_A B$; and $A \to \prod A_{a_i}$ is faithfully flat, hence a universally 2-chain morphism. Hence, by Corollary 2.18, f satisfies GGU. □

References

[1] D. E. Dobbs, *On going-down for simple overrings, II*, Comm. Algebra **1** (1974), 439–458.

[2] D. E. Dobbs, *Going-down rings with zero-divisors*, Houston J. Math **23** (1997), 1–12.

[3] D. E. Dobbs, *A going-up theorem for arbitrary chains of prime ideals*, Comm. Algebra **27** (1999), 3887–3894. Corrigendum **28** (2000), 1653–1654.

[4] D.E. Dobbs, M. Fontana and G. Picavet, *Generalized going-down homomorphisms of commutative rings*, submitted for publication.

[5] G. De Marco and A. Orsatti, *Commutative rings in which every prime ideal is contained in a unique maximal ideal*, Proc. Amer. Math. Soc. **30** (1971), 459–466.

[6] D. E. Dobbs and I. J. Papick, *On going-down for simple overrings, III*, Proc. Amer. Math. Soc. **54** (1976), 35–38.

[7] R. Gilmer, Mutiplicative Ideal Theory, Dekker, New York, 1972.

[8] A. Grothendieck and J. Dieudonné, Eléments de Géométrie Algébrique, Springer-Verlag, Berlin, 1971.

[9] I. Kaplansky, Commutative Rings, rev. ed., Univ. Chicago Press, Chicago, 1974.

[10] G. Picavet, *Submersion et descente*, J. Algebra **103** (1986), 527–591.

[11] G. Picavet, *Parties sondables d'un spectre et profondeur*, Boll. U.M.I. (7) **8-B** (1994), 677–730.

Parameter-Like Sequences and Extensions of Tight Closure

MELVIN HOCHSTER

DEPARTMENT OF MATHEMATICS
UNIVERSITY OF MICHIGAN
ANN ARBOR, MI 48109–1109
USA
E-MAIL:
hochster@umich.edu

0. INTRODUCTION

We introduce here a notion of closure for ideals (*parameter tight closure*) in arbitrary Noetherian rings, including rings of mixed characteristic, that we hope will have properties parallel to those enjoyed by tight closure in characteristic p. We are not able to prove that the definition proposed here has all the properties that tight closure does: this remains an open question. But we can show that it agrees with tight closure in prime characteristic $p > 0$, that it is, in general, contained in the solid closure introduced in [Ho8] and [Ho9] (which is known to be "too large"), and it appears *very* likely that in many cases it is smaller than solid closure: cf. Discussion (3.6) and Theorem (3.7). We also show that, quite generally, including in mixed characteristic, it captures elements of a domain which are in the expansion of an ideal to an integral extension (sse Theorem (2.5)), and that, in equal characteristic 0, it has so-called "colon-capturing" properties analogous to those of tight closure (see Theorem (3.2)). The case of complete local domains suffices for applications, and so, for simplicity, we often restrict to that case in the sequel.

In fact, closure operations of the kind we have in mind are determined by their behavior over complete local domains. We describe briefly how the definition goes in that case. The underlying idea is to define a notion of "parameter-like" sequence in an algebra, not necessarily Noetherian, over a complete local domain R. The definition is made in terms of

---

The author was supported in part by a grant from the National Science Foundation.
Version of November 18, 2001.

annihilation properties of certain local cohomology modules. See (2.2) for details. It is, in fact, the case that a system of parameters of a complete local domain R is parameter-like. We next define the notion of a "parameter-preserving" algebra S over the complete local domain R: an R-algebra S is *parameter-preserving* if and only if every system of parameters in R is parameter-like in S. We then define the *parameter tight closure* $I^{\natural}$ of $I \subseteq R$ to be the smallest ideal J of R containing I such that for every parameter preserving-algebra S over R, the contraction of JS to R is J.

If R is a complete local domain, it turns out that a parameter-preserving algebra is solid, which makes the new closure *a priori* contained in the solid closure defined in [Ho9]. In §3 we show that the solid algebra Roberts uses in [Ro6] to prove that the solid closure of an ideal in a regular ring of equal characteristic zero can be strictly larger than the ideal is not parameter-preserving. Thus, it appears to be possible that the parameter tight closure of any ideal in a regular ring is equal to the ideal, although we cannot prove this. Moreover, we do not know whether parameter tight closure agrees with any of the equal characteristic zero notions of tight closure introduced in [HH10]: but we can show that it contains the largest of them, the big equational tight closure defined there. See Theorem (3.3).

We shall show that this parameter tight closure has many of the properties that we want a tight closure theory to have. However, we do not know whether every ideal of a regular ring is parameter tightly closed, neither in equal characteristic zero nor in mixed characteristic. If the mixed characteristic case could be established, the direct summand conjecture would follow.

Of course, we were led to study the notion of parameter-like sequences because of the possibility of settling many long-standing open questions in mixed characteristic (cf. [Ho1,7], [PS1,2], [Ro1-5,7], [Ho2,3,5,6], [Du1,2], [DHM], [EvG], and [Rang]) via the construction of a suitable analogue of tight closure theory. However, we feel that parameter-like sequences are worthy of study in their own right even without the potential for this application: their behavior appears to be subtle even in finitely generated algebras over a complete local domain.

Other notions, defined quite differently, that generalize tight closure have been explored in [Heit1–3] and [HoV]: these are related to ideas from [Sm] and [HH5]. In particular, in a tremendous breakthrough, [Heit3] resolves the direct summand conjecture in dimension 3. In [Ho11] a main result of [Heit3] is used to prove the existence (in a weakly functorial sense) of big Cohen-Macaulay algebras in dimension 3, even in mixed characteristic.

1. TIGHT CLOSURE IN POSITIVE CHARACTERISTIC

For the theory of tight closure in characteristic p we refer the reader to [Ho8], [HH1–4, 6–12], [Hu], [Sm], and [Bru]. The equal characteristic zero theory is described in [Ho9] and [Ho10], and developed in complete detail in [HH10].

For the moment we shall be working over a Noetherian ring R of positive prime characteristic p. In this situation we shall always let e denote an element of the nonnegative integers $\mathbb{N}$, and write q as an abbreviation for p^e. Thus "for all sufficient large q" means "for all sufficiently large integers q of the form p^e," and so forth.

Recall that a module M over a local ring (R, m, K) is a *balanced big Cohen-Macaulay module* (cf. [Sh]) for R if $M \neq mM$ and every system of parameters for R is a regular sequence on M. If M is also an R-algebra it is called a *big Cohen-Macaulay algebra* for R (i.e., in the context of algebras we shall always assume "balanced" but we omit the word).

We also recall that if R is a domain then an R-module M is called *solid* if there exists a nonzero R-linear map from M to R, i.e., $\mathrm{Hom}_R(M, R) \neq 0$. If (R, M, K) is a complete local domain of dimension d then it turns out that M is solid if and only if $H^d_m(M) \neq 0$. An R-algebra is called *solid* if it is solid when considered as an R-module.

In order to explain, in part, why we are led to consider the notion of *parameter tight closure* we first consider four characterizations of tight closure in the characteristic $p > 0$ case. For simplicity, we consider only the case of ideals, and when it simplifies matters, we assume that the Noetherian ring R of prime characteristic $p > 0$ is a complete local domain. The first characterization given below is actually the definition of tight closure in positive characteristic. The characterizations (2)and (3) below are consequences of Theorems (11.1) and (8.6b), respectively, of [Ho9]. The characterization (4) is a consequence of

Theorem (8.17) of [HH4] and the results of [Mo] on Hilbert-Kunz multiplicities. See also the discussion in [Ho8], p. 179.

(1.1) Characterizations of of tight closure. Let R be a Noetherian ring of prime characteristic $p > 0$. Let $u \in R$ and let $I \subseteq R$ be an ideal. Let I^* denote the tight closure of I.

(1) (Definition) $u \in I^*$ precisely if there exists c not in any minimal prime of R such that $cu^q \in I^{[q]}$ for all nonnegative integers e, where $I^{[q]}$ is the ideal generated by all q *th* powers of elements of I. (When R is a domain, the condition on c is simply that it not be 0.)

(2) Let R be a complete local domain. $u \in I^*$ if and only if $u \in IS \cap R$ for some big Cohen-Macaulay R-algebra S.

(3) Let R be a complete local domain. $u \in I^*$ if and only if $u \in IS \cap R$ for some solid R-algebra S.

(4) Let R be a complete local domain. Assume also that I is m-primary. With $J = I + uR$, we have that $u \in I^*$ if and only if $\lim_{e \to \infty} \dfrac{\ell(R/J^{[p^e]})}{\ell(R/I^{[p^e]})} = 1$. (Here, "$\ell$" indicates length.)

We present these characterizations because every characterization of tight closure in prime characteristic $p > 0$ gives a potential method for generalizing the theory to mixed characteristic. We want to discuss briefly the difficulties that arise from using these characterizations to help motivate the definitions of the next section.

We first note that an analogue of (1) can be defined in equal characteristic zero by reduction to characteristic p. This idea gives a very good extension of tight closure theory to the equal characteristic zero case (cf. [Ho8], [Ho10], [HH10]), but this definition does not seem to lead to any highly useful notion in mixed characteristic.

Condition (2) might lead to a notion that is a good notion in all characteristics, but at this time this idea does little good in mixed characteristic, because big Cohen-Macaulay algebras are not known to exist in mixed characteristic.

Condition (3) leads to a notion that is explored in the author's paper [Ho9], but an example [Ro6] of Paul Roberts shows that solid closure is too big in equal characteristic zero (ideals in regular rings of dimension 3 are not always solidly closed). The situation in mixed characteristic is unresolved, but there it is difficult to prove anything and Roberts' example is discouraging. Solid closure does give some information, but not enough to settle, for example, the direct summand conjecture. For further information about this and related conjectures, we refer the the reader to [Ho1,7], [PS1,2], [Ro1-5,7], [Ho2,3,5,6], [Du1,2], [DHM], [EvG], and [Rang].

In connection with all of these conditions, we note that if R is essentially of finite type over a field or even over an excellent local ring, and has prime characteristic $p > 0$, then $u \in I^*$ if and only if the image of u is in $(ID)^*$ (working over D) for every complete local domain D to which R maps.[1] Thus, under mild conditions on the ring, tight closure theory in prime characteristic $p > 0$ is determined by its behavior for complete local domains.

Condition (4) merits some further comment. Note that in a complete local domain (R, m), the tight closure of I is the intersection of the tight closures of the m-primary ideals containing I, and so tight closure is determined by its behavior on m-primary ideals. The intriguing condition (4) can be rephrased slightly as follows: for m-primary ideals $I \subseteq J$ in a complete local domain (R, m) of prime characteristic p, $J \in I^*$ if and only if I and J have the same Hilbert-Kunz multiplicity. (It is known that, with $d = \dim R$, $\ell(R/I^{[q]}) = \gamma(q^d) + O(q^{d-1})$, where γ, the Hilbert-Kunz multiplicity, is a positive real constant (conjectured, but not known, to be rational) and the term $O(q^{d-1})$ is bounded in absolute value by a constant times q^{d-1}.) Cf. [Mo] for the basic theory, and see [HaMo] for some surprising examples. This exciting tie-in between tight closure and Hilbert-Kunz multiplicities has not, so far, led to any possible extensions of tight closure theory to mixed characteristic.

We want to come back to the conditions (2) and (3). Evidently, if one has a class of R-algebras contained in the solid R-algebras and containing the big Cohen-Macaulay R-algebras, one can use it to define a notion of closure that will agree with tight closure in prime characteristic $p > 0$ and may give a good notion in equal characteristic 0 and in mixed characteristic. In the next section we define a class of algebras, the *parameter-*

[1]One may use [HH7], Prop. (6.23) and Thm. (6.24) to show "only if". To prove "if" one may use that the rings considered have completely stable test elements. One can reduce to looking at the completions of their local rings and then the quotients of those by minimal primes by [HH4], Prop. (6.25).

preserving algebras, and prove that it lies between the class of big Cohen-Macaulay algebras and the class of solid algebras: it is not obvious that parameter-preserving algebras are solid, but that is the content of Theorem (2.7). We shall show in the sequel that this class of algebras gives a notion with many of the properties we want, and, so far as we know, it may have all of the properties that we want. In §3 we show that the algebra that Roberts uses in [Ro6] to show that solid closure is too big (in the sense that not every ideal of a regular ring is solidly closed) is not parameter-preserving. Thus, there is no known "obstruction" to prevent this notion from being a good one in equal characteristic 0 and in mixed characteristic. But whether it has all the properties one would like remains an open question.

2. PARAMETER-LIKE SEQUENCES AND PARAMETER-PRESERVING ALGEBRAS

As discussed earlier, we want to explore here the possibility of defining a closure operation that is provably useful in all characteristics along the following lines: we first define a property, *parameter-preservation*, of algebras that is stronger than being solid but weaker than being a big Cohen-Macaulay algebra. We then define u to be *immediately in the parameter tight closure* of I if $u \in IS \cap R$ for some algebra S having the specified property. We then take the *parameter tight closure* $I^{\natural}$ of I to be the smallest ideal of R containing I that is closed under immediate parameter tight closure. We can do something similar for modules. The detailed definition is given in (2.2).

Although we are primarily interested in complete local domains, it will be convenient to allow complete local rings of pure dimension as well: recall that R has *pure dimension* d if (0) has no embedded prime ideals, and for every minimal prime P of R, the dimension of R/P is d. This is equivalent to the statement that every nonzero submodule of R has dimension d. Likewise, we say that an R-module has *pure dimension* d if it and all of its nonzero submodules have dimension d.

Much of the sequel depends on the facts (b) and (c) about local cohomology in the following:

(2.1) Lemma. *Let R be a complete local ring of pure dimension d. Then:*

(a) *Given any system of parameters $x_1, \ldots, x_d$ for R, R is module-finite over a complete local ring $A \subseteq R$ such that $x_1, \ldots, x_d \in A$ and A is either regular (A can always be chosen to be regular in the equal characteristic case), or A is a hypersurface in a complete regular local ring.*

(b) *For $i \neq d$, the module $H^i_m(R)$ is annihilated by an ideal of height at least 2 in R. (The unit ideal has height $+\infty$, and so the condition is satisfied if the local cohomology module vanishes.)*

(c) *Let M be any R-module, not necessarily Noetherian. Let $\mathfrak{B} \subseteq R$ denote the annihilator of $H^d_m(M)$. Then if $H^d_m(M) \neq 0$, $\dim R/\mathfrak{B} = d$.*

Proof. (a) This is quite standard if R contains a field: it has a coefficient field K and one may choose $D = K[[x_1, \ldots, x_d]]$. In the mixed characteristic case, choose a coefficient ring B for R. This means that $B \subseteq R$ with $m_B \subseteq m = m_R$, that $B/m_B \to R/m$ is an isomorphism, and that for some mixed characteristic discrete valuation ring V with residual characteristic p such that $m_V = pV$, either $B = V$ or $B = V/(p^t)$. Choose a map ϕ of $V[[X]] = V[[X_1, \ldots, X_d]]$ to R so that the X_i map to the x_i, the given system of parameters for R. Then image A of $V[[X]]$ in R has pure dimension d, since $R \supseteq A$ is module-finite, and so $\operatorname{Ker}\phi \subseteq V[[X]]$ is a pure height one ideal of the unique factorization domain $V[[X]]$. But then $\operatorname{Ker}\phi = (f)$ is principal, and R is module-finite over $A \cong V[[X]]/(f)$ as required.

(b) Choose A as in part (a). Then, by local duality over the Gorenstein ring A, the Matlis dual of $H^i_m(R)$ is $\operatorname{Ext}^{d-i}_A(R, A)$, $0 \leq i \leq d-1$, and so it suffices to see that the Ext has an annihilator of height two or more in A: this ideal will expand to an ideal of height two or more in R. Therefore it suffices to see that for every height one prime Q of A, $\operatorname{Ext}^j_{A_Q}(R_Q, A_Q) = 0$ for $j = d - i \neq 0$. But since A_Q is a one-dimensional Gorenstein ring, the Matlis dual of the localized Ext is $H^{1-j}_{QA_Q}(R_Q)$, which is 0 if $j > 1$, or $j < 0$, clearly, and vanishes when $j = 1$ because R_Q is of pure dimensional one over A_Q, and this implies that it is Cohen-Macaulay.

(c) If $\dim R/\mathfrak{B} < d$, then $\mathfrak{B}$ contains an element x_1 that is part of a system of parameters. Hence, we can choose $A \subseteq R$ as in part (a) such that $x_1 \in A$. Since R is module-finite over A, the maximal ideal of A expands to an m-primary ideal of R. Thus, if we think of M is an A-module, the local cohomology does not change, and we still have that $H^d_{m_A}(M) \neq 0$ but that this module is killed by the parameter x_1. We may therefore assume that $R = A$ is a hypersurface. Let $E = E_R(K) \cong H^d_m(R)$. Then $H^d_m(M) \cong M \otimes_R E$ is nonzero and killed by x_1. But $\mathrm{Hom}_R(\_, E)$ is faithfully exact, and so we get that $\mathrm{Hom}_R(H^d_m(M), E) \cong \mathrm{Hom}_R(M \otimes_R E, E)$ is nonzero and killed by x_1, and by the adjointness of tensor and Hom this may be identified with $\mathrm{Hom}_R(M, Hom_R(E, E)) \cong \mathrm{Hom}_R(M, R)$, since R is complete. Since the Hom is nonzero and killed by x_1, there exists a nonzero map $M \to R$ that is killed by x_1. But this means that the image of the map is an ideal of R killed by x_1, and x_1 is not a zerodivisor in R. This is a contradiction. □

If R is a local ring and J is any ideal of R, we define J^{unmx} to be the intersection of the primary components of J corresponding to minimal primes Q of J such that $\dim R/Q = \dim R$. (We have restricted this definition to the local case to avoid difficulties that arise from rings having maximal ideals with differing heights.) Note that if R is any local ring of dimension d, then $R/(0)^{\mathrm{unmx}}$ has pure dimension d. In fact, $(0)^{\mathrm{unmx}}$ is the largest ideal I of R such that the dimension of I as an R-module is smaller than d: it consists of all elements of $u \in R$ such that $\dim Ru < d$.

(2.2) Definitions: parameter-like sequences, parameter-preserving algebras, and parameter tight closure. Although we are primarily interested in complete local domains, it will be convenient to allow complete local rings of pure dimension d as well in certain definitions. Thus, let R be a complete local ring of pure dimension d and let S be an R-algebra. Let $x_1, \ldots, x_d$ be a system of parameters for R. Let $T_0 = \mathcal{T}_0(S)$ be the quotient of S by the ideal of all elements that have an annihilator of positive height in R, and, recursively, if $T_i = \mathcal{T}_i(S)$ has been defined for $i < d$ let T_{i+1} be the quotient of $T_i/(x_{i+1}T_i)$ by the ideal of all elements u such $\dim Ru < d-(i+1)$. (Note that $Ru \in T_{i+1}$)

is killed by $(x_1, \ldots, x_{i+1})$ and that $\dim R/(x_1, \ldots, x_{i+1}) = d - (i+1)$.) If we need to make explicit the dependence of $T_i(S)$ on the choices of R and $\underline{x} = x_1, \ldots, x_d$, we shall write $T_i^R(\underline{x}; S)$, but we shall usually omit either or both of R, $\underline{x}$.

Call a system of parameters $x_1, \ldots, x_d$ *parameter-like* in S if $T_d \neq 0$, and for all i, $0 \leq i \leq d-1$, the height of the annihilator $\mathfrak{A}_i$ of $H_m^{d-1-i}(T_i)$ in R is at least $i+2$. We note again that we are making the usual convention that the height of the unit ideal is $+\infty$, and so the condition is satisfied whenever $H_m^{d-1-i}(T_i)$ vanishes. Since R was assumed to have pure dimension, it is equivalent to assert that for every i, $0 \leq i \leq d-1$, either $H^{d-1-i}(T_i) = 0$, or else $\dim R/\mathfrak{A}_i \leq d - i - 2$.

Call S a *parameter-preserving* R-algebra if every system of parameters $x_1, \ldots, x_d$ of R is parameter-like in S.

Given $N \subseteq M$, finitely generated R-modules, define $u \in M$ to be in the *immediate parameter tight closure* of N in M if there exists a parameter-preserving R-algebra S such that $1 \otimes u$ is in the image of $S \otimes_R N$ in $S \otimes_R M$ (if $M = R$ and $N = I$ is an ideal, this just says that the image of u in S is in IS). Define the *first parameter tight closure of N in M* to be the submodule of M generated by the elements in the immediate tight closure of N. The first parameter tight closure will be a submodule of M containing N. Iterating this process, we obtain an ascending chain of submodules of M that must stabilize. We define the stable submodule in this chain to be the *parameter tight closure* of N in M, and denote it $N_M^\natural$ or simply $N^\natural$. When N is an ideal of R, M is understood to be R unless otherwise specified.

Alternatively, $N^\natural$ is the smallest submodule of M containing N that has the property that for any element $u \in M$ and any parameter preserving-algebra S over R, if $1 \otimes u$ is in the image of $S \otimes_R N^\natural$ in $S \otimes_R M$, then $u \in N^\natural$.

The definition of parameter-like is rather technical. The results that follow will help explain why it was chosen. The key points that will be established are:

(1) A system of parameters in a complete local domain is parameter-like, and module-finite extensions of complete local domains are parameter-preserving. (Cf. (2.3).)

(2) A big Cohen-Macaulay algebra over R is parameter-preserving. (Cf. (2.6).)

(3) A parameter-preserving algebra is solid. (Cf. (2.7).)

(4) The algebra that is used in [Ro6] to show that not every ideal in a regular ring is solidly closed in equal characteristic 0 is *not* parameter-preserving. (Cf. (3.6) and (3.7).)

These conditions imply many good properties for this closure operation, the majority of which are discussed in §3. We now proceed to the proofs.

(2.3) Theorem. *If R is a complete local domain then every sequence of elements that is part of a system of parameters is parameter-like in R and in every module-finite extension domain of S of R. Hence, every module-finite extension domain of R is parameter-preserving, including, of course, R itself.*

Proof. Fix part of a system of parameters in R: it will also be a system of parameters for S. One sees by induction on i that T_i is a local ring module-finite over $R/(x_1, \dots, x_i)$, that T_i has pure dimension $d - i$, using the remark following Fact (2.1), and that it is a quotient of S by a proper ideal. Thus, all the T_i are nonzero. The fact that the annihilators of the local cohomology modules are as stated now follows from Lemma (2.1b). □

We next observe:

(2.4) Lemma. *Let (R, m) be a complete local ring of pure dimension d, and let S be an R-algebra. Let $\underline{x} = x_1, \dots, x_d$ be a system of parameters in R. Let $R_i = \mathcal{T}_i^R(\underline{x};\, R)$ (cf. Definition (2.2)), and $T_i = \mathcal{T}_i^R(\underline{x};\, S)$. Then:*

(a) *$R_0 = R$ and for every $i \leq d$, R_i is a homomorphic image of $R/(x_1, \dots, x_i)R$ that has pure dimension $d - i$. Moreover, S_i is an R_i module. Let y_j be the image of x_j in R_i for $i + 1 \leq j \leq d$. Then for $i \leq j \leq d$, $\mathcal{T}_{j-i}^{R_i}(y_{i+1}, \dots, y_d;\, S_i) = S_j$. Thus, $x_1, \dots, x_d$ is parameter-like in S if and only if $H^{d-1}(T_0)$ has an annihilator of height at least two in R, and the images of $x_2, \dots, x_d$ are parameter-like in T_1 over R_1.*

(b) *If S' is flat over S, then for $1 \leq i \leq d$, $\mathcal{T}_i(S') \cong S' \otimes_S T_i(S)$, and, for all j, $H_m^j(\mathcal{T}_i(S')) \cong S' \otimes_S H_m^j(\mathcal{T}_i(S))$. Thus, if $x_1, \dots, x_d$ is parameter-like in S, then it is parameter-like in S' if and only if $S' \otimes_S T_d \neq 0$. In particular, if S' is faithfully flat over S, and $x_1, \dots, x_d$ is parameter-like in S then it is parameter-like in S'. Likewise, if $S' = W^{-1}S$, where W is a multiplicative system in S, then for all i, $0 \leq i \leq d$, for* *all j, we have that $\mathcal{T}_i(W^{-1}S) \cong W^{-1}\mathcal{T}_i$, that $H^j_m(\mathcal{T}_i(W^{-1}S)) \cong W^{-1}H^j_m(\mathcal{T}_i)$, and, if $x_1, \ldots, x_d$ is parameter-like in S, then it is parameter-like in $W^{-1}S$ if and only if $W^{-1}\mathcal{T}_d \neq 0$.*

(c) *If $x_1, \ldots, x_d$ is parameter-like in S and Q is any prime in the support of $\mathcal{T}_d$; S_Q is parameter-preserving. Thus, in testing immediate parameter tight closure it suffices to consider quasilocal R-algebras (S, Q) over R such that m maps into Q.*

Proof. (a) $R_0 = R$ since R is assumed to have pure dimension, and it is clear that R_i is a homomorphic image of $R/(x_1, \ldots, x_i)$, and that $\mathcal{T}_i$ is a module over $R/(x_1, \ldots, x_i)$. The statement that R_i has pure dimension $d-i$ is immediate by induction on i.

Let $u \in R/(x_1, \ldots, x_i)$ be such that Ru has dimension $< d-i$ in $R/(x_1, \ldots, x_i)$. If $v \in \mathcal{T}_i$ then the cyclic module Ruv is a homomorphic image of Ru, and so also has dimension $< d-i$. It follows that uv is killed in $\mathcal{T}_i$, and so u kills $\mathcal{T}_i$. Thus, $\mathcal{T}_i$ is an R_i-module. Once we know this, we have at once from the definitions that for $i \leq j \leq d$, $\mathcal{T}^{R_i}_{j-i}(y_{i+1}, \ldots, y_d; S_i) = S_j$, and the final statement in part (a) is then clear.

For part (b), first note that this holds when $i = 0$. The ideal of S' that we must kill to form $\mathcal{T}_0(S')$ is the union of the annihilators in S' of the positive height ideals of R. For any such ideal I, the annihilator of I in S' is the expansion of its annihilator from S, and so the union is the expansion of the union of the annihilators in S. We may then proceed by induction on i. Killing x_{i+1} times the algebra commutes with tensoring with S' over S, and the next step is like the formation of $\mathcal{T}_0$, but working with $S_i/x_{i+1}S_i$ and R_{i+1} instead of S and R. The statement that local cohomology commutes with tensoring with S' over S is obvious from the Cech complex method of defining local cohomology, and the final statement follows at once.

Part (c) is implied at once by part (b). $\square$

(2.5) Theorem. *If R is a complete local domain, I is an ideal of R, and S is a module-finite extension of R then $IS \cap R \subseteq I^\natural$. Hence, if $I = I^\natural$ for every ideal of R, then R is a direct summand of every module-finite extension.*

Proof. We can replace S by a quotient by a minimal prime of S disjoint from the domain R, and then the first statement is immediate from (2.3) and the definition of parameter tight closure. The second statement then follows from the main result of [Ho4]. □

(2.6) Theorem. *If S is a balanced big Cohen-Macaulay algebra for R then S is parameter-preserving.*

Proof. Note that $S_0 = S$, since any height one (or more) ideal of R will contain an element that is part of a system of parameters, and so the annihilator is 0. By a trivial induction on i, we have $T_i = S/(x_1, \ldots, x_i)S$, $1 \le i \le d$, which is a big Cohen-Macaulay algebra over R_i. The Cohen-Macaulay condition on S_i implies that it has depth $d-i$ on the maximal ideal of m, and so all the $H^j_m(S_i) = 0$ for $j < d-i$, and, in particular, for $j = d-1-i$. □

(2.7) Theorem. *If R is a complete local domain and S is a parameter-preserving R-algebra, then S is a solid R-algebra. In fact, if $\dim R = d$, $x_1, \ldots, x_d$ is a system of parameters for R, and the T_i are as in Definition (2.2), then we have that for all i, $0 \le i \le d$, $H^{d-i}_m(T_i) \ne 0$. (In particular, this holds when $i = 0$, which yields the fact that S itself is solid.)*

Proof. We use reverse induction on i to show that all the $H^{d-i}_m(T_i) \ne 0$, $i = d, d-1, \ldots, 0$. When $i = d$ we have that T_d is a nonzero module killed by $x_1, \ldots, x_d$ and, hence, by a power of m. Thus, $H^0_m(T_d) \ne 0$. Now suppose that we have shown that a certain $H^{d-i}_m(T_i) \ne 0$, $1 \le i \le d$. We must show that $H^{d+1-i}_m(T_{i-1}) \ne 0$. Now $x = x_i$ is a nonzerodivisor on T_{i-1} by the construction for T_{i-1}, and so we have a short exact sequence

$$(*) \qquad 0 \to T_{i-1} \xrightarrow{x} T_{i-1} \to T_{i-1}/xT_{i-1} \to 0.$$

Also, we have a short exact sequence $0 \to I \to T_{i-1}/xT_{i-1} \to T_i \to 0$ where I is an ideal of T_{i-1}/xT_{i-1} consisting of elements that are killed by an ideal of positive height in R_i. This means that every finitely generated R_i-submodule of I has dimension $< d-i$ as an R_i-module. We can conclude that $H^{d-i}_m(I) = H^{d-i+1}_m(I) = 0$, and so, from the long exact sequence for local cohomology, $H^{d-i}(T_{i-1}/xT_{i-1}) \cong H^{d-i}(T_i) \ne 0$, by the induction hypothesis.

On the other hand, the short exact sequence $(*)$ displayed above yields a long exact sequence of local cohomology modules part of which is

$$H^{d-i}_m(T_{i-1}) \xrightarrow{x} H^{d-i}_m(T_{i-1}) \to H^{d-i}_m(T_{i-1}/xT_{i-1}) \to H^{d-i+1}_m(T_{i-1})$$

We assume that the last term is 0, and get a contradiction. If the last term is zero, we have a surjection:

$$H_m^{d-i}(T_{i-1})/xH_m^{d-i}(T_{i-1}) \twoheadrightarrow H_m^{d-i}(T_{i-1}/xT_{i-1})$$

Since $H_m^{d-i}(T_{i-1}/xT_{i-1}) \neq 0$, we know that $H_m^{d-i}(T_{i-1}) \neq 0$. By Definition (2.2), since $x_1, \dots, x_d$ is parameter-like in S, if $\mathfrak{A}$ is the annihilator of $H_m^{d-i}(T_{i-1})$ in R, we have that $\dim R/\mathfrak{A} \leq d-(i-1)-2 = d-1-i$. But $\mathfrak{A}$ annihilates $H_m^{d-i}(T_{i-1}/xT_{i-1})$ as well, and so if $\mathfrak{B}$ is the annihilator of $H_m^{d-i}(T_{i-1}/xT_{i-1})$ we have that $R/\mathfrak{B} \leq d-1-i$. If we think of T_{i-1}/xT_{i-1} as a module over R_i (which has pure dimension $d-i$) we see that we have a contradiction, by Lemma (2.1c). □

3. THE NEW CLOSURE OPERATION

(3.1) Theorem. *For a complete local domain R of prime characteristic $p>0$, parameter tight closure is the same as the tight closure.*

Proof. Let $N \subseteq M$ be finitely generated R-modules. To show that $N^\natural \subseteq M^*$, it suffices to show that if $u \in M$ is in the immediate parameter tight closure of N in M, then $u \in M^*$. This is immediate from the Theorem (8.6) of [Ho9]: since any parameter-preserving algebra is solid, by Theorem (2.7), one has that u is in the solid closure of N in M, and then by it is in N^*, by [Ho9, Thm. (8.6)].

The converse follows from Theorem (11.1) of [Ho9]: if u is in N^*, then there exists a big Cohen-Macaulay R-algebra S such that $1 \otimes u$ is in the image of $S \otimes_R N$ in $S \otimes_R M$, and S is parameter-preserving by Theorem (2.6). □

(3.2) Theorem. *Let R be a complete local domain of equal characteristic, or a complete local domain of mixed characteristic and dimension at most three.*

(a) *(Colon capturing property) Let $x_1, \dots, x_d$ be a system of parameters for R. Then for $1 \leq i \leq d-1$, if $I=(x_1, \dots, x_i)R$, then $I : x_{i+1} \subseteq I^\natural$.*

(b) *(Analogue of phantom acyclicity) Let $G_\bullet$ denote a finite complex of finitely generated free modules over R that satisfies the standard conditions on rank and height.*[2] *Then for each $i \geq 1$, the module of cycles $Z_i \in G_i$ is in the parameter tight closure in G_i of the module of boundaries B_i.*

Proof. Of course, in the prime characteristic $p > 0$ case both parts follow from the fact that parameter tight closure agrees with tight closure, for which the statements in this theorem are standard.

However, the proof that we give for equal characteristic 0 also handles the positive characteristic case. By the main results of [HH5] (for characteristic $p > 0$), [HH9] (for equal characteristic 0), and [Ho11] (for mixed characteristic and dimension at most 3), R has a big Cohen-Macaulay algebra S. In S, $x_1, \dots, x_d$ is a regular sequences and so $I :_R x_{i_1} \subseteq I :_S x_{i+1} \subseteq IS$. Part (b) follows similarly, because when we apply $S \otimes_R \_$, the complex $S \otimes_R G_\bullet$ becomes acyclic over S. □

(3.3) Theorem. *Let $N \subseteq M$ be finitely generated modules over a complete local domain R of equal characteristic 0. Then $N^\natural \supseteq N^{*\mathrm{EQ}}$, the big equational tight closure of N in the sense of [HH10].*

Proof. Theorem (11.4) of [Ho9] shows that for any element u of $N^{*\mathrm{EQ}}$ there is a big Cohen-Macaulay algebra S for R such $1 \otimes u$ is in the image of $S \otimes_R N$ in $S \otimes_R M$. □

(3.4) Theorem. *Let R be a complete local domain of dimension at most two. Let $N \subseteq M$ be finitely generated R-modules. Then $N^\natural$ is the same as the solid closure of N in M, and $u \in N^\natural$ if and only if there is a big Cohen-Macaulay algebra S for R such that $1 \otimes u$ is in the image of $S \otimes_R N$ in $S \otimes_R M$.*

Proof. By Proposition (12.3) and Theorem (12.5) of [Ho9], in the dimension two case, an algebra over R is solid if and only if it can be mapped further to a big Cohen-Macaulay algebra. The parameter tight closure is always contained in the solid closure because parameter-preserving algebras are solid. In dimension two, the converse holds because any

---

[2] This means that the sum of the determinantal ranks of the maps to and from G_i is the rank of G_i, and that the ideal generated by the rank size minors of a matrix of the map $G_i \to G_{i-1}$ has height $\geq i$.

solid algebra can be mapped further to a big Cohen-Macaulay algebra, and big Cohen-Macaulay algebras are parameter-preserving. □

(3.5) Corollary. *Over a complete regular local ring of dimension at most two, every submodule of every finitely generated module is parameter tightly closed.*

Proof. It suffices to check that the immediate parameter tight closure of a submodule is equal to the submodule. But an element is in it if and only if it gets into the expanded submodule after tensoring with a big Cohen-Macaulay algebra, by Theorem (3.4). But a big Cohen-Macaulay algebra over a regular ring is faithfully flat over the regular ring (cf. the parenthetical argument in 6.7 on p. 77 of [HH5]). □

(3.6) Discussion. Let $R = K[[x_1, x_2, x_3]]$, where K is the field of rational numbers or any other field of characteristic 0, and let $S = R[y_1, y_2, y_3]/(F)$, where $F = x_1^2x_2^2x_3^2 - \sum_{j=1}^{3} y_j x_j^3$. Then S is solid by a result of Paul Roberts [Ro6]: this shows that the ideal (x_1^3, x_2^3, x_3^3) is not solidly closed in $K[[x_1, x_2, x_3]]$. As an indication that parameter tight closure is likely to behave better than solid closure in equal characteristic 0, we want to prove that x_1, x_2, x_3 is not a parameter-like sequence in S (which is an example of what is called a *forcing algebra* in [Ho9]). The following result handles a much larger class of forcing algebras, showing that none of them is parameter-preserving. We restrict attention to dimension ≥ 3, since we already know that every ideal is parameter tightly closed in complete regular domains of dimension at most 2.

(3.7) Theorem. *Let (V, x_1V) be a complete discrete valuation ring with residue class field K (which may or may not be of equal characteristic), and let $R = V[[x_2, \ldots, x_d]]$, $d \geq 3$, so that R is a complete regular local domain of dimension d with regular system of parameters $x_1, \ldots, x_d$. Let $S = R[y_1, \ldots, y_d]/(F)$ where $y_1, \ldots, y_d$ are indeterminates over R and*

$$F = (y_1 \cdots y_d)^{t-1} - \sum_{j=1}^{d} y_i x_i^t$$

for some fixed integer $t \geq 1$. Then S is not parameter-preserving over R. Specifically, $x_1, \ldots, x_d$ is not parameter-like in S: in fact, $T_{d-2} = S/(x_1, \ldots, x_{d-2})S$ is such that $H_m^1(T_{d-2})$ is not killed by an ideal of height two or more in $R_{d-2} = K[[x_{d-1}, x_d]]$.

Proof. Killing an initial segment of $x_1, \ldots, x_{d-2}$ in S produces a domain, from which it follows that $T_i = S/(x_1, \ldots, x_i)S$ for $0 \leq d-2$, and

$$T_{d-2} \cong K[[x_{d-1}, x_d]][y_1, \ldots, y_d]/(y_{d-1}x_{d-1}^t + y_d x_d^t),$$

which is a polynomial ring in $y_1, \ldots, y_{d-2}$ over $B = K[[x_{d-1}, x_d]][y_{d-1}, y_d]/(G)$, where $G = y_{d-1}x_{d-1}^t + y_d x_d^t$. The definition of parameter-like for $x_1, \ldots, x_d$ requires that $H_m^1(T_{d-2})$ be 0 or else be killed by a height two ideal of $R_{d-2} = K[[x_{d-1}, x_d]]$. Since $H_m^1(T_{d-2})$ may be identified as the polynomials in $y_1, \ldots, y_{d-2}$ over $H_m^1(B)$, it suffices to see that this fails for $H_m^1(B)$. Let $u = x_{d-1}^t$ and $v = -x_d^t$, so that $B = K[[x_{d-1}, x_d]][y_{d-1}, y_d]/(y_{d-1}u - y_d v)$. Since $K[[x_{d-1}, x_d]][y_{d-1}, y_d]$ is a finitely-generated free module over $K[[u, v,]][y_{d-1}, y_d]$, we have that B is module-finite and free over $C = K[[u, v,]][y_{d-1}, y_d]/(y_{d-1}u - y_d v)$. Then

$$H_m^1(B) \cong H_{(x_{d-1}, x_d)}^1(B) \cong H_{(u,v)}^1(B) \cong B \otimes_C H_{(u,v)}^1(C),$$

and so it will certainly suffice to show that $H_{(u,v)}^1(C)$ is a faithful C-module: if it were annihilated by an ideal of R_{d-2} primary to the maximal ideal, it would be annihilated by an ideal of C primary to the maximal ideal.

Let $A = K[[u, v]]$, and let z be an indeterminate over A. Then the A-algebra surjection $A[[u, v]][y_{d-1}, y_d] \to A[uz, vz]$ sending y_{d-1} to vz and y_d to uz is easily seen to have $(y_{d-1}u - y_d v)$ as its kernel, so that

$$C \cong A[uz, vz] = A \oplus (u, v)Az \oplus (u, v)^2 Az^2 \oplus \cdots,$$

the Rees ring, where the direct sum is over A. Let $Q = (u, v)A$, the maximal ideal of A. Thus,

$$H_{(u,v)}^1(C) \cong \bigoplus_{j=0}^{\infty} H_Q^1(Q^j).$$

From the short exact sequence $0 \to Q^j \to A \to A/Q^j \to 0$ and the corresponding long exact sequence for $H_Q^\bullet(\_)$, we have an exact sequence

$$\cdots \to H_Q^0(A) \to H_Q^0(A/Q^j) \to H_Q^1(Q^j) \to H_Q^1(A) \to \cdots$$

Since A has depth 2 on Q, $H_Q^0(A) = H_Q^1(A) = 0$, and so $H_Q^1(Q^j) \cong H_Q^0(A/Q^j) = A/Q^j$. Thus, $H_{(u,v)}^1(C) \cong \bigoplus_{j=0}^{\infty} A/Q^j$, so that the annihilator is $\subseteq \bigcap_j Q^j = (0)$, as required. □

4. A GALOIS CONJECTURE

In [Rang] ideas involving the interaction of group cohomology for Galois groups and local cohomology, as well techniques from number theory, are used to prove certain cases of the direct summand conjecture. The question that we mention here is related to the ideas of [Rang] but a bit different, and can be presented in a reasonably elementary way. An affirmative answer would be sufficient to prove the direct summand conjecture. The conjecture is true both in equal characteristic $p > 0$ and in equal characteristic 0, although the reasons why it is true in those two cases are completely different.

Let V be a complete discrete valuation ring, which may be either equal characteristic or mixed characteristic. In the mixed characteristic case assume that the residual characteristic p is the generator of the maximal ideal. In either case, denote the generator of the maximal ideal by $x = x_1$. Let $A = V[[x_2, \dots, x_d]]$ be a formal power series ring over V. If D is any domain we denote by D^+ an *absolute integral closure* of D, i.e., the integral closure of D an algebraic closure of its fraction field (cf. [Ar]). D^+ is unique up to non-unique isomorphism. Let $\mathcal{F}$ denote the fraction field of A, and then the fraction field of A^+ is an algebraic closure of $\mathcal{F}$, which we denote $\overline{\mathcal{F}}$, although the notation $\mathcal{F}^+$ would also be appropriate. We shall write G for the group of $\mathcal{F}$-automorphisms of $\overline{\mathcal{F}}$, which also acts on A^+. Note that ${A^+}^G = A$ when $\mathcal{F}$ has characteristic zero, which includes the case where A has equal characteristic zero and the case where A has mixed characteristic.

We shall write E for $H^d_m(A)$, the highest (in fact, the only) nonzero local cohomology module of A with support in $m = m_A$, since it is also an injective hull $E_A(K)$ for the residue field $K = A/m$ of A over A. We write $M^\vee$ for $\mathrm{Hom}_A(M, E)$. If (C, n, L) is any complete local ring, we shall call a C-module W *small* if $E_C(L)$, the injective hull of $L = C/n$ over C, cannot be injected into W. Note that if $E_C(L)$ is a submodule of W, then it is actually a direct summand of W, since $E_C(L)$ is an injective module. The condition that a module be small is not a strong restriction.

The result of [Ho7, Thm. (6.1)] implies that in order to prove the direct summand conjecture, it suffices to show that the modules $H^d_m(A^+)$ are not zero. Now $x = x_1$ is a regular parameter in A, and we have a short exact sequence $0 \to A^+ \xrightarrow{x} A^+ \to A^+/xA^+ \to 0$. If we contradict the direct summand conjecture and assume that $H^d_m(A^+) = 0$, part of the corresponding long exact sequence for local cohomology gives:

$$\cdots \to H^{d-1}_m(A^+) \xrightarrow{x} H^{d-1}_m(A^+) \to H^{d-1}_m(A^+/xA^+) \to 0.$$

This implies an isomorphism $H^{d-1}_m(A^+/xA^+) \cong H^{d-1}_m(A^+)/xH^{d-1}_m(A^+)$. The regular ring A/xA injects into A^+/xA^+ (because A is normal, the principal ideal xA is contracted from A^+). If A provides a counterexample to the direct summand conjecture of smallest dimension (or if A has mixed characteristic, provides a counterexample, and $x = p$), then A/xA is a direct summand of A^+/xA^+ as an (A/xA)-module, and it follows that $H^{d-1}_m(A/xA)$ injects into $H^{d-1}_m(A^+/xA^+)$. Evidently, since G acts on A^+, since m is contained in the ring of invariants of this action, and since x is an invariant, G acts on $H^{d-1}_m(A^+)/xH^{d-1}_m(A^+)$, and it is clear that $H^{d-1}_m(A/xA)$ injects into $\left(H^{d-1}_m(A^+)/xH^{d-1}_m(A^+)\right)^G \subseteq H^{d-1}_m(A^+)/xH^{d-1}_m(A^+)$.

We therefore will have a contradiction that establishes the direct summand conjecture if we can prove the following:

(4.1) Galois Conjecture. *Let (A, m, K) be a complete regular local ring of dimension d with fraction field $\mathcal{F}$, let G be the automorphism group of the algebraic closure $\overline{\mathcal{F}}$ over $\mathcal{F}$, and let x be a regular parameter in A. Then $\left(H^{d-1}_m(A^+)/xH^{d-1}_m(A^+)\right)^G$ is a small (A/xA)-module.*

(4.2) Theorem. *The Galois Conjecture (4.1) holds if $\dim A \le 2$ or if A contains a field. In fact, in all of these cases $\left(H^{d-1}_m(A^+)/xH^{d-1}_m(A^+)\right)^G = 0$.*

Proof. The explanation when A contains a field is quite different depending on whether the field has characteristic 0 or positive characteristic. In the first case, it turns out that ^{G} is an exact functor here, so that what we have is $\left(H^{d-1}_m(A^{+G})/xH^{d-1}_m(A^{+G})\right)$, and since $A^{+G} = A$, this is $H^{d-1}_m(A)/xH^{d-1}_m(A)$, and $H^{d-1}_m(A) = 0$. In the positive characteristic case we know from the main result of [HH5] that A^+ is a big Cohen-Macaulay algebra, so that $H^{d-1}_m(A^+) = 0$, and the result follows again. The same argument shows that the conjecture is true when A has dimension at most two. □

From the discussion above, we have the following:

(4.3) Theorem. *If the Galois Conjecture is true whenever A is a formal power series ring $V[[x_2, \ldots, x_d]]$ over a complete discrete valuation domain (V, pV, K) of mixed characteristic and residual characteristic $p > 0$, then the direct summand conjecture is true.* □

5. QUESTIONS

Of course, many open questions remain. We mention some of the most important among these.

Question 1. *In a complete regular local ring containing the rationals, is every ideal parameter tightly closed?*

Question 2. *In an arbitrary complete regular ring, is every ideal parameter tightly closed?*

An affirmative answer to this question would yield the direct summand conjecture in the general case.

Question 3. *Over a complete local domain of equal characteristic 0, does parameter tight closure agree with big equational tight closure?*

Of course, an affirmative answer to Question 3 would yield an affirmative answer for Question 1, since it is known that every ideal of an equicharacteristic zero regular ring is tightly closed if one uses big equational tight closure as the operation.

Note that Theorem (3.3) shows that the parameter tight closure contains the big equational tight closure: it is the converse that is problematic.

Question 4. *Do colon-capturing and an analogue of phantom acyclicity hold for parameter tight closure in mixed characteristic local domains?*

Question 5. *Can one characterize parameter-preserving finitely generated algebras over a complete local domain or parameter-preserving complete local extensions of a complete local domain in a simpler way?*

Even when S is restricted in this way, the problem does not seem easy. Evidently, questions about parameter tight closure are abundant.

BIBLIOGRAPHY

[Ar] M. Artin, *On the joins of Hensel rings*, Advances in Math. **7** (1971), 282–296.

[Bru] W. Bruns, *Tight closure*, Bull. Amer. Math. Soc. **33** (1996), 447–458.

[Du1] S. P. Dutta, *Frobenius and multiplicities*, J. Algebra **85** (1983), 424–448.

[Du2] ———, *On the canonical element conjecture*, Trans. Amer. Math. Soc. **299** (1987), 803–811.

[DHM] S. P. Dutta, M. Hochster, and J. E. McLaughlin, *Modules of finite projective dimension with negative intersection multiplicities*, Invent. Math. **79** (1985), 253–291.

[EvG] G. Evans and P. Griffith, *The syzygy problem*, Annals of Math. **114** (1981), 323-333.

[HaMo] C. Han and P. Monsky, *Some surprising Hilbert-Kunz functions*, Math. Z. **214** (1993), 119–135.

[Heit1] R. Heitman, *The plus closure in mixed characteristic*, J. of Algebra **193** (1997), 688–708.

[Heit2] R. Heitman, J. of Algebra **238** (2001), 801–826.

[Heit3] ———, *The direct summand conjecture in dimension three*, preprint.

[Ho1] M. Hochster, *Contracted ideals from integral extensions of regular rings*, Nagoya Math. J. **51** (1973), 25–43.

[Ho2] ———, *Topics in the homological theory of modules over commutative rings*, C.B.M.S. Regional Conf. Ser. in Math. No. **24**, Amer. Math. Soc., Providence, R.I., 1975.

[Ho3] ———, *Big Cohen-Macaulay modules and algebras and embeddability in rings of Witt vectors*, in *Proceedings of the Queen's University Commutative Algebra Conference*, Queen's Papers in Pure and Applied Math. **42**, 1975, pp. 106–195.

[Ho4] ———, *Cyclic purity versus purity in excellent Noetherian rings*, Trans. Amer. Math. Soc. **231** (1977), 463–488.

[Ho5] ———, *Some applications of the Frobenius in characteristic 0*, Bull. Amer. Math. Soc. **84** (1978), 886–912.

[Ho6] ———, *Cohen-Macaulay rings and modules*, in *Proc. of the International Congress of Mathematicians* (Helsinki, Finland, 1978), Vol. I, Academia Scientarium Fennica, 1980, pp. 291–298.

[Ho7] ———, *Canonical elements in local cohomology modules and the direct summand conjecture*, J. of Algebra **84** (1983), 503–553.

[Ho8] ———, *Tight closure in equal characteristic, big Cohen-Macaulay algebras, and solid closure*, in *Commutative Algebra: Syzygies, Multiplicities and Birational Algebra*, Contemp. Math. **159**, Amer. Math. Soc., Providence, R. I., 1994, pp. 173–196.

[Ho9] ———, *Solid closure*, in *Commutative Algebra: Syzygies, Multiplicities and Birational Algebra*, Contemp. Math. **159**, Amer. Math. Soc., Providence, R. I., 1994, pp. 103–172.

[Ho10] ———, *The notion of tight closure in equal characteristic zero*, in *Proc. of the CBMS Conference on Tight Closure and Its Applications* (Fargo, North Dakota, July, 1995), Appendix to the notes on the main lectures by Craig Huneke, C.B.M.S. Regional Conference Series, Amer. Math. Soc., Providence, R.I., 1996.

[Ho11] ———, *Big Cohen-Macaulay algebras in dimension three via Hietmann's theorem*, preprint.

[HoV] M. Hochster and J. D. Vélez, *Diamond closure*, preprint.

[HH1] M. Hochster and C. Huneke, *Tightly closed ideals*, Bull. Amer. Math. Soc. **18** (1988), 45–48.

[HH2] ———, *Tight closure*, in *Commutative Algebra*, Math. Sci. Research Inst. Publ. **15**, Springer-Verlag, New York-Berlin-Heidelberg, 1989, pp. 305–324.

[HH3] ———, *Tight closure and strong F-regularity*, Mémoires de la Société Mathématique de France, numéro **38** (1989), 119–133.

[HH4] ———, *Tight closure, invariant theory, and the Briançon-Skoda theorem*, J. Amer. Math. Soc. **3** (1990), 31–116.

[HH5] ———, *Infinite integral extensions and big Cohen-Macaulay algebras*, Annals of Math. **135** (1992), 53–89.

[HH6] ———, *Phantom homology*, Memoirs Amer. Math. Soc., Vol. **103**, No. 490 (1993), 1–91.

[HH7] ———, *F-regularity, test elements, and smooth base change*, Trans. Amer. Math. Soc. **346** (1994), 1–62.

[HH8] ———, *Tight closure of parameter ideals and splitting in module-finite extensions*, J. of Algebraic Geometry **3** (1994), 599–670.

[HH9] ———, *Applications of the existence of big Cohen-Macaulay algebras*, Advances in Math. **113** (1995), 45–117.

[HH10] ———, *Tight closure in equal characteristic zero*, preprint.

[HH11] ———, *Test exponents and localization of tight closure*, Michigan Math. J. **48** (2000), 305–329.

[HH12] ———, *Comparison of symbolic and ordinary powers of ideals*, to appear, Inventiones Math.

[Hu] ———, *Tight Closure and Its Applications*, Proc. of the CBMS Conference held at Fargo, North Dakota, July, 1995, C.B.M.S. Regional Conference Series, Amer. Math. Soc., Providence, R.I., 1996.

[Mo] P. Monsky, *The Hilbert-Kunz function*, Math. Ann. **263** (198), 43-49.

[PS1] Peskine, C. and L. Szpiro, *Dimension projective finie et cohomologie locale*, I.H.E.S. Publ. Math. **42** (Paris) (1973), 323-395.

[PS2] ———, *Syzygies et multiplicités*, C. R. Acad. Sci. Paris Sér. A **278** (1974), 1421-1424.

[Rang] N. Ranganathan, *Splitting in module-finite extension rings and the vanishing conjecture for maps of Tor*, Thesis, University of Michigan, 2000.

[Ro1] Roberts, P., *Two applications of dualizing complexes over local rings*, Ann. Sci. Ec. Norm. Sup. **9** (1976), 103–106.

[Ro2] ———, *Cohen-Macaulay complexes and an analytic proof of the new intersection conjecture*, J. of Algebra **66** (1980), 225–230.

[Ro3] ———, *The vanishing of intersection multiplicities of perfect complexes*, Bull. Amer. Math. Soc. **13** (1985), 127–130.

[Ro4] ———, *Le théorème d'intersection*, C. R. Acad. Sc. Paris Sér. I **304** (1987), 177–180.

[Ro5] ———, *Intersection theorems*, in Commutative Algebra, Math. Sci. Research Inst. Publ. **15**, Springer-Verlag, New York · Berlin · Heidelberg, 1989, pp. 417–436.

[Ro6] ———, *A computation of local cohomology*, in *Commutative Algebra: Syzygies, Multiplicities and Birational Algebra*, Contemp. Math. **159**, Amer. Math. Soc., Providence, R. I., 1994.

[Ro7] ———, *Multiplicities and Chern classes in local algebra*, Cambridge Tracts in Mathematics, vol. 133, Cambridge University Press, Cambridge, England, 1998.

[Sh] R. Y. Sharp, *Cohen-Macaulay properties for balanced big Cohen-Macaulay modules*, Math. Proc. Cambridge Philos. Soc. **90** (1981), 229–238.

[Sm] K. E. Smith, *Tight closure of parameter ideals*, Inventiones Math. **115** (1994), 41–60.

The Tor Game

CRAIG HUNEKE[1] Department of Mathematics, University of Kansas, Lawrence, Kansas 66045, U. S. A., huneke@math.ukans.edu

ROGER WIEGAND[1] Department of Mathematics and Statistics, University of Nebraska, Lincoln, Nebraska 68588-0323, U. S. A., rwiegand@math.unl.edu

0 INTRODUCTION

Let M and N be non-zero finitely generated modules over a local (Noetherian) ring $(R,\mathfrak{m})$. Assume that $\operatorname{Tor}_i^R(M,N)=0$ for $i>>0$, and let $q=q_R(M,N)$ be the largest integer i for which $\operatorname{Tor}_i^R(M,N)\neq 0$. For the case $q=0$, Auslander [2, Theorem 1.2] found a useful formula relating the depths of M, N and R. In view of the Auslander-Buchsbaum formula [4, Theorem 1.3.3] Auslander's theorem goes as follows:

$$\begin{gathered}\operatorname{depth}(M)+\operatorname{depth}(N)-\operatorname{depth}(R)=\operatorname{depth}(M\otimes_R N)\\ \text{provided } M \text{ has finite projective dimension and } q(M,N)=0.\end{gathered}\tag{1}$$

For larger values of q, Auslander proved the following result (still assuming that M has finite projective dimension):

$$\operatorname{depth}(M)+\operatorname{depth}(N)-\operatorname{depth}(R)=\operatorname{depth}(\operatorname{Tor}_q(M,N))-q,\tag{2}$$

provided $\operatorname{depth}(\operatorname{Tor}_q^R(M,N))\leq 1$.

The past decade has seen a renewed interest in these formulas. The authors showed in [7] that (1) holds for complete intersections, even without the assumption that M have finite projective dimension. Indeed, there seem to be many situations where homological properties of a module hold if *either* (a) the module has finite projective dimension *or* (b) the ring is a complete interesection. This phenomenon led Avramov, Gasharov and Peeva [3] to investigate a condition (see §1 for the

[1]The research of both authors was partially supported by the National Science Foundation.

definition) called "finite complete intersection dimension" (finite CI dimension for short), a condition that generalizes both (a) and (b). In [1, Theorem 2.5] Araya and Yoshino obtained formulas (1) and (2) in this general context. They also showed by example that (2) can fail if $\operatorname{Tor}_q^R(M,N)$ has depth 2. Jorgensen [9, Theorem 2.2] showed, assuming M has finite CI dimension, that $\min\{\operatorname{depth}(M_p)+\operatorname{depth}(N_p)-\operatorname{depth}(R_p) \mid p \in \operatorname{Spec}(R)\} = -q$; and Theorem 3 of [5] shows that the minimum value $-q$ is achieved precisely at those primes p for which $\operatorname{depth}(\operatorname{Tor}_q^R(M,N)_p) = 0$. In particular, one has

$$\operatorname{depth}(M)+\operatorname{depth}(N)-\operatorname{depth}(R) \geq -q, \tag{3}$$

if M has finite CI dimension. Choi and Iyengar [5] raised the interesting question of whether there is always *some* value j between 0 and q such that

$$\operatorname{depth}(M)+\operatorname{depth}(N)-\operatorname{depth}(R) = \operatorname{depth}(\operatorname{Tor}_j^R(M,N)) - j. \tag{4}$$

They gave examples showing that this too can fail.

In this paper we describe a "game" whose goal is to prove (4) and to identify the "winning" subscript j. This approach indicates rather clearly the obstruction to winning the game, that is, the reason formula (4) can fail.

To state our main theorem we adopt some notation. Throughout this paper R is a local ring with maximal ideal $\mathfrak{m}$. Our convention is that local rings are always Noetherian. Let M and N be finitely generated non-zero R-modules, and put $q_R(M,N) = \max\{i \mid \operatorname{Tor}_i^R(M,N) \neq 0\}$. (This invariant is denoted by $\operatorname{fd}(M,N)$ in [5].) For each $i \in \mathbb{Z}$ we put $d_i(M,N) = \operatorname{depth}(\operatorname{Tor}_i^R(M,N))$. (We define the depth of the zero module to be ∞. Thus $d_i = \infty$ if $i > q$ or $i < 0$.) Finally, we let

$$D_R(M,N) = \operatorname{depth}(M)+\operatorname{depth}(N)-\operatorname{depth}(R).$$

Our main result is Theorem 2.4, reproduced here for convenience:

MAIN THEOREM 0.1. *Let M and N be non-zero finitely generated modules over a local ring R. Assume M has finite CI dimension and $q_R(M,N) < \infty$. Put $m = m(M,N) = \min\{d_i - i \mid i \in \mathbb{Z}\}$ and $j = j(M,N) = \max\{i \mid d_i - i = m\}$. Assume*

$$d_i - i \geq m+2 \text{ for } i > j. \tag{†}$$

Then $D(M,N) = m$.

The definition of j forces

$$d_i - i \geq m+1 \text{ for } i > j.$$

The assumption of the theorem is simply that the next possible value for $d_i - i$, namely $d_j - j + 1 = m+1$, is not attained. Theorem 2.4 recovers the earlier results mentioned above. For example, suppose that the depth of the last non-zero Tor is 0 or 1. Then $d_q - q$ is either $-q$ or $1-q$, while all other $d_i - i$ are $\geq 1-q$. Theorem 2.4 immediately gives that $D(M,N) = d_q - q$ in this case (which recovers formula (2) above).

1 PRELIMINARIES

We recall the following definition from [3]:

DEFINITION 1.1. A finitely generated module M over a local ring R has *finite CI dimension* provided there exist a local ring $(S,\mathfrak{n})$, a regular sequence $(x_1,\dots,x_c)$ in $\mathfrak{n}$, and a flat local homomorphism $R \to R' := S/(x_1,\dots,x_c)$ such that $\operatorname{pd}_S(M \otimes_R R') < \infty$.

We will need the following result, proved by Auslander [2, Proposition 1.1] in the case of finite projective dimension:

PROPOSITION 1.2. *Let M and N be non-zero finitely generated modules over a local ring $(R,\mathfrak{m})$. Assume M has finite CI dimension and that $q = q_R(M,N) < \infty$. If* $\operatorname{depth} N = 0$ *then* $d_q(M,N) = 0$.

Proof. With the notation of (1.1), we may assume that $R = R'$. For $0 \le j \le c$, put $S_j = S/(x_1,\dots,x_{c-j})$ (so that $S_0 = R$ and $S_c = S$), and put $T_i^{S_j} = \operatorname{Tor}_i^{S_j}(M,N)$. For $i \ge 1$ and $0 \le j \le c-1$, there is an exact sequence

$$\mathrm{T}_{i+1}^{S_j} \to \mathrm{T}_{i-1}^{S_j} \to \mathrm{T}_i^{S_{j+1}} \to \mathrm{T}_i^{S_j}. \tag{1.2.1}$$

(See, for example, [7, (2.1)].) We see that

$$0 \ne \mathrm{T}_q^{S_0} \cong \mathrm{T}_{q+1}^{S_1} \cong \cdots \cong \mathrm{T}_{q+c}^{S_c} \tag{1.2.2}$$

and that $T_i^{S_j} = 0$ for all $i > q+j$. Since M has finite projective dimension over S, Auslander's result [2, Proposition 1.1] $\mathrm{T}_{q+c}^{S_c}$ has depth 0. By (1.2.2), then, $\mathrm{T}_q^{S_0}$ also has depth 0, as desired. □

DEFINITION 1.3. Let M be a finitely generated module over a local ring $(R,\mathfrak{m})$. An element $x \in \mathfrak{m}$ is *general with respect to M*, provided x is a non-zerodivisor on $M/\mathrm{H}_{\mathfrak{m}}^0(M)$.

LEMMA 1.4. *Let M be a finitely generated module over a local ring $(R,\mathfrak{m})$, and let $x \in \mathfrak{m}$ be general with respect to M.*

(1) $\operatorname{Ann}_M(x)$ *has finite length.*

(2) If $\operatorname{depth}(M) = 0$, *then* $\operatorname{depth}(M/xM) = 0$.

Proof. If $z \in \operatorname{Ann}_M(x)$, then $xz = 0$, and since x is general this forces $z \in \mathrm{H}_{\mathfrak{m}}^0(M)$. Thus $\operatorname{Ann}_M(x) \subseteq \mathrm{H}_{\mathfrak{m}}^0(M)$, and (1) follows. Suppose now that M has depth 0. Then $\operatorname{Ann}_M(x) \ne 0$, and $\operatorname{Ann}_M(x)$ has finite length by part (1). By [8, Lemma 6.2], $\operatorname{depth}(M/xM) = 0$. □

2 THE TOR GAME

The idea of the Tor Game is to choose a general element x, replace N by $\bar{N} := N/xN$, and compute the depths of the new modules $\mathrm{Tor}_i^R(M,\bar{N})$. Repeating this procedure, we eventually decrease d_q to 0 or 1, at which point we can invoke formula (2) in the introduction. Then, with a little luck, we can backtrack and identify the winning index j for which $D(M,N) = d_j(M,N) - j$. The next theorem gives the recipe for computing the new depths, except for one annoying siutation—when $d_i = 2$ and $d_{i-1} = 0$. In a Tor Game where this situation never occurs at any stage, we win.

RULES OF THE TOR GAME 2.1. Let M and N be non-zero finitely generated modules over a local ring R. Assume that $q(M,N) < \infty$, and define $d_i = d_i(M,N)$ as above. Assume that N has positive depth, and let $x \in \mathfrak{m}$ be general with respect to N and with respect to each $\mathrm{Tor}_i^R(M,N), 0 \le i \le q(M,N)$. Let $\bar{N} = N/xN$ and put $\bar{d_i} = \mathrm{depth}(\mathrm{Tor}_i^R(M,\bar{N}))$. In the following rules, we declare the zero module to have depth ∞ and apply the usual rules of arithmetic involving ∞, e.g., $\infty - 1 = \infty$. Also, we take $\mathrm{Tor}_i^R(M,N) = 0$ if $i < 0$.

(R1) If $d_i > 0$ and $d_{i-1} > 0$, then $\bar{d_i} = d_i - 1$.
(R2) If $d_i = 0$ then $\bar{d_i} = 0$.
(R3) If $d_i \neq 2$ and $d_{i-1} = 0$, then $\bar{d_i} = 0$.
(R4) If $d_i < \infty$ then $\bar{d_i} < \infty$.

Proof. We put $\bar{N} = N/xN$, and tensor the short exact sequence $0 \to N \xrightarrow{x} N \to \bar{N} \to 0$ with M. The resulting long exact sequence provides, for each $i \in \mathbb{Z}$, a short exact sequence

$$0 \to \frac{\mathrm{Tor}_i^R(M,N)}{x\mathrm{Tor}_i^R(M,N)} \to \mathrm{Tor}_i^R(M,\bar{N}) \to \mathrm{Ann}_{\mathrm{Tor}_{i-1}^R(M,N)}(x) \to 0.$$

For ease of notation, we rewrite this exact sequence as

$$0 \to U \to V \to W \to 0.$$

If $d_{i-1} > 0$, then $W = 0$. Therefore $U \cong V$, and $\bar{d_i} = \mathrm{depth}(U)$. If also $d_i > 0$, then $\mathrm{depth}(U) = d_i - 1$, since x is a non-zerodivisor on $\mathrm{Tor}_i^R(M,N)$ (even if $d_i = \infty$). This proves (R1).

To prove (R2), suppose $d_i = 0$. Then $\mathrm{depth}(U) = 0$ by Lemma 1.4. Since V contains U, $\bar{d_i} = \mathrm{depth}(V) = 0$.

For (R3) we assume that $d_{i-1} = 0$ and $\bar{d_i} > 0$. Since x is general, W has finite length by (1) of Lemma 1.4. Also, $W \neq 0$ because $d_{i-1} = 0$. Therefore $\mathrm{depth}(W) = 0$. Since $\mathrm{depth}(V) = \bar{d_i} > 0$, the "Depth Lemma" [6, Lemma 1.1] implies that $\mathrm{depth}(U) = 1$. Now $d_i > 0$ by (R2), and since x is general x is a non-zero divisor on $\mathrm{Tor}_i^R(M,N)$. Therefore $d_i = 1 + \mathrm{depth}(U) = 2$, as desired.

Finally, we note that if $d_i < \infty$ then $U \neq 0$ by Nakayama's Lemma. Therefore $V \neq 0$ and we have (R4). □

Before stating the main theorem, we show how to use the Tor Game to prove an inequality obtained by Choi and Iyengar.

PROPOSITION 2.2. *[5, Remark 7] Let M and N be non-zero finitely generated modules over a local ring R. Assume that M has finite CI dimension and that $q_R(M,N) < \infty$. Then $D(M,N) \geq m := \min\{d_i - i \mid i \in \mathbb{Z}\}$.*

Proof. If $\operatorname{depth}(N) = 0$, then $d_q = 0$ by Proposition 1.2. Therefore $m = -q$, which equals $D(M,N)$ by formula (2) (proved by Araya and Yoshino [1] in the present context). We assume now that $\operatorname{depth}(N) > 0$ and proceed by induction on $\operatorname{depth}(N)$. We play the Tor Game, noting that $D(M,\bar{N}) = D(M,N) - 1$. It will suffice to show that $\bar{d}_i - i \geq m - 1$ for each i, for then we will have by induction that $D(M,\bar{N}) \geq m - 1$.

If $d_{i-1} > 0$, then $\bar{d}_i \geq d_i - 1$ (by (R1) and (R2)), and $\bar{d}_i - i \geq d_i - i - 1 \geq m - 1$. If, on the other hand, $d_{i-1} = 0$, then $\bar{d}_i - i \geq -i = d_{i-1} - (i-1) - 1 \geq m - 1$. □

Note that inequality (3) in the introduction follows easily from (2.2). In fact, we get the following fact, proved by Choi and Iyengar:

PROPOSITION 2.3. *[5, Theorem 3] Let M and N be non-zero finitely generated modules over a local ring R. Assume M has finite CI dimension and $q := q_R(M,N) < \infty$. Then $D(M,N) \geq -q$, with equality if and only if $d_q = 0$.*

Proof. If $d_q = 0$ we have equality by formula (2), as observed in the proof of (2.2). Assume now that $d_q > 0$, and let $d_j - j = m := \min\{d_i - i \mid i \in \mathbb{Z}\}$. Then $D(M,N) \geq d_j - j$ by (2.2). If $j = q$, then $D(M,N) \geq d_q - q > -q$; and if $j < q$ then $D(M,N) \geq -j > -q$. □

Now we come to our main theorem, which gives a sufficient condition for winning the Tor Game.

MAIN THEOREM 2.4. *Let M and N be non-zero finitely generated modules over a local ring R. Assume M has finite CI dimension and $q_R(M,N) < \infty$. Put $m = m(M,N) = \min\{d_i - i \mid i \in \mathbb{Z}\}$ and $j = j(M,N) = \max\{i \mid d_i - i = m\}$. Assume*

$$d_i - i \geq m + 2 \text{ for } i > j. \tag{†}$$

Then $D(M,N) = m$.

The proof will be deferred to Section 5 of the paper. As we shall see in Example 3.2, (†) is not a necessary condition for winning the Tor Game. It is, however, both necessary and sufficient if we play strictly by the rules (see §5).

An immediate corollary is worth stating separately.

COROLLARY 2.5. *Let M and N be non-zero finitely generated modules over a local ring R. Assume M has finite CI dimension and $q_R(M,N) < \infty$. Put $m = m(M,N) = \min\{d_i - i \mid i \in \mathbb{Z}\}$ Suppose that $m = d_q - q$. Then $D(M,N) = m$.*

Proof. This follows immediately from Theorem 2.4 since the main condition needed in that theorem is vacuously satisfied. □

3 EXAMPLES

Here we give some examples to indicate some of the possibilities that can occur in the Tor Game.

EXAMPLE 3.1. **A losing Tor Game.** Let $R = k[[X,Y,U,V]]$, where k is a field, and put $f = XY - UV$. Let $M = (X,U)/(f)$ and $N = (Y,V)/(f)$. Both M and N have depth 3 (cf. [7, Example 4.1]), whence $D(M,N) = 2$. As modules over $R/(f)$, both M and N are free on the punctured spectrum of $R/(f)$ (cf. [8, Example 1.8]). If d_0 were positive, $M \otimes_R N$ would satisfy Serre's condition (S_1) as an $R/(f)$-module and therefore be torsion-free as an $R/(f)$-module. But then $M \otimes_R N$, which requires 4 generators, would be isomorphic to the ideal product $(x,u)(y,v)$ in $R/(f)$. But $(x,u)(y,v) = (xy,xv,uy,uv) = (xy,xv,uy)$, contradiction. Thus $d_0 = 0$.

To compute d_1, we choose an exact sequence $0 \to F \to G \to M \to 0$, with F and G free R-modules. Tensoring this with N, we get an exact sequence

$$0 \to \operatorname{Tor}_1^R(M,N) \to F \otimes_R N \to G \otimes_R N \to M \otimes_R N \to 0.$$

Since the two middle terms have depth 3 and $M \otimes_R N$ has depth 0, it follows from the Depth Lemma that $d_1 = 2$.

To summarize, we have $q = 1, D(M,N) = 2, d_0 = 0$ and $d_1 = 2$. Since neither d_0 nor $d_1 - 1$ is equal to $D(M,N)$, we lose the Tor Game.

EXAMPLE 3.2. **A more general losing Tor Game.** Let R be a Cohen-Macaulay ring of dimension d and let p be a height-one prime ideal of R. Assume that R/p is not Cohen-Macaulay, and set its depth equal to s. Choose a non-zerodivisor $a \in p$ and write $p = (a : b)$, where b is also a non-zerodivisor. Set $M = R/Ra$ and $N = R/Rb$. Observe that

$$D(M,N) = (d-1) + (d-1) - d = d - 2,$$

and both M and N have projective dimension 1. We compute the depths d_0 and d_1. Since $\operatorname{Tor}_1^R(M,N) \cong p/(a)$, the short exact sequence

$$0 \to p/(a) \to R/(a) \to R/p \to 0,$$

together with the fact that $\operatorname{depth}(R/p) = s < d - 1 = \operatorname{depth}(R/(a))$, shows that $d_1 = s + 1$. Likewise, since $\operatorname{Tor}_0^R(M,N) \cong R/(a,b)$, the short exact sequence

$$0 \to R/p \to R/(a) \to R/(a,b) \to 0$$

proves that $d_0 = s - 1$. Hence the minimum value of $d_1 - 1$ and $d_0 - 0$ is $s - 1$ which is not equal to $d - 2 = D(M,N)$. This example shows very clearly that in general there is no chance of proving a relationship between $D(M,N)$ and the values $d_i - i$. The gap of 2 between d_1 and d_0 is fatal. Our main result in some sense shows that this is only way the Tor Game fails.

EXAMPLE 3.3. **Cheating in the Tor Game.** Let $R = k[[X,Y,Z]]$ and put $f = XY + Z^2$. Let $M = N = (X,Z)/(f)$. As in the first example, it is helpful to view M as an ideal of the hypersurface $R/(f)$. Certainly $M \otimes_R N$ has non-zero torsion as an $R/(f)$-module (since it needs four generators and the ideal $(x,z)^2$ of $R/(f)$ needs only 3). As shown in [7, Example 4.2], M is a reflexive $R/(f)$-module and therefore has depth 2. Therefore $D_R(M,N) = 1$ and $q(M,N) \leq 1$. Since $R/(f)$ is an isolated singularity M is free on the punctured spectrum of $R/(f)$, and it follows, essentially as in Example 4.1, that $d_0 = 0$ and $d_1 = 2$. We see that $m(M,N) = 0$, $j(M,N) = 0$, but (†) fails because $d_1 - 1 = m + 1$. In a sense we have won the Tor Game, since $D(M,N) = d_1 - 1$, but we have not followed the rules, and the answer is not the one "$D(M,N) = m(M,N)$" given by the theorem. In the next section we will describe a slightly modified game that keeps the player from having to think and prevents "accidental" wins like this.

EXAMPLE 3.4. **Fluctuating depths.** One might hope that in general the d_i are monotone, as a kind of strong rigidity of Tor. However, this is not the case as the following example (pointed out to us by Bernd Ulrich) shows. Let R be a regular local ring of dimension d and I a perfect height two ideal. Assume that I is not generated by a regular sequence. Set $M = N = R/I$. Then $\operatorname{Tor}_0^R(M,N) = R/I$ has depth $d-2$. The module $\operatorname{Tor}_1^R(M,N) \cong I/I^2$, and this module has depth $d-3$ (see [10]). A resolution of R/I is given by

$$0 \to R^{n-1} \to R^n \to R \to R/I \to 0.$$

Tensoring with R/I shows that the following is exact:

$$0 \to \operatorname{Tor}_2^R(R/I,R/I) \to (R/I)^{n-1} \to (R/I)^n \to I/I^2 \to 0.$$

This proves that $\operatorname{Tor}_2^R(R/I,R/I)$ is Cohen-Macaulay having depth $d-2$, strictly greater than $\operatorname{depth}(\operatorname{Tor}_1^R(R/I,R/I))$, which in turn is strictly less than the depth of $\operatorname{Tor}_0^R(R/I,R/I)$. Hence the depths fluctuate. In this case, however, the conditions of Theorem 2.4 are met and we win the Tor game: $D(R/I,R/I) = (d-2)+(d-2)-d = d-4$. Then $(d_0 - 0, d_1 - 1, d_2 - 2) = (d-2, d-4, d-4)$. Since the minimum value occurs at the last non-zero Tor, we automatically win according to Corollary 2.5.

EXAMPLE 3.5. **Another losing game.** We refer to [5, Prop. 11, Example 12] which constructs a losing Tor game over a regular local ring. Specifically, they construct two cyclic modules M and N over a regular local ring of dimension 5 such that $q_R(M,N) = 1$, $D(M,N) = 2$, $d_0 = 0$ and $d_1 = 2$. Hence $\{d_0 - 0, d_1 - 1\} = \{0,1\}$ and $D(M,N)$ is not in this set.

4 THE VIRTUAL TOR GAME

In order to expand the market for our game, we have developed a version that is much easier to play. It requires very little skill, and there is no strategy whatsoever. You just follow the rules and hope for the best.

In the Virtual Tor Game, Player I chooses a sequence $d_0, \ldots, d_q$, where each d_i is either a non-negative integer, the symbol ∞, or the symbol $\star$ (indicating an "unknown" depth). It is required that $d_0 \in \{0,1,2,\ldots\}$ and $d_q \neq \infty$. Player I understands that $d_i = \infty$ if $i < 0$. Values of d_i for $i > q$ are irrelevant. The symbol $\star$ is incomparable with integers and with ∞. Thus, if we hypothesize, for example, that $d_i > 0$, it is assumed that $d_i \neq \star$.

If $d_q \in \{0, \star\}$, the game is over (see below). Otherwise Player II computes the sequence $\bar{d}_0, \ldots, \bar{d}_q$ according to the rules (which we shall describe shortly), and, if the game is not over, repeats the process on the new sequence (computing $\bar{\bar{d}}_0, \ldots, \bar{\bar{d}}_q$) and so on. A consequence of the Rules is that at each stage one has $d_{-1} = \infty$, and Player I can simply be assured of this at the outset.

The rules of the Virtual Tor Game are the same as in the actual Tor Game, except that we have introduced the symbol $\star$ to deal with the undecided case $d_i = 2$, $d_{i-1} = 0$. Also, we allow some of the original data to include unknown entries.

The rules of the Virtual Tor Game are the following:

(VR1) If $d_i > 0$ and $d_{i-1} > 0$, then $\bar{d}_i = d_i - 1$.
(VR2) If $d_i \leq 1$ then $\bar{d}_i = 0$.
(VR3) If $d_i > 2$ and $d_{i-1} = 0$, then $\bar{d}_i = 0$.
(VR4) If $d_i = \star$ then $\bar{d}_i = \star$.
(VR5) If $d_i = 2$ and $d_{i-1} = 0$, then $\bar{d}_i = \star$.
(VR6) If $d_i \geq 2$ and $d_{i-1} = \star$, then $\bar{d}_i = \star$.

Player II wins if, after a finite sequence of iterations, d_q becomes 0. Player II loses if he eventually gets $d_q = \star$. It is easy to see that Player II either wins or loses. (Of course, Player I is expected to make an interesting choice of the orginal d_i, so that neither a win nor a loss is obvious from the start!)

We will give some examples of winning and losing Virtual Tor Games. In deference to Macaulay 2, we will list the d_i in the order $d_0\ d_1\ \ldots\ d_q$. This list appears in the second line of each display, the first line (in bold) being just the indices $0\ 1\ \ldots\ q$. The third line will be $\bar{d}_0\ \bar{d}_1\ \ldots\ \bar{d}_q$, the result of one move of Virtual Tor Game, and so on.

EXAMPLE 4.1. **An inauspicious beginning, but we win in the end.**

i	:	**0**	**1**	**2**	**3**	**4**	**5**	**6**
d_i	:	0	2	3	5	4	7	9
$\bar{d}_i$	:	0	$\star$	2	4	3	6	8
$\bar{\bar{d}}_i$	:	0	$\star$	$\star$	3	2	5	7
·		0	$\star$	$\star$	$\star$	1	4	6
·		0	$\star$	$\star$	$\star$	0	3	5
·		0	$\star$	$\star$	$\star$	0	0	4
·		0	$\star$	$\star$	$\star$	0	0	0

Here we get stars right from the start, and they move ominously to the right, but eventually they are blocked by 0's. In this example we have $m = 0$ and $j = 4$. Note that condition (†) is satisfied.

EXAMPLE 4.2. **A promising beginning with a devastating finale.**

i	:	**0**	**1**	**2**	**3**	**4**	**5**	**6**	**7**
d_i	:	2	4	4	7	8	10	9	11
$\bar{d}_i$	:	1	3	3	6	7	9	8	10
$\bar{\bar{d}}_i$	:	0	2	2	5	6	8	7	9
·		0	$\star$	1	4	5	7	6	8
·		0	$\star$	0	3	4	6	5	7
·		0	$\star$	0	0	3	5	4	6
·		0	$\star$	0	0	0	4	3	5
·		0	$\star$	0	0	0	0	2	4
·		0	$\star$	0	0	0	0	$\star$	3
·		0	$\star$	0	0	0	0	$\star$	$\star$

Here stars crop up early but are quickly blocked. Trouble lies ahead, however, and the culprit is d_6: Note that $m = 2$ and $j = 2$; condition (†) fails, since $d_6 - 6 = m + 1$.

The configuration $d_{i+1} = d_i + 2$ will often, but not always, lead to a losing Virtual Tor Game. In fact, one can lose the virtual game even when this configuration is not present at the outset.

EXAMPLE 4.3. **Another losing Tor Game.**

i	:	**0**	**1**	**2**	**3**	**4**	**5**
d_i	:	1	1	4	4	5	6
$\bar{d}_i$	:	0	0	3	3	4	5
$\bar{\bar{d}}_i$	:	0	0	0	2	3	4
·		0	0	0	$\star$	2	3
·		0	0	0	$\star$	$\star$	2
·		0	0	0	$\star$	$\star$	$\star$

5 PROOF OF THE MAIN THEOREM

In this section we will show that condition (†) of (1.4), slightly modified to allow some unknown depths in the original data, is necessary and sufficient for winning the Virtual Tor Game. The Main Theorem will follow as a special case.

NOTATION AND ASSUMPTIONS 5.1. Suppose that we are given a list $d_0, \ldots, d_q$, with each $d_i \in \{0, 1, 2, \ldots, \infty, \star\}$. We assume always that d_0 is an integer, that $d_q \neq \infty$, and that $d_i = \infty$ if $i < 0$ or $i > q$. We put $b = \sup\{i \mid d_i = \star\}$. (If $\star$ does not occur among the d_i we put $b = -\infty$.) Let $m = \inf\{d_i - i \mid i > b\}$ and $j = \sup\{i > b \mid d_i - i = m\}$. (Note that $m = j = \infty$ if $b = q$; otherwise $m, j < \infty$.) Assuming that $d_q \notin \{0, \star\}$, we compute $\bar{d_0}, \ldots, \bar{d_q}$ according to the rules of the Virtual Tor Game in Section 4. Note that we still have $\bar{d_0} \in \{0, 1, 2, \ldots\}$, $\bar{d_q} \neq \infty$, and $\bar{d_i} = \infty$ if $i < 0$ or $i > q$. (Informally, $\bar{q} = q$.) We define $\bar{b}$, $\bar{m}$ and $\bar{j}$ in the obvious way. We consider the following condition, which amounts to (†) in the case where there are no stars ($b = -\infty$):

$$\begin{aligned} &\text{(a) } d_i - i \geq m + 2 \text{ for each } i > j, \\ &\text{(b) } d_q \neq \star, \text{ and} \\ &\text{(c) } j \geq b + d_j. \end{aligned} \qquad (\ddagger)$$

Part (b) guarantees that j and d_j are finite, and (c) ensures that j is far enough to the right of the last star (if there are any).

THEOREM 5.2. *Assume $d_q \notin \{0, \star\}$.*

(1) Condition (‡) holds for the d_i if and only if (‡) holds for the $\bar{d_i}$.

(2) If (‡) holds and $d_j > 0$, then $\bar{j} = j$, $\bar{d_j} = d_j - 1$ and $\bar{m} = m - 1$.

(3) If (‡) holds and $d_j = 0$, then $\bar{j} = j + 1$, $\bar{d}_{j+1} = 0$ and $\bar{m} = m - 1$.

Proof. Suppose first that $d_j > 0$. We claim that if (c) of (‡) holds for $d_0, \ldots, d_q$, then $\bar{j} = j$, $\bar{m} = m - 1$, and both (b) and (c) hold for $\bar{d_0}, \ldots, \bar{d_q}$. To see this, let $i \geq j - d_j + 1$. Since $j - d_j \geq b$, we have $d_i - i \geq d_j - j$, whence $d_i \geq d_j - j + i \geq d_j - j + (j - d_j + 1) = 1$. We have shown that $d_i \geq 1$ for all $i \geq j - d_j + 1$, and it follows that

$$\bar{d_i} = d_i - 1 \text{ for all } i \geq j - d_j + 2. \qquad (5.2.1)$$

From (5.2.1) and (VR2) we see that $\bar{d_j} = j - 1$. To complete the proof of the claim, we just need to verify that $\bar{d_i} - i \geq \bar{d_j} - j$ for each $i > \bar{b}$. Let $i > \bar{b}$. Then $i > b$ (by VR6)), and hence $d_i - i \geq d_j - j$. If $\bar{d_i} - i < \bar{d_j} - j$ $(= d_j - j - 1)$ then $\bar{d_i} \leq d_i - 2$, which forces $d_{i-1} = 0$. Therefore $i - 1 > b$, and we have $\bar{d_i} - i \geq -i = d_{i-1} - (i-1) - 1 \geq d_j - j - 1 = \bar{d_j} - j$. This proves the claim. Assertion (2) and the "only if" part of (1) in the case $d_j > 0$ follow easily.

Still assuming $d_j > 0$, suppose now that (‡) fails for the d_i. If (c) holds for the d_i, then (a) must fail. By the claim in the paragraph above, (a) fails also for $\bar{d_0}, \ldots, \bar{d_q}$. Assume now that (c) fails, that is, $j - d_j < b$. Then, for each $i > b$ we have $d_i - i \geq d_j - j > -b$, whence $d_i > i - b \geq 1$. In particular, $d_{b+1} \geq 2$, and now

(VR6) implies that $\bar{d}_{b+1} = \star$. Moreover, $\bar{d}_i = d_i - 1$ for each $i > b+1$, and we see that $\bar{b} = b+1$ and $\bar{j} = j$. Now $j - \bar{d}_j = j - d_j + 1 < b + 1 = \bar{b}$, and (‡) fails for $\bar{d}_0, \ldots, \bar{d}_q$.

This completes the proof in the case $d_j > 0$. Assume now that $d_j = 0$. Then (c) is automatically satisfied for $d_0, \ldots, d_q$ (and (b) is part of the hypotheses of the theorem). If $i > j$ we have $d_i - i > d_j - j = -j$. Therefore

$$d_i \geq 2 \text{ for each } i > j. \tag{5.2.2}$$

If $d_{j+1} \geq 3$, then, by (5.2.2), $\bar{d}_{j+1} = 0$ and $\bar{d}_i = d_i - 1$ for each $i > j+1$. Assertions (1) and (3) now follow easily. Therefore we assume that $d_{j+1} = 2$, in which case (‡) fails for $\bar{d}_0, \ldots, \bar{d}_q$. To complete the proof, we need only verify that (‡) fails for $d_0, \ldots, d_q$. We may assume that $\bar{d}_q \neq \star$. Since $\bar{d}_{j+1} = \star$ by (VR5), we have $j+1 \leq \bar{b} < \bar{j} \leq q$. Noting that $\bar{d}_{\bar{j}} = d_{\bar{j}} - 1$ by (5.2.2), we have $\bar{b} + \bar{d}_{\bar{j}} \geq j + 1 + d_{\bar{j}} - 1 = j + \bar{j} + d_{\bar{j}} - \bar{j} > j + \bar{j} + d_j - j = \bar{j}$. This shows that (c) fails for $\bar{d}_0, \ldots, \bar{d}_q$, and the proof is complete. □

COROLLARY 5.3. *Given the initial data $d_0, \ldots, d_q \in \{0, 1, 2, \ldots, \infty, \star\}$, with $d_0 \in \mathbb{Z}$ and $d_q \neq \infty$, we win the Virtual Tor Game if and only if (‡) is true.*

Proof. The rules of the Virtual Tor Game dictate that if $d_i \neq \star$, then either $\bar{d}_i < d_i$ or $\bar{d}_i = \star$. Thus, after a finite number of iterations of the Tor Game, d_q becomes either 0 or $\star$. By (1) of (5.1), this happens precisely when (‡) is, respectively, is not true. □

Proof of the Main Theorem (2.4) Put $b = -\infty$, so that (†) and (‡) say the same thing. We play the Tor Game until d_q drops to 0. By (1.2), N has positive depth at the beginning of each play; therefore $D(M,N)$ drops by 1 at each stage. But also m drops by 1 at each stage, by (2) and (3) of (5.2). Since the Theorem is true when $d_q = 0$ (by formula (2) of the introduction—proved in the context of finite CI dimension by Araya and Yoshino [1]), it must be true in general. □

References

[1] T. Araya and Y. Yoshino, *Remarks on a depth formula, a grade inequality and a conjecture of Auslander*, Comm. Algebra **26** (1998), pp. 3793–3806.

[2] M. Auslander, *Modules over unramified regular local rings*, Illinois J. Math. **5** (1961), pp. 631–647.

[3] L. Avramov, V. Gasharov and I. Peeva, *Complete intersection dimension*, Publ. Math. I.H.E.S. **86** (1997), pp. 67–114.

[4] W. Bruns and J. Herzog, Cohen-Macaulay Rings, Cambridge University Press, Cambridge, 1993.

[5] S. Choi and S. Iyengar, *On a depth formula for modules over local rings*, Comm. Algebra **29** (2001), pp. 3135–3143.

[6] E. G. Evans and P. Griffith, Syzygies, London Math. Soc. Lecture Note Ser. **106**, Cambridge, 1985.

[7] C. Huneke and R. Wiegand, *Tensor products of modules and the rigidity of Tor*, Math. Ann. **299** (1994), pp. 449-476.

[8] C. Huneke and R. Wiegand, *Tensor products of modules, rigidity and local cohomology*, Math. Scand. **81** (1997), pp. 161–183.

[9] D. A. Jorgensen, *A generalization of the Auslander-Buchsbaum formula*, J. Pure Appl. Algebra **144** (1999), pp. 145–155.

[10] J. Weyman, *Resolutions of the exterior and symmetric powers of a module*, J. Algebra **58** (1979), pp. 333-341.

Trivial Extensions of Local Rings and a Conjecture of Costa

S. Kabbaj*,[1] and N. Mahdou[2]

[1] Department of Mathematics, KFUPM, P.O.Box 5046, Dhahran 31261, Saudi Arabia
[2] Department of Mathematics, FST Fez-Saïss, B.P. 2202, University of Fez, Fez, Morocco

Abstract. This paper partly settles a conjecture of Costa on (n, d)-rings, i.e., rings in which n-presented modules have projective dimension at most d. For this purpose, a theorem studies the transfer of the (n, d)-property to trivial extensions of local rings by their residue fields. It concludes with a brief discussion -backed by original examples- of the scopes and limits of our results.

* This project has been funded by King Fahd University of Petroleum & minerals under Project # MS/TRIVIAL/224

0. Introduction

All rings considered in this paper are commutative with identity elements and all modules are unital. For a nonnegative integer n, an R-module E is n-presented if there is an exact sequence $F_n \to F_{n-1} \to ... \to F_0 \to E \to 0$ in which each F_i is a finitely generated free R-module (In [1], such E is said to have an n-presentation). In particular, "0-presented" means finitely generated and "1-presented" means finitely presented. Also, $pd_R E$ will denote the projective dimension of E as an R-module.

In 1994, Costa [2] introduced a doubly filtered set of classes of rings throwing a brighter light on the structures of non-Noetherian rings. Namely, for nonnegative integers n and d, a ring R is an (n, d)-ring if every n-presented R-module has projective dimension at most d. The Noetherianness deflates the (n, d)-property to the notion of regular ring. However, outside Noetherian settings, the richness of this classification resides in its ability to unify classic concepts such as von Neumann regular, hereditary/Dedekind, and semi-hereditary/Prüfer rings. Costa was motivated by the sake of a deeper understanding of what makes a Prüfer domain Prüfer. In this context, he asked "what happens if we assume only that every finitely presented (instead of generated) sub-module of a finitely generated free module is projective?" It turned out that a non-Prüfer domain having this property exists, i.e., (In the (n, d)-jargon) a $(2, 1)$-domain which is not a $(1, 1)$-domain. This gave rise to the theory of (n, d)-rings. Throughout, we assume familiarity with n-presentation, coherence, and basics of the (n, d)-theory as in [1, 2, 3, 6, 7, 8, 10].

Costa's paper [2] concludes with a number of open problems and conjectures, including the existence of (n, d)-rings, specifically whether: "*There are examples of (n, d)-rings which are neither $(n, d-1)$-rings nor $(n-1, d)$-rings, for all nonnegative integers n and d*". Some limitations are immediate; for instance, there are no (n, 0)-domains which are not fields. Also, for $d = 0$ or $n = 0$ the conjecture reduces to "$(n, 0)$-ring not $(n-1, 0)$-ring" or "$(0, d)$-ring not $(0, d-1)$-ring", respectively.

Let's summarize the current situation. So far, solely the cases $n \leq 2$ and d arbitrary were gradually solved in [2], [3], and [14]. These partial results were obtained using various pullbacks. For obvious reasons, these were no longer useful for the specific case $d = 0$. Therefore, in [14], the author appealed to trivial extensions of fields k by infinite-dimensional k-vector spaces, and hence constructed a $(2,0)$-ring (also called 2-von Neumann regular ring) which is not a $(1,0)$-ring (i.e., not von Neumann regular). This encouraged further work for other trivial extension contexts.

Let A be a ring and E an A-module. The trivial ring extension of A by E is the ring $R = A \propto E$ whose underlying group is $A \times E$ with multiplication given by $(a,e)(a',e') = (aa', ae' + a'e)$. An ideal J of R has the form $J = I \propto E'$, where I is an ideal of A and E' is an A-submodule of E such that $IE \subseteq E'$. Considerable work, part of it summarized in Glaz's book [10] and Huckaba's book [11], has been concerned with trivial ring extensions. These have proven to be useful in solving many open problems and conjectures for various contexts in (commutative and non-commutative) ring theory. See for instance [4, 5, 9, 12, 13, 15, 16, 17].

Costa's conjecture is still elusively outstanding. A complete solution (i.e., for all nonnegative integers n and d) would very likely appeal to new techniques and constructions. Our aim in this paper is much more modest. We shall resolve the case "$n = 3$ and d arbitrary". For this purpose, Section 1 investigates the transfer of the (n,d)-property to trivial extensions of local (not necessarily Noetherian) rings by their residue fields. A surprising result establishes such a transfer and hence enables us to construct a class of $(3,d)$-rings which are neither $(3,d-1)$-rings nor $(2,d)$-rings, for d arbitrary. Section 2 is merely an attempt to show that Theorem 1.1 and hence Example 1.3 are the best results one can get out of trivial extensions of local rings by their residue fields.

1. Result and Example

This section develops a result on the transfer of the (n,d)-property for a particular context of trivial ring extensions, namely, those issued from local (not necessarily Noetherian) rings by their residue fields. This will enable us to construct a class of $(3,d)$-rings which are neither $(3,d-1)$-rings nor

$(2, d)$-rings, for d arbitrary.

The next theorem not only serves as a prelude to the contruction of examples, but also contributes to the study of the homological algebra of trivial ring extensions.

Theorem 1.1. *Let (A, M) be a local ring and let $R = A \propto A/M$ be the trivial ring extension of A by A/M. Then*

1) R is a $(3, 0)$-ring provided M is not finitely generated.

2) R is not a $(2, d)$-ring, for each integer $d \geq 0$, provided M contains a regular element.

The proof of this theorem requires the next preliminary.

Lemma 1.2. *Let A be a ring, I a proper ideal of A, and R the trivial ring extension of A by A/I. Then $pd_R(I \propto A/I)$ and hence $pd_R(0 \propto A/I)$ are infinite.*

Proof. Consider the exact sequence of R-modules

$$0 \to I \propto A/I \to R \to R/(I \propto A/I) \to 0$$

We claim that $R/(I \propto A/I)$ is not projective. Deny. Then the sequence splits. Hence, $I \propto A/I$ is generated by an idempotent element $(a, e) = (a, e)(a, e) = (a^2, 0)$. So $I \propto A/I = R(a, 0) = Aa \propto 0$, the desired contradiction (since $A/I \neq 0$). It follows from the above sequence that

$$pd_R(R/(I \propto A/I)) = 1 + pd_R(I \propto A/I). \tag{1}$$

Let $(x_i)_{i \in \Delta}$ be a set of generators of I and let $R^{(\Delta)}$ be a free R-module. Consider the exact sequence of R-modules

$$0 \to Ker(u) \to R^{(\Delta)} \oplus R \overset{u}{\to} I \propto A/I \to 0$$

where

$$u((a_i, e_i)_{i \in \Delta}, (a_0, e_0)) = \sum_{i \in \Delta}(a_i, e_i)(x_i, 0) + (a_0, e_0)(0, 1) = (\sum_{i \in \Delta} a_i x_i, a_0)$$

since $x_i \in I$ for each $i \in \Delta$. Hence,

$$Ker(u) = (U \propto (A/I)^{(\Delta)}) \oplus (I \propto A/I)$$

where $U = \{(a_i)_{i\in\Delta} \in A^{(\Delta)} / \sum_{i\in\Delta} a_i x_i = 0\}$. Therefore, we have the isomorphism of R-modules $I \propto A/I \cong (R^{(\Delta)}/(U \propto (A/I)^{(\Delta)})) \oplus (R/(I \propto A/I))$. It follows that

$$pd_R(R/(I \propto A/I)) \leq pd_R(I \propto A/I). \tag{2}$$

Clearly, (1) and (2) force $pd_R(I \propto A/I)$ to be infinite.

Now the exact sequence of R-modules

$$0 \to I \propto A/I \to R \xrightarrow{v} 0 \propto A/I \to 0,$$

where $v(a,e) = (a,e)(0,1) = (0,a)$, easily yields $pd_R(0 \propto A/I) = \infty$, completing the proof of Lemma 1.2. ◇

Proof of Theorem 1.1. 1) Suppose M is not finitely generated. Let $H_0 (\neq 0)$ be a 3-presented R-module and let $(z_i)_{i=1,\ldots,n}$ be a minimal set of generators of H_0 (for some positive integer n). Consider the exact sequence of R-modules

$$0 \to H_1 := Ker(u_0) \to R^n \xrightarrow{u_0} H_0 \to 0$$

where $u_0((r_i)_{i=1,\ldots,n}) = \sum_{i=1}^{n} r_i z_i$. Throughout this proof we identify R^n with $A^n \propto (A/M)^n$. Our aim is to prove that $H_1 = 0$. Deny. By the above exact sequence, H_1 is a 2-presented R-module. Let $(x_i, y_i)_{i=1,\ldots,m}$ be a minimal set of generators of H_1 (for some positive integer m). The minimality of $(z_i)_{i=1,\ldots,n}$ implies that $H_1 \subseteq M^n \propto (A/M)^n$, whence $x_i \in M^n$ (and $y_i \in (A/M)^n$) for $i = 1, \ldots, m$. Consider the exact sequence of R-modules

$$0 \to H_2 := Ker(u_1) \to R^m \xrightarrow{u_1} H_1 \to 0$$

where $u_1((a_i, e_i)_i) = \sum_{i=1}^{m} (a_i, e_i)(x_i, y_i) = \sum_{i=1}^{m} (a_i x_i, a_i y_i)$, since $x_i \in M^n$ for each i. Then, $H_2 = U \propto (A/M)^m$, where $U = \{(a_i)_{i=1,\ldots,m} \in A^m / \sum_{i=1}^{m} a_i x_i =$

0 and $\sum_{i=1}^{m} a_i y_i = 0\}$. By the above exact sequence, H_2 is a finitely presented (hence generated) R-module, so that (via [11, Theorem 25.1]) U is a finitely generated A-module. Further, the minimality of $(x_i, y_i)_{i=1,\ldots,m}$ yields $U \subseteq M^m$. Let $(t_i)_{i=1,\ldots,p}$ be a set of generators of U and let $(f_i)_{i=p+1,\ldots,p+m}$ be a basis of the (A/M)-vector space $(A/M)^m$. Consider the exact sequence of R-modules

$$0 \to H_3 := Ker(u_2) \to R^{p+m} \overset{u_2}{\to} H_2 \to 0$$

where

$$u_2((a_i, e_i)_i) = \sum_{i=1}^{p}(a_i, e_i)(t_i, 0) + \sum_{i=p+1}^{p+m} (a_i, e_i)(0, f_i) = (\sum_{i=1}^{p} a_i t_i, \sum_{i=p+1}^{p+m} a_i f_i),$$

since $t_i \in M^m$ for each $i = 1, \ldots, p$ and $(f_i)_i$ is a basis of the (A/M)-vector space $(A/M)^m$. It follows that $H_3 \cong (V \propto (A/M)^p) \oplus (M^m \propto (A/M)^m)$, where $V = \{(a_i)_{i=1,\ldots,p} \in A^p / \sum_{i=1}^{p} a_i t_i = 0\}$. By the above sequence, H_3 is a finitely generated R-module. Hence $M \propto A/M$ is a finitely generated ideal of R, so M is a finitely generated ideal of A by [11, Theorem 25.1], the desired contradiction.

Consequently, $H_1 = 0$, forcing H_0 to be a free R-module. Therefore, every 3-presented R-module is projective (i.e., R is a $(3,0)$-ring).

2) Assume that M contains a regular element m. We must show that R is not a $(2,d)$-ring, for each integer $d \geq 0$. Let $J = R(m, 0)$ and consider the exact sequence of R-modules

$$0 \to Ker(v) \to R \overset{v}{\to} J \to 0$$

where $v(a, e) = (a, e)(m, 0) = (am, 0)$. Clearly, $Ker(v) = 0 \propto (A/M) = R(0, 1)$, since m is a regular element. Therefore, $Ker(v)$ is a finitely generated ideal of R and hence J is a finitely presented ideal of R. On the other hand, $pd_R(Ker(v)) = pd_R(0 \propto A/M) = \infty$ by Lemma 1.2, so $pd_R(J) = \infty$. Finally, the exact sequence

$$0 \to J \to R \to R/J \to 0$$

yields a 2-presented R-module, namely R/J, with infinite projective dimension (i.e., R is not a $(2,d)$-ring, for each $d \geq 0$), completing the proof. $\diamondsuit$

We are now able to construct a class of $(3,d)$-rings which are neither $(3,d-1)$-rings nor $(2,d)$-rings, for d arbitrary. In order to do this, we first recall from [14] an interesting result establishing the transfer of the (n,d)-property to finite direct sums.

Theorem 1.3 ([14, Theorem 2.4]). *A finite direct sum $\bigoplus_{1 \leq i \leq n} A_i$ is an (n,d)-ring if and only if so is each A_i.* $\diamondsuit$

Example 1.4. Let d be a nonnegative integer and B a Noetherian ring of global dimension d. Let (A_0, M) be a nondiscrete valuation domain and let $A = A_0 \propto (A_0/M)$ be the trivial ring extension of A_0 by A_0/M. Let $R = A \times B$ be the direct product of A and B. Then R is a $(3,d)$-ring which is neither a $(3,d-1)$-ring nor a $(2,d)$-ring, for d arbitrary (The case $d = 0$ reduces to "$(3,0)$-ring not $(2,0)$-ring").

Indeed, by Theorem 1.1, A is a $(3,0)$-ring (also called 3-Von Neumann regular ring) which is not a $(2,d')$-ring for each nonnegative integer d'. Moreover, R is a $(3,d)$-ring by [14, Theorem (2.4)] since both A and B are $(3,d)$-rings (by gnomonic theorems of Costa [2]). Further, R is not a $(2,d)$-ring by [14, Theorem (2.4)] (since A is not a $(2,d)$-ring). Finally, we claim that R is not a $(3,d-1)$-ring. Deny. Then B is a $(3,d-1)$-ring by [14, Theorem (2.4)]. Hence, by [2, Theorem 2.4] B is a $(0,d-1)$-ring since B is Noetherian (i.e., 0-coherent). So that $gldim(B) \leq d-1$, the desired contradiction. $\diamondsuit$

2. Discussion

This section consists of a brief discussion of the scopes and limits of our findings. This merely is an attempt to show that Theorem 1.1 and hence Example 1.3 are the best results one can get out of trivial extensions of local rings by their residue fields.

Remark 2.1. In Theorem 1.1, the (n,d)-property holds for a trivial ring extension of a local ring (A, M) by its residue field sans any (n,d)-hypothesis on the basic ring A. This is the first surprise. The second one resides in the narrow scope revealed by this (strong) result, namely $n = 3$ and $d = 0$. Thus, the two assertions of Theorem 1.1, put together with Costa's gnomonic theorems, restrict the scope of a possible example to $n = 3$ and d arbitrary.

Furthermore, since in Theorem 1.1 the upshot is controlled solely by restrictions on M, the next two examples clearly illustrate its failure in case one denies these restrictions, namely, "*M is not finitely generated*" and "*M contains a regular element*", respectively.

Example 2.2. Let K be a field and let $A = K[[X]] = K + M$, where $M = XA$. We claim that the trivial ring extension R of A by $A/M(= K)$ is not an (n, d)-ring, for any integers $n, d \geq 0$.

Let's first show that R is Noetherian. Let $J = I \propto E$ be a proper ideal of R, where I is a proper ideal of A and E is a submodule of the simple A-module A/M (i.e., $E = 0$ or $E = A/M$). Since A is a Noetherian valuation ring, $I = Aa$ for some $a \in M$. Let $f \in A$ such that $(a, \bar{f}) \in J$. Without loss of generality, suppose $J \neq R(a, \bar{f})$. Let $(c, \bar{g}) \in J \setminus R(a, \bar{f})$, where $c, g \in A$, and let $c = \lambda a$, for some $\lambda \in A$. Then $(0, \bar{g} - \lambda\bar{f}) = (c, \bar{g}) - (a, \bar{f})(\lambda, \bar{0}) \in J \setminus R(a, \bar{f})$, so that we may assume $c = 0$ and $\bar{g} \neq \bar{0}$, i.e., g is invertible in A. It follows that $(0, \bar{1}) = (0, \bar{g})(g^{-1}, \bar{0}) \in J$ (hence $E = A/M$) and $(a, \bar{0}) = (a, \bar{f}) - (0, \bar{g})(g^{-1}f, \bar{0}) \in J$. Consequently, $J = (a, \bar{0})R + (0, \bar{1})R$, whence J is finitely generated, as desired.

Now, by Lemma 1.2, $pd_R(0 \propto A/M) = pd_R R(0, 1) = \infty$, whence $gldim(R) = \infty$; then an application of [2, Theorem 1.3(ix)] completes the proof. ◇

Example 2.3. Let K be a field and E be a K-vector space with infinite rank. Let $A = K \propto E$ be the trivial ring extension of K by E. The ring A is a local $(2, 0)$-ring by [14, Theorem 3.4]. Clearly, its maximal ideal $M = 0 \propto E$ is not finitely generated and consists entirely of zero-divisors since $(0, e)M = 0$, for each $e \in E$. Let $R = A \propto (A/M)$ be the trivial ring extension of A by $A/M(\cong K)$. Then R is a $(2, 0)$-ring (and hence Theorem 1.1(2) fails because of the gnomonic property).

Indeed, let H be a 2-presented R-module and let $(x_1, \ldots, x_n)$ be a minimal set of generators of H. Our aim is to show that H is a projective R-module. Consider the exact sequence of R-modules

$$0 \to Ker(u) \to R^n \overset{u}{\to} H \to 0$$

where $u((r_i)_{i=1,\ldots,n}) = \sum_{i=1}^{n} r_i x_i$. So, $Ker(u)$ is a finitely presented R-module with $Ker(u) = U \propto E'$, where U is a submodule of A^n and E' is a K-vector subspace of K^n. We claim that $Ker(u) = 0$. Deny. The minimality of $(x_1, \ldots, x_n)$ yields

$$Ker(u) = U \propto E' \subseteq (M \propto A/M)R^n = (M \propto A/M)^n$$

since R is local with maximal ideal $M \propto A/M$. Let $(y_i, f_i)_{i=1,\ldots,p}$ be a minimal set of generators of $Ker(u)$, where $y_i \in M^n$ and $f_i \in K^n$. Consider the exact sequence of R-modules

$$0 \to Ker(v) \to R^p \overset{v}{\to} Ker(u)(= U \propto E') \to 0$$

where $v((a_i, e_i)_{i=1,\ldots,p}) = \sum_{i=1}^{p} (a_i, e_i)(y_i, f_i) = (\sum_{i=1}^{p} a_i y_i, \sum_{i=1}^{p} a_i f_i)$. Here too the minimality of $(y_i, f_i)_{i=1,\ldots,p}$ yields $Ker(v) \subseteq (M \propto A/M)^p$; whence, $Ker(v) = V \propto (A/M)^p$, where $V = \{(a_i)_{i=1,\ldots,p} \in A^p / \sum_{i=1}^{p} a_i y_i = 0\} (\subseteq M^p)$. By the above exact sequence, $Ker(v)$ is a finitely generated R-module, so that V is a finitely generated A-module [11, Theorem 25.1]. Now, by the exact sequence

$$0 \to V \to A^p \overset{w}{\to} U \to 0$$

where $w((a_i)_{i=1,\ldots,p}) = \sum_{i=1}^{p} a_i y_i$, U is a finitely presented A-module (since U is generated by $(y_i)_{i=1,\ldots,p}$). Further, U is an A-submodule of A^n and A is a $(2,0)$-ring, then U is projective. In addition, A is local, it follows that U is a finitely generated free A-module. On the other hand, $U \subseteq M^n = (0 \propto E)^n$, so $(0, e)U = 0$ for each $e \in E$, the desired contradiction (since U has a basis). ◊

References

[1] N. Bourbaki, Algèbre commutative, chapitres 1-4, Masson, Paris, 1985.

[2] D.L. Costa, Parameterizing families of non-Noetherian rings, Comm. Algebra 22 (1994) 3997-4011.

[3] D.L. Costa, S. Kabbaj, Classes of $D + M$ rings defined by homological conditions, Comm. Algebra 24 (1996) 891-906.

[4] R. Damiano, J. Shapiro, Commutative torsion stable rings, J. Pure Appl. Algebra 32 (1984) 21-32.

[5] F. Dischinger, W. Müller, Left PF is not right PF, Comm. Algebra 14 (1986) 1223-1227.

[6] D.E. Dobbs, On the global dimensions of $D+M$, Canad. Math. Bull. 18 (1975) 657-660.

[7] D.E. Dobbs, S. Kabbaj, N. Mahdou, M. Sobrani, When is $D + M$ n-coherent and an (n, d)-domain?, Lecture Notes in Pure and Appl. Math., Dekker, 205 (1999), 257-270.

[8] D.E. Dobbs, I.J. Papick, When is $D + M$ coherent?, Proc. Amer. Math. Soc. 56 (1976) 51-54.

[9] R. Fossum, Commutative extensions by canonical modules are Gorenstein rings, Proc. Amer. Math. Soc. 40 (1973) 395-400.

[10] S. Glaz, Commutative coherent rings, Lecture Notes in Mathematics, Vol. 1371, Springer-Verlag, Berlin, 1989.

[11] J.A. Huckaba, Commutative rings with zero divisors, Marcel Dekker, New York, 1988.

[12] G. Levin, Modules and Golod homomorphisms, J. Pure Appl. Algebra 38 (1985) 299-304.

[13] C. Löfwall, On the homotopy Lie algebra of a local ring, J. Pure Appl. Algebra 38 (1985) 305-312.

[14] N. Mahdou, On Costa's conjecture, Comm. Algebra 29 (7) (2001) 2775-2785.

[15] I. Palmér, J.-E. Roos, Explicit formulae for the global homological dimensions of trivial extensions of rings, J. Algebra 27 (1973) 380-413.

[16] D. Popescu, General Néron desingularization, Nagoya Math. J. 100 (1985) 97-126.

[17] J.-E. Roos, Finiteness conditions in commutative algebra and solution to a problem of Vasconcelos, Commutative algebra: Durham 1981 (Durham, 1981), pp. 179-203, London Math. Soc. Lecture Note Ser., 72, Cambridge Univ. Press, Cambridge-New York, 1982.

On the t–Dimension of Integral Domains

MOHAMMED KHALIS Department of Mathematics, Faculté des Sciences et Techniques, P.O. Box 577, Settat, Morocco, khalis@uh1.ac.ma

ABSTRACT

This paper aims at computing the t-dimension for two classes of integral domains, namely, v-coherent domains (of the form $D+M$) and power series rings over certain integral domains. As an application, we obtain the following: If A is an n-dimensional Prüfer domain which satisfies the SFT-property, or a PVD issued from an n-dimensional discrete valuation domain, then $t\text{-dim}(A[[X]]) = t\text{-dim}(A) = \dim(A)$.

1 INTRODUCTION

All rings considered below are integral domains and, unless otherwise specified, are assumed to be finite-dimensional. Throughout, $\dim(A)$ will denote the Krull dimension of an integral domain A and $\text{qf}(A)$ its quotient field. For the convenience of the reader, we review some terminology related to the $v-$ and $t-$operations . Let A be an integral domain with quotient field K and I be a nonzero fractional ideal of A. We denote $(A:I) = \{u \in K/uI \subseteq A\}$ by I^{-1} and $(I^{-1})^{-1}$ by I_v. We say that I is divisorial (or a v-ideal) if $I_v = I$. The divisorial ideal I is v-finite if $I = J_v$ for some finitely generated fractional ideal J of A. Also, we define $I_t = \cup\{J_v \mid J$ is a finitely generated subideal of $I\}$. The ideal I is called a t-ideal if $I_t = I$. For more details about these notions, see [11].

By analogy with known results on the prime ideal structure, significant work has been concerned with the structure of prime t-ideals (also called t-primes), see [12], [2], [14], [13], and [7]. Noteworthy is the fact that every t-ideal is contained in a maximal t-ideal (which is necessarily a prime ideal). Also any prime ideal minimal over a t-ideal is a t-prime, and hence so is any height-one prime ideal. Finally, recall that the t-dimension of an integral domain A, denoted $t\text{-dim}(A)$, is the supremum of the lengths of chains of t-primes of A (For the purpose of this definition, the zero ideal is considered as a t-prime, though it is not); we also define the t-height of a prime P to be the supremum of the lengths of the chains of t-primes contained in P. Recall, at this point, that (nontrivial) Krull domains have t-dimension one. We shall use t-Spec(A) to denote the set of t-primes of A and $\text{Spec}^+(A)$ to denote the set of nonzero prime ideals of A.

A domain A is v-coherent if the intersection of any two v-finite ideals is v-finite. The class of v-coherent domains naturally arose as a general context for the validity of Nagata's theorem for the class group (originally stated for Krull domains)(Cf. [9]). Its interest also resides in its ability to unify large classes of domains such as those of Mori domains, quasi-coherent domains, and PVMDs (Cf. [17]). In [10], Gabelli and Houston have investigated the transfer of v-coherence to $D+M$ domains. In the second section of this paper, we use their result and other previous investigations on this classical pullback to establish a satisfactory analogue to a well known result on the Krull dimension, Theorem 2.4.(1), stating that: "*if* $R = D+M$ *is a v-coherent domain issued from* $T = K+M$ *such that* $\text{qf}(D) = K$ *and* M *is a t-ideal of* T, *then* $t\text{-dim}(R) = \max\ \{t\text{-ht}_T(M) + t\text{-dim}(D)\ ,\ t\text{-dim}(T)\}$". As an application, one may compute, via this theorem, the t-dimension for large families of integral domains (See Example 2.7 and Corollary 2.9).

On the other hand, in [7], Dobbs and Houston studied the t-spectrum of power series rings over an integral domain A. They developed several results enlightening the interplay between the t-prime ideal structures of A and $A[[X]]$. Their work provided a natural starting point to the problem of expressing $t\text{-dim}(A[[X]])$ in terms of numerical invariants of A such as $t\text{-dim}(A)$. The third section of this paper is devoted to this problem. Here, our main result, Theorem 3.3, asserts that "*if A is an integral domain (which is not a field) such that* $\dim(A[[X]]) = \dim(A) + 1$ *and* $t\text{-Spec}(A) = \text{Spec}^+(A)$, *then* $t\text{-dim}(A[[X]]) = t\text{-dim}(A)$". This allows us to resolve entirely the problem for the class of Prüfer domains satisfying the *SFT*-property and for the class of PVDs issued from discrete valuation rings as well.

All along, corollaries and examples illustrate the scopes or the limits of our main results.

2 THE CLASS OF v-COHERENT DOMAINS

This section examines the t-prime structure for (some classes of) v-coherent domains. We place however more emphasis on those of the form $R = D+M$ issued from $T = K+M$. We aim at expressing t-dim(R) in terms of numerical invariants such as t-dim(D) and t-dim(T). This may allow one to determine the t-dimension for large families of integral domains.

According to [8], a domain A is v-coherent if the intersection of each pair of v-finite ideals is v-finite, or equivalently, if I^{-1} is v-finite for each finitely generated ideal I of A. Note, for convenience, that v-coherence coincides with the property P^* introduced in [17], where it was shown that the class of v-coherent domains properly contains the union of the classes of Mori domains, quasi-coherent domains, and PVMDs.

Let's recall the following result which we will be using frequently in the sequel.

LEMMA 2.1. *Let $A \subseteq B$ be an extension of integral domains such that* $\mathrm{qf}(A) = \mathrm{qf}(B)$. *Then*

- *If $A \hookrightarrow B$ is flat, then for each t-ideal I of B, $I \cap A$ is a t-ideal of A.*
- *If A is v-coherent and S is a multiplicatively closed subset of A, then for each t-prime P of A such that $P \cap S = \emptyset$, $P(S^{-1}A)$ is a t-prime of $S^{-1}A$.*

Proof. For the first assertion, see [8, Proposition 0.7]. For the second assertion, see [17, Lemme 2.3]. □

For v-coherent domains, Lemma 2.1 reduces the study of the t-dimension to the local case. Specifically, we have:

PROPOSITION 2.2. *Let A be a v-coherent domain. Then*

$$t\text{-}\dim(A) = \sup\,\{t\text{-}\dim(A_P)\ /\ P \in t\text{-}\mathrm{Spec}(A)\}.$$

Next, we examine the t-primes for some particular classes of v-coherent domains. Notice first that Krull domains have t-dimension equal to 1, since the Krull hypothesis deflates the notion of prime t-ideal to that of height-one prime ideal.

COROLLARY 2.3. *Let A be a PVMD. Then*

$$t\text{-}\dim(A) = \sup\,\{\dim(A_P)\ /\ P \in t\text{-}\mathrm{Spec}(A)\}.$$

Moreover, if A is a Prüfer domain, then $t\text{-}\dim(A) = \dim(A)$.

Proof. Since a PVMD is v-coherent, the first statement follows immediately from Proposition 2.2. The "moreover" assertion is straightforward. □

It turns out from the above preliminaries that a v-coherent domain may have arbitrary t-dimension. We now turn our attention to v-coherent domains issued from the classical $D+M$ constructions. Below, Theorem 2.4.(1) is a satisfactory analogue of a well-known result on the Krull dimension due to Brewer and Rutter [5, Corollary 9].

THEOREM 2.4. *Let T be a v-coherent domain of the form $K+M$, where K is a field and M is a maximal ideal of T. Let $R = D+M$, where D is a v-coherent subring of K.*

1) Assume $\mathrm{qf}(D) = K$ *and M is a t-ideal of T. Then*

$$t\text{-}\dim(R) = \max\,\{t\text{-}\mathrm{ht}_T(M) + t\text{-}\dim(D)\,,\ t\text{-}\dim(T)\}.$$

2) Assume $dim(T) \leq 2$, $\mathrm{qf}(D) \subset K$, *and either M is not a t-ideal ideal of T or it is a v-finite ideal of T. Then*

$$t\text{-}\dim(R) = \dim(T) + t\text{-}\dim(D).$$

Proof. 1) First, recall that (under the above assumptions) R is v-coherent [10, Theorem 3.4]. Note that, without any hypothesis on T and D, for each nonzero ideal I of D, $I+M = IR$ and so $(I+M)_v = (IR)_v = I_vR = I_v + M$. Hence $(I+M)_t = I_t + M$ [2]. Therefore, for each nonzero prime ideal P of D, P is a t-prime of D if and only if $P+M$ is a t-prime of R.

Let $S := D \setminus \{0\}$. Then $T = S^{-1}R$. Let $n = t\text{-ht}_T(M)$ and choose a strictly increasing saturated chain

$$Q_1 \subset \ldots \subset Q_n = M$$

of t-primes of T contained in M. Consider any strictly increasing chain

$$P_1 \subset \ldots \subset P_s$$

of t-primes of D. Then

$$Q_1 \cap R \subset \ldots \subset Q_{n-1} \cap R \subset M \subset P_1 + M \subset \ldots \subset P_s + M,$$

is a chain of t-primes of R by Lemma 2.1. It follows that

$$t\text{-}\dim(R) \geq t\text{-ht}_T(M) + t\text{-}\dim(D).$$

On the other hand, since $R \hookrightarrow T$ is flat, every chain of t-primes of T contracts into a chain of t-primes of R, hence $t\text{-}\dim(R) \geq t\text{-}\dim(T)$.

Conversely, consider a maximal chain of t-primes of R

$$P_1 \subset P_2 \subset \ldots \subset P_r \qquad (C)$$

There are two cases:

- $P_r \cap D = (0)$. Since R is v-coherent and $T = S^{-1}R$, then

$$P_1T \subset P_2T \subset \ldots \subset P_rT$$

is a chain of t-primes of T.

- $P_r \cap D \neq (0)$. Let i_o be minimal, $1 \leq i_o \leq r$, such that $P_{i_o} \cap D \neq (0)$. Then, for each $i \geq i_o$,

$$P_i = (P_i \cap D) + M = p_i + M,$$

where $p_i \in \mathrm{Spec}(D)$. We have $P_{i_o-1} \subseteq M$. Indeed, let $x \in P_{i_o-1} \subset P_{i_o} = p_{i_o} + M$. Then $x = \alpha + \beta$, where $\alpha \in p_{i_o}$ and $\beta \in M$. Suppose that $\alpha \neq 0$. Then

$$\alpha^{-1}x = 1 + \alpha^{-1}\beta \in P_{i_o-1}T \cap R = P_{i_o-1} \subset p_{i_o} + M,$$

hence $1 \in p_{i_o}$, (since $\alpha^{-1} \in T$), contradiction. Thus, $\alpha = 0$ and so $P_{i_o-1} \subseteq M$. The maximality of the chain (C) yields that $P_{i_o-1} = M$. Thus,

$$r=(i_o-1)+(r-i_o+1)\leq t\text{-ht}_T(M)+\ t\text{-dim}(D).$$

2) Set $T_1=k+M$, where $k=\text{qf}(D)$. Then, by [10, Theorem 3.5], T_1 is v-coherent. Moreover, M is a t-ideal of T_1. It follows that R is v-coherent and by (1),

$$t\text{-dim}(R)=\max\ \{t\text{-ht}_{T_1}(M)+t\text{-dim}(D)\ ,\ t\text{-dim}(T_1)\}$$

Since M is a t-ideal of T_1 and $\dim(T_1)=n\leq 2$, we have

$$t\text{-ht}_{T_1}(M)=t\text{-dim}(T_1)=n.$$

Hence, $t\text{-dim}(R)=n+t\text{-dim}(D)$. □

Next, we present two examples which illustrate the fact that the two hypotheses in Theorem 2.4.(1) cannot be deleted. Precisely, in Example 2.5, the maximal ideal M is not a t-ideal of T, $\text{qf}(D)=K$, and hence R is not v-coherent. In Example 2.6, R is v-coherent, but M is not a t-ideal of T and $\text{qf}(D)\neq K$.

EXAMPLE 2.5. Let $T=\mathbb{Q}[[X,Y]]=\mathbb{Q}+(X,Y)\mathbb{Q}[[X,Y]]$ be the power series ring in two indeterminates over $\mathbb{Q}$ and $R=\mathbb{Z}+M$, where $M=(X,Y)\mathbb{Q}[[X,Y]]$. Since T is a Krull domain, M is not a t-ideal of T. So R is not v-coherent [10, Theorem 3.4]. However M is a t-ideal of R and $\text{ht}_R(M)=2$. Necessarily, M contains a height-one t-prime. Moreover, for each nonzero prime ideal p of $\mathbb{Z}$, $p+M$ is a t-prime of R. Thus, there is a chain $Q\subset M\subset p+M$ of t-primes of R, and since $\dim(R)=3$, we have

$$t\text{-dim}(R)=3\neq\max\ \{t\text{-ht}_T(M)+t\text{-dim}(\mathbb{Z})\ ,\ t\text{-dim}(T)\}=2.$$

EXAMPLE 2.6. Let $T=\mathbb{Q}(\sqrt{2})[X,Y]=\mathbb{Q}(\sqrt{2})+(X,Y)\mathbb{Q}(\sqrt{2})[X,Y]$ be the polynomials ring in two indeterminates over $\mathbb{Q}(\sqrt{2})$ and put $R=\mathbb{Q}+M$, where $M=(X,Y)\mathbb{Q}(\sqrt{2})[X,Y]$. Since T is a Krull domain, $t\text{-dim}(T)=1$. Further, R is v-coherent (since Noetherian) and M is a height-two maximal t-ideal of R. Therefore,

$$t\text{-dim}(R)=2\neq\max\ \{t\text{-ht}_T(M)+t\text{-dim}(\mathbb{Q})\ ,\ t\text{-dim}(T)\}=1.$$

Corollary 2.3 asserts that $t\text{-dim}(A)=\dim(A)$, for any Prüfer domain A. The following example displays, via Theorem 2.4.(1), other families of integral domains satisfying the same property.

EXAMPLE 2.7. Consider $D=\overline{\mathbb{Q}}+Y\mathbb{C}[[Y]]$, where $\overline{\mathbb{Q}}$ is the integral closure of $\mathbb{Q}$ in $\mathbb{C}$, and let K be the quotient field of D. Let $n\geq 1$ be an integer, $T=K+M$ be a Prüfer domain with maximal ideal M such that $\text{ht}_T(M)=\dim(T)=n-1$, and $R=D+M$. Clearly, $\dim(R)=n$. On the other hand, D is a v-coherent domain of t-dimension 1, (Actually, D is a Mori domain which in not a PVMD). Moreover, M is a t-ideal of T, hence R is v-coherent [10, Theorem 3.4]. Finally, Theorem 2.4.(1) yields that $t\text{-dim}(R)=t\text{-ht}_T(M)+t\text{-dim}(D)=n=\dim(R)$.

In contrast with Krull domains and Prüfer domains, PVMDs may have arbitrary t-dimensions. The following example illustrates, via Theorem 2.4, this fact.

EXAMPLE 2.8. Let $T = \mathbb{Q}(X)[[Y]]$ and $R = \mathbb{Z}[X] + Y\mathbb{Q}(X)[[Y]]$. Then R is a PVMD (by [2, Theorem 4.1]) satisfying $1 < t\text{-dim}(R) < \dim(R)$. More precisely, $t\text{-dim}(R) = 2$ (by Theorem 2.4) while $\dim(R) = 3$ (by [5, Corollary 9]).

We close this section with a satisfactory analogue of a well-known result on the famous construction $T = D + XK[X]$ of D. Costa, J.L. Mott, and M. Zafrullah (Cf. [6, Corollary 2.10]).

COROLLARY 2.9. *Let K be a field, D a subring of K such that* $\mathrm{qf}(D) = k \subset K$, *and $R = D + XK[X]$. Then*

$$t\text{-dim}(R) = 1 + t\text{-dim}(D).$$

Proof. $T = K[X]$ is a v-coherent domain and $M = XK[X]$ is a principal ideal of T, hence v-finite. Thus, $R = D + M$ is v-coherent if and only if D is v-coherent. In case D is v-coherent, Theorem 2.4.(2) yields $t\text{-dim}(R) = 1 + t\text{-dim}(D)$. Even if D is not v-coherent, however, this equality holds. To see this, note that M is a t-prime ideal of R ($\mathrm{ht}_R(M) = 1$), and, for each prime ideal p of D, p is a t-prime of D if and only if $p + M$ is a t-prime of R, (see the proof of Theorem 2.4.). It follows that we have $t\text{-dim}(R) = 1 + t\text{-dim}(D)$. □

3 POWER SERIES RINGS OVER INTEGRAL DOMAINS

First, it is convenient to recall from [3] that a ring A is an SFT-ring if, for each ideal I of A , there is a finitely generated ideal $J \subseteq I$ and an integer $k \geq 1$ such that $x^k \in J$, for each $x \in I$. A well-known result [3, Theorem 1] is that if A is not an SFT-ring, then $\dim(A[[X]]) = \infty$.

In [7], it is proved that for an integral domain A, if I is an ideal of A, then $(IA[[X]])_t \subseteq I_t[[X]]$. Hence, for each t-prime p of A, $p[[X]]$ contains a t-prime which contracts into p in A. Precisely,

PROPOSITION 3.1. [7, Corollary 2.6]. *If A is an SFT-ring and p is a t-prime of A, then $p[[X]]$ is a t-prime of $A[[X]]$. In particular,* $t\text{-dim}(A[[X]]) \geq t\text{-dim}(A)$.

Moreover, it is also noted in [7] that the inequality in the previous Proposition can be strict [7, Example 4.2]. However, there are several contexts in which the inequality reduces to equality. In contrast with the Krull dimension case, the next example displays a non-SFT-ring A such that $t\text{-dim}(A[[X]]) = t\text{-dim}(A) = 1$.

EXAMPLE 3.2. Let k be a field and Y_1, Y_2, ..., Y_n, ... be infinitely many indeterminates over k. It is known that $A = k[\{Y_i\}_{i=1}^{\infty}] = \bigcup_{n=1}^{\infty} k[Y_1, \ldots, Y_n]$ is a Krull domain (actually, a UFD). We claim that A is not an SFT-ring. Indeed, suppose that $M = (Y_1, Y_2, \ldots, Y_n, \ldots)$ is an SFT-ideal of A. Then, there exists a finitely generated ideal $J = (f_1, \ldots, f_s)A$ of A and an integer $\beta \geq 1$ such that $J \subseteq M$ and $g^{\beta} \in J$, for each $g \in M$. Since $f_i \in A$, for each $i = 1,2,...,s$, then there exists an integer $n_o \geq 1$ such that $f_i \in k[Y_1,...,Y_{n_o}]$, for each $i \leq s$. Let $J_o = (f_1,...,f_s)k[Y_1,...,Y_{n_o}]$ be the ideal of $k[Y_1,...,Y_{n_o}]$ generated by the polynomials f_i and consider a prime ideal p such that $J_o \subseteq p \subset k[Y_1,...,Y_{n_o}]$. Set $P = p[Y_{n_o+1}, Y_{n_o+2},...] = pk[\{Y_i\}_{i=n_o+1}^{\infty}]$. Then P is a prime ideal of A containing J. Let $g \in M$. Then $g^{\beta} \in J = (f_1,...,f_s)A \subseteq P$, hence $g^{\beta} \in P$, whence $g \in P$. It follows that $M \subseteq P$, and hence $P = M$ (since M is maximal), the desired contradiction (since P is an extended prime). Now, A is a Krull domain and so is $A[[X]]$, hence t-dim$(A[[X]]) = t$-dim$(A) = 1$.

THEOREM 3.3. *Let A be an integral domain which is an SFT-ring (and which is not a field). Then*

1) t-dim$(A) \leq t$-dim$(A[[X]]) \leq$ dim$(A[[X]]) - 1$.

2) *If* dim$(A[[X]]) =$ dim$(A) + 1$ *and* t-Spec$(A) =$ Spec$^+(A)$, *then*

$$t\text{-dim}(A[[X]]) = t\text{-dim}(A).$$

Proof. 1) First, observe that if p is a nonzero prime t-ideal of A, then $P = p + XA[[X]]$ is not a t-ideal of $A[[X]]$. Indeed, for each nonzero element $a \in p$, we have $(a,X)^{-1} = a^{-1}A[[X]] \cap X^{-1}A[[X]]$. Let $g \in (a,X)^{-1}$. Then $g = a^{-1}h_1 = X^{-1}h_2$, where h_1, $h_2 \in A[[X]]$. By an easy order argument on the series $Xh_1 = ah_2$, we get X divides h_2, whence $g \in A[[X]]$. It follows that $(a,X)^{-1} = A[[X]]$. In particular, every maximal ideal M of $A[[X]]$ is of the form $M = m + XA[[X]]$, where m is a maximal ideal of A, and hence is not a t-ideal of $A[[X]]$. Thus, t-dim$(A[[X]]) \leq$ dim$(A[[X]]) - 1$. Finally, since A is an *SFT*-ring, Proposition 3.1 completes the proof of (1).

2) Follows from (1) and [3, Theorem 1]. □

COROLLARY 3.4. *Let $n \geq 1$ and A be an n-dimensional Prüfer domain with the SFT-property. Then*

$$t\text{-dim}(A[[X]]) = t\text{-dim}(A) = n.$$

Proof. Follows from [4, Theorem 3.8], Corollary 2.3, and Theorem 3.3. □

REMARK 3.5. Corollary 3.4 recovers [7, Proposition 4.1] which deals with discrete valuation domains. As for (finite-dimensional) nondiscrete valuation domains V, Kang and Park proved, in [15, Theorem 3.8], that t-dim$(V[[X]]) = \infty$.

COROLLARY 3.6. *Let $n \geq 1$ and R be a PVD issued from an n-dimensional discrete valuation domain. Then*

$$t\text{-dim}(R[[X]]) = t\text{-dim}(R) = n.$$

Proof. Let V be the discrete valuation domain associated to R and M its maximal ideal. Let $k = R/M$ and $K = V/M$. Then $R = \varphi^{-1}(k)$ is the pullback of the canonical surjection $\varphi : V \longrightarrow K$. It follows by [1, Proposition 2.3] that R is an *SFT*-ring and $\dim(R[[X]]) = \dim(R) + 1$ (see also [16, Théorèmes 2.4 and 3.9]). On the other hand, for each prime ideal p of R, pR_p is a t-prime of R_p (since pR_p is the divided maximal ideal, $pR_p = pV_p$, hence divisorial). Moreover, the flatness of the extension $R \hookrightarrow R_p$, yields $p = pR_p \cap R$ is a prime t-ideal of R. Thus, $t\text{-Spec}(R) = \text{Spec}^+(R)$. Theorem 3.3 completes the proof. □

It is worth noticing that Theorem 3.3 may allow one to determine the t-dimension of power series rings $A[[X]]$ for large families of integral domains A, beyond those issued from Corollaries 3.4 and 3.6. The next example illustrates this fact.

EXAMPLE 3.7. Let $n \geq 1$, $V = K + M$ be a discrete valuation domain of rank $n-1$, and D be any one-dimensional Noetherian domain such that $\text{qf}(D) = K$. Then $R = D + M$ is an *SFT*-ring and $\dim(R[[X]]) = \dim(V[[X]]) + \dim(D[[X]]) - 1 = n + 1 = \dim(R) + 1$ (Cf. [16]). Further, $t\text{-dim}(R) = t\text{-ht}_V(M) + t\text{-dim}(D) = n = \dim(R)$ by Theorem 2.4. Moreover, for every nonzero prime ideal P of R, we have: If $P \subseteq M$, then P is a t-prime of $V = S^{-1}R$, where $S = D \setminus (0)$. Thus P is a t-prime of R by Lemma 2.1. If $M \subset P$, then $P \cap D \neq (0)$ and $P = (P \cap D) + M$ is a t-prime of R, since $P \cap D$ is a height-one prime ideal of D. It follows that $t\text{-Spec}(R) = \text{Spec}^+(R)$. By Theorem 3.3, $t\text{-dim}(R[[X]]) = t\text{-dim}(R) = n$.

References

[1] D. F. Anderson, D. E. Dobbs, M. Fontana and M. Khalis, *Catenarity of formal power series over a Pullback*, J. Pure Appl. Algebra, **78** (1992), 109–122.

[2] D. F. Anderson and A. Ryckaert, *The class group of $D+M$*, J. Pure Appl. Algebra, **52** (1988), 199–212.

[3] J. T. Arnold, *Krull dimension in power series rings*, Tran. Amer.Math. Soc., **177** (1973), 299–304.

[4] J. T. Arnold, *Power series rings over a Prüfer domain*, Pacific J. Math., **44** (1973), 1–11.

[5] J. W. Brewer and E. A. Rutter, *"$D+M$" Construction with general overings*, Michigan Math. J., **23** (1976), 33–42.

[6] D. Costa, J.L. Mott and M. Zafrullah, *The construction $D + X.D_S[X]$*, J. Algebra, **53** (1978), 423–439.

[7] D. E. Dobbs and E. G. Houston, *On t-Spec($R[[X]]$)*, Canad. Math. Bull., **38**(2) (1995), 187–195.

[8] M. Fontana and S. Gabelli, *On the class group and the local class group of a Pullback*, J. Algebra, **181**(3) (1996), 803–835.

[9] S. Gabelli, *On Nagata's theorem for the class group, II*, Lecture Notes in Pure and Appl. Math., **206**, Dekker, New York (1999), pp.117–142.

[10] S. Gabelli and E. G. Houston, *Coherent-like conditions in Pullbacks*, Michigan Math. J., **44**(1) (1997), 99–123.

[11] R. Gilmer, Multiplicative Ideal Theory, Dekker, New York, 1972.

[12] J. Hedstrom and E. G. Houston, *Some remarks on star-operation*, J. Pure Appl. Algebra, **18** (1980), 37–44.

[13] E. G. Houston, *Prime t-ideals in* $R[X]$, in Commutative ring theory, Lecture Notes in Pure and Appl. Math., **153**, Dekker, New York (1994), pp.163–170.

[14] E. G. Houston and M. Zafrullah, *On t-invertibility II*, Comm. Algebra, **17** (1989), 1955–1969.

[15] B. G. Kang and M. H. Park, *A localization of a power series ring over a valuation domain*, J. Pure Appl. Algebra, **140**(2) (1999), 107–124.

[16] M. Khalis, *Dimension de Krull des anneaux de séries formelles sur un produit fibré*, Rend.Circ.Mat.Palermo, **39** (1990), 395–411.

[17] D. Nour El Abidine, *Groupe des classes de certains anneaux intègres et idéaux transformés*, Thèse de Doctorat, Université Claude Bernard, F-Lyon 1, 1992.

On Some Annihilator Conditions Over Commutative Rings

FARID KOURKI, Département de Mathématiques et Informatique, Université Abdelmalek Essaâdi, Faculté des Sciences de Tétouan, B.P. 2121, Tétouan, Morocco

ABSTRACT. It is proved that if a commutative ring R has finite Goldie dimension then $T(R)$, the total ring of quotient of R, is semilocal. It is also shown that a ring R is CS if and only if $Ann I + Ann J = R$ for any ideals I and J such that $I \cap J = 0$ (H). As a consequence, we characterize completely CS rings. A ring is pseudo–PIF if it satisfies Property (H) for any principal ideals I and J. Several examples are given and completely pseudo–PIF rings are characterized as arithmetical rings. Finally, we prove that if R is pseudo–PIF with acc on annihilators, then $T(R)$ is quasi–Frobenius, extending known results on the subject.

1. INTRODUCTION

All rings considered are commutative with unity. A ring R is called Kasch if every proper ideal of R has nonzero annihilator. Rings R having semilocal Kasch total ring of quotient $T(R)$ were studied by Faith in [9]. This work was motivated by the following result stated by Faith [9, Theorem 4.3]: If R is semiGoldie (i.e. no infinite direct sum of nonzero ideals imbeds in R) and zip (i.e. every faithful ideal of R contains a finitely generated faithful ideal) then $T(R)$ is semilocal and Kasch. One may ask whether this is true if we drop the zip condition. We prove that the semilocality of $T(R)$ is still preserved (Proposition 1) but $T(R)$ is not always Kasch (see example after Theorem 16). However $T(R)$ enjoys an annihilator property close to the Kasch property: Every finitely generated proper ideal of $T(R)$ has a nonzero annihilator. As an application, we show that if R is semiGoldie and if F is a free

R–module, then any two maximal linearly independent subsets of F have the same cardinality (Corollary 2). Rings with Krull dimension (in the sense of Gabriel and Rentschler, see [11]) form an interesting class of semiGoldie rings. We prove that if R has Krull dimension then $T(R)$ is semilocal Kasch (Proposition 3). We use this to give another proof of a result of Armendariz and Park (Corollary 4). In Section 3 CS rings are considered. We give a new characterization of CS rings involving annihilators (Theorem 6). As an application, we characterize completely CS rings (rings R such that R/I is CS for every ideal I of R). A theorem on decomposition of CS rings is also given (Theorem 8). In Section 4 pseudo–PIF rings are introduced as generalizations of CS rings. We see that arithmetical rings characterize completely pseudo–PIF rings (Theorem 15). We also show that if R is pseudo–PIF and has acc on annihilators then $T(R)$ is quasi–Frobenius (Theorem 19). Several applications are given including those which extend known sufficient conditions that a ring be quasi–Frobenius.

Throughout, for a ring R, $T(R)$ will denote the total ring of quotient of R and $Max(R)$ the set of maximal ideals of R. If M is an R–module, let $AnnM = \{a \in R/\ aM = 0\}$ and if I is an ideal of R, let $Ann_M I = \{x \in M/\ Ix = 0\}$. If N is an essential submodule of M, we write $N \subseteq^{ess} M$.

2. SEMIGOLDIE RINGS

A ring R has finite (Goldie) dimension if R contains no infinite direct sums of nonzero ideals. In this case, the injective hull $E(R)$ of R is a finite direct sum $E(R) = E(U_1) \oplus \ldots \oplus E(U_n)$ of indecomposable injective modules $E(U_i)$, the injective hull of uniform ideals U_i, $i = 1, \ldots, n$. By the Krull Schmidt theorem n is a unique integer called (Goldie) dimension of R and denoted by $dimR$. Any uniform ring, e.g. an integral domain, has dimension 1, as has any chain (=valuation) ring. A ring is semiGoldie if it has finite dimension. A ring R is finitely embedded if it has a finite essential socle; equivalently, $E(R) = E(S_1) \oplus \ldots \oplus E(S_n)$ for some minimal ideals S_i, $i = 1, \ldots, n$.

If M is an R–module then $\mathcal{Z}(M)$ denotes the set of zero divisors of M; that is, $\mathcal{Z}(M) = \{a \in R/\ ax = 0 \text{ for some } 0 \neq x \in M\}$. In general $\mathcal{Z}(M)$ is not always an ideal of R. When $\mathcal{Z}(M)$ is an ideal, then it is prime. It is not difficult to see that if M and N are R–modules then $\mathcal{Z}(M) = \mathcal{Z}(E(M))$ and $\mathcal{Z}(M \oplus N) = \mathcal{Z}(M) \cup \mathcal{Z}(N)$. A ring R is said to be McCoy (see [9]) or "sondable" (see [19]) provided that every finitely generated faithful ideal contains a regular element. This is equivalent to the statement that every finitely generated proper ideal of $T(R)$ has a nonzero annihilator. Picavet G. shows that any semiGoldie ring is McCoy [19, Theorème 6.21]. Using the ideas of Picavet we show that any semiGoldie ring has a semilocal total ring of quotient. The following remark is the key to our result: If R is any commutative ring and if U is a uniform R–module then $\mathcal{Z}(U)$ is a prime ideal.

Indeed, let a and b be two elements of $\mathcal{Z}(U)$. So $Ann_U a \neq 0$ and $Ann_U b \neq 0$. Since U is uniform, $0 \neq Ann_U a \cap Ann_U b \subseteq Ann_U(a+b)$. So $a+b \in \mathcal{Z}(U)$. Therefore $\mathcal{Z}(U)$ is an ideal of R and consequently it is prime. A ring R is called semilocal if $Max(R)$ is finite and, in this case, the number of its maximal ideals is denoted by $\mid Max(R) \mid$.

PROPOSITION 1. If R is a semiGoldie ring, then R is McCoy, $T(R)$ is semilocal and we have $\mid Max(T(R)) \mid \leq dimR$.

Proof. First we suppose that $T(R) = R$. Let $E(R) = E(U_1) \oplus \ldots \oplus E(U_n)$ for some uniform ideals $U_1, \ldots, U_n$ $(dimR = n)$. So $\mathcal{Z}(R) = \mathcal{Z}(U_1) \cup \ldots \cup \mathcal{Z}(U_n)$ and, as noted above, each $\mathcal{Z}(U_i)$ is a prime ideal. If $I = a_1R + \ldots + a_mR$ is a finitely generated ideal contained in $\mathcal{Z}(R)$, then $I \subseteq \mathcal{Z}(U_j)$ for some $j \in \{1, \ldots, n\}$. As $a_k \in \mathcal{Z}(U_j)$ we have $Ann_{U_j} a_k \neq 0$ for all $k \in \{1, \ldots, m\}$. Since U_j is uniform, $\bigcap_{k=1}^{m} Ann_{U_j} a_k \neq 0$. Take a nonzero element t from this intersection, then $tI = 0$ and hence R is McCoy. Now we proceed to show that $R = T(R)$ is semilocal and $\mid Max(R) \mid \leq dimR$. Let $\mathcal{B}_i = \mathcal{Z}(U_i)$ so that $\mathcal{Z}(R) = \mathcal{B}_1 \cup \ldots \cup \mathcal{B}_n$. We show that the maximal ideals of R are exactly the maximal elements of $\{\mathcal{B}_1, \ldots, \mathcal{B}_n\}$. Let $\mathcal{M}$ be a maximal ideal of R. Since $T(R) = R$, $\mathcal{M} \subseteq \mathcal{Z}(R) = \mathcal{B}_1 \cup \ldots \cup \mathcal{B}_n$ and so $\mathcal{M} \subseteq \mathcal{B}_j$ for some $j \in \{1, \ldots, n\}$. Therefore $\mathcal{M} = \mathcal{B}_j$ by the maximality of $\mathcal{M}$. Now let $\mathcal{B}_l$ be a maximal element of $\{\mathcal{B}_1, \ldots, \mathcal{B}_n\}$ and $\mathcal{M}$ a maximal ideal of R such that $\mathcal{B}_l \subseteq \mathcal{M}$. We have $\mathcal{M} \subseteq \mathcal{Z}(\mathcal{R})$, and so $\mathcal{M} \subseteq \mathcal{B}_k$ for some $k \in \{1, \ldots, n\}$. Therefore $\mathcal{B}_l \subseteq \mathcal{B}_k$ and hence $\mathcal{B}_l = \mathcal{B}_k = \mathcal{M}$. So R is semilocal and $\mid Max(R) \mid \leq dimR$. To complete the proof of Proposition 1 we use the following observations: (1) A ring R is McCoy if and only if $T(R)$ is McCoy; (2) A ring R is semiGoldie if and only if $T(R)$ is semiGoldie and, in this case, $dimR = dimT(R)$. ◇

Let R be a ring. Following Lazarus [15], we say that an R–module M satisfies Property (P) if any two maximal linearly independent subsets of M have the same cardinality. If R is noetherian or if R is finitely embedded, then every free R–module has Property (P) (see [2, p. 136]). With the help of Proposition 1 we obtain a common generalization of these two results.

COROLLARY 2. If R is semiGoldie then every free R–module has Property (P).

Proof. To prove that every free R–module has Property (P), it is sufficent to show that $T(R)$ is weakly semi–Steinitz [2, Corollary 1]. Since $T(R)$ is semilocal by Proposition 1, $T(R)$ is Hermite [13, Theorem 3.7]. Also every finitely generated proper ideal of $T(R)$ has nonzero annihilator since R is McCoy. The later two properties of $T(R)$ guarantee that it is weakly semi–Steinitz [16, Theorem 2.2]. ◇

A ring R is called Kasch if every simple module embeds in R; equivalently, $Ann\mathcal{M} \neq 0$ for every maximal ideal $\mathcal{M}$ of R. Examples of rings R such that $T(R)$ is semilocal Kasch include rings with ascending chain condition on annihilators

(= $acc \perp$), finitely embedded rings and perfect rings (see for example [9]). The next proposition affords a new class of rings satisfying this property.

PROPOSITION 3. If R is a ring with Krull dimension (in the sense of Gabriel and Rentschler, see [11]) then $T(R)$ is semilocal Kasch.

Proof. From [11, Proposition 1.4], R is a semiGoldie ring; therefore there exist some uniform ideals $U_1, \ldots, U_n$ such that $E(R) = E(U_1) \oplus \ldots \oplus E(U_n)$. For each $i \in \{1, \ldots, n\}$, we may find an ideal $0 \neq U_i' \subseteq U_i$ such that $\mathcal{Z}(U_i') = AnnU_i'$ [11, Theorem 8.3]. Since U_i is uniform, $\mathcal{Z}(x_iR) = \mathcal{Z}(U_i') = \mathcal{Z}(U_i)$ for every $0 \neq x_i \in U_i'$. We have $Annx_i \subseteq \mathcal{Z}(x_iR) = AnnU_i'$ and $AnnU_i' \subseteq Annx_i$, so $\mathcal{Z}(U_i) = \mathcal{Z}(U_i') = Annx_i$. If $\mathcal{Z}(R) = \mathcal{Z}(U_1) \cup \ldots \cup \mathcal{Z}(U_n) = Annx_1 \cup \ldots \cup Annx_n$ contains an ideal I, then $I \subseteq Annx_j$ for some $j \in \{1, \ldots, n\}$, which leads to $x_jI = 0$. Therefore $T(R)$ is Kasch. The semilocality of $T(R)$ comes from Proposition 1. ◇

A ring R is called pseudo–Frobenius (PF) if every faithful R–module is a generator; equivalently, R is self–injective and Kasch. If R is a self–injective ring with Krull dimension then $T(R) = R$ is Kasch by the preceding proposition. So we get the following result.

COROLLARY 4. [1, Proposition 10] If R is self–injective with Krull dimension then R is PF.

3. CS RINGS

A ring R is called Baer if $AnnI$ is a direct summand of R for every ideal I. A ring is said to be reduced if it has no nonzero nilpotent elements. A characterization of Baer rings is given in the following proposition.

PROPOSITION 5. The following are equivalent for a ring R:
(1) R is Baer.
(2) The condition $IJ = 0$ implies that $AnnI + AnnJ = R$ for any ideals I and J of R.

Proof. (1) ⇒ (2). Let I and J be ideals of R such that $IJ = 0$. This condition implies that $AnnAnnI \subseteq AnnJ$. But $AnnI \oplus AnnAnnI = R$ since $AnnI$ is a summand of R. Therefore $AnnI + AnnJ = R$.

(2) ⇒ (1). First we show that R is reduced. To see this, suppose that there exist $0 \neq a \in R$ and an integer $m \geq 2$ minimal with respect to $a^m = 0$. By hypothesis $R = Anna + Anna^{m-1} = Anna^{m-1}$. This implies that $a^{m-1} = 0$, a contradiction. Let I be an ideal of R. Since $IAnnI = 0$ then $AnnI + AnnAnnI = R$ and, since R is reduced, we have also $AnnI \cap AnnAnnI = 0$. So $AnnI \oplus AnnAnnI = R$. ◇

For convenience, we call a ring R pseudo–Baer if $AnnI + AnnJ = R$ for every ideals I and J such that $I \cap J = 0$. Clearly every Baer ring is pseudo–Baer as is any uniform ring. Here are some other examples.

EXAMPLES. (1) A ring R is called Ikeda–Nakayama (IN) if for any ideals I and J of R we have $Ann(I \cap J) = AnnI + AnnJ$. These rings were recently investigated by Camillo, Nicholson and Yousif in [4]. Clearly every IN ring is pseudo–Baer.

(2) A ring R is called FPF if every finitely generated faithful R–module generates the R–modules. Every FPF ring is pseudo–Baer. For if I and J are ideals of R such that $I \cap J = 0$, then $R/I \oplus R/J$ is a finitely generated faithful R–module and so it generates the R–modules. Therefore its trace ideal $AnnI + AnnJ$ is R.

A ring R is called CS if every ideal of R is essential in a direct summand of R. Every self–injective ring is CS. It turns out that commutative CS rings are exactly pseudo–Baer rings.

THEOREM 6. A commutative ring R is CS if and only if $AnnI + AnnJ = R$ for every ideals I and J such that $I \cap J = 0$.

Proof. Assume that R is pseudo–Baer. In [4, Theorem 3] it is shown that if R is IN then R is CS. But we remark that the proof of this result still works only assuming that R is pseudo–Baer. Conversely, let R be a CS ring and let I and J be ideals of R such that $I \cap J = 0$. Then $I \subseteq^{ess} Re$ and $J \subseteq^{ess} Rf$ where e and f are idempotents of R. Therefore $eR \cap fR = 0$. As $Anne = (1-e)R$ and $Annf = (1-f)R$ and also $(1-e)f + (1-f) = 1$, we have $Anne + Annf = R$. But $Anne \subseteq AnnI$ and $Annf \subseteq AnnJ$, so $AnnI + AnnJ = R$. ◇

REMARKS (1) Call (a not necessarily commutative ring) R right pseudo–Baer if $l(I) + l(J) = R$ ($l(I)$ is the left annihilator of I) for any right ideals I and J of R such that $I \cap J = 0$. Every right pseudo–Baer ring is right CS (the proof is the same as the proof of Theorem 3 of [4]) but the converse is not always true. Let $R = \begin{bmatrix} F & F \\ 0 & F \end{bmatrix}$; then R is left and right CS. Take $x = \begin{bmatrix} 0 & 0 \\ 0 & 1 \end{bmatrix}$ and $y = \begin{bmatrix} 0 & 1 \\ 0 & 1 \end{bmatrix}$, then $xR \cap yR = 0$ but $l(x) + l(y) \subseteq \begin{bmatrix} F & F \\ 0 & 0 \end{bmatrix} \neq R$ (see [4, p. 1005]).

(2) Camillo, Nicholson and Yousif have given an example of a commutative local and uniform ring which is not a IN ring [4, Example 6]. We deduce that a CS ring is not always IN.

As a consequence of Theorem 6 we obtain one of Faith's results.

COROLLARY 7. ([7]) A commutative FPF ring is CS.

For a reduced ring R, the condition $IJ = 0$ is equivalent to $I \cap J = 0$, for every ideals I and J of R. We conclude that if R is reduced then R is CS if and only if

R is Baer (Proposition 5 and Theorem 6). For the nonreduced case we have the following decomposition of CS rings (we use the ideas of the proof of [6, Theorem 10]).

THEOREM 8. If R is a CS ring then $R = R_1 \times R_2$ where R_1 is a Baer ring and $nil(R) = nil(R_2) \subseteq^{ess} R_2$.

Proof. First we claim that $I \subseteq^{ess} AnnAnnI$ for every ideal I of R. Indeed, let J be an ideal contained in $AnnAnnI$ such that $I \cap J = 0$. Then $AnnI + AnnJ = R$ (Theorem 6). Moreover $AnnI \subseteq AnnJ$ ($J \subseteq AnnAnnI$), so $AnnJ = R$ proving that $J = 0$. Now, let $N = nil(R)$, the nilradical of R, and let K be an ideal maximal with respect to $K \cap N = 0$. Let $L = AnnK$. Since $(K \cap L)^2 = 0$, then $K \cap L \subseteq K \cap N = 0$, so $AnnK + AnnL = L + AnnL = R$ (Theorem 6). The condition $K \cap L = 0$ also implies that $L \cap AnnL = L \cap AnnAnnK = 0$ as $K \subseteq^{ess} AnnAnnK$. Hence $R = R_1 \times R_2$ where $R_1 = AnnL$ and $R_2 = L$. As $N \cap K = 0$, we have $N \subseteq AnnK = R_2$, so $nil(R_2) = N$ and R_1 is reduced. Also, it is not difficult to see that R_1 is also CS and so it is Baer as it is noted just before this theorem. If I is any ideal of R_2 and if $I \neq 0$, then $(K + I) \cap N \neq 0$. So if $0 \neq x = b + y$ with $x \in N$, $b \in K$, $y \in I$, then $b = x - y \in K \cap L = 0$. So $0 \neq x = y \in N \cap I$. Hence $N = nil(R_2)$ is essential in R_2. ◇

A ring R called CF if every R-module M which is a direct sum of finitely many cyclic R-modules has a canonical form, i.e. $M \simeq R/I_1 \oplus \cdots \oplus R/I_n$ where $I_1 \subseteq I_2 \subseteq \cdots \subseteq I_n \neq R$. If A and B are ideals of a ring R, let $(A : B) = \{a \in R \ / \ aB \subseteq A\}$. A ring R is CF if and only if for every ideals A and B of R we have $(A \cap B : A) + (A \cap B : B) = R$ [21, Corollary 1.6]. We see that CF rings characterize completely CS rings as the following theorem shows.

THEOREM 9. The following are equivalent for a commutative ring R:
(1) R/I is CS for every ideal I of R.
(2) R is CF.
(3) R is a finite direct product of valuation rings, h–local Prüfer domains and torch rings.

Proof. (2) ⇔ (3). See Theorem 3.12 of [21].

(1) ⇒ (2). We have $(A/A \cap B) \cap (B/A \cap B) = 0$ for every ideals A and B of R. Therefore $Ann(A/A \cap B) + Ann(B/A \cap B) = R/A \cap B$ since $R/A \cap B$ is a CS ring (Theorem 6). Thus $(A \cap B : A) + (A \cap B : B) = R$ and consequently R is a CF ring [21, Corollary 1.6].

(2) ⇒ (1). Let I be an ideal of R. It is not difficult to see that $R' = R/I$ is also a CF ring. Let A and B are ideals of R' such that $A \cap B = 0$. By [21, Corollary 1.6] we have $(A \cap B : A) + (A \cap B : B) = R'$. So $AnnA + AnnB = R'$ and therefore R' is CS (Theorem 6). ◇

4. PSEUDO–PIF RINGS

In this section we are interested in a weaker condition than the condition on annihilators satisfied by CS rings (Theorem 6).

PROPOSITION 10. The following are equivalent for a commutative ring R:
(1) Every principal ideal of R is flat.
(2) $R_{\mathcal{M}}$ is a domain for every maximal ideal $\mathcal{M}$ of R.
(3) If $ab = 0$ then $Anna + Annb = R$, for every elements a and b of R.

Proof. (1) $\Leftrightarrow$ (2) is proved in [10, Theorem 4.2.2] and (2) $\Leftrightarrow$ (3) is by [12, Exercice 33 p. 65].

When these equivalent conditions hold, R is called PIF (principal ideals are flat). Any PP ring (principal ideals are projective) is PIF. The preceding proposition suggests the following definition.

DEFINITION 11. A ring R is called pseudo–PIF if $Anna + Annb = R$ for every elements a and b of R such that $aR \cap bR = 0$.

EXAMPLES. (1) If R is PIF then R is pseudo–PIF; and if R is reduced, then these two conditions are equivalent.

(2) If R is CS then R is pseudo–PIF (Theorem 6).

(3) A ring R is said to be FP^2F if every finitely presented faithful R–module generates the category of R–modules. Let a and b be elements of such a ring R satisfying $aR \cap bR = 0$. Then the R–module $R/aR \oplus R/bR$ is finitely presented and faithful, so it generates the R–modules. Therefore its trace ideal $Anna + Annb$ is R. Consequently every FP^2F ring is pseudo–PIF.

(4) If $R_{\mathcal{M}}$ is pseudo–PIF for all $\mathcal{M} \in Max(R)$ then R is pseudo–PIF. Indeed, let a and b be elements of R such that $aR \cap bR = 0$. If $Anna + Annb \neq R$ then $Anna + Annb$ is contained in a maximal ideal $\mathcal{M}$. Since $R_{\mathcal{M}}$ is pseudo–PIF, $aR_{\mathcal{M}} \cap bR_{\mathcal{M}} \neq 0$, a contradiction. We deduce that every arithmetical ring R (i.e. $R_{\mathcal{M}}$ is a valuation ring for all $\mathcal{M} \in Max(R)$) is pseudo–PIF.

A ring R is called principally injective (p–injective) if every homomorphism $\alpha : aR \to R$, $a \in R$, extends to R; equivalently, $AnnAnna = aR$ for every $a \in R$. Clearly every self–injective ring is p–injective as is any Von Neumann regular ring.

LEMMA 12. If R is p–injective then for every elements a and b of R we have:

$$(Ann(aR \cap bR))^2 \subseteq Anna + Annb \subseteq Ann(aR \cap bR)$$

In particular $\sqrt{Ann(aR \cap bR)} = \sqrt{Anna + Annb}$.

Proof. The inclusion $Anna + Annb \subseteq Ann(aR \cap bR)$ is true in any ring.

Let β and γ be two elements of $Ann(aR \cap bR)$. If α is an element of R such that $(a+b)\alpha = 0$, then $\alpha a \in aR \cap bR$ and hence $\beta\alpha a = 0$. Therefore the R–homomorphism f from $(a+b)R$ to R such that $f((a+b)\alpha) = \beta\alpha a$ is well defined. Since R is p–injective, there exists $c \in R$ such that $f(x) = cx$ for every $x \in R$. So $\beta a = c(a+b)$ and hence $(\beta - c)a = cb \in aR \cap bR$, whence $\gamma(\beta - c)a = \gamma cb = 0$. Thus $(\gamma\beta - \gamma c) \in Anna$ and $\gamma c \in Annb$ and $(\gamma\beta - \gamma c) + \gamma c = \gamma\beta \in Anna + Annb$. As β and γ were arbitrary, $(Ann(aR \cap bR))^2 \subseteq Anna + Annb$. ◇

COROLLARY 13. Every p–injective ring is pseudo–PIF.

A ring R is called mininjective if every R–homomorphism from a simple ideal to R extends to R; equivalently, $AnnAnna = aR$ for every $a \in R$ such that aR is simple. Every p–injective ring is mininjective. A ring R is said to have a square–free socle if the socle of R contains no more than one copy of each simple R–module. A commutative ring is mininjective if and only if it has a square–free socle [18, Corollary 2.11]. We remark that a ring R is mininjective if and only if $Anna + Annb = R$ for every simple ideals aR and bR such that $aR \cap bR = 0$ (see [18, p. 460]).

LEMMA 14. If R is pseudo–PIF then R is mininjective.

THEOREM 15. Let R be a ring. The following conditions are equivalent:
(1) R/I is pseudo–PIF for every ideal I of R.
(2) $T(R/I)$ is pseudo–PIF for every ideal I of R.
(3) R/I is mininjective for every ideal I of R.
(4) $T(R/I)$ is mininjective for every ideal I of R.
(5) R/I is FP^2F for every ideal I of R.
(6) $T(R/I)$ is FP^2F for every ideal I of R.
(7) R is arithmetical.

Proof. By Lemma 14 (1) ⇒ (3). By [18, Theorem 2.12] (3) ⇔ (7) and by [6, Corollary 5E] (5) ⇔ (7).

If R is arithmetical then R/I is arithmetical for every ideal I of R. So we have the implication (7) ⇒ (1). Thus (1) ⇔ (3) ⇔ (5) ⇔ (7).

The proof of (4) ⇒ (7) is the same as the proof of Theorem 1 of [5].

If R is a FP^2F ring then $T(R)$ is FP^2F. Indeed, let M be a finitely presented faithful $T(R)$–module. If

$$T(R)^n \xrightarrow{\left(\frac{a_{ij}}{s_{ij}}\right)} T(R)^m \longrightarrow M \longrightarrow 0$$

is a presentation of M, let M' be an R–module such that

$$R^n \xrightarrow{(a_{ij})} R^m \longrightarrow M' \longrightarrow 0$$

Then M' is a finitely presented faithful R-module such that $S^{-1}M' = M$ (where $S = R - \mathcal{Z}(R)$). Since R is FP^2F, R is a generator of Mod–R; so $M' \cong R \oplus L$. Therefore $M \cong T(R) \oplus S^{-1}L$ and hence M generates Mod–R. So (5) $\Rightarrow$ (6). Thus (7) $\Leftrightarrow$ (5) $\Rightarrow$ (6) $\Rightarrow$ (2) $\Rightarrow$ (4) $\Rightarrow$ (7). This completes the proof. $\diamondsuit$

REMARKS. (1) If $R[X]$ is pseudo–PIF then R is pseudo–PIF. Assume that $R[X]$ is pseudo–PIF and let a and b be elements of R such that $aR \cap bR = 0$. Then we have $aR[X] \cap bR[X] = 0$. Hence $Ann_{R[X]}a + Ann_{R[X]}b = R[X]$. If $f(X) \in Ann_{R[X]}a$ and $g(X) \in Ann_{R[X]}b$ such that $f(X) + g(X) = 1$, then $f(0) + g(0) = 1$ and $af(0) = bg(0) = 0$. Therefore $Ann_R a + Ann_R b = R$. If R is a ring which is not pseudo–PIF (see [14, Example p. 352]), then $R[X]$ is mininjective [14, Remark 1.1] but $R[X]$ is not pseudo–PIF.

(2) If a ring R is pseudo–PIF so is $T(R)$, but the converse is not true in general. To find a counter example, first we recall that a ring R is PP if and only if R is PIF and $T(R)$ is Von Neumann regular [10, p. 121]. Let R be a ring such that $T(R)$ is Von Neumann regular and R is not PP (see [10, p. 122]). Obviously $T(R)$ is pseudo–PIF. But, since R is not PIF and R is reduced, R is not pseudo–PIF.

Turning to semiGoldie rings, we have:

THEOREM 16. Let R be a pseudo–PIF ring such that $T(R) = R$. Then:
(1) R is semiGoldie if and only if R is a semilocal ring and, in this case, we have $\mid Max(R) \mid = dimR$.
(2) R is finitely embedded if and only if R is semilocal Kasch.

Proof. (1) Proposition 1 implies that R is semilocal whenever it is semiGoldie. Conversely, assume that R is a semilocal ring and that $\mid Max(R) \mid = n$, say $Max(R) = \{\mathcal{M}_1, \ldots, \mathcal{M}_n\}$. Suppose that $a_1, \ldots, a_{n+1}$ are elements of R such that $a_1R \oplus \ldots \oplus a_{n+1}R$ is direct. We have $Anna_1 \neq R$, so $Anna_1$ is contained in a maximal ideal $\mathcal{M}_1$ (after relabeling the $\mathcal{M}_i$). Since $Anna_1 + Anna_2 = R$, we have $Anna_2$ is contained in a maximal ideal different from $\mathcal{M}_1$, say $Anna_2 \subseteq \mathcal{M}_2$ (after relabeling the $\mathcal{M}_i$). We continue this process to get $Anna_i \subseteq \mathcal{M}_i$ for $1 \leq i \leq n$. But $Anna_{n+1} \subseteq \mathcal{M}_t$ for some $t = 1, \ldots, n$, contrary to the claim that $Anna_t + Anna_{n+1} = R$ (since R is pseudo–PIF). This proof also shows that $dimR \leq \mid Max(R) \mid$. But $\mid Max(R) \mid \leq dimR$ by Proposition 1. So we have the equality.

(2) By [9, Corollary 1.1] if R is a finitely embedded ring then $T(R)$ is semilocal Kasch. Conversely, assume that R is semilocal Kasch. By (1) R is semiGoldie, say $E(R) = E(U_1) \oplus \ldots \oplus E(U_n)$ where U_i are uniform ideals ($dimR = n$). The maximal ideals of R are among $\mathcal{Z}(U_i)$, $i = 1, \ldots, n$ (see the proof of Proposition 1). By (1) $\mid Max(R) \mid = dimR$, so $Max(R) = \{\mathcal{Z}(U_1), \ldots, \mathcal{Z}(U_n)\}$. Let i be an element of $\{1, \ldots, n\}$. As R is Kasch, we have $\mathcal{Z}(U_i) = Annx_i$ where $0 \neq x_i \in R$. We have $R/Annx_i \hookrightarrow R$, so $E(R/Annx_i)$ is an indecomposable direct summand

of $E(R)$. Therefore, by the Azumaya theorem, $E(R/Annx_i) \cong E(U_j)$ for some $j \in \{1, \ldots, n\}$. So $\mathcal{Z}(E(R/Annx_i)) = \mathcal{Z}(E(U_j))$ and hence $Annx_i = \mathcal{Z}(U_j)$, thus $i = j$. Therefore $E(R) \cong E(R/Annx_1) \oplus \ldots \oplus E(R/Annx_n)$ where $R/Annx_i$ are simple R–modules. This means that R is finitely embedded. ◇

Let K be a field and $R = K[X_1, X_2, \ldots]$ such that $X_1^2 = 0$ and $X_n^2 = X_{n-1}$ for all $n \geq 2$. Then R is local and uniform such that $T(R) = R$. So R is pseudo–PIF and $\mid Max(R) \mid= dimR = 1$. However R is not Kasch (see [22, page 635]).

A result of Nicholson and Yousif [17, Corollary 3.2] asserts that if R is a commutative p–injective ring with Krull dimension then R has an essential socle. Here is a generalization of this result.

COROLLARY 17. Let R be a pseudo–PIF ring such that $T(R) = R$. If R has Krull dimension then R is finitely embedded.

Proof. By Proposition 3 $T(R) = R$ is semilocal Kasch, so Theorem 16 applies. ◇

COROLLARY 18. A ring R is perfect and arithmetical if and only if R is principal and artinian.

Proof. Assume that R is perfect and arithmetical. Let I be an ideal of R. By Theorem 15, R/I is pseudo–PIF . Since R is perfect, R/I is perfect and hence R/I is semilocal Kasch. Therefore R/I is finitely embedded (Theorem 16). As I is arbitrary R is artinian [23, Proposition 2*]. From the fact that R is arithmetical and semilocal we have R is Bezout, i.e. every finitely generated ideal is principal (see for example [3, Proposition 3.8]). But R is artinian, so it is noetherian; hence R is principal. The converse is clear. ◇

Now we come to a fundamental fact about pseudo–PIF rings:

THEOREM 19. Let R be a pseudo–PIF ring such that $T(R) = R$. Then R satisfies acc on annihilators if and only if R is QF.

Proof. If R satisfies acc on annihilators, then $T(R) = R$ is semilocal Kasch [9, Corollary 3.7]. So R is finitely embedded by Theorem 16. But R is also mininjective (Lemma 14), hence R is QF [8, Corollary 2]. The converse is clear. ◇

COROLLARY 20. [20, Corollary 1]. If R is p–injective with acc on annihilators then R is QF.

Proof. R is pseudo–PIF by Corollary 13. Since $AnnAnna = aR$ for every $a \in R$, $T(R) = R$. So Theorem 19 applies. ◇

COROLLARY 21. If R is FP^2F with acc on annihilators then $T(R)$ is QF.

Proof. R is pseudo–PIF and so is T(R). Also, the acc on annihilators is inherited by $T(R)$. Now use Theorem 19. ◇

ACKNOWLEDGMENT

I would like to thank the referee for his helpful comments. Particularly, he directed my attention to the following two papers (dealing with non-commutative contexts) authored by G. F. Birkenmeier: "Decompositions of Baer-like rings", Acta Math. Hungarica 59 (3-4) (1992) 319-326; and "Baer rings and quasi-continuous rings have MDSN", Pacific J. Math. 97 (2) (1981) 283-292. These are relevant to the present work since Theorem 6 can be obtained from Lemma 2.1 of the first paper, while Theorem 8 follows from another theorem of the second paper.

References

[1] E.P. Armendariz and J.K. Park, Self-injective rings with restricted chain conditions, Arch. Math. **58** (1992) 24-33.

[2] A. Bouanane and F. Kourki, On weakly semi-Steinitz rings, Lecture Notes Pure Appl. Math., Marcel Dekker, **185** (1997) 131-139.

[3] W. Brandal, Commutative rings whose finitely generated modules decompose, Lecture Notes in Math. **723**, Springer–Verlag, 1979.

[4] V. Camillo, W.K. Nicholson and M.F. Yousif, Ikeda-Nakayama rings, J. Algebra **226** (2000) 1001-1010.

[5] A. Facchini, C. Faith, FP-injective quotient rings and elementary divisors rings, Lecture Notes Pure Appl. Math., Marcel Dekker, **185** (1997) 293-302.

[6] C. Faith, Injective quotient rings of commutative rings, Lecture Notes Pure Appl. Math., Marcel Dekker, **700** (1979) 151-203.

[7] C. Faith, Injective modules and injective quotient rings, Lecture Notes Pure Appl. Math., Marcel Dekker, **72** (1982) 71-105.

[8] C. Faith, Finitely embedded commutative rings, Proc. Amer. Math. Soc. **112** (3) (1991) 657-659.

[9] C. Faith, Annihilator ideals, associated primes and Kasch-McCoy commutative rings, Comm. Algebra **19** (7) (1994) 1867-1892.

[10] S. Glaz, Commutative coherent rings, Lecture Notes in Math. **1371**, Springer–Verlag, 1989.

[11] R. Gordon and J.C. Robson, Krull dimension, Amer. Math. Soc. **133**, Providence, 1973.

[12] A. Jebli, Algèbre commutative, Scientifika, 1993.

[13] T.Y. Lam, Serre's conjecture, Lecture Notes in Math. **635**, Springer–Verlag, 1978.

[14] T.Y. Lam, Lectures on modules and rings, Graduate Texts in Math. **189**, Springer–Verlag, 1999.

[15] M. Lazarus, Les familles libres maximales d'un module ont-elles le même cardinal? Publ. Sém. Math. Rennes **4** (1973) 677-730.

[16] B. Nashier and W. Nichols, On Steinitz properties, Arch. Math **57** (1991) 247-253.

[17] W.K. Nicholson and M.F. Yousif, Principally injective rings, J. Algebra **174** (1995) 77-93.

[18] W.K. Nicholson and M.F. Yousif, Mininjective rings, J. Algebra **187** (1997) 548-578.

[19] G. Picavet, Parties sondables d'un spectre et profondeur, Boll. Unione Mat. Ital. **7** (8–B) (1994) 677-730.

[20] E.A. Rutter, Rings with the principal extension property, Comm. Algebra **3** (3) (1975) 203-212.

[21] T.S. Shores and R. Wiegand, Rings whose finitely generated modules are direct sums of cyclics, J. Algebra **32** (1974) 152-172.

[22] H. Storrer, On Goldman's primary decomposition, Lecture Notes in Math. **246** (1972) 617-661.

[23] P. Vamos, The dual of the notion of "finitely generated", J. London Math. Soc. **43** (1968) 643-646.

On Projective Modules Over Polynomial Rings

Yves Lequain[(*),(†)]
Instituto de Matemática Pura e Aplicada
Estrada Dona Castorina 110
22.460-320, Rio de Janeiro-RJ
Brazil
e-mail: ylequain@impa.br

Najib Mahdou
Département de Mathématiques et Informatique
Faculté des Sciences et Techniques Fez-Saïss
Université S.M. Ben Abdellah
B.P. 2202, Fez
Morocco
e-mail: mahdou@hotmail.com

Summary: We show that if A is a domain obtained by a certain rather general type of pullback from Prüfer and Bézout domains, then it satisfies the following property: for every $n \geq 0$, all finitely generated projective $A[X_1, \ldots, X_n]$-modules are extended from A. We construct a wide class of domains of Krull dimension one and arbitrary valuative dimension that satisfy the above property.

1 INTRODUCTION

D. Quillen and A. Suslin have independently proved Serre's conjecture: if A is a Dedekind domain, then A satisfies the following property (see [7] or [5]):

$$\begin{array}{l}\text{For every } n \geq 0, \text{ all finitely generated} \\ \text{projective } A[X_1, \ldots, X_n]\text{-modules are extended from } A.\end{array} \tag{1}$$

[(*)]Partially supported by a grant from the group ALGA-PRONEX/MCT
[(†)] The research was done while the first named author was visiting Florida State University.

Y. Lequain and A. Simis have given a dimension-free generalization of Quillen-Suslin's result, showing that if A is a Prüfer domain of arbitrary dimension, then A satisfies Property (1) (see [6]).

In the context of domains of Krull dimension one, the junction of results of Schanuel, Gilmer-Heitmann and Brewer-Costa gives the following theorem (see [1] and [3]), the second part of which is a generalization of Quillen-Suslin's result:

"Let A be a domain of Krull dimension one.

(a) If A satisfies Property (1), then A is (2,3)-closed;

(b) Conversely, if A is (2,3)-closed and if A has valuative dimension one, then A satisfies Property (1)".

In the proof of (b), Serre's theorem is an important ingredient used to reduce the study of the finitely generated projective $A[X]$-modules to the study of the invertible ideals of $A[X]$. If the valuative dimension of A is ≥ 2, this reduction is not possible anymore and nothing is known about whether A may, or may not, satisfy Property (1).

In this paper, we do not call on Serre's theorem. In our main result (Theorem 3), we show that if A is a domain obtained by a rather general type of pullback from domains that satisfy Property (1), then A itself satisfies Property (1). When coupled with Lequain-Simis' result on Prüfer and Bézout domains, this allows us to fill in part the gap existing in the theory of domains of Krull dimension one: in Example 10, we construct a wide class of (necessarily $(2,3)$-closed) domains of Krull dimension one and arbitrary valuative dimension (even infinite) that satisfy Property (1). We are lead to make the conjecture that every $(2,3)$-closed domain of Krull dimension one satisfies Property (1).

2 NOTATIONS AND DEFINITIONS

If A is a ring, $\mathcal{P}(A)$ will stand for the set of finitely generated projective A-modules.

If $A \overset{i}{\hookrightarrow} B$ is an extension of rings and if M is a B-module, then M is *extended* from A if there exists an A-module N such that $M \simeq N \underset{A}{\otimes} B$. Observe that if there exists a retraction $\varphi\colon B \to A$ (i.e., a ring homomorphism φ such that $\varphi \circ i = Id_A$), and if such an A-module exists, then it is unique, up to isomorphism, and equal to $M \underset{B}{\otimes} A$. If furthermore M is finitely generated and projective (respectively, free), then N is also finitely generated and projective (respectively, free).

We shall often use the symbol $\underline{X}$ to denote the set of indeterminates $\{X_1, \ldots, X_n\}$ where n is an integer ≥ 0 that will be specified in the context. If A is a ring, the notation $A[\underline{X}]$ will therefore stand for the polynomial ring $A[X_1, \ldots, X_n]$. If $n = 0$, $\underline{X}$ will stand for the empty set and $A[\underline{X}]$ for the ring A.

3 PROJECTIVE MODULES OVER $R[X_1, \ldots, X_n]$, R A PULLBACK

In this section, we are working in the following framework:

- B is a ring of the type $B = L \oplus I$ where L is a subring of B and I is an ideal of B.
- R is a subring of B of the type $R = D \oplus I$ where D is a subring of L.

[As an example of such a situation, we can give: $B = \mathbb{C}[Y] = \mathbb{C} \oplus Y\mathbb{C}[Y]$, $D = \mathbb{Z}$, $R = \mathbb{Z} \oplus Y\mathbb{C}[Y]$]

The objective of the section is to study, for two given integers $n \geq 0$, $r \geq 1$ and a set of indeterminates $\underline{X} := \{X_1, \ldots, X_n\}$, the relation between the following two properties:

1) Every $P \in \mathcal{P}(R[\underline{X}])$ of rank r is extended from R.

2) Every $Q \in \mathcal{P}(D[\underline{X}])$ of rank r is extended from D.

The following diagram might help to keep track of our modules:

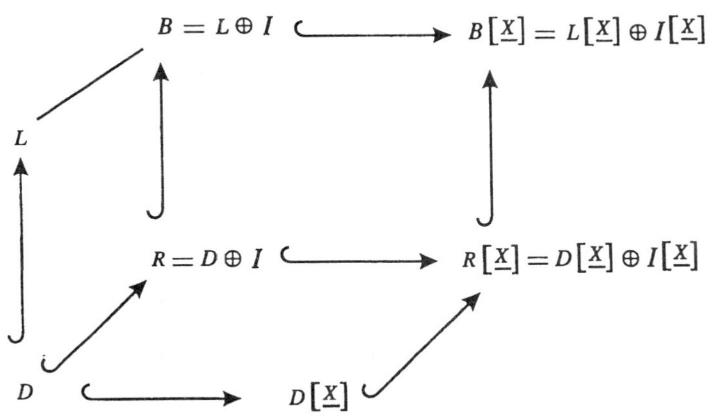

(*)

Of course, we also have the natural retractions from R to D, $B[\underline{X}]$ to B, $R[\underline{X}]$ to R, $R[\underline{X}]$ to $D[\underline{X}]$, $D[\underline{X}]$ to D and $R[\underline{X}]$ to D.

THEOREM 1. *Let R be a ring of the type $R = D \oplus I$, where D is a subring of R and I an ideal of R. Let $n \geq 0$, $r \geq 1$ be two integers and $\underline{X} = \{X_1, \dots, X_n\}$ some indeterminates.*

(a) *If every $P \in \mathcal{P}(R[\underline{X}])$ of rank r is extended from R, then every $Q \in \mathcal{P}(D[\underline{X}])$ of rank r is extended from D.*

(b) *If every $P \in \mathcal{P}(R[\underline{X}])$ of rank r is free, then every $Q \in \mathcal{P}(D[\underline{X}])$ of rank r is free.*

(c) *If R satisfies Property (1), then D satisfies Property (1).*

We first establish a result valid for any (not necessarily projective) module:

LEMMA 2. *Let $R, D, I, n, \underline{X}$ be as in Theorem 1. Let M be an $R[\underline{X}]$-module.*

(a) *The following statements are equivalent:*

(i) *There exist an R-module N_1 and a $D[\underline{X}]$-module N_2 such that $M \simeq N_1 \otimes_R R[\underline{X}]$ and $M \simeq N_2 \otimes_{D[\underline{X}]} R[\underline{X}]$ (i.e., M is extended from both R and $D[\underline{X}]$).*

(ii) *There exists a D-module N_0 such that $M \simeq N_0 \otimes_D R[\underline{X}]$ (i.e., M is extended from D).*

(b) *When (i)–(ii) are satisfied, then N_0, N_1, N_2 are uniquely determined up to an isomorphism, and*

$$N_1 \simeq M \otimes_{R[\underline{X}]} R, \quad N_2 \simeq M \otimes_{R[\underline{X}]} D[\underline{X}], \quad N_0 \simeq M \otimes_{R[\underline{X}]} D$$

$$N_1 \simeq N_0 \otimes_D R, \quad N_2 \simeq N_0 \otimes_D D[\underline{X}], \quad N_0 \simeq N_1 \otimes_R D \simeq N_2 \otimes_{D[\underline{X}]} D.$$

In particular, both N_1 and N_2 are extended from D.

PROOF. (a) (ii) $\Rightarrow$ (i). Let N_0 be a D-module such that $M \simeq N_0 \otimes_D R[\underline{X}]$. It is clear that, taking $N_1 := N_0 \otimes_D R$ and $N_2 := N_0 \otimes_D D[\underline{X}]$, one has $M \simeq N_1 \otimes_R R[\underline{X}]$ and $M \simeq N_2 \otimes_{D[\underline{X}]} R[\underline{X}]$.

(i) $\Rightarrow$ (ii). Let N_1 be an R-module and N_2 a $D[\underline{X}]$-module such that

$$N_1 \underset{R}{\otimes} R[\underline{X}] \simeq M \simeq N_2 \underset{D[\underline{X}]}{\otimes} R[\underline{X}]. \tag{2}$$

Observe that the following diagram, where φ_1 and φ_2 are the natural retractions, is commutative:

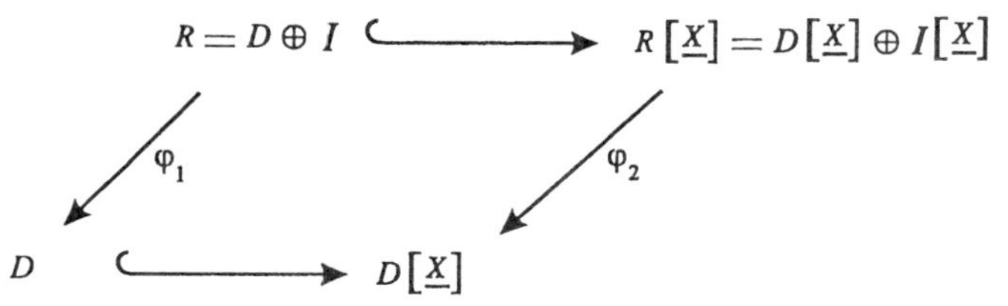

(**)

Then, we have:

$$\begin{aligned}
M &\simeq N_2 \underset{D[\underline{X}]}{\otimes} R[\underline{X}] \quad \text{by (2)} \\
&\simeq \left(N_2 \underset{D[\underline{X}]}{\otimes} R[\underline{X}] \underset{R[\underline{X}]}{\otimes} D[\underline{X}]\right) \underset{D[\underline{X}]}{\otimes} R[\underline{X}] \text{ because } \varphi_2 \text{ is a retraction} \\
&\simeq \left(N_1 \underset{R}{\otimes} R[\underline{X}] \underset{R[\underline{X}]}{\otimes} D[\underline{X}]\right) \underset{D[\underline{X}]}{\otimes} R[\underline{X}] \quad \text{by (2)} \\
&\simeq N_1 \underset{R}{\otimes} D \underset{D}{\otimes} D[\underline{X}] \underset{D[\underline{X}]}{\otimes} R[\underline{X}] \text{ because the diagram (**) is commutative} \\
&\simeq N_0 \underset{D}{\otimes} R[\underline{X}] \quad \text{with} \quad N_0 := N_1 \underset{R}{\otimes} D.
\end{aligned}$$

(b) Since there exist retractions $R[\underline{X}] \to D$, $R[\underline{X}] \to R$, $R[\underline{X}] \to D[\underline{X}]$, then N_0, N_1, N_2 are uniquely determined up to isomorphisms, and $N_0 \simeq M \underset{R[\underline{X}]}{\otimes} D$, $N_1 \simeq M \underset{R[\underline{X}]}{\otimes} R \simeq N_0 \underset{D}{\otimes} R$, $N_2 \simeq M \underset{R[\underline{X}]}{\otimes} D[\underline{X}] \simeq N_0 \underset{D}{\otimes} D[\underline{X}]$. Finally, since there exist retractions $R \to D$, $D[\underline{X}] \to D$, then $N_1 \underset{R}{\otimes} D \simeq N_0 \simeq N_2 \underset{D[\underline{X}]}{\otimes} D$. □

Proof of Theorem 1. (a) Let $Q \in \mathcal{P}(D[\underline{X}])$ of rank r. The $R[\underline{X}]$-module $P := Q \underset{D[\underline{X}]}{\otimes} R[\underline{X}]$ is projective of rank r, hence is extended from R by hypothesis. Furthermore, by definition itself, P is extended from $R[\underline{X}]$. Thus, by Lemma 2(b), Q is extended from D.

(b) Let $Q \in \mathcal{P}(D[\underline{X}])$ of rank r. The $R[\underline{X}]$-module $P := Q \underset{D[\underline{X}]}{\otimes} R[\underline{X}]$ is projective of rank r, hence is free by hypothesis. Since there exists a retraction $R[\underline{X}] \to D[\underline{X}]$, then $Q \simeq P \underset{R[\underline{X}]}{\otimes} D[\underline{X}]$ and Q is free. □

(c) Is a consequence of (a).

Next, we establish a kind of converse for Theorem 1.

THEOREM 3. *Let B be a ring of the type $B = L \oplus I$ where L is a subring of B and I an ideal of B. Let R be a subring of B of the type $R = D \oplus I$ where D is a subring of L. Let $n \geq 0$ and $r \geq 1$ be two integers, $\underline{X} := \{X_1, \ldots, X_n\}$ some indeterminates and suppose that every $M \in \mathcal{P}(B[\underline{X}])$ of rank r is free.*

(a) *If every $Q \in \mathcal{P}(D[\underline{X}])$ of rank r is extended from D, then every $P \in \mathcal{P}(R[\underline{X}])$ of rank r is extended from D (hence also from R).*

(b) *If every $Q \in \mathcal{P}(D[\underline{X}])$ of rank r is free, then every $P \in \mathcal{P}(R[\underline{X}])$ of rank r is free.*

(c) *R satisfies Property (1) if and only if D satisfies Property (1).*

Theorem 3 will be obtained as a consequence of the following lemma.

LEMMA 4. *Let $B, L, I, R, D, n, \underline{X}$ be as in Theorem 3. Let $P \in \mathcal{P}(R[\underline{X}])$ such that $P \underset{R[\underline{X}]}{\otimes} B[\underline{X}]$ is free.*

(a) *If $P \underset{R[\underline{X}]}{\otimes} D[\underline{X}]$ is extended from D, then P is extended from D (hence also from R).*

(b) *If $P \underset{R[\underline{X}]}{\otimes} D[\underline{X}]$ is free, then P is free.*

PROOF. (a) Let $m_1, \ldots, m_t \in P \underset{R[\underline{X}]}{\otimes} B[\underline{X}]$ be a $B[\underline{X}]$-basis of $P \underset{R[\underline{X}]}{\otimes} B[\underline{X}]$. Then,

$$P \underset{R[\underline{X}]}{\otimes} B[\underline{X}] = (L[\underline{X}]m_1 \oplus \cdots \oplus L[\underline{X}]m_t) \oplus (I[\underline{X}]m_1 \oplus \cdots \oplus I[\underline{X}]m_t) \quad (3)$$

$$I[\underline{X}]m_1 \oplus \cdots \oplus I[\underline{X}]m_t = (I[\underline{X}]P) \underset{R[\underline{X}]}{\otimes} 1_{B[\underline{X}]} \quad (4)$$

$$I[\underline{X}]m_1 \oplus \cdots \oplus I[\underline{X}]m_t \subseteq P \underset{R[\underline{X}]}{\otimes} 1_{B[\underline{X}]} \subseteq P \underset{R[\underline{X}]}{\otimes} B[\underline{X}]. \quad (5)$$

¿From (3) and (5), we obtain

$$P \underset{R[\underline{X}]}{\otimes} 1_{B[\underline{X}]} = G \oplus (I[\underline{X}]m_1 \oplus \cdots \oplus I[\underline{X}]m_t) \quad (6)$$

where

$$G = (L[\underline{X}]m_1 \oplus \cdots \oplus L[\underline{X}]m_t) \cap (P \underset{R[\underline{X}]}{\otimes} 1_{B[\underline{X}]}). \quad (7)$$

Note that G is a $D[\underline{X}]$-module. We claim that

$$G \simeq P \underset{R[\underline{X}]}{\otimes} D[\underline{X}]. \tag{8}$$

Indeed, on one hand, by (6) and (4) we have

$$G \simeq \left(P \underset{R[\underline{X}]}{\otimes} 1_{B[\underline{X}]}\right) \Big/ \left(I[\underline{X}]P \underset{R[\underline{X}]}{\otimes} 1_{B[\underline{X}]}\right) \tag{9}$$

and, on the other hand, we clearly have

$$P \underset{R[\underline{X}]}{\otimes} D[\underline{X}] \simeq P/I[\underline{X}]P. \tag{10}$$

Then, since P is a projective $R[\underline{X}]$-module and since $R[\underline{X}] \hookrightarrow B[\underline{X}]$ is an injective map of $R[\underline{X}]$-modules, the right hand sides of (9) and (10) are isomorphic $D[\underline{X}]$-modules; this proves (8).

Since $G \simeq P \underset{R[\underline{X}]}{\otimes} D[\underline{X}]$, then G is a finitely generated projective $D[\underline{X}]$-module. By hypothesis $P \underset{R[\underline{X}]}{\otimes} D[\underline{X}]$ is extended from D; let G_0 be a D-module such that

$$G \simeq G_0 \underset{D}{\otimes} D[\underline{X}]. \tag{11}$$

Since there exists a retraction $D[\underline{X}] \to D$, then

$$G_0 \simeq G \underset{D[\underline{X}]}{\otimes} D \tag{12}$$

and, since G is a finitely generated projective $D[\underline{X}]$-module, then G_0 is a finitely generated projective D-module.

Note that $L[\underline{X}]$ is not an $R[\underline{X}]$-module, and hence $P \underset{R[\underline{X}]}{\otimes} L[\underline{X}]$ does not make sense. However, $B[\underline{X}]$ is an $R[\underline{X}]$-module, hence $P \underset{R[\underline{X}]}{\otimes} B[\underline{X}]$ makes sense, and we can consider the subset $\left(P \underset{R[\underline{X}]}{\otimes} 1_{B[\underline{X}]}\right) L[\underline{X}]$, or more generally the subsets HC where H is any additive subgroup of $P \underset{R[\underline{X}]}{\otimes} 1_{B[\underline{X}]}$ and C any subring of $B[\underline{X}]$. We claim that

$$P \underset{R[\underline{X}]}{\otimes} B[\underline{X}] = \left(P \underset{R[\underline{X}]}{\otimes} 1_{B[\underline{X}]}\right) L. \tag{13}$$

Indeed,

$$\begin{aligned} P \underset{R[\underline{X}]}{\otimes} B[\underline{X}] &= P \underset{R[\underline{X}]}{\otimes} (L[\underline{X}] \oplus I[\underline{X}]) \\ &= \left(P \underset{R[\underline{X}]}{\otimes} 1_{B[\underline{X}]}\right) L[\underline{X}] + \left(P \underset{R[\underline{X}]}{\otimes} 1_{B[\underline{X}]}\right) I[\underline{X}] \\ &= \left(P \underset{R[\underline{X}]}{\otimes} 1_{B[\underline{X}]}\right) L. \end{aligned}$$

Now, we claim that

$$G_0 \underset{D}{\otimes} I[\underline{X}] \simeq GI = I[\underline{X}]m_1 \oplus \cdots \oplus I[\underline{X}]m_t. \tag{14}$$

Indeed, since G_0 is a finitely generated projective D-module and since $I[\underline{X}] \hookrightarrow B[\underline{X}]$ is an injective map of D-modules, then the canonical map $G_0 \underset{D}{\otimes} I[\underline{X}] \to G_0 \underset{D}{\otimes} B[\underline{X}]$ is injective. Hence $G_0 \underset{D}{\otimes} I[\underline{X}] \simeq (G_0 \underset{D}{\otimes} 1_{B[\underline{X}]})I[\underline{X}] = (G_0 \underset{D}{\otimes} 1_{B[\underline{X}]})D[\underline{X}]I$. Since G_0 is a finitely generated projective D-module and since $D[\underline{X}] \hookrightarrow B[\underline{X}]$ is an injective map of D-modules, then the canonical map $G_0 \underset{D}{\otimes} D[\underline{X}] \to G_0 \underset{D}{\otimes} B[\underline{X}]$ is injective, hence $G_0 \underset{D}{\otimes} D[\underline{X}] \simeq (G_0 \underset{D}{\otimes} 1_{B[\underline{X}]})D[\underline{X}]$. We therefore obtain that $G_0 \underset{D}{\otimes} I[\underline{X}] \simeq (G_0 \underset{D}{\otimes} D[\underline{X}])I \simeq GI$, which is the first part of (14).

Now, we want to prove the second part of (14). By (3), (13) and (6), we have

$$\begin{aligned}(L[\underline{X}]m_1 + \cdots + L[\underline{X}]m_t) &\oplus (I[\underline{X}]m_1 + \cdots + I[\underline{X}]m_t) \\ &= [G \oplus (I[\underline{X}]m_1 + \cdots + I[\underline{X}]m_t)]L,\end{aligned}$$

hence

$$\begin{aligned}(L[\underline{X}]m_1 + \cdots + L[\underline{X}]m_t) &\oplus (I[\underline{X}]m_1 + \cdots + I[\underline{X}]m_t) \\ &= GL + (I[\underline{X}]m_1 + \cdots + I[\underline{X}]m_t).\end{aligned} \tag{15}$$

Furthermore, by (7), we know that $G \subseteq L[\underline{X}]m_1 + \cdots + L[\underline{X}]m_t$, hence $GL \subseteq L[\underline{X}]m_1 + \cdots + L[\underline{X}]m_t$. Together with (15), this implies that $GL = L[\underline{X}]m_1 + \cdots + L[\underline{X}]m_t$ and therefore that $GI = GLI = (L[\underline{X}]m_1 + \cdots + L[\underline{X}]m_t)I = I[\underline{X}]m_1 + \cdots + I[\underline{X}]m_t$. This terminates the proof of (14).

Finally, we have

$$P \simeq G_0 \underset{D}{\otimes} R[\underline{X}]. \tag{16}$$

Indeed,

$$\begin{aligned}G_0 \underset{D}{\otimes} R[\underline{X}] &= G_0 \underset{D}{\otimes} (D[\underline{X}] \oplus I[\underline{X}]) \simeq (G_0 \underset{D}{\otimes} D[\underline{X}]) \oplus (G_0 \underset{D}{\otimes} I[\underline{X}]) \\ &\simeq G \oplus (I[\underline{X}]m_1 + \cdots + I[\underline{X}]m_t) \quad \text{by (11) and (14)} \\ &= P \underset{R[\underline{X}]}{\otimes} 1_{B[\underline{X}]} \quad \text{by (6)} \\ &\simeq P \quad \text{since} \quad P \in \mathcal{P}(R[\underline{X}]).\end{aligned}$$

Thus, P is extended from D.

(b) By (12) and (8) we have $G_0 \simeq (P \underset{R[\underline{X}]}{\otimes} D[\underline{X}]) \underset{D[\underline{X}]}{\otimes} D$. Thus G_0 is free if $P \underset{R[\underline{X}]}{\otimes} D[\underline{X}]$ is free. In that case, G is also free since $P \simeq G_0 \underset{D}{\otimes} R[\underline{X}]$ by (16). □

PROOF OF THEOREM 3. (a) and (b) are consequences of Lemma 4.

(c) Is a consequence of (a) and Theorem 1. □

4 APPLICATIONS

If D is a domain that satisfies Property (1), then it is clear that $D[\underline{Y}]$ also satisfies Property (1). We shall need the converse:

PROPOSITION 5. *Let D be a domain, $m \geq 0$ and $n \geq 1$ two integers, $\underline{Y} := \{Y_1, \ldots, Y_m\}$ and $\underline{X} := \{X_1, \ldots, X_n\}$ some indeterminates.*

(a) *If every $P \in \mathcal{P}(D[\underline{Y}, \underline{X}])$ of rank r is extended from $D[\underline{Y}]$, then every $Q \in \mathcal{P}(D[\underline{X}])$ of rank r is extended from D.*

(b) *D satisfies Property (1) if (and only if) $D[\underline{Y}]$ satisfies Property (1).*

(c) *If M is a $D[\underline{Y}, \underline{X}]$-module, then M is extended from D if (and only if) M is extended from both $D[\underline{Y}]$ and $D[\underline{X}]$.*

(d) *If every $P \in \mathcal{P}(D[\underline{Y}, \underline{X}])$ of rank r is extended from $D[\underline{Y}]$, then every such P is extended from D.*

PROOF. Let $R := D[\underline{Y}]$, $I := \underline{Y}D[\underline{Y}]$. Evidently, we have $R = D \oplus I$.

(a) By hypothesis, every $P \in \mathcal{P}(R[\underline{X}])$ of rank r is extended from R. Then, by Theorem 1(a), every $Q \in \mathcal{P}(D[\underline{X}])$ of rank r is extended from D.

(b) Is a consequence of (a).

(c) Apply Lemma 2(a).

(d) Let $P_0 \in \mathcal{P}(D[\underline{Y}, \underline{X}]$ of rank r. We want to show that P_0 is extended from D. We proceed by induction on $m =$ the number of Y_i's. If $m = 0$, there is nothing to do. Thus, we can suppose $m \geq 1$.

If $m = 1$, let $Y = Y_1$. By hypothesis, P_0 is extended from $D[Y]$; thus there exists $Q_0 \in \mathcal{P}(D[Y])$ (necessarily of rank r) such that $P_0 \simeq Q_0 \underset{D[Y]}{\otimes} D[Y, \underline{X}]$. We will show that Q_0 is extended from D.

Let P' be any element of $\mathcal{P}(D[Y, \underline{X}])$ of rank r. If $\varphi\colon D[Y, \underline{X}] \to D[Y, \underline{X}]$ is the D-isomorphism defined by $\varphi(Y) = X_1$, $\varphi(X_1) = Y$ and $\varphi(X_i) = X_i$ for $i \geq 2$, then there exists $P \in \mathcal{P}(D[Y, \underline{X}])$ of rank r such that $P' \simeq P \underset{D[Y,\underline{X}]}{\otimes} {}^{\varphi}D[Y, \underline{X}]$. By hypothesis, there exists $N \in \mathcal{P}(D[Y])$ of rank r such that $P \simeq N \underset{D[Y]}{\otimes} D[Y, \underline{X}]$. Then, setting $\varphi' :=$ the restriction of φ to $D[Y]$, we have

$$\begin{aligned} P' &\simeq \left(N \underset{D[Y]}{\otimes} D[Y, \underline{X}]\right) \underset{D[Y,\underline{X}]}{\otimes} {}^{\varphi}D[Y, \underline{X}] \\ &\simeq \left(N \underset{D[Y]}{\otimes} {}^{\varphi'}D[X_1]\right) \underset{D[X_1]}{\otimes} D[\underline{X}, Y] \end{aligned}$$

since the following diagram commutes:

$$\begin{array}{ccc} D[Y] & \longrightarrow & D[Y,\underline{X}] \\ \varphi' \downarrow & & \downarrow \varphi \\ D[X_1] & \longrightarrow & D[Y,\underline{X}] \end{array}$$

(here, the horizontal arrows are the natural embeddings). Thus, P' is extended from $D[X_1]$ (from the $D[X_1]$-module $N \underset{D[Y]}{\otimes} {}^{\varphi'} D[X_1]$), hence a fortiori from $D[\underline{X}]$.

Thus, every $P' \in \mathcal{P}(D[Y,\underline{X}])$ of rank r is extended from $D[\underline{X}]$. Then, by (a), every $Q' \in \mathcal{P}(D[Y])$ of rank r is extended from D. In particular, Q_0 is extended from D as we wanted, and this finishes the proof for $m = 1$.

Now, suppose that $m \geq 2$. By the case $m = 1$ applied to the domain $D[Y_2, \dots, Y_m]$, P_0 is extended from $D[Y_2, \dots, Y_m]$. By the case $m-1$ applied to the domain $D[Y_1]$, P_0 is extended from $D[Y_1]$, hence a fortiori from $D[Y_1, \underline{X}]$. Being extended from both $D[Y_2, \dots, Y_n]$ and from $D[Y_1, \underline{X}]$, then by Lemma 2(a), P_0 is extended from D. □

REMARK 6. 1) Let D be a domain and X, Y two indeterminates. Let $P \in \mathcal{P}(D[X,Y])$, with P extended from $D[Y]$. In view of Proposition 5(d), one could ask whether P is necessarily extended from D. The answer is negative: indeed, if k is a field, T an indeterminate and $D := k[T^2, T^3]$, then Schanuel has shown that $J := (T^2, 1+TY)D[X,Y]$ is a projective $D[X,Y]$-ideal that is clearly extended from $D[Y]$, but that is not extended from $D[X]$ (see [1, p.209]); then, of course, J is not extended from D either.

2) When $m = n$, Part (d) of Proposition 5 could be obtained as a consequence of Part (a). But, when $m \neq n$, it requires a proof.

Theorem 3 allows us to give a big class of domains that satisfy Property (1). Before doing that, we recall some definitions.

DEFINITIONS. A domain D is a *Prüfer domain* if $D_{\mathfrak{p}}$ is a valuation domain for every prime ideal $\mathfrak{p}$ or, equivalently, if every finitely generated ideal of D is invertible. The domain D is a *Bézout domain* if every finitely generated ideal of D is principal.

A domain D with quotient field K and integral closure $\bar{D}$ is (2,3)-*closed* if it satisfies the following property:

$$\xi \in K, \quad \xi^2 \in D, \quad \xi^3 \in D \Rightarrow \xi \in D.$$

An extension $D \hookrightarrow A$ of domains is *geometrically unibranched* if for every prime ideal $\mathfrak{p}$ of D, there exists a unique prime ideal $\mathfrak{p}'$ of A lying over $\mathfrak{p}$ and the residue fields of $D_{\mathfrak{p}}$ and $A_{\mathfrak{p}'}$ are equal. The domain D is *seminormal* if it satisfies the following property:

A domain, $D \subseteq A \subseteq \bar{D}$, $D \hookrightarrow A$ geometrically unibranched $\Rightarrow D = A$.

It is known that a domain D is $(2,3)$-closed if and only if D is seminormal (see [1, Theorem 1 p. 209]).

COROLLARY 7. *Let $E \subseteq F$ be an extension of rings with F a Bézout domain. Let $u \geq 0$, $v \geq 0$, $r \geq 0$, $n \geq 0$ be integers and $\underline{Z} := \{Z_1, \dots, Z_u\}$, $\underline{T} := \{T_1, \dots, T_v\}$, $\underline{Y} := \{Y_1, \dots, Y_r\}$, $\underline{X} := \{X_1, \dots, X_n\}$ some finite sets of indeterminates. Let $R := E[\underline{Z}] + (\underline{Y})F[\underline{Z}, \underline{T}, \underline{Y}]$. Then,*

(a) *R satisfies Property (1) if and only if E satisfies Property (1).*

(b) *R satisfies Property (1) if E is a Prüfer domain.*

(c) *In case E is a domain of Krull dimension one and valuative dimension one, R satisfies Property (1) if and only if E is (2,3)-closed.*

(d) *For every integer $n \geq 0$, every $P \in \mathcal{P}(R[\underline{X}])$ is free if E is a Bézout domain.*

PROOF. (a) Set $L := F[\underline{Z}, \underline{T}]$, $I := (\underline{Y})F[\underline{Z}, \underline{T}, \underline{Y}]$, $B := F[\underline{Z}, \underline{T}, \underline{Y}]$, $D := E[\underline{Z}]$. Evidently, we have $D \subset L$, $R = D \oplus I$, $B = L \oplus I$. If R satisfies Property (1), then by Theorem 1(a), $E[\underline{Z}]$ satisfies Property (1) and by Proposition 5(b), E satisfies Property (1).

Now, we look at the converse. Since F is a Bézout domain, then by [6, Theorem b, p. 166], for any integer $n \geq 0$, every $M \in \mathcal{P}(F[\underline{Z}, \underline{T}, \underline{Y}, \underline{X}])$ is free. Since E satisfies Property (1), then clearly so does $E[\underline{Z}]$. Hence, by Theorem 3(a), R satisfies Property (1).

(b) If E is a Prüfer domain, then by [6, Theorem b, p. 166], E satisfies Property (1) and, by (a), R satisfies Property (1).

(c) If E is a domain of Krull dimension one and valuative dimension one, then by [4, (25.13) p. 354] the integral closure of E is a Prüfer domain and by [1, Theorem 2 and Theorem 1], E satisfies Property (1) if and only if E is (2,3)-closed. Then, by (a), R satisfies Property (1) if and only if E is (2,3)-closed.

(d) If E is a Bézout domain, then by [6, Theorem b, p. 166], for any integer $n \geq 0$, every $Q \in \mathcal{P}(E[\underline{Z}, \underline{X}])$ is free. Thus, by Theorem 3(b), every $P \in \mathcal{P}(R[\underline{X}])$ is free. □

Wanting some specific examples out of Corollary 7, we can give the following:

EXAMPLE 8. Let $u, v, r, n, \underline{Z}, \underline{T}, \underline{Y}, \underline{X}$ be as in Corollary 7. Then,

(a) $R_1 := \mathbb{Z}[\underline{Z}] + (\underline{Y})\mathbb{Q}[\underline{Z}, \underline{T}, \underline{Y}]$ is a domain such that, for every $n \geq 0$, every $P \in \mathcal{P}(R_1[\underline{X}])$ is free.

(b) $R_2 := \mathbb{Z}[e^{i2\pi/23}][\underline{Z}] + (\underline{Y})\mathbb{Q}(e^{i2\pi/23})[\underline{Z}, \underline{T}, \underline{Y}]$ satisfies Property (1) but possesses some projective ideals that are not free. Indeed, $\mathbb{Z}[e^{i2\pi/23}]$ is a Dedekind domain (hence a Prüfer domain) that is not a PID, hence that is not a Bézout domain.

We now turn our attention to domains of Krull dimension one. Before that, we observe that in this context of relating Property (1) with (2, 3)-closeness of a domain A, we can restrict ourselves to the study of the quasi-local domains (i.e., domains that have only one maximal ideal). Indeed, if A is a domain that satisfies Property (1), then $A_{\mathcal{M}}$ satisfies Property (1) for every maximal ideal $\mathcal{M}$ by [9, Proposition 2 p. 54]; conversely, if $A_{\mathcal{M}}$ satisfies Property (1) for every maximal ideal $\mathcal{M}$, then A satisfies Property (1) by Quillen's localization theorem [7, Theorem 1, p. 169]. Also, A is (2, 3)-closed if and only if $A_{\mathcal{M}}$ is (2, 3)-closed for every maximal ideal $\mathcal{M}$.

PROPOSITION 9. *Let $(A, \mathcal{M}) \subseteq (B, \mathcal{M})$ be two quasi-local domains that admit a field of representatives and that have the same maximal ideal $\mathcal{M}$. If B satisfies Property (1), then A satisfies Property (1).*

PROOF. Let k be a field of representatives for A and L a field of representatives for B. We have $B = L \oplus \mathcal{M}$ and $A = k \oplus \mathcal{M}$. Let $n \geq 0$ be any integer and $\underline{X} := \{X_1, \ldots, X_n\}$ a set of indeterminates. By hypothesis, every $M \in \mathcal{P}(B[\underline{X}])$ is extended from B; since B is quasi-local, this implies that every $M \in \mathcal{P}(B[\underline{X}])$ is free. Since k is a field, then by Quillen-Suslin's result, every $Q \in \mathcal{P}(k[\underline{X}])$ is free. Thus, by Theorem 3(b), every $P \in \mathcal{P}(A[\underline{X}])$ is free (equivalently is extended from A). Thus A satisfies Property (1). □

We now exhibit some quasi-local domains of Krull dimension one and arbitrary valuative dimension, that satisfy Property (1).

EXAMPLE 10. Let r be an integer ≥ 0 or $r = \infty$. let $L|k$ be a field extension of transcendance degree equal to r. Let B be any quasi-local Bezout domain of Krull dimension one and of the type $B = L \oplus \mathcal{M}$, where $\mathcal{M}$ is an ideal (for example, $B = L[Y]_{(Y)}$ and $\mathcal{M} = YL[Y]_{(Y)}$, or $B = L[[Y]]$ and $\mathcal{M} = YL[[Y]]$), and let $A = k \oplus \mathcal{M}$.

The domain A has Krull dimension one and valuative dimension equal to r. Furthermore, B satisfies Property (1) by [6, Theorem b, p. 166]. Thus, A satisfies Property (1) by Proposition 9.

Note that A is (2, 3)-closed and that it is integrally closed if and only if k is algebraically closed in L.

We are lead to make the following conjecture:

CONJECTURE A: Let A be a quasi-local domain of Krull dimension one. Then, A satisfies Property (1) if (and only if) A is $(2,3)$-closed.

In view of Proposition 9, we make the following weaker preliminary conjecture:

CONJECTURE B: Let A be a $(2,3)$-closed domain of Krull dimension one. Suppose that the integral closure $\bar{A}$ of A satisfies Property (1). Then, A satisfies Property (1).

REMARK 11. If A is a Prüfer domain with quotient field K, then every ring B between A and K is a Prüfer domain and therefore satisfies Property (1). In view of this result, one could ask the following question: if A is a domain that satisfies Property (1) and if B is any ring between A and its quotient field, does B satisfy Property (1)?

It is easy to see that the answer is negative, even if A is a one-dimensional noetherian local domain. Indeed, let k be a field, Y an indeterminate, A the one-dimensional local ring $k + Y(Y+1)k[Y]_{(Y)\cup(Y+1)}$ and B the ring $A[Y^{-2}, Y^{-3}]$. The integral closure $\bar{A}$ of A, which is the ring $k[Y]_{(Y)\cup(Y+1)}$, is a Bézout domain. Furthermore, it is easy to see that there is no local ring between A and $\bar{A}$ other than A, hence A is seminormal. Then, A satisfies Property (1) by [1, Theorem 3 p. 213]. On the other hand, the overring B is not $(2,3)$-closed and therefore does not satisfy Property (1).

If A is a Prüfer domain and Y is an indeterminate, then by [6, Theorem b, p. 166], we know that $A[Y]$ satisfies Property (1). One may ask whether any ring B between $A[Y]$ and its quotient field satisfies Property (1). Once again, the answer is negative. Indeed, if m is a nonzero nonunit element of A and $\mathcal{M}$ is a maximal ideal of A that contains m, then $B := A_{\mathcal{M}}[Y, Y^2/m^2, Y^3/m^3]$ is an overring that is not $(2,3)$-closed, hence that does not satisfy Property (1).

References

[1] Brewer, J.W.; Costa, D.L. Seminormality and Projective Modules over Polynomial Rings, *J. Algebra* **58** (1979), 208–216.

[2] Endo, S. Projective modules over polynomial rings, *J. Math. Soc. Japan* **15** (1963), 339–352.

[3] Gilmer, R.; Heitmann, R. On Pic $R[X]$ for R seminormal, *J. Pure Appl. Algebra* **16** (1980), 251–257.

[4] Gilmer, R. Multiplicative Ideal Theory; Queen's Papers on Pure and Applied Mathematics, Kingston (1968).

[5] Lam, T.Y. Serre's Conjecture; Lecture Notes in Mathematics, Springer-Verlag, Berlin and New York (1978).

[6] Lequain, Y.; Simis, A. Projective modules over $R[X_1, \ldots, X_n]$, R a Prüfer domain. *J. Pure Appl. Algebra* **18** (1980), 165–171.

[7] Quillen, D. Projective modules over polynomial rings, *Inv. Math.* **36** (1976), 167–171.

[8] Roitman, M. On Projective Modules over Polynomial Rings, *J. Algebra* **58** (1979), 51–63.

Rings, Conditional Expectations, and Localization

THOMAS G. LUCAS

DEPARTMENT OF MATHEMATICS, UNIVERSITY OF NORTH CAROLINA CHARLOTTE, CHARLOTTE, NC 28223

INTRODUCTION.

We start with an example. Let μ represent Lebesgue measure on the interval $[-1,1]$ and let $\mathcal{B}$ denote the corresponding set of Lebesgue measurable subsets of $[-1,1]$. Let $\mathcal{A}$ denote the subset of $\mathcal{B}$ consisting of those Lebesgue measurable subsets of $[-1,1]$ which are symmetric with the origin; i.e., $Y \in \mathcal{B}$ is in $\mathcal{A}$ if for each $x \in [-1,1]$, $x \in Y$ if and only if $-x \in Y$. Now any real valued function whose domain is symmetric with respect to the origin can be written uniquely as a sum of an even function and an odd function, one simply uses the functions $f_e(x) = (f(x) + f(-x))/2$ and $f_o(x) = (f(x) - f(-x))/2$. If f is integrable on $[-1,1]$, then we have that $\int_Y f\,d\mu = \int_Y f_e\,d\mu$ since the integral of an odd (integrable) function is zero on any measurable set which is symmetric with the origin. In such a case, f_e is said to be the "conditional expectation of f with respect to $\mathcal{A}$". We will revisit this example in the second section.

The notion of conditional expectation is valid for any pair of comparable complete sigma-finite sigma-algebras on a measurable space $\mathbf{X}$. Given $\mathbf{X}$ and two distinct comparable complete sigma-finite sigma-algebras $\mathcal{A} \subset \mathcal{B}$ with regard to some sigma-finite measure μ on $\mathbf{X}$. Then for each $p \geq 1$ including $p = \infty$ and each function $f \in L^p(\mathbf{X}, \mathcal{B}, \mu)$, there is a unique (up to differing on sets of measure zero) $\mathcal{A}$-measurable function $\mathcal{E}(f) \in L^p(\mathbf{X}, \mathcal{A}, \mu)$ such that for each $Y \in \mathcal{A}$ with $\mu(Y) < \infty$, $\int_Y f\,d\mu = \int_Y \mathcal{E}(f)\,d\mu$ [see for example [1, page 916]]. While the conditional expectation can be defined for any integrable function, we shall restrict to those functions which are essentially bounded on $\mathbf{X}$; i.e., those measurable functions f for which there exists a real number r such that the measure of the set $\{x \in \mathbf{X} \mid |f(x)| > r\}$ is zero. The primary reason for doing so here is that the set of essentially bounded functions forms a commutative ring with identity—in our case, the rings $L^\infty(\mathbf{X}, \mathcal{B}, \mu)$ and $L^\infty(\mathbf{X}, \mathcal{A}, \mu)$. Moreover, if f is essentially bounded and $\mathcal{B}$-measurable, then $\mathcal{E}(f)$ is essentially bounded (and, of course, $\mathcal{A}$-measurable). Thus $\mathcal{E}(\_)$ is a function from the ring $L^\infty(\mathbf{X}, \mathcal{B}, \mu)$ to the ring $L^\infty(\mathbf{X}, \mathcal{A}, \mu)$. Note that if $\mu(\mathbf{X}) = \infty$, such functions need not be integrable, but the conditional expectation still exists. For a $\mathcal{B}$-measurable set Z, we let $\mathcal{X}_Z$ denote its corresponding characteristic function. The function $\mathcal{E}$ has several very nice properties. Here are several that we will explore in a more general setting later.

(A) For each $f \in L^\infty(\mathbf{X}, \mathcal{B}, \mu)$ and each $g \in L^\infty(\mathbf{X}, \mathcal{A}, \mu)$, not only is $\mathcal{E}(g) = g$ but $\mathcal{E}(gf) = gE(f)$.
(B) If $f(x) \geq 0$ for all $x \in \mathbf{X}$ (up to the usual "set of measure zero" restriction), then $\mathcal{E}(f)(x) \geq 0$ for all x.
(C) For each $\mathcal{B}$ measurable set Z, the support of $\mathcal{E}(\mathcal{X}_Z)$ contains Z.
(D) Let Z denote the support of f. If $f(x) > 0$ for each $x \in Z$, then $\mathcal{E}(f)$ is positive on its support. Moreover, $\mathcal{E}(f)$ and $\mathcal{E}(\mathcal{X}_Z)$ have the same support.
(E) If $\mathcal{E}(f\mathcal{X}_Y) = 0$ for each $Y \in \mathcal{B}$, then $f = 0$.
(F) If $Y \in \mathcal{B}$ contains the support of f and $\mathcal{E}(f)\mathcal{X}_Y = 0$, then $\mathcal{E}(f) = 0$.

In terms of measurable sets we have the following.

(G) For each set $Y \in \mathcal{B}$, there are unique sets $Y^\#$ and $Y_\#$ in $\mathcal{A}$ such that (i) $Y_\# \subseteq Y \subseteq Y^\#$, (ii) if $Z \in \mathcal{A}$ is such that $Z \subseteq Y$, then $Z \subseteq Y_\#$, and (iii) if $W \in \mathcal{A}$ is such that $Y \subseteq W$, then $Y^\# \subseteq W$.
(H) For each $Y \in \mathcal{B}$, $Y^\#$ is the support of $\mathcal{E}(\mathcal{X}_Y)$ and $Y_\#$ is the complement of the support of $\mathcal{E}(1 - \mathcal{X}_Y)$.

The set $Y^\#$ is the intersection of all sets in $\mathcal{A}$ that contain Y and the set $Y_\#$ is the union of all sets in $\mathcal{A}$ that are contained in Y. A good reference for the topic is the book by M.M. Rao, [7]. In particular, see [7, Chapter 7] for statements and proofs matching up with the properties listed above.

From here on we drop the reference to $\mathbf{X}$ and μ and simply write $L^\infty(\mathcal{B})$ and $L^\infty(\mathcal{A})$ in place of $L^\infty(\mathbf{X}, \mathcal{B}, \mu)$ and $L^\infty(\mathbf{X}, \mathcal{A}, \mu)$.

One of the nice ring theoretic properties of the ring $L^\infty(\mathcal{B})$ is that its total quotient ring is von Neumann regular. Another is that each finitely generated ideal of $L^\infty(\mathcal{B})$ is principal. Suppose $I = (f, g)$ is a two-generated ideal. Set $h(x) = max\{|f(x)|, |g(x)|\}$. Since $f, g \in L^\infty(\mathcal{B})$, so is h. Let $Y = \{x \in \mathbf{X} \mid |f(x)| \geq |g(x)|\}$. Then $h = sgn(f)\mathcal{X}_Y f + sgn(g)(1 - \mathcal{X}_Y)g \in I$. We also have that h divides both f and g. For those x for which $h(x)$ is not zero, we have $h(x) \geq |f(x)|$ and $h(x) \geq |g(x)|$ so that $-1 \leq f(x)/h(x) \leq 1$ and $-1 \leq g(x)/h(x) \leq 1$. So in this case, $(f/h)(x) = f(x)/h(x)$ and $(g/h)(x) = g(x)/h(x)$. As $h(x) = 0$ implies both $f(x)$ and $g(x)$ are zero, we can simply declare $(f/h)(x) = (g/h)(x) = 0$ when $h(x) = 0$. Thus both f/h and g/h denote elements of $L^\infty(\mathcal{B})$ and $I = (h)$. Note that $h(x) \geq 0$ for each x.

Now let $R \subseteq S$ be a pair of reduced rings with the same identity such that the total quotient rings, $T(R) \subseteq T(S)$, are von Neumann regular. Further assume that R contains each idempotent of $T(R)$ and that S contains each idempotent of $T(S)$. Denote the set of idempotents of R by E_R and of S by

E_S. We say that an R-module homomorphism $\mathcal{E} : S \to R$ is a *conditional ring expectation on S* if $\mathcal{E}$ satisfies the following:

(I) $\mathcal{E}(r) = r$ for each $r \in R$ (equivalently, $\mathcal{E}(1) = 1$).
(II) For $s \in S$, $\mathcal{E}(se) = 0$ for each $e \in E_S$ implies $s = 0$.
(III) For $f \in E_S$, $\mathcal{E}(f) = 0$ if and only if $f = 0$.

What we want to see is how many of the properties of conditional expectation for the rings of L^∞ functions carry over to this general ring theoretic setting.

First note that since both $T(R)$ and $T(S)$ are von Neumann regular and we have required that R and S contain all possible idempotents we have that for each $r \in R$ ($s \in S$), there is an idempotent $e \in E_R$ ($f \in E_S$) and a regular element $t \in R$ ($v \in S$) such that $r = re = te$ ($s = sf = vf$). For e, we use the fact that $T(R)$ ($T(S)$) is von Neumann regular to produce an element $a \in T(R)$ ($b \in T(S)$) such that $r^2a = r$ ($s^2b = s$). Then $e = ra$ ($f = sb$) is idempotent with $r(1-e) = 0$ ($s(1-f) = 0$) and $t = r + (1-e)$ ($v = s + (1-f)$) is a regular element for which $r = te$ ($s = vf$). We will refer to e (f) as the *support idempotent* of r (s).

Examples exist to show that properties (II) and (III) of a conditional ring expectation are independent [4, Remark 2.1 (b) and (d)]. In particular, note that because of condition (II), we cannot have a conditional ring expectation from $S = R[X]$ to R since the only idempotent elements of $R[X]$ are the idempotents of R [see for example [2]]. We of course have the rather trivial R-module homomorphism from $R[X]$ to R that simply maps each polynomial to its constant term and therefore does satisfy (I) and (III).

CONDITIONAL RING EXPECTATIONS.

Our first theorem involves just the idempotents. It matches up with properties (G) and (H) above. It appeared first in [4]. We provide a proof to introduce some of the techniques that will be used later.

THEOREM 1. Let $f \in E_S$. Then there exist unique idempotents $f^\#$ and $f_\#$ in E_R such that

(i) $f_\# f = f_\#$, $f^\# f = f$
(ii) If $g \in E_R$ and $fg = f$, then $f^\# g = f^\#$
(iii) If $h \in E_R$ and $fh = h$, then $f_\# h = h$

PROOF. Let $g \in E_R$ be such that $\mathcal{E}(f)g = \mathcal{E}(f)$ and $\mathcal{E}(f)(1-g) = 0$. Since $1-g$ is in R, $\mathcal{E}(f)(1-g) = 0$ implies $\mathcal{E}(f(1-g)) = 0$. As a product of idempotents is idempotent, we have $f(1-g) = 0$ by property (III) of a conditional ring expectation. Hence $f = fg$. Now let $f^\# \in E_R$ be the idempotent for which $\mathcal{E}(f) = f^\# t$ for some regular element $t \in R$. Obviously, $\mathcal{E}(f)f^\# = \mathcal{E}(f)$. Hence $ff^\# = f$. We also have $gf^\# t = f^\# t$. As t is regular $gf^\# = f^\#$.

For $f_{\#}$ we apply what we have so far to $1-f$ and obtain the idempotent $(1-f)^{\#}$ for which $(1-f)(1-f)^{\#} = 1-f$ and $(1-f)^{\#}e = (1-f)^{\#}$ for each idempotent $e \in E_R$ such that $(1-f)e = (1-f)$. Set $f_{\#} = 1-(1-f)^{\#}$. From $(1-f)(1-f)^{\#} = 1-f$ we first get $(1-f)^{\#} - f(1-f)^{\#} = 1-f$. Then by rearranging we get $f - f(1-f)^{\#} = 1-(1-f)^{\#}$ and thus $ff_{\#} = f_{\#}$. To complete the proof let $h \in E_R$ be such that $fh = h$. Then we have $h(1-f) = 0$ in which case $(1-h)(1-f) = 1-f$. But then $(1-f)^{\#}(1-h) = (1-f)^{\#}$ which implies that $h(1-f)^{\#} = 0$ and therefore $hf_{\#} = h$.

The uniqueness of $f^{\#}$ and $f_{\#}$ is taken care of by conditions (ii) and (iii). If $e \in E_R$ is such that $ef = f$ and for each $g \in E_R$ with $gf = f$ we have $eg = e$. Then we simply apply condition (ii) to e with $g = f^{\#}$ to have $f^{\#}e = f^{\#}$ and then to $f^{\#}$ with $g = e$ to have $f^{\#}e = e$. A similar proof shows that $f_{\#}$ is unique.♦

Note that for idempotents e and f, we can set $e \leq f$ if $ef = e$. With this notion (i)-(iii) in Theorem 1 match up with the subset properties in the L^∞ setting. Also note that once we know we have a conditional ring expectation from S to R, the idempotents $f^{\#}$ and $f_{\#}$ are independent of the expectation (assuming there is more than one). Essentially $f_{\#}$ is the unique largest idempotent of R that is less than or equal to f and $f^{\#}$ is the unique smallest idempotent of R that is greater than or equal to f. Moreover, if $\mathcal{E}' : S \to R$ is a conditional ring expectation on S, then $\mathcal{E}'(f) = f^{\#}\mathcal{E}'(f)$ and $r = (1-f^{\#}) + \mathcal{E}'(f)$ is a regular element of R. We will give an example of how to construct various conditional ring expectations for a particular pair of rings $R \subset S$ [see the remarks after Examples 10 and 11].

Our next theorem corresponds to property (F) in the L^∞ setting. Theorem 1 required only properties (I) and (III) of a conditional (ring) expectation and our assumption that R and S are saturated with respect to idempotents. The same can be said about Theorem 2.

THEOREM 2. For $s \in S$ and $f \in E_S$, if $sf = s$ and $\mathcal{E}(s)f = 0$, then $\mathcal{E}(s) = 0$.

PROOF Obviously, if $f \in E_R$ and $sf = s$, then $\mathcal{E}(s) = \mathcal{E}(sf) = \mathcal{E}(s)f$. Thus we may assume f is not in R. We also have $sf^{\#} = s$ since $ff^{\#} = f$. Hence $\mathcal{E}(s) = \mathcal{E}(s)f^{\#}$. Consider $\mathcal{E}(\mathcal{E}(s)f)$. Since $\mathcal{E}(s) \in R$, we have $0 = \mathcal{E}(\mathcal{E}(s)f) = \mathcal{E}(s)\mathcal{E}(f) = \mathcal{E}(s)f^{\#}v$ for some regular element $v \in S$. Hence $\mathcal{E}(s) = \mathcal{E}(s)f^{\#} = 0$.♦

Our next result is specific to the L^∞ functions, we have not been able to find a natural way to extend it to the general setting. The basic notions expressed go back at least to S.-T. Moy, [6], where they are developed for L^1 functions on a probability space.

THEOREM 3. Let $R = L^\infty(\mathcal{A}) \subset S = L^\infty(\mathcal{B})$ with $\mathcal{A} \subset \mathcal{B}$ complete sigma-finite sigma-algebras and let $f, g \in S$.

(1) If $g(x) \geq f(x) \geq 0$ for each $x \in \mathbf{X}$, then g divides f and $\mathcal{E}(g)$ divides $\mathcal{E}(f)$.
(2) If $g(x) > 0$ on its support and g divides f, then $\mathcal{E}(g)$ divides $\mathcal{E}(f)$.

PROOF. That g divides f is essentially what makes each finitely generated ideal of S (and of R) principal. As we did above, simply set $s(x) = f(x)/g(x)$ whenever $g(x) \neq 0$ and set $s(x) = 0$ when $g(x) = 0$. Then for each x, $0 \leq s(x) \leq 1$. Hence $s \in S$ and we have $f = sg$.

Since $\mathcal{E}$ is an R-module homomorphism, $\mathcal{E}(g - f) = \mathcal{E}(g) - \mathcal{E}(f)$. As $g(x) \geq f(x) \geq 0$, $\mathcal{E}(g)(x) \geq 0$ and $\mathcal{E}(f)(x) \geq 0$ and $\mathcal{E}(g)(x) - \mathcal{E}(f)(x) = \mathcal{E}(g - f)(x) \geq 0$. Thus we have $\mathcal{E}(g)(x) \geq \mathcal{E}(f)(x) \geq 0$. As in S, this is enough to guarantee that $\mathcal{E}(g)$ divides $\mathcal{E}(f)$ as elements of R.

Now assume g divides f and $g(x) > 0$ on it support, Y. Obviously, the support of f must be contained in the support of g and we may assume f is not the zero function. We first consider the case where $f(x) \geq 0$ on Y, so it is positive on some set of positive measure. Write $f = sg$ for some $s \in S$. We may assume the support of s is contained in Y, in which case f and s have the same support. Let $b = ||s||_\infty$. Then $||f||_\infty = ||sg||_\infty \leq ||s||_\infty ||g||_\infty = b||g||_\infty$. Consider the function f/b. Since b is a constant $||f/b||_\infty = ||f||_\infty/b \leq ||g||_\infty$. Hence $f/b \in S$. Now $b \geq s(x) \geq 0$ for all x. Hence $f(x)/b \leq f(x)/s(x) = g(x)$ whenever $s(x)$ is not zero. Thus $0 \leq f(x)/b \leq g(x)$ for all x. By (1), g divides f/b and $\mathcal{E}(g)$ divides $\mathcal{E}(f/b)$. As b is a positive constant, we have $\mathcal{E}(f/b) = \mathcal{E}(f)/b$. Write $\mathcal{E}(f)/b = r\mathcal{E}(g)$ for some $r \in R$. Then we have $\mathcal{E}(f) = br\mathcal{E}(g)$ with $br \in R$ since b is a constant.

For arbitrary f, simply note that we can split f into its positive and negative parts, f^+ and f^- respectively. Let W denote the support of f^+ and let Z denote the support of f^-. Then $f^+ = \mathcal{X}_W f$ and $f^- = \mathcal{X}_Z f$. So g divides both f^+ and f^-. From above we have that $\mathcal{E}(g)$ divides both $\mathcal{E}(f^+)$ and $\mathcal{E}(-f^-) = -\mathcal{E}(f^-)$. It follows that $\mathcal{E}(g)$ divides $\mathcal{E}(f) = \mathcal{E}(f^+) + \mathcal{E}(f^-)$.♦

An easy corollary to Theorem 3 is the following result about principal ideals in the L^∞ setting. In general it is not known under what circumstances this result can be extended to conditional ring expectations. Later we give an example to show that the image of a principal ideal need not be principal.

COROLLARY 4. Let $R = L^\infty(\mathcal{A}) \subset S = L^\infty(\mathcal{B})$ with $\mathcal{A} \subset \mathcal{B}$ complete sigma-finite sigma-algebras. If I is a principal ideal of S, then there is an element $b \in I$ such that $bS = I$ and $\mathcal{E}(I) = \mathcal{E}(b)R$.

PROOF. Let $b \in I$ be such that $bS = I$. Then $sgn(b)$ is in S and we have $sgn(b)b(x) \geq 0$ for each x. Obviously $(sgn(b))^2 b = b$. Hence $sgn(b)bS = bS = I$. Thus we may assume $b(x) \geq 0$ for each x. By Theorem 3, each element $f \in I$ is such that $\mathcal{E}(b)$ divides $\mathcal{E}(f)$. Hence, we have $\mathcal{E}(I) = \mathcal{E}(b)R$.♦

LOCALIZING SETS AND IDEMPOTENTS.

We revisit out original example of a conditional expectation. Recall the setting: μ represents Lebesgue measure on the interval $[-1,1]$, $\mathcal{B}$ is the corresponding set of Lebesgue measurable subsets of $[-1,1]$ and $\mathcal{A}$ is the subset of $\mathcal{B}$ consisting of those Lebesgue measurable subsets of $[-1,1]$ which are symmetric with the origin. For $f \in L^\infty(\mathcal{B})$, $\mathcal{E}(f) = f_e$, the even part of f. Consider the set $Y = [0,1]$. For each $f \in L^\infty(\mathcal{A})$, if we know the value of $f(x)$ for each $x \in [0,1]$, then we know it for each $x \in [-1,1]$ since each function in $L^\infty(\mathcal{A})$ is even. Moreover, ignoring the value of $f(0)$ (which is perfectly okay since finite sets have measure zero), we know how to build a $\mathcal{B}$-measurable function g to get an (essentially) odd function for which $g(x) = f(x)$ when $x > 0$ and $g(x) = -f(x)$ when $x < 0$. Thus for each $h \in L^\infty(\mathcal{B})$, with $h(x) = 0$ for each $x \in [-1,0]$, there is a function $f \in L^\infty(\mathcal{A})$ for which $f(x) = h(x)$ for each $x \in [0,1]$ and there is a function $j \in L^\infty(\mathcal{B})$ with $j(x) = h(x)$ for each $x \in [0,1]$ and $j(x) = -h(-x)$ for each $x \in [-1,0)$. Hence when we consider the product of an arbitrary function r in $L^\infty(\mathcal{B})$ with the characteristic function for the set $[0,1]$, it is impossible to tell whether r is in $L^\infty(\mathcal{A})$ or in the kernel of $\mathcal{E}$, or in neither. In this setting the set $[0,1]$ is referred to as a "localizing set" for the pair $\mathcal{A}$ and $\mathcal{B}$, or is simply said to "localize" $\mathcal{A}$ [3, page 112]. This example is due to A. Lambert and B. Weinstock [5], a similar one can be found in [1].

More generally we have the following for a comparable pair of complete sigma-finite sigma-algebras $\mathcal{A} \subset \mathcal{B}$, a nonempty subset Y in $\mathcal{B}$ is said to be a *localizing set* for the pair $\mathcal{A}$ and $\mathcal{B}$ if for each set $Z \in \mathcal{B}$, there is a set $W \in \mathcal{A}$ such that $Y \bigcap Z = Y \bigcap W$ [3, page 112]. Obviously, one may restrict the definition to only considering subsets of Y and simply say that Y is a localizing set if for each $Z \subseteq Y$ there is a set $W \in \mathcal{A}$ such that $Z = Y \bigcap W$. In [3]. Lambert gives several ways of characterizing when a set is a localizing set, several of these were later generalized in [1]. We collect some of those characterizations in the next theorem, and give a further generalization to the L^∞ case for spaces of infinite measure. Note that all five of the statements in Theorem 5 are known to be equivalent for spaces of finite measure (see [3] and [1]).

THEOREM 5. Let $R = L^\infty(\mathcal{A}) \subseteq S = L^\infty(\mathcal{B})$ with $\mathcal{A} \subseteq \mathcal{B}$ complete sigma-finite sigma-algebras and let Y be a nonempty member of $\mathcal{B}$. Then the following are equivalent.

(1) Y is a localizing set for $\mathcal{A}$ and $\mathcal{B}$.
(2) For each $Z \subseteq Y$, $Z^\# \bigcap Y = Z$.
(3) Each subset $Z \in \mathcal{B}$ of Y with positive measure is a localizing set for $\mathcal{A}$ and $\mathcal{B}$.
(4) For each function $f \in S$ for which $f\mathcal{X}_Y = f$, $\mathcal{E}(f)\mathcal{X}_Y = f\mathcal{E}(\mathcal{X}_Y)$.
(5) $\mathcal{X}_Y S = \mathcal{X}_Y R$.
(6) For each $f \in S$ there is a function $g \in R$ such that $f\mathcal{X}_Y = g\mathcal{X}_Y$.

PROOF. The equivalence of (1) through (3) is due to Lambert [3, Proposition 1.4]. Lambert also proved that Y is a localizing set if and only if for each function $f \in \bigcup_{1 \leq p < \infty} L^p(\mathcal{B})$, $\mathcal{E}(f)\mathcal{X}_Y = f\mathcal{E}(\mathcal{X}_Y)$. Obviously, (5) and (6) are equivalent. Also (6) implies (1) since for each set $Y \in \mathcal{B}$, $\mathcal{X}_Z\mathcal{X}_Y = g\mathcal{X}_Y$ for some $g \in R$ implies that the support of g must intersect Y in the set $Z \bigcap Y$. Corollary 1.6 of [1] establishes that (1) implies (6) in the (somewhat) restricted case that $f \in S$ is also integrable on the set Y.

[(1) $\rightarrow$ (4)] Assume Y is a localizing set for $\mathcal{A}$ and $\mathcal{B}$ and let $f \in S$ be such that $f\mathcal{X}_Y = f$; i.e., $f \in \mathcal{X}_Y S$. By Theorem 3, $\mathcal{E}(\mathcal{X}_Y)$ divides $\mathcal{E}(f)$. Thus we may write $\mathcal{E}(f) = r\mathcal{E}(\mathcal{X}_Y) = \mathcal{E}(r\mathcal{X}_Y)$ for some $r \in R$. Let Z be the support of f. Then $Z^\#$ is the support of $\mathcal{E}(\mathcal{X}_Z)$, $\mathcal{E}(f)\mathcal{X}_{Z\#} = \mathcal{E}(f\mathcal{X}_{Z\#}) = \mathcal{E}(f)$ (since $Z \subseteq Z^\#$) and $\mathcal{X}_Y\mathcal{X}_{Z\#} = \mathcal{X}_Z$ (since Y is a localizing set). It follows that $\mathcal{E}(f) = r\mathcal{E}(\mathcal{X}_Y)\mathcal{X}_{Z\#} = r\mathcal{E}(\mathcal{X}_Y\mathcal{X}_{Z\#}) = r\mathcal{E}(\mathcal{X}_Z) = \mathcal{E}(r\mathcal{X}_Z)$. Thus $\mathcal{E}(f)\mathcal{X}_Y = r\mathcal{E}(\mathcal{X}_Z)\mathcal{X}_Y = r\mathcal{X}_Z\mathcal{E}(\mathcal{X}_Y)$. It suffices to show that $r\mathcal{X}_Z = f$.

We at least have $\mathcal{E}(f - r\mathcal{X}_Z) = 0$. By property (C) of a conditional expectation, we can show equality of f and $r\mathcal{X}_Z$ by simply showing that for each characteristic function $\mathcal{X}_W$ with $W \in \mathcal{B}$, $\mathcal{E}((f - r\mathcal{X}_Z)\mathcal{X}_W) = 0$. Of course, as the support (if any) of $f - r\mathcal{X}_Z$ is contained in Z, we need only show this for those W which are subsets of Z. So let $W \in \mathcal{B}$ be a subset of Z and consider $\mathcal{E}((f - r\mathcal{X}_Z)\mathcal{X}_W)$. By (3), Z is also a localizing set so we have $f\mathcal{X}_W = fZ\mathcal{X}_{W\#} = f\mathcal{X}_{W\#}$ and $r\mathcal{X}_Z\mathcal{X}_W = r\mathcal{X}_Z\mathcal{X}_{W\#}$. Hence $\mathcal{E}((f - r\mathcal{X}_Z)\mathcal{X}_W) = \mathcal{E}(f\mathcal{X}_{W\#} - r\mathcal{X}_Z\mathcal{X}_{W\#}) = \mathcal{E}(f - r\mathcal{X}_Z)\mathcal{X}_{W\#} = 0$ as desired. Therefore $\mathcal{E}(f)\mathcal{X}_Y = f\mathcal{E}(\mathcal{X}_Y)$.

To complete the proof we show that (4) implies (6).

[(4) $\rightarrow$ (6) (& (5))] Assume the statement in (4) holds and let $f \in S$. Since we only need $f\mathcal{X}_Y = g\mathcal{X}_Y$ for some $g \in R$, we may assume $f\mathcal{X}_Y = f$, Thus we have $\mathcal{E}(f)\mathcal{X}_Y = f\mathcal{E}(\mathcal{X}_Y)$. By Theorem 3, we also have $\mathcal{E}(f) = r\mathcal{E}(\mathcal{X}_Y)$ for some $r \in R$. Since the support of $\mathcal{E}(f)$ is contained in $Y^\#$, $\mathcal{E}(f)\mathcal{X}_{Y\#} = \mathcal{E}(f)$. Moreover, without loss of generality we may assume $r\mathcal{X}_{Y\#} = r$, and therefore $r(1 - \mathcal{X}_{Y\#}) = 0$. Let $t = (1 - \mathcal{X}_{Y\#}) + \mathcal{E}(\mathcal{X}_Y)$. This is a regular element of R and we have $\mathcal{E}(f) = rt$. Moreover, $f\mathcal{E}(\mathcal{X}_Y) = ft$. Thus $\mathcal{E}(f)/t = r \in R$ and $r\mathcal{X}_Y = f$.♦

Returning to the interval $[-1, 1]$ it is easy to see that any set W of positive measure which has the property that $x \in W$ implies $-x$ is not in W (for $x \neq 0$, of course) will be a localizing set. There are fairly simple examples of pairs $\mathcal{A} \subset \mathcal{B}$ which have no localizing sets. One such example is to let $\mathcal{B}$ denote the (two-dimensional) Lebesgue measurable subsets of the unit square $[0, 1] \times [0, 1]$ and let $\mathcal{A}$ denote the sets of the form $A \times [0, 1]$ for (one-dimensional) Lebesgue measurable subsets A of $[0, 1]$. In this case, $Y^\#$ is the set consisting of the union of the "vertical stripes" $\{(x, y) \mid 0 \leq y \leq 1\}$ where x is such that the set $\{z \in [0, 1] \mid (x, z) \in Y\}$ has positive (one-dimensional) Lebesgue measure [see [1, Example 2.1]].

The name "localizing set" turns out to be linked to the ring theoretic notion of localization. We will define it in terms of idempotents in such a way to match up with the set theoretic definition in terms of intersections given above in the measure theoretic setting. Specifically, we say that a nonzero idempotent $f \in E_S$ *localizes* R and S if for each idempotent $g \in E_S$ there is an idempotent $e \in E_R$ such that $gf = ef$. As above, it suffices to restrict to those idempotents $g \in E_S$ for which $gf = g$ (so $g \leq f$).

We need to set a little notation before we give a ring theoretic characterization of localizing idempotents.

For an idempotent $f \in E_S$, let $t = (1 - f^{\#}) + \mathcal{E}(f)$, the "canonical" regular element of R for which $\mathcal{E}(f) = tf^{\#}$. Set $F = \{1, t, t^2, \dots\}$ and form the rings R_F and S_F. Now define a map $\mathcal{E}_f : fS_F \to fR_F$ by setting $\mathcal{E}_f(s/t^n) = f\mathcal{E}(s)/t^{n+1}$ [for $n \geq 0$].

THEOREM 6. For $f \in E_S$, the map $\mathcal{E}_f$ is an R-module homomorphism such that $\mathcal{E}_f^2 = \mathcal{E}_f$. Moreover, $\mathcal{E}_f(s/t^n) = 0$ if and only if $\mathcal{E}(s) = 0$.

PROOF. Let s/t^n and u/t^m be elements of fS_F and let a and b be elements of R. That $\mathcal{E}_f$ is well defined follows from the fact that t is a regular element of R. [So, in particular, $\mathcal{E}_f(st^k/t^n) = f\mathcal{E}(st^k)/t^{n+1} = ft^k\mathcal{E}(s)/t^{n+1} = f\mathcal{E}(s)/t^{n-k+1} = \mathcal{E}_f(s/t^{n-k})$ for $0 \leq k \leq n$.] Consider $\mathcal{E}_f((as/t^n)+(bu/t^m))$. Without loss of generality we may assume $0 \leq n \leq m$. Thus $(as/t^n) + (bu/t^m) = (ast^{m-n} + bu)/t^m$ and we have $\mathcal{E}_f((as/t^n) + (bu/t^m)) = \mathcal{E}_f((ast^{m-n} + bu)/t^m) = f\mathcal{E}(ast^{m-n} + bu)/t^{m+1} = (fat^{n-m}\mathcal{E}(s) + fb\mathcal{E}(u))/t^{m+1} = fa\mathcal{E}_f(s/t^n) + fb\mathcal{E}_f(b/t^m)$. Hence $\mathcal{E}_f$ is an R-module homomorphism.

To see that $E_f^2 = E_f$, start with $\mathcal{E}_f^2(s/t^n) = \mathcal{E}_f(\mathcal{E}_f(s/t^n))$. Now, by the definition of $\mathcal{E}_f$, we have $\mathcal{E}_f(\mathcal{E}_f(s/t^n)) = \mathcal{E}_f(f\mathcal{E}(s)/t^{n+1}) = f\mathcal{E}(f\mathcal{E}(s))/t^{n+2} = f\mathcal{E}(s)\mathcal{E}(f)/t^{n+2} = f\mathcal{E}(s)/t^{n+1}$ since $t = (1-f^{\#})+\mathcal{E}(f)$ and $\mathcal{E}(f)(1-f^{\#}) = 0$. Thus $\mathcal{E}_f^2 = \mathcal{E}_f$.

For the final conclusion simply recall that one of the properties of a conditional ring expectation is that if an element $v \in S$ is such that $ev = v$ and $e\mathcal{E}(v) = 0$ for some idempotent $e \in E_S$, then $\mathcal{E}(v) = 0$.♦

For each idempotent $g \in E_S$, let $J_g = \{x \in ker(\mathcal{E}) | xg = 0\}$ and $I_g = \{x \in ker(\mathcal{E}) | \mathcal{E}(xg) = 0\}$. Obviously, if $x \in J_g$ implies $x \in I_g$. For localizing idempotents we can say even more. Many of these statements are analogous to statements in the L^∞ and/or L^p setting [in particular, see [1]], but several appear here for the first time.

THEOREM 7. Let R, S, $\mathcal{E}$, f, F and $\mathcal{E}_f$ be as above. Then the following are equivalent:

(1) f localizes R and S.
(2) For each $g \in E_S$, $gf = g$ implies $g = g^{\#}f$.
(3) For each $g \in E_S$, $gf = g$ implies $\mathcal{E}(g) = g^{\#}\mathcal{E}(f)$.

(4) For each $g \in E_S$, $gf = g$ implies $\mathcal{E}(g)\mathcal{E}(f) = \mathcal{E}(g)^2$.
(5) $J_{1-f} = (0)$
(6) $J_f = I_f$
(7) For each pair $s, u \in fS$, $\mathcal{E}(s)u = s\mathcal{E}(u)$.
(8) For each $s \in fS$, $\mathcal{E}(s)f = s\mathcal{E}(f)$.
(9) For each pair of elements $s, u \in fS$, $\mathcal{E}(su)\mathcal{E}(f) = \mathcal{E}(s)\mathcal{E}(u)$.
(10) For each $s \in fS$, $\mathcal{E}(s)^2 = \mathcal{E}(s^2)\mathcal{E}(f)$.
(11) For each $s \in fS$ and each integer $n \geq 2$, $\mathcal{E}(s)^n = \mathcal{E}(s^n)\mathcal{E}(f)^{n-1}$.
(12) For each integer $n > 1$ and each product $\prod s_i$ of n elements of fS, $\prod \mathcal{E}(s_i) = \mathcal{E}(\prod s_i)\mathcal{E}(f)^{n-1}$.
(13) For each integer $n > 0$ and each $s \in fS$, $s\mathcal{E}(s)^n = f\mathcal{E}(s^{n+1})\mathcal{E}(f)^{n-1}$, with $\mathcal{E}(f)^0 = f^{\#}$.
(14) $fS_F = fR_F$.
(15) $\mathcal{E}_f$ injective.
(16) For each $x \in fS_F$, $\mathcal{E}_f(x)f = x\mathcal{E}_f(f)$.
(17) The map $\alpha : f^{\#}R_F \to fS_F$ defined by $\alpha(f^{\#}r) = fr$ is surjective (moreover, it is a ring isomorphism)

PROOF. With 17 statements, the scheme for the proof is a bit complicated. Basically what we will do is show (1) through (4) are equivalent, establish a large loop of successive implications (4) $\to$ (3) $\to$ (5) $\to$ (6) $\to$ (7) $\to$ (8) $\to$ (9) $\to$ (12) $\to$ (11) $\to$ (10) $\to$ (4) (not quite in this order, for example, we actually show (5) $\leftarrow$ (6) and (5) $\to$ (7)), two small loops of implications (8) $\to$ (13) $\to$ (3) $\to$ (8) and (8) $\to$ (14) $\to$ (15) $\to$ (5) $\to$ (8) (the latter implication in each small loop by way of the large loop), and two simple equivalences (8) $\leftarrow$ (16) and (15) $\leftarrow$ (17). In some instances an equivalence will be verified rather than only an implication in one direction.

A few of the implications are obvious. In particular (7) implies (8), (12) implies (11), (11) implies (10), (10) implies (4), and (9) implies (10). All but (9) directly implies (10) will be used in our proof.

[(1) $\leftarrow$ (2)] Let $g \in E_S$. Obviously, $gf = ef$ for some idempotent $e \in E_R$ if and only if $(gf)f = ef$. Hence the statement in (1) is equivalent to the statement that for each idempotent $g \in fS$ there is an idempotent $e \in E_R$ such that $g = ef$. Thus we have (2) implies (1). For the reverse implication we simply make use of Theorem 1. Write $g = ef$ for some idempotent $e \in E_R$. Then $g = ef = efg = eg$. Thus we must have $eg^{\#} = g^{\#}$. The result now follows from the fact that $gg^{\#} = g$.

[(2) $\to$ (3) $\to$ (4) $\to$ (3) $\to$ (2)] If $g = g^{\#}f$ for each $g \leq f$, then we also have $\mathcal{E}(g) = \mathcal{E}(g^{\#}f) = g^{\#}(f)$. So (2) implies (3). Now multiply both sides of this equation by g to $g\mathcal{E}(g) = gg^{\#}\mathcal{E}(f)$ and apply $\mathcal{E}$ to both sides. As $\mathcal{E}(g)$ and $\mathcal{E}(f)$ are in R and $gg^{\#} = g$, we obtain $\mathcal{E}(g)^2 = \mathcal{E}(g)\mathcal{E}(f)$. So (3) implies (4). In this equation we may replace $\mathcal{E}(g)$ by $rg^{\#}$ for some regular element $r \in R$ to obtain $rg^{\#}\mathcal{E}(g) = rg^{\#}\mathcal{E}(f)$. As r is regular we may cancel it from both sides to have $\mathcal{E}(g) = g^{\#}\mathcal{E}(g) = g^{\#}\mathcal{E}(f)$. Hence (4) implies

(3). Finally we start with the equation $\mathcal{E}(g) = g^{\#}\mathcal{E}(f)$ and consider the idempotent $e = g^{\#}f - g$. Now $\mathcal{E}(e) = \mathcal{E}(g^{\#}f - g) = g^{\#}\mathcal{E}(f) - \mathcal{E}(g) = 0$ and therefore $e = 0$ by condition (III) of a conditional ring expectation.

[(3) $\to$ (5)] Assume $g = g^{\#}f$ for each idempotent $g \leq f$ and let $s \in J_{1-f}$. Thus $\mathcal{E}(s) = 0$ and $s(1-f) = 0$, so $s = sf$. Since $s = sf$ it suffices to show that $\mathcal{E}(gs) = 0$ for each idempotent $g \leq f$. For such an idempotent g we have $sg = sfg^{\#} = sg^{\#}$. Hence $\mathcal{E}(sg) = g^{\#}\mathcal{E}(s) = 0$ since $g^{\#}$ is in R. Therefore $s = 0$ by condition (II) of a conditional ring expectation.

[(5) $\leftarrow$ (6)] First note that $J_g \subseteq I_g$ for each idempotent $g \in E_S$. If $J_{1-f} = 0$ and $s \in I_f$, we have $\mathcal{E}(s) = 0$ and $\mathcal{E}(sf) = 0$. As $sf(1-f) = 0$, sf must be in J_{1-f}. But then $sf = 0$ which implies $s \in J_f$. On the other hand, if $I_f = J_f$ and $u \in J_{1-f}$ we have $\mathcal{E}(u) = 0$ and $u(1-f) = 0$. But the latter implies that $u = uf$ and we have $u \in I_f$. Hence $u = uf = 0$ as desired.

[(5) $\to$ (7)] Assume $J_{1-f} = (0)$ and let $s, u \in fS$. Consider the element $\mathcal{E}(s)u - s\mathcal{E}(u)$. This is obviously an element of fS and therefore it is annihilated by $(1-f)$. But since $\mathcal{E}(a\mathcal{E}(b)) = \mathcal{E}(a)\mathcal{E}(b)$ for each pair $a, b \in S$, we also have that $\mathcal{E}(\mathcal{E}(s)u - s\mathcal{E}(u)) = 0$. Hence $\mathcal{E}(s)u - s\mathcal{E}(u)$ is in J_{1-f} and therefore $\mathcal{E}(s)u = s\mathcal{E}(u)$.

[(8) $\to$ (9)] Assume that $\mathcal{E}(s)f = s\mathcal{E}(f)$ for each $s \in fS$ and let $s, u \in fS$. Then we have $\mathcal{E}(s)f = s\mathcal{E}(f)$. Now simply multiply both sides by u and apply $\mathcal{E}$ to get $\mathcal{E}(s)\mathcal{E}(u) = \mathcal{E}(su)\mathcal{E}(f)$ since $\mathcal{E}(\mathcal{E}(s)u) = \mathcal{E}(s)\mathcal{E}(u)$ and $uf = u$.

[(9) $\to$ (12)] Assume $\mathcal{E}(su)\mathcal{E}(f) = \mathcal{E}(s)\mathcal{E}(u)$ for each pair $s, u \in fS$. By way of induction assume that the statement in (12) holds for all products of m elements of fS when $2 \leq m \leq n-1$ and let $s_1, s_2, \ldots s_n \in fS$ be n not necessarily distinct elements of fS. We have $\mathcal{E}(s_1(s_2s_3\cdots s_n))\mathcal{E}(f) = \mathcal{E}(s_1)\mathcal{E}(s_2s_3\cdots s_n)$. Replace $\mathcal{E}(s_2s_3\cdots s_n)$ by $\mathcal{E}(s_2)\mathcal{E}(s_3)\cdots\mathcal{E}(s_n)\mathcal{E}(f)^{n-2}$ to obtain the desired equality $\mathcal{E}(s_1s_2\cdots s_n)\mathcal{E}(f)^{n-1} = \mathcal{E}(s_1)\mathcal{E}(s_2)\cdots\mathcal{E}(s_n)$. Combined with the "obvious" implications given above, this completes the "large" loop.

[(8) $\to$ (13)] Let $s \in fS$ and assume $\mathcal{E}(u)f = u\mathcal{E}(f)$. for each $u \in fS$. First we show $s\mathcal{E}(u) = f\mathcal{E}(su)$ for each $u \in fS$. We have $\mathcal{E}(u)f = u\mathcal{E}(f)$ and $\mathcal{E}(su)f = su\mathcal{E}(f)$. Multiply both sides of the former by s and note that $sf = s$ so we have $s\mathcal{E}(u) = s\mathcal{E}(u)f = su\mathcal{E}(f) = \mathcal{E}(su)f$. This takes care of the case $n = 1$. We complete the proof using induction on n. Assume the result holds for each positive integer $m < n$. Then for $m = n-1$ we have $s\mathcal{E}(s)^{n-1} = f\mathcal{E}(s^n)\mathcal{E}(f)^{n-2}$. Applying $\mathcal{E}$ to both sides yields $\mathcal{E}(s)^n = \mathcal{E}(s^n)\mathcal{E}(f)^{n-1}$. Now multiply both sides by s and apply the general rule above that $s\mathcal{E}(u) = f\mathcal{E}(su)$ with $u = s^n$ to obtain the desired equality $s\mathcal{E}(s)^n = s\mathcal{E}(s^n)\mathcal{E}(f)^{n-1} = f\mathcal{E}(s^{n+1})\mathcal{E}(f)^{n-1}$.

[(13) $\to$ (3)&(4)] Assume that for each positive integer n and each element $s \in fS$, $s\mathcal{E}(s)^n = f\mathcal{E}(s^{n+1})\mathcal{E}(f)^{n-1}$ (with $\mathcal{E}(f)^0 = f^{\#}$). Let $g \in fS$ be idempotent. Then we have $g\mathcal{E}(g) = f\mathcal{E}(g^2) = f\mathcal{E}(g)$. Now either apply $\mathcal{E}$ and rewrite to obtain the equation in statement (4) or do as in the proof of (4) implies (3) and replace $E(g)$ by $rg^{\#}$ for some regular element $r \in R$, then cancel r which leaves us with $g = gg^{\#} = fg^{\#}$. This is enough to close the first small loop.

[(8) $\to$ (14)] Assume $\mathcal{E}(u)f = u\mathcal{E}(f)$ for each $u \in fS$ and let s/t^n be an element of fS_f. Since $\mathcal{E}(f) = tf^{\#}$, we have $\mathcal{E}(s)f/t = sf^{\#} = sff^{\#} = sf = s$. Thus $s/t^n = \mathcal{E}(s)f/t^{n+1} \in fR_F$.

[(14) $\to$ (15)] Assume $fS_F = fR_F$. Then for each $s/t^n \in fS_F$ there is an element $r \in R$ and a nonnegative integer m such that $s/t^n = rf/t^m$. As elements of R factor out of $\mathcal{E}_f$, we have $\mathcal{E}_f(s/t^n) = \mathcal{E}_f(rf/t^m) = rf\mathcal{E}(f)/t^{m+1} = rftf^{\#}/t^{m+1} = rff^{\#}/t^m = rf/t^m = s/t^n$, in other words $\mathcal{E}_f$ is the identity function and so it is (trivially) injective.

[(15) $\leftarrow$ (5)] If $\mathcal{E}_f$ is injective and $s \in J_{1-f}$. Then $\mathcal{E}(s) = 0$ and $s(1-f) = 0$, so $s = sf$. But this implies $\mathcal{E}_f(s) = 0$ and therefore $s = 0$. On the other hand, if $J_{1-f} = (0)$ and $s/t^n \in fS_F$ is in the kernel of $\mathcal{E}_f$. Then we have $s(1-f) = 0$ and $\mathcal{E}(s) = 0$ which together imply $s = 0$. We now have the second small loop closed, so all that are left are the equivalences.

[(8) $\to$ (16)] Assume $\mathcal{E}(u)f = u\mathcal{E}(f)$ for each $u \in fS$ and let $x = s/t^n$ be an element of fS_F. Then $\mathcal{E}(s)f = s\mathcal{E}(f)$ and therefore $\mathcal{E}(s)f/t^{n+1} = s\mathcal{E}(f)/t^{n+1} = sf\mathcal{E}(f)/t^{n+1}$. The left hand side of this equation is $\mathcal{E}_f(x)f$ and the right hand side is $x\mathcal{E}_f(f)$.

[(16) $\to$ (8)] Assume $\mathcal{E}_f(x)f = x\mathcal{E}_f(f)$ for each $x \in fS_F$ and let $s \in fS$. Then we have $\mathcal{E}_f(s)f = s\mathcal{E}_f(f)$, or in terms of values of $\mathcal{E}$, $f\mathcal{E}(s)/t = s\mathcal{E}(f)/t$. As t is a regular element of R, we have $\mathcal{E}(s)f = s\mathcal{E}(f)$.

[(15) $\leftarrow$ (17)] First note that no matter whether f is a localizing idempotent or not, α is simply the function from $f^{\#}R_F$ into fS_F that is defined by multiplication by f. Thus the image of α is simply fR_F and α is a ring homomorphism from the ring $f^{\#}R_F$ (whose identity element is $f^{\#}$) onto the ring fR_F (whose identity is f). Hence α is surjective if and only if $fS_F = fR_F$. Moreover, no matter whether α is surjective or not, it is always injective for if $\alpha(f^{\#}u) = 0$ for some $u = r/t^n \in R$, then $rf = 0$. But then we have $0 = \mathcal{E}(rf) = r\mathcal{E}(f) = rf^{\#}t$. As t is a regular element of R, $0 = rf^{\#} = uf^{\#}$.♦

Recall from above that principal ideals in the L^∞ setting get mapped to principal ideals. Thus we have $\mathcal{E}(\mathcal{X}_Y)L^\infty(\mathcal{A}) = \mathcal{E}(\mathcal{X}_Y L^\infty(\mathcal{B}))$ for each $Y \in \mathcal{B}$. From this we see that $\mathcal{E}(\mathcal{X}_Y L^\infty(\mathcal{B}))^2 = \mathcal{E}(\mathcal{X}_Y)\mathcal{E}(\mathcal{X}_Y L^\infty(\mathcal{B}))$ no matter whether Y is a localizing set for $\mathcal{A}$ and $\mathcal{B}$ or not. In the general setting we do no know whether $\mathcal{E}(f)R = \mathcal{E}(fS)$ for each idempotent $f \in E_S$, but it is the case that $\mathcal{E}(fS)^2$ is equal to $\mathcal{E}(f)\mathcal{E}(fS)$ if f is a localizing idempotent for R and S. This most easily follows from statement (9) in Theorem 7 using the fact that $\mathcal{E}(fS)^2$ is generated by the elements of the form $\mathcal{E}(s)\mathcal{E}(u)$ for pairs of elements $s, u \in fS$.

As $f^{\#}$ is defined independently from any particular conditional ring expectation from S to R, all that is needed is the existence of at least one such map $\mathcal{E}$ to have all of the conditions above pertaining to $\mathcal{E}$ hold for each conditional ring expectation from S to R.

COROLLARY 8. For $f \in E_S$, if there is exists a conditional expectation $\mathcal{E} : S \to R$ such that $\mathcal{E}(s)f = s\mathcal{E}(f)$ for all $s = sf \in S$, then $\mathcal{E}'(s)f = s\mathcal{E}'(f)$ for all conditional expectations $\mathcal{E}' : S \to R$.

COROLLARY 9. For a nonzero idempotent $f \in E_S$, $fS = fR$ if and only if $\mathcal{E}(fS) = \mathcal{E}(f)R$ and f is a localizing idempotent for the pair $R \subset S$.

PROOF. Obviously, if $fS = fR$, then we have $fS_F = fR_F$ and $\mathcal{E}(fS) = \mathcal{E}(fR) = \mathcal{E}(f)R$. On the other hand, if $\mathcal{E}(fS) = \mathcal{E}(f)R$ and f is a localizing idempotent, then for each $s \in fS$ we have $\mathcal{E}(s)f = s\mathcal{E}(f)$ and $\mathcal{E}(s) = \mathcal{E}(f)r$ for some $r \in R$. Simply replace $\mathcal{E}(s)$ by $\mathcal{E}(f)r$ to obtain $s\mathcal{E}(f) = r\mathcal{E}(f)$. We may substitute $tf^{\#}$ for $\mathcal{E}(f)$ (with $t = (1 - f^{\#}) + \mathcal{E}(f)$ and then cancel the ts since to have $sf^{\#} = rf^{\#}$. As $s = sf$ and $ff^{\#} = f$, $rf^{\#} = rf$ and we have $s = rf$. ♦

EXAMPLES.

For the L^∞ setting, if a localizing set exists for $\mathcal{A}$ and $\mathcal{B}$, then maximal localizing sets exist. Essentially this is due to the fact that the union of a countable ascending chain of localizing sets turns out to be a localizing set which is in $\mathcal{B}$.

Our first two examples are based on comparable pairs of subrings of the rings $Q = \prod Q_n$ and $T = \prod R_n$ where $Q_n = \mathbb{R}^n$ and R_n is the diagonal set of Q_n. Elements will be represented in the form $q = (q_1, q_2, q_3, \ldots)$ with $q_n = (q_{n,1}, q_{n,2}, \ldots, q_{n,n}) \in Q_n$ for each n. Define $\mathcal{E} : S \to R$ by $\mathcal{E}(s) = r$ where $r_{n,j} = \sum s_{n,i}/n$ for each n and $1 \le j \le n$. In both examples, principal ideals get mapped to principal ideals. Each localizing idempotent is dominated by a maximal localizing idempotent in the first while in the second there are no maximal localizing idempotents at all.

EXAMPLE 10. Let $Q = \prod Q_n$ and $T = \prod R_n$ where $Q_n = \mathbb{R}^n$ and R_n is the diagonal set of Q_n. Finally let S and R be the bounded sequences in Q and T, respectively. Then

(a) Localizing idempotents exist, and each is dominated by a maximal localizing idempotent.
(b) If f is a localizing idempotent, then $fS = fR$.
(c) For each principal ideal I of S there is a generator s such that $\mathcal{E}(I) = \mathcal{E}(s)R$.

PROOF. Since $\mathcal{E}$ is the composite of the simple averaging function from each Q_n to the corresponding R_n, it will satisfy the properties of a conditional ring expectation. Moreover, if $s_{n,j} \ge 0$ for each pair (n, j), then the same is true for $\mathcal{E}(s)_{n,j}$. If, in addition, some $s_{m,i}$ is positive for some fixed, but arbitrary, pair (m, i), then $\mathcal{E}(s)_{m,k} > 0$ for each pair (m, k).

Let $f = (f_1, f_2, f_3, \dots) \in S$ be an idempotent. The corresponding idempotent $f^{\#}$ is simply the idempotent of R where the nth component $(f^{\#})_n$ is $(1, 1, \dots, 1)$ if some $f_{n,i}$ is nonzero and is $(0, 0, \dots, 0)$ if each $f_{n,i}$ is zero. Suppose some f_n has more than one component that is not zero, say it has $k+1(>1)$ nonzero components. Without loss of generality we may assume $fn, 1 = 1$. Consider the element $s_n = (-1, 1/k, 1/k, \dots, 1/k)f_n \in Q_n$ and let $s = (s_1, s_2, \dots)$ be such that $s_m = (0, 0, \dots, 0)$ for each $m \neq n$. Obviously, $s(1-f) = 0$ and $\mathcal{E}(s) = 0$ but $s \neq 0$. Thus for f to be a localizing idempotent it must be that at most one component of each f_n is nonzero. Due to the structure of $g^{\#}$ for each idempotent $g \in S$, it is easy to see that $g = g^{\#} f$ for each $g \leq f$ when f has this form. If no f_n is $(0, 0, \dots, 0)$, then f is a maximal localizing idempotent. Otherwise, simply replace each $f_n = (0, 0, 0 \dots, 0)$ with $(1, 0, 0 \dots, 0)$ and leave the other components of f alone to find a maximal localizing idempotent that dominates f.

Now assume we have a localizing idempotent f and let $s \in fS$ with $r = \mathcal{E}(s)$. Let f_{n,k_n} be the nonzero component of f_n (if f_n has one). Then s_{n,k_n} can be nonzero but $s_{n,i} = 0$ for $i \neq k_n$. No matter whether s_{n,k_n} is nonzero or not, $r_n = (s_{n,k_n}/n, \dots, s_{n,k_n}/n) = (s_{n,k_n}, \dots, s_{n,k_n})(1/n, \dots, 1/n)$. It follows that $\mathcal{E}(s) = r'\mathcal{E}(f)$ where $r'_n = (s_{n,k_n}, s_{n,k_n}, \dots, s_{n,k_n})$. Since $s \in S$, the sequence $\{s_{n,k_n}\}$ is bounded and therefore r' is in R. Hence $\mathcal{E}(fS) = \mathcal{E}(f)R$. Moreover, we have $fS = fR$.

Let sS be a principal ideal of S and let e be the support idempotent of s (i.e., $es = s$ and $s + (1-e)$ is a regular element of S). As with real valued functions, we may split each element of S into its zero part and its positive and negative parts. Split e into e^+ and e^- where $e^+_{n,k} = 1$ if and only if $s_{n,k} > 0$ and $e^-_{n,k} = 1$ if and only if $s_{n,k} < 0$. Obviously, $e^+s - e^-s$ has all components nonnegative and generates sS. Thus we may assume that $s_{n,k}$ is nonnegative for each pair (n, k). Let $b \in sS$ and let b^+ be the positive part of b and b^- denote the negative part of b. Both b^+ and b^- are multiples of s since b is, thus it suffices to simply show that if $b_{n,k} \geq 0$ for each pair (n, k), then $\mathcal{E}(s)$ divides $\mathcal{E}(b)$. Write $b = as$ and let a' denote the supremum of the set $\{a_{n,k} \mid n \geq 0, 1 \leq k \leq n\}$. We must have $eb = b$, so the support idempotent f of b is such that $ef = f$. Hence, f is the support idempotent of a. Let $c = a'(1, (1, 1), (1, 1, 1), \dots)$. Then c is a unit of both S and R. Moreover we have $0 < b_{n,k}/a' \leq b_{n,k}/a_{n,k} = s_{n,k}$ for each pair (n, k) where $b_{n,k}$ is not zero and, of course, $b_{n,k}/a' = 0$ whenever $b_{n,k} = 0$, but then $a_{n,k} = 0$ as well. Consider the element b/c. Since c is a unit of S, b/c is not only an element of S, but it is in sS as well. Moreover, for each pair (n, k), $0 \leq (b/c)_{n,k} \leq s_{n,k}$. Thus $\mathcal{E}(s - b/c)_{n,k} \geq 0$ for each pair (n, k). Hence $\mathcal{E}(s)_{n,k} \geq \mathcal{E}(b/c)_{n,k} \geq 0$. It follows that whenever $\mathcal{E}(s)_{n,k}$ is positive, we have $0 \leq \mathcal{E}(b/c)_{n,k} \leq \mathcal{E}(s)_{n,k} \leq 1$ and when

$\mathcal{E}(s)_{n,k} = 0$ then so is $\mathcal{E}(b/c)_{n,k}$. For fixed n, $\mathcal{E}(s)_{n,k}$ and $\mathcal{E}(b/c)_{n,k}$ are independent of k. Thus we may define an element $r \in R$ by setting $r_n = (\mathcal{E}(b/c)_{n,1}/\mathcal{E}(s)_{n,1}, \ldots, \mathcal{E}(b/c)_{n,1}/\mathcal{E}(s)_{n,1})$ whenever $\mathcal{E}(b/c)_{n,1}$ is not zero and setting $r_n = (0, 0, \ldots, 0)$ otherwise. Obviously $\mathcal{E}(b/c) = r\mathcal{E}(s)$. As c is a unit of R, cr is in R and we have $\mathcal{E}(b) = cr\mathcal{E}(s)$. Therefore $\mathcal{E}(sS) = \mathcal{E}(s)R$.♦

EXAMPLE 11. Let $Q = \prod Q_n$ and $T = \prod R_n$ where $Q_n = \mathbb{R}^n$ and R_n is the diagonal set of Q_n. Finally let S_c and R_c be the bounded sequences in Q and T, respectively, which are eventually constant. Then

(a) Localizing idempotents exist, but no maximal localizing idempotents exist.
(b) If f is a localizing idempotent, then $fS = fR$.
(c) For each principal ideal I of S there is a generator s such that $\mathcal{E}(I) = \mathcal{E}(s)R$.

PROOF. The proofs for (b) and (c) are the same as those given for the corresponding statements in Example 10.

Let $f = (f_1, f_2, \ldots)$ be an idempotent of S. Then there is a positive integer m such that either $f_n = (1, 1, \ldots, 1)$ for each $n \geq m$ or $f_n(0, 0, \ldots, 0)$ for each $n \geq m$. If f is of the latter form, it is a localizing idempotent if and only if each f_k for $k < m$ has at most one nonzero component. Otherwise, f is not a localizing idempotent. It is easiest to see this by considering an idempotent g where $g_{m+2} = (1, 1, 0, 0, \ldots, 0)$ where m is such that $f_n = (1, 1, \ldots, 1)$ for each $n \geq m$. Then $(g^{\#})_{m+2} = (1, 1, 1, \ldots, 1)$ and therefore $(fg^{\#})_{m+2} \neq g_{m+2}$. Hence by statement (2) of Theorem 7 such an idempotent f cannot a localizing idempotent.♦

To obtain a different conditional ring expectation $\mathcal{E}'$ for the rings above, simply take for each $n > 1$, some fixed set of n positive rational numbers $a_{n,1}, a_{n,2}, \ldots a_{n,n}$ with sum $a_{n,1} + a_{n,2} + \cdots + a_{n,n} = 1$ (with at least one $a_{n,k}$ not $1/n$) and then set $\mathcal{E}'(s) = r$ where each $r_{n,k}$ is the sum $a_{n,1}s_{n,1} + a_{n,2}s_{n,2} + \cdots + a_{n,n}s_{n,n}$ for each pair (n, k).

In our last example we show that principal ideals need not be mapped to principal ideals by conditional ring expectations. In this particular example the image of a principal idempotent ideal will be principal and generated by the image of the idempotent. Thus by Corollary 9, an idempotent f of S is localizing if and only if $fS = fR$.

EXAMPLE 12. Let $Q = \prod Q_n$ and $T = \prod T_n$ where for each $n \geq 1$, $Q_n = \mathbb{R}[x]^n$ and T_n is the diagonal set of Q_n. Let R and S be the sequences in T and Q, respectively, which are bounded by the size of the coefficients. Write the elements of Q in the form $q = (q_1, q_2, \ldots)$ where in this case $q_n = (q_{n,1}, q_{n,2}, \ldots, q_{n,n})$ is an n-tuple of polynomials. Define a function $\mathcal{E} : S \to R$ as in Examples 10 and 11, by simply averaging the components of each s_n in $s = (s_1, s_2, \ldots)$ to obtain $\mathcal{E}(s) = r$ where $r_n = \sum s_{n,k}/n$.

(1) $\mathcal{E}$ is a conditional ring expectation.

(2) For each idempotent f, $\mathcal{E}(fS) = \mathcal{E}(f)R$.

(3) There are principal ideals sS for which $\mathcal{E}(sS)$ is principal but cannot be generated by the image of a generator of sS and there are principal ideals qS for which $\mathcal{E}(qS)$ is not principal.

PROOF. As above, it is a simple exercise to show that $\mathcal{E}$ is a conditional ring expectation.

For statement (2), let $f = (f_1, f_2, \ldots)$ be an idempotent of S and let $s \in fS$. As above we denote $\mathcal{E}(f)$ as $tf^\#$ where $t = (1 - f^\#) + \mathcal{E}(f)$. Let $r = \mathcal{E}(s)$ and let $e^\# = (e_1, e_2, \ldots)$ denote the support idempotent for r. Note that $e^\# \leq f^\#$. Of course, we need only consider the nonzero e_n, each of which is simply the n-tuple $(1, 1, \ldots, 1)$. For such an e_n, r_n is the n-tuple all of whose components are $\sum s_{n,k}/n$ and $t_n = (k_n/n, k_n/n, \ldots, k_n/n)$ where k_n is the number of 1's that appear in the n-tuple for f_n. Thus the components of r_n/t_n are simply $\sum s_{n,k}/k_n$. As $k_n \geq 1$, the coefficients of $\sum s_{n,k}/k_n$ are bounded by the same universal bound as on s. Thus t divides r in R. As $f^\# e^\# = e^\#$ we have $\mathcal{E}(s) = (r/t)tf^\# \in \mathcal{E}(fS)$.

It is actually quite easy to give an example of a principal ideal of sS whose image in R while principal, is not generated by the image of a generator of sS. Let $s = (0, (x, x^2), (0,0,0), \ldots)$. Then $\mathcal{E}(sS) = (0, ((x + x^2)/2, (x+x^2)/2), (0,0,0), \ldots)$. But we also have $a = (0, (0, x^2), (0,0,0), \ldots)$ and $b = (0, (x, 0), (0,0,0), \ldots)$ in sS. For these two elements we have $\mathcal{E}(a) = (0, (x^2/2, x^2/2), (0,0,0), \ldots)$ and $\mathcal{E}(b) = (0, (x/2, x/2), (0,0,0), \ldots)$. An arbitrary element of sS has the form $u = (0, (xp, x^2 q), (0,0,0), \ldots)$ with $\mathcal{E}(u) = (0, ((xp + x^2 q)/2, (xp + x^2 q)/2), (0,0,0), \ldots)$. With this we have $E(u) = (0, (x/2, x/2), (0,0,0), \ldots)((0, (p+xq, p+xq), (0,0,0), \ldots) \in \mathcal{E}(b)R$. Hence $\mathcal{E}(sS)$ is principal, but the only generators of sS are the elements of the form $(0, (mx, nx^2), (0,0,0), \ldots)$ where both m and n are nonzero real numbers. The image of such an element will always involve both x and x^2 so it cannot divide $\mathcal{E}(b)$.

Now let $q = (q_1, q_2, q_3, \ldots)$ where $q_1 = 1$, $q_2 = (1 + x, x^2/4)$ and $q_n = (1+x, x^2/n^2, 0, \ldots, 0)$ for $n \geq 3$. Thus $a = (1, (1+x, 0), (1+x, 0, 0), \ldots)$ and $b = (1, (0, x^2/4), (0, x^2/9, 0), \ldots)$ are in qS, but $c = (1, (0, x^2), (0, x^2, 0), \ldots)$ is not in qS. Let $r = \mathcal{E}(a)$ and $v = \mathcal{E}(b)$. Then $r_n = ((1 + x)/n, (1 + x)/n, \ldots, (1 + x)/n)$ and $v_n = (x^2/n^3, x^2/n^3, \ldots)$ for each $n \geq 2$. By way of contradiction assume $p = (p_1, p_2, \ldots)$ generates $\mathcal{E}(qS)$. Since the gcd of $1 + x$ and x^2 is 1, each p_n is of the form $(d_n, d_n, \ldots, d_n)$ for some (positive) constant d_n. For p to be in R the sup of $\{|d_n|\}$ must be finite. Let h_n and k_n be polynomials for which $h_n(1 + x)/n + k_n x^2/n^3 = d_n$. Then $h = (h_1, (h_2, 0), (h_3, 0, 0), \ldots)$ and $k = (k_1, (0, k_2), (0, k_3, 0), \ldots)$ are such that $\mathcal{E}(hq + kq) = p$. We will show that either at least one of h and k is not in S, or p does not divide at least one of $\mathcal{E}(a)$ or $\mathcal{E}(b)$. It is clear that $(1 + x)(1 - x) + n^2(x^2/n^2) = 1$, so we have $[(1 + x)/n](d_n - d_n x) +$

$d_n n^2(x^2/n^3) = d_n$. Since the gcd of $1+x$ and x^2 is 1, any other combination of $1+x$ and x^2/n^2 that yields d_n must be of a form $[(1+x)/n](d_n - d_n x + y_n(x^2/n^2)) + (d_n n^2 - y_n(1+x))(x^2/n^3)$ for some polynomial y_n. Write $y_n = y_{n,0} + y_{n,1}x + \cdots + y_{n,i_n}x^{i_n}$. To have both $d_n - d_n x + y_n(x^2/n^2)$ and $d_n n^2 - y_n(1+x)$ as the entries of an element of R, we must have a single uniform upper bound for all of the sets $\{d_n, y_{n,0}/n^2, \ldots, y_{n,i_n}/n^2\}$ and $\{d_n n^2 - y_{n,0}, y_{n,0}, y_{n,1}, \ldots, y_{n,i_n}\}$. Thus all we need is a single uniform bound for the set $\{d_n n^2, y_{n,0}, \ldots, y_{n,i_n}\}$. On the other hand, it is easy to see that $\mathcal{E}(a)$ factors uniquely as $(d_1, d_2, d_3, \ldots)(r_1/d_1, r_2/d_2, \ldots)$ where the components of r_n/d_n are $(1+x)/nd_n$ when $n \geq 2$. Thus there must be a uniform upper bound for $\{1/nd_n\}$. For $\mathcal{E}(b)$ we have the unique factorization as $(d_1, d_2, \ldots)(v_1/d_1, v_2/d_2, \ldots)$ where the components of v_n/d_n are $x^2/n^3 d_n$. Obviously, any bound that works for $\{1/nd_n\}$ will work for $\{1/n^3 d_n\}$. As we are free to choose any y_n provided $(y_1, y_2, \ldots)$ is in R, we may concentrate on the d_ns. What we must have is a single uniform bound for the sets $\{d_n, d_n n^2, 1/nd_n\}$. This is impossible for if $M \geq n^2 d_n$ and $M \geq 1/nd_n$, then $M \geq n^2 d_n \geq n^2/nM = n/M$ for each positive integer n. Thus the ideal $\mathcal{E}(qS)$ is not principal. ♦

QUESTIONS AND OPEN PROBLEMS.

We end with a question and a series of open problems. All are based on a conditional ring expectation $\mathcal{E} : S \to R$.

Q_1: If $f \in E_S$ localizes R, is $fS = fR$?

Q_2: Characterize those pairs of rings $R \subset S$ for which each localizing idempotent of S is dominated by a maximal localizing idempotent (assuming localizing idempotents exist for the pair $R \subset S$).

Q_3: Characterize those pairs of rings $R \subset S$ for which each increasing sequence of localizing idempotents is dominated by a maximal localizing idempotent (again assuming localizing idempotents exist for the pair $R \subset S$).

Q_4: Characterize those pairs of rings $R \subset S$ for which $\mathcal{E}(sS)$ is a principal ideal of R for each $s \in S$.

Q_5: Characterize those pairs of rings $R \subset S$ for which not only is $\mathcal{E}(sS)$ principal for each $s \in S$, but can be generated by an element of the form $\mathcal{E}(b)$ where $bS = sS$.

Q_6: Characterize those pairs of rings $R \subset S$ for which each principal idempotent ideal of S is "well-preserved" by the map $\mathcal{E}$; i.e., $\mathcal{E}(eS) = \mathcal{E}(e)R$ for each $e \in E_S$.

Q_7: Characterize those pairs of rings $R \subset S$ with conditional ring expectation $\mathcal{E}$ which have the property that each nonzero principal ideal of S can be generated by an element s for which $\mathcal{E}(se) = 0$ implies $se = 0$.

$\mathcal{Q}_8$: Characterize those pairs of rings $R \subset S$ with conditional ring expectation $\mathcal{E}$ for which $\mathcal{E}$ can be extended to a conditional ring expectation $\mathcal{E}' : T(S) \to T(R)$.

With regard to $\mathcal{Q}_8$, consider the rings $R = \mathbb{R}$ and $Q = \prod R_n$ where each $R_n = \mathbb{R}$. Both R and Q are von Neumann regular. Let $S \subset Q$ be the set of bounded sequences in Q and let $\sum a_n = 1$ be a convergent series where each a_n is positive. Define $\mathcal{E} : S \to R$ by $\mathcal{E}(s) = \sum s_n a_n$. Since the sequence $\{s_n\}$ is bounded, the series is absolutely convergent. Moreover, if $s_n \geq 0$ for each n, then $\mathcal{E}(s) \geq 0$ with equality if and only if each s_n is zero. There is no way to extend $\mathcal{E}$ to an R-module homomorphism from Q to $T(R) = R$. Note that R can be embedded in Q as the diagonal set.

ACKNOWLEDGEMENTS.

The author expresses his sincere thanks to Alan Lambert for introducing him to conditional expectations, localizing sets and Rao's book, [7].

REFERENCES

(1) J. Campbell, A. Lambert and B. Weinstock, Localizing sets and the structure of sigma-algebras, Indiana Univ. Math. J., 47 (1998), 913–938.
(2) R. Gilmer and T. Parker, Semigroup rings as Prüfer rings, Duke Math. J. 41 (1974), 219–230.
(3) A. Lambert, Localising sets for sigma-algebras and related point transformations, Proc. Royal Soc. Edinburgh 118A (1991), 111-118.
(4) A. Lambert and T. Lucas, Nagata's Principle of Idealization in relation to module homomorphisms and conditional expectations, Kyungpook Math. J. 40 (2000), 327–337.
(5) A. Lambert and B. Weinstock, A class of operator algebras related to probabilistic conditional expectations, Michigan Math. J. 40 (1993), 359–376.
(6) S.-T. C. Moy, Characterizations of conditional expectations as a transformation on function spaces, Pac. J. Math. 4 (1954), 47–63.
(7) M. M. Rao, Conditional Measures and Applications, *Pure and Applied Mathematics*, 177, Dekker, New York, 1993.

Errata: "Pullbacks and Coherent-Like Properties"

[In: Lect. Notes Pure Appl. Math., Dekker, 205 (1999) 437-459]

ABDESLAM MIMOUNI Department of Mathematics, Faculty of Sciences "Dhar Al-Mehraz", University of Fez, Fez, Morocco

In Theorem 4, Corollary 6, and Corollary 7, we add the hypothesis $Z(T/I) \cap D = (0)$, but without any changes in the proofs . However, in Proposition 8 (old Corollary 8), we give a new proof in which the hypothesis $Z(T/I) \cap D = (0)$ is not needed.

THEOREM 4. For the diagram (Δ), assume that $Z(T/I) \cap D = (0)$:
1) If $T/I \not\subseteq qf(D)$, then $I = J^{-1}$ for some finitely generated fractional ideal J of R with $J \subseteq T$, so I is a v-ideal of R.
2) If $T/I \subseteq qf(D)$, then $I = P$ is a prime ideal of T and $\chi(P) = qf(D)$. If P is not a v-ideal of R, then $(T : P) = P^{-1} = (P : P)$, P is a prime ideal of $(T : P)$ and for each overring B of R, P is not a maximal ideal of B.

COROLLARY 6. For the diagram (Δ), assume that $Z(T/I) \cap D = (0)$. If R is a v-domain, then $I = P$ is a prime ideal of T and $\chi(P) = qf(D)$.

COROLLARY 7. For the diagram (Δ_1), assume that $Z(V/I) \cap D = (0)$. Then R is a v-domain if and only if D is a v-domain, $I = P$ is a prime ideal of T and $\chi(P) = qf(D)$.

PROPOSITION 8. For the diagram (Δ_1), R is a $PVMD$ (resp., Prüfer), if and only if D is a $PVMD$ (resp., Prüfer), $I = P$ is a prime ideal of V and $\chi(P) = qf(D)$.

Proof. $\Longrightarrow$). Assume that R is a $PVMD$. By Theorem 1, I is a divisorial ideal of R. If I is t-maximal, by [10, Proposition 2.1], I is t-invertible. So $R \subset V \subseteq (I : I) \subseteq ((II^{-1})_t : (II^{-1})_t) = R$, absurd. By [10, Proposition 1.2], I is a prime ideal of $(I : I)$, and therefore a prime ideal of V. Now, it suffices to consider the diagram where $I = IV_I$ is a maximal ideal of V_I and conclude by [9, Theorem 4.13].
$\Longleftarrow$) Follows from [9, Theorem 4.13].

Ultraproducts of Commutative Rings

BRUCE OLBERDING, Department of Mathematics, The University of Louisiana at Monroe, Monroe, LA 71209, maolberding@ulm.edu

SERPIL SAYDAM, Department of Mathematics, The University of Louisiana at Monroe, Monroe, LA 71209, masaydam@ulm.edu

1. INTRODUCTION

The focus of this paper is the structure of ultraproducts of commutative rings, and in particular, ultraproducts of integral domains. Ultraproducts of certain classes of integral domains, such as orders in algebraic number fields, have been well-studied but often from a model-theoretic point of view. The model-theoretic view is a natural one to take given the importance of the ultraproduct construction in the theory of elementary models. In particular, the Keisler-Shelah Theorem states that two models have the property that any sentence in first-order logic satisfied by one is satisfied by the other if and only if these two models have isomorphic ultrapowers. In this same vein is the Ax-Kochen-Ershov Theorem, which implies that two henselian valuation domains have isomorphic ultrapowers if and only if their value groups and residue fields have isomorphic ultrapowers. Hence an understanding of arbitrary ultrapowers of a given class of commutative rings leads to a better understanding of the first order theory of the class of rings.

Our goal in this paper is to give some new examples of how one can construct interesting rings using ultraproducts, then focus on the ring-theoretic properties of the ultraproducts of integral domains. Section 2 contains some simple observations regarding first order properties of commutative integral domains. In Section 3, we analyze the n-generator property and coherency and use this analysis to construct a GCD domain that is not coherent, the existence of which was an open question until recently. In Section 4, we turn to some of the main structural results of the paper. These focus, for example, on the prime spectrum of an ultraproduct of integral domains and the representation of an ultraproduct of domains as an intersection of its localizations.

An ultraproduct of commutative rings is a certain homomorphic image of a cartesian product of these same rings. The elements of the ultraproduct can be interpreted as equivalence classes of elements of the cartesian product with respect to an "ultrafilter," a maximal filter on the index set of the product.

Let $\mathcal{I}$ be a set and $\mathcal{F}$ be a collection of subsets of $\mathcal{I}$. Then $\mathcal{F}$ is a *filter* on $\mathcal{I}$ if (a) for all $\mathcal{A},\mathcal{B} \in \mathcal{F}$, $\mathcal{A} \cap \mathcal{B} \in \mathcal{F}$, and (b) for all $\mathcal{A} \in \mathcal{F}$, if $\mathcal{A} \subseteq \mathcal{B} \subseteq \mathcal{I}$, then $\mathcal{B} \in \mathcal{F}$. A filter $\mathcal{F}$ is an *ultrafilter* on $\mathcal{I}$ if for all $\mathcal{A} \subset \mathcal{I}$, either $\mathcal{A} \in \mathcal{F}$ or $\mathcal{I} \backslash \mathcal{A} \in \mathcal{F}$. It is not hard to see that if an ultrafilter $\mathcal{F}$ contains a finite set, then it contains a singleton set, say $\{i\}$, and i is an element of every element of $\mathcal{F}$. In this case we say $\mathcal{F}$ is a *principal* ultrafilter.

Let $\{R_i\}_{i\in\mathcal{I}}$ be a collection of commutative rings for some index set $\mathcal{I}$. If $\mathcal{D}$ is an ultrafilter on $\mathcal{I}$, then we write $R^* = \prod_{\mathcal{D}} R_i$ for the *ultraproduct* of the R_i's with respect to the ultrafilter $\mathcal{D}$. An element of $\prod_{\mathcal{D}} R_i$ is an equivalence class of elements of $\prod_{i\in\mathcal{I}} R_i$ defined by:

$$(r_i)_{i\in\mathcal{I}} \sim (s_i)_{i\in\mathcal{I}} \Leftrightarrow \{i \in \mathcal{I} : r_i = s_i\} \in \mathcal{D}.$$

If $(r_i)_{i\in\mathcal{I}}$ is an element of the equivalence class $x \in \prod_{\mathcal{D}} R_i$, then we write $x = (r_i)_{\mathcal{D}}$. If there is a ring R such that $R_i = R$ for all $i \in \mathcal{I}$, then $R^{\mathcal{D}} := R^*$ is the *ultrapower* of R with respect to $\mathcal{D}$. If for each i, S_i is a subset of R_i, then we write $(S_i)_{\mathcal{D}}$ for the subset of R^* consisting of elements of the form $(s_i)_{\mathcal{D}}$, $s_i \in S_i$.

If $\mathcal{D}$ is a principal ultrafilter, say $\{i\} \in \mathcal{D}$, then it is not hard to see that $R^* \cong R_i$. Thus we are only interested in the case where $\mathcal{I}$ is an infinite set and $\mathcal{D}$ is a non-principal ultrafilter. (Note the standing hypotheses below.)

Although we will not touch on the following articles, we reference them here because they are relevant to the topics in the present paper. The article [16] contains an interesting application of ultraproducts to the construction of non-Noetherian Prüfer domains having specified Picard group. In [22], ultraproducts are used to construct a quasilocal commutative ring R such that every regular ideal of R is stable and R is one-dimensional but not Noetherian. The articles [11] and [12] deal with infinite products of commutative rings and show how ultrafilters arise naturally in the consideration of prime spectra of such rings. These articles are especially relevant to some of the issues in Sections 4-6 of the present paper, where the rather unwieldly behavior of prime spectra and Krull dimension are considered. In particular, R. Gilmer and W. Heinzer give an example in [12] of an infinite product R of zero-dimensional local rings such that for every $d > 0$, there is a homomorphic image of R of Krull dimension d.

Standing hypotheses. *Throughout this paper, $\{R_i\}_{i\in\mathcal{I}}$ is a collection of commutative rings indexed by an infinite set $\mathcal{I}$, and $\mathcal{D}$ is a non-principal ultrafilter. The total quotient ring of each R_i is denoted by F_i. The total quotient ring of R^* is denoted by F^*, and we identify F^* and $\prod_{\mathcal{D}} F_i$.*

2. BASIC PROPERTIES

We refer frequently to "first order" properties in the language $\mathcal{L} = \{+, \cdot, 0, 1\}$ of commutative rings. Roughly speaking, these properties involve quantifications only over elements (rather than, say, ideals) of commutative rings. For a more formal treatment, see [3, Section 1.3]. Our use of model theory is very modest here, but the notion of a first order sentence or formula will be helpful in proving our results since one of the most fundamental properties of ultraproducts is that they preserve first order properties of their constituent parts. This is part of the content of the following theorem. We say that a property $\mathcal{P}$ *holds for $\mathcal{D}$-many i* if the set of all i such that R_i satisifies $\mathcal{P}$ is an element of $\mathcal{D}$.

ŁOS'S THEOREM. [3] *Let ϕ be a sentence in the language of commutative rings. Then R^* satisfies ϕ if and only if, for $\mathcal{D}$-many i, R_i satisfies ϕ.*

It follows from Łos's Theorem that R^* is an integral domain if and only if R_i is an integral domain for $\mathcal{D}$-many i. This is because the theory of integral domains consists of finitely many axioms, all of which can easily be expressed in the first order language of rings with identity. (See [3], for example.) Similarly, R^* is a field if and only if R_i is a field for $\mathcal{D}$-many i. We will make use of both of these facts without further reference.

An ideal I of a commutative ring R is *definable* in R if there exists a first order formula $\phi(x, y_1, y_2, \dots, y_n)$ in the language of commutative rings and $r_1, \dots, r_n \in R$ such that

$$r \in I \Leftrightarrow \phi(r, r_1, \dots, r_n) \text{ is true in } R.$$

A finitely generated ideal I and its dual I^{-1} are definable, as is the maximal ideal of any quasilocal ring. For example, if $I := (r_1, r_2, \dots, r_n)$ is a finitely generated ideal of R, then I is defined by the formula

$$\phi(x, y_1, y_2, \dots, y_n) : \quad \exists w_1, w_2, \dots, w_n(x = w_1y_1 + w_2y_2 + \cdots + w_ny_n),$$

since $r \in I$ if and only if $\phi(r, r_1, r_2, \dots, r_n)$ is true in R.

An ideal I of R^* is *induced* if $I = (I_i)_{\mathcal{D}}$ for ideals I_i of R_i, $i \in \mathcal{I}$. Lemma 2.1 records some useful facts about the algebra of induced ideals. Recall that an ideal I of a commutative ring is *n-generated* if I can be generated by n elements.

LEMMA 2.1 *Suppose $I := (I_i)_{\mathcal{D}}$, $J := (J_i)_{\mathcal{D}}$ and $K := (K_i)_{\mathcal{D}}$ are induced ideals of R^*.*

(i) *$I \subseteq J$ if and only if $I_i \subseteq J_i$ for $\mathcal{D}$-many i.*
(ii) *I is a prime (maximal) ideal if and only if I_i is a prime (maximal) ideal of R_i for D-many i.*
(iii) *There exists $n > 0$ such that I is an n-generated ideal if and only if I_i is an n-generated ideal for D-many i.*
(iv) *$I \cap J = K$ if and only if $I_i \cap J_i = K_i$ for $\mathcal{D}$-many i.*

Moreover, if R^* is a domain, then:

(v) *I is an invertible ideal of R^* if and only if there exists $n > 0$ such that I_i is an n-generated invertible ideal of R_i for $\mathcal{D}$-many i.*
(vi) *I is a divisorial ideal of R^* if and only if I_i is a divisorial ideal of R_i for $\mathcal{D}$-many i.*

Proof. Statement (i) is a direct application of the relevant definitions. To prove statement (ii), observe that $R^*/I \cong \prod_{\mathcal{D}} R_i/I_i$. Thus I is a prime ideal of R^* if and only if R_i/I_i is a domain for $\mathcal{D}$-many i; if and only if I_i is a prime ideal of $\mathcal{D}$-many i. A similar argument shows I is maximal if and only if I_i is a maximal ideal of R_i for $\mathcal{D}$-many i. For statement (iii), observe that I is n-generated if and only if $I = (L_i)_{\mathcal{D}}$ for some n-generated ideals L_i of R_i. Thus if I is n-generated, by (i), $L_i = I_i$ for $\mathcal{D}$-many i. The converse is clear. Statement (iv) asserts $I \cap J = (I_i \cap J_i)_{\mathcal{D}}$, and this is easily verified by direct application of definitions. Now suppose R^* is a domain. In order to prove (v), note that if I_i is an n-generated invertible ideal of R_i for $\mathcal{D}$-many i, then $I_iI_i^{-1} = R_i$ for $\mathcal{D}$-many i. Since I_i is n-generated for $\mathcal{D}$-many i, it follows that $(I_iI_i^{-1})_{\mathcal{D}} = (I_i)_{\mathcal{D}}(I_i^{-1})_{\mathcal{D}}$. Thus $(I_i)_{\mathcal{D}}$ is an invertible ideal of R^*. The converse follows from (i) and the easily verified fact that if $L := (L_i)_{\mathcal{D}}$ is an induced ideal of R^*, then $L^{-1} = (L_i^{-1})_{\mathcal{D}}$. Statement (vi) also from from this observation. □

The following properties are all more or less well-known, so we give only cursory proofs using Łos's Theorem and Lemma 2.1. Recall that an integral domain R is a *Prüfer domain* if every finitely generated ideal is invertible, and that R is a *Bezout domain* if every finitely generated ideal is principal.

PROPOSITION 2.2 *R^* is a commutative ring that satisfies the following statements.*

(i) *R^* is quasilocal if and only if R_i is quasilocal for $\mathcal{D}$-many i.*
(ii) *R^* is a valuation domain if and only if R_i is a valuation domain for $\mathcal{D}$-many i.*
(iii) *R^* is a Prüfer (Bezout) domain if and only if R_i is a Prüfer (resp. Bezout) domain for $\mathcal{D}$-many i.*
(iv) *R^* is an integrally closed domain if and only if R_i is an integrally closed domain for $\mathcal{D}$-many i.*

Proof. By Łos's Theorem, it is enough to show each assumed property is a first order property in the language of rings. For (i), use the characterization of a quasilocal ring as a ring in which the non-units are closed under addition. For statement (ii), encode the statement that valuation domains are characterized by the property that given any two elements, one divides the other. To prove the Prüfer case of (iii), recall that a domain is Prüfer if and only if every ideal generated by 2 elements is invertible [10]. By Lemma 2.1, the statement *every ideal generated by two elements is invertible* is a first order sentence in the language of commutative rings. Hence the Prüfer case of (iii) follows from Łos's Theorem. The proof of the Bezout case of (iii) is similar. For statement (iv), observe first that a domain R is integrally closed if and only if for all $n > 0$, R satisfies the sentence ϕ_n that asserts that if $a, b \in R$ and $f(a,b) = 0$ for some degree n form $f(x,y) := x^n + r_{n-1}x^{n-1}y + \cdots + r_0y^n \in R[x,y]$, then b divides a in R. Thus (iv) is an application of Łos's Theorem. □

In the article [13], R. Gilmer, W. Heinzer and M. Roitman construct a quasilocal domain R with maximal M such that M cannot be generated by finitely many elements but M^2 can be generated by 3 elements. They obtain such examples in every dimension greater than two and show that the maximal ideal in their examples has maximum possible height in the ring [13, Example 3.2]. By their results, a one-dimensional example is impossible: They show that an integral domain is Noetherian if and only if every prime ideal has some power that is finitely generated [13, Theorem 1.17]. Thus a one dimensional quasilocal domain whose maximal ideal has a finitely generated power is Noetherian.

They also raise the problem of whether an even more extreme example can be obtained, that of a quasilocal domain with non-finitely generated maximal ideal whose square is 2-generated. By a result of J. Sally in [25], such a ring cannot exist: If the square of a maximal ideal in a quasilocal domain is 2-generated, then every power of the maximal ideal, including the first power, must be 2-generated. Thus the number 3 is optimal.

To conclude this section, we give another approach to examples of non-finitely generated prime ideals whose squares are finitely generated.

PROPOSITION 2.3 *Suppose $\mathcal{I} = \mathbb{N}$, and for each $i \in \mathbb{N}$, P_i is a prime ideal of R_i. Suppose there exists $k > 0$ such that:*

(i) *for $\mathcal{D}$-many i, P_i can be generated by no fewer than i generators, and*
(ii) *for $\mathcal{D}$-many i, P_i^2 can be generated by k elements of the form $p_i q_i$, with $p_i, q_i \in P_i$.*

Then R^* is a commutative ring with non-finitely generated prime ideal $(P_i)_\mathcal{D}$ whose square is k-generated.

Proof. By Lemma 2.1, $P := (P_i)_\mathcal{D}$ is a prime ideal that is not finitely generated. Also by Lemma 2.1, $(P_i^2)_\mathcal{D}$ is k-generated. Clearly, $P^2 \subseteq (P_i^2)_\mathcal{D}$. Also, $(P_i^2)_\mathcal{D}$ is generated by k elements of the form $(p_i q_i)_\mathcal{D}$, where $p_i, q_i \in P_i$ for $\mathcal{D}$-many i. Thus $(P_i^2)_\mathcal{D} \subseteq P^2$, and the claim follows. □

Sally's method for constructing quasilocal domains such that the squares of the prime ideals are 3-generated is as follows. Let J be an ideal of some quasilocal domain A and let y be an indeterminate. Set $B = A[[y^3, y^4, Jy^5]]$ and let $P = (y^3, y^4, Jy^5)$. Then $B/P \cong A$ and P is a prime ideal of B that cannot be generated by fewer than the minimal number of generators of J while $P^2 = (y^6, y^7, y^8)$ is 3-generated. By varying the choice of A one may vary the Krull dimension of the ring B. In this way one can construct a collection of local domains $\{R_i : i \in \mathbb{N}\}$ with prime ideals $\{P_i : i \in \mathbb{N}\}$ such that the P_i satisify the hypotheses of Proposition 2.3.

In [13], the question is posed of whether an integrally closed quasilocal domain can have a non-finitely generated maximal ideal whose square is 3-generated, and it is shown that the answer is yes if one omits the requirement that the domain be integrally closed. In light of this result and Proposition 2.3, we ask: *Does there exist a collection of quasilocal domains $\{R_i\}_{i \in \mathbb{N}}$ such that for each $i > 0$, the maximal ideal M_i of R_i is finitely generated but not i-generated and M_i^2 is 3-generated?* By Propositions 2.2 and 2.3, an affirmative answer to this question yields another example of a quasilocal domain with non-finitely generated maximal ideal whose square is 3-generated. An affirmative answer to the integrally closed version of this question yields an affirmative answer to the question in [13].

3. COHERENCY AND THE n-GENERATOR PROPERTY

In this section we examine coherency for ultraproducts. Recall that a commutative ring R has the *n-generator property* if every finitely generated ideal of R is n-generated.

PROPOSITION 3.1 *The following statements hold for R^*.*

(i) *R^* is quasilocal with n-generated maximal ideal if and only if, for $\mathcal{D}$-many i, R_i is quasilocal with n-generated maximal ideal.*
(ii) *R^* has the n-generator property if and only if R_i has the n-generator property for $\mathcal{D}$-many i.*
(iii) *Let $n, m > 0$. Intersections of pairs of m-generated ideals of R^* are n-generated if and only if for $\mathcal{D}$-many i, intersections of m-generated ideals of R_i are n-generated.*

Proof. (i) By Proposition 2.2, R^* is quasilocal if and only if the set $\mathcal{A}$ of elements $i \in \mathcal{I}$ such that R_i is quasilocal is an element of $\mathcal{D}$. Let ψ be the sentence that asserts that there exist $x_1, x_2, \ldots, x_n$ such that for each non-unit y there exist $y_1, y_2, \ldots, y_n$ such that

$$y = y_1 x_1 + y_2 x_2 + \cdots + y_n x_n.$$

Then by Łos's Theorem, R satisfies ψ if and only if there is a set $\mathcal{B} \in \mathcal{D}$ such that R_i satisfies ψ for all $i \in \mathcal{B}$. Thus (i) follows from the fact that since $\mathcal{D}$ is a filter, $\mathcal{A} \cap \mathcal{B} \in \mathcal{D}$.

(ii) To prove a ring has the n-generator property it is enough to show that every $(n+1)$-generated ideal can be generated by n elements. This is a first order property, so Łos's Theorem yields (ii).

(iii) It is not hard to see that given $n, m > 0$, the property that *the intersection of all pairs of m-generated ideals is n-generated* can be encoded into a first order sentence. Hence Łos's Theorem implies (iii).

□

A commutative ring R is *coherent* if for all $n > 0$, every non-zero homomorphism $R^n \to R$ has a finitely generated kernel. A commutative ring R is *uniformly coherent* if there exists a map $\phi : \mathbb{N} \to \mathbb{N}$ such that for all $n > 0$ and non-zero homomorphisms $f : R^n \to R$, the kernel of f can be generated by $\phi(n)$ elements. The map ϕ is the *uniformity map* for R. Examples of uniformly coherent domains include local (Noetherian) domains of Krull dimension less than 3 and, more generally, domains of global dimension less than 3 for which every finitely generated projective module is free. By contrast, a Noetherian domain of Krull dimension greater than 2 is coherent but not uniformly coherent. See [15, Section 6.1] for a discussion of these results.

There is a more intrinsic characterization of coherency when no zero-divisors are present: An integral domain R is coherent if and only if intersections of pairs of finitely generated ideals are finitely generated [15, Theorem 2.3.2]. Lemma 3.2 contains an analogous statement for uniformly coherent domains. The lemma is probably well-known, but for lack of a reference we include a proof.

LEMMA 3.2 *An integral domain R is uniformly coherent if and only if there exists a function $\phi : \mathbb{N} \to \mathbb{N}$ such that for all $n > 0$, the intersection of any two n-generated ideals of R can be generated by $\phi(n)$ elements.*

Proof. Suppose ψ is a uniformity map for R. Let I and J be n-generated ideals of R. There is an R-submodule K of $R^n \oplus R^n$ that yields the commutative diagram:

$$\begin{array}{ccccccccc}
0 & \longrightarrow & K & \longrightarrow & R^n \oplus R^n & \longrightarrow & I+J & \longrightarrow & 0 \\
 & & {\scriptstyle f}\downarrow & & \downarrow & & \downarrow & & \\
0 & \longrightarrow & I \cap J & \longrightarrow & I \oplus J & \longrightarrow & I+J & \longrightarrow & 0
\end{array}$$

By assumption, K can be generated by $\psi(2n)$ elements. Since f is surjective, $I \cap J$ can be generated by $\psi(2n)$ elements. Thus if $\phi : \mathbb{N} \to \mathbb{N}$ is defined by $\phi(n) := \psi(2n)$ for all $n > 0$, then ϕ is the desired mapping.

Conversely, suppose that there exists a map $\phi : \mathbb{N} \to \mathbb{N}$ such that for all $n > 0$, the intersection of any two n-generated ideals of R can be generated by $\phi(n)$ elements. We shall construct a uniformity map $\psi : \mathbb{N} \to \mathbb{N}$ for R by induction on $n \in \mathbb{N}$. Fix

$n > 1$, and suppose that for all $k < n$, there exists a number $\psi(k)$ such that every homomorphism $f : R^k \to R$ has a kernel that can be generated by $\psi(k)$ elements. Let I be an n-generated ideal of R, and let J be an ideal of R that can be generated by $n-1$ elements and such that $Ra + J = I$. There is a commutative diagram:

$$\begin{array}{ccccccccc} 0 & \longrightarrow & K & \longrightarrow & R \oplus R^{n-1} & \longrightarrow & I & \longrightarrow & 0 \\ & & {\scriptstyle f}\downarrow & & {\scriptstyle g}\downarrow & & \downarrow & & \\ 0 & \longrightarrow & Ra \cap J & \longrightarrow & Ra \oplus J & \longrightarrow & I & \longrightarrow & 0 \end{array}$$

By the induction hypothesis, the kernel of g can be generated by $\psi(n-1)$ elements. Since Ker f is isomorphic to Ker g and $Ra \cap J$ can, by assumption, be generated by $\phi(n-1)$ elements, it follows that K can be generated by $\psi(n-1) + \phi(n-1)$ elements. Thus we define $\psi(n) := \psi(n-1) + \phi(n-1)$ for all $n \in \mathbb{N}$ and conclude the mapping ψ is a uniformity map for R. □

For most of this article we deal with a generic non-principal ultrafilter $\mathcal{D}$ on an infinite set $\mathcal{I}$. However, in Theorem 3.3 and later in Propositions 3.7 and 6.2, we place an additional restriction on our ultrafilter $\mathcal{D}$. If κ is a cardinal number, then a filter $\mathcal{F}$ on $\mathcal{I}$ is *κ-complete* if $\cap\mathcal{G} \in \mathcal{F}$ whenever $\mathcal{G} \subseteq \mathcal{F}$ and $|\mathcal{G}| < \kappa$. A filter is *countably incomplete* if it is not ω^+-complete. Observe that $\mathcal{F}$ is a countably incomplete ultrafilter if and only if there exists a collection $\{\mathcal{A}_k : k > 0\}$ of subsets of $\mathcal{I}$ such that $\cup_{k>0}\mathcal{A}_k \in \mathcal{F}$ but $\mathcal{A}_k \notin \mathcal{F}$ for all $k > 0$. Thus (and this is the relevance of countable incompleteness in our context) an ultrafilter $\mathcal{F}$ is countably incomplete if and only if there exists a function $\nu : \mathcal{I} \to \mathbb{N}$ such that for all $k > 0$, $\{i : \nu(i) > k\} \in \mathcal{F}$. Notice that if $\mathcal{I}$ is a countably infinite set, then every ultrafilter on $\mathcal{I}$ is countably incomplete.

THEOREM 3.3 *Suppose R is a commutative ring and $R_i = R$ for $\mathcal{D}$-many i. If $\mathcal{D}$ is countably incomplete, then the following statements are equivalent.*

(1) *R is a uniformly coherent domain.*
(2) *R^* is a uniformly coherent domain.*
(3) *R^* is a coherent domain.*

Proof. To prove (1) ⇒ (2), suppose that R is a uniformly coherent domain. By Lemma 3.2, there exists a function $\phi : \mathbb{N} \to \mathbb{N}$ such that for all $n > 0$, the intersection of any two n-generated ideals of R is $\phi(n)$-generated. By Lemma 3.1, the intersection of any two n-generated ideals of R^* is also $\phi(n)$-generated. Hence R^* is uniformly coherent. It is clear that (2) ⇒ (3), so it remains to show (3) ⇒ (1). Suppose R^* is a coherent domain but that R is not uniformly coherent. Then there exists $n > 0$ such that for all $k > 0$, there exists a pair of n-generated ideals I_k and J_k such that $I_k \cap J_k$ cannot be generated by k elements. Define $\mathcal{A} := \{i \in \mathcal{I} : R_i = R\}$. Since $\mathcal{D}$ is countably incomplete, there exists a function $\nu : \mathcal{I} \to \mathbb{N}$ such that for all $k > 0$, $\{i \in \mathcal{I} : \nu(i) > k\} \in \mathcal{D}$. For each $i \in \mathcal{A}$, define $U_i := I_{\nu(i)}$ and $V_i := J_{\nu(i)}$. If $i \in \mathcal{I} \backslash \mathcal{A}$, set $U_i = R$ and $V_i = R$. Then $(U_i)_{\mathcal{D}}$ and $(V_i)_{\mathcal{D}}$ are n-generated ideals of R^*. However, by Lemma 2.1, the ideal $(U_i)_{\mathcal{D}} \cap (V_i)_{\mathcal{D}} = (U_i \cap V_i)_{\mathcal{D}}$ of R^* cannot be finitely generated, contradicting (3). Therefore, R is a uniformly coherent domain. □

As a corollary, we obtain a result about uniformly coherent domains that does not refer to ultraproducts in its formulation. M. Roitman shows in [23, Theorem 1.8] that if R is a coherent domain and P is a prime ideal having some power that is finitely generated, then P is finitely generated. Corollary 3.4 shows that when R is uniformly coherent, one can assert something stronger. This corollary can be proved using an extension of an argument like that of Roitman in the proof of [23, Theorem 1.8], but we include here an alternate proof via ultraproducts.

COROLLARY 3.4 *If R is a uniformly coherent domain, then for all $k > 0$, there exists a function $\phi_k : \mathbb{N} \to \mathbb{N}$ such that whenever P is a finitely generated prime ideal of R and P^k is an n-generated ideal of R for some $n > 0$, then P can be generated by $\phi_k(n)$ elements.*

Proof. Suppose R is a uniformly coherent domain. Define $\mathcal{I} = \mathbb{N}$ and set $R_i = R$ for all $i \in \mathcal{I}$. Then by Theorem 3.3, R^* is coherent. Suppose there exists $\ell, k > 0$ such that for each $n > 0$, there exists a prime ideal P_n with the property that P_n cannot be generated by n elements but P_n^k can be generated by ℓ elements. Define $P := (P_n)_{\mathcal{D}}$ and observe P^k can be generated by ℓ elements but P is not finitely generated. This contradicts [23, Theorem 1.8], which states that prime ideals of coherent domains having a finitely generated power are finitely generated. Hence for each $k > 0$, there exists a map ϕ_k as in the statement of the theorem. □

An integral domain R is a *finite conductor domain* if the intersection of any two principal ideals is finitely generated. Every coherent domain is a finite conductor domain, but whether the converse is true remained an open question until recently, when S. Glaz gave the first examples of finite conductor domains that are not coherent [14]. Later we show that non-coherent finite conductor domains arise naturally in the context of ultraproducts of Noetherian rings. First, in Example 3.5, we observe that an ultraproduct of coherent domains need not be coherent. In fact, ultraproducts of one dimensional local (Noetherian) domains do not have to be coherent, nor even finite conductor domains. Examples are easy to come by.

EXAMPLE 3.5 *Let $\mathcal{I} = \mathbb{N}$. For each $i > 0$, let K_i be a degree i algebraic extension of $\mathbb{Q}$. Set $R_i = \mathbb{Q} + XK_i[X]$, where X is an indeterminate. Then R^* is an ultraproduct of Noetherian (hence coherent) domains that is not a finite conductor domain.*

Proof. For each $i > 0$, the maximal ideal $M_i = XK_i[X]$ of R_i cannot be generated by fewer than i elements. Also, $M_i^{-1} \neq R_i$ and it follows that M_i is a divisorial ideal of R_i. Since M_i is a maximal ideal of R_i, M_i is the intersection of two principal fractional ideals, one of which can be chosen to be R_i. Write $M_i = R_i \cap R_i x_i$ for some x_i in the quotient field of R_i. Then the maximal ideal $M := (M_i)_{\mathcal{D}}$ of $R^* = \prod_{\mathcal{D}} R_i$ is the intersection of two principal fractional ideals, $R^* \cap R^*(x_i)_{\mathcal{D}}$. If R^* is a finite conductor domain, then M is finitely generated and by Proposition 3.1, there exists $n > 0$ such that M_i is n-generated for $\mathcal{D}$-many i. Since $\mathcal{D}$ contains the cofinite sets and M_i is n-generated for only finitely many i, this is a contradiction. □

More generally, if each R_i, $i \in \mathbb{N}$, is a quasilocal domain that has a divisorial maximal ideal that cannot be generated by less than i elements, then our argument shows that the ring R^* is not a finite conductor domain.

Examination of this example shows that coherency is violated because as $i \to \infty$, the number of generators of M_i grows without bound. Thus to preserve coherency, one needs some kind of boundedness condition such as the existence of a uniformity map as in Theorem 3.3. The n-generator property also suffices:

PROPOSITION 3.6 *R^* is a coherent domain with the n-generator property if and only if R_i is a coherent domain with the n-generator property for $\mathcal{D}$-many i.*

Proof. Apply Lemma 2.1 and Proposition 3.1. □

We turn now to the problem of generating examples of finite conductor domains that are not coherent. The basic idea behind our construction is that height two prime ideals in $K[[x,y,z]]$, K a field, have arbitrarily large generating sets. The non-analytic version of this assertion is classical and is due to Macaulay. As discussed in [25, p. 58], Macaulay's primes do not remain prime under completion, so, since we wish to work in the setting of local rings, we appeal instead to Moh's construction of large primes in $K[[x,y,z]]$. Sally shows that this construction implies the existence of large primes in any three dimensional regular local ring that contains a coefficient field [25, pp. 61-62]. Thus if R is a regular local ring of dimension greater than two that contains a coefficient field, then R has arbitrarily large primes of height two since R can be localized at a height three prime ideal to form a three dimensional regular local ring that contains a coefficient field.

PROPOSITION 3.7 *Let R be a regular local domain of Krull dimension greater than two that contains a coefficient field. Let $\mathcal{I}$ be an infinite set and $\mathcal{D}$ be a countably incomplete non-principal ultrafilter on $\mathcal{I}$. Then $R^* = \prod_{\mathcal{D}} R$ is a non-coherent finite conductor domain. In particular, R^* is a GCD domain with elements $a, b, c \in R^*$ such that $(a) \cap (b, c)$ is not finitely generated.*

Proof. Since $\mathcal{D}$ is countably incomplete, there exists a function $\nu : \mathcal{I} \to \mathbb{N}$ such that for all $k > 0$, $\{i \in \mathcal{I} : \nu(i) > k\} \in \mathcal{D}$. For each $i \in \mathcal{I}$, let P_i be a height two prime ideal of R that cannot be generated by fewer than $\nu(i)$ elements. For each i, let (a_i, b_i) be a regular sequence in R that is contained in P_i. Choose $c_i \in R$ such that $P_i = ((a_i, b_i) :_R c_i)$. Then $c_i P_i = Rc_i \cap (a_i, b_i)$. Set $a = (a_i)_{\mathcal{D}}, b = (b_i)_{\mathcal{D}}$, and $c = (c_i)_{\mathcal{D}}$. Define $P := (P_i)_{\mathcal{D}}$. Then $cP = R^*c \cap (R^*a + R^*b)$ but if P is m-generated, then Lemma 2.1 implies P_i is m-generated for $\mathcal{D}$-many i, a contradiction. Thus P is not finitely generated and R^* is not coherent. By Lemma 2.1 and Łos's Theorem, R^* is a GCD, since regular local rings are UFD's. □

COROLLARY 3.8 *Let R be a regular local domain of Krull dimension greater than two that contains a coefficient field. Then R is elementarily equivalent (see* [3, Section 1.3]*) to a non-coherent GCD domain.*

Proof. This follows from Proposition 3.7 and Łos's Theorem. □

4. MAXIMAL IDEALS OF R^*

We turn now to the description of maximal ideals of R^*. The easiest case is when R^* has finitely many maximal ideals. If R^* has infinitely many maximal ideals, then the situation is considerably more complicated.

PROPOSITION 4.1 *R^* has precisely n many maximal ideals if and only if R_i has precisely n many maximal ideals for $\mathcal{D}$-many i.*

Proof. Observe that the Jacobson radical of a ring is easily seen to be definable (in the sense of Section 2) by the formula $\phi(x)$ that asserts $1-x$ is a unit, so one may construct a first order sentence that asserts a ring modulo its Jacobson radical is a product of n-many fields, namely there exist, modulo the set of all x such that $1-x$ is a unit, orthogonal idempotents $e_1, \ldots, e_n$ such that for all x, $xe_i \neq 0$ implies xe_i has a multiplicative inverse. The claim now follows from Los's Theorem. □

We will see in Theorem 4.7 that if R^* has infinitely many maximal ideals, then R^* has infinitely many non-induced maximal ideals, but first we use the terminology of G. Cherlin to note there is a one-to-one correspondence between maximal ideals of R^* and ultrafilters on the set of induced maximal ideals of R^* [4]. The following notation will be helpful. If R^* is a domain and J is an induced ideal of R^*, then we define S_J to be the set of all induced maximal ideals of R^* that contain J. For convenience, we write S_r for $S_{R^* r}$, where $r \in R^*$. If K is an ideal of R^*, then $\mathcal{E}(K)$ is the filter generated by the set $\{S_J : J \subseteq K$, where J is an induced ideal of $R^*\}$. Note that the set $\mathcal{E}(K)$ is closed under finite intersections, since if $S_{J_1}, S_{J_2} \in \mathcal{E}(K)$, then $S_{J_1} \cap S_{J_2} = S_{J_1+J_2} \in \mathcal{E}(K)$. Thus to every induced ideal is associated a filter on the set of induced maximal ideals. Conversely, if $\mathcal{E}$ is a filter on the set of induced maximal ideals of R^*, then we define $M(\mathcal{E}) = \{r \in R^* : S_r \in \mathcal{E}\}$. Since $\mathcal{E}$ is a filter, $M(\mathcal{E})$ is an ideal of R^*.

Following Cherlin, we say that an ultrafilter on the set of induced maximal ideals of R^* is *bounded* if it contains an element of the form S_J for some finitely generated ideal J of R^*. The following lemma, the proof of which extends directly from Cherlin's context, gives a general description of the maximal ideals of R^*.

PROPOSITION 4.2 *There is a one-to-one correspondence between maximal ideals of R^* and bounded ultrafilters on the set of induced maximal ideals of R^* given by*

$$M \mapsto \mathcal{E}(M)$$

and

$$\mathcal{E} \mapsto M(\mathcal{E}).$$

The principal ultrafilters correspond to induced maximal ideals, and vice versa.

Proof. If M is a maximal ideal, then $M \subseteq M(\mathcal{E}(M)) \subset R$, so $M = M(\mathcal{E}(M))$. Similarly, if $\mathcal{E}$ is a bounded ultrafilter, then $\mathcal{E} \subseteq \mathcal{E}(M(\mathcal{E}))$. Hence $\mathcal{E} = \mathcal{E}(M(\mathcal{E}))$, since $\mathcal{E}$ and $\mathcal{E}(M(\mathcal{E}))$ are proper ultrafilters. The last assertion is clear. □

In the terminology of [2], Proposition 4.2 states that if R^* has infinitely many maximal ideals, then every maximal ideal of R^* is a "limit" of the induced maximal ideals with respect to an ultrafilter on the set of induced maximal ideals.

We now indicate a second, more constructive method for exhibiting maximal ideals of R^*. Let $\mathcal{J}$ be the set of induced ideals $J = (J_i)_\mathcal{D}$ of R^* such that for $\mathcal{D}$-many i, $J_i = R_i$ or J_i is the intersection of at most finitely many maximal ideals of R_i. We view $\mathcal{J}$ as a partially ordered set under reverse inclusion, namely, if $J, K \in \mathcal{J}$, then we write $J \leq K$ if $K \subseteq J$. Notice that $R^* \in \mathcal{J}$ and $R^* \leq J$ for all $J \in \mathcal{J}$. We define the following operations on $\mathcal{J}$:

- $J \vee K := J \cap K$
- $J \wedge K := (\sqrt{J_i + K_i})_\mathcal{D}$
- $J \backslash K := (L_i)_\mathcal{D}$, where L_i is the intersection of the maximal ideals of R_i that contain J_i but not K_i ($L_i = R_i$ if this intersection is empty).

These notions are well-defined since $\mathcal{D}$ is an ultrafilter. We conclude that $(\mathcal{J}, \vee, \wedge)$ is a complemented distributive lattice with least element R^*. Thus, if $J \in \mathcal{J}$, then the set $\mathcal{B}(J) := \{K \in \mathcal{J} : K \leq J\}$ is a Boolean algebra since this set has a greatest element, namely J. Therefore, every proper filter on $\mathcal{B}(J)$ extends to an ultrafilter [17, p. 76, Corollary 3]. We show that for each $J \in \mathcal{J}$, these ultrafilters on $\mathcal{B}(J)$ are in one-to-one correspondence with the maximal ideals of R^* that contain J. If $J \in \mathcal{J}$ and $\mathcal{E}$ is an ultrafilter on $\mathcal{B}(J)$, we define $M_J(\mathcal{E}) := \sum_{K \in \mathcal{E}} K$. If M is a maximal ideal of R^* that contains J, then $\mathcal{E}_J(M) := \{K \in \mathcal{J} : J \subseteq K \subseteq M\}$.

THEOREM 4.3 *Let $J \in \mathcal{J}$. There is a bijective mapping between the set of ultrafilters $\mathcal{E}$ on $\mathcal{B}(J)$ and the maximal ideals M of R^* that contain J given by*

$$\mathcal{E} \mapsto M_J(\mathcal{E})$$

and

$$M \mapsto \mathcal{E}_J(M).$$

Moreover, the non-principal ultrafilters on $\mathcal{B}(J)$ correspond precisely to the non-induced maximal ideals of R^* that contain J.

Proof. Let $\mathcal{E}$ be an ultrafilter on $\mathcal{B}(J)$ and suppose $r := (r_i)_\mathcal{D} \in R^* \backslash M_J(\mathcal{E})$. For each i, let $K_i = \sqrt{J_i + R_i r_i}$, with the convention that $K_i = R_i$ if $J_i + R_i r_i = R_i$. Then $K := (K_i)_\mathcal{D} \in \mathcal{B}(J)$, and by assumption $K \notin \mathcal{E}$. Thus $L := J \backslash K \in \mathcal{E}$. However, if $L := (L_i)_\mathcal{D}$, then $L_i + R_i r_i = R_i$ for $\mathcal{D}$-many i, so $R^* r + M_J(\mathcal{E}) = R^*$, and it follows that $M_J(\mathcal{E})$ is a maximal ideal of R^*. Conversely, suppose M is a maximal ideal of R^* that contains J. Clearly, $\mathcal{E} := \mathcal{E}_J(M)$ is a filter on $\mathcal{B}(J)$ with respect to the ordering $\leq$, so $\mathcal{E}$ extends to an ultrafilter $\mathcal{E}'$. We show that $M = M_J(\mathcal{E}')$. If $K \in \mathcal{B}(J) \backslash \mathcal{E}$, then $K + M = R^*$, since M is a maximal ideal of R^*. Suppose $r := (r_i)_\mathcal{D} \in M$. Then $J + Rr$ is contained in $L := (\sqrt{J_i + R_i r_i})_\mathcal{D}$ and $L \in \mathcal{B}(J)$. Thus $L \subseteq M$ and $L \in \mathcal{E} \subseteq \mathcal{E}'$, so $r \in L \subseteq M_J(\mathcal{E}')$. It follows that $M \subseteq M_J(\mathcal{E}')$, so, since M is a maximal ideal of R^*, $M = M_J(\mathcal{E}')$. Thus $\mathcal{E} = \mathcal{E}'$. The last assertion of the theorem is now an easy consequence of the definition of $\mathcal{B}(J)$. ☐

The construction of non-induced maximal ideals in Theorem 4.3, when applied to the case where $R_i = \mathbb{Z}$ for $\mathcal{D}$-many i, can be translated into the terminology of [19], where a similar theorem is proved for infinite products of $\mathbb{Z}$.

LEMMA 4.4 *For each $i \in I$, let A_i be a collection of ideals of R_i and let $K_i = \cap_{K \in A_i} K$. Then $(K_i)_D = \cap\{J : J = (J_i)_D$ with $J_i \in A_i\}$.*

Proof. Set $J = \cap\{J : J = (J_i)_D \text{ with } J_i \in A_i\}$. Clearly $(K_i)_D \subseteq J$, so what must be shown is that $J \subseteq (K_i)_D$. Let $(r_i)_D \in R^*$ such that $(r_i)_D \notin (K_i)_D$. Then the set $\{i \in I : r_i \in K_i\} \notin D$. Hence, since D is an ultrafilter, $\{i \in I : r_i \notin K_i\} \in D$. In particular, for D-many i, there exists an ideal L_i of R_i with $L_i \in A_i$ but $r_i \notin L_i$. Then $(r_i)_D \notin (K_i)_D$, so $(r_i)_D \notin J$. Therefore, $J \subseteq (K_i)_D$. □

LEMMA 4.5 *If R^* has infinitely many maximal ideals, then R^* has infinitely many induced maximal ideals.*

Proof. Suppose R^* has infinitely many maximal ideals but only finitely many induced maximal ideals. Then there exists a non-induced maximal ideal N of R^* and we may choose $r \in N$ such that r is not contained in any induced maximal ideals. However, this is impossible since $r := (r_i)_{\mathcal{D}}$, for some $r_i \in R_i$, and r must thus be contained in an induced maximal ideal of R^*. □

LEMMA 4.6 *If R^* has infinitely many maximal ideals, there exists a non-zero ideal $J \in \mathcal{J}$ that is contained in infinitely many induced maximal ideals of R^*.*

Proof. For each $k > 0$, define $A_k := \{i \in \mathcal{I} : R_i$ has at least k many maximal ideals$\}$. Then by Proposition 4.1 and the assumption that R^* has infinitely many maximal ideals, each $A_k \in \mathcal{D}$ and $A_{k+1} \subseteq A_k$ for all $k > 0$. Define $B_k := A_k \backslash A_{k+1}$ for each $k > 0$. For each $i \in B_k$, let J_i be the intersection of k maximal ideals that contain I_i. Then for each $k > 0$, $\{i \in \mathcal{I} : R_i/J_i$ has at least k maximal ideals$\} = A_k \in \mathcal{D}$. It follows from Proposition 4.1 that $J := (J_i)_{\mathcal{D}}$ is contained in infinitely many maximal ideals of R^* (apply the proposition to $R^*/J \cong \prod_{\mathcal{D}} R_i/J_i$) and that $J \in \mathcal{J}$. □

THEOREM 4.7 *If I is an induced ideal of R^* that is contained in infinitely many maximal ideals of R^*, then I is contained in at least $2^{2^{\omega}}$ many non-induced maximal ideals of R^*.*

Proof. Since I is contained in infinitely many maximal ideals, Lemma 4.6 (applied to $R^*/I \cong \prod_{\mathcal{D}} R_i/I_i$) guarantees there exists $J \in \mathcal{J}$ such that $I \subseteq J$ and J is contained in infinitely many maximal ideals of R^*. By Lemma 4.5, J is contained in infinitely many induced maximal ideals of R^*. It follows that $\mathcal{B}(J)$ is an infinite set. If the set of induced maximal ideals containing J has cardinality α then the cardinality β of $\mathcal{B}(J)$ is at most 2^{α}. The elements of $\mathcal{B}(J)$ correspond to the principal ultrafilters on $\mathcal{B}(J)$, and by [5, Corollary 7.4], there are at least $2^{2^{\beta}}$ ultrafilters on $\mathcal{B}(J)$. Thus there are at least $2^{2^{\beta}}$ many non-principal ultrafilters on $\mathcal{B}(J)$ and the claim follows. □

COROLLARY 4.8 *Every maximal ideal of R^* is an induced ideal if and only if R^* has finitely many maximal ideals.*

Proof. If R^* has finitely many maximal ideals and N is a a maximal ideal of R^*, then there exists $r \in N$ such that N is the only maximal ideal of R^* containing r. However, r is contained in an induced maximal ideal of R^*, and this forces N to be induced. Conversely, suppose every maximal ideal of R^* is induced but that R^* has infinitely many maximal ideals. Then by Lemma 4.5, R^* has infinitely many induced maximal ideals. Since R^* has infinitely many maximal ideals, it follows from Lemma 4.6 that there is an ideal $J \in \mathcal{J}$ such that R^*/J has infinitely many maximal ideals. By Theorem 4.7, there are infinitely many non-induced maximal ideals, contrary to assumption. Thus R^* has finitely many maximal ideals. □

A commutative ring R has *finite character* if every non-zero element of R is contained in at most finitely many maximal ideals of R. For example, one-dimensional Noetherian domains have finite character, so the introduction of the finite character property in our context is relevant to the study of Peano rings. Note however that an ultraproduct of finite character rings need not have finite character, as is evidenced by the case $R_i = \mathbb{Z}$ for all $i \in \mathbb{N}$. (The element $r := (r_i)_\mathcal{D}$, where for each $i > 0$, r_i is the product of the first i primes, is contained in infinitely many induced maximal ideals of R^*. This follows from Proposition 4.1 applied to R^*/R^*r.) Our strongest result regarding the description of maximal ideals of R^* is obtained under the finite character assumption:

THEOREM 4.9 *If R_i has finite character for $\mathcal{D}$-many i, then every maximal ideal of R^* is of the form $M_J(\mathcal{E})$ for some $J \in \mathcal{J}$ and ultrafilter $\mathcal{E}$ on $\mathcal{B}(J)$.*

Proof. Let M be a maximal ideal of R^*, and let $I := (I_i)_\mathcal{D}$ be the largest induced ideal of R^* contained in N (such an ideal exists by Zorn's Lemma). For each i, define J_i to be the intersection of the maximal ideals of R_i that contain I_i. Then by Lemma 4.4, $J := (J_i)_\mathcal{D}$ is the intersection of the induced maximal ideals that contain I. If $J \not\subseteq N$, then $J + R^*(s_i)_\mathcal{D} = R^*$ for some $(s_i)_\mathcal{D} \in N$. Thus, for $\mathcal{D}$-many i, $J_i + Rs_i = R_i$ and s_i is not contained in any maximal ideal containing I_i. By design, J_i is contained in every maximal ideal of R_i that contains I_i, so $I_i + R_i s_i = R_i$ for $\mathcal{D}$-many i. Thus $R^* = I + R^*(s_i)_\mathcal{D} \subseteq N$, a contradiction that implies $J \subseteq N$. This forces $I = J$. Since R_i has finite character for $\mathcal{D}$-many i, $I \in \mathcal{J}$. Now $\mathcal{E}_I(M)$ is an ultrafilter on $\mathcal{B}(I)$, so by Theorem 4.3, $M = M_I(\mathcal{E}_I(M))$. □

5. REPRESENTATIONS OF R^*

We now study representations of R^* as intersections of localizations of R^* at induced maximal ideals. The first proposition indicates a sense in which the non-induced maximal ideals are superfluous in such representations. Theorem 5.3 then characterizes when this representation is unique. This uniqueness of representation is an attractive property that one finds, for example, in studies of rings of continuous functions and nonstandard models of orders in algebraic number fields.

PROPOSITION 5.1 *If R^* is an integral domain, then $R^* = \cap_M R^*_M$, where M ranges over the set of all induced maximal ideals of R^*.*

Proof. It is easy to see that an element q of the quotient field of a domain R is in R_M for some maximal ideal M of R if and only if $R \cap Rq^{-1} \not\subseteq M$. Thus to prove the proposition, it suffices to show every proper ideal of the form $R^* \cap R^*q$, $q \in F^*$, of R^* is contained in some induced maximal ideal of R^*. Let $q := (q_i)_\mathcal{D} \in F^*$ be such that $R^* \cap R^*q \subset R^*$ (recall that F^* is the quotient field of R^*). Then for $\mathcal{D}$-many i, $R_i \cap R_i q_i \subset R_i$, so for each such i, there is a maximal ideal M_i such that $R_i \cap R_i q_i \subseteq M_i$. If for some $i \in \mathcal{I}$, $R_i \cap R_i q_i = R_i$, then set $M_i = R_i$. It follows from Lemma 2.1 that $R^* \cap R^*q \subseteq (M_i)_\mathcal{D}$ and $(M_i)_\mathcal{D}$ is a maximal ideal of R^*. □

We are interested in when the representation in Proposition 5.1 is unique. If R is an integral domain and A is a set of maximal ideals of R such that $R = \cap_{M \in A} R_M$ but for any maximal ideal N in A, $R \neq \cap_{M \in A \setminus \{N\}} R_M$, then we say the representation of R as an intersection of the R_M, $M \in A$, is *irredundant.* Condition (#) provides the strongest form of an irredundant representation:

> *An integral domain R satisfies (#) if and only if for every subset A of maximal ideals of R, $R = \cap_{M \in A} R_M$ implies A is the set of all maximal ideals of R.*

Thus if R satisfies (#), every localization of R at a maximal ideal is essential in the representation of R as the intersection of its localizations.

If R is a domain and I is an ideal of R, then the *J-radical* of I is the intersection of all maximal ideals of R that contain I. It can be shown that if A is a collection of maximal ideals of R, then a representation of R as an intersection of the R_M, $M \in A$, is irredundant if and only if each maximal ideal $M \in A$ is the J-radical of a divisorial ideal (see [21, Proposition 2.2], for example).

LEMMA 5.2 *If R^* is a domain, then M is the J-radical of a divisorial ideal of R^* if and only if $M = (M_i)_\mathcal{D}$ for maximal ideals M_i of R_i such that for $\mathcal{D}$-many i, M_i is the J-radical of a divisorial ideal of R_i.*

Proof. Suppose M is the J-radical of a divisorial ideal of R^*. Then M is the J-radical of an ideal of the form $R^* \cap R^*(q_i)_\mathcal{D}$, where $q := (q_i)_\mathcal{D} \in F^*$. For each $i \in \mathcal{I}$, set $K_i = R_i \cap R_i q_i$. Now Lemma 2.1 implies that for $\mathcal{D}$-many i, $K_i \subseteq M_i$ for some maximal ideal M_i of R_i. Let $\mathcal{B}$ be the set of all i such that M_i is not the J-radical of K_i. If $\mathcal{B} \in \mathcal{D}$, then for each i in $\mathcal{B}$, R_i/K_i is not a quasilocal ring. Thus, since by Lemma 2.1(iv) $K := (K_i)_\mathcal{D} = R^* \cap R^*q$, it follows from Proposition 4.1 that R^*/K has at least two maximal ideals, contrary to assumption. Thus M_i is the J-radical of $R_i \cap R_i q_i$ for $\mathcal{D}$-many i and $(M_i)_\mathcal{D}$ is the J-radical of K in R^*. This forces $M = (M_i)_\mathcal{D}$.

Conversely, suppose $\mathcal{A}$ is the set of all $i \in \mathcal{I}$ such that M_i is the J-radical of a divisorial ideal of R_i and that $\mathcal{A} \in \mathcal{D}$. Then there exists for each $i \in \mathcal{A}$, $q_i \in F_i$ such that the J-radical of $R_i \cap R_i q_i$ is M_i. If $i \notin \mathcal{A}$, set $q_i = 1$. Define $q = (q_i)_\mathcal{D}$ and set $K = R^* \cap R^*q$. Then $K \subseteq (M_i)_\mathcal{D}$. If R^*/K has more than one maximal ideal, then there exist $a, b \in R^*$ such that $a + K$ and $b + K$ are non-units in R^*/K but $a + b + K$ is a unit in R^*/K. Write $a = (a_i)_\mathcal{D}, b = (b_i)_\mathcal{D}$, and $K = (K_i)_\mathcal{D}$. Then for $\mathcal{D}$-many i, $a_i + K_i$ and $b_i + K_i$ are non-units but $a_i + b_i + K_i$ is a unit of R_i/K_i. Hence, for $\mathcal{D}$-many i, R_i/K_i has at least two maximal ideals, contrary to assumption. We conclude $(M_i)_\mathcal{D}$ is the only maximal ideal of R^* that contains K. □

THEOREM 5.3 *If R^* is a domain, the representation $R^* = \cap_M R^*_M$, where M ranges over all induced maximal ideals of R, is irredundant if and only if R_i satisfies (#) for $\mathcal{D}$-many i.*

Proof. If R_i satisfies (#) for $\mathcal{D}$-many i, then for $\mathcal{D}$-many i, each maximal ideal of R_i is the J-radical of a divisorial ideal. Hence by Lemma 5.2, each induced maximal ideal of R^* is the J-radical of a divisorial ideal, and the representation $R^* = \cap_M R_M$, where M ranges over the induced maximal ideals of R^*, is irredundant. Conversely, suppose the representation $\cap_M R^*_M$, M an induced maximal ideal of R^*, is irredundant but that $\mathcal{D}$ does not contain the set of elements $i \in \mathcal{I}$ such that R_i satisfies (#). Then for $\mathcal{D}$-many i, each R_i has a maximal ideal M_i such that M_i is not the J-radical of a divisorial ideal of R^*. Hence, by Lemma 5.2, $M = (M_i)_\mathcal{D}$ is not the J-radical of a divisorial ideal of R^*. Since $\cap_{N \neq M} R^*_N \not\subseteq R^*_M$, where N ranges over the induced maximal ideals of R^* distinct from M, there is an element $q := (q_i)_\mathcal{D} \in F$ such that $J := R^* \cap R^* q \subseteq M$ but $J \not\subseteq N$ for all induced maximal ideals $N \neq M$. By assumption there exists for $\mathcal{D}$-many i, a maximal ideal $N_i \neq M_i$ of R_i such that $J_i \subseteq N_i \cap M_i$. Thus, if $N := (N_i)_\mathcal{D}$, $J \subseteq N \cap M$, and since $N \neq M$, this contradicts the choice of J. Hence R_i satisfies (#) for $\mathcal{D}$-many i. □

COROLLARY 5.4 *R^* satisfies (#) if and only if there exists $n > 0$ such that for $\mathcal{D}$-many i, R_i has precisely n many maximal ideals.*

Proof. If R^* satisfies (#), then every maximal ideal of R^* is the J-radical of a divisorial ideal and by Lemma 5.2 is induced by maximal ideals of the R_i. By Corollary 4.6, this means R^* has finitely many maximal ideals and the claim follows from Proposition 4.1. The converse is a consequence of Proposition 4.1 and the fact that domains with only finitely many maximal ideals satisify (#). □

6. NON-MAXIMAL PRIME IDEALS OF R^*

In the present section we examine the non-maximal prime ideals of R^*. We first show that in some important cases (e.g. $\mathcal{I} = \mathbb{N}$) R^* has infinite Krull dimension, then we characterize when every non-zero prime ideal of R^* is contained in a unique maximal ideal of R^*.

LEMMA 6.1 *If R is an integral domain and there exists a non-unit $r \in R$ such that $0 \neq \cap_{k>0} Rr^k$, then R has Krull dimension at least two.*

Proof. Let M be a maximal ideal of R containing r. By localizing R at M, we may assume that R is quasilocal. If R is one-dimensional, then the radical of Rr is M. Moreover, if $0 \neq s \in \cap_{k>0} Rr^k$, then the radical of Rs is also M. But this implies there exists $n > 0$ such that $r^n \in Rs \subseteq \cap_{k>0} Rr^k$, a contradiction that forces the Krull dimension of R to be at least two. □

In Proposition 6.2, we use the notion of countable incompleteness from Section 3.

PROPOSITION 6.2 *If R^* is a domain that is not a field and $\mathcal{D}$ is a countably incomplete ultrafilter, then each non-zero prime ideal of R^* has infinite height. Consequently, R^* is not Noetherian when $\mathcal{D}$ is countably incomplete.*

Proof. Let $r := (r_i)_{\mathcal{D}} \in R^*$ be a non-unit. Since $\mathcal{D}$ is countably incomplete, there exists a function $n : \mathcal{I} \to \mathbb{N}$ such that for all $k > 0$, $\{i : n(i) > k\} \in \mathcal{D}$. Set $s = (r_i^{n(i)})_{\mathcal{D}}$. Then for all $k > 0$, $\frac{1}{r^k}s \in R^*$. Hence $0 \neq s \in \cap_{k>0} R^* r^k$. Let P be a prime ideal of R^*. Then by Lemma 6.1, R^*_P has a nonzero non-maximal prime ideal, say P_1. Localizing R^*_P at P_1 and applying Lemma 6.1 again implies P_1 contains a non-zero non-maximal prime ideal P_2 of R^*_P. Continuing in this manner, we conclude that P has infinite height. □

The referee has shown us an interesting connection between "fragmented" domains and the property of ultraproducts that the intersection of powers of a principal ideal is non-zero. An integral domain R is *fragmented* if for every nonunit $r \in R$, there is a nonunit $s \in R$ such that $r \in \cap_{n>0} Rs^n$. An integral domain R that is fragmented has infinite Krull dimension [7]; compare to Lemma 6.1 and Proposition 6.2.

In the important example of Peano rings, those ultrapowers of orders in algebraic number fields, the ultraproduct has what we shall follow [6, 8] and term **PM**: *each non-zero non-maximal prime ideal is contained in a unique maximal ideal.* This is a familiar phenomenon in the prime spectra of rings of continuous functions of various sorts (e.g. the ring of entire functions [9, Theorem 8.1.11], the ring of continuous integer-valued functions [1], and N. Schwartz's real closed rings [26]). We characterize when this property arises in the ultraproduct construction. Our approach is to use the somewhat surprising fact that this ostensibly second order property of prime ideals can be encoded into first order language. This gives a general approach to some results in the study of the aforementioned Peano rings. That Peano rings are PM domains has been proved previously using model-theoretic and topological methods [4, 19]. However, these proofs rely heavily on properties of Dedekind domains. Theorem 6.6 offers a more transparent explanation of these results. If R is a domain, max(R) denotes the space of maximal ideals of R, where the closed sets of max(R) are those sets $V \subseteq \max(R)$ such that there exists an ideal I of R with the property that the set of maximal ideals containing I is precisely V.

LEMMA 6.3 *Let R be an integral domain with quotient field F. If V_1 and V_2 are disjoint closed subsets of max(R), $S_1 := R \setminus \cup_{M \in V_1} M$ and $S_2 := R \setminus \cup_{N \in V_2} N$, then $R_{S_1} R_{S_2} = F$ if and only if $P \not\subseteq \cup_{N \in V_2} N$ for every non-zero prime ideal P of R such that $P \subseteq \cup_{M \in V_1} M$.*

Proof. The proof is a simple modification of an argument of Matlis in [20, Theorem 20]. Observe that $R_{S_1} R_{S_2} = F$ if and only if $(R_{S_1})_{S_2} = F$; if and only if every non-zero prime ideal of R_{S_1} meets S_2; if and only if every non-zero prime ideal of R that does not meet S_1 must meet S_2. Observe that a non-zero prime ideal P of R does not meet S_1 if and only if $P \subseteq \cup_{M \in V_1} M$. On the other hand, if P meets S_2, then $P \not\subseteq \cup_{N \in V_2} N$. The claim now follows. □

LEMMA 6.4 *Let R be an integral domain and suppose $\{M_\alpha\}$ is a collection of maximal ideals of R. If M is a maximal ideal of R and $M \subseteq \cup_\alpha M_\alpha$, then $\cap_\alpha M_\alpha \subseteq M$.*

Proof. If $M \subseteq \cup_\alpha M_\alpha$, then $M + \cap_\alpha M_\alpha \subseteq \cup_\alpha M_\alpha$ and the claim follows. □

LEMMA 6.5 *An integral domain R is a PM domain if and only if for each closed subset V of $max(R)$ and non-zero prime ideal P of R such that $P \subseteq \cup_{N \in V} N$, it must be that $M \subseteq \cup_{N \in V} N$ for all maximal ideals M of R that contain P.*

Proof. Suppose R is a PM domain. Let V be a closed subset of $\max(R)$ and P be a non-zero prime ideal of R such that $P \subseteq A := \cup_{N \in V} N$. By Zorn's Lemma, there exists a largest ideal J containing P such that $J \subseteq A$. If $K := \cap_{N \in V} N$, then $J + K \subseteq A$; hence $K \subseteq J$. Thus $K \subseteq M$, where M is the unique maximal ideal of R containing P, and since K is closed, $M \in V$. In particular, $M \subseteq A$, proving the claim. The converse is clear. □

THEOREM 6.6 *Let R be an integral domain. Then R is a PM domain if and only if for all non-units $x, y, z \in R$ such that $Rx + Ry = R$, there exists $x', y' \in R$ such that $Rx + Rx' = R$, $Ry + Ry' = R$ and $x'y' \in Rz$.*

Proof. Suppose R is a PM domain. Let x, y, z be nonunits in R such that $Rx+Ry = R$. Let V_1 be the set of maximal ideals of R containing x and V_2 be the set of maximal ideals of R containing y. By assumption, V_1 and V_2 are disjoint. For $i = 1, 2$, define $A_i = \cup_{M \in V_i} M$ and $S_i = R \backslash A_i$. Using Lemma 6.3, we show first that $R_{S_1} R_{S_2} = F$. Let P be a prime ideal of R such that $P \subseteq A_1$, and let M be a maximal ideal of R that contains P. Then by our assumption on R and Lemma 6.5, $M \subseteq A_1$. If $M \subseteq A_2$, then by Lemma 6.4, $\cap_{N \in V_2} N \subseteq M$. But V_2 is a closed subset of $\max(R)$, so this forces $M \in V_2$. Similarly, since $M \subseteq A_1$, $M \in V_1$. But V_1 and V_2 are disjoint subsets of $\max(R)$, so $M \not\subseteq A_2$. Furthermore, $P \not\subseteq A_2$, since by Lemma 6.5 this would force $M \subseteq A_2$. Thus by Lemma 6.3, $R_{S_1} R_{S_2} = F$. Hence $\frac{1}{z} = \frac{r}{x'} \frac{1}{y'}$ for some $r \in R$, $x' \in S_1$ and $y' \in S_2$. In particular, $rz = x'y'$, $Ry + Ry' = R$, and $Rx + Rx' = R$.

Conversely, to prove the stated property implies R is a PM domain, we will use the characterization of PM domains in Lemma 6.5. Let M be a maximal ideal of R and V be a closed subset of $\max(R)$ not containing M. Define $J := \cap_{M \in V} M$. Suppose $M \not\subseteq A := \cup_{N \in V} N$. We show that $R_M R_S = F$, where $S = R \backslash A$. Lemma 6.3 then substantiates the claim. Let $0 \neq z \in R$, $x \in M \backslash A$ and $y \in J \backslash M$ (recall that $M \notin V$). Then $Rx + Ry = R$, so by assumption, there exist $x', y', w \in R$ such that $Rx + Rx' = R$, $Ry + Ry' = R$ and $x'y' = wz$. Thus $\frac{1}{z} = \frac{w}{x_2} \frac{1}{y_2}$. Since $Rx + Rx' = R$, $x' \notin M$. Similarly, $y' \notin A$ since y is contained in every maximal ideal of R contained in V. Thus $\frac{1}{z} \in R_M R_S \subseteq F$. The choice of z was arbitrary, so it follows that $F \subseteq R_M R_S$. Hence $R_M R_S = F$. It follows that R is a PM domain. □

COROLLARY 6.7 *R^* has the property that each non-zero prime ideal of R^* is contained in a unique maximal ideal of R^* if and only if for $\mathcal{D}$-many i, R_i has the property that each non-zero prime ideal of R_i is contained in a unique maximal ideal of R_i.*

Proof. Apply Theorem 6.6 and Łos's Theorem. □

REFERENCES

[1] N. Alling, Rings of continuous integer-valued functions and non-standard arithmetic, Trans. Amer. Math. Soc. **118** (1965), 498-525.

[2] P. J. Cahen, K. A. Loper and F. Tartarone, Integer-valued polynomials and Prüfer v-multiplication domains, J. Algebra **226** (2000), no. 2, 765-787.
[3] C. Chang and H. Keisler, *Model Theory*, North-Holland, 1990.
[4] G. Cherlin, Ideals of integers in nonstandard number fields, *Model theory and algebra*, Lecture Notes in Math. **498**, Springer-Verlag, 1975, 60-90.
[5] W. Comfort and S. Negrepontis, *The theory of ultrafilters*, Springer-Verlag, New York, 1974.
[6] M. Contessa, On pm-rings, Comm. Algebra **10** (1982), 93-108.
[7] J. Coykendall and D. Dobbs, Fragmented domains have infinite Krull dimension, Rend. Circ. Mat. Palermo (2) **50** (2001), no. 2, 377-388.
[8] G. De Marco and A. Orsatti, Commutative rings in which every ideal is contained in a unique maximal ideal, Proc. Amer. Math. Soc **30** (1971), 459-466.
[9] M. Fontana, J. Huckaba and I. Papick, *Prüfer Domains*; Marcel Dekker, New York, 1997.
[10] R. Gilmer, *Multiplicative Ideal Theory*, Queen's Papers in Pure and Applied Mathematics, No. 90, Queen's University, Kingston, Ontario, 1992.
[11] R. Gilmer and W. Heinzer, Imbeddability of a commutative ring in a finite-dimensional ring, manuscripta math. **84** (1994), 401-414.
[12] R. Gilmer and W. Heinzer, Infinite products of zero-dimensional commutative rings, Houston J. Math. **21** (1995), 247-259.
[13] R. Gilmer, W. Heinzer, M. Roitman, Finite generation of powers of ideals, Proc. Amer. Math. Soc. **127** (1999), 3141-3151.
[14] S. Glaz, Finite conductor rings, Proc. Amer. Math. Soc., to appear.
[15] S. Glaz, *Commutative coherent rings*, LNM 1371, Springer Verlag, New York, 1989.
[16] R. Heitmann and L. Levy, $1\frac{1}{2}$ and 2 generator ideals in Prüfer domains, Rocky Mountain J. Math. **5** (1975), 361-373.
[17] N. Jacobson, *Basic algebra II*, W. H. Freeman and Co., 1989.
[18] C. U. Jensen, Peano Rings of abritrary global dimension, J. London Math. Soc. **21** (1980), 39-44.
[19] R. Levy, P. Loustauna and J. Shapiro, The prime spectrum of an infinite product of copies of $\mathbb{Z}$, Fund. Math. **138** (1991), 155-164.
[20] E. Matlis, *Torsion-Free Modules*, University of Chicago Press, 1973.
[21] B. Olberding, Globalizing local properties of Prüfer domains, J. Algebra **205** (1998), 480-504.
[22] B. Olberding, On the classification of stable domains, J. Algebra **243** (2001), 177-197.
[23] M. Roitman, On finite generation of powers of ideals, J. Pure Appl. Algebra **161** (2001), 327-340.
[24] J. D. Sally, On the number of generators of powers of an ideal, Proc. Amer. Math. Soc. **53** (1975), 24-26.
[25] J. D. Sally, *Numbers of generators of ideals in local rings*, Marcel Dekker, 1978.
[26] N. Schwartz, Real closed rings, Algebra and order (Luminy-Marseille, 1984), 175–194.

Geometric Subsets of a Spectrum

GABRIEL PICAVET Laboratoire de Mathématiques Pures, Université de Clermont II, 63177 Aubière Cedex, France
Gabriel.Picavet@math.univ-bpclermont.fr

ABSTRACT

It is well known that a spectrum is not sufficient to recover a ring. To clarify the situation, we are aiming to show that some subsets of a spectrum are solutions to universal problems. Namely, if $X \subset \mathrm{Spec}(A)$ where A is a ring, then X is called geometric if there is a ring morphism $A \to \Delta$ with spectral image contained in X and universal for this property. The similar problem has always a solution in the category of locally ringed spaces. We examine when points are geometric. We show that when A is locally Noetherian or an almost multiplication ring, then a prime ideal P is geometric if and only if P is a minimal prime ideal. Geometric points are characterized for Prüfer domains. An arbitrary geometric subset is stable under formal generizations and a solution is an epimorphism focussing on X. The geometric property is local on the spectrum, universal and is descended by algebraically pure morphisms. Moreover, local morphisms induced by completions are isomorphisms. We give a complete characterization of geometric subsets which are either closed or quasi-compact and stable under generizations. The akin problem of quasi-geometric subsets is examined.

0 INTRODUCTION AND NOTATION

As most Algebraic Geometry textbooks explain, a spectrum is not sufficient to recover a ring. For instance, all fields have homeomorphic spectra. Thus to remedy, Algebraic Geometry uses sheaves. Our notation and definitions are those of N. Bourbaki and E.G.A. of A. Grothendieck and J. Dieudonné. In this paper, we are aiming to show that some subsets of spectra determine ring morphisms. More precisely, let A be a ring and $X \subset \mathrm{Spec}(A)$. We say that X is geometric if there exists a ring morphism $\delta : A \to \Delta$ such that, firstly ${}^a\delta(\mathrm{Spec}(\Delta)) \subset X$ and secondly, each ring morphism $f . A \to B$ such that ${}^a f(\mathrm{Spec}(B)) \subset X$ can be factored uniquely through δ. Hence δ is a solution of a universal problem. We call it the geometric problem associated to X and δ is called a solution. It turns out that in this case ${}^a\delta(\mathrm{Spec}(\Delta)) = X$ so that X determines a ring morphism within an isomorphism. A variant of this problem is gotten when one restricts to the category of reduced rings, in which case we consider quasi-geometric subsets. Obviously, if X is geometric with solution $A \to \Delta$, then X is quasi-geometric with solution $A \to \Delta_{\mathrm{red}}$.

As above recalled, sheaves are the right tool to use in this kind of problem. So in Section 1, the problem is studied inside the category of ringed spaces and the category of locally ringed spaces. Let $(Y, \mathcal{F})$ be a ringed (respectively, locally ringed) space and X a subset of Y. Then consider the induced ringed space (resp. locally ringed space) $(X, \mathcal{F}_{|X})$. We show that $(X, \mathcal{F}_{|X}) \to (Y, \mathcal{F})$ is a monomorphism of the considered categories. Moreover, each morphism $(\mu, \theta) : (Z, \mathcal{G}) \to (Y, \mathcal{F})$ of the category, such that $\mu(Z) \subset X$, can be factored through $(X, \mathcal{F}_{|X}) \to (Y, \mathcal{F})$. Thus the geometric problem associated to X has always a solution in these categories. This is not really surprising. But things are going wrong when we consider the category of schemes. If $(Y, \mathcal{O}_Y)$ is a scheme and $X \subset Y$, then unless $(X, \mathcal{O}_{Y|X})$ be a scheme, the geometric problem associated to X need not have a solution. Indeed, in Section 3 and 4 we consider geometric problems in the category of affine schemes (equivalently, the category of commutative rings). A subset of a spectrum has a solution in this category, namely is geometric, if and only if it has a solution in the category of schemes and we show that most of time, a point is not geometric. We give conditions for a subscheme to be geometric and we show that the spectral image of a flat epimorphism is geometric. Section 1 ends with some considerations about a functor from the category of ringed spaces to the category of locally ringed spaces and solutions to a geometric problem. J. Malgoire and C. Voisin introduced this functor [15].

In Section 2, we introduce some materials and present some results used in the next sections. They have their own interest. Indeed, they give factorization properties of ring morphisms, linked to the subsets of a spectrum. Let A be a ring and $X \subset \mathrm{Spec}(A)$. Denote by $(\mathrm{Spec}(A), \widetilde{A})$ the associated affine scheme. In view of Section 1, it is clear that the ring of sections $\widetilde{A}(X)$ on X of the affine scheme has a prominent part since an arbitrary ring morphism $f : A \to B$ such that ${}^a f(\mathrm{Spec}(B)) \subset X$ can be factored $A \to \widetilde{A}(X) \to B$. Unfortunately, in general neither the factorization is unique nor the spectral image of $A \to \widetilde{A}(X)$ is contained in X. Actually X^g is contained in this image. If X is a patch, we give an estimate of this image and recall a calculation given by M. Raynaud of $\widetilde{A}(X^g)$ as a direct limit. Then we give expressions of $\widetilde{A}(X)$ in some cases. For instance, if A is an integral domain,

$\widetilde{A}(X) = \cap\,[A_P\ ;\ P \in X]$. Then we consider the Gabriel's localization associated to X. Let $f : A \to B$ be a ring morphism with spectral image X. The set $\mathcal{F}$ of all ideals I of A such that $X \subset \mathrm{D}(I)$ is a site and I belongs to $\mathcal{F}$ if and only if $IB = B$. We show that there is a factorization $A \to A_{\mathcal{F}} = \widetilde{A}(X^g) \to \widetilde{A}(X) \to B$. Next we consider the ring $A(\mathcal{F})$ introduced by C. Năstăsescu and N. Popescu. An element $b \in B$ belongs to $A(\mathcal{F})$ if and only if $A :_A b \in \mathcal{F}$. There is a factorization $A \to A_{\mathcal{F}} \to A(\mathcal{F}) \to B$. If A is an integral domain, then $A_{\mathcal{F}} = \widetilde{A}(X)$ while $A_{\mathcal{F}} = A(\mathcal{F})$ if f is flat or injective. The main interest of $A(\mathcal{F})$ is that a factorization $A \to A(\mathcal{F}) \to D$ for a ring morphism $A \to D$ is unique as soon as the spectral image of $A \to D$ is contained in X. Next, the ring $A(\mathcal{F})$ is contained in the dominiom of f (see 0.2 below). We give a generalization of an unpublished result of D. Ferrand. If $f : A \to B$ is a flat morphism and $X = \mathrm{Im}({}^a f)$, then $\widetilde{A}(X) = A(\mathcal{F}) = \mathrm{Dom}_A(B)$. We showed that a quasi-compact subset X of $\mathrm{Spec}(A)$, stable under generizations, is the spectral image of an explicit flat morphism [21]. Thus $\widetilde{A}(X)$ can be calculated for an arbitrary quasi-compact and stable under generizations subset. Ferrand's result deals with a quasi-compact open subset. Other factorization results in the same vein are given.

Section 3 gives elementary properties of geometric subsets. A geometric subset X of $\mathrm{Spec}(A)$ is a patch (see 0.1 below) and is stable under formal generizations that is, if P belongs to X then the spectral image of $A \to \widehat{A_P}$ is contained in X (here, $\widehat{A_P}$ denotes the PA_P-adic completion of A_P). A particular attention is paid to geometric points. A minimal prime ideal P of A is geometric with solution $A \to A_P$. The converse is true if PA_P is finitely generated or A is an almost multiplication ring. Now if V is a valuation domain with an idempotent maximal ideal M, then M is geometric with solution $A \to A/M$. Solutions in these examples are epimorphisms. This is a general fact (see Section 4). But one of them is flat while the other is not. We show that $\sqrt{\cap_n P^{(n)}} = P$ if P is geometric. The converse is true if A is a Prüfer domain. We show also that $\mathrm{Min}(A)$ is geometric if A is a reduced ring and $\mathrm{Min}(A)$ is compact, a solution being the maximal flat epimorphic extension.

In Section 4, we give a key result. If $X \subset \mathrm{Spec}(A)$ is geometric with solution $A \to \Delta(X)$ and $f : A \to B$ is a ring morphism, then ${}^a f^{-1}(X)$ is geometric with solution $B \to B \bigotimes_A \Delta(X)$. It follows that a solution $A \to \Delta(X)$ associated to a geometric subset is an epimorphism. A characterization of geometric subsets using strict morphisms is deduced. We show also that an intersection of geometric subsets is a geometric subset. Moreover, the property of being a geometric subset is local on the spectrum that is to say X is geometric if and only if $X \cap \mathrm{D}(f_1), \ldots, X \cap \mathrm{D}(f_n)$ are geometric whenever $\mathrm{Spec}(A) = \mathrm{D}(f_1) \cup \cdots \cup \mathrm{D}(f_n)$. We do not know whether $X \cap P^g$ is geometric for each $P \in \mathrm{Spec}(A)$ involves X is geometric. This is linked to the descent of the geometric property by pure morphisms. We are only able to show that algebraic purity descends the geometric property. A solution $A \to \Delta(X)$ associated to a geometric subset X verifies akin properties of flat epimorphism as follows. The canonical morphism $A_P \to \Delta(X)_Q$ is an isomorphism for each $Q \in \mathrm{Spec}(\Delta(X))$ lying over P in A and such that $P^g \subset X$ (for a flat epimorphism, we only need $P \in X$). There is a similar result with respect to factor rings. Moreover, $\widehat{A_P} \to \widehat{\Delta(X)_Q}$ is an isomorphism. Hence, a solution of a geometric problem is a formally flat epimorphism. Unfortunately, we do not know of any converse. Nevertheless, we are in position to prove a main result. If X is a patch stable under generizations, then X is geometric if and only if X is the spectral image of a flat epimorphism $A \to B$. This last morphism is then a solution which identifies with $A \to \widetilde{A}(X)$. We deduce from this result that if A is a locally Noetherian ring, then X is geometric if and only if X is the spectral image of a flat epimorphism. Moreover, a quasi-compact open subset is geometric if and only if it is affine. We also characterize geometric closed subsets. Let $X = \mathrm{V}(I)$ be a closed subset of $\mathrm{Spec}(A)$, then X is geometric if and only if $I \subset \sqrt{J}$ implies $I \subset J$ for each ideal J of A. This last property is verified by pure ideals. Ideals verifying this property are idempotent. Conversely, an idempotent ideal I of a Prüfer domain A defines a geometric subset $\mathrm{V}(I)$. The end of the section is dedicated to quasi-geometric subsets. Similar results are gotten. But there are more examples. For instance, a closed subset is quasi-geometric as well as an arbitrary point of the spectrum. We show that a pre-flat morphism has a quasi-geometric spectral image.

We give below some recalls and results used in the paper. Rings are assumed to be commutative (with unit).

0.1. The patch topology (in French, topologie constructible) on the spectrum of a ring A is a compact topology on $\mathrm{Spec}(A)$ finer than the Zariski topology [10], [8]. Its closed sets (patches or proconstructible subsets) are the subsets ${}^a f(\mathrm{Spec}(B))$ where $f : A \to B$ is a ring morphism. If X is a subset of $\mathrm{Spec}(A)$ and $P \in \mathrm{Spec}(A)$, then P belongs to the patch closure X^c of X if and only if $X \cap \mathrm{V}(I) \cap \mathrm{D}(a) \neq \emptyset$ for every finitely generated ideal I of A and $a \in A$ such that $P \in \mathrm{V}(I) \cap \mathrm{D}(a)$. The generization of a subset $Z \in \mathrm{Spec}(A)$ is denoted by Z^g, its specialization by Z^s and its Zariski closure by $\overline{Z}$. If Z is a patch, then $\overline{Z} = Z^s$. Let $f : A \to B$ be a ring morphism with spectral image X, then $\overline{X} = \mathrm{V}(\mathrm{Ker}(f))$ which gives $\cap\, [P\, ; P \in X] = \sqrt{\mathrm{Ker}(f)}$.

0.2. In this paper, a ring epimorphism is an epimorphism of the category of commutative rings with units. For general results on epimorphisms, the reader is referred to the Samuel's Seminar on epimorphism 1967-1968 [28]. The dominiom $\mathrm{Dom}_A(B)$ of a ring morphism is the set of all $b \in B$ such that $b \otimes 1 = 1 \otimes b$ in $B \bigotimes_A B$. Then f is an epimorphism if and only if

$\mathrm{Dom}_A(B) = B$ or also, if and only if $B \bigotimes_A (B/f(A)) = 0$. We give here some new results used in Section 4.

Let $g : A \to B$ be a ring epimorphism and $Q \in \mathrm{Spec}(B)$ lying over P in A. Therefore, there is a ring epimorphism $f : R = A_P \to B_Q = S$. We set $M = PA_P$ and $N = QA_Q$.

• **(a)** We claim that $MS = N$.

Indeed, $R/M \to S/MS$ is a ring epimorphism and R/M is a field. It follows from [13, IV.1.3] that $R/M \to S/MS$ is an isomorphism because $S/MS \neq 0$. Hence, MS is a maximal ideal contained in N.

Now consider the M-adic topology on R and the N-adic topology on S and let $\widehat{R}$ and $\widehat{S}$ be the associated completions. There is a local ring morphism $\widehat{R} \to \widehat{S}$ and we can consider that $\widehat{S}$ is the MS-adic completion of S.

• **(b)** We claim that $\widehat{R} \to \widehat{S}$ is a formal ring epimorphism, that is to say $u = v$ for every commutative diagram $\widehat{R} \to \widehat{S} \underset{v}{\overset{u}{\rightrightarrows}} T$ of the category of commutative topological Hausdorff rings.

Indeed, $R \to S$ is an epimorphism and the image of S in $\widehat{S}$ is everywhere dense.

• **(c)** It seems that in general, $\widehat{R} \to \widehat{S}$ is not an epimorphism. What follows aims to examine the situation. Set $I = \cap_n M^n$ and $J = \cap_n N^n$, the induced ring morphism $R/I \to S/J$ is an epimorphism and we can replace f by this morphism. The tensor product topology of the N-adic topologies on $S \bigotimes_R S$ is for instance defined in [8, 0.7.7]. Let n be an arbitrary integer, then the image of $N^n \bigotimes_R S$ and $S \bigotimes_R N^n$ in $S \bigotimes_R S$ are $M^n (S \bigotimes_R S)$ because $M^n S = N^n$. Therefore, the tensor product topology is the M-adic topology. Denote by $\nabla : S \bigotimes_R S \to S$ the canonical ring morphism. Since f is an epimorphism, ∇ is an isomorphism so that the tensor product topology is Hausdorff. Since the N-adic topology on S is also the M-adic topology, there is an isomorphism $\widehat{S} \widehat{\bigotimes}_{\widehat{R}} \widehat{S} \to \widehat{S}$. But $\widehat{S} \widehat{\bigotimes}_{\widehat{R}} \widehat{S}$ is canonically isomorphic to the completion of $\widehat{S} \bigotimes_{\widehat{R}} \widehat{S}$. If the tensor product topology on $\widehat{S} \bigotimes_{\widehat{R}} \widehat{S}$ is Hausdorff, then the composite $\widehat{S} \bigotimes_{\widehat{R}} \widehat{S} \to \widehat{S} \widehat{\bigotimes}_{\widehat{R}} \widehat{S} \to \widehat{S}$ is injective. In this case, $\widehat{R} \to \widehat{S}$ is an epimorphism.

0.3. We recall some well known facts about pure ring morphisms and strict monomorphisms.

• **(a)** There is a general categorical definition of strict monomorphisms. In the category of commutative rings with units, an injective ring morphism $f : A \to B$ is strict if and only if $A = \mathrm{Dom}_A(B)$. Therefore, an arbitrary ring morphism $A \to B$ can be factored $A \to \mathrm{Dom}_A(B) \to B$ where the last morphism is strict. A strict ring epimorphism is an isomorphism. Let

$$\begin{array}{ccc} A & \longrightarrow & B \\ \downarrow & & \downarrow \\ C & \longrightarrow & D \end{array}$$

be a pullback diagram. If $C \to D$ is strict, so is $A \to B$.

• **(b)** A pure morphism is a universally injective ring morphism $f : A \to B$. A pure morphism is spectrally surjective and strict. Pure morphisms descend many properties (see [19]).

0.4. Let $f : A \to B$ and $g : A \to C$ be ring morphisms. Then the canonical map $\mathrm{Spec}(B \bigotimes_A C) \to \mathrm{Spec}(B) \times_{\mathrm{Spec}(A)} \mathrm{Spec}(C)$ is surjective and the spectral image of $A \to B \bigotimes_A C$ is ${}^a f(\mathrm{Spec}(B)) \cap {}^a g(\mathrm{Spec}(C))$ [8, Section 3].

1 THE CATEGORY OF RINGED SPACES

In this section, we intend to show that the geometric problem has a solution in the category of ringed spaces. For the reader's convenience, we begin by recalling some basic facts which can be found in [8, 0.3]. They are also summarized in [9]. We denote by $(X, \mathcal{F})$ or $\mathcal{F}$ a sheaf of commutative rings (a ringed space) on a topological space X and by $\mathrm{Hom}_X(\mathcal{F}, \mathcal{G})$ the set of all morphisms of ringed spaces $\mathcal{F} \to \mathcal{G}$ on X. When $\psi : X \to Y$ is a continuous map, the direct image ringed space $(Y, \psi_\star(\mathcal{F}))$ of $(X, \mathcal{F})$ is defined by $\psi_\star(\mathcal{F})(V) = \mathcal{F}(\psi^{-1}(V))$ for any open set $V \subset Y$. Obviously, $\psi_\star$ is a functor and $(\varphi \circ \psi)_\star = \varphi_\star \circ \psi_\star$.

Now a morphism of ringed spaces $(X, \mathcal{F}) \to (Y, \mathcal{G})$ is a pair (ψ, θ) where $\psi : X \to Y$ is a continuous map and $\theta : \mathcal{G} \to \psi_\star(\mathcal{F})$ is a morphism of sheaves on Y. The composite $(X, \mathcal{F}) \xrightarrow{(\psi,\theta)} (Y, \mathcal{G}) \xrightarrow{(\phi,\sigma)} (Z, \mathcal{H})$ is given by $(\phi \circ \psi, \phi_\star(\theta) \circ \sigma)$.

The category of ringed spaces is denoted by RS.

Let $\psi : X \to Y$ be a continuous map between topological spaces and $(Y, \mathcal{G})$ a ringed space. The inverse image ringed space $\psi^{-1}(\mathcal{G})$ of $\mathcal{G}$ on X is the sheaf associated to the presheaf $\mathcal{G}[\psi]$ on X defined by $U \mapsto \varinjlim_{V \supset \psi(U)} \mathcal{G}(V)$ where U is any open subset of X and the limit is taken over all open sets $V \supset \psi(U)$. When X is a subspace of Y and $j : X \to Y$ is the canonical injection, we get the induced ringed space $\mathcal{G}_{|X} = j^{-1}(\mathcal{G})$ on X.

There is a functorial isomorphism $v \mapsto v^\flat$ with respect to ringed spaces $(X, \mathcal{F})$

$$\mathrm{Hom}_X(\psi^{-1}(\mathcal{G}), \mathcal{F}) \to \mathrm{Hom}_Y(\mathcal{G}, \psi_\star(\mathcal{F})) \quad (\dagger)$$

Its inverse is denoted by $u \mapsto u^\sharp$. For each morphism of ringed spaces $w \in \mathrm{Hom}_X(\mathcal{F}_1, \mathcal{F}_2)$, we have $(w \circ v)^\flat = \psi_\star(w) \circ v^\flat$. We will have to consider $\rho_\mathcal{G} = \left(\mathrm{Id}_{\psi^{-1}(\mathcal{G})}\right)^\flat : \mathcal{G} \to \psi_\star\left(\psi^{-1}(\mathcal{G})\right)$ inducing a morphism of ringed spaces $(\psi, \rho_\mathcal{G}) : \left(X, \psi^{-1}(\mathcal{G})\right) \to (Y, \mathcal{G})$. Moreover, the following relation holds

$$v^\flat = \psi_\star(v) \circ \rho_\mathcal{G} \quad (\ddagger)$$

PROPOSITION 1.1. *Let $(Y, \mathcal{G})$ be a ringed space and $X \subset Y$ a subspace. Denote by $j : X \to Y$ the canonical continuous injective map.*

(1) *There is a monomorphism of ringed spaces $(j, \rho_\mathcal{G}) : \left(X, \mathcal{G}_{|X}\right) \to (Y, \mathcal{G})$.*
(2) *A morphism of ringed spaces $(\mu, \theta) : (Z, \mathcal{H}) \to (Y, \mathcal{G})$ such that $\mu(Z) \subset X$ can be factored by a unique morphism of ringed spaces $(Z, \mathcal{H}) \to \left(X, \mathcal{G}_{|X}\right)$.*

Proof. We first show that $(j, \rho_\mathcal{G}) : \left(X, \mathcal{G}_{|X}\right) \to (Y, \mathcal{G})$ is a monomorphism of the category of ringed spaces. Assume that there is a commutative diagram

$$(Z, \mathcal{H}) \underset{(\alpha_2, \theta_2)}{\overset{(\alpha_1, \theta_1)}{\rightrightarrows}} \left(X, \mathcal{G}_{|X}\right) \to (Y, \mathcal{G})$$

Since $\alpha_1 = \alpha_2$, we set $\alpha = \alpha_1$. Then from $j_\star(\theta_1) \circ \rho_\mathcal{G} = j_\star(\theta_2) \circ \rho_\mathcal{G}$, we deduce that $\theta_1^\flat = \theta_2^\flat$ by $(\ddagger)$, so that $\theta_1 = \theta_2$ in view of $(\dagger)$. Therefore, $\left(X, \mathcal{G}_{|X}\right) \to (Y, \mathcal{G})$ is a monomorphism. Now consider a morphism of ringed space $(\mu, \theta) : (Z, \mathcal{H}) \to (Y, \mathcal{G})$ such that $\mu(Z) \subset X$. Denote by α the continuous restriction of μ to X so that $\mu = j \circ \alpha$. Then θ is nothing but $\mathcal{G} \to j_\star\left(\alpha_\star(\mathcal{H})\right)$. Thanks to the functorial isomorphism $(\dagger)$, there is a morphism of ringed spaces $\theta^\sharp : j^{-1}(\mathcal{G}) \to \alpha_\star(\mathcal{H})$ providing a morphism $\left(\alpha, \theta^\sharp\right) : (Z, \mathcal{H}) \to \left(X, \mathcal{G}_{|X}\right)$. As above we get $j_\star\left(\theta^\sharp\right) \circ \rho_\mathcal{G} = \left(\theta^\sharp\right)^\flat = \theta$. Thus the proof is complete. □

The monomorphism exhibited in 1.1,(1) is a solution of a universal problem. We call it a solution in the category of ringed spaces of the geometric problem associated to X. The same definition is valid for subcategories of RS.

Recall that for a ringed space $\mathcal{F}$ on X and $x \in X$, the stalk $\mathcal{F}_x$ of $\mathcal{F}$ is the ring $\varinjlim_{U \ni x} \mathcal{F}(U)$ where the limit is taken over all open subsets $U \ni x$. Let $(X, \mathcal{F})$ be a ringed space and $\psi : X \to Y$ a continuous map. There is a ring morphism $\psi_x : \left(\psi_\star(\mathcal{F})\right)_{\psi(x)} \to \mathcal{F}_x$ for every $x \in X$ [8, 0.3.4.4].

$(*)$ When ψ induces a homeomorphism $X \to \psi(X)$, the ring morphism ψ_x is an isomorphism [8, 0.3.4.5]. This property holds for a subspace X of Y.

Now if $(\psi, \theta) : (X, \mathcal{F}) \to (Y, \mathcal{G})$ is a morphism of ringed spaces, there is a stalk ring morphism $\mathcal{G}_{\psi(x)} \to \mathcal{F}_x$ which can be factored for $x \in X$,

$$\mathcal{G}_{\psi(x)} \to (\psi_\star(\mathcal{F}))_{\psi(x)} \to \mathcal{F}_x.$$

(∗∗) Hence, for a ringed space $(Y, \mathcal{G})$ and a continuous map $\psi : X \to Y$, there is a stalk ring isomorphism $\psi_x \circ (\rho_{\mathcal{G}})_{\psi(x)} : \mathcal{G}_{\psi(x)} \to \left(\psi^{-1}(\mathcal{G})\right)_x$ for $x \in X$ [8, 0.3.7.2].

Notice that $\mathcal{G}[\psi] \to \psi^{-1}(\mathcal{G})$ induces an isomorphism $\mathcal{G}[\psi]_x \to \psi^{-1}(\mathcal{G})_x$ for every $x \in X$.

A locally ringed space is a ringed space, the stalks of which are local rings, while a morphism of locally ringed spaces is a morphism of ringed spaces with local stalk ring morphisms. The category of locally ringed spaces is denoted by LRS.

When $(Y, \mathcal{G})$ is a locally ringed space and X is a subspace of Y, the above results show that $(X, \mathcal{G}_{|X}) \to (Y, \mathcal{G})$ is a monomorphism of LRS with bijective stalk ring morphisms.

Although LRS is not a full subcategory of RS, the previous result holds in LRS.

PROPOSITION 1.2. *Let $(Y, \mathcal{G})$ be a locally ringed space and $X \subset Y$ a subspace. Denote by $j : X \to Y$ the canonical continuous injective map.*

(1) *$(j, \rho_{\mathcal{G}}) : (X, \mathcal{G}_{|X}) \to (Y, \mathcal{G})$ is a monomorphism of locally ringed spaces.*

(2) *A morphism of locally ringed spaces $(\mu, \theta) : (Z, \mathcal{H}) \to (Y, \mathcal{G})$ such that $\mu(Z) \subset X$ can be factored by a unique morphism of locally ringed spaces $(Z, \mathcal{H}) \to (X, \mathcal{G}_{|X})$.*

Proof. We denote by $\alpha : Z \to X$ the restriction of μ and $j : X \to Y$ is the canonical injection. There is a factorization $(Z, \mathcal{F}) \to (X, \mathcal{G}_{|X}) \to (Y, \mathcal{G})$ by 1.1. For $z \in Z$, there are ring morphisms $G_{j(\alpha(z))} \to (\mathcal{G}_{|X})_{\alpha(z)} \to \mathcal{F}_z$ where the composite is local. As the first morphism is an isomorphism by (∗∗), the second is local. Therefore, $(Z, \mathcal{F}) \to (X, \mathcal{G}_{|X})$ is a morphism of locally ringed spaces as well as the monomorphism $(X, \mathcal{G}_{|X}) \to (Y, \mathcal{G})$. □

COROLLARY 1.3. *Let $(Y, \mathcal{O}_Y)$ be a scheme and $X \subset Y$. Assume that X equipped with the induced ringed space is a scheme, for instance an open subset of Y. Then $X \to Y$ is a flat monomorphism of schemes and a solution to the geometric problem associated to X in the category of schemes.*

We give a converse in the category of affine schemes. If A is a ring, the associated affine scheme is denoted by $\left(\mathrm{Spec}(A), \widetilde{A}\right)$.

PROPOSITION 1.4. *Let $f : A \to B$ be a flat epimorphism of rings with spectral image X. Then the morphism of ringed spaces $(\mathrm{Spec}(B), \widetilde{B}) \to (X, \widetilde{A}_{|X})$ is an isomorphism of locally ringed spaces. Hence $(X, \widetilde{A}_{|X})$ is an affine scheme and $(\mathrm{Spec}(B), \widetilde{B}) \to (\mathrm{Spec}(A), \widetilde{A})$ is a solution to the geometric problem associated to X in the category of schemes.*

Proof. By [13, IV.2.2], the restriction $\alpha : \mathrm{Spec}(B) \to X$ of ${}^a f$ is a homeomorphism and $A_P \to B_Q$ is an isomorphism for every prime ideal Q of B lying over P in A [27]. We set $j : X \to \mathrm{Spec}(A)$ for the canonical injection so that ${}^a f = j \circ \alpha$. Since ${}^a f(\mathrm{Spec}(B)) = X$, there is a composite morphism of ringed spaces $(\mathrm{Spec}(B), \widetilde{B}) \to (X, (\widetilde{A})_{|X}) \to (\mathrm{Spec}(A), \widetilde{A})$ providing us a factorization

$$\widetilde{A}_{j(\alpha(Q))} \to ((\widetilde{A})_{|X})_{\alpha(Q)} \to (\alpha_\star(\widetilde{B}))_{\alpha(Q)} \to \widetilde{B}_Q$$

for $Q \in \mathrm{Spec}(B)$, where the first and the last morphisms are isomorphisms by (**) and (*). Therefore, $((\widetilde{A})_{|X})_{\alpha(Q)} \to (\alpha_\star(\widetilde{B}))_{\alpha(Q)}$ is an isomorphism. It follows that $(\widetilde{A})_{|X} \to \alpha_\star(\widetilde{B})$ is an isomorphism because α is surjective. □

For the two following results, we use [8, I.4.1]. A subscheme X of a scheme Y is locally closed in X. But in general, there is only a surjective morphism of ringed spaces $\omega : (\mathcal{O}_Y)_{|X} \to \mathcal{O}_X$. Moreover, there is a morphism of schemes $j_X = (j, \omega^\flat) : X \to Y$ which is a monomorphism in LRS. Then [8, I.4.1.6] can be rephrased as follows.

PROPOSITION 1.5. *Let X be a subscheme of a scheme Y. Then a morphism of schemes $f : Z \to Y$ such that $f(Z) \subset X$ can be factored $Z \to X \to Y$ if and only if $\mathcal{O}_{Y,f(z)} \to \mathcal{O}_{Z,z}$ can be factored $\mathcal{O}_{Y,f(z)} \to \mathcal{O}_{X,f(z)} \to \mathcal{O}_{Z,z}$ for every $z \in Z$ or equivalently,* $\mathrm{Ker}\left(\mathcal{O}_{Y,f(z)} \to \mathcal{O}_{X,f(z)}\right) \subset \mathrm{Ker}\left(\mathcal{O}_{Y,f(z)} \to \mathcal{O}_{Z,z}\right)$.

PROPOSITION 1.6. *Let X be a subscheme of a scheme Y and $U = (Y \setminus \overline{X}) \cup X$ the largest open subset of Y, containing X and such that $X = U \cap \overline{X}$. Let $\mathcal{I}$ be the quasi-coherent ideal of $\mathcal{F} = (\mathcal{O}_Y)_{|U}$ defining the closed subscheme $(X, \mathcal{O}_X)$ of $(U, \mathcal{F})$. Assume that $\mathcal{I}_x = 0$ for every $x \in X$.*

(1) *$\mathcal{I}$ is locally trivial whence a flat $\mathcal{F}$-module such that $\mathcal{I} = \mathcal{I}^2$.*
(2) *$X \to Y$ is a solution to the geometric problem associated to X in the category of schemes.*

Proof. We get $\mathcal{I}_z = \mathcal{F}_z$ for every $z \in U \setminus X$ because X is the support of $\mathcal{F}/\mathcal{I}$. Hence $\mathcal{I}$ is locally trivial. Let $f : Z \to Y$ be a morphism of schemes such that $f(Z) \subset X = \overline{X} \cap U$. By 1.3 there is a factorization $Z \xrightarrow{g} U \xrightarrow{i} Y$ and $g(Z) \subset X$. Thus we can assume that X is a closed subscheme of Y. In this case, $\mathcal{O}_{Y,x} \to \mathcal{O}_{X,x}$ for $x \in X$ can be identified to $\mathcal{O}_{Y,x} \to \mathcal{O}_{Y,x}/\mathcal{I}_x$ and is therefore injective. Thus we can use 1.5 to conclude. □

EXAMPLE 1.7. Let X be a totally disconnected compact topological space and K a field. Let A be the ring of locally constant functions from X to K. This ring is absolutely flat. Let $x \in X$ be a non-isolated point. Consider the maximal ideal M of all $f \in A$ such that $f(a) = 0$. Then the morphism of schemes $\mathrm{Spec}(A/M) \to \mathrm{Spec}(A)$ is such that the stalk morphisms for every $x \in X$ are bijective [8, I.4.2.3]. Therefore, the geometric problem associated to $\{M\}$ has a solution in the category of schemes. Notice that $\{M\}$ is not open.

To end this section we give some considerations about ringed space spectra, introduced in [15]. The following result summarizes the main properties.

PROPOSITION 1.8. *Let $(Z, \mathcal{H})$ be a ringed space.*

(1) *There is a morphism of ringed spaces $(p, \tilde{p}) : (Z', \mathcal{H}') \to (Z, \mathcal{H})$ where $(Z', \mathcal{H}')$ is a locally ringed space.*

(2) *For every morphism of ringed spaces $(f, \tilde{f}) : (Y, \mathcal{G}) \to (Z, \mathcal{H})$ where $(Y, \mathcal{G})$ is a locally ringed space, there is a unique factorization*

$$(Y, \mathcal{G}) \to (Z', \mathcal{H}') \to (Z, \mathcal{H}) \text{ of } (Y, \mathcal{G}) \to (Z, \mathcal{H})$$

where $(Y, \mathcal{G}) \to (Z', \mathcal{H}')$ is a morphism of locally ringed spaces.

The locally ringed space $(Z', \mathcal{H}')$ is called the spectrum of $(Z, \mathcal{H})$.

If $(Z, \mathcal{H})$ is a locally ringed space, there is a factorization $(Z, \mathcal{H}) \to (Z', \mathcal{H}') \to (Z, \mathcal{H})$ where the first morphism lies in $\underline{\mathrm{LRS}}$.

We get a covariant functor $F : \underline{\mathrm{RS}} \to \underline{\mathrm{LRS}}$ defined by $F(X, \mathcal{F}) = (X', \mathcal{F}')$ and such that for every morphism of ringed spaces $(f, \tilde{f}) : (X, \mathcal{F}) \to (Y, \mathcal{G})$ there is a commutative diagram:

$$\begin{array}{ccc} F(X, \mathcal{F}) & \longrightarrow & F(Y, \mathcal{G}) \\ \downarrow & & \downarrow \\ (X, \mathcal{F}) & \longrightarrow & (Y, \mathcal{G}) \end{array}$$

LEMMA 1.9. *Let $(Z, \mathcal{H})$ be a ringed space and $(p, \tilde{p}) : (Z', \mathcal{H}') \to (Z, \mathcal{H})$ the ringed space morphism associated to the spectrum of $(Z, \mathcal{H})$. Then $(Z', \mathcal{H}') \to (Z, \mathcal{H})$ is an epimorphism of the category of ringed spaces and a monomorphism with respect to the subcategory $\underline{\mathrm{LRS}}$.*

Proof. In view of [15, 1.3], the continuous map $p : Z' \to Z$ is surjective and proper and $\tilde{p} : \mathcal{H} \to p_\star(\mathcal{H}')$ is an isomorphism. Therefore, $\varphi_\star(\mathcal{H}) \to \varphi_\star(p_\star(\mathcal{H}'))$ is an isomorphism for every continuous map $\varphi : Z \to X$, because $\varphi_\star$ is a functor. □

PROPOSITION 1.10. *Let* $(Y, \mathcal{G})$ *be a ringed space and* $X \subset Y$. *Set* $\mathcal{F} = \mathcal{G}_{|X}$. *Let* $(j', \sigma') = (j, \rho_{\mathcal{G}})' : (X', \mathcal{F}') \to (Y', \mathcal{G}')$ *be the associated morphism of locally ringed spaces. Then a solution to the geometric problem associated to* $j'(X')$ *in the category* <u>LRS</u> *is given by the monomorphism* $(j, \rho_{\mathcal{G}})' : (X', \mathcal{F}') \to (Y', \mathcal{G}')$.

Proof. Denote by $(q, \tilde{q}) : (X', \mathcal{F}') \to (X, \mathcal{F})$ and by $(p, \tilde{p}) : (Y', \mathcal{G}') \to (Y, \mathcal{G})$ the canonical morphisms. Consider a morphism $(\mu, \theta) : (Z, \mathcal{H}) \to (Y', \mathcal{G}')$ of <u>LRS</u> such that $\mu(Z) \subset j'(X')$. Because q is surjective, we get $p(\mu(Z)) \subset j(X) = X$. Therefore, there is a unique morphism $(Z, \mathcal{H}) \to (X, \mathcal{F})$ of <u>RS</u> factorizing $(Z, \mathcal{H}) \to (Y, \mathcal{G})$. By the universal property of $(X', \mathcal{F}')$, there is a unique morphism of locally ringed spaces $(\nu, \tau) : (Z, \mathcal{H}) \to (X', \mathcal{F}')$ factorizing $(Z, \mathcal{H}) \to (X, \mathcal{F})$. Then the morphisms of locally ringed spaces $(j', \sigma') \circ (\nu, \tau)$ and (μ, θ) are equal in view of 1.9, because they are equalized by $(p, \tilde{p})$. It is easy to prove that the factorization is unique. □

2 RINGS OF SECTIONS AND GABRIEL'S LOCALIZATION

In the next sections, we consider solutions in the category of commutative rings of geometric problems associated to subsets of a spectrum. These subsets are necessarily patches. In this section we intend to exhibit unique factorizations linked to the subsets of a spectrum. But they do not give in general solutions to geometric problems. Nevertheless, the material of this section will be used in the next section and the gotten results have their own interest.

In the following, we consider a ring A and a subset X of $\mathrm{Spec}(A)$.

There is a monomorphism of locally ringed spaces with bijective stalk morphisms $(j, \rho_{\mathcal{G}}) : (X, \widetilde{A}_{|X}) \to (\mathrm{Spec}(A), \widetilde{A})$. We set $\widetilde{A}(X) = \widetilde{A}_{|X}(X)$. Hence there is a ring morphism $f_X : A \to \widetilde{A}(X)$. When $X = \mathrm{D}(I)$ is an open subset, $\widetilde{A}(X)$ is nothing but the ring of sections over $\mathrm{D}(I)$. If I is an ideal of A, the spectral image of $A \to \widetilde{A}(\mathrm{D}(I))$ is denoted by $\mathrm{W}(I)$. We denote by $\mathcal{QO}$ the familly of all quasi-compact open subsets V of $\mathrm{Spec}(A)$ (such that $V = \mathrm{D}(I)$ where I is a finitely generated ideal of A).

LEMMA 2.1. *Let* A *be a ring and* $X \subset \mathrm{Spec}(A)$.

(1) $X^g \subset \mathrm{Im}({}^a f_X) \subset \cap [\, \mathrm{W}(I) \; ; \; X \subset \mathrm{D}(I) \,] = \mathrm{Im}\,({}^a f_{X^g})$.

(2) *If* X *is a patch, then* $X^g = \cap [\mathrm{D}(I) \; ; \; X \subset \mathrm{D}(I), \; \mathrm{D}(I) \in \mathcal{QO} \;]$ *is a patch and* $\widetilde{A}(X^g) = \varinjlim_{V \supset X} \widetilde{A}(V)$ *where the limit is taken over all open subsets* $V \supset X$ *in* $\mathcal{QO}$.

(3) *For each ring morphism* $g : B \to C$ *such that* ${}^a g(\mathrm{Spec}(B)) \subset X$, *there is a factorization* $A \to \widetilde{A}(X) \to B$.

Proof. The statement (2) follows from [27, 2.1 and 2.3]. We show (1). In view of (**) in Section 1, there is a factorization $A \to \widetilde{A}(X) \to A_P$ for every $P \in X$. The first inclusion of (1) follows. Moreover, denoting by $j : X \to \mathrm{Spec}(A)$ the canonical map, $\widetilde{A}_{|X}$ is the sheaf associated to the presheaf $\widetilde{A}[j]$ (see Section 1). Hence there is a factorization $A \to \varinjlim_{V \supset X} \widetilde{A}(V) \to \widetilde{A}(X)$ where the limit is taken over all open sets $V \supset X$. The second inclusion follows from [EGA, I.3.4.10] while the equality is a consequence of (2). Now (3) is a consequence of 1.2. □

Next we examine the case of an integral domain A with quotient field K. In that case $\widetilde{A}(\mathrm{D}(I)) = \cap[\ A_P\ ;\ P \in \mathrm{D}(I)\ ] \subset K$ for an ideal I of A. If k belongs to K, we denote by $I(k)$ its denominator ideal $A :_A k$. If P is a prime ideal of A, then $P \in \mathrm{D}(I(k))$ is equivalent to $k \in A_P$.

LEMMA 2.2. *Let A be an integral domain and $X \subset \mathrm{Spec}(A)$. Then $\widetilde{A}(X)$ is equal to $\cap[\ A_P\ ;\ P \in X\ ]$ so that $\widetilde{A}(X) = \widetilde{A}(X^g) = \{k \in K\ /\ X \subset \mathrm{D}(I(k))\}$.*

Proof. Denoting by $j : X \to \mathrm{Spec}(A)$ the canonical map, we show that the presheaf $\mathcal{F} = \widetilde{A}[j]$ is a sheaf. Let $W = X \cap U$ be an open subset of X where U is an open subset of $\mathrm{Spec}(A)$, then $\mathcal{F}(X \cap U) = \cup\left[\widetilde{A}(V)\,;\, V \supset W\right]$ by Section 1. Clearly $\widetilde{A}(V)$ is contained in $B = \cap[\,A_P\,;\,P \in X \cap U\ ]$ and so is $\mathcal{F}(W)$. Now let $k \in B$ with denominator ideal I. If P belongs to $X \cap U$, we get $P \in \mathrm{D}(I)$ so that $W \subset \mathrm{D}(I)$. Moreover, $Q \in \mathrm{D}(I)$ implies $k \in A_Q$. In short, k belongs to B. Therefore, $\mathcal{F}(W) = \cap[\,A_P\,;\,P \in W\ ]$. Since rings of sections $\mathcal{F}(W)$ are contained in K, it is easy to show that $\mathcal{F}$ is a sheaf. Indeed, if $U = \cup_i U_i$ is an open covering, then $\mathcal{F}(X \cap U) = \cap_i\,\mathcal{F}\,(X \cap U_i\ )$. It follows that $\widetilde{A}(X) = \cap[\,A_P\,;\,P \in X\ ]$. To complete the proof, it is enough to observe that $X \subset \mathrm{D}(I)$ is equivalent to $X^g \subset \mathrm{D}(I)$. □

We introduce a relevant ring associated to a patch X. Its definition generalizes the above expression of $\widetilde{A}(X)$.

For a general theory of localizations with respect to Gabriel topologies (called here sites), we refer the reader to the Stenström's book [29]. Let $f : A \to B$ be a ring morphism and $\mathcal{F}$ the set of all ideals I of A such that $B = IB$. Then $\mathcal{F}$ is a site on A. We consider the subring $A(\mathcal{F})$ of all elements $b \in B$ for which there is some I in $\mathcal{F}$ such that $f(I)b \subset f(A)$ [16, Section 4]. For $b \in B$, we denote by $A :_A b$ the ideal of all elements $a \in A$ such that $f(a)b \in f(A)$. Thus b belongs to $A(\mathcal{F})$ if and only if $A :_A b \in \mathcal{F}$ because a site is stable under overideals. If M is an A-module, then $\mathcal{F}M$ is the submodule of all $x \in M$ such that $0 :_A x \in \mathcal{F}$. We denote as usual by $A_{\mathcal{F}}$ the localization of A with respect to $\mathcal{F}$, that is to say $A_{\mathcal{F}} = \varinjlim_{I \in \mathcal{F}} \mathrm{Hom}_A\,(I, A/\mathcal{F}A\ )$. When

M is an A-module, the localization $M_{\mathcal{F}}$ of M with respect to $\mathcal{F}$ is defined in a similar way. If we consider the sheaf $\widetilde{M}$, we set $\widetilde{M}(X) = \widetilde{M}_{|X}(X)$.

LEMMA 2.3. *Let* $f : A \to B$ *be a ring morphism,* $\mathcal{F}$ *the associated site,* $X \subset \mathrm{Spec}(A)$ *such that* $\mathrm{Im}\,({}^a f\,) \subset X$ *and* I *an ideal of* A.

(1) $X \subset \mathrm{D}(I)$ *implies* $IB = B$.

(2) *If* $X = \mathrm{Im}\,({}^a f\,)$, *then* $X \subset \mathrm{D}(I)$ *is equivalent to* $B = IB$.

Therefore, when $X = \mathrm{Im}\,({}^a f\,)$, *an ideal* I *of* A *belongs to* $\mathcal{F}$ *if and only if* $X \subset \mathrm{D}(I)$ *and a prime ideal* P *of* A *belongs to* X^g *if and only if* $P \notin \mathcal{F}$.

Proof. Obvious. □

Let A be a ring, $X \subset \mathrm{Spec}(A)$ and define $\mathcal{F}$ to be the set of all ideals in A such that $X \subset \mathrm{D}(I)$. Then $\mathcal{F}$ is a site. R. Bkouche proved the following facts [3, unpublished]. For every A-module M, there is an injective morphism $\zeta(M) : M_{\mathcal{F}} \to \widetilde{M}(X)$ (defining a functorial monomorphism ζ) such that $M \to \widetilde{M}(X) = M \to M_{\mathcal{F}} \to \widetilde{M}(X)$. Moreover, $\widetilde{M}(X) \to \widetilde{M}(X)_{\mathcal{F}}$ is an isomorphism. A subset X of $\mathrm{Spec}(A)$ is called agreeable if ζ is an isomorphism. If X is quasi-compact and stable under generizations, then X is agreeable.

A prime ideal P of a ring A is called flabby if $A \to A_P$ is surjective. Let $N(P)$ be the kernel of $A \to A_P$ where P is an arbitrary prime ideal, then $\overline{P^g} = \mathrm{V}(N(P))$. Thus when P is flabby, P^g is closed, $N(P)$ is a pure ideal and A/P is a local ring. A ring A is called flabby if every maximal ideal is flabby. For instance, a local ring is flabby and a ring of continuous numerical functions $\mathcal{C}(E, \mathbb{R})$ is flabby. When A is flabby, an arbitrary subset X of $\mathrm{Max}(A)$ is agreeable [3].

LEMMA 2.4. *Let* $f : A \to B$ *be a ring morphism,* $\mathcal{F}$ *the associated site and* $X = \mathrm{Im}({}^a f)$.

(1) $A_{\mathcal{F}} = \widetilde{A}(X^g)$ *and there is a factorization* $A \to \widetilde{A}(X^g) \to \widetilde{A}(X) \to B$ *where* $\widetilde{A}(X^g) \to \widetilde{A}(X)$ *is injective.*

(2) *If* $X \subset \mathrm{Max}(A)$, *each* $P \in X$ *is flabby and* $\cap\,[P\ ;\ P \in X] = 0$, *then* $\widetilde{A}(X^g) \to \widetilde{A}(X)$ *is bijective* [2, 2.6.1].

(3) *There is a factorization* $A \to A_{\mathcal{F}} \to A(\mathcal{F}) \to B$ *so that* $\mathcal{F}A \subset \mathrm{Ker}(f)$.

Proof. (1) is a consequence of the above Bkouche's results and 2.1 since X is a patch. We prove (3). Consider the factorization $A \to f(A) \to B$ which gives a factorization $A_{\mathcal{F}} \to f(A)_{\mathcal{F}} \to B_{\mathcal{F}}$. Set $\mathcal{G} = f(\mathcal{F}) = \{IB \ /\ I \in \mathcal{F}\} = \{B\}$ so that $B_{\mathcal{F}} = B_{\mathcal{G}} = B$. In view of [16, 4.1], we have $f(A)_{\mathcal{F}} = A(\mathcal{F})$. To complete the proof, observe that $\mathcal{F}A = \mathrm{Ker}\,(A \to A_{\mathcal{F}}\,)$. □

Under some hypotheses, the rings appearing in the above factorizations identify.

LEMMA 2.5. *Let* $f : A \to B$ *be a ring morphism,* $\mathcal{F}$ *the associated site and* $X = \mathrm{Im}\,({}^a f\,)$.

(1) *If* A *is an integral domain, then* $A_{\mathcal{F}} = \widetilde{A}(X)$.
(2) *If* $\mathcal{F}A = \mathrm{Ker}(f)$, *then* $A_{\mathcal{F}} = A(\mathcal{F})$. *This is the case when* $A \to B$ *is flat or injective.*

Proof. If A is an integral domain, use [16, 2.3] and 2.3 and observe that $P \in X^g$ is equivalent to $P \notin \mathcal{F}$. Assume that $A \to B$ is flat. We have $\mathcal{F}A = \mathrm{Ker}(f)$. Indeed, $\mathcal{F}A \subset \mathrm{Ker}(f)$ is always true. Conversely, assume that $f(a) = 0$; then we have $(0 : a)B = 0 : f(a)B = B$ by flatness so that $a \in \mathcal{F}A$. Assume that $\mathcal{F}A = \mathrm{Ker}(f)$, so that $A/\mathcal{F}A = f(A)$. To complete the proof, by using the definition of the localization with respect to $\mathcal{F}$, it is enough to observe that $\mathcal{F}(A/\mathcal{F}A) = 0$. □

The dominiom $\mathrm{Dom}_A(B)$ of a ring morphism $f : A \to B$ is defined in 0.2.

PROPOSITION 2.6. *Let* $f : A \to B$ *be a ring morphism,* $\mathcal{F}$ *the associated site and* $X = \mathrm{Im}\,({}^a f\,)$.

(1) $A(\mathcal{F})$ *is the set of all* $b \in B$ *such that* $X \subset \mathrm{D}\,(A :_A b\,)$.
(2) *There is a factorization* $A \xrightarrow{\mu} A(\mathcal{F}) \to \mathrm{Dom}_A(B) \to B$ *so that* $X \subset \mathrm{Im}\,({}^a\mu\,)$.
(3) $C \to C \bigotimes_A A(\mathcal{F})$ *is surjective for every ring morphism* $g : A \to C$ *such that* $\mathrm{Im}\,({}^a g\,) \subset X$.
(4) *If* X *is stable under generizations,* ${}^a\mu$ *defines a bijection* ${}^a\mu^{-1}\,(X\,) \to X$.
(5) *If* $A \to A(\mathcal{F}) \overset{u}{\underset{v}{\rightrightarrows}} D$ *is a commutative diagram such that the spectral image of* $g = u \circ \mu = v \circ \mu$ *is contained in* X, *then* $u = v$.

Proof. (1) is a consequence of 2.3. We show (2). Let $b \in A(\mathcal{F})$ and $I \in \mathcal{F}$ such that $f(I)b \subset f(A)$. From $I(b \otimes 1 - 1 \otimes b)$ and $IB = B$, we deduce that $b \otimes 1 = 1 \otimes b$. The hypotheses of (3) being granted, consider $z = \sum c_k \otimes b_k \in C \bigotimes_A A(\mathcal{F})$. Since $\mathcal{F}$ is stable under products, there is some $I \in \mathcal{F}$ such that $f(I)b_k \in f(A)$ for each k. Thanks to 2.3,(1), we have $IC = C$. Pick some $a_j \in I$ and $\gamma_j \in C$ such that $1 = \sum g\,(a_j\,)\,\gamma_j$ and set $f\,(a_j\,)\,b_k = f\,(\alpha_{j,k}\,)$. We get $(g\,(a_j\,) \otimes 1\,)\,z = (1 \otimes f\,(a_j\,)\,)\,z = \sum c_k \otimes f\,(\alpha_{j,k}\,) = \delta_j \otimes 1$ where $\delta_j \in C$. It follows easily that $z = c \otimes 1$. In short, $C \to C \bigotimes_A A(\mathcal{F})$ is surjective. To prove (4), consider $P \in X$ and $R \in \mathrm{Spec}(B)$ lying over P and over $Q \in \mathrm{Spec}\,(A(\mathcal{F})\,)$. If I belongs to $\mathcal{F}$, then $IA(\mathcal{F}) \not\subset Q$ follows from $I \not\subset P$ by 2.3. Thus Q is lying over P and belongs to $\cap_{I \in \mathcal{F}} \mathrm{D}\,(IA(\mathcal{F})\,) = {}^a\mu^{-1}\,(X\,)$. Now take Q_1 and Q_2 in ${}^a\mu^{-1}\,(X\,)$ lying over P in A. Let $b \in Q_1$, there is some $I \in \mathcal{F}$ such that $f(I)b \subset f(A) \cap Q_1 = \mu(A) \cap Q_1 = \mu(P) \subset Q_2$. As Q_2 does not contain $\mu(I)$ by 2.3, we get $Q_1 \subset Q_2$. It follows that $Q_1 = Q_2$. To end, assume that the hypotheses of (5) hold and let $b \in A(\mathcal{F})$. There is some

$I \in \mathcal{F}$ such that $f(I)b \subset f(A)$. Then for $a \in I$, we get $u(\mu(a)b) = v(\mu(a)b)$ so that $g(a)(u(b) - v(b)) = 0$. Thanks to 2.3,(1), we have $ID = D$ from which it follows that $u(b) = v(b)$. $\square$

We present now a generalization of an unpublished Ferrand's result which asserts that $\widetilde{A}(\mathrm{D}(I))$ is a dominiom for a quasi-compact open subset $\mathrm{D}(I)$.

THEOREM 2.7. *Let $f : A \to B$ be a flat ring morphism, $\mathcal{F}$ the associated site and $X = \mathrm{Im}\,({}^a f\ )$, then $A(\mathcal{F}) = \mathrm{Dom}_A(B)$. It follows then that $\widetilde{A}(X) = \mathrm{Dom}_A(B)$.*

Proof. In view of 2.6, $A(\mathcal{F}) \subset \mathrm{Dom}_A(B)$ holds. Now consider the canonical morphism $\varphi : B/f(A) \to B/f(A) \bigotimes_A B$. Then Lazard and Huet proved that $\mathrm{Dom}_A(B)/f(A) = \mathrm{Ker}(\varphi)$ [14, 2.2]. For $b \in \mathrm{Dom}_A(B)$, define a morphism of A-modules $u : A \to B/f(A)$ by $u(a) = \overline{f(a)b}$ where $\bar{x}$ is the class of $x \in B$ in $B/f(A)$. Its kernel is $A :_A b$. Now consider $v = u \otimes 1_B : B \to B/f(A) \bigotimes_A B$. For $c \in B$, we get $v(c) = u(1) \otimes c = \bar{b} \otimes c = 0$ because $b \in \mathrm{Dom}_A(B)$. Since f is flat, $(A :_A b)B = B$ follows and b belongs to $A(\mathcal{F})$. Then use 2.4,(1) and 2.5,(2), observing that $X = X^g$. $\square$

REMARK 2.8.

(1) Let $I = (a_1, \dots , a_n\ )$ be a finitely generated ideal of A. The canonical morphism $\psi_I = A \to A\,[X_1, \dots , X_n\ ]/\,(a_1X_1 + \cdots + a_nX_n - 1\ ) = B_I$ is a flat morphism such that ${}^a\psi_I\,(\mathrm{Spec}\,(B_I\ )\ ) = \mathrm{D}(I)$. Then Ferrand's result is $\widetilde{A}(\mathrm{D}(I)) = \mathrm{Dom}_A\,(B_I\ )$. Moreover, $\mathrm{Dom}_A(B_I) \to B_I$ is an injective flat ring morphism (use [13, IV.3.1]).

(2) Now if $X \subset \mathrm{Spec}(A)$ is quasi-compact and stable under generizations, X is the intersection of all $\mathrm{D}(I)$ where I is a finitely generated ideal such that $X \subset \mathrm{D}(I)$. Set $B_X = \bigotimes_A B_I$ where I varies in the preceding set of ideals and consider $\psi_X : A \to B_X$, then $\widetilde{A}(X) = \mathrm{Dom}_A\,(B_X\ )$.

(3) We can use an alternative morphism. Let X be a subset of $\mathrm{Spec}(A)$. The following results may be found in [21]. Consider the subset Σ_X of all $p(t) \in A\,[t\ ]$ such that $X \subset \mathrm{D}(C(p(t)))$ where $C(p(t))$ is the content ideal of $p(t)$. Then Σ_X is a multiplicative subset of $\mathrm{Spec}(A)$. Let $\varphi_X : A \to A\,[t\ ]_{\Sigma_X} = X(A)$ be the canonical flat morphism. The intersection of all quasi-compact open subsets of $\mathrm{Spec}(A)$ containing X is ${}^a\varphi_X(\mathrm{Spec}(X(A)))$ [21]. Therefore, if X is quasi-compact and stable under generizations, then $\widetilde{A}(X) = \mathrm{Dom}_A(X(A))$.

(4) We recall now some notation introduced in [23, Section 1]. Let A be a ring and $X \subset \mathrm{Spec}(A)$. The size of X is $\mathcal{U}(X) = \cup\ [P\ ;\ P \in X]$ and its radical is $\mathcal{R}(X) = \cap\ [\ P\ ;\ \in X]$. We denote by S_X the subset of all elements $a \in A$ such that $X \subset \mathrm{D}(a)$. Then S_X is a multiplicative subset of A and we set $A_X = A_{S_X}$; clearly, $A_X = A_{X^g}$. The ring A_X is

called the localization of A at X (obviously, $A_{\{P\}} = A_P$). Denote by X^u the set of all prime ideals P of A such that $P \subset \mathcal{U}(X)$ and by $u_X : A \to A_X$ the canonical morphism. Then $X^u = {}^a u_X(\mathrm{Spec}(A_X))$ is a patch stable under generizations. Now let $f : A \to B$ be a ring morphism such that ${}^a f(\mathrm{Spec}(B)) \subset X^g$. Since each element of $f(S_X)$ is a unit in B, we get a factorization $A \to A_X \to B$ of f.

We will need the following result.

PROPOSITION 2.9. *Let $f : A \to B$ be a ring morphism, $\mathcal{F}$ the associated site and $X = \mathrm{Im}({}^a f)$. Then f is called pre-flat (or immersing) if one of the following equivalent statements holds* [18] *and* [16].

(1) *$A_P \to B_P$ is surjective for every $P \in X$ (resp. $P \in X^g$).*
(2) *$A(\mathcal{F}) = B$.*
(3) *$A/\mathrm{Ker}(f) \to B$ is a flat epimorphism.*

Proof. For (1) ⇔ (2) see [16, 3.4 and 4.2] and for (1) ⇔ (3) see [18, II.4]. □

A pre-flat morphism is an epimorphism. A flat epimorphism is pre-flat and a pre-flat injective morphism is a flat epimorphism [16, Section 3].

The next result could be used in Section 4 but we prefer to work with different methods. It is related to the preceding results.

PROPOSITION 2.10. *Let $f : A \to B$ be a ring morphism, $\mathcal{F}$ the associated site and $X = \mathrm{Im}({}^a f)$. Let $g : A \to C$ be a ring morphism such that ${}^a g(\mathrm{Spec}(C)) \subset X$. Then there is a unique factorization $A \to C = A \to A_{\mathcal{F}} \to C$.*

Proof. First observe that $C_{\mathcal{F}} = C_{\mathcal{G}}$ where $\mathcal{G} = g(\mathcal{F}) = \{C\}$. From $C_{\mathcal{G}} = C$ we deduce that the factorization exists. Denote by $j : A \to A_{\mathcal{F}}$ the canonical morphism and assume that there are two ring morphisms $u, v : A_{\mathcal{F}} \to C$ such that $u \circ j = g = v \circ j$. Now let q be in $A_{\mathcal{F}}$, there is some $I \in \mathcal{F}$ such that $j(I)q \subset j(A)$. It follows that $g(I)(u(q) - v(q)) = 0$. From $IC = C$ we deduce that $u(q) = v(q)$. Therefore the factorization is unique. □

3 THE CATEGORY OF RINGS

DEFINITION 3.1. Let A be a ring and $X \subset \mathrm{Spec}(A)$. Then X is called a geometric subset of $\mathrm{Spec}(A)$ if the following statements hold.

(1) There is a ring morphism $\delta : A \to \Delta(X)$ such that ${}^a\delta(\mathrm{Spec}(\Delta(X))) \subset X$.
(2) For every ring morphism $f : A \to B$ such that ${}^a f(\mathrm{Spec}(B)) \subset X$, there is a unique ring morphism $g : \Delta(X) \to B$ such that $f = g \circ \delta$.

In that case δ is called a solution of the geometric (universal) problem associated to X.

Clearly, the subset $\mathrm{Spec}(A)$ has a solution $A \to \Delta(X) = A$.

The hypotheses of 2.10 being granted, X is geometric if the spectral image Y of $A \to A_{\mathcal{F}}$ is contained in X. Actually, Gabriel's localization theory shows that prime ideals $P \notin \mathcal{F}$ belong to Y. Thanks to 2.3, we get that $X^g \subset Y$. Therefore, the above condition says that X is quasi-compact and stable under generizations. We will give a complete answer for such subsets.

LEMMA 3.2. *Let A be a ring and X a geometric subset of* $\mathrm{Spec}(A)$. *Then we have* ${}^a\delta(\mathrm{Spec}(\Delta(X))) = X$ *so that X is a patch.*

Proof. It is enough to show that $X \subset {}^a\delta(\mathrm{Spec}(\Delta(X)))$. Let $P \in X$, there is a ring morphism $A \to \mathrm{k}(P)$ providing a factorization $A \to \Delta(X) \to \mathrm{k}(P)$. □

We say that a ring morphism $f : A \to B$ focuses on a subset X of $\mathrm{Spec}(A)$ if ${}^af(\mathrm{Spec}(B)) = X$ and that f focuses on a prime ideal P of A if ${}^af(\mathrm{Spec}(B)) = \{P\}$. Trivial examples are the ring morphisms $A \to A_P/P^nA_P = \mathrm{k}_n(P)$ for each integer $n > 0$.

REMARK 3.3. If a ring morphism $f : A \to B$ focuses on $X \subset \mathrm{Spec}(A)$, then

(1) $\sqrt{\mathrm{Ker}(f)} = \mathcal{R}(X)$ since $\overline{{}^af(\mathrm{Spec}(B))} = \mathrm{V}(\mathrm{Ker}(f))$ by 0.1.

(2) There is a factorization $A \to A_{1+\mathcal{R}(X)} \to B$ of f. Indeed, $f(\mathcal{R}(X)) \subset \mathrm{Nil}(B)$ forces each element of $1 + \mathcal{R}(X)$ to be a unit in B.

LEMMA 3.4. *Let $f : A \to B$ be a ring morphism and X a patch of* $\mathrm{Spec}(A)$. *If f focuses on X then $B \to B_X = B \bigotimes_A A_X$ is an isomorphism.*

Proof. There is a ring morphism $\theta_X : A \to \prod[\mathrm{k}(P)\,;\, P \in X] = P(X)$ which focuses on X because X is a patch [17, Lemme 6]. Hence, there is a factorization $A \to A_X \to P(X)$ by 2.8,(4) from which we deduce a factorization $B \to B_X \to P(X) \bigotimes_A B$. By 0.4, $B \to P(X) \bigotimes_A B$ is surjective on the spectrum and so is $B \to B_X$. This last morphism is therefore a faithfully flat epimorphism whence an isomorphism [13, IV.1.2]. □

COROLLARY 3.5. *Let $f : A \to B$ be a ring morphism and* $P \in \mathrm{Spec}(A)$. *Then f focuses on P if and only if $B \to B_P$ is an isomorphism and $PB \subset$* $\mathrm{Nil}(B)$. *In particular, if f is an epimorphism focusing on P, then* $\mathrm{Spec}(B) =$ *$\{PB\}$ and B_{red} is isomorphic to* $\mathrm{k}(P)$.

Proof. One implication is given by 3.4 and 3.3. The reverse implication is obvious. Assume in addition that f is an epimorphism, then $\mathrm{k}(P) \to B_P/PB_P = B/PB$ is an isomorphism so that PB_P is a maximal ideal. □

We pause to examine when a prime ideal P of A is geometric (*i. e.* $X = \{P\}$ is geometric). A minimal prime ideal P is geometric, a solution being $A \to A_P$. To see this, use 2.8,(4) and observe that $A \to A_P$ is a flat epimorphism.

For informations on almost multiplication rings, the reader is referred to the book of Larsen and McCarthy [11].

PROPOSITION 3.6. *Let A be a ring and P a prime ideal of A. Then P is geometric if and only if P is a minimal prime ideal when one of the following conditions holds:*

1. *PA_P is a finitely generated ideal of A_P (for instance, if A is locally Noetherian).*
2. *A is an almost multiplication ring.*

Proof. Assume that P is geometric with a solution $\delta : A \to \Delta$. Thanks to 2.8,(4), there is a factorization $A \to A_P \xrightarrow{g} \Delta$ where the last morphism g focuses on $PA_P = Q$. In view of 3.3, we get $Q = \sqrt{\mathrm{Ker}(g)}$. If Q is finitely generated, there is some integer n such that $Q^n \subset \mathrm{Ker}(g)$. Now assume that A is an almost multiplication ring. We have $P = \sqrt{\mathrm{Ker}(\delta)}$ since δ focuses on P. Let $N(P)$ be the kernel of $A \to A_P$. If $N(P)$ is not a prime ideal, then by [11, 9.25], P is a minimal prime ideal. If $N(P) = P$, then A_P is a field so that P is a minimal prime ideal. If $N(P)$ is a prime ideal and $N(P) \neq P$, then thanks to [11, 9.26], $\mathrm{Ker}(\delta) = P^n$ for some integer $n > 0$ and A_P is a discrete rank one valuation ring. It follows that $Q^n \subset \mathrm{Ker}(g)$. Therefore, we are reduced to consider that in both cases $Q^n \subset \mathrm{Ker}(g)$ and Q is finitely generated. If so, there is a factorization $A \to \mathrm{k}_n(P) \to \Delta$ of δ. By definition, there is a factorization $A \to \Delta \to \mathrm{k}_{n+1}(P)$. Since $A \to \mathrm{k}_{n+1}(P)$ is an epimorphism, the composite morphism $\mathrm{k}_{n+1}(P) \to \mathrm{k}_n(P) \to \Delta \to \mathrm{k}_{n+1}(P)$ is the identity. It follows that $\mathrm{k}_{n+1}(P) \to \mathrm{k}_n(P)$ is injective so that $Q^n = Q^{n+1}$. Now $Q = 0$ by Nakayama's lemma because Q is finitely generated. Hence P is a minimal prime ideal. □

REMARK 3.7. Let V be a valuation domain with maximal ideal M such that $M^2 = M$. Then M is geometric, a solution being $V \to V/M$. Indeed, let $f : V \to B$ be a ring morphism focussing on M. By 3.3,(1), $M = \sqrt{\mathrm{Ker}(f)}$. Then by [7, 17.1], there is some integer n such that $M^n \subset \mathrm{Ker}(f)$ unless $\mathrm{Ker}(f) = V$ in which case $B = 0$. It follows that $M = \mathrm{Ker}(f)$ for every ring morphism $f : V \to B$ focussing on M.

We will see in Section 4 that a solution $A \to \Delta(X)$ associated to a geometric subset X is an epimorphism. It follows from 3.5 that when P is geometric, $\mathrm{Spec}(\Delta(P)) = \{P\Delta(P)\}$ and that $\Delta(P)_{\mathrm{red}}$ is isomorphic to $\mathrm{k}(P)$.

Observe that 3.6 provides us examples of solutions which are flat epimorphisms while in 3.7, the solution $V \to V/M$ is a non-flat epimorphism because if not, $\mathrm{V}(M) = \{M\}$ is stable under generizations.

Geometric subsets are almost stable under generizations. To see this, we need a definition. If P is a prime ideal of A, we can consider the ring morphism $\gamma_P : A \to A_P \to \widehat{A_P}$ where $\widehat{A_P}$ is the separated completion of A_P

with respect to the PA_P-adic topology. Setting $P^{fg} = {}^a\gamma_P(\mathrm{Spec}(\widehat{A_P}))$, we have $P^{fg} \subset P^g$.

DEFINITION 3.8. Let X be a subset of Spec(A). We say that X is stable under formal generizations if $P^{fg} \subset X$ for each $P \in X$.

If X is stable under generizations, then X is stable under formal generizations. The converse is true when A is locally Noetherian. Indeed, $A_P \to \widehat{A_P}$ is faithfully flat since A_P is Noetherian.

PROPOSITION 3.9. *Let X be a geometric subset of* Spec(A), *then X is stable under formal generizations. Therefore, X is stable under generizations when A is locally Noetherian.*

Proof. Recall that $A_P \to \widehat{A_P}$ identifies to $A_P \to \varprojlim \mathrm{k}_n(P)$. Now let $P \in X$, we have a factorization $A \to \Delta(X) \xrightarrow{u_n} \mathrm{k}_n(P)$. Then $\{u_n\}_n$ is an inverse system of ring morphisms because δ is an epimorphism (see 4.2). Therefore, there exists a factorization $A \to \Delta(X) \to \widehat{A_P}$. □

COROLLARY 3.10. *Let P be a prime ideal of a ring A.*

(1) *If P is geometric, then $\sqrt{\cap_n P^n A_P} = PA_P$ (equivalently, $\sqrt{\cap_n P^{(n)}} = P$).*

(2) *If $\sqrt{\cap_n P^n A_P} = PA_P$ and A is a Prüfer domain, then P is geometric with solution $A \to A_P \to A_P/\cap_n P^n A_P$.*

It follows that a prime ideal P of a Prüfer domain is geometric if and only if PA_P is idempotent. In this case, a solution is given by $A \to \mathrm{k}(P)$.

Proof. Since $\cap_n P^n A_P = \mathrm{Ker}(A_P \to \widehat{A_P})$, a minimal prime ideal Q of this ideal can be lifted in $\mathrm{Spec}\left(\widehat{A_P}\right)$. Assume that P is geometric. From $P^{fg} = \{P\}$, we deduce that $Q = PA_P$. Thus (1) is proved. If the hypotheses of (2) hold, let $f : A \to B$ be a ring morphism focussing on P. We know that there is a factorization $A \to A_P \xrightarrow{g} B$ by 2.8,(4) so that g focuses on PA_P. By 0.1, $\sqrt{\mathrm{Ker}(g)} = PA_P$. Since A_P is a valuation ring, there is some integer n such that $P^n A_P \subset \mathrm{Ker}(g)$ [7, 17.1]. Setting $I = \cap_n P^n A_P$, we get $I \subset \mathrm{Ker}(g)$. Therefore, g can be factored $A_P \to A_P/I \to B$. In short, there is a factorization $A \to A_P/I \to B$ where the first morphism is an epimorphism with spectral image $\{P\}$ because $\sqrt{I} = PA_P$. To complete the proof, use [1, 2.7] which asserts that $I = \sqrt{I}$. □

REMARK 3.11. Let P be a prime ideal of a ring A with total quotient ring T and $N(P) = \mathrm{Ker}(A \to A_P)$. Denote by $\overline{T}$ the total quotient ring of $A/N(P)$. Then $A \to A/N(P)$ may be continued to a ring morphism $T \to \overline{T}$ and A_P may be considered as an overring of $A/N(P)$. Assume that P is quasi-invertible, that is $P < PP^{-1}$. By reworking the proof of [4, Theorem], we find that PA_P is invertible so that this ideal is finitely generated and $0 = \cap_n P^n A_P$ is a prime ideal by [4, Lemma]. Therefore, a quasi-invertible prime ideal P is geometric if and only if P is a minimal prime ideal.

PROPOSITION 3.12. *Let A be a reduced ring such that* $\mathrm{Min}(A)$ *is compact. Then* $\mathrm{Min}(A)$ *is geometric with solution* $m : A \to M(A)$, *the maximal flat epimorphic extension of A.*

Proof. Thanks to [17, II,Proposition 7], $\mathrm{Min}(A)$ is compact if and only if $M(A)$ is absolutely flat where $A \to M(A)$ is the maximal flat epimorphic extension of A (see for instance [13, IV.3]). In this case m focuses on $\mathrm{Min}(A)$. Now let $f : A \to B$ be a ring morphism such that ${}^a f(\mathrm{Spec}(B)) \subset \mathrm{Min}(A)$. Then the canonical morphism $B \to B \bigotimes_A M(A)$ is a spectrally surjective flat epimorphism by 0.4, whence an isomorphism because a faithfully flat epimorphism is an isomorphism [13, IV.1.2]. □

4 PROPERTIES OF GEOMETRIC SUBSETS

The first proposition is a key result.

PROPOSITION 4.1. *Let* $X \subset \mathrm{Spec}(A)$ *be a geometric subset. Let* $f : A \to B$ *be a ring morphism and* $Y = {}^a f^{-1}(X)$. *Then Y is geometric and a solution is given by* $B \to \Delta(Y) = B \bigotimes_A \Delta(X)$.

Proof. Let $g : B \to C$ be a ring morphism such that ${}^a g(\mathrm{Spec}(C)) \subset Y$. It follows that ${}^a(g \circ f)(\mathrm{Spec}(C)) \subset X$. Then there is a unique morphism $h : \Delta(X) \to C$ such that $h \circ \delta = g \circ f$. By the universal property of tensor products, $B \to C$ can be factored by $B \bigotimes_A \Delta(X) \to C$. Moreover, the spectral image of $B \bigotimes_A \Delta(X)$ in $\mathrm{Spec}(A)$ is ${}^a f(\mathrm{Spec}(B)) \cap {}^a\delta(\mathrm{Spec}(\Delta(X))) \subset X$ by 0.4 so that the spectral image of $B \bigotimes_A \Delta(X)$ in $\mathrm{Spec}(B)$ is contained in Y. The proof of uniqueness is straightforward. □

REMARK. Let $X \subset \mathrm{Spec}(A)$ be a geometric subset. Let $f : A \to B$ be an epimorphism factoring δ and let $g : B \to \Delta(X)$. Then $g : B \to \Delta(X)$ is a solution for the geometric subset ${}^a f^{-1}(X)$. This can be shown by using the definition of a geometric subset. This result is also a consequence of 4.1. Indeed, there is a factorization $\Delta(X) \to B \bigotimes_A \Delta(X) \to \Delta(X)$ where the first morphism is a pure epimorphism whence an isomorphism (see 0.3).

PROPOSITION 4.2. *The following conditions are equivalent for* $X \subset \mathrm{Spec}(A)$:

(1) *X is geometric.*
(2) *There exists an epimorphism* $\delta : A \to \Delta$ *such that*
    (a) ${}^a\delta(\mathrm{Spec}(\Delta)) = X$.

(b) For each ring morphism $h : A \to B$ *such that* ${}^a h(\mathrm{Spec}(B)) \subset X$*, there is a commutative diagram of ring morphisms*

$$\begin{array}{ccc} A & \longrightarrow & \Delta \\ \downarrow & & \downarrow \\ B & \longrightarrow & C \end{array}$$

where $B \to C$ *is strict (respectively pure,* B *is a retract of* C*).*

Therefore, $\delta : A \to \Delta(X)$ *is an epimorphism when* X *is geometric. In this case, let* $A \to S \to \Delta(X)$ *be a factorization, then* $S \to \Delta(X)$ *is surjective whenever the image of* $\mathrm{Spec}(S) \to \mathrm{Spec}(A)$ *is contained in* X *(for instance, when* $S \to \Delta(X)$ *is spectrally surjective).*

Proof. Assume that (1) holds. Then $A \to \Delta(X) = \Delta$ is an epimorphism. To see this, take $f = \delta$ in 4.1, so that $Y = \mathrm{Spec}(\Delta(X))$ by 3.2. Then in view of 4.1, $\Delta(X) \to \Delta(X) \bigotimes_A \Delta(X)$ provides us a solution for $\mathrm{Spec}(\Delta(X))$. The remark following 3.1 shows that this last morphism is an isomorphism. Thus δ is an epimorphism [13, IV.1.0], such that ${}^a\delta(\mathrm{Spec}(\Delta)) = X$. To show (b), let $h : A \to B$ be as in (b). There is a factorization $A \to \Delta \to B$ and Id_B can be factored $B \to B \bigotimes_A \Delta \to B$. Hence it is enough to take $C = B \bigotimes_A \Delta$. Therefore (2) is proved. Conversely, assume that the hypotheses of (2) hold and let $h : A \to B$ be a ring morphism such that ${}^a h(\mathrm{Spec}(B)) \subset X$. There is a commutative diagram as in (b). Set $S = \Delta \times_C B$, then $S \to \Delta$ is strict by 0.3 and an epimorphism whence an isomorphism. It follows that $A \to B$ can be factored through δ and (1) is proved. To complete the proof, assume that δ is factored by a ring morphism $f : A \to S$ such that ${}^a f(\mathrm{Spec}(S)) \subset X$. Denote by g the ring morphism $S \to B$ so that $g \circ f = \delta$. Then f can be factored $f = h \circ g \circ f$ where $h : \Delta \to C$ is a ring morphism. It follows that $g \circ h \circ \delta = g \circ f = \delta$. Since δ is an epimorphism, $g \circ h = \mathrm{Id}_\Delta$ so that g is surjective. □

When X is a geometric subset of $\mathrm{Spec}(A)$, the monomorphism of affine schemes $\Delta = \left(\mathrm{Spec}(\Delta(X)), \widetilde{\Delta(X)}\right) \to \left(\mathrm{Spec}(A), \widetilde{A}\right) = Y$ is a solution of the geometric problem associated to X in the category of schemes. Indeed, first observe that this morphism is obviously a solution in the category of affine schemes. Then consider a morphism of schemes $f : Z \to Y$ such that $f(Z) \subset X$. The affine open subsets of Z are a base of the topology of the underlying set of Z. Let $V \subset U$ be affine open subsets of Z, defining morphisms of schemes $V \to U \to Z$. Then $U \to Y$ and $V \to Y$ being morphisms of affine schemes can be factored $U \to \Delta \to Y$ and $V \to \Delta \to Y$. Since $\Delta \to Y$ is a monomorphism, we get a factorization $V \to U \to \Delta$. Thanks to [24, 2.12,(3)], there is a morphism of schemes $Z \to \Delta$ such that

$U \to Z \to \Delta \to Y = U \to Z \to Y$ for each affine open subset U. Then the uniqueness asserted in [24, 2.12,(3)] implies that $Z \to Y = Z \dashrightarrow \Delta \to Y$.

Clearly, there is a factorization $\Delta \to \left(X, \widetilde{A}_{|X}\right) \to Y$. But the first morphism may not be an isomorphism for if not, $A_P \to \Delta(X)_P$ is an isomorphism for each $P \in X$ by Section 1 (∗∗), that is $A \to \Delta(X)$ is a flat epimorphism [13, IV.2.4]. This denies 3.7.

Anyway, there is a factorization $A \to \widetilde{A}(X) \to \Delta(X)$ (see 2.1). Unlike to the locally ringed space case, there are two obstructions for $f_X : A \to \widetilde{A}(X)$ to be a solution of the geometric problem associated to X: its spectral image contains X^g and f_X may not be an epimorphism. By 2.1, X is geometric when f_X is an epimorphism and $\operatorname{Im}({}^a f_X) = X$. In this case, X is a patch stable under generizations. We will give a complete answer for such subsets X.

REMARK 4.3. If $X \subset \operatorname{Spec}(A)$ is a patch, the family of all epimorphisms $f : A \to B$ such that $\operatorname{Im}({}^a f) \subset X$ has a final object. Indeed, Let $t : A \to \mathrm{T}(A)$ be the ring morphism of A into its associated universal absolutely flat ring. Then t is an epimorphism, ${}^a t$ is a homeomorphism for the Zariski topology on $\operatorname{Spec}(\mathrm{T}(A))$ and the patch topology on $\operatorname{Spec}(A)$ (see Olivier's paper [17]). Thus $F = {}^a t^{-1}(X)$ is a Zariski closed set $\mathrm{V}(I)$ where I is a radical ideal. Then $\theta : A \to \mathrm{T}(A)/I = E$ is an epimorphism with an absolutely flat range such that $\operatorname{Im}({}^a\theta) = X$. Let $f : A \to B$ be an epimorphism such that $\operatorname{Im}({}^a f) \subset X$. Then $E \to E \bigotimes_A B$ is a flat epimorphism. Now if Q is a prime ideal of E lying above $P \in X$, there is some prime ideal R in B lying over P. By 0.4, there is a prime ideal S in $E \bigotimes_A B$ lying over Q. Hence $E \to E \bigotimes_A B$ is a faithfully flat epimorphism whence an isomorphism [13, IV.1.2]. Therefore, there is a factorization $A \to B \to E$. If X is geometric, the above family has an initial object.

PROPOSITION 4.4. *Let $\{X_i\}_{i\in I}$ be a family of geometric subsets of* $\operatorname{Spec}(A)$ *admitting a solution $A \to \Delta(X_i)$. Then $X = \cap_{i\in I} X_i$ is a geometric subset with solution $A \to \bigotimes_{i\in I} \Delta(X_i)$.*

Proof. Let $f : A \to B$ such that ${}^a f(\operatorname{Spec}(B)) \subset X \subset X_i$ for every $i \in I$. There is a unique factorization $A \to \Delta(X_i) \to B$. Thus we get a unique factorization $A \to \bigotimes_{i\in I} \Delta(X_i) \to B$. Moreover, the spectral image of $\bigotimes_{i\in I} \Delta(X_i)$ in $\operatorname{Spec}(A)$ is X [8, I.3.4.10]. □

Therefore, for an arbitrary subset X of $\operatorname{Spec}(A)$, there is a smallest geometric subset X' containing X. It follows from 2.8,(4) and 4.7 below that $X' \subset X^u$.

LEMMA 4.5. *Let $X \subset Y$ be geometric subsets of* $\operatorname{Spec}(A)$. *There is a factorization $A \to \Delta(Y) \to \Delta(X)$.*

Proof. Obvious. □

REMARK 4.6. The previous result 4.4 can be refined. Let X be a patch in $\mathrm{Spec}(A)$, we proved that X is the intersection of a family $\{X_i\}_{i\in I}$ where I is an ordered set, each X_i is constructible and $X_j \subset X_i$ when $i \leq j$ [20, I.2,Lemme 4]. Assume that each X_i is geometric. For $i \leq j$, there is a ring morphism $\delta_{j,i} : \Delta(X_i) \to \Delta(X_j)$. Since each $A \to \Delta(X_i)$ is an epimorphism, we get a direct system of A-algebras $\{\delta_{j,i}, \Delta(X_i)\}$ with limit an epimorphism $\delta : A \to \Delta$, focussing on X [8, I.3.4.10]. Then a ring morphism $f : A \to B$ such that ${}^a f(\mathrm{Spec}(B)) \subset X$ can be factored uniquely $A \to \Delta \to B$ because each $A \to \Delta(X_i)$ is an epimorphism. Therefore, X is geometric with solution δ. We do not know whether an arbitrary geometric subset can be gotten in this way.

PROPOSITION 4.7. *Let $f : A \to B$ be a flat epimorphism with spectral image X. Then X is geometric, $A \to B$ is a solution which identifies with $A \to \widetilde{A}(X)$. In particular, if S is a multiplicative subset of A, then $X = \cap\ [\mathrm{D}(s)\ ;\ s \in S]$ is geometric with solution $A \to A_S$.*

Proof. Let $g : A \to C$ be a ring morphism such that ${}^a g(\mathrm{Spec}(C)) \subset X$. Then $h : C \to C \bigotimes_A B$ is a flat epimorphism. Moreover, ${}^a h$ is surjective because $\mathrm{Spec}\,(C \bigotimes_A B\ ) \to \mathrm{Spec}(B) \times_{\mathrm{Spec}(A)} \mathrm{Spec}(C)$ is surjective by 0.4. Thus h is a faithfully flat epimorphism, whence an isomorphism [13, IV.1.2]. The factorization is unique because $A \to B$ is an epimorphism. Then $A \to B$ identifies with $A \to \widetilde{A}(X)$ by 1.4. □

Let X be a geometric subset of $\mathrm{Spec}(A)$ and $f \in A$. Then $X_f = X \cap \mathrm{D}(f)$ is geometric by 4.6 and 4.7, a solution being given by $A \to \Delta(X)_f$. We intend to show a converse. To this end, consider the following situation. Let $\{M_i\}$ be a finite family of A-modules where $i = 1, \ldots, n$ and $\{M_{\{i,j\}}\}$ another family of A-modules for $i, j = 1, \ldots, n$, together with morphisms of A-modules $f_{i,j} : M_i \to M_{\{i,j\}}$ and $f_{j,i} : M_j \to M_{\{i,j\}}$. Denote by $p_i : M = \prod_{k=1}^n M_k \to M_i$ the canonical surjection and by K the set of all $(x_1, \ldots, x_n) \in M$ such that $f_{i,j}(x_i) = f_{j,i}(x_j)$ for each $\{i,j\}$. Then K is the kernel of the morphism of A-modules $f : M \to \prod_{\{i,j\}} M_{\{i,j\}}$ defined by $f(x_1, \ldots, x_n) = (f_{i,j} \circ p_i(x_1, \ldots, x_n) - f_{j,i} \circ p_j(x_1, \ldots, x_n)\ )$. Now if P is a flat A-module, then $K \bigotimes_A P$ is the kernel associated to the A-modules $M_i \bigotimes_A P$, $M_{\{i,j\}} \bigotimes_A P$ and the morphisms $f_{i,j} \otimes P$. When the modules are A-algebras, so is K.

PROPOSITION 4.8. *The geometric property on a subset is local on the spectrum. In other words, let X be a subset of* $\mathrm{Spec}(A)$ *and $f_1, \ldots, f_n \in A$ generating the ideal A. Then X is geometric if and only if $X_{f_1}, \ldots, X_{f_n}$ are geometric.*

Proof. Assume that each X_{f_i} is geometric and let K be the associated kernel. We set $\Delta_i = \Delta(X_{f_i})$ and $\Delta_{\{i,j\}} = \Delta(X_{f_i f_j})$. The pievious results show that there are A-algebras morphisms $f_{i,j} : \Delta_i \to \Delta_{\{i,j\}}$. Let $g : A \to B$ be a ring morphism such that ${}^a g(\mathrm{Spec}(B)) \subset X$, then ${}^a g_i(\mathrm{Spec}(B_{f_i})) \subset X_{f_i}$ holds for $g_i : A \to B_{f_i}$. Hence, there exist morphisms of A-algebras $\Delta_i \to B_{f_i}$ and $\Delta_{\{i,j\}} \to B_{f_i f_j}$ such that the following diagrams are commutative:

$$\begin{array}{ccc} \Delta_i & \longrightarrow & \Delta_{\{i,j\}} \\ \downarrow & & \downarrow \\ B_{f_i} & \longrightarrow & B_{f_i f_j} \end{array}$$

because each $A \to \Delta_i$ is an epimorphism. Observe that B is the kernel associated to the families $\{B_{f_i}\}, \{B_{f_i f_j}\}$ because $\widetilde{B}$ is a sheaf and the ideal generated by $g(f_1), \ldots, g(f_n)$ is B. It follows that there is a unique factorization $A \to K \to B$. It remains to show that $\delta : A \to K$ focuses on X. Tensoring the exact sequence $K \to \prod \Delta_i \to \prod \Delta_{\{i,j\}}$ by the flat module A_{f_1} provides us an exact sequence $K_{f_1} \to \prod_i (\Delta_1)_{f_i} \to \prod_{i,j} (\Delta_1)_{f_i f_j}$ because $(\Delta_1)_{f_j} = \Delta(X \cap \mathrm{D}(f_1) \cap \mathrm{D}(f_j)) = (\Delta_j)_{f_1}$. Now the images of $f_1, \ldots, f_n$ generate Δ_1. Because $\widetilde{\Delta_1}$ is a sheaf, $K_{f_1} = \Delta_1$ follows. But $K \to \prod K_{f_i}$ is faithfully flat whence spectrally surjective. It follows that $X = \cup X_{f_i} = {}^a\delta(\mathrm{Spec}(K))$. □

Take $n = 2$ in the above result. Then K is nothing but the pullback associated to the ring morphisms $\Delta_1 \to \Delta_{1,2}$ and $\Delta_2 \to \Delta_{1,2}$.

In view of the preceding result, it may be asked whether pure morphisms descend geometric subsets, that is, if $f : A \to B$ is pure and $X \subset \mathrm{Spec}(A)$ such that ${}^a f^{-1}(X)$ is geometric, then X is geometric. This is true if ${}^a f^{-1}(X)$ is the spectral image of a flat epimorphism (hence geometric by 4.7) [22, 4.8]. We do not know the answer for an arbitrary pure morphism, but for algebraically pure morphisms introduced by D. Popescu [26], the result is valid. Algebraic purity is defined in the same way as purity, linear relations being replaced by polynomial relations.

PROPOSITION 4.9. *Let $f : A \to B$ be a ring morphism focussing on X in* $\mathrm{Spec}(A)$ *and $\varphi : A \to A'$ an algebraically pure morphism. If ${}^a\varphi^{-1}(X)$ is geometric with solution $A' \to A' \bigotimes_A B$, then X is geometric with solution $f : A \to B$.*

Proof. By definition, $A' \to A' \bigotimes_A C$ is factored by $A' \to A' \bigotimes_A B$. Hence $A \to C$ is factored by $A \to B$ [25, 1.17]. Then the uniqueness of the factorization follows from purity. □

Let X be a geometric subset of $\mathrm{Spec}(A)$ and $f : A \to B$ a ring morphism such that ${}^a f(\mathrm{Spec}(B)) \subset X$. We deduce from 4.1 and 3.1 that $B \to B \bigotimes_A \Delta(X)$ is an isomorphism because ${}^a f^{-1}(X) = \mathrm{Spec}(B)$. Thus we get the following result.

PROPOSITION 4.10. *Let $X \subset \mathrm{Spec}(A)$ be a geometric subset and $P \in \mathrm{Spec}(A)$.*

(1) *$A_P \to \Delta(X)_P$ is an isomorphism if and only if $P^g \subset X$.*
(2) *$A/P \to \Delta(X)/P\Delta(X)$ is an isomorphism if and only if $P^s \subset X$.*

But we can prove a more precise result.

THEOREM 4.11. *Let X be a geometric subset of $\mathrm{Spec}(A)$ and $Q \in \mathrm{Spec}(\Delta(X))$ lying over P in A.*

(1) *$A_P \to \Delta(X)_Q$ is an isomorphism provided $P^g \subset X$.*
(2) *$A/P \to \Delta(X)/Q$ is an isomorphism provided $P^s \subset X$.*
(3) *$\widehat{A_P} \to \widehat{\Delta(X)_Q}$ is an isomorphism.*

Proof. There is a commutative diagram

$$\begin{array}{ccc} A & \xrightarrow{\delta} & \Delta(X) \\ \varphi\downarrow & & \downarrow\psi \\ A_P & \xrightarrow{\varepsilon} & \Delta(X)_Q \end{array}$$

Assume that $P^g \subset X$. There is a ring epimorphism $m : \Delta(X) \to A_P$ such that $\varphi = m \circ \delta$. Then $\varepsilon \circ m \circ \delta = \psi \circ \delta$ implies $\psi = \varepsilon \circ m$ because δ is an epimorphism. Now let $b \in \Delta(X) \setminus Q$, then $m(b) \in PA_P$ implies $\psi(b) \in Q\Delta(X)_Q$ which is absurd. Thus m can be factored $\Delta(X) \xrightarrow{\psi} \Delta(X)_Q \xrightarrow{n} A_P$. Hence we get $n \circ \varepsilon \circ m = n \circ \psi = m$ so that $n \circ \varepsilon = \mathrm{Id}$ because m is an epimorphism. Therefore, ε is a pure epimorphism whence an isomorphism. The proof of (2) is analogous. Indeed, considering the injective ring morphism $\varepsilon : A/P \to \Delta(X)/Q$, we get that $\mathrm{Ker}(m) = Q$ from which we deduce the factorization $\Delta(X) \to \Delta(X)/Q \to A/P$. Next we show (3). We give some notation for the following canonical morphisms: $\alpha_P : A \to A_P$, $c_P : A_P \to \widehat{A_P}$, $d_P = c_P \circ \alpha_P$ (similarly for Q), $\delta' : A_P \to \Delta(X)_Q$, $\hat{\delta} : \widehat{A_P} \to \widehat{\Delta(X)_Q}$. Thanks to 3.9, the ring morphism $d_P : A \to \widehat{A_P}$ can be factored $d_P = \varphi \circ \delta$ where $\varphi : B \to \widehat{A_P}$ is a ring morphism. From $\hat{\delta} \circ \varphi \circ \delta = \hat{\delta} \circ d_P = d_Q \circ \delta$, we get $\hat{\delta} \circ \varphi = d_Q$ because δ is an epimorphism. Now let σ be in $\Delta(X) \setminus Q$ and assume that $\varphi(s)$ is not a unit. It follows that $\varphi(s) \in \widehat{P}$ so that $d_Q(s) = \hat{\delta}(\varphi(s)) \in \widehat{Q}$ and s belongs to Q. This is absurd. Therefore, there is a ring morphism $\psi : \Delta(X)_Q \to \widehat{A_P}$ providing us a factorization $\varphi = \psi \circ \alpha_Q$. We get $\psi \circ \delta' \circ \alpha_P = \psi \circ \alpha_Q \circ \delta = \varphi \circ \delta = c_P \circ \alpha_P$. Thus $\psi \circ \delta' = c_P$ follows because

α_P is an epimorphism. This implies that $\psi(Q\Delta(X)_Q) = \psi(PB_Q) \subset \widehat{P}$ by 0.2,(a). Therefore, ψ induces a ring morphism $\rho : \widehat{\Delta(X)_Q} \to \widehat{A_P}$ such that $\rho \circ c_Q = \psi$. Now we have $\rho \circ \hat{\delta} \circ d_P = \rho \circ c_Q \circ \delta' \circ \alpha_P = \psi \circ \delta' \circ \alpha_P = \psi \circ \alpha_Q \circ \delta = \varphi \circ \delta = c_P \circ \alpha_P$ from which we deduce $\rho \circ \hat{\delta} \circ c_P = c_P = \mathrm{Id}_{\widehat{A_P}} \circ c_P$ because α_P is an epimorphism. Since the image of c_P is everywhere dense, we get $\rho \circ \hat{\delta} = \mathrm{Id}_{\widehat{A_P}}$. Therefore, $e = \hat{\delta} \circ \rho$ is an idempotent endomorphism of $\widehat{B_Q}$, so that $e^2 = \mathrm{Id}_{\widehat{B_Q}} \circ e$. Since ρ is surjective and $\hat{\delta}$ is a formal epimorphism by 0.2,(b), e is a formal epimorphism. It follows that e is the identity of $\widehat{B_Q}$ and $\hat{\delta}$ is an isomorphism. □

The previous result shows that a ring morphism which is a solution of a geometric problem is not so far from being a flat epimorphism. Statement (3) says that δ is a formally flat epimorphism. It would be interesting to know whether (3) has a converse.

THEOREM 4.12. *Let $X \subset \mathrm{Spec}(A)$ be a patch stable under generizations. Then X is geometric if and only if there is a flat ring epimorphism $f : A \to B$ such that ${}^a f(\mathrm{Spec}(B)) = X$, or also if and only if $(X, \widetilde{A}_{|X})$ is an affine scheme. In that case, $A \to \widetilde{A}(X)$ is a solution and $\widetilde{A}(X) = \mathrm{Dom}_A(X(A))$.*

Proof. Assume that X is geometric and a patch stable under generizations. Then $A \to \Delta(X)$ is an epimorphism and is flat thanks to 4.11. To complete the proof, use 4.7 and 2.8,(3). □

REMARK. We can get part of 4.12 as follows. Let $f : A \to B$ be a ring morphism and $X \subset \mathrm{Spec}(A)$. Assume that f is an isomorphism along X, that is $A_P \to B_P$ is an isomorphism for every $P \in X$. Let $g : A \to C$ such that ${}^a g(\mathrm{Spec}(C)) \subset X$ and $Q \in \mathrm{Spec}(C)$ lying over $P \in X$. There is a cocartesian square

$$\begin{array}{ccc} A_P & \longrightarrow & B_P \\ \downarrow & & \downarrow \\ C_Q & \longrightarrow & (C \bigotimes_A B)_Q \end{array}$$

It follows that $C_Q \to (C \bigotimes_A B)_Q$ is an isomorphism for each $Q \in \mathrm{Spec}(B)$ and so is $C \to C \bigotimes_A B$. Therefore, there is a factorization $A \to B \to C$. Now if in addition ${}^a f(\mathrm{Spec}(B)) \subset X$, then f is a flat epimorphism by [13, IV.2.4,(iii)].

COROLLARY 4.13. *Let A be a locally Noetherian ring and $X \subset \mathrm{Spec}(A)$. Then X is geometric if and only if there is a flat ring epimorphism $f : A \to B$ such that ${}^a f(\mathrm{Spec}(B)) = X$.*

Proof. Use 3.9. □

EXAMPLE 4.14. We use Ferrand's thesis [6, 5.4 & 5.5]. Let A be a one-dimensional Noetherian integral domain and $\{(B_i, P_i)\}_{i\in I}$ a family of local subrings of the quotient field K of A. Set $B = \cap_{i\in I} B_i$, $N_i = B \cap P_i$, $M_i = A \cap P_i$ and assume that $M_i \neq M_j$ for $i \neq j$. Then the ring $B = \cap_{i\in I} B_i$ is a one-dimensional Noetherian integral domain, $\mathrm{Max}(B) = \{N_i\}_{i\in I}$ and $B_i = B_{N_i} = B_{M_i}$. Now take a subset X of $\mathrm{Max}(A)$. The ring B associated to the family $\{A_M\}_{M\in X}$ is $\widetilde{A}(X) = \widetilde{A}(X^g)$ by 2.2. We get that $\mathrm{Im}(f_X) = X \cup \{0\} = X^g$ is proconstructible and stable under generizations. Therefore, $A \to \widetilde{A}(X^g)$ is a solution of the geometric problem associated to X^g if and only if this morphism is flat (indeed, f_X flat implies f_X is an epimorphism by [13, IV.3.2] applied to the factorization $A \to B \to K$).

EXAMPLE 4.15. Let I be an ideal of a ring A. Then $X = \mathrm{V}(I)^g$ is geometric. To see this, observe that $X = \mathrm{Spec}\,(A_{1+I})$ [12, 3.1] and that $A \to A_{1+I}$ is a flat epimorphism.

We use the notation of Section 2, in particular 2.8,(1). Some parts of 4.16 are well known. Moreover, Proposition 4.16 is related to Theorem 2.4 of the paper by M. Fontana and N. Popescu "Universal property of the Kaplansky ideal transform and affineness of open subsets" J. Pure Appl. Algebra (to appear) where the authors characterize, with different techniques, when an open subspace of the prime spectrum of an integral domain is geometric.

PROPOSITION 4.16. *Let A be a ring and $I = (a_1, \dots, a_n)$ a finitely generated ideal of A. The following statements are equivalent:*

(1) $A \to \widetilde{A}(\mathrm{D}(I))$ *is an epimorphism and* $\mathrm{W}(I) = \mathrm{D}(I)$.
(2) $A \to \widetilde{A}(\mathrm{D}(I))$ *is a flat epimorphism.*
(3) $\mathrm{D}(I)$ *is an affine open subset of* $\mathrm{Spec}(A)$.
(4) $\mathrm{D}(I)$ *is geometric.*

If one of the preceding statements holds, then $A \to \mathrm{Dom}\,(B_I)$ *is of finite presentation.*

Proof. Assume that $A \to \mathrm{Dom}_A\,(B_I) = \widetilde{A}(\mathrm{D}(I))$ is an epimorphism and $\mathrm{W}(I) = \mathrm{D}(I)$. In this case $\mathrm{Dom}_A\,(B_I) \to B_I$ has a surjective spectral map because the spectral map of an epimorphism is injective. Therefore, $\mathrm{Dom}_A\,(B_I) \to B_I$ is faithfully flat by 2.8,(1), from which it follows that $A \to \mathrm{Dom}_A\,(B_I)$ is flat of finite presentation. Assume that $A \to \mathrm{Dom}_A\,(B_I)$ is a flat epimorphism. The spectral image of $A \to \mathrm{Dom}_A\,(B_I)$ is $\mathrm{D}(I)$ by the same argumentation as above. It follows from 1.4 that $\mathrm{D}(I)$ is an affine open subset. Now if $\mathrm{D}(I)$ is an affine open subset, the morphism of schemes $\left(\mathrm{D}(I), \widetilde{A}_{|\mathrm{D}(I)}\right) \to (\mathrm{Spec}(A), \widetilde{A})$ is an open immersion so that $A \to \widetilde{A}(\mathrm{D}(I))$ is a flat epimorphism of finite presentation. □

Next we characterize geometric closed subsets. Let X be a closed subset of Spec(A), then X defines many structures of subschemes of $\left(\operatorname{Spec}(A), \widetilde{A}\right)$ depending on the representation $X = \mathrm{V}(I)$ where I is an ideal of A.

THEOREM 4.17. *Let* $X = \mathrm{V}(I)$ *be a closed subset of* Spec(A). *The following conditions are equivalent:*

(1) X *is geometric.*
(2) *If* $f : A \to B$ *is a ring morphism such that* ${}^a f(\operatorname{Spec}(B)) \subset X$, *then* $IB = 0$.
(3) I *is contained in each ideal* J *of* A *such that* $I \subset \sqrt{J}$.

If one of the preceding conditions holds, a solution is given by $A \to A/I$.

Proof. Assume that X is geometric and let $f : A \to B$ be a ring morphism such that ${}^a f(\operatorname{Spec}(B)) \subset X$. Let $Q \in \operatorname{Spec}(B)$ lying over P in A. Thanks to 1.5, we get $I_P \subset \operatorname{Ker}(A_P \to B_Q)$ so that $(IB)_Q = 0$. It follows that $IB = 0$. Assume that (2) holds, then clearly $\mathrm{V}(I)$ is geometric with solution $A \to A/I$. Assuming again that (2) holds, let J be an ideal such that $I \subset \sqrt{J}$ so that $\mathrm{V}(J) \subset \mathrm{V}(I)$. It follows that $I(A/J) = 0$ and then $I \subset J$. Hence (3) is proved. If (3) holds, let $f : A \to B$ be a ring morphism such that ${}^a f(\operatorname{Spec}(B)) \subset X$. It follows that $\mathrm{V}(\operatorname{Ker}(f)) \subset \mathrm{V}(I)$ by 0.1 so that $I \subset \operatorname{Ker}(f)$. Therefore, there is a factorization $A \to A/I \to B$ and (1) is proved. $\square$

REMARKS 4.18. We give some informations on ideals verifying (3).

(1) A pure ideal I (such that $A \to A/I$ is flat) verifies (3). Recall that I is pure if and only if for each $a \in I$ there is some $b \in I$ such that $a = ba$. Now if $I \subset \sqrt{J}$ where J is an ideal, for each a, $b \in I$ such that $ab = a$, there is some integer $n > 0$ such that $b^n \in J$. It follows that $a = ab^n \in J$ and we get $I \subset J$. We recover a special case of 4.12.
(2) An ideal I verifying (3) is idempotent, because $I \subset \sqrt{I^2}$. Conversely, an idempotent ideal of a Prüfer domain verifies (3). Indeed, let J be an ideal such that $I \subset \sqrt{J}$ and P a prime ideal of A. Then from $IA_P \subset \sqrt{JA_P}$, we get $IA_P = (IA_P)^n \subset JA_P$ for some integer $n > 0$ because A_P is a valuation ring [7, 17.1] so that $I \subset J$.
(3) We recover 3.7 where we considered a valuation ring V with an idempotent maximal ideal M. This example shows that an ideal verifying (3) is not necessarily pure.

Now we intend to give some results about geometric problems with respect to reduced rings. In the category of reduced rings, things are somewhat easier.

DEFINITION 4.19. Let A be a ring and $X \subset \operatorname{Spec}(A)$. Then X is called a quasi-geometric subset of Spec(A) if the following statements hold:

(1) There is a ring morphism $\omega : A \to \Omega(X)$ such that ${}^a\omega(\operatorname{Spec}(\Omega(X))) \subset X$ and $\Omega(X)$ is reduced.

(2) For every ring morphism $f : A \to B$ where B is reduced and such that ${}^a f(\mathrm{Spec}(B)) \subset X$, there is a unique ring morphism $g : \Omega(X) \to B$ such that $f = g \circ \omega$.

When (1) and (2) hold, we say that ω is a solution of the quasi-geometric (universal) problem associated to X. In this case, ${}^a\omega(\mathrm{Spec}(\Omega(X))) = X$ shows that X is a patch (let P be in X, there is a ring morphism $A \to \mathrm{k}(P)$ giving a factorization $A \to \Omega(X) \to \mathrm{k}(P)$).

Clearly, the subset $\mathrm{Spec}(A)$ has a solution $A \to \Omega(X) = A_{\mathrm{red}}$.

Obviously, if $X \subset \mathrm{Spec}(A)$ is geometric, then X is quasi-geometric. A solution is given by $A \to \Delta(X)_{\mathrm{red}}$.

Recall that a ring morphism $f : A \to B$ is called radicial if f is universally spectrally injective [8, I.3.7.2]. An equivalent condition is $B_{\mathrm{red}} \to (B \bigotimes_A B)_{\mathrm{red}}$ is an isomorphism. Mimicking the proofs of 4.1, 4.2 and 4.4, we get the following results:

(1) If X is quasi-geometric, then $A \to \Omega(X)$ is radicial.

(2) If X is quasi-geometric and $f : A \to B$ is a ring morphism, then ${}^a f^{-1}(X)$ is quasi-geometric.

(3) If $\{X_i\}_{i \in I}$ is a family of quasi-geometric subsets, then so is $\bigcap_{i \in I} X_i$.

To show the last property, it is enough to use two well known facts. Firstly, a direct limit of reduced rings is a reduced ring and secondly, the spectrum of a direct limit of rings is homeomorphic to the inverse limit of the spectra.

EXAMPLE 4.20. Consider a ring A and an ideal I of A. Let $f : A \to B$ be a ring morphism such that ${}^a f(\mathrm{Spec}(B)) \subset \mathrm{V}(I)$ and B is reduced. Since the Zariski closure of ${}^a f(\mathrm{Spec}(B))$ is $\mathrm{V}(\mathrm{Ker}(f))$ by 0.1, we get a unique factorization $A \to A/\sqrt{I} \to B$. Therefore, $\mathrm{V}(I)$ is quasi-geometric.

For the next result, we use the notation and results of Section 2.

THEOREM 4.21. *Let $f : A \to B$ be a ring morphism and $X = \mathrm{Im}\,({}^a f\,)$. For every ring morphism $g : A \to C$ such that $\mathrm{Im}\,({}^a g\,) \subset X$, there is a unique factorization $A \to A(\mathcal{F}) \to C_{\mathrm{red}}$. Moreover, if f is pre-flat then $A(\mathcal{F}) = B$ and X is quasi-geometric.*

Proof. We can assume that C is reduced. In view of 2.6,(3), there is a surjection $\nu : C \to C \bigotimes_A A(\mathcal{F})$. Its spectral map is surjective. Indeed, let $Q \in \mathrm{Spec}(C)$ lying over $P \in X$. By 2.6,(2), there is a prime ideal R in $\mathrm{Spec}\,(A(\mathcal{F})\,)$ lying over P. By 0.4, there is a prime ideal of $C \bigotimes_A A(\mathcal{F})$ lying over Q. Therefore, $\mathrm{Ker}(\nu)$ is contained in $\mathrm{Nil}(C) = 0$. Hence the factorization is proved. Its uniqueness follows from 2.6,(5). If $A \to B$ is pre-flat, then $A(\mathcal{F}) = B$ by [16, 4.2] and X is quasi-geometric with solution $A \to B_{\mathrm{red}}$. $\square$

EXAMPLE 4.22. Let X be a subset of $\mathrm{Spec}(A)$. Then consider the ring morphism $A \to A_X \bigotimes_A A/\mathcal{R}(X)$ with spectral image $Y = X^u \cap \overline{X}$. This morphism is pre-flat by 2.9, its kernel being $\mathcal{R}(X)$. It follows that Y is quasi-geometric. Indeed, a ring morphism $f : A \to B$ with reduced range and verifying ${}^a f(\mathrm{Spec}(B)) \subset X$ is such that $\mathcal{R}(X) \subset \mathrm{Ker}(f)$ by 0.1. It follows that $A \to B$ can be factored $A \to A/\mathcal{R}(X) \to B$ and $A \to A_X \to B$ by 2.8,(4). Hence an arbitrary prime ideal is quasi-geometric.

Recall that a ring A is a going-down ring if A/P is a going-down domain for each $P \in \mathrm{Spec}(A)$. Morevover, a going-down domain is an integral domain A such that $A \to B$ has going-down for each overring B of A.

EXAMPLE 4.23. Let A be a going-down ring and X a maximal chain of $\mathrm{Spec}(A)$. Then X is quasi-geometric with solution $A \to (A_X \bigotimes_A A/\mathcal{R}(X))_{\mathrm{red}}$.
In view of [5, 2.6], a maximal chain of prime ideals is stable under unions and intersections. Therefore, $\mathcal{U}(X)$ and $\mathcal{R}(X)$ belong to X. The spectral image of $A \to A_X \bigotimes_A A/\mathcal{R}(X)$ is $Y = X^u \cap \mathrm{V}(\mathcal{R}(X))$. Now let P be in Y so that $\mathcal{R}(X) \subset P \subset \mathcal{U}(X)$. There is a valuation overring V of $\overline{A} = A/\mathcal{R}(X)$ such that the spectral image of $\overline{A} \to V$ is $\overline{X} = \{P/\mathcal{R}(X)\ ;\ P \in X\}$ [5]. Since $\overline{A}$ is a going-down domain, $\overline{A} \to V$ is going-down and hence P belongs to X. Therefore, $Y = X$ and X is quasi-geometric by 4.22.

EXAMPLE 4.24. Let $f : A \to B$ be a pre-flat morphism, $X = \mathrm{Im}({}^a f)$ and I an ideal of A such that $\overline{X} \subset \mathrm{D}(I)$. From $X \subset \mathrm{D}(I)$ follows $B/IB = 0$. Let the ring morphism $g : A \to B \times A/I = C$. If $P \in X$, then $A_P \to C_P$ identifies to $A_P \to B_P$ since $P \notin \mathrm{V}(I)$ and hence is surjective. Now $A_P \to C_P$ identifies to $A_P \to (A/I)_P$ for $P \in \mathrm{V}(I)$. Indeed, $B_P = 0$ for if not, $X \cap P^g \neq \emptyset$ implies $P \in X^s = \overline{X}$ (see 0.1) which is absurd. Since $Y = \mathrm{Im}({}^a g) = X \cup \mathrm{V}(I)$, we see that $A \to C$ is pre-flat and Y is quasi-geometric.

REFERENCES

1. J. T. Arnold and R. Gilmer, *Idempotent ideals and union of nets of Prüfer domains*, J. Sci. Hiroshima Univ. **Ser. A-I, 31** (1967), 131–145.
2. R. Bkouche, *Couples spectraux et faisceaux associés. Applications aux anneaux de fonctions*, Bull. Soc. Math. France **98** (1970), 253–295.
3. R. Bkouche, *Parties agréables d'un spectre d'anneau, géométrie pure et mollesse*, Unpublished.
4. H. S. Butts, *Quasi-invertible prime ideals*, Proc. Amer. Math. Soc **16** (1965), 291–292.
5. D. E. Dobbs, M. Fontana and G. Picavet, *Generalized going-down homomorphisms of commutative rings*, Submitted to publication (2001).
6. D. Ferrand, *Monomorphismes et morphismes absolument plats*, Bull. Soc. Math. France **100** (1972), 97–128.
7. R. Gilmer, *Multiplicative ideal theory*, Dekker, New York, 1972.

8. A. Grothendieck and J. Dieudonné, *Eléments de Géométrie Algébrique*, Spinger Verlag, Berlin, 1971.
9. R. Hartshorne, *Algebraic Geometry*, Springer Verlag, Berlin, 1977.
10. M. Hochster, *Prime ideal structure in commutative rings*, Trans. Amer. Math. Soc. **142** (1969), 43–60.
11. M. D. Larsen and P. J. McCarthy, *Multiplicative theory of ideals*, Academic Press, New York, London, 1971.
12. D. Lazard, *Disconnexités des spectres d'anneaux et de préschémas*, Bull. Soc. Math. France **95** (1967), 95 –108.
13. D. Lazard, *Autour de la platitude*, Bull. Soc. Math. France **97** (1969), 81–128.
14. D. Lazard and P. Huet, *Dominions des anneaux commutatifs*, Bull. Sc. Math. **94, Série 2** (1970), 193–199.
15. J. Malgoire and C. Voisin, *Spectre d'un espace annelé et dualité de Stone généralisée*, C. R. Acad. Sc. Paris **289 - Série A** (1979), 449–451.
16. C. Năstăsescu and N. Popescu, *On the localization ring of a ring*, J. Algebra **15** (1970), 41–56.
17. J. P. Olivier, *Anneaux absoluments plats et épimorphismes à buts réduits*, Séminaire Samuel sur les épimorphismes, Secrétariat Mathématique, Paris, 1967-1968, pp. Section 6.
18. J. P. Olivier, *Morphismes immergeants de Ann*, Université des Sciences et Techniques du Languedoc, Montpellier **106** (1971).
19. J. P. Olivier, *Descente de quelques propriétés élémentaires par morphismes purs*, An. Acad. Brasil. Ciênc. **45** (1973), 2–33.
20. G. Picavet, *Submersion et descente*, J. Algebra **103** (1985), 527–591.
21. G. Picavet, *Propriétés et applications de la notion de contenu*, Comm. Algebra **13** (1985), 2231–2265.
22. G. Picavet, *Pureté, rigidité et morphismes entiers*, Trans. Amer. Math. Soc **323** (1991), 283–313.
23. G. Picavet, *Parties sondables d'un spectre et profondeur*, Boll. U.M.I. **8-B** (1994), 677–730.
24. G. Picavet, *Seminormal or t-closed schemes and Rees rings*, Algebra and Representation theory **1** (1998), 255–309.
25. G. Picavet, *Algebraically flat or projective morphisms*, Submitted to publication.
26. D. Popescu, *Algebraically pure morphisms*, Rev. Roum. Math. Pures et Appl. **24** (1979), 947–977.
27. M. Raynaud, *Un critère d'effectivité de descente*, Séminaire Samuel sur les épimorphismes, Secrétariat Mathématique, Paris, 1967-1968 Sec. 5.
28. *Séminaire Samuel sur les épimorphismes*, Secrétariat Mathématique, Paris, 1967-1968.
29. B. Stenström, *Rings of Quotients*, Springer Verlag, Berlin, 1975.

Trigonometric Polynomial Rings

GABRIEL PICAVET and MARTINE PICAVET-L'HERMITTE
Laboratoire de Mathématiques Pures, Université de Clermont II,
63177 Aubière Cedex, France
Gabriel.Picavet@math.univ-bpclermont.fr
Martine.Picavet@math.univ-bpclermont.fr

ABSTRACT

Let T (resp. T') be the ring of real (resp. complex) trigonometric polynomials. Then T' is a Euclidean domain while T is a half-factorial Dedekind domain. We characterize irreducible elements of T, the behavior of maximal ideals of T with respect to irreducible elements of T and the correspondence between maximal ideals of T and T'. Moreover, we consider the embedding of T into the ring of continuous numerical functions $\mathcal{C}(\mathbb{R})$ and give the link between prime ideals of T and $\mathcal{C}(\mathbb{R})$. As a by-product, we find that T can be equipped with total orders associated to ultrafilters on the field of real numbers.

1 INTRODUCTION

The theory of factorization in polynomial rings has been investigated for a long time and a lot of properties can be found in the literature.

In 1927, J.F. Ritt obtained the following theorem for factorization of exponential polynomials [5, Theorem].

THEOREM 1.1 [J.F. Ritt]. *If* $1+a_1e^{\alpha_1 x}+\cdots+a_ne^{\alpha_n x}$ *is divisible by* $1+b_1e^{\beta_1 x}+\cdots+b_re^{\beta_r x}$ *with no b equal to zero, then every* β *is a linear combination of* $\alpha_1,\ldots,\alpha_n$ *with rational coefficients.*

When the α_k are of the form im, with $m \in \mathbb{Z}$, we obtain trigonometric polynomials. In this paper, we are interested in factorization properties of the two following sets of trigonometric polynomials:

$$T' = \left\{ \sum_{k=0}^{n} (a_k \cos kx + b_k \sin kx) \mid n \in \mathbb{N},\ a_k, b_k \in \mathbb{C} \right\}$$

and

$$T = \left\{ \sum_{k=0}^{n} (a_k \cos kx + b_k \sin kx) \mid n \in \mathbb{N},\ a_k, b_k \in \mathbb{R} \right\}$$

Indeed, $\sin^2 x = (1 - \cos x)(1 + \cos x)$ shows that two different nonassociated irreducible factorizations of the same element may appear.

In the following, we denote by $\cos kx$ and $\sin kx$ the two functions $x \mapsto \cos kx$ and $x \mapsto \sin kx$ (defined over $\mathbb{R}$). It is well known that for each $n \in \mathbb{N}^*$, we get $\cos nx$ as a polynomial in $\cos x$ with degree n and $\sin nx$ as the product of $\sin x$ by a polynomial in $\cos x$ with degree $n-1$. Conversely, linearization formulas show that any product $\cos^n x \sin^p x$ can be written as $\sum_{k=0}^{q} (a_k \cos kx + b_k \sin kx)$, $q \in \mathbb{N}$, $a_k, b_k \in \mathbb{Q}$. It follows that $T = \mathbb{R}[\cos x, \sin x]$ and $T' = \mathbb{C}[\cos x, \sin x]$.

We first recall some needed definitions.

(1) An integral domain is called **atomic** if each nonzero nonunit is a finite product of irreducible elements (**atoms**).
(2) An integral domain R is said to be a **half-factorial domain (HFD)** if R is atomic and whenever $x_1 \cdots x_m = y_1 \cdots y_n$ with $x_i, y_j \in R$ irreducible, then $m = n$ (Zaks [8]).

If the first ring T', studied in section 2, is a Euclidean domain (isomorphic to $\mathbb{C}[X]_X$), the second T is in fact a half-factorial Dedekind domain where irreducible elements are trigonometric polynomials of degree one. Section 3 is devoted to T, where we consider the behavior of maximal ideals of T with respect to irreducible elements, and the correspondence of maximal ideals between T and T'.

In section 4, we consider T as a subring of rings of continuous numerical functions $\mathcal{C}(\mathbb{R})$ and $\mathcal{C}([0, 2\pi])$, with emphasis on the correspondence between maximal ideals. As a by-product, we get that T can be equipped with total orders inducing the usual total order on $\mathbb{R}$. Each ultrafilter on $\mathbb{R}$ gives such a total order.

Let $z = \sum_{k=0}^{n} (a_k \cos kx + b_k \sin kx)$, $n \in \mathbb{N}$, $a_k, b_k \in \mathbb{C}$, be an element of T'.

We denote by $\bar{z}$ the element $\sum_{k=0}^{n} (\bar{a}_k \cos kx + \bar{b}_k \sin kx)$, where $\bar{a}$ denotes the conjugate of $a \in \mathbb{C}$.

Notice that an element $z = \sum_{k=0}^{n}(a_k \cos kx + b_k \sin kx)$ of T or T' determines uniquely the elements a_k and b_k thanks to the theory of Fourier's series.

For a ring R we denote by $\mathcal{U}(R)$ the group of units of R.

2 ON THE STRUCTURE OF $\mathbb{C}[\cos x, \sin x]$

Relations $\cos x = \frac{e^{ix}+e^{-ix}}{2}$ and $\sin x = \frac{e^{ix}-e^{-ix}}{2i}$ show that an arbitrary element $z \in T' = \mathbb{C}[\cos x, \sin x]$ is of the following form

$$z = e^{-inx}P(e^{ix}),\ n \in \mathbb{N},\ \text{where}\ P(X) \in \mathbb{C}[X]\ \text{and}\ \deg(P) = 2n\ (*)$$

Conversely, in view of $e^{ix} = \cos x + i \sin x$, an element of the form $e^{-inx}P(e^{ix})$, $n \in \mathbb{N}$, $P(X) \in \mathbb{C}[X]$ is in T'. So there is an isomorphism $f : \mathbb{C}[X]_X \to T'$ through the substitution morphism $X \mapsto e^{ix}$. The following theorem results

THEOREM 2.1. *$T' = \mathbb{C}[\cos x, \sin x]$ is a Euclidean domain with quotient field $K' = \mathbb{C}(\cos x)[\sin x]$. The irreducible elements of T' are, up to units, trigonometric polynomials of the form $\cos x + i \sin x - a$, $a \in \mathbb{C}^*$.*

Notice that each element z of $\mathbb{C}[X]_X$ can be be written uniquely $X^k P(X)$ where $k \in \mathbb{Z}$, $P(X) \in \mathbb{C}[X]$ and $P(0) \neq 0$. It is well known that the algorithm φ defining the Euclidean domain $\mathbb{C}[X]_X$ is given by $\varphi(z) = \deg(P)$ with the previous notation (see for instance [6, Proposition 7]).

The following corollary provides us a generalization of Ritt's factorization theorem.

COROLLARY 2.2. *Let $z = \sum_{k=0}^{n}(a_k \cos kx + b_k \sin kx)$, $n \in \mathbb{N}^*$, $a_k, b_k \in \mathbb{C}$ with $(a_n, b_n) \neq (0,0)$. Let d be a common divisor of the integers k such that $(a_k, b_k) \neq (0,0)$. Then z has a unique factorization*

$$z = \lambda(\cos nx - i \sin nx) \prod_{j=1}^{\frac{2n}{d}} (\cos dx + i \sin dx - \alpha_j)$$

where $\lambda, \alpha_j \in \mathbb{C}^*$.

Proof. Setting $y = dx$, we have

$$z = \sum_{k'=0}^{\frac{n}{d}} (a_{dk'} \cos k'y + b_{dk'} \sin k'y) = e^{-i\frac{n}{d}y}P(e^{iy})$$

where $P(X) \in \mathbb{C}[X]$ and $\deg(P) = 2\frac{n}{d}$ thanks to $(*)$. Use now the factoriality of $\mathbb{C}[\cos y, \sin y]$ with its irreducible elements $\cos y + i \sin y - a$, $a \in \mathbb{C}^*$. $\square$

3 ON THE STRUCTURE OF $\mathbb{R}[\cos x, \sin x]$

(a) $\mathbb{R}[\cos x, \sin x]$ IS HALF-FACTORIAL

$T = \mathbb{R}[\cos x, \sin x]$ is isomorphic to $\mathbb{R}[X,Y]/(X^2+Y^2-1)$ through the substitution morphism $g : \mathbb{R}[X,Y] \to T$ defined by $g(X) = \cos x,\ g(Y) = \sin x$.

THEOREM 3.1. $\mathbb{R}[\cos x, \sin x]$ *is a Dedekind half-factorial domain.*

Proof. R. Fossum showed that the class group of $\mathbb{R}[X,Y]/(X^2+Y^2-1)$ is isomorphic to $\mathbb{Z}/2\mathbb{Z}$ [3, Proposition 11.8].

Since $\mathbb{R}[X,Y]/(X^2+Y^2-1) = \mathbb{R}[X][Y]/(Y^2-(1-X^2))$ with $\mathbb{R}[X]$ factorial and $1-X^2$ square-free, $\mathbb{R}[X,Y]/(X^2+Y^2-1)$ is a one-dimensional Noetherian integrally closed integral domain [3, Lemma 11.1]. Therefore, T is a Dedekind domain, and so a Krull domain with class group isomorphic to $\mathbb{Z}/2\mathbb{Z}$. Thus T is half-factorial by Zaks [8, Theorem 1.4]. □

REMARKS. (1) T is a free $\mathbb{R}[\cos x]$-module with basis $\{1, \sin x\}$ and T' is a free T-module with basis $\{1, i\}$.

(2) $\mathbb{R}[\cos x]$ is a Euclidean domain because isomorphic to $\mathbb{R}[X]$. It follows that the half-factorial domain T is wedged in between the two Euclidean domains $\mathbb{R}[\cos x]$ and $\mathbb{C}[\cos x, \sin x]$.

(3) The quotient field of $\mathbb{R}[\cos x, \sin x]$ is $K = \mathbb{R}(\cos x)[\sin x]$ and the quotient field of $\mathbb{C}[\cos x, \sin x]$ is $K' = \mathbb{C}(\cos x)[\sin x]$.

PROPOSITION 3.2. $\mathbb{C}[\cos x, \sin x]$ *is the integral closure of* $\mathbb{R}[\cos x, \sin x]$ *in the quotient field* K' *of* T'.

Proof. $K' = \mathbb{C}(\cos x)[\sin x]$ is the quadratic extension $K[i]$ where $K = \mathbb{R}(\cos x)[\sin x]$ and any element of K' integral over T is in T'. The reverse inclusion is obvious. □

REMARK. We saw that $T' = \mathbb{C}[\cos x, \sin x]$ is a Euclidean domain with algorithm φ given by $\varphi(z) = \deg(P)$ where $P(X) \in \mathbb{C}[X]$ is such that $z = f[X^k P(X)]$ with $k \in \mathbb{Z}$ and $P(0) \neq 0$, for any $z \in T' \setminus \{0\}$.

$\mathbb{R}[\cos x]$ is a Euclidean domain with the degree of a polynomial in $\cos x$ as algorithm. The restriction of φ to $\mathbb{R}[\cos x]$ does not give the degree but twice the degree.

Let $P(\cos x) = \sum_{k=0}^{n} a_k \cos^k x,\ n \in \mathbb{N}^*,\ a_k \in \mathbb{R},\ a_n \neq 0$. Then $P(\cos x) =$

$$\sum_{k=0}^{n} a_k \left(\frac{e^{ix}+e^{-ix}}{2}\right)^k = f\left(\sum_{k=0}^{n} a_k \left(\frac{X^2+1}{2X}\right)^k\right) = f\left(X^{-n}Q(X)\right),$$ where

$Q(X) = 2^{-n}\sum_{k=0}^{n} a_k \left(X^2+1\right)^k (2X)^{n-k}$ and $Q(0) = 2^{-n}a_n \neq 0$. It follows that $\varphi[P(\cos x)] = 2n = 2\deg(P)$.

(b) IRREDUCIBLE ELEMENTS OF $\mathbb{R}[\cos x, \sin x]$

DEFINITION 3.3. Let $P = \sum_{k=0}^{n} a_k \cos^k x + \left(\sum_{j=0}^{p} b_j \cos^j x \right) \sin x, \; a_k, b_j \in \mathbb{R}$. The **degree** of P is defined by $\delta(P) = \sup\{k, j+1 \mid a_k, b_j \neq 0\}$ if $P \neq 0$ and $\delta(P) = -\infty$ if $P = 0$.

This definition makes sense because the family $\{\cos^k x, \sin x \cos^k x\}$ is a basis of T over $\mathbb{R}$.

This implies the following formula $\delta(PQ) = \delta(P) + \delta(Q)$ for any $P, Q \in T$. In particular, trigonometric polynomials of degree one are irreducible and $\mathbb{R}^* = \mathcal{U}(T)$.

THEOREM 3.4. *The irreducible elements of* $\mathbb{R}[\cos x, \sin x]$ *are of the form*

$$a \cos x + b \sin x + c, \; (a, b, c) \in \mathbb{R}^3, \; (a, b) \neq (0, 0)$$

Proof. The first part of the proof is given above by degree considerations.

Let $z \in T$ with $\delta(z) = k > 0$. Then z is not a unit and $z \in T'$. By the relation $(*)$ of Section 2, we can write $z = \lambda e^{-ikx} \prod_{j=1}^{2k} (e^{ix} + a_j), \; \lambda, a_j \in \mathbb{C}^*$. But $z \in T$ implies $z = \bar{z}$, so that $z^2 = z\bar{z} = \lambda\bar{\lambda} \prod_{j=1}^{2k} (e^{ix} + a_j) \prod_{j=1}^{2k} (e^{-ix} + \overline{a_j})$, with $\lambda\bar{\lambda} \in \mathbb{R}_+^*$. Moreover, for $a \in \mathbb{C}^*$, we have

$$(e^{ix} + a)(e^{-ix} + \bar{a}) = 1 + a\bar{a} + 2(\alpha \cos x + \beta \sin x)$$

where $a = \alpha + i\beta$, $(\alpha, \beta) \in \mathbb{R}^2 \setminus \{(0,0)\}$ since $a \neq 0$. So we get $z^2 = z_1 \cdots z_{2k}$, where the z_j are irreducible elements of T of degree 1.

In particular, if z is irreducible, we get $2 = 2\delta(z)$ since T is an HFD, so that $\delta(z) = 1$. □

COROLLARY 3.5. *Let* $P = \sum_{k=0}^{n} (a_k \cos kx + b_k \sin kx), \; n \in \mathbb{N}^*, \; a_k, b_k \in \mathbb{R}$ *with* $(a_n, b_n) \neq (0, 0)$. *Let* d *be a common divisor of the integers* k *such that* $(a_k, b_k) \neq (0, 0)$. *Then* P *is a product of* n/d *elements of the form* $a \cos dx + b \sin dx + c$ *where* $(a, b, c) \in \mathbb{R}^3$, $(a, b) \neq (0, 0)$.

Proof. The same as the proof of Corollary 2.2. □

COROLLARY 3.6. *For any nonzero nonunit* $z \in \mathbb{R}[\cos x, \sin x]$, *there exists a unique factorization, up to order and associates, of the form* $z^2 = uz_1 \cdots z_n$, *where* $u \in \mathbb{R}_+^*$, *the* z_j *are irreducible elements of the form*

$$a_j \cos x + b_j \sin x + 1 + \frac{1}{4}(a_j^2 + b_j^2), (a_j, b_j) \in \mathbb{R}^2 \setminus \{(0,0)\}$$

and $n = 2\delta(z)$.

Proof. Existence of such a factorization follows from the proof of the previous theorem. As $\delta(z_j) = 1$, we get $n = 2\delta(z)$.
Consider two factorizations $z^2 = uz_1 \cdots z_n = u'z_1' \cdots z_n'$ $(*)$, with $u, u' \in \mathbb{R}_+^*$,
$z_j = a_j \cos x + b_j \sin x + 1 + \frac{1}{4}(a_j^2 + b_j^2)$,
$z_j' = a_j' \cos x + b_j' \sin x + 1 + \frac{1}{4}({a_j'}^2 + {b_j'}^2)$,
$(a_j, b_j), (a_j', b_j') \in \mathbb{R}^2 \setminus \{(0,0)\}$. Set $c_j = \frac{1}{2}(a_j + ib_j)$ and $c_j' = \frac{1}{2}(a_j' + ib_j')$.
Equality $(*)$ in T' implies $u \prod_{j=1}^{n} (e^{ix} + c_j)(e^{-ix} + \bar{c}_j) = u' \prod_{j=1}^{n} (e^{ix} + c_j')(e^{-ix} + \bar{c}_j')$
which is equivalent to $u \prod_{j=1}^{n} (X + c_j)(1 + X\bar{c}_j) = u' \prod_{j=1}^{n} (X + c_j')(1 + X\bar{c}_j')$.
Since $\mathbb{C}[X]$ is factorial, we get that for any $j \in \{1, \dots, n\}$, there exists $j' \in \{1, \dots, n\}$ such that $c_j = c_{j'}'$ or $1 - c_j \bar{c}_{j'}' = 0$.

- $c_j = c_{j'}'$ gives $z_j = z_{j'}'$.
- $1 - c_j \bar{c}_{j'}' = 0$ gives $c_{j'}' = (\bar{c}_j)^{-1} = \frac{c_j}{c_j \bar{c}_j} = \frac{4(a_j + ib_j)}{2(a_j^2 + b_j^2)}$, so that $a_{j'}' = \frac{4a_j}{a_j^2 + b_j^2}$

and $b_{j'}' = \frac{4b_j}{a_j^2 + b_j^2}$. It follows that $z_{j'}' = \frac{4a_j}{a_j^2 + b_j^2} \cos x + \frac{4b_j}{a_j^2 + b_j^2} \sin x + \frac{4 + a_j^2 + b_j^2}{a_j^2 + b_j^2}$. At last, we obtain $z_j' = \frac{4z_j}{a_j^2 + b_j^2}$ which implies that z_j and z_j' are associated. The uniqueness is gotten. □

REMARK. We recover here a result of S.T. Chapman and U. Krause concerning Cale monoids. Recall that a monoid M is a Cale monoid with basis Q if for every nonunit $x \in M$ there exists a positive integer n such that x^n factors uniquely up to order and units as elements from $Q \subset M \setminus \mathcal{U}(M)$. Then, [2, Theorem 3.9], a Krull monoid M is a Cale monoid if and only if the divisor class group of M is a torsion group. In our situation we consider the multiplicative monoid of all nonzero elements of T, which is a Cale monoid. The following corollary gives the basis of the Cale monoid $T \setminus \{0\}$.

COROLLARY 3.7. $\mathbb{R}[\cos x, \sin x] \setminus \{0\}$ *is a Cale monoid with basis the elements of the form* $a \cos x + b \sin x + 1 + \frac{1}{4}(a^2 + b^2)$, $a, b \in \mathbb{R}$, $a^2 + b^2 \geq 4$.

Proof. Corollary 3.6 shows that the square of a nonzero nonunit has a unique factorization into irreducibles of the form $a \cos x + b \sin x + 1 + \frac{1}{4}(a^2 + b^2)$ with $(a, b) \in \mathbb{R}^2 \setminus \{(0,0)\}$, up to order and units.

Consider two associated elements $z = a\cos x + b\sin x + 1 + \frac{1}{4}(a^2+b^2)$ and $z' = a'\cos x + b'\sin x + 1 + \frac{1}{4}(a'^2 + b'^2)$, $a, b, a', b' \in \mathbb{R}$, $a^2+b^2, a'^2+b'^2 \geq 4$ so that $z' = cz$ where $c \in \mathbb{R}^*$.

It follows that $a' = ca$, $b' = cb$ and $1 + \frac{1}{4}(a'^2+b'^2) = c\left[1 + \frac{1}{4}(a^2+b^2)\right]$. This relation implies $\frac{c}{4}(c-1)(a^2+b^2) = c-1$ so that $c = 1$ or $c = \frac{4}{a^2+b^2}$. But in this case $a'^2 + b'^2 = \frac{16}{a^2+b^2} \leq 4$, a contradiction, except when $a^2+b^2 = 4$ where we recover $c = 1$. □

REMARK. As M. Zafrullah told us at the Fez Conference (2001), T is also an almost Bézout domain (resp. almost GCD-domain): for $t, z \in T \setminus \{0\}$, there exists $n \in \mathbb{N}^*$ with (t^n, z^n) (resp. $(t^n) \cap (z^n)$) principal. Indeed, T is a Prüfer domain with torsion class group and every nonzero ideal of T is divisorial [1, Theorem 4.7].

(c) MAXIMAL IDEALS OF $\mathbb{R}[\cos x, \sin x]$

Since $\mathbb{R}[\cos x, \sin x]$ is a Dedekind domain with class number 2 (Theorem 3.1), any maximal ideal is either a principal ideal or of order 2. Moreover, if M and M' are of order 2, we get that M^2 and MM' are principal ideals generated by irreducible elements [7, Proposition 1 and Corollary]. Conversely, if $z \in T$ is irreducible, three cases may occur: (z) is a maximal ideal, $(z) = M^2$, where M is a maximal ideal, $(z) = MM'$, where M and M' are distinct maximal ideals.

Irreducible elements are given by Theorem 3.4. We are going to study the primary decomposition of (z) for an irreducible $z \in T$.

THEOREM 3.8. *Let $z = a\cos x + b\sin x + c \in \mathbb{R}[\cos x, \sin x]$ with $(a, b) \neq (0, 0)$. Then*

(1) *(z) is a maximal ideal if and only if $c^2 > a^2 + b^2$.*
(2) *(z) is the square of a maximal ideal if and only if $c^2 = a^2 + b^2$.*
(3) *(z) is the product of two maximal ideals if and only if $c^2 < a^2 + b^2$.*

Proof. Since $(a, b) \neq (0, 0)$, we get $a^2 + b^2 > 0$. Set $a' = \frac{a}{\sqrt{a^2+b^2}}$ and $b' = \frac{b}{\sqrt{a^2+b^2}}$. Then z is associated to $z' = a'\cos x + b'\sin x + k$, where $k = \frac{c}{\sqrt{a^2+b^2}}$. Moreover, $a'^2 + b'^2 = 1$ implies that there exists $\alpha \in \mathbb{R}$ such that $a' = \sin\alpha$ and $b' = \cos\alpha$, which gives $z' = \sin(x+\alpha) + k$. So we are led to study the principal ideal generated by $\sin(x+\alpha) + k$.

Define the substitution morphism $h : T \to T$ by $h(\cos x) = \cos(x+\alpha)$ and $h(\sin x) = \sin(x+\alpha)$, which is an isomorphism.
Then $h((\sin x + k)) = (\sin(x+\alpha) + k)$ and it follows that the primary decompositions of $(\sin(x+\alpha)+k)$ and $(\sin x + k)$ are of the same form.

Now define the substitution morphism $g' : \mathbb{R}[X] \to T$ by $g'(X) = \cos x$. Let $t = \sin x + k$, $k \in \mathbb{R}$. We are going to calculate ${g'}^{-1}((t))$.

Let $P(X) \in \mathbb{R}[X]$. We have
$P(X) \in g'^{-1}((t)) \Leftrightarrow$ there exist $Q(X), R(X) \in \mathbb{R}[X]$ such that

$$\begin{aligned} P(\cos x) &= (\sin x + k)[Q(\cos x) + \sin x R(\cos x)] \\ &= [kQ(\cos x) + (1-\cos^2 x)R(\cos x)] + \sin x[Q(\cos x) + kR(\cos x)] \end{aligned}$$

$$\Rightarrow \begin{cases} kQ(\cos x) + (1-\cos^2 x)R(\cos x) & = P(\cos x) \\ Q(\cos x) + kR(\cos x) & = 0 \end{cases}$$

$$\Rightarrow \begin{cases} kQ(X) + (1-X^2)R(X) & = P(X) \\ Q(X) + kR(X) & = 0 \end{cases}$$

so that $P(X) = (1-k^2-X^2)R(X)$. Moreover,
$g'(1-k^2-X^2) = 1-k^2-\cos^2 x = \sin^2 x - k^2 = (\sin x - k)(\sin x + k) \in (t)$.
It results that $g'^{-1}((t)) = (1-k^2-X^2)\mathbb{R}[X]$. This gives the following commutative diagram:

$$\begin{array}{ccc} \mathbb{R}[X] & \xrightarrow{g'} & T \\ \downarrow & & \downarrow \\ \mathbb{R}[X]/(1-k^2-X^2) & \xrightarrow{\bar{g}'} & T/(t) \end{array}$$

where $\bar{g}'$ is injective. But $\bar{g}'$ is surjective since $\overline{\sin x} = -\bar{k}$ in $T/(t)$ so that $\overline{P(\cos x) + \sin x Q(\cos x)} = \bar{g}'\left[\overline{P(X) - kQ(X)}\right]$.
Then $T/(t) \simeq \mathbb{R}[X]/(1-k^2-X^2)$ allows a dicussion with respect to (t).
(a) (t) is a maximal ideal $\Leftrightarrow 1-k^2-X^2$ is irreducible in $\mathbb{R}[X] \Leftrightarrow k^2 > 1$.
(b) (t) is the square of a maximal ideal $\Leftrightarrow (1-k^2-X^2)$ is the square of a maximal ideal $\Leftrightarrow 1-k^2-X^2$ is a square $\Leftrightarrow k^2 = 1$.
(c) (t) is the product of two distinct maximal ideals $\Leftrightarrow k^2 < 1$. □

Now we are able to characterize the maximal ideals of $\mathbb{R}[\cos x, \sin x]$ by means of their generators.

COROLLARY 3.9. *Let M be a maximal ideal of $T = \mathbb{R}[\cos x, \sin x]$.*

(1) *If M is a principal ideal, there exist $\alpha, k \in \mathbb{R}$, $k > 1$ such that $M = (\sin(x+\alpha) + k)$ and T/M is isomorphic to $\mathbb{C}$.*
(2) *If M is not a principal ideal, there exists $\alpha \in \mathbb{R}$ such that $M = (\sin(x+\alpha)+1, \cos(x+\alpha))$, $M^2 = (\sin(x+\alpha)+1)$ and $\mathbb{R} \to T/M$ is an isomorphism.*

Conversely, such ideals are maximal ideals in $\mathbb{R}[\cos x, \sin x]$.

Proof. (1) The form of principal maximal ideals is deduced from Theorem 3.8 (1). If $k < 0$, it is enough to take $\alpha + \pi$ instead of α.

(2) Assume that M is not a principal ideal. Since the class group of T is isomorphic to $\mathbb{Z}/2\mathbb{Z}$, M^2 is a principal ideal generated by an irreducible element by [7] quoted above and of the form $t = \sin(x+\alpha)+1$. Indeed, M^2 is generated by an element of the form $\sin(x+\alpha)+k$, $k \in \mathbb{R}$ where $k^2 = 1$ by Theorem 3.8 (2) (if $k=-1$, take $\alpha+\pi$ instead of α).
But $\sin(x+\alpha)+1 \in M$ implies $\cos^2(x+\alpha) = [1-\sin(x+\alpha)][1+\sin(x+\alpha)] \in M$ and so $\cos(x+\alpha) \in M$. Hence we have $I = (\sin(x+\alpha)+1, \cos(x+\alpha)) \subset M$. The substitution morphism $h: T \to T$ defined by $h(\cos x) = \cos(x+\alpha)$ and $h(\sin x) = \sin(x+\alpha)$ gives the isomorphism $T/I \simeq \mathbb{R}$, and I is a maximal ideal. Therefore, $M = (\sin(x+\alpha)+1, \cos(x+\alpha))$. □

REMARK. This last result allows us to confirm a remark of S.T. Chapman and U. Krause concerning Cale monoids [2, Introduction]. The elements occuring in the basis of the Cale monoid $\mathbb{R}[\cos x, \sin x] \setminus \{0\}$ are primary elements (generating a primary principal ideal). Corollary 3.7 said that these elements are of the form

$$z = a\cos x + b\sin x + 1 + \frac{1}{4}(a^2+b^2),\ (a,b) \in \mathbb{R}^2,\ a^2+b^2 \geq 4$$

But $[1+\frac{1}{4}(a^2+b^2)]^2 - (a^2+b^2) = [1-\frac{1}{4}(a^2+b^2)]^2 \geq 0$ gives that z is a primary element (see Theorem 3.8,(1),(2)).

It follows from Theorem 3.8 that the principal ideal generated by an element $a\cos x + b\sin x + c$ of T with $(a,b) \neq (0,0)$ and such that $c^2 < a^2+b^2$ is the product of two distinct maximal nonprincipal ideals. We intend to determine these maximal ideals and conversely, to find a generator for a product of two distinct maximal nonprincipal ideals.

PROPOSITION 3.10. *Product of two nonprincipal maximal ideals of T.*

(1) *If* $M = (\sin(x+\alpha)+1, \cos(x+\alpha))$, $M' = (\sin(x+\beta)+1, \cos(x+\beta))$ *with* $\alpha \not\equiv \beta \pmod{2\pi}$ *are two maximal non principal ideals of* T, *then* MM' *is a principal ideal generated by the irreducible element* $\sin(\frac{\alpha+\beta}{2})\cos x + \cos(\frac{\alpha+\beta}{2})\sin x + \cos(\frac{\alpha-\beta}{2})$.

(2) *Conversely, let* $z = a\cos x + b\sin x + c$, $a, b, c \in \mathbb{R}$ *be an irreducible element of* $\mathbb{R}[\cos x, \sin x]$ *such that* $c^2 < a^2+b^2$.

Then (z) *is the product of two distinct maximal nonprincipal ideals* $M = (\sin(x+\alpha)+1, \cos(x+\alpha))$ *and* $M' = (\sin(x+\beta)+1, \cos(x+\beta))$ *such that there exists some* $\theta \in \mathbb{R}$ *satisfying* $\sin\theta = \frac{a}{\sqrt{a^2+b^2}}$, $\cos\theta = \frac{b}{\sqrt{a^2+b^2}}$, $\cos(\theta-\beta) = \frac{c}{\sqrt{a^2+b^2}}$ *and* $\alpha = 2\theta - \beta$.

Proof. (1) Under the hypotheses of (1) we get $MM' = (z)$ where $z = a\cos x + b\sin x + c$ with $c^2 < a^2+b^2$ by Theorem 3.8 and $M^2M'^2 = (z)^2$. But

M^2 and M'^2 are also principal ideals generated by the irreducible elements $\sin(x+\alpha)+1$ and $\sin(x+\beta)+1$ by Corollary 3.9. It follows that

$$(\sin(x+\alpha)+1)(\sin(x+\beta)+1) = (a\cos x + b\sin x + c)^2$$

which gives, after identification and up to a common factor of a, b, c in $\mathbb{R}^*$, the following system:

$$\begin{cases} b^2 - a^2 = \cos(\alpha+\beta) \\ 2ab = \sin(\alpha+\beta) \\ 2ac = \sin\alpha + \sin\beta \\ 2bc = \cos\alpha + \cos\beta \\ b^2 + c^2 = 1 + \cos\alpha\cos\beta \end{cases}$$

with solution, up to the sign, $a = \sin\frac{\alpha+\beta}{2}$, $b = \cos\frac{\alpha+\beta}{2}$, $c = \cos\frac{\alpha-\beta}{2}$.
(2) Conversely, let $z = a\cos x + b\sin x + c$ be such that $c^2 < a^2 + b^2$. Since $a^2 + b^2 > 0$, set $a' = \frac{a}{\sqrt{a^2+b^2}}$, $b' = \frac{b}{\sqrt{a^2+b^2}}$ and $c' = \frac{c}{\sqrt{a^2+b^2}}$.

Then $z' = a'\cos x + b'\sin x + c'$ generates the same principal ideal $(z) = (z')$ with $|c'| < 1$. It follows that there exist $\theta \in \mathbb{R}$ such that $a' = \sin\theta$, $b' = \cos\theta$ and $\beta \not\equiv \theta \pmod{\pi}$ such that $c' = \cos(\theta - \beta)$. Set $\alpha = 2\theta - \beta$ which implies $\alpha \not\equiv \beta \pmod{2\pi}$. After identification, the previous calculation leads to

$$(\sin(x+\alpha)+1)(\sin(x+\beta)+1) = (a'\cos x + b'\sin x + c')^2$$

Moreover, $M = (\sin(x+\alpha)+1, \cos(x+\alpha))$ and $M' = (\sin(x+\beta)+1, \cos(x+\beta))$ are two different maximal ideals such that $MM' = (z)$. □

Now, we are going to look at the correspondence between maximal ideals of $T = \mathbb{R}[\cos x, \sin x]$ and $T' = \mathbb{C}[\cos x, \sin x]$.

PROPOSITION 3.11. *Let M be a maximal ideal of T.*

(1) *If M is a principal ideal, then M is of the form $(p\cos x + q\sin x + r)$ with $4r = 4 + p^2 + q^2$, $r \neq 2$ and $T'M$ is the product of two distinct maximal ideals M' and M'' of T' with $M' = (\cos x + i\sin x + a)$, $M'' = (\cos x + i\sin x + \bar{a}^{-1})$ and $2a = p + iq$.*
(2) *If M is a nonprincipal ideal, there exists α such that M is of the form $(\sin(x+\alpha)+1, \cos(x+\alpha))$ and $T'M = (\cos x + i\sin x + a)$ is a maximal ideal of T', where $a = ie^{-i\alpha}$.*

Let $M' = (\cos x + i\sin x + a)$, $a \in \mathbb{C}^*$, be a maximal ideal in T'. Then $M' \cap T$ is a principal ideal if and only if $|a| \neq 1$. When $M' \cap T$ is a principal ideal, the other maximal ideal M'' lying over $M' \cap T$ is $(\cos x + i\sin x + \bar{a}^{-1})$.

Proof. Let M be a maximal ideal of T. Then $T'M = \prod_{i=1}^{n} M_i'^{e_i}$, where the M_i' are maximal ideals of T' and $\sum_{i=1}^{n} e_i f_i = 2$ $(*)$, with $f_i = [T'/M_i' : T/M]$

thanks to Proposition 3.2 and [9, Corollary p. 287]. For each maximal ideal M' of T', we have $T'/M' \simeq \mathbb{C}$.

(1) If M is a principal ideal of T, then M is generated by an irreducible element z of the form $p\cos x + q\sin x + r$ with $(p,q) \neq (0,0)$ and $r^2 > p^2 + q^2$ by Theorem 3.8. But in this case $z = \lambda e^{-ix}(e^{ix} - \alpha)(e^{ix} - \beta)$, $\lambda, \alpha, \beta \in \mathbb{C}^*$ with $\alpha \neq \beta$ since $p^2 + q^2 - r^2 \neq 0$. It follows that z belongs to two distinct maximal ideals of T' and $n = 2, e_i = f_i = 1$ for $i = 1, 2$ by $(*)$.

(2) If M is a nonprincipal ideal, there exists α such that M is of the form $M = (\sin(x+\alpha) + 1, \cos(x+\alpha))$.

In this case, $T/M \simeq \mathbb{R}$, so that $[T'/M'_i : T/M] = 2$ for any maximal ideal M'_i lying over M. It follows that $f_i = 2$ and $(*)$ gives $n = e_i = 1$.

Consider in T' the maximal ideal $M' = (z)$ where $z = \cos x + i\sin x + a$, $a \in \mathbb{C}^*$ and set $M = M' \cap T$. Assume that M is not a principal ideal, then M is of the form $M = (\sin(x+\alpha) + 1, \cos(x+\alpha))$, $\alpha \in \mathbb{R}$ and $e^{i(x+\alpha)} + i \in M'$ gives $a = ie^{-i\alpha}$, which implies $|a| = 1$.

Conversely, let $a \in \mathbb{C}$ be such that $|a| = 1$. There exists $\alpha \in \mathbb{R}$ such that $a = ie^{-i\alpha}$. Then $t = \frac{e^{i(\alpha - x)}}{2i} z^2 = \sin(x+\alpha) + 1$ is an irreducible element of T and generates the square of the maximal ideal $M = M' \cap T$ of T (Theorem 3.8). Thus M is not a principal ideal.

Let $M' = (\cos x + i\sin x + a)$, $a \in \mathbb{C}^*$ be a maximal ideal in T' such that $|a| \neq 1$. Then $M = M' \cap T$ is a principal ideal. Let M'' be the other maximal ideal of T' lying over M. Set $z = \cos x + i\sin x + a$. Then $t = z\bar{z}$ is a generator of M. But $\bar{z} = \cos x - i\sin x + \bar{a} = \bar{a}e^{-ix}(\cos x + i\sin x + \bar{a}^{-1}) \notin M'$ since $\bar{a}^{-1} \neq a$. It follows that $\cos x + i\sin x + \bar{a}^{-1}$ generates M''. Moreover, we saw in the proof of Theorem 3.4 that $z\bar{z} = 1 + a\bar{a} + 2(\alpha\cos x + \beta\sin x)$, where $a = \alpha + i\beta$. Set $p = 2\alpha$, $q = 2\beta$ and $r = 1 + a\bar{a} = 1 + \frac{1}{4}(p^2 + q^2)$ so that $p\cos x + q\sin x + r$ generates M. Since M is a maximal ideal in T, we get $r^2 > p^2 + q^2 = 4r - 4$ and $r \neq 2$. $\square$

REMARK. We can observe that no maximal ideal M of T is ramified in T' (such that $T'M = M'^2$, M' a maximal ideal of T').

In fact, we have the two following situations of maximal ideals of T' lying over a maximal ideal M of T

$$M \quad \text{nonprincipal} \quad \rightarrow \quad M' = (\cos x + i\sin x + a), \qquad a \in \mathbb{C}^*,\ |a| = 1$$

$$M \quad \text{principal} \quad \begin{array}{l} \nearrow \ M' = (\cos x + i\sin x + a) \\ \searrow \ M'' = (\cos x + i\sin x + \bar{a}^{-1}) \end{array} \qquad a \in \mathbb{C}^*,\ |a| \neq 1$$

Here are some examples of factorizations.

EXAMPLE 1. Consider the equality $\sin^2 x = (1 - \cos x)(1 + \cos x)$ $(*)$. We have two maximal ideals $M_1 = (1 - \cos x, \sin x)$ and $M_2 = (1 + \cos x, \sin x)$.

The three irreducible elements $\sin x$, $1-\cos x$ and $1+\cos x$ occuring in $(*)$ generate the following products M_1M_2, M_1^2 and M_2^2.

EXAMPLE 2. Consider the three following nonassociated irreducible factorizations of $\cos 2x$

$$\begin{aligned}\cos 2x &= (\sqrt{2}\cos x - 1)(\sqrt{2}\cos x + 1)\\ &= (\cos x - \sin x)(\cos x + \sin x)\\ &= (1+\sqrt{2}\sin x)(1-\sqrt{2}\sin x)\end{aligned}$$

All these irreducible elements generate a product of two distinct maximal ideals among the four following:

$$\begin{aligned}M_1 &= \left(1+\sin\left(x+\frac{\pi}{4}\right), \cos\left(x+\frac{\pi}{4}\right)\right)\\ M_2 &= \left(1+\sin\left(x-\frac{\pi}{4}\right), \cos\left(x-\frac{\pi}{4}\right)\right)\\ M_3 &= \left(1+\sin\left(x+\frac{3\pi}{4}\right), \cos\left(x+\frac{3\pi}{4}\right)\right)\\ M_4 &= \left(1+\sin\left(x-\frac{3\pi}{4}\right), \cos\left(x-\frac{3\pi}{4}\right)\right)\end{aligned}$$

such that
$(\sqrt{2}\cos x - 1) = M_2M_4$, $(\sqrt{2}\cos x + 1) = M_1M_3$, $(\cos x - \sin x) = M_1M_4$, $(\cos x + \sin x) = M_2M_3$, $(1+\sqrt{2}\sin x) = M_1M_2$, $(1-\sqrt{2}\sin x) = M_3M_4$ providing $(\cos 2x) = M_1M_2M_3M_4$.

EXAMPLE 3. Let $n \in \mathbb{N}^*$. We are going to calculate the number of distinct nonassociated irreducible factorizations of $\sin nx$. Considering first unique factorization in T', we get

$$\begin{aligned}\sin nx &= -\frac{ie^{-inx}}{2}(e^{2inx}-1)\\ &= -\frac{ie^{-inx}}{2}\prod_{k=1}^{2n}\left(e^{ix}-e^{\frac{2ik\pi}{2n}}\right)\\ &= -\frac{ie^{-inx}}{2}(e^{ix}+1)(e^{ix}-1)\prod_{k=1}^{n-1}\left(e^{2ix}-2e^{ix}\cos\frac{2k\pi}{2n}+1\right)\\ &= 2^{n-1}\sin x\prod_{k=1}^{n-1}\left(\cos x-\cos\frac{2k\pi}{2n}\right)\end{aligned}$$

It follows that we have in T the following product of ideals

$$(\sin nx) = (\sin x)\prod_{k=1}^{n-1}\left(\cos x - \cos\frac{2k\pi}{2n}\right)$$

But, for $k = 1, \dots, n-1$, we have
$\left(\cos x - \cos \frac{2k\pi}{2n}\right) =$
$\left(1 - \cos\left(x + \frac{k\pi}{n}\right), \sin\left(x + \frac{k\pi}{n}\right)\right)\left(1 - \cos\left(x - \frac{k\pi}{n}\right), \sin\left(x - \frac{k\pi}{n}\right)\right)$
and $(\sin x) = (1 + \cos x, \sin x)(1 - \cos x, \sin x)$ so that $(\sin nx)$ is the product of $2n$ distinct nonprincipal maximal ideals.

By Proposition 3.10, the number of distinct nonassociated irreducible factorizations is the number of different ways of getting pairs of these maximal ideals. Hence, $\sin nx$ has $(2n-1) \times (2n-3) \times \cdots \times 3 \times 1 = \frac{(2n)!}{2^n n!}$ distinct nonassociated irreducible factorizations.

4 PROPERTIES RELATED TO RINGS OF CONTINUOUS NUMERICAL FUNCTIONS

Clearly, T is a subring of three rings of continuous numerical functions, that is $\mathcal{C} = \mathcal{C}(\mathbb{R})$, $\mathcal{D} = \mathcal{C}([0, 2\pi])$ and $\mathcal{D}' = \mathcal{C}([0, 2\pi[)$. We use results which may be found in the book of L. Gillman and M. Jerison [4]. Such rings have a property which is contrary to Dedekind domains properties, namely $P^2 = P$ for each prime ideal [4, 2B]. If X is a completely regular topological space as $\mathbb{R}$, $[0, 2\pi]$ and $[0, 2\pi[$, a class of maximal ideals in $\mathcal{C}(X)$ is given by all the fixed maximal ideals. A fixed maximal ideal is of the form $M_x = \{f \in \mathcal{C}(X) \mid f(x) = 0\}$ where $x \in X$ and then $\mathbb{R} \to \mathcal{C}(X)/M_x$ is an isomorphism [4, 4.6]. Moreover, $x \neq y$ implies $M_x \neq M_y$. We denote by $\mathrm{Max}_F(\mathcal{C}(X))$ the set of all fixed maximal ideals of $\mathcal{C}(X)$. When X is compact, each maximal ideal of $\mathcal{C}(X)$ is fixed [4, 4.9].

PROPOSITION 4.1. *Let M be a maximal ideal of T.*

(1) *If M is principal of the form $(\sin(x + \alpha) + k)$ where $k > 1$, there is no prime ideal in $\mathcal{C}$ (resp. in $\mathcal{D}$) lying over M.*
(2) *If M is a nonprincipal ideal of the form $(\sin(x + \alpha) + 1, \cos(x + \alpha))$, a fixed maximal ideal in $\mathcal{C}$ lying over M is of the form M_u, where $u = \frac{3\pi}{2} - \alpha + 2k\pi$, $k \in \mathbb{Z}$. Moreover, $\mathrm{Max}_F(\mathcal{C}) \to \mathrm{Spec}(T)$ and $\mathrm{Max}(\mathcal{D}) \to \mathrm{Spec}(T)$ are surjections onto the set of all nonprincipal maximal ideals of T and $\mathrm{Max}_F(\mathcal{D}') \to \mathrm{Spec}(T)$ gives a bijection onto the set of all nonprincipal maximal ideals of T.*

Proof. If M is principal and generated by $p = \sin(x + \alpha) + k$ where $k > 1$, then p is a unit in $\mathcal{C}$ and in $\mathcal{D}$. Now assume that M is nonprincipal and generated by $p = \sin(x + \alpha) + 1$ and $q = \cos(x + \alpha)$. Then p and q belongs to M_u where $u = \frac{3\pi}{2} - \alpha$ so that M_u is lying over M. To complete the proof, observe that an arbirary element $u \in \mathbb{R}$ can be written $u = \frac{3\pi}{2} - \alpha$. $\square$

PROPOSITION 4.2. *Let N be a minimal prime ideal of $\mathcal{C}$, $\mathcal{D}$ or $\mathcal{D}'$. Then N is lying over* 0 *in T. It follows that T is a totally ordered ring inducing on $\mathbb{R}$ the usual total order.*

Proof. Assume that N is lying over a maximal ideal M of T. Then by Proposition 4.1, M is generated by $p = \sin(x+\alpha)+1$ and $q = \cos(x+\alpha)$. Thus there is some $f \notin N$ such that $fq = 0$ because the rings are reduced and N is minimal. It follows that $f = 0$ because f is continuous, a contradiction. Therefore, N is lying over 0. Now recall that an arbitrary factor integral domain A of $\mathcal{C}$ or $\mathcal{D}$ is totally ordered [4, 5.5] and $\mathbb{R} \to A$ is an order-preserving ring morphism. To complete the proof, it is enough to consider the composite of injective ring morphisms $\mathbb{R} \to T \to A$. □

For $f \in \mathcal{C}$, we denote by $\mathrm{Z}(f)$ the zero-set of f, that is the set of all $x \in \mathbb{R}$ such that $f(x) = 0$.

REMARK 4.3. The ring $\mathcal{C}$ can be considered as a subring of the absolutely flat ring $R = \mathbb{R}^{\mathbb{R}}$. Therefore, for each minimal prime ideal N of $\mathcal{C}$ there is a minimal prime ideal M of R lying over N. Now the minimal prime ideals (equivalently, maximal ideals) of R are well known. For each minimal prime ideal M of R there is a unique ultrafilter $\mathcal{F}$ on $\mathbb{R}$ such that $M = \{f \in R \mid \mathrm{Z}(f) \in \mathcal{F}\}$. By [4, 5.2], an element $\bar{f} \in \mathcal{C}/N$ verifies $\bar{f} \geq 0$ if and only if there is some $g \geq 0$ in $\mathcal{C}$ such that $f - g \in N$. Therefore, for a given minimal prime ideal N of $\mathcal{C}$ lying over 0 in T, there is some ultrafilter $\mathcal{F}$ defining a total ordering $\geq$ on T as follows. An element $t \in T$ is ≥ 0 if there is some $f \geq 0$ in $\mathcal{C}$ such that $\mathrm{Z}(t - f) \in \mathcal{F}$. The same proof works for $\mathcal{D}$. Notice that $\mathcal{F}$ is necessarily a free ultrafilter (an ultrafilter which is not the set of all subsets containing a fixed element of $\mathbb{R}$). Deny, in that case there is some $u \in \mathbb{R}$ such that $N = M_u$, contradicting Proposition 4.1. The total order on T associated to the minimal prime ideal N of $\mathcal{C}$ can also be described in the following way. An element $t \in T$ is ≥ 0 if and only if there is some $f \geq 0$ in $\mathcal{C}$ and some $s \in \mathcal{C} \setminus N$ such that $s(t - f) = 0$. Now there is a natural question, since we have associated a total ordering on T to each minimal prime ideal of $\mathcal{C}$. It may be asked whether these total orders on T are equal. We do not know the answer but we conjecture that the answer is negative. Indeed, let A be a totally ordered commutative ring, then $A[X]$ can be equipped with two total orders. Let $p(X) \neq 0$ be a polynomial with coefficients $a_i \in A$, then a first total order is defined by $p(X) > 0$ if and only if $a_n > 0$ where n is the degree of $p(X)$. A second total order is defined by $p(X) > 0$ if and only if $a_{n-k} > 0$ where $n - k$ is the valuation of $p(X)$. Now observe that T is an extension of $\mathbb{R}[\cos x]$.

REFERENCES

1. D.D. Anderson and M. Zafrullah, *Almost Bézout domains*, J. Algebra **142** (1991), 285–309.
2. S.T. Chapman and U. Krause, *On the Cale property in integral domains and monoids, preprint.*
3. R.F. Fossum, *The Divisor Class group of a Krull Domain*, Springer-Verlag, Berlin, Heidelberg, New York, 1973.
4. L. Gillman and M. Jerison, *Rings of continuous functions*, Van Nostrand, Toronto, 1960.
5. J.F. Ritt, *A factorization theory for functions* $\sum_{i=1}^{n} a_i e^{\alpha_i x}$, Trans. Amer. Math. Soc. **29** (1927), 584 – 596.
6. P. Samuel, *About Euclidean Rings*, J. Algebra **19** (1971), 282–301.
7. R.J. Valenza, *Elasticity of Factorization in Number fields*, J. Number Theory **36** (1990), 212–218.
8. A. Zaks, *Half-factorial domains*, Israel J. Math. **37** (1980), 281–302.
9. O. Zariski and P. Samuel, *Commutative Algebra I*, Van Nostrand, Toronto, 1963.

The First Mayr–Meyer Ideal

IRENA SWANSON

New Mexico State University - Department of Mathematical Sciences, Las Cruces, New Mexico 88003-8001, USA. E-mail: iswanson@nmsu.edu.

Summary. This paper gives a complete primary decomposition of the first, that is, the smallest, Mayr-Meyer ideal, its radical, and the intersection of its minimal components. The particular membership problem which makes the Mayr-Meyer ideals' complexity doubly exponential in the number of variables is here examined also for the radical and the intersection of the minimal components. It is proved that for the first Mayr-Meyer ideal the complexity of this membership problem is the same as for its radical. This problem was motivated by a question of Bayer, Huneke and Stillman.

Grete Hermann proved in [H] that for any ideal I in an n-dimensional polynomial ring over the field of rational numbers, if I is generated by polynomials $f_1, \ldots, f_k$ of degree at most d, then it is possible to write $f = \sum r_i f_i$, where each r_i has degree at most $\deg f + (kd)^{(2^n)}$. Mayr and Meyer in [MM] found ideals $J(n,d)$ for which a doubly exponential bound in n is indeed achieved. Bayer and Stillman [BS] showed that for these same ideals also any minimal generating set of syzygies has elements of degree which is doubly exponential in n. Koh [K] modified the original ideal to obtain homogeneous quadric ideals with doubly exponential degrees of syzygies and ideal membership coefficients.

Bayer, Huneke and Stillman have raised questions about the structure of these Mayr-Meyer ideals: is the doubly exponential behavior due to the number of minimal primes, to the number of associated primes, or to the structure of one of them? This paper, together with [S], is an attempt at answering these questions. More precisely, the Mayr-Meyer ideal $J(n,d)$ is an ideal in a polynomial ring in $10n + 2$ variables whose generators have degree at most $d + 2$. This paper analyzes the case $n = 1$ and shows that in this base case the embedded components do not play a role.

Theorem 1 of this paper gives a complete primary decomposition of $J(1,d)$, after which the intersection of the minimal components and the radical come as easy corollaries. The last proposition shows that the complexity of the particular membership problem from

[MM, BS, K] for the radical of $J(1,d)$ is the same as the complexity of the membership problem for $J(1,d)$. Thus at least for the case $n = 1$, neither the embedded components nor the non-reducedness play a role in the complexity.

In a developing paper "Primary decomposition of the Mayr-Meyer ideal" [S], partial primary decompositions are determined for the Mayr-Meyer ideals $J(n,d)$ for all $n \geq 2$, $d \geq 1$. Under the assumption that the characteristic of the field does not divide d, for $n \geq 2$, the number of minimal primes is exactly $nd^2 + 20$, and the number of embedded primes likewise depends on n and d. However, a precise number of embedded components is not known. The case $n = 1$ is very different from the case $n \geq 2$. For example, under the same assumption on the characteristic of the field, the number of minimal primes of the first Mayr-Meyer ideal is $d + 4$, and there is exactly one embedded prime. For understanding the asymptotic behavior of the Mayr-Meyer ideals $J(n,d)$, the case $n = 1$ may not seem interesting, however, it is a basis of the induction arguments for the behavior of the other $J(n,d)$. Furthermore, the case $n = 1$ is computationally and notationally more accessible.

All results of this paper were verified for specific low values of d on Macaulay2.

Acknowledgement. This research was done during 2000/01 at University of Kansas, supported by the NSF-POWRE grant. I thank Craig Huneke for introducing me to the Mayr-Meyer ideals, for all the conversations about them, and for his enthusiasm for this project. I also thank the NSF for partial support on DMS-9970566.

The first Mayr-Meyer ideal $J(1,d)$ is defined as follows. Let K be a field, and d a positive integer. In case the characteristic of K is a positive prime p, write $d = d'i$, where i is a power of p, and d' and p are relatively prime integers. In case the characteristic of K is zero, let $d' = d$, $i = 1$. For notational simplicity we assume that K contains all the $\frac{d}{i}$th roots of unity. Let $s, f, s_1, f_1, c_1, \ldots, c_4, b_1, \ldots, b_4$ be indeterminates over K, and $R = K[s, f, s_1, f_1, c_1, \ldots, c_4, b_1, \ldots, b_4]$. Note that R has dimension 12. The Mayr-Meyer ideal for $n = 1$ is the ideal in R with the generators as follows:

$$J = J(1,d) = (s_1 - sc_1, f_1 - sc_4) + \left(c_i \left(s - fb_i^d\right) | i = 1,2,3,4\right) + \left(fc_1 - sc_2, fc_4 - sc_3, s\left(c_3 - c_2\right), f\left(c_2b_1 - c_3b_4\right), fc_2\left(b_2 - b_3\right)\right).$$

THEOREM 1: *A minimal primary decomposition of $J = J(1,d)$ is as follows:*

$$J = (s_1 - sc_1, f_1 - sc_4, c_1, c_2, c_3, c_4)$$

$$\bigcap_{\alpha}(s_1 - sc_1, f_1 - sc_4, c_4 - c_1, c_3 - c_2, c_1 - c_2b_1^d, s - fb_1^d, b_1 - b_4, b_2 - b_3, b_1^i - \alpha b_2^i)$$
$$\cap\,(s_1 - sc_1, f_1 - sc_4, s, f)$$
$$\cap\,(s_1 - sc_1, f_1 - sc_4, s, c_1, c_2, c_4, b_3^d, b_4)$$
$$\cap\,(s_1 - sc_1, f_1 - sc_4, s, c_1, c_4, b_3^d, b_2 - b_3, c_2b_1 - c_3b_4)$$
$$\cap\,(s_1 - sc_1, f_1 - sc_4, s^2, f^2, c_4(s - fb_4^d), c_3(s - fb_3^d), sc_3 - fc_4, c_3^2, c_4^2,$$
$$c_1 - c_4, c_2 - c_3, b_2 - b_3, b_1 - b_4)\,,$$

where the α vary over the $\frac{d}{i}$th roots of unity in K.

It is easy to verify that $J = J(1,d)$ is contained in the intersection, and that all but the last ideal on the right-hand side of the equality are primary. The following lemma proves that the last ideal is primary as well:

LEMMA 2: *The last ideal in the intersection in Theorem 1 is primary.*

Proof: Here is a simple fact: let $x_1, \ldots, x_n$ be variables over a ring A, $S = A[x_1, \ldots, x_n]$, and I an ideal in A. Then I is primary (respectively, prime) if and only if for any $f_1, \ldots, f_n \in A$, $IS + (x_1 - f_1, \ldots, x_n - f_n)S$ is a primary (respectively, prime) ideal in S.

By this fact it suffices to prove that the ideal

$$L = (s^2, f^2, c_4(s - fb_4^d), c_3(s - fb_3^d), sc_3 - fc_4, c_3^2, c_4^2)$$

is primary. Note that $\sqrt{L} = (s, f, c_3, c_4)$ is a prime ideal. It suffices to prove that the set of associated primes of L is $\{\sqrt{L}\}$. It is an easy fact that for any $x \in R$,

$$\operatorname{Ass}\left(\frac{R}{L}\right) \subseteq \operatorname{Ass}\left(\frac{R}{L : x}\right) \cup \operatorname{Ass}\left(\frac{R}{L + (x)}\right).$$

In particular, when $x = f$, $L + (f) = (s^2, f, sc_4, sc_3, c_3^2, c_4^2)$ is clearly primary to $\sqrt{L}$. Thus it suffices to prove that $L : f$ is primary to $\sqrt{L}$.

We fix the monomial lexicographic ordering $s > f > c_4 > c_3 > b_4 > b_3$. Clearly $L : f$ contains $(s^2, f, c_4 - c_3b_3^d, sc_3, c_3^2)$. If $r \in L : f$, then the leading term of r times f is contained in the ideal of leading terms of L, namely in $(s^2, f^2, sc_4, sc_3, c_3^2, c_4^2, fc_4)$, so that the leading term of r lies in $(s^2, f, c_4, sc_3, c_3^2)$. This proves that $L : f = (s^2, f, c_4 - c_3b_3^d, sc_3, c_3^2)$. This ideal is clearly primary to $\sqrt{L}$, which proves the lemma. ∎

We next prove that the intersection of ideals in Theorem 1 equals $J = J(1,d)$. Note that it suffices to prove the shortened equality:

$$
\begin{aligned}
&(c_1, c_2, c_3, c_4)\\
&\quad \bigcap_\alpha \left(c_4 - c_1, c_3 - c_2, c_1 - c_2 b_1^d, s - f b_1^d, b_1 - b_4, b_2 - b_3, b_1^i - \alpha b_2^i\right)\\
&\quad \cap (s, f)\\
&\quad \cap \left(s, c_1, c_2, c_4, b_3^d, b_4\right)\\
&\quad \cap \left(s, c_1, c_4, b_3^d, b_2 - b_3, c_2 b_1 - c_3 b_4\right)\\
&\quad \cap \left(s^2, f^2, c_4(s - f b_4^d), c_3(s - f b_3^d), s c_3 - f c_4, c_3^2, c_4^2, c_1 - c_4, c_2 - c_3, b_2 - b_3, b_1 - b_4\right)\\
&= \left(c_i \left(s - f b_i^d\right), f c_1 - s c_2, f c_4 - s c_3, s\left(c_3 - c_2\right), f\left(c_2 b_1 - c_3 b_4\right), f c_2\left(b_2 - b_3\right)\right).
\end{aligned}
$$

The intersection of the first two rows equals:

$$
\begin{aligned}
&(c_1, c_2, c_3, c_4) \cap \left(c_4 - c_1, c_3 - c_2, c_1 - c_2 b_1^d, s - f b_1^d, b_1 - b_4, b_2 - b_3, b_1^d - b_2^d\right)\\
&= \left(c_4 - c_1, c_3 - c_2, c_1 - c_2 b_1^d\right) + (c_1, c_2, c_3, c_4) \cdot \left(s - f b_1^d, b_1 - b_4, b_2 - b_3, b_1^d - b_2^d\right)\\
&= J + \left(c_4 - c_1, c_3 - c_2, c_1 - c_2 b_1^d\right) + c_2 \cdot \left(b_1 - b_4, b_2 - b_3, b_1^d - b_2^d\right).
\end{aligned}
$$

This intersected with the third row, namely with (s, f), equals

$$
J + (s, f)\left(c_4 - c_1, c_3 - c_2, c_1 - c_2 b_1^d\right) + c_2 (s, f)\left(b_1 - b_4, b_2 - b_3, b_1^d - b_2^d\right).
$$

Modulo J,

$$
\begin{aligned}
s c_1 &\equiv f c_1 b_1^d \equiv s c_2 b_1^d \equiv f c_2 b_1^d b_2^d \equiv f c_2 b_1^d b_3^d\\
&\equiv f c_3 b_1^{d-1} b_4 b_3^d \equiv s c_3 b_1^{d-1} b_4 \equiv s c_2 b_1^{d-1} b_4 \equiv f c_2 b_1^{d-1} b_4 b_2^d \equiv f c_2 b_1^{d-1} b_4 b_3^d\\
&\equiv f c_3 b_1^{d-2} b_4^2 b_3^d \equiv \cdots \equiv f c_3 b_1^0 b_4^d b_3^d \equiv s c_3 b_4^d \equiv f c_4 b_4^d \equiv s c_4,
\end{aligned}
$$

so that $s(c_1 - c_4) \in J$. Also it is clear that $f(c_1 - c_4), s(c_3 - c_2), s(c_1 - c_2 b_1^d) \in J$, that $s c_2 \in J + (f c_2)$, and that $f c_2 (b_2 - b_3) \in J$. Thus the intersection of the ideals in the first three rows of Theorem 1 simplifies to $J + f\left(c_3 - c_2, c_1 - c_2 b_1^d\right) + f c_2 \left(b_1 - b_4, b_1^d - b_2^d\right)$. Furthermore, $f c_2 (b_1 - b_4) \in (f(c_3 - c_2)) + J$ and modulo J, $f(c_1 - c_2 b_1^d) \equiv c_2 (s - f b_1^d) \equiv f c_2 (b_2^d - b_1^d)$, so that finally the intersection of the first three rows simplifies to

$$
J + \left(f(c_3 - c_2), f c_2 (b_1^d - b_2^d)\right).
$$

We intersect this with the (shortened) ideal in the fourth row of Theorem 1, namely with $(s, c_1, c_2, c_4, b_3^d, b_4)$, to get

$$\begin{aligned} &J + (fc_2(b_1^d - b_2^d)) + (f(c_3 - c_2)) \cap (s, c_1, c_2, c_4, b_3^d, b_4) \\ &= J + (fc_2(b_1^d - b_2^d)) + f(c_3 - c_2) \cdot (s, c_1, c_2, c_4, b_3^d, b_4) \\ &= J + (fc_2(b_1^d - b_2^d)) + f(c_3 - c_2) \cdot (c_2, b_3^d, b_4). \end{aligned}$$

As modulo J, $f(c_3 - c_2)b_4 \equiv fc_2(b_1 - b_4)$, and $f(c_3 - c_2)b_3^d \equiv fc_3b_3^d - fc_2b_2^d \equiv sc_3 - sc_2 \equiv 0$, the intersection of the first four rows simplifies to

$$J + fc_2\left(b_1^d - b_2^d, c_3 - c_2, b_1 - b_4\right).$$

Next we intersect this with the ideal in the fifth row (of Theorem 1) namely with $(s, c_1, c_4, b_3^d, b_2 - b_3, c_2b_1 - c_3b_4)$, to get:

$$\begin{aligned} &J + \left(fc_2\left(b_1^d - b_2^d, c_3 - c_2, b_1 - b_4\right)\right) \cap \left(s, c_1, c_4, b_3^d, b_2 - b_3, c_2b_1 - c_3b_4\right) \\ &= J + fc_2\left(\left(b_1^d - b_2^d, c_3 - c_2, b_1 - b_4\right) \cap \left(s, c_1, c_4, b_3^d, b_2 - b_3, c_2b_1 - c_3b_4\right)\right) \\ &= J + fc_2\left((c_2b_1 - c_3b_4) + \left(b_1^d - b_2^d, c_3 - c_2, b_1 - b_4\right) \cap \left(s, c_1, c_4, b_3^d, b_2 - b_3\right)\right) \\ &= J + fc_2\left((c_2b_1 - c_3b_4) + \left(b_1^d - b_2^d, c_3 - c_2, b_1 - b_4\right) \cdot \left(s, c_1, c_4, b_3^d, b_2 - b_3\right)\right) \\ &= J + fc_2\left(b_1^d - b_2^d, c_3 - c_2, b_1 - b_4\right) \cdot \left(s, c_1, c_4, b_3^d\right). \end{aligned}$$

As modulo J,

$$\begin{aligned} sc_2(b_1 - b_4) &\equiv fc_3b_3^d(b_1 - b_4) \equiv fb_3^d(c_3 - c_2)b_1 \equiv (sc_3 - fb_2^dc_2)b_1 \equiv s(c_3 - c_2)b_1 \equiv 0, \\ sfc_2(b_1^d - b_2^d) &\equiv f^2c_1b_1^d - s^2c_2 \equiv sfc_1 - sfc_1 = 0, \\ fc_1 &\equiv fc_4, \end{aligned}$$

the intersection of the ideals in the first five rows simplifies to

$$\begin{aligned} &J + fc_2\left(b_1^d - b_2^d, c_3 - c_2, b_1 - b_4\right) \cdot \left(c_1, b_3^d\right) \\ &= J + sc_2\left(b_1^d - b_2^d, c_3 - c_2, b_1 - b_4\right) \\ &= J + sc_2\left(b_1^d - b_2^d\right). \end{aligned}$$

Finally we intersect this intersection of the ideals in the first five rows in the statement of Theorem 1 with the (shortened) last ideal there, namely with $L = (s^2, f^2, c_4(s - fb_4^d), c_3(s - fb_3^d), sc_3 - fc_4, c_3^2, c_4^2, c_1 - c_4, c_2 - c_3, b_2 - b_3, b_1 - b_4)$, to get:

$$J + sc_2\left(b_1^d - b_2^d\right) \cap L.$$

It is easy to see that $L : sc_2$ contains $\sqrt{L}$. As sc_2 is not in L, then $L : sc_2 = \sqrt{L}$, so that the intersection of all the ideals in Theorem 1 equals

$$\begin{aligned} J+&sc_2\left(b_1^d - b_2^d\right)\left(L : sc_2\left(b_1^d - b_2^d\right)\right) = J + sc_2\left(b_1^d - b_2^d\right)\left(\sqrt{L} : \left(b_1^d - b_2^d\right)\right) \\ &= J + sc_2\left(b_1^d - b_2^d\right)\sqrt{L} \\ &= J + sc_2\left(b_1^d - b_2^d\right)(s, f, c_1, c_2, c_3, c_4, b_2 - b_3, b_1 - b_4) \\ &= J + sc_2\left(b_1^d - b_2^d\right)(f, c_2). \end{aligned}$$

It has been proved that $sfc_2\left(b_1^d - b_2^d\right) \in J$, and similarly $sc_2^2\left(b_1^d - b_2^d\right) \in J$. This proves that the intersection of all the listed ideals in Theorem 1 does equal J. ∎

In order to finish the proof of Theorem 1, it remains to prove that none of the listed components is redundant. The last component is primary to a non-minimal prime, whereas there are no inclusion relations among the rest of the primes. Thus the first $d' + 4$ listed components belong to minimal primes and are not redundant. With this it suffices to prove that J has an embedded prime:

LEMMA 3: *When $n = 1$, $c_4(s - fb_3^d)$ is in every minimal component but not in J. Thus there exists an embedded component.*

Proof: It has been established that $c_4(s - fb_3^d)$ is in every minimal component. Suppose that $c_4(s - fb_3^d)$ is in J. Then

$$\begin{aligned} c_4(s - fb_3^d) = \sum_{i=1}^{4} r_i c_i\left(s - fb_i^d\right) + r_5(fc_1 - sc_2) + r_6(fc_4 - sc_3) + r_7 s\,(c_3 - c_2) \\ + r_8 f\,(c_2 b_1 - c_3 b_4) + r_9 f c_2\,(b_2 - b_3)\,, \end{aligned}$$

for some elements r_i in the ring. By the homogeneity of all elements in the two sets of variables $\{s, f\}$ and $\{c_1, c_2, c_3, c_4\}$, without loss of generality each r_i is an element of $K[b_i | i = 1, 2, 3, 4]$. Therefore the coefficients of the fc_i, sc_i yield the following equations:

$$\begin{aligned} sc_4 &: \quad 1 = r_4, \\ fc_4 &: \quad -b_3^d = -r_4 b_4^d + r_6, \text{ so } r_6 = b_4^d - b_3^d, \\ fc_3 &: \quad 0 = r_3 b_3^d + r_8 b_4, \text{ so } r_3 = rb_4, r_8 = -rb_3^d \text{ for some } r \in R, \\ sc_3 &: \quad 0 = b_4^d - b_3^d - r_3 - r_7, \text{ so } r_7 = b_4^d - b_3^d - rb_4, \end{aligned}$$

$$\begin{aligned}
sc_1 &: \quad 0 = r_1, \\
fc_1 &: \quad 0 = -r_1 b_1^d + r_5, \text{ so } r_5 = 0, \\
sc_2 &: \quad 0 = r_2 - b_4^d + b_3^d + r b_4, \text{ so } r_2 = b_4^d - b_3^d - r b_4, \\
fc_2 &: \quad 0 = -r_2 b_2^d - r b_3^d b_1 + r_9(b_2 - b_3).
\end{aligned}$$

After expanding r_2 in the last equation, $0 = -(b_4^d - b_3^d - rb_4)b_2^d - rb_3^d b_1 + r_9(b_2 - b_3)$, so that $b_2^d b_3^d \in (b_1, b_4, b_2 - b_3)$, which is a contradiction. ■

As one embedded component has been established, this proves the Theorem. Thus in the case $n = 1$, the Mayr-Meyer ideal $J(1,d)$ has $d' + 4$ minimal primes and one embedded one, and these associated prime ideals are as follows (α varies over the d' th roots of unity):

associated prime ideal	height
$(s_1 - sc_1, f_1 - sc_4, c_1, c_2, c_3, c_4)$	6
$(s_1 - sc_1, f_1 - sc_4, c_4 - c_1, c_3 - c_2, c_1 - c_2 b_1^d, s - f b_1^d, b_1 - b_4, b_2 - b_3, b_1 - \alpha b_2)$	9
$(s_1 - sc_1, f_1 - sc_4, s, f)$	4
$(s_1 - sc_1, f_1 - sc_4, s, c_1, c_2, c_4, b_3, b_4)$	8
$(s_1 - sc_1, f_1 - sc_4, s, c_1, c_4, b_2, b_3, c_2 b_1 - c_3 b_4)$	8
$(s_1 - sc_1, f_1 - sc_4, s, f, c_1, c_2, c_3, c_4, b_2 - b_3, b_1 - b_4)$	10

The proof of the Theorem also explicitly computes the intersection of the first five rows of the primary decomposition, so that:

PROPOSITION 4: *The intersection of all the minimal components of $J(1,d)$ equals $J + (sc_2(b_1^d - b_2^d))$.* ■

Furthermore, it is straightforward to compute the radical of $J(1,d)$:

PROPOSITION 5: *The radical of $J(1,d)$ equals $J(1,d') + fb_3(c_3 - c_2, c_2(b_1^{d'} - b_2^{d'}))$.*

Proof: It is straightforward to compute the radical of each component. Note that as in the previous proof it suffices to compute the shortened intersection:

$$(c_1, c_2, c_3, c_4)$$

$$
\begin{aligned}
&\bigcap_{\alpha} \left(c_4 - c_1, c_3 - c_2, c_1 - c_2 b_1^d, s - f b_1^d, b_1 - b_4, b_2 - b_3, b_1 - \alpha b_2\right) \\
&\cap (s, f) \\
&\cap (s, c_1, c_2, c_4, b_3, b_4) \\
&\cap (s, c_1, c_4, b_2, b_3, c_2 b_1 - c_3 b_4) \\
= &\left(c_i \left(s - f b_i^d\right), f c_1 - s c_2, f c_4 - s c_3, s\left(c_3 - c_2\right), f\left(c_2 b_1 - c_3 b_4\right)\right) \\
&+ \left(f c_2 \left(b_2 - b_3\right), f b_3 (c_3 - c_2), f b_3 c_2 (b_1^{d'} - b_2^{d'})\right).
\end{aligned}
$$

As in the proof of the Theorem, the intersection of the first three rows equals $J(1, d') + \left(f(c_3 - c_2), f c_2(b_1^{d'} - b_2^{d'})\right)$. Intersection with the ideal in the fourth row, namely with $(s, c_1, c_2, c_4, b_3, b_4)$, equals

$$
\begin{aligned}
J(1, d') &+ \left(f c_2(b_1^{d'} - b_2^{d'})\right) + (f(c_3 - c_2)) \cap (s, c_1, c_2, c_4, b_3, b_4) \\
&= J(1, d') + \left(f c_2(b_1^{d'} - b_2^{d'})\right) + (f(c_3 - c_2)) \cdot (s, c_1, c_2, c_4, b_3, b_4) \\
&= J(1, d') + \left(f c_2(b_1^{d'} - b_2^{d'})\right) + (f(c_3 - c_2)) \cdot (c_2, b_3, b_4).
\end{aligned}
$$

When this is intersected with the ideal in the fifth row, namely with $(s, c_1, c_4, b_2, b_3, c_2 b_1 - c_3 b_4)$, the resulting radical of $J(1, d)$ equals

$$
\begin{aligned}
J(1, d') &+ (f b_3(c_3 - c_2)) \\
&+ \left(f c_2(b_1^{d'} - b_2^{d'}), f c_2(c_3 - c_2), f b_4(c_3 - c_2)\right) \cap (s, c_1, c_4, b_2, b_3, c_2 b_1 - c_3 b_4) \\
&= J(1, d') + (f b_3(c_3 - c_2)) \\
&\quad + f c_2 \left(\left(b_1^{d'} - b_2^{d'}, c_3 - c_2, b_1 - b_4\right) \cap (s, c_1, c_4, b_2, b_3, c_2 b_1 - c_3 b_4)\right) \\
&= J(1, d') + (f b_3(c_3 - c_2)) + f c_2 \left(\left(b_1^{d'} - b_2^{d'}, c_3 - c_2, b_1 - b_4\right) \cdot (s, c_1, c_4, b_2, b_3)\right),
\end{aligned}
$$

and by previous computations this simplifies to

$$
\begin{aligned}
J(1, d') &+ (f b_3(c_3 - c_2)) + f c_2 b_3 \left(b_1^{d'} - b_2^{d'}, c_3 - c_2, b_1 - b_4\right) \\
&= J(1, d') + f b_3(c_3 - c_2, c_2(b_1^{d'} - b_2^{d'})). \quad \blacksquare
\end{aligned}
$$

Mayr and Meyer [MM] observed that whenever the element $s(c_4 - c_1)$ of $J(1, d)$ is expressed as an R-linear combination of the given generators of $J(1, d)$, at least one of the coefficients has degree at least d. In fact, as the proposition below proves, the degree of at least one of the coefficients is at least $2d-1$, and this lower bound is achieved. (See also the proof of Theorem showing that $sc_4 \equiv sc_1$ modulo $J(1,d)$.) Mayr and Meyer also showed the analogues for $n \geq 1$, with degrees of the coefficients depending on $n-1$ doubly exponentially.

Bayer, Huneke and Stillman questioned how much this doubly exponential growth depends on the existence of embedded primes of $J(n,d)$, or on the structure of the components. The proposition below shows that at least for $n=1$, the facts that $J(1,d)$ has an embedded prime and that the minimal components are not radical, do not seem to be crucial for this property:

PROPOSITION 6: *Let I be any ideal between $J(1,d)$ and its radical. Then whenever the element $s(c_4-c_1)$ is expressed as an R-linear combination of the minimal generators of I which include all the given generators of $J(1,d)$, at least one of the coefficients has degree at least $2d-1$.*

Proof: All the cases can be deduced from the case of I being the radical of $J(1,d)$. To simplify the notation it suffices to replace d by d', so that $I = J(1,d) + fb_3(c_3-c_2, c_2(b_1^d - b_2^d))$. Write

$$s(c_4-c_1) = \sum_{i=1}^{4} r_i c_i \left(s - fb_i^d\right) + r_5(fc_1 - sc_2) + r_6(fc_4 - sc_3) + r_7 s\,(c_3-c_2)$$
$$+ r_8 f\,(c_2b_1 - c_3b_4) + r_9 fc_2\,(b_2-b_3) + r_{10} fb_3(c_3-c_2) + r_{11} fb_3c_2(b_1^d - b_2^d)$$

for some elements r_i in the ring. Note that each of the explicit elements of I is homogeneous in the two sets of variables $\{s,f\}$ and $\{c_1,c_2,c_3,c_4\}$. Thus it suffices to prove that in degrees 1 in each of the two sets of variables, one of the coefficients has degree at least $2d-1$. So without loss of generality each r_i is an element of $K[b_i|i=1,2,3,4]$. Therefore the coefficients of the sc_i, fc_i yield

$$\begin{aligned}
sc_4 &: \quad 1 = r_4,\\
fc_4 &: \quad 0 = -r_4b_4^d + r_6, \text{ so } r_6 = b_4^d,\\
sc_1 &: \quad -1 = r_1,\\
fc_1 &: \quad 0 = -r_1b_1^d + r_5, \text{ so } r_5 = -b_1^d,\\
sc_2 &: \quad 0 = r_2 + b_1^d - r_7, \text{ so } r_7 = r_2 + b_1^d,
\end{aligned}$$

$$fc_2: \quad 0 = -r_2 b_2^d + r_8 b_1 + r_9 (b_2 - b_3) - r_{10} b_3 + r_{11} b_3 (b_1^d - b_2^d),$$
$$sc_3: \quad 0 = r_3 - b_4^d + r_7, \text{ so } r_3 = b_4^d - b_1^d - r_2,$$
$$fc_3: \quad 0 = -r_3 b_3^d - r_8 b_4 + r_{10} b_3.$$

The last equation implies that $r_8 = r_8' b_3$, so that the equations for the coefficients of fc_3 and fc_2 can be rewritten as

$$r_{10} = r_3 b_3^{d-1} + r_8' b_4 = (b_4^d - b_1^d) b_3^{d-1} - r_2 b_3^{d-1} + r_8' b_4,$$
$$0 = -r_2 b_2^d + r_8' b_3 b_1 + r_9 (b_2 - b_3) - (b_4^d - b_1^d) b_3^d + r_2 b_3^d - r_8' b_4 b_3 + r_{11} b_3 (b_1^d - b_2^d),$$
$$= -r_2 (b_2^d - b_3^d) + r_8' b_3 (b_1 - b_4) + r_9 (b_2 - b_3) - (b_4^d - b_1^d) b_3^d + r_{11} b_3 (b_1^d - b_2^d).$$

Thus $r_8' b_3 (b_1 - b_4) \in (b_2 - b_3, b_3^d (b_1^d - b_4^d), b_1^d - b_2^d)$, so that $r_8' \in (b_2 - b_3, b_3^{d-1} \frac{b_1^d - b_4^d}{b_1 - b_4}, b_1^d - b_2^d)$. If $r_8' \in (b_2 - b_3, b_1^d - b_2^d)$. Then $(b_4^d - b_1^d) b_3^d \in (b_2 - b_3, b_1^d - b_2^d)$, which is a contradiction. Thus r_8' has a multiple of $b_3^{d-1} \frac{b_1^d - b_4^d}{b_1 - b_4}$ as a summand, so r_8' has degree at least $2d - 2$, so that r_8 has degree at least $2d - 1$. In fact, by setting all the free variables r_2, r_{11} to zero, the maximum degree of the coefficients r_i is $2d - 1$. ■

Note that in the proof above it is possible to have both $r_{10} = r_{11} = 0$ and the degrees of the r_i still at most $2d - 1$, with $2d - 1$ attained on some r_i. (Lemma 2.3 of [BS] erroneously claims that the degree of some r_i is at least $2d$.)

References

[BS] Bayer, D.; Stillman, M., On the complexity of computing syzygies, *J. Symbolic Comput.* **6**, (1988), 135-147.

[GS] Grayson, D.; Stillman, M., Macaulay2. 1996. A system for computation in algebraic geometry and commutative algebra, available via anonymous `ftp` from `math.uiuc.edu`.

[H] Herrmann, G., Die Frage der endlich vielen Schritte in der Theorie der Polynomideale, *Math. Ann.* **95**, (1926), 736-788.

[K] Koh, J., Ideals generated by quadrics exhibiting double exponential degrees, *J. Algebra* **200**, (1998), 225-245.

[MM] Mayr, E.; Meyer, A., The complexity of the word problems for commutative semigroups and polynomial ideals, *Adv. Math.* **46**, (1982), 305-329.

[S] Swanson, I., Primary decomposition of the Mayr-Meyer ideals, in preparation, 2001.

Facets on Rings Between $D[X]$ and $K[X]$

MUHAMMAD ZAFRULLAH Department of Mathematics, Idaho State University, Pocatello, ID 83209-8085 E-mail: mzafrullah@usa.net

ABSTRACT: Since the circulation, in 1974, of the first draft of "The construction $D + XD_S[X]$, J. Algebra 53 (1978), 423-439" a number of variations of this construction have appeared. Some of these are: The generalized $D + M$ construction, the $A + (X)B[X]$ construction, with X a single variable or a set of variables, and the $D + I$ construction (with I not necessarily prime). These constructions have proved their worth not only in providing numerous examples and counter examples in commutative ring theory, but also in providing statements that often turn out to be forerunners of results on general pullbacks. The aim of this paper will be to discuss these constructions and the remarkable uses they have been put to. I will concentrate more on the $A + XB[X]$ construction, its basic properties and examples arising from it.

INTRODUCTION

Let A be a subring of an integral domain B, and let X be an indeterminate over B. The set $\{f(X) \in B[X] : f(0) \in A\}$ is a ring denoted by $A + XB[X]$. This article is an attempt at a survey of the polynomial ring constructions of the form $A + XB[X]$ that have come into vogue in recent years. The article consists of a slightly modified version of the talk that I gave at the Fez Conference held in the year 2001 and a number of appendices or supplements. In the talk, I briefly surveyed the history of the $A + XB[X]$ construction and the various constructions that seem to have risen from similar considerations. The talk is the first part of the article. In the appendices, I study topics that are either essential to the understanding of the $A + XB[X]$ construction or are ones that give rise to examples that, in my opinion, are useful. In the first appendix, which is part 2, I study the prime ideal structure of $A + XB[X]$ construction. Then in part 3, I indicate how the study of Spec($A + XB[X]$) led to the construction of various examples and in part 4, I indicate how useful examples can be constructed from the $D+XD_S[X]$ construction. Part 5 is a wish list, i.e., I briefly go over topics that I hoped to write on but could not because that will make the article a bit too long. The necessary terminology is explained where needed, and any terminology that has not been explained can be found either in Gilmer [1] and/or Kaplansky [2].

PART 1 (THE TALK)

Let D be an integral domain with quotient field F and let K be a field extension of F. We know a great deal about $D[X]$ and almost everything about $K[X]$, where X is an indeterminate over K. About rings between $D[X]$ and $K[X]$ we have just begun to learn. Let us call the rings between $D[X]$ and $K[X]$ the intermediate rings. At present we can split these rings into two main types: (1) Intermediate rings that are composite, i.e. are of the form $A + XB[X] = \{a_0 + \sum_{i=1}^{n} a_i X^i \mid a_0 \in A$ and $a_i \in B\}$ where $D \subseteq A \subseteq B \subseteq K$ are ring extensions, and (2) the intermediate rings that are not of the form $A + XB[X]$.

The only, very, well known rings of the latter kind are the rings of integer-valued polynomials. These rings are very well known and the best I can do is refer the readers to the wonderful book by Cahen and Chabert [3]. There are other less well known though equally important examples, of such rings due to Eakin and Heinzer [4]. Indeed rings that are not of the form $A + XB[X]$ have a composite cover as indicated by D.D. Anderson, D.F. Anderson and myself in [5]. This composite cover, very often, determines some of the properties of these rings.

If you have not got the drift yet, then let me tell you, I intend to spend more time on the composites, i.e., rings of the form $A+XB[X]$. My reasons for choosing this course of action are the following:

(a) The composites, over the past few years, have provided directly constructible examples of rings which were once very hard to construct.

(b) The composites have given rise to new notions and new constructions which make it easier to bring in new concepts and study them.

(c) The composites are pullbacks and it has become customary to prove a statement for a special kind of a composite, then for a general composite, and then for a general pullback. Consequently the appreciation of pullbacks has increased.

My plan is essentially to give a brief description of what it (i.e. the $A+XB[X]$ construction) is, then brag about what it does, and then indicate several of the variants of this construction that have appeared recently. Recently, Tom Lucas [6] has written a survey on examples of pullbacks using the $A + XB[X]$ construction. I will try not to repeat those examples and will cover the material that Tom left out because of space restrictions.

Basic properties of the intermediate rings.

Given a ring R between $D[X]$ and $K[X]$, we can split R into two parts as follows: $M_R = \{f(x) \in R \mid f(0) = 0\}$ and $S_R = \{f(X) \in R \mid f(0) \neq 0\}$. (Let $S_R(0) = \{f(0) : f \in S_R\}$.) Of these M_R is a prime ideal and S_R is a multiplicative set with the property that $S_R(0) \cup \{0\} = \{f(0) \mid f(x) \in R\} = R_0$. Now R_0 is a subring of K, though R_0 may or may not be a subring of R, but there are some direct observations that can be made:

(i) The map $\pi : R \longrightarrow R_0$ defined by $\pi(f) = f(0)$ is a ring epimorphism with $\ker \pi = M_R$. Thus $R_0 \cong R/M_R$.

(ii) Every unit of R is a unit of R_0, but the converse may not hold. ($Q[X] \subseteq R = Q[X][\sqrt{2}X + \sqrt{3}] \subset \mathcal{R}[X]$, where Q is the rationals and $\mathcal{R}$ is the reals. Clearly $R_0 = Q[\sqrt{3}]$.)

(iii) If $R_0 \subseteq R$ then $R = R_0 + M_R$. Using the fact that if B is a ring I an ideal of B and A a subring of B then $A + I$ is a subring of B, we can construct for each subring A of $R \cap R_0$ a subring $A + M_R$ of R. Now as $R \supseteq D[X]$, we have $XA[X] \subseteq M_R$, and so $A + M_R$ contains $D[X]$, whenever A contains D.

(iv) If it so happens that $R_0 \subseteq R$ and $M_R = XR_1[X]$ for some R_0-subalgebra R_1 of K, then we have $R = R_0 + XR_1[X]$ a ring of the form $A + XB[X]$. Indeed, $M_R = XR_1[X]$ for some R_1 if and only if $\sum_{i=1}^{n} a_i X^i \in R$ implies $a_i X \in R$ for all i such that $1 \leq i \leq n$. In this case, $R_1 = \{a \mid aX \in R\}$ [5].

(v) Now we have seen that $R_0 \cong R/M_R$. For some of us it is grounds enough to set up, for each subring R' of R_0, the following diagram of canonical homomorphisms:

$$\begin{array}{ccc} S = \pi^{-1}(R') & \hookrightarrow & R \\ \downarrow & & \downarrow \pi \\ R' & \hookrightarrow & R/M_R \end{array} .$$

In the usual terminology S is called a pullback. Indeed, if $R' \subseteq R$, then $S = R' + M_R$. Thus the composites $A + XB[X]$ are pullbacks. Yet, while the pullback would give you $\pi^{-1}(R_0) = R$, the composites will go a step further. By [5, Cor. 2.4] associated with an intermediate ring R, there is a unique composite $S = R_0 + XS_1[X]$, where S_1 is a subring of K generated by $\bigcup_{f \in R} A_f$ and A_f is the D-submodule of K generated by the coefficients of f. Let us call this unique composite the composite cover of R.

Now, what is so special about the composite cover of an intermediate ring R? For one thing, R is integral over $D[X]$ if and only if the composite cover of R is integral over $D[X]$ ([5, Prop. 2.6]). On its own, $A + XB[X]$ is integrally closed if and only if B is integrally closed and A is integrally closed in B.

(vi) The name composite fits because $A+XB[X]$ is the composite of A and $B[X]$ over the ideal $XB[X]$. Mott and Schexnayder's paper [7] gives a good description of composites of several kinds.

Special types of composites.

Enough of the basic properties. Let us now see the different types of composites. The current wave of study of composites started with the circulation of an earlier version of my paper with Costa and Mott [8]. The construction to be studied was given in the title, "The construction $D + XD_S[X]$", where D is an integral domain of your choice, S a multiplicative set in D and X an indeterminate. The immediate special case: the then well known $D + XK[X]$, where K is the quotient field of D, which I was using to produce examples of GCD domains each proper principal ideal of which has finitely many minimal primes. I called these GCD domains the unique representation domains (URD's). Indeed, if D is a URD, then so is $D + XK[X]$. At Paul Cohn's suggestion I started looking into $D^{(S)} = D + XD_S[X]$. It turned out that $D + XD_S[X]$ is a GCD domain if and only if D is a GCD domain and for each $d \in D$, $\mathrm{GCD}(d, X)$ exists [8]. It turned out also that if $D^{(S)}$ is a GCD domain then $D^{(S)}$ is a URD iff D is [9]. Of interest in a GCD domain D are the PF-primes, say the primes P such that D_P is a valuation domain. The PF dimension of a GCD domain can be defined in the same way as the Krull dimension or the valuative dimension of a domain is defined. Sheldon [10] had studied the PF primes and had conjectured that a GCD domain D with PF-dim(D) = Krull-dim(D) should be Bézout (f.g. ideals are principal).

Next, as $D^{(S)}$ is a polynomial ring construction, it was natural to ask questions about Krull-dim($D^{(S)}$) with reference to Krull-dim(D); it meant a study of Spec($D^{(S)}$). I found out that if P is a prime ideal of $D^{(S)}$ with $P \cap S \neq \phi$, then $P = P \cap D + XD_S[X]$, and if $P \cap S = \phi$, then $P = PD_S[X] \cap D^{(S)}$.

Adding a little bit here and a little bit there, my paper was complete. When I was writing my thesis, Paul (Cohn) told me to send whatever I produced to Robert Gilmer, who was and of course is a leading expert in Multiplicative Ideal Theory. So, out of habit, I mailed a copy to Robert who had indeed been very kind to me. This way I got in contact with Joe Mott and Doug Costa who had done some similar things. Between us, we added results including results such as: If D is Noetherian, then $D^{(S)}$ is a coherent ring; $D + XK[X]$ is a PVMD if and only if D is a PVMD. Now this needs a little bit of introduction. A function $*$ on the set $F(D)$ of nonzero fractional ideals is said to be a star operation if for all $a \in K\backslash\{0\}, A, B \in F(D)$, we have (1) $(a)^* = (a)$, $(aA)^* = aA^*$ (2) $A \subseteq A^*$ and $A \subseteq B \Rightarrow A^* \subseteq B^*$ and (3) $(A^*)^* = A^*$. If $*$ is a star operation, we can also define $*$-multiplication as $A \times_* B = (AB)^* = (A^*B)^* = (A^*B^*)^*$. An ideal $A \in F(D)$ is a $*$-ideal if $A = A^*$. The operation defined, on $F(D)$, by $A \longrightarrow A_v = (A^{-1})^{-1}$ is called the v-operation, and the one defined by $A_t = \bigcup F_v$, where F ranges over nonzero finitely generated subideals of A, is called the t-operation. Now D is a PVMD if for all finitely generated $A \in F(D)$, A is t-invertible, i.e., $(AA^{-1})_t = D$.

The message of [8], like any other construction involving two rings, was: see how, and under what conditions, some properties of D get transferred to $D^{(S)}$. This brought up the question: If D is a PVMD, for what S should $D^{(S)}$ be a PVMD? That meant knowing all the maximal t-ideals of $D^{(S)}$. An integral ideal P that is maximal w.r.t. being a t-ideal is a prime ideal, and Griffin [11] had shown that D is a PVMD if and only if for each maximal t-ideal P of D, D_P is a valuation domain. The main hurdle was that the prime ideals of $D^{(S)}$ that are disjoint with S are contractions of prime ideals of $D_S[X]$. To see how this problem could be resolved, I decided to study the $D + XD_S[X]$ construction from GCD domains [12]. (A GCD domain is a PVMD.) It did not give me what I wanted, but it brought simple examples of locally GCD domains that were not GCD, and the fact that if P is a prime t-ideal of D then PD_P may not be a t-ideal of D_P. I also discovered that if D is GCD, then $D^{(S)}$ is GCD if and only if S is a splitting set, i.e., each $d \in D\backslash\{0\}$ is expressible as $d = sd_1$, where $s \in S$ and $(d_1) \cap (t) = (td_1)$ for all $t \in S$. This rediscovery of the splitting sets of [7] led not only to a lot of activity from the factorization point of view, but also to the ultimate solution of my problem. The solution turned out to be: If D is a PVMD, $D^{(S)}$ is a PVMD if and only if S is a t-splitting set of D. Now, a saturated multiplicative set is a t-splitting set if for every element $d \in D\backslash\{0\}$ we have $(d) = (AB)_t$, where A and B are integral t-ideals such that $A \cap S \neq \phi$ and $B \cap (t) = Bt$ for all $t \in S$. This result has appeared in a paper of mine with the Anderson brothers [13].

Apparently the contents of [8] had started taking effect before it was published. Brewer and Rutter[14] came up with the idea: if $R = k + M$, where k is a field, then for every subring D of k you have a subring $D + M$ of R. They called it the generalized $D + M$ construction, generalized because it generalized the celebrated $D + M$ construction, greatly popularized by Gilmer, which required $R = k + M$ to be a valuation domain. Using this construction they were able to recover all that was proved about $D + XK[X]$ in [8], here $K = qf(D)$ and, on top of that, their construction allowed results on subrings of $k[[X]]$, where k is a field. While this was happening, Malik and Mott [15] had come up with their study of strong S-domains. D is a S(eidenberg) domain if for each height one prime P of D, $PD[X]$ is of height one, and D is strong S if for each prime P, D/P is an S-domain. They showed that the $D + XK[X]$ construction is a strong S-ring if and only if D is; yet they pointed out that if $D[X]$ is strong S, it is not necessary that $(D + M)[X]$ should be strong S. Now, [14] had encouraged me to go general. I wrote up a piece on the overrings of $D + XL[X]$, where L is an extension of the quotient field of D. Then Costa and Mott gave it the language of generalized $D + M$, and we had another paper [16]. (Later Joe and I [17] showed that if D is a Noetherian Hilbert domain and L is an extension of the quotient field of D, then $D + XL[X]$ is a non-Noetherian Hilbert domain whose maximal ideals are finitely generated. Constructing such a domain by conventional means was quite difficult, as shown by Gilmer and Heinzer [18].)

The beauty of the $D + XD_S[X]$ construction seems to lie in the fact that it is so close to well known examples of pullbacks and composites, yet so open to reinterpretation and so easy to work with. Marco Fontana wrote a longish article in Italian [19] and sent a copy to me in Libya. (Marco tells me that those were his seminar notes.) He had treated all the constructions and composites that I have talked about above with reference to [7], including $D^{(S)}$, showing how the spectral space of a pullback can be shown to be connected with the spectral spaces of the constituents of the pullback. I think that Marco's interest in the $D + XD_S[X]$ construction had a profound effect on the development of polynomial ring constructions.

In trying to get some examples in a completely different context, I had found out that if $k \subseteq K$ is an extension of fields and if X is an indeterminate over K, then $k + XK[X]$ satisfies ACC on principal ideals [20]. When, in 1986, I went to Lyon, (France) I gave, among other talks, a talk on an earlier version of [5]. There I met several young men, Salah-Eddine Kabbaj included, who were eager to learn and ready to experiment with new ideas and techniques. From these young men issued forth a barrage of papers containing all sorts of variations of $A + XB[X]$ construction. Strong S-domains and Jaffard domains were in vogue. Jaffard domains are domains D such that valuative $\dim(D) =$ Krull $\dim(D)$ (symbolically ($\dim_v(D) =$ K-$\dim(D)$). Anderson, Bouvier, Dobbs, Fontana and Kabbaj wrote papers, [21] and [22]. [21], using various pullbacks to construct examples, and [22] showing that if D is Jaffard, then so is $D + XD_S[X]$. Then Fontana and Kabbaj [23] studied the Krull and valuative dimensions of $D + (X_1, X_2,.., X_n)D_S[X_1, X_2,.., X_n] = D^{(S,n)}$. It turned out that $\dim_v(D^{(S,n)}) = \dim_v(D) + n$, and that D is a Jaffard domain if and only if so is $D^{(S,n)}$. Next, they prove that $D^{(S,n)}$ is a strong S ring if and only if both D and $D_S[X_1, X_2,.., X_n]$ are. To top it all, they showed that $D[X]$ is an S-domain for any D. Later, Fontana, Izelgue, and Kabbaj [24] studied the

Krull and valuative dimensions of the $A+XB[X]$ construction and showed that the results are different, especially when the quotient field of B is a proper extension of the quotient field of A. Recently, Anderson and Nour-el-Abidine [25] have studied the $A+XB[X]$ and $A+XB[[X]]$ constructions from GCD domains.

Current trends.

The sole purpose of studying the $D+XD_S[X]$ construction was to get examples of domains that did not satisfy ACC on principal ideals. But we could not ignore the possibility of ACCP holding for an intermediate ring. In [5], we came up with: Let R be an intermediate ring, then R has ACCP if and only if any ascending chain of principal ideals generated by polynomials of R, of the same degree, terminates. As a demonstration of this, we proved a proposition for $D[X] \subseteq R \subseteq K[X]$, where K is the quotient field of D. Of course we thrashed the case of $A+XB[X]$ for A and B fields, but Barucci, Izelgue, and Kabbaj [26] came up with the somewhat remarkable discovery that if A is a field then $A+XB[X]$ has ACCP no matter what kind of integral domain B is. (I recall having written a good review of this paper but, apparently, what I sent was hard to understand, for reasons I am trying to explain to myself. Possibly some part of the review got deleted!) This remarkable short note had some other gems that started off a lot of activity in the study of factorization properties of $A+XB[X]$ and $A+XB[[X]]$ constructions. The names to mention in this connection are Nathalie Gonzalez [27] and [28] [29] , David Anderson and Nour-el-Abidine [30], Dumitrescu, Radu, Salihi and Shah [31].

Let $T(D)$ denote the set of t-invertible t-ideals of D. Then $T(D)$ is a group under t-product and when we quotient it by its subgroup $P(D)$ of principal fractional ideals we get what I call the t-class group $Cl_t(D) = T(D)/P(D)$. This class group was introduced by Bouvier in [32]. Anderson and Ryckaert [33] studied the t-class group of the generalized $D+M$ construction. The fact that $Cl_t(D+XK[X]) \cong Cl_t(D)$ came to the fore in a strange way in a paper of mine with Bouvier [34]. Then as the Fontana factor grew, a lot of the above questions were considered for pullbacks. Fontana and Gabelli's [35], and independently of them Khalis and Nour El-Abidine [36], considered the t-class group of a pullback. Yet the class group of $A+XB[X]$ has also been studied by Anderson, El-Baghdadi and Kabbaj [37]. In [37] the main question studied is: Under what conditions is $Cl_t(A+XB[X]) \cong Cl_t(A)$. The same authors go on to study other forms of the t-class group of $A+XB[X]$; a good source for their work is El-Baghdadi's thesis [38]. Coming back to the pullbacks, the hot questions these days are something like: when is a pullback...? For example, see Houston, Kabbaj, Lucas, and Mimouni, [39]. Also, see coherent-like conditions in pullbacks as in [40].

PART 2 (KRULL DIMENSION OF A+XB[X])

Being honest to goodness polynomial ring constructions, $A+XB[X]$ domains qualify for a comparative study of their Krull dimensions with the Krull dimensions of $A, A[X]$ and $B[X]$. Of course, so do the general pullbacks, but in the case of the $A+XB[X]$ construction, we can get a somewhat better picture. This picture becomes clearer for some special cases of this construction. Now Fontana, Izelgue, and Kabbaj [41], [24] (one of these two is a translation of the other) and [42] took good care of this need both for speakers of English and French. In the following, I will try to give an idea of what they produced in this connection for readers who are interested, but not too interested.

Let us start with the observation that for each prime ideal P of $R = A + XB[X]$, either $X \in P$ or $X \notin P$. Clearly if $X \in P$, then $XR \subseteq P$, and so $X^2B[X] \subseteq XR \subseteq P$. But as $X^2B[X] = (XB[X])^2$, we have $XB[X] \subseteq P$. This means that all the prime ideals that contain X are of the form $p+XB[X]$, where p is a prime ideal of A. Now the prime ideals Q that do not contain X are hidden in places that may be hard to reach. Let us fix some notation to make the task a little lighter. Let $\mathcal{L} = \{P \in Spec(R) : X \in P\}$ and $\mathcal{M} = \{P \in Spec(R) : X \notin P\}$. Then $\mathcal{L} = \{p + XB[X] : p \in Spec(A)\}$. Now let $l = sup\{ht_R(P) : P \in \mathcal{L}\} = sup\{ht_R(p + XB[X] : p \in Spec(A)\} = sup\{ht_A(P) + ht_R(XB[X] : P \in \mathcal{L}) = dim(A) + ht_R(XB[X])$. Next, let m be the supremum of lengths of chains in $\mathcal{M}$. Then, if $S = \{X^n : \text{where } n \in N\}$, there is a one-to-one order-preserving correspondence between the primes in $\mathcal{M}$ and the primes in $R_S = (A+XB[X])_S = B[X]_S = B[X, X^{-1}]$; so $m = dim(B[X, X^{-1}])$. Now it has been established that $dim(B[X, X^{-1}] = dim(B[X])$, see for instance [21, Proposition 1.14]. Thus $m = dim(B[X])$. Since both these sets of primes come from $Spec(R)$, we have $dim(R) \geq max\{l, m\} = max\{dim(A) + ht_R(XB[X])$, $dim(B[X])\}$. Let us record this for future reference as an observation.

OBSERVATION 2.0. Let $R = A + XB[X]$, where $A \subseteq B$ is an extension of domains and X is an indeterminate over B. Then $dim(R) \geq max\{dim(A) + ht_R(XB[X])$, $dim(B[X])\}$.

Now we must find out the answers to the obvious questions. That is, what is $ht_R(XB[X])$?, is there an upper bound for $dim(R)$?, etc. Besides, even though the sets $\mathcal{L}$ and $\mathcal{M}$ are disjoint, some members of $\mathcal{M}$ may be contained in some members of $\mathcal{L}$. Since the Krull dimension is nothing but the supremum of lengths of chains of prime ideals, we may have to consider the case when, after taking the longest chain in $\mathcal{L}$, we are faced with the possibility that there is a sizeable chain of prime ideals of R contained in $XB[X]$, and obviously each of those prime ideals is coming from $\mathcal{M}$. So let us find out what kind of prime ideals of R will be contained in $XB[X]$.

LEMMA 2.1. Let $R = A + XB[X]$, where $A \subseteq B$ is an extension of domains and X is an indeterminate over B. If P is a prime ideal of R such that $P \subsetneq XB[X]$, then the following hold.

(1) no power of X is contained in P.

(2) $X^{-1}P \cap A = (0)$. (Consequently, if there is a nonzero prime ideal $P \subsetneq XB[X]$, then $X^{-1}P \cap A = (0)$).

(3) $X^{-1}P$ does not contain a polynomial $f(X)$ such that $f(0) \in A \backslash \{0\}$.

(4) $X^{-1}P$ is a prime ideal of $B[X]$.

Proof. (1) Since P is a prime, any power of X in P means $X \in P$. But then $XR \subseteq P$, which means that $X^2B[X] \subseteq P$. But as we have already observed, $X^2B[X] = (XB[X])^2$ we have $XB[X] \subseteq P$, and a contradiction. (2) Clearly $X^{-1}P$ is an ideal of $B[X]$. Suppose on the contrary that $X^{-1}P \cap A = \alpha \neq (0)$. Then $X^{-1}P \supseteq \alpha[X]$. Select $f(X) \in \alpha[X]$ such that $f(0) \neq 0$. Now $f(X) \in A[X] \subseteq A + XB[X], X \in A+XB[X], f(X) \notin P$ because $f(X) \notin XB[X]$ and $X \notin P$ by (1). Yet $Xf(X) \in P$, because $X\alpha[X] \subseteq P$, contradicting the primality of P. (3) The proof of (2) can be modified to take care of this. (4) Let $f(X)g(X) \in X^{-1}P$ where $f, g \in B[X]$. Then $Xf(X)g(X) \in P$, and hence $X^2f(X)g(X) = Xf(X)(Xg(X)) \in P$, which forces $Xf(X) \in P$ or $Xg(X) \in P$. That is, $f(X) \in X^{-1}P$ or $g(X) \in X^{-1}P$.

LEMMA 2.2. Given that A, B, X and R are as in Lemma 2.1. If p is a prime ideal of B such that $p \cap A = (0)$ then $Xp[X]$ is a prime ideal of R contained in $XB[X]$.

Proof. It is enough to note that $p[X] \cap (A + XB[X]) = \{f(X) \in p[X] : f(0) = 0$ and every coefficient of f comes from $\{Xg(X) : g(X) \in p[X]\}$ $Xp[X]$. That $Xp[X]$ is a prime ideal of R follows from the fact that it is a contraction of a prime ideal.

LEMMA 2.3. Let $A \subseteq B$ be an extension of domains, let X be an indeterminate over B, and let $R = A + XB[X]$. Next let P be an ideal of B that is maximal w.r.t. the property that $P \cap A = (0)$. Then P is a prime ideal such that for any prime ideal Q strictly containing $P[X]$, either $X \in Q$ or there is a polynomial $f \in Q$ such that $f(0) \in A$. Consequently, if P has the above stated property, there is no prime ideal strictly between $XB[X]$ and $P[X] \cap R = XP[X]$.

Proof. Note that $A \backslash \{0\}$ is a multiplicatively closed set in B. So, P being maximal w.r.t. $P \cap A = (0)$ means P is maximal w.r.t. being disjoint from $A \backslash (0)$. This makes P a prime ideal. Next, suppose that Q is a prime ideal such that $P[X] \subsetneqq Q$ and $X \notin Q$. Then there is a polynomial $f(X) \in Q \backslash P[X]$ such that no coefficient of f is in P. Since $X \notin Q$, we can arrange $f(X)$ so that $f(0) \neq 0$. But then, due to the maximality of P w.r.t. disjointness from $A \backslash (0)$, we have $rf(0) + p \in A$ for some $r \in B$ and $p \in P$. Next note that as $P \cap A = (0)$, we have $P[X] \cap R = XP[X] \subseteq XB[X]$. Now if there were a prime ideal H strictly between $XB[X]$ and $XP[X]$, then by (1) of Lemma 2.1, $X^n \notin H$ for any n. Yet, as $H \subseteq XB[X]$, every element of H is of the form $h = Xg(X)$ where $g(X) \in B[X]$. Let $h = Xg(X) \in H \backslash XP[X]$. Then $g(X) \in X^{-1}H \backslash P[X]$. This means that $X^{-1}H \supsetneqq P[X]$ and as $X \notin X^{-1}H$ by the first part there is $f(X) \in X^{-1}H$ such that $f(0) \in A \backslash \{0\}$. But this contradicts (3) of Lemma 2.1. Hence there is no prime ideal H strictly between $XB[X]$ and $P[X] \cap R = XP[X]$.

Before we go any further, a word about $dim(B[X])$. Let $C : P_n \supsetneqq P_{n-1} \supsetneqq \ldots \supsetneqq P_1 \supsetneqq (0)$ be a chain of prime ideals in $B[X]$. Jaffard [43] calls C a special chain if $P_i \in C$ implies $(P_i \cap B)[X] \in C$. In [43], it was shown that $dim(B[X])$ can be realized as the length of a special chain. Let B be finite dimensional and let C described above be a chain that realizes $dim(B[X])$. Then $P_n \supsetneqq (P_n \cap B)[X]$. For if $P_n = (P_n \cap B)[X]$, then $P_{n+1} = (P_n, X) = (P_n \cap B) + XB[X]$, and so there is a longer (special) chain of prime ideals in $B[X]$, a contradiction. Next we note that $(P_n \cap B)[X] = P_{n-1}$. For if not then, say $(P_n \cap B)[X] = P_{n-i}$ because C is a special chain, which gives $P_n \cap B \supseteq P_{n-1} \cap B \supseteq P_{n-i} \cap B = P_n \cap B$. This forces three distinct prime ideals of $B[X]$ to contract to the same prime ideal of B, which is impossible. Finally we note that $P_n \cap B$ must be a maximal ideal of B, because if not, then say a prime ideal $Q \supsetneqq (P_n \cap B)$, and then $Q[X] \supsetneqq (P_n \cap B)[X] = P_{n-1}$ and we end up, again, with a longer chain $Q + XB[X] \supsetneqq Q[X] \supsetneqq P_{n-1} \ldots$. (A reader who is seriously interested in dimension theory of polynomial rings may want to read [1, p.366] and chase the references given there.) Now, having made these notes, we can make the following statement.

OBSERVATION 2.4. For an integral domain B, $dim(B[X]) = 1 + max\{ht_{B[X]}(P[X]) : P \in Spec(B)\} = 1 + max\{ht_{B[X]}(P[X]) : P$ ranges over maximal ideals of $B\}$.

LEMMA 2.5. Let $R = A + XB[X]$, where A is a field. Then $dim(R) = dim(B[X])$ and so $dim(R) = 1$ if and only if B is a field.

Proof. In this case, $\mathcal{L} = \{XB[X]\}$ and so $l = 1$ and $\mathcal{M} = \{P \in Spec(R) : X \notin P\}$, and as before $m = dim(B[X]) = 1 + max\{ht_{B[X]}(P[X]) : P \in Spec(B)\}$. Now for each $P \in Spec(B), P \cap A = (0)$, and so $XB[X]$ properly contains $XP[X]$ for each $P \in Spec(B)$. Thus $ht_R(XB[X]) \geq 1 + max\{ht_{B[X]}(P[X]) : P \in Spec(B)\} = dim(B[X])$. On the other hand, each prime ideal properly contained in $XB[X]$ corresponds to a prime ideal properly contained in some member of $Spec(B[X])$ by Lemma 2.3. Thus $ht_R(XB[X]) \leq dim(B[X])$.

Now we are in a position to find out the height of $XB[X]$ in R.

LEMMA 2.6. (Cf. Lemma 1.3 of [24].) Let A, B, X, and R be as in Lemma 2.1 and let $S = A^* = A \backslash \{0\}$. Then $ht_R(XB[X]) = 1 + max\{ht_B(p[X]) : p \in Spec(B)$ such that $p \cap A = (0)\} = dim(B_S[X]) \leq dim(B[X])$.

Proof. Indeed, if for every nonzero prime ideal p of B we have $p \cap A \neq (0)$, then $qf(A) = qf(B)$, and so $ht_R(XB[X]) \leq dim(A + XB[X])_S = dim(K[X]) = 1$ because $XB[X] \cap A = 0$. Now as $ht_R(XB[X]) \geq 1$, we have $ht_R(XB[X]) = 1$. This establishes the lemma for the case when $\{p \in Spec(B) \backslash \{(0)\} : p \cap A = 0\} = \phi$. Now suppose that $\{p \in Spec(B) \backslash \{(0)\} : p \cap A = 0\} \neq \phi$. Then, as $XB[X] \cap A = (0)$, we have $ht_R(XB[X]) = ht_{R_{A^*}}(XB_S[X]) = dim(B_S[X])$ because $R_S = A_S + XB_S[X]$ meets the requirements of Lemma 2.5. The inequality is self evident.

Now let us take a chain of prime ideals $C = P_n \supsetneq P_{n-1} \supsetneq \cdots \supsetneq P_r \supsetneq P_{r-1} \supsetneq \cdots \supsetneq P_1 \supsetneq P_0 = (0)$ in $A + XB[X]$, and let us use our trick of spotting X. If X does not belong to any of the P_i, then all the P_i are in $\mathcal{M}$, and so $n \leq m = dim(B[X])$. If X belongs to some, but not all of the P_i, then we reason as follows. If $X \in P_r$, then $P_i = (P_i \cap A) + XB[X]$ for all i such that $r \leq i \leq n$. Now if $P_r = XB[X]$, the largest value that $n - r$ can take corresponds to the longest chain of prime ideals in A. So $n - r \leq dim(A)$. That is, $n \leq dim(A) + r$. But, by Lemma 2.6, $r \leq dim(B[X])$. So we have $dim(R) \leq dimA + dim(B[X]$. Next, according to Lemma 2.6, $ht_R(XB[X]) = dim(B_S[X])$. Combining this information with Observations 0, we have that $dim(R) \geq max\{dim(A) + dim(B_S[X]), dim(B[X])\}$ This completes the proof of the following theorem.

THEOREM 2.7. Let $A \subseteq B$ be an extension of domains, X an indeterminate over B and let $R = A + XB[X]$. Then $max\{dim(A) + dim(B_S[X]), dim(B[X])\} \leq dim(R) \leq dimA + dim(B[X])$.

Now the usual questions. Can these bounds be attained? How do these observations link up with earlier work? First of all, note that if $dim(B_S[X]) = dim(B[X])$, then the inequalities are replaced by equalities, that is $dim(R) = dimA + dim(B[X])$. What are the circumstances under which this can happen? Of course one possibility is when $qf(A) \subseteq B$. That is, if B is a field or B is a $qf(A)$-algebra. It would be interesting to know if there is an example of a domain B, where $qf(A) \nsubseteq B$, and still $dim(R) = dimA + dim(B[X])$. It may be noted however that $qf(A) \subseteq B$ if and only if for each $P \in Spec(B)$, $P \cap A = (0)$. Now for the one-ended (lower) limits. Let A

be a one-dimensional domain such that $dim(A[X]) = 3$. Then $R = A + XA[X]$ gets the lower limit and understandably misses the upper limit. All we need now is an example of $R = A + XB[X]$ such that $max\{dim(A) + dim(B_S[X]), dim(B[X])\} < dim(R) < dimA + dim(B[X])$. Such an example was constructed in [24, Example 3.1]. I will mention the example below and what it does, and let an interested reader look for proofs in [24].

EXAMPLE 2.8. Let K be a field, and let X, X_1, X_2, X_3, X_4 be indeterminates over K. Set $A = K[X_1]_{(X_1)} + X_4K(X_1, X_2, X_3)[X_4]_{(X_4)}$, $B = K(X_1)[X_2]_{(X_2)} + X_3K(X_1, X_2)[X_3]_{(X_3)} + X_4K(X_1, X_2, X_3)[X_4]_{(X_4)}$, $R = A + XB[X]$, and $S = A^*$. Then $max\{dim(A) + dim(B_S[X]), dim(B[X])\} < dim(R) < dimA + dim(B[X])$. In the illustration of this example, it is also shown that in this case $dim(R) > dim(A[X])$. This is important to know because in [8, Theorem 2.6] it was shown that if S is a multiplicative set in A such that $B = A_S$ and $R = A + XA_S[X]$, then $dim(A_S[X]) \leq dim(R) \leq dim(A[X])$. So, this example also serves to show that there was a need for the study of Krull dimension of $A + XB[X]$, and that the general $A + XB[X]$ construction can behave differently from the $A + XA_S[X]$ construction.

I do not know if it has occurred to anyone, but I feel that the inequalities appearing in Theorem 2.7 can be used as a forcing tool, to draw conclusions about $A + XB[X]$ constructions having certain properties. However, the construction being too general, the forcing that I suggest may have very limited scope. For instance, Theorem 2.7 provides the following estimates for $A + XA_S[X] : max\{dim(A) + 1, dim(A_S[X])\} \leq dim(R) \leq dimA + dim(A_S[X])$, and the upper bound may turn out to be somewhat higher than the corresponding inequalities given in [8, Corollary 2.9] which says $max\{dim(A) + 1, dim(A_S[X])\} \leq dim(R) \leq S - dimA + dim(A_S[X])$. Here $S - dim(A)$ represents the maximum length of the chain of prime ideals $P_n \supsetneq P_{n-1} \supsetneq ..... \supsetneq P_1 = P$ such that $P_i \cap S \neq \phi$.

I must record here the fact that a study of the Krull dimension, nearly on the same lines as [24], has been carried out by Cahen [44] and Ayache [45] for the $A+I$ construction. Indeed, the first application of their work is $A + XB[X]$, though it is more useful in the situation when $A + I$ is a subring of $K[X_1, .., X_n]$. (Rings of this kind were studied by Visweswaran [46].) Moreover, the general study of subrings of the form $D + I$ of a domain R has been carried out in [47]. The study of Krull dimension or of chains of prime ideals has also gone on in some other directions in Dobbs and Khalis's joint work [48]. In this paper, they also have a construction of the form $A + XA_S[[X]]$.

PART 3 (CONSTRUCTING INTERESTING EXAMPLES 1)

It is good to have a proof that something exists, but if there is a simple example to support a claim, we would do well to use it. An example at hand may well pave the way to better understanding. After this "philosophical" statement, I should come up with some really interesting constructions. I hope to do just that, but I have to let some excess material from the previous part flow in.

Let us talk a little about the valuative dimension of $R = A+XB[X]$. Recall that if A is an integral domain, then the supremum of $dim(V)$ for all valuation overrings

V of A is called the valuative dimension of A and is denoted by $dim_v(A)$. The notion of the valuative dimension was introduced by Jaffard in [49]. However, Gilmer [1] has given a good basic treatment to this topic, and the following remarks can be traced back to [1, Section 30]. Indeed, for an integral domain A, $dim(A) \leq dim_v(A)$. Now what is so important about the valuative dimension is the result that $dim_v(A[X]) = dim_v(A) + 1$. Following [21], we may call an integral domain A a Jaffard domain if $dim_v(A) = dim(A)$. Pulling out two of the several equivalent conditions of Theorem 30.9 of [1], we have that $dim_v(A) = n$ is the same as $dim(A[X_1, X_2, .., X_n]) = 2n$. So, Noetherian domains and Prüfer domains are Jaffard domains, along with a host of other examples of Jaffard domains mentioned in [21]. Coming back to the business at hand, we have the following result, in connection with the $A + XB[X]$ construction, to report from [24]. Here, let us recall that if $A \subseteq B$ is an extension of domains, then the degree of transcendence of $qf(B)$ over $qf(A)$ is called the degree of transcendence of B over A, denoted by $tr.deg(B/A)$.

THEOREM 3.1. $dim_v(A + XB[X]) = dim_v(A) + tr.deg(B/A) + 1$.

Now if you are interested in the proof, look up [24]. However, I would be more interested in a proof that is based on the observation that every valuation overring of $R = A + XB[X]$ is the ring of a valuation on $qf(R)$ that is an extension of a valuation on $qf(A)$.

Now let us see what Theorem 3.1 has to offer. Indeed, if B is algebraic over A, then $tr.deg(B/A) = 0$. So, if B is algebraic over A, in particular if B is an overring of A, then $dim_v(A + XB[X]) = dim_v(A) + 1$. This obviously takes care of the case when B is a quotient ring of A. Now what is the use?

COROLLARY 3.2. Suppose that A is a Jaffard domain. Then $dim_v(A + XB[X]) = dim(A) + tr.deg(B/A) + 1$.

Thus if $tr.deg(L/K) = \infty$, where L is a field extension of K, then $R = K + XL[X]$ is a one-dimensional domain whose valuative dimension is infinite. There is a wealth of results on valuative dimensions of pullbacks and generalized $D + M$ constructions in [21]. One may wonder about the need to write [24] if pullbacks are so perfect. My response, as usual, is that $A + XB[X]$ constructions, crude though they may look, do provide valuable information which may be hard to glean from pullbacks.

Recall that an integral domain A is an S-domain (S for Seidenberg) if for each height-one prime ideal P of A we have that $P[X]$ is a height-one prime ideal of $A[X]$. Let us also recall from Kaplansky [2], who is responsible for this terminology, that A is a strong S-ring if A/P is an S-domain for each prime ideal P of A. Clearly if A is a strong S-ring, then so are the homomorphic images of A. The terminology, whose motivation can in part be traced back to Seidenberg [50, Theorem 3], seemed to provide a useful tool for recognizing integral domains that behaved like Noetherian domains in that they satisfied $dim(A[X]) = dim(A) + 1$.

THEOREM 3.3. ([2, Theorem 39]). Let A be a strong S-ring, let X be an indeterminate over A, and let P be a prime ideal of A. Then $ht_{A[X]}(P[X]) = ht_A(P)$. Moreover, if Q is a prime ideal of $A[X]$ such that $Q \cap A = P$ and $Q \supsetneq P[X]$, then $ht_{A[X]}(Q) = ht_A(P) + 1$.

Then in a later section, he shows that a valuation ring is a strong S-ring [2, Theorem 68]. Now this is where Malik and Mott [15] picked up the strand and started pulling, of course in a multiplicative sort of way. Their results were of the type:

PROPOSITION 3.4. ([15, 2.1 and 2.2]) A domain A is an S-domain if and only if A_T is an S-domain for each multiplicative set T, if and only if A_M is an S-domain for each maximal ideal M.

On the strong S-property, they proved, likewise, the following statement.

PROPOSITION 3.5. ([15, 2.3,2.4]). A ring A is a strong S-ring if and only if A_T is a strong S-ring for each multiplicative set T, if and only if A_M is a strong S-ring for each maximal ideal M.

Now, coupling Proposition 3.5 with Kaplansky's Theorem 68 mentioned above, and adding some more work they stated the following result.

PROPOSITION 3.6 ([15, 2.5]). A Prüfer domain is a strong S-domain.

Using their criteria, they came up with the following scheme.

PROPOSITION 3.7. ([15, 3.1, 3.2]) For A an S-domain (a strong S-domain), $A[X]$ is an S-domain (resp., a strong S-domain) if and only if $A_P[X]$ is an S-domain (resp., a strong S-domain) for each prime ideal P of R. (Here X denotes a finite set of indeterminates.)

There were several other interesting statements in section 3 of [15]. All this culminated in a beautiful result and that can be stated as follows.

THEOREM 3.8. ([15,3.5]) Let A be a Prüfer domain and let $X = \{X_1, X_2, .., X_n\}$ be a finite set of indeterminates over A. Then $A[X]$ is a strong S-domain.

Next, using Kaplansky's Theorem 39 (Theorem 3.3 here) it is easy to observe that a strong S-domain is a Jaffard domain. This observation was made in [15], along with an example of a Jaffard domain that is not a strong S-ring [15, 3.11]. It appears that no one has tried to find a minimal set of conditions under which a Jaffard domain should be a strong S-domain. The paper ([15]) goes on to display other goodies, but I must leave the rest for the interested readers and hasten to answer the question that has by now started popping up in every reader's mind, "Where is the $A + XB[X]$?" Let me take you to Salah Kabbaj's earlier work. He picked up where [15] had left off. I found in an earlier version of his thesis the following statement which stayed as it was in the final version ([51, Théorème 0.8, Chap. II]). Let A be an S-domain. Then $A[X_1, X_2, .., X_n]$ is an S-domain for all $n \geq 1$. Looked like a pretty result, it was a considerable improvement on [15, 3.1] (which is a part of Proposition 3.7 here), so I started playing with it. The first thing that came to my mind was, "Where is he using the fact that A is an S-domain?" The answer came out, "Only at one place, and that could be avoided." I made the suggestion, in my report on his thesis, indicating how his proof can be modified to prove the following statement.

THEOREM 3.9. If A is an integral domain and X an indeterminate over A, then $A[X]$ is an S-domain.

Now, for some reason, this suggestion was not taken and I was hopping mad, (I so wanted Kabbaj to have this result!). I was working on [5] and I mentioned the result to Dan. He agreed to include the result, but at a price, as usual, he would write his own proof. He did give a pretty proof though. Possibly simultaneously, Kabbaj and Fontana [23] did prove Theorem 3.9 and many more interesting results. Now my trouble is that I like both very much. For this reason, I have decided to give a proof that has the flavor of both.

LEMMA 3.10. ([5]) For an integral domain A the following statements are equivalent.

(1) A is an S-domain.

(2) For each height-one prime ideal P of A, A_P is an S-domain.

(3) For each height-one prime ideal P of A, A_P (the integral closure of A_P) is a Prüfer domain.

Proof. (1)$\Rightarrow$ (2) by Proposition 3.4 above. (2)$\Rightarrow$ (3) By (2), $A_P[X]$ is two-dimensional and by [1, 30.14], A_P is Prüfer. (3)$\Rightarrow$ (1). Suppose that for each height-one prime ideal P, A_P is one-dimensional Prüfer. Then $A_P[X] = A_P[X]$ is two-dimensional, which requires that $A_P[X]$ is two dimensional, which means that $PA_P[X] = P(A[X])_{(A\setminus P)}$ is of height one. This indeed means that $PA[X] = P[X]$ is of height one. Now recall that P is any height-one prime ideal of A.

This lemma provides a neat characterization of S-domains. Now before we start proving the theorem, let me digress a little. Call A a stably strong S-domain if $A[X_1, .., X_n]$ is a strong S-domain for all $n \geq 1$. (A being a homomorphic image of $A[X_1, .., X_n]$, stably strong S implies strong S.) So, a Prüfer domain, by Theorem 3.8 above, is a (stably) strong S-domain. Of the various equivalences on one-dimensional domains in [21, Theorem 1.10], we recall that for a one-dimensional domain A the properties: (a) S-domain, (b) strong S-domain, and (c) stably strong S-domain, are all equivalent.

Proof of Theorem 3.9. By Lemma 3.10, all we need is to show that $A[X]_P$ is an S-domain for each height-one prime P of $A[X]$. There are two possibilities: (i) $P \cap A = (0)$, (ii) $P \cap A = p \neq (0)$. In the first case, it is well known that $(A[X])_P$ is a valuation domain and hence an S-domain. In the second case, $P = p[X]$, where p is of height one. Now $PA[X]_P = pA[X]_{p[X]}$ is of height one. But $A[X]_{p[X]} = A_p(X) = (A_p[X])_{pA_p[X]}$. So $pA_p[X]$ is of height one. Whence the one dimensional A_p is an S-domain. But then, by the remarks prior to the proof, A_p is a stably strong S-domain. This means that $A_p[X]$ is an S-domain. But then so is every quotient ring of $A_p[X]$ by Proposition 3.4 above. Whence $A_p(X) = A[X]_{p[X]} = A[X]_P$ is an S-domain, and this completes the proof.

An immediate corollary is the following statement.

COROLLARY 3.11. ([5, 3.3]) $D + XD_S[X]$ is an S-domain for every multiplicative set S of D.

As already mentioned in part 1, in [23] the authors study a construction defined as

$$D^{(S,r)} = D + (X_1, .., X_r)D_S[X_1, .., X_r]$$

and show that $D^{(S,r)}$ is an S-domain. Now the question is, what form Theorem 3.9 will take for $R = A + XB[X]$? The answer comes from [42]. Yet before we quote from [42], let us see what we can do with what we have established so far. The following statement can be regarded as a corollary to Lemma 3.10 and Theorem 3.9.

PROPOSITION 3.12. Let $A \subseteq B$ be an extension of domains such that B is a subring of the quotient field of A. Then $R = A + XB[X]$ is an S-domain.

Proof. We show that R_P is an S-domain for each height-one prime P of R. Note that $ht_R(XB[X]) = 1$ and that $R_{XB[X]} = (A + XB[X])_{XB[X]} = (K[X])_{XK[X]}$ is a valuation domain, and hence is an S-domain. For height-one primes $P \neq XB[X]$, let $S = \{X^n : n \geq 1\}$ and note that $R_P = (R_S)_{P_S} = B[X, X^{-1}]_{P_S}$. But then R_P is a quotient ring of $B[X]$ which is an S-domain, and we know that every quotient ring of an S-domain is again an S-domain (Proposition 3.4).

Now comes the promised result.

PROPOSITION 3.13. ([42, Theorem 1.1]) $R = A + XB[X]$ is an S-domain if and only if $ht_R(XB[X]) > 1$ or B is algebraic over A.

What is the use of this? For one thing, it takes care of Corollary 3.11 and Proposition 3.12. Moreover, you can construct composite examples of S-domains and non S-domains with interesting properties. For instance, using Proposition 3.13, if $\mathcal{Z}$, $\mathcal{Q}$ and $\mathcal{R}$ represent the integers, rationals, and reals, respectively then $\mathcal{Z}+X\mathcal{R}[X]$ is not an S-domain, while $\mathcal{Z} + X\mathcal{Q}(i)[X]$ is an S-domain. Of course, these are just two very simple examples.

In [42], the reader can find a wealth of examples on strong S-domains. For this survey, we select a number of results that could intrigue the reader enough to want to prove some direct results in this direction. Let us recall first that the extension of domains $A \subseteq B$ is said to be incomparable (INC) if two distinct primes P, Q of B contract to the same prime p of A, then P and Q are incomparable. ([2, Section1-6] is a good source for a study of INC and related notions.) For example, $A \subseteq A[X]$, where X is an indeterminate, does not have INC because $(0) \subseteq (X)$ both contract to (0) in A. So if $A \subseteq B$ is incomparable, then every nonzero prime of B contracts to a nonzero prime of A. Next, $A \subseteq B$ is said to be residually algebraic if for each prime P of B, we have B/P algebraic over $A/(P \cap A)$. For this notion the reader may look up [52].

THEOREM 3.14. ([42, Théorème 1.7]) Let $A \subseteq B$ be an extension of domains. Then the following are equivalent:

(1) $R = A + XB[X]$ is a strong S-domain and $A \subseteq B$ is an incomparable extension.

(2) A and $B[X]$ are strong S-domains and $A \subseteq B$ is a residually algebraic extension.

As a direct consequence of the above equivalence, we have the following statement.

COROLLARY 3.15. Let D be an integral domain, S a multiplicative set of D and let K be a field containing D as a subring. Then the following hold.

(1) $D + XK[X]$ is a strong S-domain if and only if D is strong S-domain and K is algebraic over the quotient field of D.

(2) $D + XD_S[X]$ is a strong S-domain if and only if D and $D_S[X]$ are strong S-domains.

Now this little corollary gives us a host of examples of strong S-domains, including ones that have terrible and totally unaccommodating properties elsewhere and ones that serve as examples of beautiful new notions. I would mention only two here.

EXAMPLE 3.16. Let D be a Prüfer domain and let S be a multiplicative set of D. Then $D^{(S)} = D + XD_S[X]$ is a strong S-domain.

In fact, $D^{(S)}$ is a stably strong S-domain, i.e., for any set $\{Y_1, Y_2, .., Y_m\}$ of indeterminates $D^{(S)}[Y_1, Y_2, .., Y_m]$ is a strong S-domain. The main reason is that, by Theorem 3.8, $D[X]$ and $D_S[X]$ are both strong S-domains for any set X of indeterminates and $D_S[X] = (D[X])_S$.

Now $D^{(S)}$ is Prüfer if and only if $S = D^*$ [8], so Example 3.16 affords an example of a non Prüfer domain that is a strong S-domain. For reasons of organization, I will not go too deep into this example and will refer to it later.

EXAMPLE 3.17. Let $K \subseteq L$ be an extension of fields. Then $R = K + XL[X]$ is one-dimensional and according to [21], R is an S-domain $\Leftrightarrow$ R is a strong S-domain $\Leftrightarrow$ R the integral closure is a Prüfer domain $\Leftrightarrow$ R is a stably strong S-domain. Now R is Prüfer if and only if L is algebraic over K. Next, let K be of characteristic $p \neq 0$ and let L be a purely inseparable extension of K such that $L^p \subseteq K$. (See [2, Theorem 100] for an example with $p = 2$.) Then, apart from being a stably strong S-domain, $R = K + XL[X]$ has the added property that for each pair $f, g \in R$ we have $f^p, g^p \in K[X]$, which is a PID, and so $(f^p, g^p)K[X] = hK[X]$. Now as $K[X]R = R$, we have $(f^p, g^p)R = hR$. From this it also follows that $f^pR \cap g^pR$ is principal.

This example is Example 2.13 of [53]. Let me use this example to introduce an interesting set of concepts.

An integral domain D is called an almost GCD domain (AGCD domain) if for each pair $x, y \in D$ there is a natural number $n = n(x, y)$ such that $x^nD \cap y^nD$ is principal. (Indeed, if for all $x, y \in D$ we have $n(x, y) = 1$, we get a GCD domain.) Apart from the above example, there are other well known examples, such as almost factorial domains of Storch. These are Krull domains with torsion divisor class group (see, for instance, Fossum [54]). (The reader can look up [54] to check that an integral domain D is Krull if for each height-one prime P, D_P is a discrete rank one valuation domain, and D is a locally finite intersection of localizations at height-one primes.) For other examples of AGCD domains, that use the $A + XB[X]$ construction, the reader may consult [55] when it appears. Next, D is called an almost Bézout domain if for each pair $x, y \in D$ there is a natural number $n = n(x, y)$ such that $x^nD + y^nD$ is principal. Example 3.17 above may serve as an example again. For more examples of almost Bézout domains, you may consult [56]. Now, the integral closure of an almost Bézout domain is a Prüfer domain with torsion ideal class group [56, Corollary 4.8]. Indeed, as we already know from [21] that a one-dimensional almost Bézout domain is stably strong S-domain, but the general case of almost Bézout domains is not quite clear.

Now, coming back to our task at hand, a ring A is called a Hilbert ring if every prime ideal of A is expressible as an intersection of maximal ideals containing it. So, a one-dimensional domain A is a Hilbert ring if and only if there is a set $\{M_\alpha\}$ of maximal ideals of A such that $(0) = \cap M_\alpha$. Now, because everything in a Hilbert ring seems to be in terms of maximal ideals, it is fair to ask if Cohen's criterion for a Noetherian ring (R is Noetherian if and only if every prime ideal of R is finitely generated) can be relaxed for Hilbert rings to: A Hilbert ring A is Noetherian if and only if every maximal ideal of A is finitely generated? A.V. Geramita asked Robert Gilmer and/or William Heinzer this question and they came up with a non-Noetherian example of a Hilbert domain whose maximal ideals are all finitely generated [18]. Later, Joe Mott and I [17] came up with the following theorem.

THEOREM 3.18. Let D be a Hilbert domain and let L be a field containing D. Then $D + XL[X]$ is a Hilbert domain.

The proof is straight-forward and short, and if D is a Hilbert PID with quotient field L then $D + XL[X]$ is a two dimensional Bézout domain, with each maximal ideal principal. This example is uncannily similar to the one constructed in [18]. Obviously, taking any Noetherian Hilbert domain for D in Theorem 3.18, you can construct a Hilbert domain whose maximal ideals are finitely generated and which is not Noetherian. Theorem 3.18 also pre-empts the obvious question about a Hilbert domain being an S-domain. The answer of course is, "Not in general". (Strangely, [18] is still on my recommended reading list because of its useful auxiliary results.) In [5], we proved something in a slightly different direction. The result can be stated as follows.

THEOREM 3.19. Let D be an integral domain and S a multiplicative set of D such that each prime P of D that intersects S, intersects S in detail, that is for each nonzero prime $Q \subseteq P$, $Q \cap S \neq \phi$. Then $R = D + XD_S[X]$ is a Hilbert domain if and only if both D and D_S are Hilbert domains.

It is remarkable that unruly Hilbert domains of [17] did not seem to have as much effect as Theorem 3.19 had. There was a renewed interest in Hilbert domains and out came a paper by Anderson, Dobbs, and Fontana [57] on Hilbert rings arising as pullbacks. In this paper, they discuss as applications the events of $D + (X_1, .., X_n)D_S[X_1, ..X_n]$, $A + XB[X]$, and $D + M$ constructions being Hilbert. In particular, it was shown in [57] that, $A + XB[X]$ is a Hilbert domain if and only if A and B are Hilbert domains. This includes Theorems 3.18 and 3.19 above. It appears that someone else was interested in showing when $A + XB[X]$ is a Hilbert domain. On reading this survey Lahoucine Izelgue sent me some of his old work on this topic. I have decided to include it here because it is simple and it is efficient.

Constructions Cachées en Algèbre Abstraite: Le Principe Local-Global

Henri Lombardi ([1]), Claude Quitté ([2])

Résumé

Nous appliquons une forme constructive de principes local-global en algèbre commutative pour décrypter, cachées dans des théorèmes d'algèbre abstraite, des constructions de matrices inversibles dans des anneaux de polynomes. Ceci nous donne une nouvelle preuve constructive de la conjecture de Serre (théorème de Quillen-Suslin) et une preuve constructive du théorème de stabilité de Suslin.

Abstract

Wa apply a constructive form of local-global principles in commutative algebra in order to decipher some constructions of invertible polynomial matrices hidden in theorems of abstract algebra. This leads us to a new constructive proof of Serre's conjecture (Quillen-Suslin theorem). We get also a constructive proof of Suslin's stability theorem.

---

[1] Equipe de Mathématiques, UMR CNRS 6623, UFR des Sciences et Techniques, Université de Franche-Comté, 25030 BESANCON cedex, FRANCE, email: lombardi@math.univ-fcomte.fr

[2] Laboratoire de Mathématiques, SP2MI, Boulevard 3, Teleport 2, BP 179, 86960 FUTUROSCOPE Cedex, FRANCE, email: quitte@mathlabo.univ-poitiers.fr

Introduction

Nous nous situons dans la philosophie développée dans les articles [4, 5, 17, 18, 19, 20, 21, 22, 23]. Il s'agit de débusquer un contenu constructif caché dans des preuves abstraites de théorèmes concrets.

La méthode générale consiste à remplacer certains objets abstraits idéaux qui n'existent qu'en vertu du principe du tiers exclu et de l'axiome du choix, par des spécifications incomplètes de ces mêmes objets.

Dans cet article nous nous attaquons à la méthode abstraite qui utilise des principes du type local-global. Un résultat est démontré vrai après localisation en n'importe quel idéal premier. On déduit ensuite qu'il est vrai globalement par un argument adéquat.

Notre but n'est en aucun cas de donner des algorithmes performants, mais de montrer qu'il n'y a pas de miracle en mathématiques : si une preuve abstraite donne un résultat concret, le calcul concret du résultat doit d'une manière ou d'une autre être caché dans la preuve abstraite (sauf à croire en la réalité de l'Univers Cantorien censé officiellement justifier la preuve abstraite).

Ou encore pour le dire autrement. Nos preuves explicites ont ceci de particulier qu'elles sont obtenues par un simple décryptage des arguments contenus dans une preuve abstraite. Par contre ces algorithmes sont a priori peu efficaces. Ils n'ont pas pour but de reposer la machine, mais de reposer le concepteur d'algorithmes. Et surtout d'annoncer une bonne nouvelle : les méthodes abstraites en algèbre sont, en fait, constructives.

Nous pensons engager ainsi un début de réalisation du programme de Hilbert pour ce qui concerne les méthodes de l'algèbre abstraite.

Dans son esprit, notre méthode est à rapprocher de celle de Kreisel lorsqu'il "déroule" (unwind) des preuves classiques pour en faire des preuves constructives "sans introduire de nouvelles idées" (cf. la description du programme de Kreisel par Feferman dans [7]). Mais nous utilisons des moyens purement algébriques, relativement élémentaires, tandis que Kreisel met en oeuvre une artillerie métamathématique assez impressionnante (cf. [6]).

Dans la section 1 nous expliquons la machinerie de relecture constructive grâce à laquelle nous remplaçons "la localisation en tous les idéaux premiers" par des localisations en des monoïdes convenables, décrits explicitement en termes finis à partir d'une lecture attentive de la preuve abstraite. En pratique les idéaux premiers "purement idéaux" qui interviennent dans la preuve abstraite sont remplacés par certaines spécifications incomplètes d'idéaux premiers, qui suffisent à faire fonctionner la preuve, et qui la font fonctionner de manière constructive. Notre procédé de relecture automatique transforme la preuve du théorème local en celle d'un théorème que nous appelons quasi-global.

Quant à la preuve que la version quasi-globale implique la version globale, elle est en général déjà dans la littérature classique, sous la forme d'un *lemme de propagation* (pas toujours énoncé sous forme d'un lemme séparé), qui est au coeur de la preuve du principe local-global concret abstrait correspondant. Nous préférons quant à nous énoncer le lemme de propagation sous forme d'un principe local-global concret, car cette terminologie nous paraît plus parlante.

Dans les sections 2 et 3 nous donnons deux exemples de théorèmes célèbres pour lesquels nous appliquons cette méthode. Le théorème de Quillen-Suslin (conjecture de Serre) et le théorème de stabilité de Suslin. Dans les deux cas nous nous limitons au cas des corps (il y a des versions plus générales que nous ne traitons pas ici).

Pour ces théorèmes (dans le cas des corps) d'autres preuves constructives basées sur des idées différentes sont déjà connues.

Tous les anneaux considérés sont commutatifs unitaires. Soit $\mathbf{A}$ un tel anneau. Un vecteur $f = {}^t(f_1, \ldots, f_n)$ de $\mathbf{A}^{n\times 1}$ est dit *unimodulaire* lorsque l'idéal $\mathcal{I}(f_1, \ldots, f_n)$ contient 1. On dit encore que les éléments $f_1, \ldots, f_n$ de $\mathbf{A}$ sont *comaximaux*. Nous notons $\mathrm{Rad}(\mathbf{A})$ le radical (de Jacobson) de $\mathbf{A}$, c.-à-d. l'ensemble des x tels que $1 + x\mathbf{A} \subset \mathbf{A}^\times$ (le groupe des unités de $\mathbf{A}$). Nous notons $\mathbf{M}_n(\mathbf{A})$ l'anneau des matrices carrées d'ordre n à coefficients dans $\mathbf{A}$, $\mathbb{SL}_n(\mathbf{A})$ le groupe des matrices de déterminant 1, $\mathbb{E}_n(\mathbf{A})$ le sous-groupe du précédent engendré par les matrices élémentaires.

La section 2 décrypte une preuve “à la Quillen” du théorème de Quillen-Suslin.

Rappelons le théorème suivant dû à Horrocks (cf. [10]).

Théorème de Horrocks local
Soit un entier $n \geq 3$, $\mathbf{A}$ un anneau local et $f(X) = {}^t(f_1(X), \ldots, f_n(X))$ un vecteur unimodulaire dans $\mathbf{A}[X]^{n\times 1}$, avec f_1 unitaire, alors il existe une matrice $H(X) \in \mathbb{E}_n(\mathbf{A}[X])$ telle que $H(X)f(X) = {}^t(1, 0, \ldots, 0)$.

Ce théorème possède une preuve constructive lorsque l'anneau local vérifie explicitement l'axiome suivant :

$$\text{“ } \forall x \in \mathbf{A} \quad x \in \mathbf{A}^\times \ \vee \ x \in \mathrm{Rad}(\mathbf{A}) \text{ ”},$$

(en mathématiques constructives, cet axiome signifie que l'anneau est local et que son corps résiduel est discret, cf. [24]).

Nous rappelons cette preuve dans la section 2.1 (nous l'avons extraite d'une preuve un peu moins explicite dans [14]). Nous nous intéressons ensuite à une version située à mi-chemin entre la version locale et la version globale.

Théorème de Horrocks quasi-global
Soit un entier $n \geq 3$, $\mathbf{A}$ un anneau et $f(X) = {}^t(f_1(X), \ldots, f_n(X))$ un vecteur unimodulaire dans $\mathbf{A}[X]^{n\times 1}$, avec f_1 unitaire, alors il existe des éléments comaximaux $a_1, \ldots, a_\ell \in \mathbf{A}$ et pour chaque $i = 1, \ldots, \ell$ une matrice $H_i(X) \in \mathbb{E}_n(\mathbf{A}[1/a_i][X])$ telle que $H_i(X)f(X) = {}^t(1, 0, \ldots, 0)$.

Nous montrons dans la section 2.1 que la preuve constructive du théorème quasi-global est cachée dans la preuve (constructive) du théorème local. Nous appliquons pour ce faire la machinerie décrite à la section 1.

Dans la section 2.2, nous établissons un principe local-global concret qui est la version constructive d'un principe local-global abstrait de Quillen.

Dans la section 2.3, nous déduisons des résultats précédents la version globale du théorème de Horrocks puis la conjecture de Serre.

Théorème de Horrocks global
Soit un entier $n \geq 3$, $\mathbf{A}$ un anneau intègre et $f(X) = {}^t(f_1(X), \ldots, f_n(X))$ un vecteur unimodulaire dans $\mathbf{A}[X]^{n\times 1}$, avec f_1 unitaire, alors il existe une matrice $H \in \mathbb{SL}_n(\mathbf{A}[X])$ telle que $H(X)f(X) = f(0)$.

La conjecture de Serre dont nous donnons ici une nouvelle preuve constructive, a été résolue indépendamment par D. Quillen et A. Suslin en 1976 [26, 27]. L'exposé classique de leurs travaux est le livre de Lam [13]. On peut également citer le livre de Kunz [12] et celui de Gupta et Murthy [9]. D'autres solutions constructives, parfois relativement efficaces, ont été proposées notamment dans [1, 2, 3, 8, 15, 16, 25]. Aucune cependant ne découle comme la nôtre du décryptage automatique d'une preuve abstraite.

Dans la section 3 nous examinons la preuve du théorème de stabilité de Suslin dans le cas des corps, telle qu'elle est donnée dans [9] en s'appuyant sur une méthode locale-globale. Pareillement, nous la décryptons en une preuve constructive selon la méthode exposée dans la section 1. Le seul véritable argument non constructif dans [9] est l'utilisation du lemme 3.6 page 46. Ce lemme est de nature locale mais il est ensuite utilisé dans un argument de type local-global. C'est le lemme suivant, dans lequel $\binom{f}{g}$ désigne le symbole de Mennicke.

Lemme 20 (local) *Soit* **A** *un anneau local et* $f, g \in \mathbf{A}[X]$ *avec* f *unitaire et* $af + bg = 1$. *Alors, on a*

$$\binom{f}{g} = \binom{f(0)}{g(0)} = 1.$$

Autrement dit la matrice

$$\begin{bmatrix} f & g & 0 \\ -b & a & 0 \\ 0 & 0 & 1 \end{bmatrix}$$

est dans $\mathbb{E}_3(\mathbf{A}[X])$.

Notre machinerie de relecture automatique de la preuve locale donne le lemme quasi-global suivant :

Lemme 21 (quasi-global) *Soit* **A** *un anneau et* $f, g \in \mathbf{A}[X]$ *avec* f *unitaire et* $af + bg = 1$. *Alors, il existe des éléments comaximaux* s_i *tels que dans chaque localisé* $\mathbf{A}[1/s_i]$ *on ait l'égalité des symboles de Mennicke suivante*

$$\binom{f}{g} = \binom{f(0)}{g(0)} = 1.$$

Et cette version quasi-globale permet de remplacer l'utilisation abstraite du lemme local par une construction explicite pour aboutir à la version constructive du théorème global suivant, qui est la clef du théorème de stabilité de Suslin.

Théorème 24 (version globale du lemme 20)
Soit **A** *un anneau et* $f, g \in \mathbf{A}[X]$ *avec* f *unitaire et* $af + bg = 1$. *Alors on a l'égalité des symboles de Mennicke suivante*

$$\binom{f}{g} = \binom{f(0)}{g(0)}.$$

1 Le principe de la méthode

Nous donnons ici quelques explications sur le fonctionnement du décryptement constructif des preuves classiques utilisant un principe local-global en algèbre commutative.

1.1 Du local au quasi-global

L'argument de localisation classique fonctionne comme suit. Lorsque l'anneau est local une certaine propriété P est vérifiée en vertu d'une preuve assez concrète. Lorsque l'anneau n'est pas local, la même propriété est encore vraie (d'un point de vue classique non constructif) car il suffit de la vérifier localement.

Nous examinons avec un peu d'attention la première preuve. Nous voyons alors apparaître certains calculs qui sont faisables en vertu du principe suivant :

$$\forall x \in \mathbf{A} \quad x \in \mathbf{A}^{\times} \ \vee \ x \in \mathrm{Rad}(\mathbf{A}),$$

Principe qui est appliqué à des éléments x provenant de la preuve elle-même. Dans le cas d'un anneau non nécessairement local, nous répétons la même preuve, en remplaçant chaque disjonction "x est une unité ou x est dans le radical", par la considération des deux anneaux $\mathbf{B}_x$ et $\mathbf{B}_{1+x\mathbf{B}}$, où $\mathbf{B}$ est la localisation "courante" de l'anneau $\mathbf{A}$ de départ, à l'endroit de la preuve où on se trouve. Lorsque la preuve initiale est ainsi déployée, on a construit à la fin un certain nombre, fini parce que la preuve est finie, de localisés $\mathbf{A}_{S_i}$, pour lesquels la propriété est vraie. En outre les ouverts de Zariski $\mathbf{U}_{S_i}$ correspondants recouvrent Spec($\mathbf{A}$) et cela implique que la propriété P est vraie avec $\mathbf{A}$, cette fois-ci de manière entièrement explicite.

Notons que cette méthode consiste pour l'essentiel à mettre à plat les calculs qui sont impliqués par la mise en oeuvre de la méthode de l'évaluation dynamique donnée dans [17].

Dans la suite, lorsqu'on parle d'un monoïde dans un anneau, on entend toujours une partie contenant 1 et stable pour la multiplication. Un monoïde S d'un anneau $\mathbf{A}$ est dit *saturé* lorsqu'on a l'implication

$$\forall s, t \in \mathbf{A} \ \ (st \in S \ \Rightarrow \ s \in S).$$

On note $\mathbf{A}_S$ le localisé $S^{-1}\mathbf{A}$ de $\mathbf{A}$ en S. Si S est engendré par $s \in \mathbf{A}$, on note $\mathbf{A}_s$ ou $\mathbf{A}[1/s]$ le localisé, qui est isomorphe à $\mathbf{A}[T]/(sT-1)$. Si on sature un monoïde, on ne change pas la localisation. Deux monoïdes sont dits *équivalents* s'ils ont même saturé.

Définition 1

(1) *Des monoïdes $S_1, \ldots, S_n$ de l'anneau* $\mathbf{A}$ *sont dits* comaximaux *si un idéal de* $\mathbf{A}$ *qui coupe chacun des S_i contient toujours* 1, *autrement dit si on a :*

$$\forall s_1 \in S_1 \ \cdots \ \forall s_n \in S_n \ \exists a_1, \ldots, a_n \in \mathbf{A} \quad \sum_{i=1}^{n} a_i s_i = 1.$$

(2) *On dit que* les monoïdes $S_1, \ldots, S_n$ de l'anneau $\mathbf{A}$ recouvrent le monoïde S *si S est contenu dans les S_i et si un idéal de* $\mathbf{A}$ *qui coupe chacun des S_i coupe toujours S, autrement dit si on a :*

$$\forall s_1 \in S_1 \ \cdots \ \forall s_n \in S_n \ \exists a_1, \ldots, a_n \in \mathbf{A} \quad \sum_{i=1}^{n} a_i s_i \in S.$$

En algèbre classique (avec l'axiome de l'idéal premier) cela revient à dire dans le premier cas que les ouverts de Zariski $\mathbf{U}_{S_i}$ recouvrent Spec($\mathbf{A}$) et dans le deuxième cas que les ouverts de Zariski $\mathbf{U}_{S_i}$ recouvrent l'ouvert $\mathbf{U}_S$. Du point de vue constructif, Spec($\mathbf{A}$) est un espace topologique connu via ses ouverts $\mathbf{U}_S$ mais dont les points sont souvent difficilement accessibles.

Un recouvrement de recouvrements est un recouvrement (calculs immédiats) :

Lemme 2 (associativité et transitivité des recouvrements)

(1) *(associativité) Si les monoïdes $S_1, \ldots, S_n$ de l'anneau* $\mathbf{A}$ *recouvrent le monoïde S et si chaque S_ℓ est recouvert par des monoïdes $S_{\ell,1}, \ldots, S_{\ell,m_\ell}$, alors les $S_{\ell,j}$ recouvrent S.*

(2) *(transitivité) Soit S un monoïde de l'anneau* $\mathbf{A}$ *et $S_1, \ldots, S_n$ des monoïdes comaximaux de l'anneau* $\mathbf{A}_S$. *Pour $\ell = 1, \ldots, n$ soit V_ℓ le monoïde de* $\mathbf{A}$ *formé par les numérateurs des éléments de S_ℓ. Alors les monoïdes $V_1, \ldots, V_n$ recouvrent S*

Définition et notation 3 *Soient I et U deux parties de $\mathbf{A}$. Nous noterons $\mathcal{M}(U)$ le monoïde engendré par U, $\mathcal{I}_{\mathbf{A}}(I)$ ou $\mathcal{I}(I)$ l'idéal engendré par I et $\mathcal{S}(I;U)$ le monoïde $\mathcal{M}(U)+\mathcal{I}(I)$. Si $I=\{a_1,\dots,a_k\}$ et $U=\{u_1,\dots,u_\ell\}$, on note respectivement $\mathcal{M}(U)$, $\mathcal{I}(I)$ et $\mathcal{S}(I;U)$ par $\mathcal{M}(u_1,\dots,u_\ell)$, $\mathcal{I}(a_1,\dots,a_k)$ et $\mathcal{S}(a_1,\dots,a_k;u_1,\dots,u_\ell)$.*

Il est clair que si u est égal au produit $u_1\cdots u_\ell$, les monoïdes $\mathcal{S}(a_1,\dots,a_k;u_1,\dots,u_\ell)$ et $\mathcal{S}(a_1,\dots,a_k;u)$ sont équivalents.

Notez que lorsqu'on localise en $S=\mathcal{S}(I;U)$, les éléments de U deviennent inversibles et ceux de I se retrouvent dans le radical de $\mathbf{A}_S$.

Notre sentiment est que la "bonne catégorie" serait celle dont les objets sont les couples $(\mathbf{A},I)$ où $\mathbf{A}$ est un anneau commutatif et I un idéal contenu dans le radical de $\mathbf{A}$, et les flèches de $(\mathbf{A},I)$ vers $(\mathbf{A}',I')$ sont les homomorphismes $f:\mathbf{A}\to\mathbf{A}'$ tels que $f(I)\subset I'$. On retrouve les anneaux usuels en prenant $I=0$ et les anneaux locaux (avec la notion de morphisme local) en prenant I égal à l'idéal maximal. Pour "localiser" un objet $(\mathbf{A},I)$ dans cette catégorie, on utilise un monoïde U et un idéal J de manière à former le nouvel objet $(\mathbf{A}_{\mathcal{S}(J_1;U)}, J_1\mathbf{A}_{\mathcal{S}(J_1;U)})$, où $J_1=I+J$.

Le lemme fondamental suivant récupère la mise constructivement lorsqu'on relit avec un anneau arbitraire une preuve donnée dans le cas d'un anneau local.

Lemme 4 *Soit U et I des parties de l'anneau $\mathbf{A}$ et $a\in\mathbf{A}$, alors les monoïdes $\mathcal{S}(I;U,a)$ et $\mathcal{S}(I,a;U)$ recouvrent le monoïde $\mathcal{S}(I;U)$.*

Preuve Pour $x\in\mathcal{S}(I;U,a)$ et $y\in\mathcal{S}(I,a;U)$ on doit trouver une combinaison linéaire $x_1x+y_1y\in\mathcal{S}(I;U)$ $(x_1,y_1\in\mathbf{A})$. On écrit $x=u_1a^k+j_1$, $y=(u_2+j_2)-(az)$ avec $u_1,u_2\in\mathcal{M}(U)$, $j_1,j_2\in\mathcal{I}(I)$, $z\in\mathbf{A}$. L'identité classique $c^k-d^k=(c-d)\times\cdots$ donne un $y_2\in\mathbf{A}$ tel que $y_2y=(u_2+j_2)^k-(az)^k=(u_2^k+j_3)-(az)^k$ et on écrit $z^kx+u_1y_2y=u_1u_2^k+u_1j_3+j_1z^k=u_4+j_4$. □

On en déduit le principe général de décryptage suivant, qui *permet d'obtenir automatiquement une version quasi-globale d'un théorème à partir de sa version locale.*

Principe général 5 *Lorsqu'on relit une preuve explicite, donnée pour le cas où l'anneau $\mathbf{A}$ est local, avec un anneau $\mathbf{A}$ arbitraire, que l'on considère au départ comme $\mathbf{A}=\mathbf{A}_{\mathcal{S}(0;1)}$ et qu'à chaque disjonction (pour un élément a qui se présente au cours du calcul dans le cas local)*

$$a\in\mathbf{A}^\times \ \vee\ a\in\mathrm{Rad}(\mathbf{A}),$$

on remplace l'anneau "en cours" $\mathbf{A}_{\mathcal{S}(I;U)}$ par les deux anneaux $\mathbf{A}_{\mathcal{S}(I;U,a)}$ et $\mathbf{A}_{\mathcal{S}(I,a;U)}$ (dans chacun desquels le calcul peut se poursuivre), on obtient à la fin de la relecture, une famille finie d'anneaux $\mathbf{A}_{\mathcal{S}(I_j;U_j)}$ avec les monoïdes $\mathcal{S}(I_j;U_j)$ comaximaux et I_j, U_j finis.

On notera que si $b=a/(u+i)$ avec $u\in\mathcal{M}(U)$ et $i\in\mathcal{I}(I)$ et si la disjonction porte sur "$b\in\mathbf{A}^\times\ \vee\ b\in\mathrm{Rad}(\mathbf{A})$", alors il faut considérer les localisés $\mathbf{A}_{\mathcal{S}(I;U,a)}$ et $\mathbf{A}_{\mathcal{S}(I,a;U)}$.

Les exemples suivants sont fréquents et résultent immédiatement des lemmes 2 et 4, sauf le premier qui se fait par un petit calcul simple.

Exemples 6 Soit $\mathbf{A}$ un anneau, U et I des parties de $\mathbf{A}$, $S=\mathcal{S}(I;U)$.

(1) Soient $s_1,\dots,s_n\in\mathbf{A}$ des éléments comaximaux (c'est-à-dire tels que $\mathcal{I}(s_1,\dots,s_n)=\mathbf{A}$). Alors les monoïdes $S_i=\mathcal{M}(s_i)$ sont comaximaux.
Plus généralement, si $t_1,\dots,t_n\in\mathbf{A}$ sont des éléments comaximaux dans $\mathbf{A}_S$, les monoïdes $\mathcal{S}(I;U,t_i)$ recouvrent le monoïde S.

(2) Soient $s_1, \ldots, s_n \in \mathbf{A}$. Les monoïdes $S_1 = \mathcal{S}(0; s_1)$, $S_2 = \mathcal{S}(s_1; s_2)$, $S_3 = \mathcal{S}(s_1, s_2; s_3)$, ..., $S_n = \mathcal{S}(s_1, \ldots, s_{n-1}; s_n)$ et $S_{n+1} = \mathcal{S}(s_1, \ldots, s_n; 1)$ sont comaximaux.
Plus généralement, les monoïdes $V_1 = \mathcal{S}(I; U, s_1)$, $V_2 = \mathcal{S}(I, s_1; U, s_2)$, $V_3 = \mathcal{S}(I, s_1, s_2; U, s_3)$, ..., $V_n = \mathcal{S}(I, s_1, \ldots, s_{n-1}; U, s_n)$ et $V_{n+1} = \mathcal{S}(I, s_1, \ldots, s_n; U)$ recouvrent le monoïde S.

(3) Si $S, S_1, \ldots, S_n \subset \mathbf{A}$ sont des monoïdes comaximaux et si $b = a/(u+i) \in \mathbf{A}_S$ alors $\mathcal{S}(I; U, a), \mathcal{S}(I, a; U), S_1, \ldots, S_n \in \mathbf{A}$ sont comaximaux.

1.2 Du quasi-global au global

Différentes variantes du principe local-global abstrait en algèbre commutative ont leur contrepartie concrète dans laquelle la localisation en tout idéal premier est remplacée par la localisation en une famille finie de monoïdes comaximaux.

Autrement dit, dans ces versions “concrètes” on affirme que certaines propriétés passent du quasi-global au global.

Citons par exemple les résultats suivants, qui sont souvent utiles pour terminer notre travail de relecture constructive.

Principe local-global concret 7 *Soient $S_1, \ldots, S_n$ des monoïdes comaximaux de $\mathbf{A}$ et soit $a, b \in \mathbf{A}$. Alors on a les équivalences suivantes :*

(1) *Recollement concret des égalités :*
$$a = b \quad \text{dans} \quad \mathbf{A} \qquad \Longleftrightarrow \qquad \forall i \in \{1, \ldots, n\} \ a/1 = b/1 \quad \text{dans} \quad \mathbf{A}_{S_i}$$

(2) *Recollement concret des non diviseurs de zéro :*
$$a \text{ est non diviseur de zéro dans } \mathbf{A} \qquad \Longleftrightarrow$$
$$\forall i \in \{1, \ldots, n\} \ a/1 \text{ est non diviseur de zéro dans } \mathbf{A}_{S_i}$$

(3) *Recollement concret des inversibles :*
$$a \text{ est inversible dans } \mathbf{A} \qquad \Longleftrightarrow$$
$$\forall i \in \{1, \ldots, n\} \ a/1 \text{ est inversible dans } \mathbf{A}_{S_i}$$

(4) *Recollement concret des solutions de systèmes linéaires : soit B une matrice $\in \mathbf{A}^{m \times p}$ et C un vecteur colonne $\in \mathbf{A}^{m \times 1}$.*
$$\text{Le système linéaire } BX = C \text{ admet une solution dans } \mathbf{A}^{p \times 1} \qquad \Longleftrightarrow$$
$$\forall i \in \{1, \ldots, n\} \text{ le système linéaire } BX = C \text{ admet une solution dans } \mathbf{A}_{S_i}^{p \times 1}$$

(5) *Recollement concret de facteurs directs : soit M un sous module de type fini d'un module de présentation finie N.*
$$M \text{ est facteur direct dans } N \qquad \Longleftrightarrow$$
$$\forall i \in \{1, \ldots, n\} \ M_{S_i} \text{ est facteur direct dans } N_{S_i}$$

Principe local-global concret 8 (recollement concret de propriétés de finitude pour les modules) *Soient $S_1, \ldots S_n$ des monoïdes comaximaux de $\mathbf{A}$ et soit M un $\mathbf{A}$-module. Alors on a les équivalences suivantes :*

(1) *M est de type fini si et seulement si chacun des M_{S_i} est un $\mathbf{A}_{S_i}$-module de type fini.*

(2) *M est de présentation finie si et seulement si chacun des M_{S_i} est un $\mathbf{A}_{S_i}$-module de présentation finie.*

(3) *M est plat si et seulement si chacun des M_{S_i} est un $\mathbf{A}_{S_i}$-module plat.*

(4) *M est projectif de type fini si et seulement si chacun des M_{S_i} est un $\mathbf{A}_{S_i}$-module projectif de type fini.*

(5) *M est projectif de rang k si et seulement si chacun des M_{S_i} est un $\mathbf{A}_{S_i}$-module projectif de rang k.*

(6) *M est cohérent si et seulement si chacun des M_{S_i} est un $\mathbf{A}_{S_i}$-module cohérent.*

(7) *M est noethérien si et seulement si chacun des M_{S_i} est un $\mathbf{A}_{S_i}$-module noethérien.*

On trouve rarement ces principes énoncés sous cette forme dans la littérature classique usuelle. Citons cependant le petit livre d'algèbre commutative de Knight [11] : le lemme 3.2.3 signale que l'anneau produit $\prod_{i=1}^{k} \mathbf{A}[1/s_i]$ est une extension fidèlement plate de $\mathbf{A}$ lorsque les s_i sont comaximaux. Un certain nombre de propriétés des extensions fidèlement plates sont par ailleurs démontrées. Mis ensemble ces résultats couvrent à peu près les principes local-global concrets 7 et 8.

Dans le style de Quillen, on voit en général énoncé le principe correspondant sous la forme abstraite (on localise en tous les idéaux premiers). Mais la preuve fait souvent intervenir un lemme crucial qui a exactement la signification du principe local-global concret correspondant. Par exemple on pourrait énoncer le principe local-global concret 8 sous la forme suivante "à la Quillen".

Lemme 9 (lemme de propagation pour certaines propriétés de finitude pour les modules) *Soit M un $\mathbf{A}$-module. Les parties I_k suivantes de $\mathbf{A}$ sont des idéaux.*

(1) $I_1 = \{\, s \in \mathbf{A} \ : M_s$ *est un* $\mathbf{A}_s$*-module de type fini* $\}$.

(2) $I_2 = \{\, s \in \mathbf{A} \ : M_s$ *est un* $\mathbf{A}_s$*-module de présentation finie* $\}$.

(3) $I_3 = \{\, s \in \mathbf{A} \ : M_s$ *est un* $\mathbf{A}_s$*-module plat* $\}$.

(4) $I_4 = \{\, s \in \mathbf{A} \ : M_s$ *est un* $\mathbf{A}_s$*-module projectif de type fini* $\}$.

(5) $I_5 = \{\, s \in \mathbf{A} \ : M_s$ *est un* $\mathbf{A}_s$*-module projectif de rang* $k \}$.

(6) $I_6 = \{\, s \in \mathbf{A} \ : M_s$ *est un* $\mathbf{A}_s$*-module cohérent* $\}$.

(7) $I_7 = \{\, s \in \mathbf{A} \ : M_s$ *est un* $\mathbf{A}_s$*-module noethérien* $\}$.

Remarque 10 De manière générale soit une propriété P qui reste vraie après localisation en un monoïde. Alors la version *principe local-global concret pour des éléments comaximaux* :

- pour tout anneau $\mathbf{A}$, si P est vraie après localisation en des éléments comaximaux de $\mathbf{A}$, alors elle est vraie dans $\mathbf{A}$,

et la version *lemme de propagation* :

- l'ensemble $I_P = \{\, s \in \mathbf{A} \ : P$ est vraie dans $\mathbf{A}_s \}$, est un idéal de $\mathbf{A}$,

sont équivalentes. D'une part la version lemme de propagation implique clairement la première. Dans l'autre sens, si $s, s' \in I_P$ et $t = s + s'$ alors $s/1$ et $s'/1$ sont des éléments comaximaux de $\mathbf{A}_t$ et P est vraie dans $(\mathbf{A}_t)_s \simeq (\mathbf{A}_s)_t \simeq \mathbf{A}_{st}$ et $(\mathbf{A}_t)_{s'} \simeq (\mathbf{A}_{s'})_t \simeq \mathbf{A}_{s't}$ donc par le principe local-global concret vraie dans $\mathbf{A}_t$.
Notons aussi qu'en général on a pour tout monoïde S l'implication suivante

- P vraie dans $\mathbf{A}_S \Rightarrow P$ vraie dans $\mathbf{A}_s$ pour un $s \in S$,

ce qui donne l'équivalence du *principe local-global concret pour les éléments comaximaux* et du *principe local-global concret pour les monoïdes comaximaux.* Ceci nous est en général indispensable car notre système de relecture (principe général 5) produit naturellement une version quasi-globale avec des monoïdes comaximaux plutôt qu'avec des éléments comaximaux.

2 Théorèmes de Horrocks, versions constructives

2.1 Preuves constructives du théorème local et du théorème quasi-global

Si $\mathbf{G}$ est un sous-groupe de $\mathbb{SL}_n(\mathbf{A})$ et $A, B \in \mathbf{A}^{n\times 1}$ nous noterons $A \overset{\mathbf{G}}{\simeq} B$ pour $\exists H \in \mathbf{G},\ HA = B$. Il est clair qu'il s'agit d'une relation d'équivalence.

Nous sommes intéressés par la possibilité de trouver dans la classe d'équivalence d'un vecteur défini sur $\mathbf{A}[X]$ un vecteur défini sur $\mathbf{A}$, en un sens convenable. La remarque banale suivante nous sera utile.

Remarque 11 Soit $f(X) \in \mathbf{A}[X]^{n\times 1}$. Alors on a :

$$f(X) \overset{\mathbb{SL}_n(\mathbf{A}[X])}{\simeq} f(0) \iff \exists g \in \mathbf{A}^{n\times 1}\ f(X) \overset{\mathbb{SL}_n(\mathbf{A}[X])}{\simeq} g.$$

En effet si $f(X) = H(X)g$ avec $H(X) \in \mathbb{SL}_n(\mathbf{A}[X])$, alors $f(0) = H(0)g$.

Nous utiliserons aussi le lemme suivant :

Lemme 12 *Soit $\mathbf{A}$ un anneau et $f(X) = {}^t(f_1(X), \ldots, f_n(X))$ un vecteur unimodulaire dans $\mathbf{A}[X]^{n\times 1}$, avec f_1 unitaire de degré d et $f_2, \ldots, f_n$ de degrés $< d$. Notons $f_{i,j}$ le coefficient de X^j dans f_i. Alors l'idéal engendré par les $f_{i,j}$ pour $i = 2, \ldots, n$ contient 1.*

Preuve du lemme Soit I cet idéal. On a : $1 \equiv u_1 f_1$ modulo I. Soit m le degré de u_1 on a $u_{1,m} \equiv 0$ modulo I puisque f_1 est unitaire. De proche en proche, on montre en descendant que tous les coefficients $u_{1,j}$ de u_1 sont dans I. Supposons qu'on l'ait déjà montré pour $j+1, \ldots, m$. Exprimons le coefficient de degré $j+d$ dans $u_1 f_1$. On trouve $0 = u_{1,j} + u_{1,j+1} f_{1,d-1} + \cdots$ ce qui donne $0 \equiv u_{1,j}$ modulo I. Donc finalement $1 \equiv u_1 f_1 \equiv 0$ modulo I. □

Théorème de Horrocks local
Soit un entier $n \geq 3$, $\mathbf{A}$ un anneau local et $f(X) = {}^t(f_1(X), \ldots, f_n(X))$ un vecteur unimodulaire dans $\mathbf{A}[X]^{n\times 1}$, avec f_1 unitaire. Alors

$$f(X) = \begin{bmatrix} f_1 \\ \vdots \\ \vdots \\ f_n \end{bmatrix} \overset{\mathbb{E}_n(\mathbf{A}[X])}{\simeq} \begin{bmatrix} 1 \\ 0 \\ \vdots \\ 0 \end{bmatrix}.$$

Preuve Soit d le degré de f_1. Par manipulations élémentaires de lignes, on ramène $f_2, \ldots, f_n$ à être de degrés $< d$. Notons $f_{i,j}$ le coefficient de X^j dans f_i. Le vecteur ${}^t(f_1(X), \ldots, f_n(X))$ reste unimodulaire. Si $d = 0$ c'est terminé. Sinon vu le lemme 12 et puisque l'anneau est local, l'un des $f_{i,j}$ pour $i = 2, \ldots, n$ est une unité. Supposons par exemple que $f_{2,k}$ est inversible. On va voir que l'on peut trouver deux polynomes v_1 et v_2 tels que le polynome $g_2 = v_1 f_1 + v_2 f_2$ soit unitaire de degré $d-1$. Si $k = d-1$ cela marche avec $v_1 = 0$ et v_2 constant. Si $k < d-1$ considérons la disjonction suivante

$$f_{2,d-1} \in \mathbf{A}^\times \ \vee\ f_{2,d-1} \in \mathrm{Rad}(\mathbf{A}).$$

Dans le premier cas, on est ramené à $k = d-1$. Dans le deuxième cas le polynome $q_2 = Xf_2 - f_{2,d-1} f_1$ est de degré $\leq d-1$ et vérifie : $q_{2,k+1}$ est une unité. On a donc gagné un cran. Il suffit donc d'itérer le processus.

Nous avons donc maintenant $g_2 = v_1 f_1 + v_2 f_2$ de degré $d-1$ et unitaire. On peut donc diviser f_3 par g_2 et on obtient $g_3 = f_3 - g_2 q$ de degré $< d-1$ ($q \in \mathbf{A}$), donc le polynome

$$h_1 = g_2 + g_3 = f_3 + g_2(1-q) = f_3 + (1-q)v_1 f_1 + (1-q)v_2 f_2$$

est unitaire de degré $d-1$. Ainsi par une manipulation élémentaire de lignes on a pu remplacer ${}^{t}(f_1, f_2, f_3)$ par ${}^{t}(f_1, f_2, h_1)$ avec h_1 unitaire de degré $d-1$.
Nous pouvons donc par une suite de transformations élémentaires de lignes ramener ${}^{t}(f_1(X), \ldots, f_n(X))$ avec f_1 unitaire de degré d à ${}^{t}(h_1(X), \ldots, h_n(X))$ avec h_1 unitaire de degré $d-1$. □

Le lemme suivant est immédiat.

Lemme 13 *Soit* $\mathbf{A}$ *un anneau,* S *un monoïde dans* $\mathbf{A}$. *Soit une matrice* $H(X) \in \mathbb{E}_n(\mathbf{A}_S[X])$, *alors il existe* $s \in S$ *tel que* $H(X) \in \mathbb{E}_n(\mathbf{A}[1/s][X])$.

Théorème de Horrocks quasi-global
Soit un entier $n \geq 3$, $\mathbf{A}$ *un anneau et* $f(X) = {}^{t}(f_1(X), \ldots, f_n(X))$ *un vecteur unimodulaire dans* $\mathbf{A}[X]^{n\times 1}$, *avec* f_1 *unitaire. Alors il existe des éléments comaximaux* $a_1, \ldots, a_\ell$ *tels que*

$$f(X) = \begin{bmatrix} f_1 \\ \vdots \\ \vdots \\ f_n \end{bmatrix} \overset{\mathbb{E}_n(\mathbf{A}[1/a_i][X])}{\simeq} \begin{bmatrix} 1 \\ 0 \\ \vdots \\ 0 \end{bmatrix}.$$

Autrement dit pour chaque $i = 1, \ldots, \ell$ il existe une matrice $H_i(X) \in \mathbb{E}_n(\mathbf{A}[1/a_i][X])$ telle que $H_i(X)f(X) = {}^{t}(1, 0, \ldots, 0)$.

Preuve En relisant la preuve du théorème local comme on l'a indiqué dans la section 1.1, on voit que, pour faire descendre le degré de d à $d-1$ il faut, après avoir rendu les degrés de $f_2, \ldots, f_n$ strictement inférieurs à d par division euclidienne, rendre inversible l'un des $f_{i,j}$ pour $i = 2, \ldots, n$. Et on sait que les $f_{i,j}$ sont comaximaux d'après le lemme 12. Ensuite, on utilise plusieurs fois (au plus $d-1$ fois) une disjonction du type

$$f_{2,d-1} \in \mathbf{A}^{\times} \;\vee\; f_{2,d-1} \in \mathrm{Rad}(\mathbf{A}),$$

(dans le cas inversible le calcul se termine sans nouvelle disjonction). Notre relecture de la preuve, pour faire descendre le degré de d à $d-1$ crée donc des localisés $\mathbf{A}_{S_j}$ (avec les S_j comaximaux) dont le nombre est majoré par $d(n-1)(d-1)$.
La mise à plat complète de la preuve crée en définitive des localisés (avec des monoïdes comaximaux) dont le nombre est majoré par

$$d(d-1)(n-1) \times \cdots \times 6(n-1) \times 2(n-1) \times (n-1) \;\leq\; (d!)^2 \times (n-1)^d < (nd^2)^d.$$

Pour chacun des localisés $\mathbf{A}_i$ on a une matrice $H_i(X) \in \mathbb{E}_n(\mathbf{A}_i[X])$ telle que $H_i(X)f(X) = {}^{t}(1, 0, \ldots, 0)$. On termine en appliquant le lemme 13. □

Remarque 14 Ce calcul peut être fait dans la situation générique où l'entier n ainsi que les degrés des f_i et les degrés des u_i sont fixés dans l'égalité polynomiale

$$u_1 f_1 + \cdots + u_n f_n = 1 \quad (*).$$

En outre on prend tous les coefficients comme des indéterminées, soumises aux seules relations données par l'égalité $(*)$.

2.2 Un principe local-global concret

Les calculs dans cette section n'ont rien de bien original, mais leur agencement et l'interprétation que nous leur donnons en termes de principe local-global concret nous semblent particulièrement éclairants.

Lemme 15 *Soit* $\mathbf{A}$ *un anneau intègre,* $b \in \mathbf{A}$ *et* $H(X) \in \mathbb{SL}_n(\mathbf{A}[1/b][X])$. *Alors pour un* $a \in \mathbf{A}$ *égal à une certaine puissance de* b *on a* $H(X+aY)H(X)^{-1} \in \mathbb{SL}_n(\mathbf{A}[X,Y])$.
De manière équivalente : si S *est un monoïde de* $\mathbf{A}$ *et* $H(X) \in \mathbb{SL}_n(\mathbf{A}_S[X])$, *alors pour un* $s \in S$ *on a* $H(X+sY)H(X)^{-1} \in \mathbb{SL}_n(\mathbf{A}[X,Y])$.

Preuve Soit $L(X) = s_1 H(X) \in \mathbf{M}_n(\mathbf{A}[X])$ avec $s_1 \in S$ et

$$s = s_1^n = \det(L(X)) = \det(L(X+Y)).$$

Soit $M(X)$ la matrice cotransposée de $L(X)$. On considère la matrice

$$B(X,Y) = H(X+Y)H(X)^{-1} = L(X+Y)L(X)^{-1} = s^{-1}L(X+Y)M(X) = s^{-1}B_1(X,Y).$$

On a $B_1(X,Y) \in \mathbf{A}[X,Y]$, $B_1(X,0) = s\mathrm{I}_n$ et donc $B_1(X,Y) = s\mathrm{I}_n + YC_1(X,Y))$ avec $C_1(X,Y) \in \mathbf{A}[X,Y]$. Donc $B_1(X,sY) = s(\mathrm{I}_n + YC_1(X,sY)$ et $H(X+sY)H(X)^{-1} = s^{-1}B_1(X,sY) = \mathrm{I}_n + YC_1(X,sY) \in \mathbf{A}[X,Y]$. □

Corollaire 16 *Soit* $\mathbf{A}$ *un anneau intègre,* S *un monoïde de* $\mathbf{A}$ *et* $f(X) \in \mathbf{A}[X]^{n\times 1}$. *Alors*

$$f(X) \overset{\mathbb{SL}_n(\mathbf{A}_S[X])}{\simeq} f(0) \implies \exists s \in S \ f(X+sY) \overset{\mathbb{SL}_n(\mathbf{A}[X,Y])}{\simeq} f(X)$$

Preuve Si $f(X) = H(X)f(0)$ avec $H(X) \in \mathbb{SL}_n(\mathbf{A}_S[X])$ alors $f(X+sY) = H(X+sY)f(0)$ et $H(X+sY)H(X)^{-1}f(X) = f(X+sY)$. Il suffit donc de prendre s comme dans le lemme 15. □

Lemme 17 *Soit* $f(X) \in \mathbf{A}[X]^{n\times 1}$, *soit*

$$I = \left\{ a \in \mathbf{A} \ : \ f(X+aY) \overset{\mathbb{SL}_n(\mathbf{A}[X,Y])}{\simeq} f(X) \right\}.$$

Alors I *est un idéal de* $\mathbf{A}$.

Preuve Si $f(X+aY) = P_a(X,Y)f(X)$ et $f(X+bY) = P_b(X,Y)f(X)$ alors,

$$f(X+(a+b)Y) = f((X+aY)+bY) = P_b(X+aY,Y)f(X+aY) = P_b(X+aY,Y)P_a(X,Y)f(X)$$

et $f(X+acY) = P_a(X,cY)f(X)$. □

Le corollaire 16 et le lemme 17 mis ensemble peuvent être énoncés sous forme d'un principe local-global concret :

Principe local-global concret 18 *Soient* $\mathbf{A}$ *un anneau intègre,* $S_1,\ldots,S_k$ *des monoïdes comaximaux et* $f(X) \in \mathbf{A}[X]^{n\times 1}$. *Alors*

$$f(X) \overset{\mathbb{SL}_n(\mathbf{A}[X])}{\simeq} f(0) \iff \bigwedge_{i=1}^{k} f(X) \overset{\mathbb{SL}_n(\mathbf{A}_{S_i}[X])}{\simeq} f(0).$$

Preuve En appliquant les résultats précédents on obtient

$$f(X+Y) \overset{\mathbb{SL}_n(\mathbf{A}[X,Y])}{\simeq} f(X)$$

c'est-à-dire $f(X+Y) = Q(X,Y)f(X)$ avec $Q(X,Y) \in \mathbb{SL}_n(\mathbf{A}[X,Y])$ et donc aussi, en faisant $X = 0$, $f(Y) = Q(0,Y)f(0)$. □

2.3 Preuve du théorème global et de la conjecture de Serre

Le principe local-global concret 18 permet de transformer le théorème de Horrocks quasi-global en sa version globale, de manière constructive.

Théorème de Horrocks global
Soit un entier $n \geq 3$, $\mathbf{A}$ *un anneau intègre et* $f(X) = {}^{t}(f_1(X), \ldots, f_n(X))$ *un vecteur unimodulaire dans* $\mathbf{A}[X]^{n\times 1}$, *avec* f_1 *unitaire, alors il existe une matrice* $H \in \mathbb{SL}_n(\mathbf{A}[X])$ *telle que* $H(X)f(X) = f(0)$.

Preuve Vue la remarque 11 on applique le théorème quasi-global puis le principe local-global concret 18. □

Remarque 19 Si l'anneau générique décrit dans la remarque 14 est intègre (ce qui semble probable), le calcul peut être fait une fois pour toutes (dans cet anneau) et se spécialise ensuite dans n'importe quel anneau, intègre ou non, ce qui permet d'enlever l'hypothèse d'intégrité dans le théorème global.

Théorème de Quillen-Suslin
Soit $\mathbf{K}$ *un corps,* $\mathbf{A} = \mathbf{K}[X_1, \ldots, X_r]$ *et dans* $\mathbf{A}^{n\times 1}$ *un vecteur unimodulaire*

$$f = {}^{t}(f_1(X_1, \ldots, X_r), \ldots, f_n(X_1, \ldots, X_r)),$$

alors il existe une matrice $H \in \mathbb{SL}_n(\mathbf{A})$ *telle que* $Hf = {}^{t}(1, 0, \ldots, 0)$.

Preuve Si $n = 1$ ou 2, le résultat est immédiat. Si $n > 2$ et $r = 1$ le résultat provient du fait que $\mathbf{A}$ est un anneau principal. Il est donné explicitement par une réduction de Smith de la matrice colonne f. Pour $r \geq 2$ on raisonne par induction sur r. En appliquant le théorème de Horrocks global à l'anneau $\mathbf{B} = \mathbf{K}[X_1, \ldots, X_{r-1}]$ on a gagné si l'un des f_i est un polynome unitaire en X_r. Si le corps a suffisamment d'éléments, on obtient cela par un changement linéaire de variable. Sinon, on fait un changement de variable à la Nagata : $Y_r = X_r$, et pour $1 \leq j < r$, $Y_j = X_j + X_r^{d^j}$, avec un entier d suffisamment grand. □

Solution de la conjecture de Serre (Quillen-Suslin)
Soit $\mathbf{K}$ *un corps,* $\mathbf{A} = \mathbf{K}[X_1, \ldots, X_r]$ *et* M *un* $\mathbf{A}$*-module projectif de type fini stablement libre, alors* M *est libre.*

Preuve On a par hypothèse un isomorphisme

$$\varphi : \mathbf{A}^k \oplus M \longrightarrow \mathbf{A}^{\ell+k}$$

pour deux entiers k et ℓ. Si $k = 0$ il n'y a rien à faire. Supposons $k > 0$. Le vecteur $f = \varphi((e_{k,1}, 0_M))$ (où $e_{k,1}$ est le premier vecteur de la base canonique de $\mathbf{A}^k$) est unimodulaire : considérer la forme linéaire λ sur $\mathbf{A}^{\ell+k}$ qui à un vecteur y fait correspondre la première coordonnée de $\varphi^{-1}(y)$. On a $\lambda(y_1, \ldots, y_{k+\ell}) = u_1y_1 + \cdots + u_{k+\ell}y_{k+\ell}$ et $\lambda(f) = 1$.
Considérons f comme un vecteur colonne. En composant φ avec l'isomorphisme donné dans le théorème de Quillen-Suslin on obtient un isomorphisme ψ qui envoie $(e_{k,1}, 0_M)$ sur $e_{k+\ell,1}$. En passant au quotient par $\mathbf{A}(e_{k,1}, 0_M)$ et $\mathbf{A}e_{k+\ell,1}$ on obtient un isomorphisme

$$\theta : \mathbf{A}^{k-1} \oplus M \longrightarrow \mathbf{A}^{\ell+k-1}.$$

□

3 Une preuve constructive d'un théorème de stabilité de Suslin

Dans cette section, nous examinons la preuve du théorème de stabilité de Suslin dans le cas des corps, telle qu'elle est donnée dans [9] en s'appuyant sur une méthode locale-globale. Nous la décryptons en une preuve constructive selon la méthode exposée dans la section 1.

3.1 Un théorème local et sa version quasi-globale

Le seul véritable argument non constructif dans [9] est *l'utilisation* du lemme II 3.6 page 46. Ce lemme est de nature locale mais il est ensuite utilisé dans un argument de type local-global. C'est le lemme suivant, dans lequel $\binom{f}{g}$ désigne le symbole de Mennicke.

Lemme 20 *(local)*
Soit $\mathbf{A}$ *un anneau local et* $f, g \in \mathbf{A}[X]$ *avec* f *unitaire et* $af + bg = 1$. *Alors on a :*

$$\binom{f}{g} = \binom{f(0)}{g(0)} = 1.$$

Preuve (cf. [9])
Notons pour commencer qu'on peut diviser b par f et qu'on obtient alors une égalité $a_1 f + b_1 g = 1$ avec $\deg(b_1) < \deg(f)$ et donc, puisque f est unitaire, $\deg(a_1) < \deg(g)$. Nous supposerons donc sans perte de généralité que $\deg(b) < \deg(f)$ et $\deg(a) < \deg(g)$.
Rappelons que $\mathbf{E}_n(\mathbf{A})$ est un sous-groupe distingué de $\mathbb{SL}_n(\mathbf{A})$ si $n \geq 3$, et que le symbole de Mennicke $\binom{f}{g}$ représente la classe d'équivalence de la matrice

$$B = \begin{bmatrix} f & g & 0 \\ -b & a & 0 \\ 0 & 0 & 1 \end{bmatrix}$$

dans le groupe quotient $\mathbb{SL}_3/\mathbb{E}_3$ (la classe d'équivalence ne dépend pas du choix de a et b), et qu'on a les propriétés suivantes (cf. Proposition II 3.5 page 44 dans [9]) :

$$\binom{u}{a} = \binom{1}{0} = 1 \text{ pour } u \in \mathbf{A}^\times, \ \binom{aa'}{b} = \binom{a}{b}\binom{a'}{b}, \ \binom{a}{b} = \binom{b}{a}, \ \binom{a+bd}{b} = \binom{a}{b}.$$

Soit r le reste de la division euclidienne de g par f. Alors $\binom{f}{g} = \binom{f}{r}$. En particulier si $\deg(f) = 0$ on a terminé. Sinon on peut supposer $\deg(g) < \deg(f)$ et on raisonne par induction sur $\deg(f)$.
Puisque $\mathbf{A}$ est local, $g(0)$ est inversible ou dans le radical $\mathcal{M}$ de $\mathbf{A}$.
Supposons tout d'abord $g(0)$ inversible. Alors

$$\binom{f}{g} = \binom{f - g(0)^{-1} f(0) g}{g}$$

si bien que nous pouvons supposer $f(0) = 0$ et $f = X f_1$. Alors

$$\binom{X f_1}{g} = \binom{X}{g}\binom{f_1}{g} = \binom{X}{g(0)}\binom{f_1}{g} = \binom{f_1}{g}$$

et la preuve est terminée par induction puisque f_1 est unitaire.
Supposons maintenant que $g(0)$ est dans $\mathcal{M}$. On note que $a(0)f(0) + b(0)g(0) = 1$, donc $a(0)f(0) \in 1 + \mathcal{M} \subseteq \mathbf{A}^\times$ et donc $a(0) \in \mathbf{A}^\times$. Or

$$\begin{bmatrix} f & g & 0 \\ -b & a & 0 \\ 0 & 0 & 1 \end{bmatrix} \equiv \begin{bmatrix} f-b & g+a & 0 \\ -b & a & 0 \\ 0 & 0 & 1 \end{bmatrix} \mod \mathbb{E}_3(\mathbf{A}[X]),$$

donc

$$\begin{pmatrix} f \\ g \end{pmatrix} = \begin{pmatrix} f-b \\ g+a \end{pmatrix}$$

avec $f - b$ unitaire de même degré que f, $\deg(g + a) < \deg(f)$ et $(g + a)(0) \in \mathbf{A}^\times + \mathcal{M} = \mathbf{A}^\times$. On est donc ramené au cas précédent. La preuve est complète. □

Notre machinerie de relecture automatique de la preuve locale donne le lemme quasi-global suivant (qui ne se trouve pas dans [9]), par application directe du principe général 5 :

Lemme 21 *(quasi-global)*
Soit $\mathbf{A}$ *un anneau et* $f, g \in \mathbf{A}[X]$ *avec* f *unitaire et* $af + bg = 1$. *Alors, il existe dans* $\mathbf{A}$ *des éléments comaximaux* s_i *tels que dans chaque localisé* $\mathbf{A}[1/s_i]$ *on ait l'égalité des symboles de Mennicke suivante*

$$\begin{pmatrix} f \\ g \end{pmatrix} = \begin{pmatrix} f(0) \\ g(0) \end{pmatrix} = 1.$$

3.2 Un principe local-global concret et la preuve constructive d'un théorème global

Maintenant nous rappelons le lemme I 5.9 page 26 dans [9].

Lemme 22 *Soit* $n \geq 3$ *et* $A \in \mathbb{SL}_n(\mathbf{A}[X])$ *avec* $A(0) = \mathrm{I}_n$. *Soit*

$$I = \{s \in \mathbf{A} \mid A \in \mathbb{E}_n(\mathbf{A}[1/s][X])\}.$$

Alors I *est un idéal de* $\mathbf{A}$.

La belle preuve constructive de ce beau lemme (dont seule la version abstraite est qualifiée de théorème) occupe les pages 22 à 26 de [9].

Ce lemme aurait pu être reformulé sous la forme du principe local-global concret suivant, qui est d'ailleurs à très peu près le lemme I 5.8 de [9] :

Principe local-global concret 23 *Soient* $n \geq 3$, $\mathbf{A}$ *un anneau,* $S_1, \ldots, S_k$ *des monoïdes comaximaux et* $A \in \mathbb{SL}_n(\mathbf{A}[X])$ *avec* $A(0) = \mathrm{I}_n$. *Alors*

$$A \in \mathbb{E}_n(\mathbf{A}[X]) \quad \Longleftrightarrow \quad \bigwedge_{i=1}^{k} A \in \mathbb{E}_n(\mathbf{A}_{S_i}[X])$$

Le principe local-global concret 23 et le lemme 21 donnent alors le théorème global suivant (corollaire II 3.8 de [9]).

Théorème 24 *(version globale du lemme 20)*
Soient $n \geq 3$, $\mathbf{A}$ *un anneau et* $f, g \in \mathbf{A}[X]$ *avec* f *unitaire et* $af + bg = 1$. *Alors on a l'égalité des symboles de Mennicke suivante*

$$\begin{pmatrix} f \\ g \end{pmatrix} = \begin{pmatrix} f(0) \\ g(0) \end{pmatrix}.$$

Preuve Soit B la matrice donnée au début de la preuve du lemme 20. L'égalité $\binom{f}{g} = \binom{f(0)}{g(0)}$ signifie : $A = BB(0)^{-1} \in \mathbb{E}_3(\mathbf{A}[X])$. On a évidemment $A(0) = \mathrm{I}_3$. Le principe local-global concret 23 nous dit qu'il suffit de vérifier l'assertion dans des anneaux localisés $\mathbf{A}_{s_i}$ pour une famille s_i d'éléments comaximaux. Et le lemme 21 nous construit cette famille. □

Enfin la preuve que ce corollaire implique le théorème de stabilité de Suslin est simple et constructive, telle que donnée dans [9].

Théorème de stabilité de Suslin (cas des corps)
Si $\mathbf{K}$ *est un corps et* $n \geq 3$, $\mathbb{SL}_n(\mathbf{K}[X_1, \ldots, X_k]) = \mathbb{E}_n(\mathbf{K}[X_1, \ldots, X_k])$.

Remarque 25 Du point de vue constructif, pour faire tourner les algorithmes correspondants au théorème précédent et au théorème de Quillen-Suslin, nous devons supposer que les opérations du corps et le test d'égalité sont explicites, c'est-à-dire dans le langage de l'algèbre constructive ([24]), que le corps est un corps discret. En fait le test d'égalité à 0 est nécessaire pour pouvoir faire les changements de variables en vue de rendre des polynômes unitaires. Enfin il reste un travail intéressant à faire pour rendre constructives des versions plus générales de ces théorèmes. Notamment les versions qui utilisent comme anneau de base, non plus un corps, mais un anneau noethérien de dimension de Krull fixée.

Une preuve de nature différente pour le dernier théorème, utilisant l'artillerie des bases de Gröbner et basée sur la connaissance des vrais idéaux maximaux de $\mathbf{K}[X_1, \ldots, X_n]$ a été donnée par Park et Woodburn dans [25].

Une preuve basée sur les mêmes idées que les nôtres, mais s'appliquant dans un cadre beaucoup plus général (anneaux noethériens de dimension de Krull majorée) nous a été signalée par I. Yengui (cf. [28]).

Références

[1] Almeida M., Blaum M., D'Alfonso L., Solernó P. *Computing bases of complete intersection rings in Noether position.* Journal of Pure and Applied Algebra **162** (2001) 127–170.

[2] Almeida M., D'Alfonso L., Solernó P. *On the degrees of bases of free modules over a polynomial ring.* Math. Zeitschrift **231** (1999), 679–706.

[3] Caniglia L., Cortinas G., Danón S., Heintz J., Krick T., Solernó P. *Algorithmic Aspects of Suslin's Proof of Serre's Conjecture.* Computational Complexity **3** (1993), 31–55.

[4] Coquand T., Lombardi H. *Hidden constructions in abstract algebra (3) Krull dimension of distributive lattices and commutative rings.* Preprint.

[5] Coste M., Lombardi H., Roy M.-F. *Dynamical method in algebra : Effective Nullstellensätze.* Annals of Pure and Applied Logic **111** (2001), 203–256.

[6] Delzell C.N. *Kreisel's unwinding of Artin's proof*, 113–245 in Kreiseliana : About and Around Georg Kreisel, ed. P. Odifreddi, A K Peters, Wellesley, MA, 1996.

[7] Feferman S. *Kreisel's "unwinding" program*, 247–273 in Kreiseliana : About and Around Georg Kreisel, ed. P. Odifreddi, A K Peters, Wellesley, MA, 1996.

[8] Fitchas N., Galligo A. *Nullstellensatz effectif et Conjecture de Serre (Théorème de Quillen-Suslin) pour le Calcul Formel.* Math. Nachr. **149** (1990), 231–253.

[9] Gupta S., Murthy M. *Suslin's work on linear groups over polynomial rings and Serre conjecture.* ISI Lecture Notes n°8. The Macmillan Company of India Limited. 1980.

[10] Horrocks G. *Projective modules over an extension of a local ring.* Proc. Lond. Math. Soc. **14** (1964), 714–718.

[11] Knight J. *Commutative Algebra.* London Mathematical Society LNS n°5. Cambridge University Press, 1971.

[12] Kunz E. *Introduction to Commutative Algebra and Algebraic Geometry.* Birkhäuser, 1991.

[13] Lam T. Y. *Serre's conjecture.* Lecture Notes in Mathematics, Vol. 635. Springer-Verlag, Berlin-New York, 1978.

[14] Lang S. *Algebra.* Addison-Wesley, troisième édition. 1993.

[15] Laubenbacher, R., Woodburn, C. *An algorithm for the Quillen-Suslin theorem for monoid rings. Algorithms for algebra (Eindhoven, 1996).* J. Pure Appl. Algebra **117/118** (1997), 395–429.

[16] Logar A., Sturmfels B. *Algorithms for the Quillen-Suslin theorem.* J. Algebra **145** no. 1, (1992), 231–239.

[17] Lombardi H. *Le contenu constructif d'un principe local-global avec une application à la structure d'un module projectif de type fini.* Publications Mathématiques de Besançon. Théorie des nombres. 1997.

[18] Lombardi H. *Relecture constructive de la théorie d'Artin-Schreier.* Annals of Pure and Applied Logic **91** (1998), 59–92.

[19] Lombardi H. *Dimension de Krull, Nullstellensätze et Évaluation dynamique.* Math. Zeitschrift, **242** (2002), 23–46.

[20] Lombardi H. *Hidden constructions in abstract algebra (1) Integral dependance relations.* Journal of Pure and Applied Algebra **167** (2002), 259–267.

[21] Lombardi H. *Platitude, localisation et anneaux de Prüfer, une approche constructive.* Publications Mathématiques de Besançon. Théorie des nombres. 2002.

[22] Lombardi H. *Constructions cachées en algèbre abstraite (4) La solution du 17ème problème de Hilbert par la théorie d'Artin-Schreier.* Publications Mathématiques de Besançon. Théorie des nombres. 2002.

[23] Lombardi H. *Constructions cachées en algèbre abstraite (5) Principe local-global de Pfister et variantes.* Preprint 2002.

[24] Mines R., Richman F., Ruitenburg W. *A Course in Constructive Algebra.* Universitext. Springer-Verlag, 1988.

[25] Park, H., Woodburn, C. *An algorithmic proof of Suslin's stability theorem for polynomial rings.* J. Algebra **178** no. 1 (1995), 277–298.

[26] Quillen D. *Projective modules over polynomial rings.* Invent. Math. **36** (1976), 167–171.

[27] Suslin A. *Projective modules over polynomial rings are free. (en Russe).* Dokl. Akad. Nauk SSSR **229** no. 5 (1976), 1063–1066.

[28] Yengui I. *An algorithmic proof of Suslin's stability theorem for Noetherian rings.* Preprint.

Hidden Constructions in Abstract Algebra: Krull Dimension of Distributive Lattices and Commutative Rings

Thierry Coquand Chalmers University, Goteborg, Sweden

Henri Lombardi Université de Franche-Comté, Besancon, France

Abstract

We present constructive versions of Krull's dimension theory for commutative rings and distributive lattices. The foundations of these constructive versions are due to Joyal, Español and the authors. We show that the notion of Krull dimension has an explicit computational content in the form of existence (or lack of existence) of some algebraic identities. We can then get an explicit computational content where abstract results about dimensions are used to show the existence of concrete elements. This can be seen as a partial realization of Hilbert's program for classical abstract commutative algebra.

Contents

Introduction

We present constructive versions of Krull's dimension theory for commutative rings and distributive lattices. The foundations of these constructive versions are due to Joyal, Español and the authors. We show that the notion of Krull dimension has an explicit computational content in the form of existence (or lack of existence) of some algebraic identities. This confirms the feeling that commutative algebra can be seen computationally as a machine that produces algebraic identities (the most famous of which being called Nullstellensatz). This can be seen as a partial realization of Hilbert's program for classical abstract commutative algebra.

Our presentation follows Bishop's style (cf. in algebra [19]). As much as possible, we kept minimal any explicit mention to logical notions. When we say that we have a constructive version of an abstract algebraic theorem, this means that we have a theorem the proof of which is constructive, which has a clear computational content, and from which we can recover the usual version of the abstract theorem by an immediate application of a well classified non-constructive principle. An abstract classical theorem can have several distinct interesting constructive versions.

In the case of abstract theorems in commutative algebra, such a non-constructive principle is the completeness theorem, which claims the existence of a model of a formally consistent propositional theory. We recall the exact formulation of this theorem in the appendix, as well as its derivation from the compactness theorem When this is used for algebraic structures of enumerable presentation (in a suitable sense) the compactness and completeness theorem can be seen as a reformulation of Bishop **LLPO** (a real number is ≥ 0 or ≤ 0).

To avoid the use of completeness theorem is not motivated by philosophical but by practical considerations. The use of this principle leads indeed to replace quite direct (but usually hidden) arguments by indirect ones which are nothing else than a double contraposition of the direct proofs, with a corresponding lack of computational content. For instance [2] the abstract proof of 17th Hilbert's problem claims : if the polynomial P is not a sum of rational fractions there is a field K in which one can find an absurdity by reading the (constructive) proof that the polynomial is everywhere positive or zero. The direct version of this abstract proof is: from the (constructive) proof that the polynomial is everywhere positive or zero, one can show (using arguments of the abstract proofs) that any attempt to build K will fail. This gives explicitly the sum of squares we are looking for. In the meantime, one has to replace the abstract result: "any real field can be ordered" by the constructive theorem: "in a field in which any attempt to build an ordering fails -1 is a sum of squares". One can go from this explicit version to the abstract one by completeness theorem, while the proof of the explicit version is hidden in the algebraic manipulations that appear in the usual classical proof of the abstract version.

Here is the content of the paper.

Distributive lattices

In this section, we present basic theorems on distributive lattices. An important simplification of proofs and computations is obtained via the systematic use of the notion of entailment relation, which has its origin in the cut rule in Gentzen's sequent calculus, with the fundamental theorem 1.7.

Dimension of distributive lattices In this section, we develop the theory of Krull dimension of distributive lattices, explaining briefly the connection with Español's developments of Joyal' s theory. We show that the property to have a Krull dimension $\leq \ell$ can be formulated as the existence of concrete equalities in the distributive lattice.

Zariski and Krull lattice In section 3 we define the Zariski lattice of a commutative ring (whose elements are radicals of finitely generated ideals), which is the constructive counterpart of Zariski spectrum : the points of Zariski spectrum are the prime ideals of Zariski lattice, and the constructible subsets of Zariski spectrum are the elements of the Boolean algebra generated by the Zariski lattice. Joyal's idea is to define Krull dimension of a commutative ring as the dimension of its Zariski lattice. This avoids any mention of prime ideals. We show the equivalence between this (constructive) point of view and the (constructive) presentation given in [14], showing that the property to have a Krull dimension $\leq \ell$ can be formulated as the existence of concrete equalities in the ring.

Conclusion
This article confirms the actual realization of Hilbert's program for a large part of abstract commutative algebra. (cf. [2, 4, 10, 11, 12, 13, 14, 15, 16, 17]). The general idea is to replace ideal abstract structures by *partial specifications* of these structures. The very short elegant abstract proof which uses these ideal objects has then a corresponding computational version at the level of the partial specifications of these objects. Most of classical results in abstract commutative algebra, the proof of which seem to require in an essential way excluded middle and Zorn's lemma, seem to have in this way a corresponding constructive version. Most importantly, the abstract proof of the classical theorem always contains, more or less implicitly, the constructive proof of the corresponding constructive version.
Finally, we should note that the explicit characterizations of Krull dimension of distributive lattices (Theorem 2.9), of spectral spaces (Theorem 2.14), and of rings (Corollary 3.6), are new.

1 Distributive lattice, Entailment relations

Elementary though it has become after successive presentations and simplifications, the theory of distributive lattices is the ideal instance of a mathematical theory, where a syntax is specified together with a complete description of all models, and what is more, a table of semantic concepts and syntactic concepts is given, together with a translation algorithm between the two kinds of concepts. Such an algorithm is a "completeness theorem" (G. C. Rota [20]).

1.1 Distributive lattices, filters and spectrum

As indicated by the quotation above, the structure of distributive lattices is fundamental in mathematics, and G.C. Rota has pointed out repeatedly its potential relevance to commutative algebra and algebraic geometry. A distributive lattice is an ordered set with finite sups and infs, a minimum element (written 0) and a maximum element (written 1). The operations sup and inf are supposed to be distributive with respect to each other. We write these operations

$\vee$ and $\wedge$. The relation $a \le b$ can then be defined by $a \vee b = b$ or, equivalently, $a \wedge b = a$. The theory of distributive lattices is then purely equational. It makes sense then to talk of distributive lattices defined by generators and relations.

A quite important rule, the *cut rule*, is the following

$$(((x \wedge a) \le b) \ \ \& \ \ (a \le (x \vee b))) \implies a \le b.$$

In order to prove this, write $x \wedge a \wedge b = x \wedge a$ and $a = a \wedge (x \vee b)$ hence

$$a = (a \wedge x) \vee (a \wedge b) = (a \wedge x \wedge b) \vee (a \wedge b) = a \wedge b.$$

A totally ordered set is a distributive lattice as soon as it has a maximum and a minimum element. We write $\mathbf{n}$ for the totally ordered set with n elements (this is a distributive lattice for $n \neq 0$.) A product of distributive lattices is a distributive lattice. Natural numbers with the divisibility relation form a distributive lattice (with minimum element 1 and maximum element 0). If L and L' are two distributive lattices, the set $\mathrm{Hom}(L, L')$ of all morphisms (*i.e.*, maps preserving sup, inf, 0 and 1) from L to L' has a natural order given by

$$\varphi \le \psi \quad \stackrel{\text{def}}{\iff} \quad \forall x \in L \ \varphi(x) \le \psi(x).$$

A map between two totally ordered distributive lattices L and S is a morphism if, and only if, it is nondecreasing and 0_L and 1_L are mapped into 0_S and 1_S.

The following proposition is direct.

Proposition 1.1 *Let L be a distributive lattice and J a subset of L. We consider the distributive lattice L' generated by L and the relations $x = 0$ for $x \in J$ (L' is a quotient of L). Then*

- *the equivalence class of 0 is the set of a such that for some finite subset J_0 of J:*

$$a \le \bigvee_{x \in J_0} x \quad \text{in } L$$

- *the equivalence class of 1 is the set of b such that for some finite subset J_0 of J:*

$$1 = \left(b \vee \bigvee_{x \in J_0} x\right) \quad \text{in } L$$

- *More generally $a \le_{L'} b$ if, and only if, for some finite subset J_0 of J:*

$$a \le \left(b \vee \bigvee_{x \in J_0} x\right)$$

In the previous proposition, the equivalence class of 0 is called an *ideal* of the lattice; it is the ideal generated by J. We write it $\langle J \rangle_L$. We can easily check that an ideal I is a subset such that:

$$0 \in I$$
$$x, y \in I \implies x \vee y \in I$$
$$x \in I,\ z \in L \implies x \wedge z \in I$$

(the last condition can be written $(x \in I,\ y \le x) \Rightarrow y \in I$).

Furthermore, for any morphim $\varphi : L_1 \to L_2$, $\varphi^{-1}(0)$ is an ideal of L_1.

A *principal ideal* is an ideal generated by one element a. We have $\langle a \rangle_L = \{x \in L \,;\, x \leq a\}$. Any finitely generated ideal is principal.

The dual notion to ideal is that of *filter*. A filter F is the inverse image of 1 by a morphism. This is a subset such that:

$$\begin{array}{rcl} & & 1 \in F \\ x, y \in F & \Longrightarrow & x \wedge y \in F \\ x \in F,\ z \in T & \Longrightarrow & x \vee z \in F \end{array}$$

Notation 1.2 We write $\mathrm{P_f}(X)$ for the set of all finite subsets of the set X. If A is a finite subset of a distributive lattice L we define

$$\bigvee A := \bigvee_{x \in A} x \qquad \text{and} \qquad \bigwedge A := \bigwedge_{x \in A} x$$

We write $A \vdash B$ or $A \vdash_L B$ for the relation defined on the set $\mathrm{P_f}(L)$:

$$A \vdash B \quad \stackrel{\text{def}}{\Longleftrightarrow} \quad \bigwedge A \leq \bigvee B$$

Note the relation $A \vdash B$ is well defined on finite subsets because of the associativity, commutativity and idempotence of the operations $\wedge$ and $\vee$. Note also $\emptyset \vdash \{x\} \Rightarrow x = 1$ and $\{y\} \vdash \emptyset \Rightarrow y = 0$. This relation satisfies the following axioms, where $A, B, A', B' \in \mathrm{P_f}(L)$; we write x for $\{x\}$ and A, B for $A \cup B$.

$$\begin{array}{rclcl} & & a \vdash a & & (R) \\ (A \vdash B)\ \&\ (A \subseteq A')\ \&\ (B \subseteq B') & \Longrightarrow & A' \vdash B' & & (M) \\ (A, x \vdash B)\ \&\ (A \vdash B, x) & \Longrightarrow & A \vdash B & & (T) \end{array}$$

We say that the relation is reflexive, monotone and transitive. The last rule is also called the*cut rule*. Let us also mention the two following rules of "distributivity":

$$\begin{array}{rcl} (A,\ x \vdash B)\ \&\ (A,\ y \vdash B) & \Longleftrightarrow & A,\ x \vee y \vdash B \\ (A \vdash B,\ x)\ \&\ (A \vdash B,\ y) & \Longleftrightarrow & A \vdash B,\ x \wedge y \end{array}$$

The following is proved in the same way as Proposition 1.1.

Proposition 1.3 *Let L be a distributive lattice and (J, U) a pair of subsets of L. We consider the distributive lattice L' generated by L and by the relations $x = 0$ for $x \in J$ and $y = 1$ for $y \in U$ (L' is a quotient of L). We have that:*

- *the equivalence class of* 0 *is the set of elements a such that:*

$$\exists J_0 \in \mathrm{P_f}(J),\ U_0 \in \mathrm{P_f}(U) \qquad a,\ U_0 \vdash_L J_0$$

- *the equivalence class of* 1 *is the set of elements b such that: vérifient:*

$$\exists J_0 \in \mathrm{P_f}(J),\ U_0 \in \mathrm{P_f}(U) \qquad U_0 \vdash_L b,\ J_0$$

- *More generally $a \leq_{L'} b$ if, and only if, there exists a finite subset J_0 of J and a finite subset U_0 of U such that, in L:*

$$a,\ U_0 \vdash_L b,\ J_0$$

We shall write $L/(J=0, U=1)$ for the quotient lattice L' described in Proposition 1.3. Let $\psi : L \to L'$ be the canonical surjection. If I is the ideal $\psi^{-1}(0)$ and F the filter $\psi^{-1}(1)$, we say that the *ideal I and the filter F are conjugate.* By the previous proposition, an ideal I and a filter F are conjugate if, and only if, we have:

$$[I_0 \in \mathrm{P_f}(I),\ F_0 \in \mathrm{P_f}(F),\ (x,\ F_0 \vdash I_0)] \implies x \in I \quad \text{and}$$
$$[I_0 \in \mathrm{P_f}(I),\ F_0 \in \mathrm{P_f}(F),\ (F_0 \vdash x,\ I_0)] \implies x \in F.$$

This can also be formulated as follows:

$$(f \in F,\ x \wedge f \in I) \Longrightarrow x \in I \quad \text{and} \quad (j \in I,\ x \vee j \in F) \Longrightarrow x \in F.$$

When an ideal I and a filter F are conjugate, we have

$$1 \in I \iff 0 \in F \iff (I, F) = (L, L).$$

We shall also write $L/(I, F)$ for $L' = T/(J=0, U=1)$. By Proposition 1.3, a homomorphism φ from L to another lattice L_1 satisfying $\varphi(J) = \{0\}$ and $\varphi(U) = \{1\}$ can be factorised in an unique way through the quotient L'.

As shown by the example of totally ordered sets a quotient of distributive lattices is not in general characterized by the equivalence classes of 0 and 1.

Classically a *prime ideal* I of a lattice is an ideal whose complement F is a filter (which is then a *prime filter*). This can be expressed by

$$1 \notin I \quad \text{and} \quad (x \wedge y) \in I \implies (x \in I \text{ or } y \in I) \qquad (*)$$

which can also be expressed by saying that I is the kernel of a morphism from L into the lattice with two elements written **2**. Constructively, at least in the case where L is discrete, it seems natural to take the definition $(*)$, where "or" is used constructively. The notion of prime filter is then defined in a dual way.

Definition 1.4 *Let L be a distributive lattice.*

- *An* idealistic prime *in L is given by a pair (J, U) of finite subsets of L. We consider this as an incomplete specification for a prime ideal P satisfying $J \subseteq P$ and $U \cap P = \emptyset$.*
- *To any idealistic prime (J, U) we can associate a pair (I, F) as described in Proposition 1.3 where I is an ideal, F is a filter and I, F are conjugate.*
- *We say that the idealistic prime (J, U)* collapses *iff we have $I = F = L$. This means that the quotient lattice $L' = T/(J=0, U=1)$ is a singleton i.e., $1 \leq_{L'} 0$, which means also $U \vdash J$.*

Theorem 1.5 (Simultaneous collapse for idealistic primes) *Let (J, U) be an idealistic prime for a lattice L and x be an element of L. If the idealistic primes $(J \cup \{x\}, U)$ and $(J, U \cup \{x\})$ collapse, then so does (J, U).*

Proof.

We have two finite subsets J_0, J_1 of J and two finite subsets U_0, U_1 of U such that

$$x,\ U_0 \vdash J_0 \quad \text{and} \quad U_1 \vdash x,\ J_1$$

hence

$$x,\ U_0,\ U_1 \vdash J_0,\ J_1 \quad \text{and} \quad U_0,\ U_1 \vdash x,\ J_0,\ J_1$$

By the cut rule

$$U_0,\ U_1 \vdash J_0,\ J_1$$

□

Notice the crucial role of the cut rule.

1.2 Distributive lattices and entailment relations

An interesting way to analyze the description of distributive lattices defined by generators and relations is to consider the relation $A \vdash B$ defined on the set $\mathrm{P_f}(L)$ of finite subsets of a lattice L. Indeed if $S \subseteq L$ generates the lattice L, then the relation $\vdash$ on $\mathrm{P_f}(S)$ is enough to characterize the lattice L, because any formula on S can be rewritten, in normal conjunctive form (inf of sups in S) and normal disjunctive form (sup of infs in S). Hence if we want to compare two elements of the lattice generated by S we write the first in normal disjunctive form, the second in normal conjunctive form, and we notice that

$$\bigvee_{i\in I}\left(\bigwedge A_i\right) \le \bigwedge_{j\in J}\left(\bigvee B_j\right) \iff \&_{(i,j)\in I\times J}\ (A_i \vdash B_j)$$

Definition 1.6 *For an arbitrary set S, a relation over $\mathrm{P_f}(S)$ which is reflexive, monotone and transitive (see page 5) is called an* entailment relation.

The notion of entailment relations goes back to Gentzen sequent calculus, where the rule (T) (the cut rule) is first explicitly stated, and plays a key role. The connection with distributive lattices has been emphasized in [3, 4]. The following result (cf. [3]) is fundamental. It says that the three properties of entailment relations are exactly the ones needed in order to have a faithfull interpretation in distributive lattices.

Theorem 1.7 (fundamental theorem of entailment relations) *Let S be a set with an entailment relation $\vdash_S$ over $\mathrm{P_f}(S)$. Let L be the lattice defined by generators and relations as follows: the generators are the elements of S and the relations are*

$$\bigwedge A \le \bigvee B$$

whenever $A \vdash_S B$. For any finite subsets A and B of S we have

$$A \vdash_L B \iff A \vdash_S B.$$

Proof.
We give an explicit possible description of the lattice L. The elements of L are represented by finite sets of finite sets of elements of S

$$X = \{A_1, \ldots, A_n\}$$

(intuitively X represents $\bigwedge A_1 \vee \cdots \vee \bigwedge A_n$). We define then inductively the relation $A \prec Y$ with $A \in \mathrm{P_f}(S)$ and $Y \in L$ (intuitively $\bigwedge A \le \bigvee_{C\in Y}(\bigwedge C)$)

- if $B \in Y$ and $B \subseteq A$ then $A \prec Y$
- if $A \vdash_S y_1, \ldots, y_m$ and $A, y_j \prec Y$ for $j = 1, \ldots, m$ then $A \prec Y$

It is easy to show that if $A \prec Y$ and $A \subseteq A'$ then we have also $A' \prec Y$. It follows that $A \prec Z$ holds whenever $A \prec Y$ and $B \prec Z$ for all $B \in Y$. We can then define $X \le Y$ by $A \prec Y$ for all $A \in X$ and one can then check that L is a distributive lattice[1] for the operations

$$0 = \emptyset, \quad 1 = \{\emptyset\}, \quad X \vee Y = X \cup Y, \quad X \wedge Y = \{A \cup B \mid A \in X,\ B \in Y\}.$$

For establishing this one first show that if $C \prec X$ and $C \prec Y$ we have $C \prec X \wedge Y$ by induction on the proofs of $C \prec X$ and $C \prec Y$. We notice then that if $A \vdash_S y_1, \ldots, y_m$ and $A, y_j \vdash_S B$ for all j then $A \vdash_S B$ using m times the cut rule. It follows that if we have $A \vdash_L B$, *i.e.*, $A \prec \{\{b\} \mid b \in B\}$, then we have also $A \vdash_S B$. □

[1] L is actually the quotient of $\mathrm{P_f}(\mathrm{P_f}(S))$ by the equivalence relation: $X \le Y$ and $Y \le X$.

As a first application, we give the description of the Boolean algebra generated by a distributive lattice. A Boolean algebra can be seen as a distributive lattice with a complement operation $x \mapsto \overline{x}$ such that $x \wedge \overline{x} = 0$ and $x \vee \overline{x} = 1$. The application $x \mapsto \overline{x}$ is then a map from the lattice to its dual.

Proposition 1.8 *Let L be a distributive lattice. There exists a free Boolean algebra generated by L. It can be described as the distributive lattice generated by the set $L_1 = L \cup \overline{L}$ ([2]) with the entailment relation $\vdash_{L_1}$ defined as follows: if A, B, A', B' are finite subsets of L we have*

$$A, \overline{B} \vdash_{L_1} A', \overline{B'} \quad \overset{\text{def}}{\Longleftrightarrow} \quad A, B' \vdash A', B \quad \text{in } L$$

If we write L_{Bool} for this lattice (which is a Boolean algebra), there is a natural embedding of L_1 in L_{Bool} and the entailment relation of L_{Bool} induces on L_1 the relation $\vdash_{L_1}$.

Proof.
See [3]. □

Notice that by Theorem 1.7 we have $x \vdash_L y$ if, and only if, $x \vdash_{L_1} y$ hence the canonical map $L \to L_1$ is one-to-one and L can be identified to a subset of L_1.

1.3 Spectrum and completeness theorem

The *spectrum* of the lattice L, written $\mathrm{Spec}(L)$ is defined as the set $\mathrm{Hom}(L, \mathbf{2})$. It is isomorphic to the ordered set of all detachable prime ideals. The order relation is then reverse inclusion. We have $\mathrm{Spec}(\mathbf{2}) \simeq \mathbf{1}$, $\mathrm{Spec}(\mathbf{3}) \simeq \mathbf{2}$, $\mathrm{Spec}(\mathbf{4}) \simeq \mathbf{3}$, etc...

Proposition 1.9 *The completeness theorem implies the following result. If (J, U) is an idealistic prime which does not collapse then there exists $\varphi \in \mathrm{Spec}(L)$ such that $J \subseteq \varphi^{-1}(0)$ and $U \subseteq \varphi^{-1}(1)$. In particular if $a \not\leq b$, there exists $\varphi \in \mathrm{Spec}(L)$ such that $\varphi(a) = 1$ and $\varphi(b) = 0$. Also, if $L \neq \mathbf{1}$, $\mathrm{Spec}(L)$ is nonempty.*

Proof.
This follows from the completeness theorem for geometric theories (see Appendix). □

A corollary is the following representation theorem (Birkhoff theorem)

Theorem 1.10 (Representation theorem) *The completeness theorem implies the following result. The map $\theta_L : L \to \mathcal{P}(\mathrm{Spec}(L))$ defined by $a \mapsto \{\varphi \in \mathrm{Spec}(L)\ ;\ \varphi(a) = 1\}$ is an injective map of distributive lattice. This means that any distributive lattice can be represented as a lattice of subsets of a set.*

Another corollary is the following proposition.

Proposition 1.11 *The completeness theorem implies the following result. Let $\varphi : L \to L'$ a map of distributive lattices; φ is injective if, and only if, $\mathrm{Spec}(\varphi) : \mathrm{Spec}(L') \to \mathrm{Spec}(L)$ is surjective.*

---

[2] $\overline{L}$ is a disjoint copy of L.

Proof.
We have the equivalence

$$a \neq b \iff a \wedge b \neq a \vee b \iff a \vee b \not\leq a \wedge b$$

Assume that Spec(φ) is surjective. If $a \neq b$ in L, take $a' = \varphi(a)$, $b' = \varphi(b)$ and let $\psi \in \mathrm{Spec}(L)$ be such that $\psi(a \vee b) = 1$ and $\psi(a \wedge b) = 0$. Since Spec(φ) is surjective there exists $\psi' \in \mathrm{Spec}(L')$ such that $\psi = \psi'\varphi$ hence $\psi'(a' \vee b') = 1$ is $\psi'(a' \wedge b') = 0$, hence $a' \vee b' \not\leq a' \wedge b'$ and $a' \neq b'$.
Suppose that φ is injective. We identify L to a sublattice of L'. If $\psi \in \mathrm{Spec}(L)$, take $I = \psi^{-1}(0)$ and $F = \psi^{-1}(1)$. By the compactness theorem (see appendix), there exists $\psi' \in \mathrm{Spec}(L')$ such that $\psi'(I) = 0$ and $\psi'(F) = 1$, which means $\psi = \psi' \circ \varphi$. □

Of course, these three last results are hard to interpret in a computational way. An intuitive interpretation is that we can proceed "as if" any distributive lattice is a lattice of subsets of a set. The goal of Hilbert's program is to give a precise meaning to this sentence, and explain what is meant by "as if" there.

2 Krull dimension of distributive lattices

2.1 Definition of $\mathrm{Kr}_\ell(L)$

To develop a suitable constructive theory of the Krull dimension of a distributive lattice we have to find a constructive counterpart of the notion of increasing chains of prime ideals.

Definition 2.1 *To any distributive lattice L and $\ell \in \mathbb{N}$ we associate a distributive lattice $\mathrm{Kr}_\ell(L)$ which is the lattice defined by the generators $\varphi_i(x)$ for $i \leq \ell$ and $x \in L$ (thus we have $\ell+1$ disjoint copies of L and we let ϕ_i be the bijection between L and the ith copy) and relations*

- $\vdash \varphi_i(1)$
- $\varphi_i(0) \vdash$
- $\varphi_i(a), \varphi_i(b) \vdash \varphi_i(a \wedge b)$
- $\varphi_i(a \vee b) \vdash \varphi_i(a), \varphi_i(b)$
- $\varphi_i(a) \vdash \varphi_i(b)$ *whenever* $a \leq b$ *in* L
- $\varphi_{i+1}(a) \vdash \varphi_i(a)$ *for* $i < \ell$

Let S be the disjoint union $\bigcup \varphi_i(L)$ and $\vdash_S$ the entailment relation generated by these relations.

From this definition, we get directly the following theorem.

Theorem 2.2 *The maps φ_i are morphisms from the lattice L to the lattice $\mathrm{Kr}_\ell(L)$. Furthermore the lattice $\mathrm{Kr}_\ell(L)$ with the maps φ_i is then a solution of the following universal problem: to find a distributive lattice K and $\ell+1$ homomorphisms $\varphi_0 \geq \varphi_1 \geq \cdots \geq \varphi_\ell$ from L to K such that, for any lattice L' and any morphism $\psi_0 \geq \psi_1 \geq \cdots \geq \psi_\ell \in \mathrm{Hom}(L, L')$ we have one and only one morphism $\eta : K \to L'$ such that $\eta\varphi_0 = \psi_0$, $\eta\varphi_1 = \psi_1$, ..., $\eta\varphi_\ell = \psi_\ell$.*

The next theorem is the main result of this paper, and uses crucially the notion of entailment relation.

Theorem 2.3 *If U_i and J_i $(i = 0, \dots, \ell)$ are finite subsets of L we have in* $\mathrm{Kr}_\ell(L)$

$$\varphi_0(U_0) \wedge \dots \wedge \varphi_\ell(U_\ell) \leq \varphi_0(J_0) \vee \dots \vee \varphi_\ell(J_\ell)$$

if, and only if,

$$\varphi_0(U_0), \dots, \varphi_\ell(U_\ell) \vdash_S \varphi_0(J_0), \dots, \varphi_\ell(J_\ell)$$

if, and only if, there exist $x_1, \dots, x_\ell \in L$ such that (where $\vdash$ is the entailment relation of L):

$$\begin{array}{rcl} x_1,\ U_0 & \vdash & J_0 \\ x_2,\ U_1 & \vdash & J_1,\ x_1 \\ \vdots & \vdots & \vdots \\ x_\ell,\ U_{\ell-1} & \vdash & J_{\ell-1},\ x_{\ell-1} \\ U_\ell & \vdash & J_\ell,\ x_\ell \end{array}$$

Proof.
The equivalence between the first and the second statement follows from Theorem 1.7.

We show next that the relation on $\mathrm{P_f}(S)$ described in the statement of the theorem is indeed an entailment relation. The only point that needs explanation is the cut rule. To simplify notations, we take $\ell = 3$. We have then 3 possible cases, and we analyze only one case, where $X, \varphi_1(z) \vdash_S Y$ and $X \vdash_S Y, \varphi_1(z)$, the other cases being similar. By hypothesis we have $x_1, x_2, x_3, y_1, y_2, y_3$ such that

$$\begin{array}{rcl} x_1,\ U_0 & \vdash & J_0 \\ x_2,\ U_1,\ z & \vdash & J_1,\ x_1 \\ x_3,\ U_2 & \vdash & J_2,\ x_2 \\ U_3 & \vdash & J_3,\ x_3 \end{array} \qquad \begin{array}{rcl} y_1,\ U_0 & \vdash & J_0 \\ y_2,\ U_1 & \vdash & J_1, y_1,\ z \\ y_3, U_2 & \vdash & J_2, y_2 \\ U_3 & \vdash & J_3,\ y_3 \end{array}$$

The two entailment relations on the second line give

$$x_2,\ y_2,\ U_1,\ z \ \vdash\ J_1,\ x_1,\ y_1 \qquad x_2,\ y_2,\ U_1 \ \vdash\ J_1,\ x_1,\ y_1,\ z$$

hence by cut

$$x_2,\ y_2,\ U_1 \ \vdash\ J_1,\ x_1,\ y_1$$

i.e.,

$$x_2 \wedge y_2,\ U_1 \ \vdash\ J_1,\ x_1 \vee y_1$$

Finally, using distributivity

$$\begin{array}{rcl} (x_1 \vee y_1),\ U_0 & \vdash & J_0 \\ (x_2 \wedge y_2),\ U_1 & \vdash & J_1,\ (x_1 \vee y_1) \\ (x_3 \wedge y_3),\ U_2 & \vdash & J_2,\ (x_2 \wedge y_2) \\ U_3 & \vdash & J_3,\ (x_3 \wedge y_3) \end{array}$$

and hence $\varphi_0(U_0), \dots, \varphi_3(U_3) \vdash_S \varphi_0(J_0), \dots, \varphi_3(J_3)$.
Finally it is left to notice that the entailment relation we have defined is clearly the least possible relation ensuring the φ_i to form a non-increasing chain of morphisms. □

Notice that the morphisms φ_i are injective: it is easily seen that for $a, b \in L$ the relation $\varphi_i(a) \vdash_S \varphi_i(b)$ implies $a \vdash b$, and hence that $\varphi_i(a) = \varphi_i(b)$ implies $a = b$.

2.2 Partially specified chains of prime ideals

Definition 2.4 *In a distributive lattice L, a* partial specification for a chain of prime ideals *(that we shall call* idealistic chain*) is defined as follows. An idealistic chain of length ℓ is a list of $\ell+1$ idealistic primes of L: $\mathcal{C} = ((J_0, U_0), \dots, (J_\ell, U_\ell))$. An idealistic chain of length 0 is nothing but an* idealistic prime.

We think of an idealistic chain of length ℓ as a partial specification of an increasing chains of prime ideals $P_0, \dots, P_\ell$ such that $J_i \subseteq P_i$, $U_i \cap P_i = \emptyset$, $(i = 0, \dots, \ell)$.

Definition 2.5 *We say that an idealist chain $((J_0, U_0), \dots, (J_\ell, U_\ell))$* collapses *if, and only if, we have in* $\mathrm{Kr}_\ell(L)$

$$\varphi_0(U_0), \dots, \varphi_\ell(U_\ell) \vdash_S \varphi_0(J_0), \dots, \varphi_\ell(J_\ell)$$

Thus an idealistic chain $((J_0, U_0), \dots, (J_\ell, U_\ell))$ collapses in L if, and only if, the idealistic prime $\mathcal{P} = (\varphi_0(J_0), \dots, \varphi_\ell(J_\ell); \varphi_0(U_0), \dots, \varphi_\ell(U_\ell))$ collapses in $\mathrm{Kr}_\ell(L)$. From the completeness theorem we deduce the following result which justifies this idea of partial specification.

Theorem 2.6 (formal Nullstellensatz for chains of prime ideals) *The completeness theorem implies the following result. Let L be a distributive lattice and $((J_0, U_0), \dots, (J_\ell, U_\ell))$ be an idealistic chain in L. The following properties are equivalent:*

(*a*) *There exist $\ell+1$ prime ideals $P_0 \subseteq \cdots \subseteq P_\ell$ such that $J_i \subseteq P_i$, $U_i \cap P_i = \emptyset$, $(i = 0, \dots, \ell)$.*

(*b*) *The idealistic chain does not collapse.*

Proof.
If (*b*) holds then the idealistic prime $\mathcal{P} = (\varphi_0(J_0), \dots, \varphi_\ell(J_\ell); \varphi_0(U_0), \dots, \varphi_\ell(U_\ell))$ does not collapse in $\mathrm{Kr}_\ell(L)$. It follows then from Proposition 1.9 that there exists $\sigma \in \mathrm{Spec}(\mathrm{Kr}_\ell(L))$ such that σ is 0 on $\varphi_0(J_0), \dots, \varphi_\ell(J_\ell)$ and 1 on $\varphi_0(U_0), \dots, \varphi_\ell(U_\ell))$. We can then take $P_i = (\sigma \circ \varphi_i)^{-1}(0)$. That (*a*) implies (*b*) is direct. □

2.3 Krull dimension of a distributive lattice

Definition 2.7

1) An elementary idealistic chain *in a distributive lattice L is an idealistic chain of the form*

$$((0, x_1), (x_1, x_2), \dots, (x_\ell, 1))$$

(with x_i in L).

2) A distributive lattice L is of dimension $\leq \ell-1$ *iff it satisfies one of the equivalent conditions*

- *Any elementary idealistic chain of length ℓ collapses.*
- *For any sequence $x_1, \dots, x_\ell \in L$ we have*

$$\varphi_0(x_1), \dots, \varphi_{\ell-1}(x_\ell) \vdash \varphi_1(x_1), \dots, \varphi_\ell(x_\ell)$$

in $\mathrm{Kr}_\ell(L)$,

The following result shows that this definition coincides with the classical definition of Krull dimension for lattices.

Theorem 2.8 *The completeness theorem implies that the Krull dimension of a lattice L is $\leq \ell - 1$ if, and only if, there is no strictly increasing chains of prime ideals of length ℓ.*

Using Theorem 2.3, we get the following characterisation.

Theorem 2.9 *A distributive lattice L is of Krull dimension $\leq \ell - 1$ if, and only if, for all $x_1, \ldots, x_\ell \in L$ there exist $a_1, \ldots, a_\ell \in L$ such that*

$$a_1 \wedge x_1 = 0, \quad a_2 \wedge x_2 \leq a_1 \vee x_1, \quad , \ldots, a_\ell \wedge x_\ell \leq a_{\ell-1} \vee x_{\ell-1}, \quad 1 = a_\ell \vee x_\ell$$

In this way we have given a concrete form of the statement that the distributive lattice L has a dimension $\leq \ell - 1$ in the form of an existence of a sequence of inequalities.

In particular the distributive lattice L is of dimension ≤ -1 if, and only if, $1 = 0$ in L, and it is of dimension ≤ 0 if, and only if, L is a Boolean algebra (any element has a complement).

We have furthermore.

Lemma 2.10 *A distributive lattice L generated by a set G is of dimension $\leq \ell - 1$ if, and only if, for any sequence $x_1, \ldots, x_\ell \in G$*

$$\varphi_0(x_1), \ldots, \varphi_{\ell-1}(x_\ell) \vdash \varphi_1(x_1), \ldots, \varphi_\ell(x_\ell)$$

in $\mathrm{Kr}_\ell(L)$.

Indeed using distributivity, one can deduce

$$a \vee a', A \vdash b \vee b', B \quad a \wedge a' \vdash b \wedge b', B$$

from $a, A \vdash b, B$ and $a', A \vdash b', B$. Furthermore any element of L is an inf of sups of elements of G.

2.4 Implicative lattice

A lattice L is said to be an *implicative lattice* [5] or *Heyting algebra* [8] if, and only if, there is a binary operation $\rightarrow$ such that

$$a \wedge b \leq c \Longleftrightarrow a \leq b \rightarrow c$$

Theorem 2.11 *If L is an implcative lattice, we have in* $\mathrm{Kr}_\ell(L)$

$$\varphi_0(U_0), \ldots, \varphi_\ell(U_\ell) \vdash_S \varphi_0(J_0), \ldots, \varphi_\ell(J_\ell)$$

if, and only if,

$$1 = u_\ell \rightarrow (j_\ell \vee (u_{\ell-1} \rightarrow (j_{\ell-1} \vee \ldots (u_0 \rightarrow j_0))))$$

where $u_j = \wedge U_j$ and $j_k = \vee J_k$.

In the case where L is an implicative lattice, we can write explicitely that L is of dimension $\leq \ell - 1$ as an identity. For instance that L is of dimension ≤ 0 is equivalent to the identity

$$1 = x \vee \neg x$$

where $\neg x = x \rightarrow 0$ and that L is of dimension ≤ 1 is equivalent to the identity

$$1 = x_2 \vee (x_2 \rightarrow (x_1 \vee \neg x_1))$$

and so on.

Corollary 2.12 *An implicative lattice L is of dimension $\leq \ell - 1$ if, and only if, for any sequence $x_1, \ldots, x_\ell$*

$$1 = x_\ell \vee (x_\ell \rightarrow \ldots (x_2 \vee (x_2 \rightarrow (x_1 \vee \neg x_1))) \ldots)$$

2.5 Decidability

To any distributive lattice L we have associated a family of distributive lattices $\mathrm{Kr}_\ell(L)$ with a complete description of their ordering. A lattice is *discrete* if, and only if, its ordering is decidable, which means intuitively that there is an algorithm to decide the ordering (or, equivalently, the equality) in this lattice. It should be intuitively clear that we could find a discrete lattice L such that $\mathrm{Kr}_1(L)$ is not discrete since, by 2.3, the ordering on $\mathrm{Kr}_1(L)$ involves an existential quantification on the set L, that may be infinite (this point is discussed in [1], with another argument). However we can use the characterization of Theorem 2.3 to give a general sufficient condition ensuring that all $\mathrm{Kr}_\ell(L)$ are discrete.

Theorem 2.13 *Suppose that the lattice L is a discrete implicative lattice then each* $\mathrm{Kr}_\ell(L)$ *is discrete.*

Proof.
This is direct from Theorem 2.11. □

2.6 Dimension of Spectral Spaces

This subsection is written from a classical point of view. Following [7], a topological space X is called a spectral space if it satisfies the following conditions: (a) X is a compact T_0-space; (b) X has a compact open basis which is closed under finite intersections; (c) each irreducible closed subspace of X has a generic point. $\mathrm{Spec}(R)$, with the Zariski topology, is spectral for any commutative ring R with identity. Similary, if we take for basic open the sets $U_a = \{\phi \in \mathrm{Spec}(L) \mid \phi(a) = 1\}$ then $\mathrm{Spec}(L)$ is spectral for any distributive lattice. The compact open subsets of a spectral space form a distributive lattice, and it is well-known [21, 8] that, if L is an arbitrary distributive lattice, then L is isomorphic to the lattice of compact open subsets of the space $\mathrm{Spec}(L)$.

If U, V are open subsets of a topological space X we define $U \to V$ to be the largest open W such that $W \cap U \subseteq V$ and $\neg\, U = U \to \emptyset$. In a classical setting a spectral space X is said to be of dimension $\leq \ell - 1$ if, and only if, there is no strictly increasing chains of length ℓ of irreducible closed subsets of X. We can reformulate Theorem 2.9 as follows.

Theorem 2.14 *A spectral space X is of dimension $\leq \ell - 1$ if, and only if, for any compact open subsets $x_1, \ldots, x_\ell$ of X*

$$X = x_\ell \vee (x_\ell \to \ldots (x_2 \vee (x_2 \to (x_1 \vee \neg x_1)))\ldots)$$

2.7 Connections with Joyal's definition

Let L be a distributive lattice, Joyal [6] gives the following definition of $\dim(L) \leq \ell$. Let $\varphi_i^\ell : L \to \mathrm{Kr}_\ell(L)$ be the $\ell + 1$ universal morphisms. By universality of $\mathrm{Kr}_{\ell+1}(L)$, we have $\ell + 1$ morphisms $\sigma_i : \mathrm{Kr}_{\ell+1}(L) \to \mathrm{Kr}_\ell(L)$ such that $\sigma_i \circ \varphi_j^{\ell+1} = \varphi_j^\ell$ if $j \leq i$ and $\sigma_i \circ \varphi_j^{\ell+1} = \varphi_{j-1}^\ell$ if $j > i$. Joyal defines then $\dim(L) \leq \ell$ to mean that $(\sigma_0, \ldots, \sigma_\ell) : \mathrm{Kr}_{\ell+1}(L) \to \mathrm{Kr}_\ell(L)^{\ell+1}$ is injective. This definition can be motivated by Proposition 1.11: the elements in the image of $Sp(\sigma_i)$ are the chains of prime ideals $(\alpha_0, \ldots, \alpha_\ell)$ with $\alpha_i = \alpha_{i+1}$, and $Sp(\sigma_0, \ldots, \sigma_\ell)$ is surjective if, and only if, for any chain $(\alpha_0, \ldots, \alpha_\ell)$ there exists $i < \ell$ such that $\alpha_i = \alpha_{i+1}$. This means exactly that there is no nontrivial chain of prime ideals of length $\ell + 1$. Using the completeness theorem, one can then see the equivalence with Definition 2.7. One could check directly this equivalence using a constructive metalanguage, but for lack of space, we shall not present here this argument. Similarly, it would be possible to establish the equivalence of our definition with the one of Español [6] (here also, this connection is clear via the completeness theorem).

3 Zariski and Krull lattice

3.1 Zariski lattice

Let R be a commutative ring. We write $\langle J\rangle$ or explicitly $\langle J\rangle_R$ for the ideal of R generated by the subset $J \subseteq R$. We write $\mathcal{M}(U)$ for the monoid (3) generated by the subset $U \subseteq R$. Given a commutative ring R the *Zariski lattice* $\mathrm{Zar}(R)$ has for elements the radicals of finitely generated ideals (the order relation being inclusion). It is well defined as a lattice. Indeed $\sqrt{I_1} = \sqrt{J_1}$ and $\sqrt{I_2} = \sqrt{J_2}$ imply $\sqrt{I_1 I_2} = \sqrt{J_1 J_2}$ (which defines $\sqrt{I_1} \wedge \sqrt{I_2}$) and $\sqrt{I_1 + I_2} = \sqrt{J_1 + J_2}$ (which defines $\sqrt{I_1} \vee \sqrt{I_2}$). The Zariski lattice of R is always distributive, but may not be discrete, even if R is discrete. Nevertheless an inclusion $\sqrt{I_1} \subseteq \sqrt{I_2}$ can always be certified in a finite way if the ring R is discrete. This lattice contains all the informations necessary for a constructive development of the abstract theory of the Zariski spectrum.
We shall write $\widetilde{a}$ for $\sqrt{\langle a\rangle}$. Given a subset S of A we write $\widetilde{S}$ for the subset of $\mathrm{Zar}(R)$ the elements of which are $\widetilde{s}$ for $s \in S$. We have $\widetilde{a_1} \vee \cdots \vee \widetilde{a_m} = \sqrt{\langle a_1, \ldots, a_m\rangle}$ and $\widetilde{a_1} \wedge \cdots \wedge \widetilde{a_m} = \widetilde{a_1 \cdots a_m}$
Let U and J be two finite subsets of R, we have

$$U \vdash_{\mathrm{Zar}(R)} J \iff \prod_{u \in U} u \in \sqrt{\langle J\rangle} \iff \mathcal{M}(U) \cap \langle J\rangle \neq \emptyset$$

This describes completely the lattice $\mathrm{Zar}(R)$. More precisely we have:

Proposition 3.1 *The lattice* $\mathrm{Zar}(R)$ *of a commutative ring* R *is (up to isomorphism) the lattice generated by* $(R, \vdash)$ *where* $\vdash$ *is the least entailment relation over* R *such that*

$$\begin{array}{lll} 0 \ \vdash & x,\, y \ \vdash \ xy & \\ \ \ \vdash \ 1 & xy \ \vdash \ x & x+y \ \vdash \ x,\, y \end{array}$$

Proof.
It is clear that the relation $U \vdash J$ defined by "$\mathcal{M}(U)$ meets $\langle J\rangle$" satisfies these axioms. It is also clear that the entailment relation generated by these axioms contains this relation. Let us show that this relation is an entailment relation. Only the cut rule is not obvious. Assume that $\mathcal{M}(U, a)$ meets $\langle J\rangle$ and that $\mathcal{M}(U)$ meets $\langle J, a\rangle$. There exist then $m_1, m_2 \in \mathcal{M}(U)$ and $k \in \mathbb{N}, x \in R$ such that $a^k m_1 \in \langle J\rangle$, $m_2 + ax \in \langle J\rangle$. Eliminating a this implies that $\mathcal{M}(U)$ intersects $\langle J\rangle$. □

We have $\widetilde{a} = \widetilde{b}$ if, and only if, a divides a power of b and b divides a power of a.

Proposition 3.2 *In a commutative ring* R *to give an ideal of the lattice* $\mathrm{Zar}(R)$ *is the same as to give a radical ideal of* R. *If* I *is a radical ideal of* R *one associates the ideal*

$$\mathcal{I} = \{J \in \mathrm{Zar}(R) \mid J \subseteq I\}$$

of $\mathrm{Zar}(R)$. *Conversely if* $\mathcal{I}$ *is an ideal of* $\mathrm{Zar}(R)$ *one can associate the ideal*

$$I = \bigcup_{J \in \mathcal{I}} J = \{x \in A \mid \widetilde{x} \in \mathcal{I}\},$$

which is a radical ideal of R. *In this bijection the prime ideals of the ring correspond to the prime ideals of the Zariski lattice.*

3 A monoid will always be multiplicative.

Proof.
We only prove the last assertion. If I is a prime ideal of R, if $J, J' \in \mathrm{Zar}(R)$ and $J \wedge J' \in \mathcal{I}$, let $a_1, \ldots, a_n \in R$ be some "generators" of J (*i.e.*, $J = \sqrt{\langle a_1, \ldots, a_n \rangle}$) and let $b_1, \ldots, b_m \in A$ be some generators of J'. We have $a_i b_j \in I$ and hence $a_i \in I$ or $b_j \in I$ for all i, j. It follows from this (constructively) that we have $a_i \in I$ for all i or $b_j \in I$ for all j. Hence $J \in \mathcal{I}$ or $J' \in \mathcal{I}$ and $\mathcal{I}$ is a prime ideal of $\mathrm{Zar}(R)$.
Conversely if $\mathcal{I}$ is a prime ideal of $\mathrm{Zar}(R)$ and if we have $\widetilde{xy} \in \mathcal{I}$ then $\widetilde{x} \wedge \widetilde{y} \in \mathcal{I}$ and hence $\widetilde{x} \in \mathcal{I}$ or $\widetilde{y} \in \mathcal{I}$. This shows that $\{x \in A \mid \widetilde{x} \in \mathcal{I}\}$ is a prime ideal of R. □

3.2 Krull lattices of a commutative ring

Definition 3.3 *We define* $\mathrm{Kru}_\ell(R) := \mathrm{Kr}_\ell(\mathrm{Zar}(R))$. *This is called the* Krull lattice of order ℓ *of the ring R. We say also that R is of Krull dimension $\leq \ell$ iff the distributive lattice* $\mathrm{Zar}(R)$ *is of dimension $\leq \ell$.*

Theorem 3.4 *The ring R is of dimension $\leq \ell - 1$ if, and only if, for any $x_1, \ldots, x_n \in R$ we have in* $\mathrm{Kru}_\ell(R)$

$$\varphi_0(\widetilde{x_1}), \ldots, \varphi_{\ell-1}(\widetilde{x_\ell}) \vdash \varphi_1(\widetilde{x_1}), \ldots, \varphi_\ell(\widetilde{x_\ell})$$

Proof.
This is a direct consequence of Lemma 2.10 and the fact that the elements $\widetilde{x}$ generates $\mathrm{Zar}(R)$.
□

Theorem 3.5 *Let $C = ((J_0, U_0), \ldots, (J_\ell, U_\ell))$ be a list of $\ell + 1$ pairs of finite subsets of R, the following properties are equivalent:*

1. *there exist $j_i \in \langle J_i \rangle$, $u_i \in \mathcal{M}(U_i)$, $(i = 0, \ldots, \ell)$, such that*

$$u_0 \cdot (u_1 \cdot (\cdots (u_\ell + j_\ell) + \cdots) + j_1) + j_0 = 0$$

2. *there exist $L_1, \ldots, L_\ell \in \mathrm{Zar}(R)$ such that in* $\mathrm{Zar}(R)$:

$$\begin{array}{rcl} L_1,\ \widetilde{U_0} & \vdash & \widetilde{J_0} \\ L_2,\ \widetilde{U_1} & \vdash & \widetilde{J_1},\ L_1 \\ \vdots & \vdots & \vdots \\ L_\ell,\ \widetilde{U_{\ell-1}} & \vdash & \widetilde{J_{\ell-1}},\ L_{\ell-1} \\ \widetilde{U_\ell} & \vdash & \widetilde{J_\ell},\ L_\ell \end{array}$$

3. *there exist $x_1, \ldots, x_\ell \in R$ such that (for the entailment relation described in Proposition 3.1):*

$$\begin{array}{rcl} x_1,\ U_0 & \vdash & J_0 \\ x_2,\ U_1 & \vdash & J_1,\ x_1 \\ \vdots & \vdots & \vdots \\ x_\ell,\ U_{\ell-1} & \vdash & J_{\ell-1},\ x_{\ell-1} \\ U_\ell & \vdash & J_\ell,\ x_\ell \end{array}$$

Proof.
It is clear that 1 entails 3: simply take

$$x_\ell = u_\ell + j_\ell,\ x_{\ell-1} = x_\ell u_{\ell-1} + j_{\ell-1}, \ldots,\ x_0 = x_1 u_0 + j_0$$

and that 3 entails 2.

Let us prove that 2 implies 3. We assume:

$$\begin{array}{rcl} L_1,\ \widetilde{U_0} & \vdash & I_0 \\ L_2,\ \widetilde{U_1} & \vdash & I_1,\ L_1 \\ \widetilde{U_2} & \vdash & I_2,\ L_2 \end{array}$$

The last line means that $\mathcal{M}(U_2)$ intersects $I_2 + L_2$ and hence $I_2 + \langle x_2 \rangle$ for some element x_2 of L_2. Hence we have $\widetilde{U}_2 \vdash I_2,\ \widetilde{x}_2$. Since $\widetilde{x}_2 \leq L_2$ in $\mathrm{Zar}(R)$ we have $\widetilde{x}_2,\ \widetilde{U}_1 \vdash I_1,\ L_1$. We have then replaced L_2 by $\widetilde{x}_2$. Reasoning as previously one sees that one can replace as well L_1 by a suitable $\widetilde{x}_1$. One gets then 3.

Finally, let us show that 3 entails 1: if we have for instance

$$\begin{array}{rcl} x_1,\ U_0 & \vdash & I_0 \\ x_2,\ U_1 & \vdash & I_1,\ x_1 \\ U_2 & \vdash & I_2,\ x_2 \end{array}$$

by the last line we know that we can find y_2 both in the monoid $M_2 = \mathcal{M}(U_2) + \langle I_2 \rangle$ and in $\langle x_2 \rangle$. Since $y_2 \vdash x_1$

$$y_2,\ U_1 \vdash I_1,\ x_1$$

and since $y_2 \in M_2$ we can find y_1 both in the monoid $M_1 = M_2\mathcal{M}(U_1) + \langle I_1 \rangle$ and in $\langle x_1 \rangle$. We have $y_1 \vdash x_1$ and hence

$$y_1, U_0 \vdash I_0$$

and since $y_1 \in M_1$ this implies $0 \in M_1\mathcal{M}(U_0) + \langle I_0 \rangle$ as desired. □

Corollary 3.6 *A ring R is of Krull dimension $\leq \ell - 1$ iff for any sequence $x_1, \ldots, x_\ell$ there exist $a_1, \ldots, a_\ell \in R$ and $m_1, \ldots, m_\ell \in \mathbb{N}$ such that*

$$x_1^{m_1}(\cdots(x_\ell^{m_\ell}(1 + a_\ell x_\ell) + \cdots) + a_1 x_1) = 0$$

Proof.

By Theorem 3.4, we have in $\mathrm{Kru}_\ell(R)$

$$\varphi_0(\widetilde{x_1}), \ldots, \varphi_{\ell-1}(\widetilde{x_\ell}) \vdash \varphi_1(\widetilde{x_1}), \ldots, \varphi_\ell(\widetilde{x_\ell})$$

we can then apply Theorem 3.5 to the elementary idealistic chain

$$((0, \widetilde{x_1}), (\widetilde{x_1}, \widetilde{x_2}), \ldots, (\widetilde{x_\ell}, 1))$$

and we get in this way $j_i \in \langle x_i \rangle$, $j_0 = 0$ and $u_i \in \mathcal{M}(x_i)$, $u_0 = 1$ such that

$$u_0 \cdot (u_1 \cdot (\cdots (u_\ell + j_\ell) + \cdots) + j_1) + j_0 = 0$$

as desired. □

This concrete characterisation of the Krull dimension of a ring can be found in [14], where it is derived using dynamical methods [2].

Lemma 3.7 *If R is coherent and noetherian then $\mathrm{Zar}(R)$ is an implicative lattice.*

Proof.
Let $L \in \operatorname{Zar}(R)$, radical of an ideal generated by elements $y_1, \ldots, y_n$ and $x \in R$, we show how to define an element $\widetilde{x} \to L \in \operatorname{Zar}(R)$ such that, for any $M \in \operatorname{Zar}(R)$

$$M \wedge \widetilde{x} \leq L \quad \Longleftrightarrow \quad M \leq \widetilde{x} \to L$$

For this, we consider the sequence of ideals

$$I_k = \{z \in R \mid zx^k \in \langle y_1, \ldots, y_n \rangle\}$$

Since R is coherent, each I_k is finitely generated. Since furthermore R is noetherian and $I_k \subseteq I_{k+1}$ the sequence I_k is stationary and $\bigcup_k I_k$ is finitely generated. We take for $\widetilde{x} \to L$ the radical of this ideal.

If $M \in \operatorname{Zar}(R)$ then M is the radical of an ideal generated by finitely many elements $x_1, \ldots, x_m$ and we can take $M \to L = (\widetilde{x_1} \to L) \wedge \ldots \wedge (\widetilde{x_m} \to L)$. □

Corollary 3.8 *If R is coherent, noetherian and strongly discrete then each lattice $\operatorname{Kr}_n(R)$ is discrete.*

Proof.
Using Theorem 2.13 and Lemma 3.7 we are left to show that $\operatorname{Zar}(R)$ is discrete. We have $M \leq L$ if, and only if, $1 = M \to L$. But to test if an element of $\operatorname{Zar}(R)$ is equal to the ideal $\langle 1 \rangle$ is decidable since R is strongly discrete. □

The hypotheses of this corollary are satisfied if R is a polynomial ring $K[X_1, \ldots, X_n]$ over a discrete field K [19].

3.3 Krull dimension of a polynomial ring over a discrete field

Let R be a commutative ring, let us say that a sequence $x_1, \ldots, x_\ell$ is *singular* if, and only if, there exists $a_1, \ldots, a_\ell \in R$ and $m_1, \ldots, m_\ell \in \mathbb{N}$ such that

$$x_1^{m_1}(\cdots(x_\ell^{m_\ell}(1 + a_\ell x_\ell) + \cdots) + a_1 x_1) = 0$$

A sequence is *pseudo regular* if, and only if, it is not singular. Corollary 3.6 can be reformulated as: a ring R is of Krull dimension $\leq \ell - 1$ if, and only if, any sequence in R of length ℓ is singular.

Proposition 3.9 *Let K be a discrete field, R a commutative K-algebra, and $x_1, \ldots, x_\ell$ in R algebraically dependent over K. The sequence $x_1, \ldots, x_\ell$ is singular.*

Proof.
Let $Q(x_1, \ldots, x_\ell) = 0$ be a algebraic dependence relation over K. Let us order the nonzero monomials of Q along the lexicographic ordering. We can suppose that the coefficient of the first monomial is 1. Let $x_1^{m_1} x_2^{m_2} \cdots x_\ell^{m_\ell}$ be this momial, it is clear that Q can be written on the form

$$Q = x_1^{m_1} \cdots x_\ell^{m_\ell} + x_1^{m_1} \cdots x_\ell^{1+m_\ell} R_\ell + x_1^{m_1} \cdots x_{\ell-1}^{1+m_{\ell-1}} R_{\ell-1} + \cdots + x_1^{m_1} x_2^{1+m_2} R_2 + x_1^{1+m_1} R_1$$

and this is the desired collapsus. □

Let us say that a ring is of dimension ℓ if it is of dimension $\leq \ell$ but not of dimension $\leq \ell - 1$. It follows that we have:

Theorem 3.10 *Let K be a discrete field. The Krull dimension of the ring $K[X_1, \ldots, X_\ell]$ is equal to ℓ.*

Proof.
Given Proposition 3.9 it is enough to check that the sequence $(X_1, \ldots, X_\ell)$ is pseudo regular, which is direct. □

Notice that we got this basic result quite directly from the characterisation of Corollary 3.6, and that our argument is of course also valid classically (with the usual definition of Krull dimension). This contradicts the current opinion that constructive arguments are necessarily more involved than classical proofs.

References

[1] Boileau, A., Joyal, A. *La logique des topos* J. Symbolic Logic **46** (1981), no. 1, 6–16

[2] Coste M., Lombardi H., Roy M.-F. *Dynamical method in algebra: Effective Nullstellensätze* Annals of Pure and Applied Logic **111**, (2001) 203–256

[3] Cederquist, Coquand T. *Entailment relations and Distributive Lattices* Logic Colloquium '98 (Prague), 127–139, Lect. Notes Log., 13. Assoc. Symbol. Logic, Urbana, (2000).

[4] Coquand T., Persson H. *Valuations and Dedekind's Prague Theorem.* J. Pure Appl. Algebra **155** (2001), no. 2-3, 121–129.

[5] Curry, H. B. *Foundations of mathematical logic* McGraw-Hill Book Co., Inc., New York-San Francisco, Calif.-Toronto-London 1963

[6] Español L. *Constructive Krull dimension of lattices.* Rev. Acad. Cienc. Zaragoza (2) **37** (1982), 5–9.

[7] Hochster M. *Prime ideal structure in commutative rings.* Trans. Amer. Math. Soc. **142** 1969 43–60.

[8] Johnstone P. *Stone Spaces.* Cambridge Studies in Advanced Mathematics, 3. Cambridge University Press, Cambridge, 1986.

[9] Joyal A. *Le théorème de Chevalley-Tarski.* Cahiers de Topologie et Géometrie Differentielle, 1975.

[10] Kuhlmann F.-V., Lombardi H. *Construction du hensélisé d'un corps valué.* Journal of Algebra **228**, (2000), 624–632.

[11] Lombardi H. *Un nouveau positivstellensatz effectif pour les corps valués.* Séminaire "Structures Ordonnées" (Paris 6-7) (18 pages, published in 96. Editeurs: F. Delon, A. Dickmann, D. Gondard)

[12] Lombardi H. *Le contenu constructif d'un principe local-global avec une application à la structure d'un module projectif de type fini* . Publications Mathématiques de Besançon. Théorie des nombres. (1997). Fascicule 94–95 & 95–96.

[13] Lombardi H. *Relecture constructive de la théorie d'Artin-Schreier.* Annals of Pure and Applied Logic, **91**, (1998), 59–92.

[14] Lombardi H. *Dimension de Krull, Nullstellensätze et Évaluation dynamique.* Math. Zeitschrift, **242**, (2002), 23–46.

[15] Lombardi H. *Platitude, localisation et anneaux de Prüfer: une approche constructive.* Preprint (1999).

[16] Lombardi H. *Hidden constructions in abstract algebra. I. Integral dependance.* J. Pure Appl. Algebra **167** (2002), 259–267.

[17] Lombardi H., Quitté C. *Constructions cachées en algèbre abstraite (2) Le principe local-global.* Preprint 1999.

[18] MacNeille H. M. *Partially ordered sets.* Trans. Amer. Math. Soc. **42** (1937), no. 3, 416–460.

[19] Mines R., Richman F., Ruitenburg W. *A Course in Constructive Algebra*. Universitext. Springer-Verlag, 1988.

[20] Rota G.C. *Indiscrete Thoughts* Birkhauser, 1995.

[21] Stone, M.H. *Topological representations of distributive lattices and Brouwerian logics*. Cas. Mat. Fys. **67**, 1-25 (1937).

Annex: Completeness, compactness theorem, LLPO and geometric theories

A.4 Theories and models

We fix a set V of *atomic propositions* or *propositional letters*. A proposition $\phi, \psi, \ldots$ is a syntactical object built from the atoms $p, q, r \in V$ with the usual logical connectives

$$0,\ 1,\ \phi \wedge \psi,\ \phi \vee \psi,\ \phi \rightarrow \psi,\ \neg\phi$$

We let P_V be the set of all propositions. Let F_2 be the Boolean algebra with two elements. A *valuation* is a function $v \in F_2^V$ that assigns a truth value to any of the atomic propositions. Such a valuation can be extended to a map $P_V \rightarrow \{0,1\}$, $\phi \longmapsto v(\phi)$ in the expected way. A *theory* T is a subset of P_V. A *model* of T is a valuation v such that $v(\phi) = 1$ for all $\phi \in T$.

More generally given a Boolean algebra B we can define B-valuation to be a function $v \in B^V$. This can be extended as well to a map $P_V \rightarrow B$, $\phi \longmapsto v(\phi)$. A *B-model* of T is a valuation v such that $v(\phi) = 1$ for all $\phi \in T$. The usual notion of model is a direct special case, taking for B the Boolean algebra F_2. For any theory there exists always a free Boolean algebra over which T is a model, the *Lindenbaum* algebra of T, which can be also be defined as the Boolean algebra generated by T, thinking of the elements of V as generators and the elements of T as relations. The theory T is *formally consistent* if, and only if, its Lindenbaum algebra is not trivial.

A.5 Completeness theorem

Theorem A.11 *(Completeness theorem) Let T be a theory. If T is formally consistent then T has a model.*

This theorem is the completeness theorem for propositional logic. Such a theorem is strongly related to Hilbert's program, which can be seen as an attempt to replace the question of existence of model of a theory by the formal fact that this theory is not contradictory.

Let B the Lindenbaum algebra of T. To prove completeness, it is enough to find a morphism $B \rightarrow F_2$ assuming that B is not trivial, wich is the same as finding a prime ideal (which is then automatically maximal) in B. Thus the completeness theorem is a consequence of the existence of prime ideal in nontrivial Boolean algebra. Notice that this existence is clear in the case where B is finite, hence that the completeness theorem is direct for finite theories.

A.6 Compactness theorem

The completeness theorem for an arbitrary theory can be seen as a corollary of the following fundamental result.

Theorem A.12 *(Compactness theorem) Let T be a theory. If all finite subsets of T have a model then so does T.*

Suppose indeed that the compactness theorem holds, and let T be a formally consistent theory. Then an arbitrary finite subset T_0 of T is also formally consistent. Furthermore, we have seen that this implies the existence of a model for T_0. It follows then from the compactness theorem that T itself has a model.

Conversely, it is clear that the compactness theorem follows from the completeness theorem, since a theory is formally consistent as soon as all its finite subsets are.

A simple general proof of the compactness theorem is to consider the product topology on $\{0,1\}^V$ and to notice that the set of models of a given subset of T is a closed subset. The theorem is then a corollary of the compactness of the space $W := \{0,1\}^V$ when compactness is expressed (in classical mathematics) as: if a family of closed subsets of W has non-void finite intersections, then its intersection is non-void.

A.7 LPO and LLPO

If V is countable (*i.e.*, discrete and enumerable) we have the following alternative argument. One writes $V = \{p_0, p_1, \ldots\}$ and builds by induction a partial valuation v_n on $\{p_i \mid i < n\}$ such that any finite subset of T has a model which extends v_n, and v_{n+1} extends v_n. To define v_{n+1} one first tries $v_{n+1}(p_n) = 0$. If this does not work, there is a finite subset of T such that any of its model v that extends v_n satisfies $v(p_n) = 1$ and one can take $v_{n+1}(p_n) = 1$.

The non-effective part of this argument is contained in the choice of $v_{n+1}(p_n)$, which demands to give a gobal answer to an infinite set of (elementary) questions.

Now let us assume also that we can enumerate the infinite set T. We can then build a sequence of finite subsets of T in a nondecreasing way $K_0 \subseteq K_1 \subseteq \ldots$ such that any finite subset of T is a subset of some K_n. Assuming we have construct v_n such that all K_j's have a model extending v_n, in order to define $v_{n+1}(p_n)$ we have to give a global answer to the questions: do all K_j's have a model extending v_{n+1} when we choose $v_{n+1}(p_n) = 1$? For each j this is an elementary question, having a clear answer. More precisely let us define $g_n : \mathbb{N} \to \{0,1\}$ in the following way: $g_n(j) = 1$ if there is a model $v_{n,j}$ of K_j extending v_n with $v_{n,j}(p_n) = 1$, else $g_n(j) = 0$. By induction hypothesis if $g_n(j) = 0$ then all K_ℓ have a model $v_{n,\ell}$ extending v_n with $v_{n,\ell}(p_n) = 1$, and all models $v_{n,\ell}$ of K_ℓ extending v_n satisfy $v_{n,\ell}(p_n) = 1$ if $\ell \geq j$. So we can "construct" inductively the infinite sequence of partial models v_n by using at each step the non-constructive Bishop's principle LPO (Least Principle of Omniscience): given a function $f : \mathbb{N} \to \{0,1\}$, either $f = 1$ or $\exists j \in \mathbb{N}\ f(j) \neq 1$. This principle is applied at step n to the function g_n.

In fact we can slightly modify the argument and use only a combination of Dependant Choice and of Bishop's principle LLPO (Lesser Limited Principle of Omniscience), which is known to be strictly weaker than LPO: given two non-increasing functions $g, h : \mathbb{N} \to \{0,1\}$ such that, for all j

$$g(j) = 1 \vee h(j) = 1$$

then we have $g = 1$ or $h = 1$. Indeed let us define $h_n : \mathbb{N} \to \{0,1\}$ in a symmetric way: $h_n(j) = 1$ if there is a model $v_{n,j}$ of K_j extending v_n with $v_{n,j}(p_n) = 0$, else $h_n(j) = 0$. Cleraly g_n and h_n are non-increasing functions. By induction hypothesis, we have for all j $g_n(j) = 1 \vee h_n(j) = 1$. So, applying LLPO, we can define $v_{n+1}(p_n) = 1$ if $g_n = 1$ and $v_{n+1}(p_n) = 1$ if $h_n = 1$. Nevertheless, we have to use dependant choice in order to make this choice inifnitely often since the answer "$g = 1$ or $h = 1$" given by the oracle LLPO may be ambiguous.

In a reverse way it is easy to see that the completeness theorem restricted to the countable case implies LLPO.

A.8 Geometric formulae and theories

What would have happened if topologies without points had been discovered before topologies with points, or if Grothendieck had known the theory of distributive lattices? (G. C. Rota [20]).

A formula is *geometric* if, and only if, it is built only with the connectives $0, 1, \phi \wedge \psi, \phi \vee \psi$ from the propositional letters in V. A theory if a (propositional) *geometric* theory iff all the formula in T are of the form $\phi \to \psi$ where ϕ and ψ are geometric formulae.

It is clear that the formulae of a geometric theory T can be seen as relations for generating a distributive lattice L_T and that the Lindenbaum algebra of T is nothing else but the free Boolean algebra generated by the lattice L_T. It follows from Proposition 1.8 that T is formally consistent if, and only if, L_T is nontrivial. Also, a model of T is nothing else but an element of $\mathrm{Spec}(L_T)$.

Theorem A.13 *(Completeness theorem for geometric theories) Let T be a geometric theory. If T generates a nontrivial distributive lattice, then T has a model.*

The general notion of geometric formula allows also existential quantification, but we restrict ourselves here to the propositional case. Even in this restricted form, the notion of geometric theory is fundamental. For instance, if R is a commutative ring, we can consider the theory with atomic propositions $D(x)$ for each $x \in R$ and with axioms

- $D(0_R) \to 0$
- $1 \to D(1_R)$
- $D(x) \wedge D(y) \to D(xy)$
- $D(xy) \to D(x)$
- $D(x + y) \to D(x) \vee D(y)$

This is a geometric theory T. The model of this theory are clearly the complement of the prime ideals. What is remarkable is that, while the existence of models of this theory is a nontrivial fact which may be dependent on set theoretic axioms (such as dependent axiom of choices) its formal consistency is completely elementary (as explained in the beginning of Section 3). This geometric theory, or the distributive lattice it generates, can be seen as a point-free description of the Zariski spectrum of the ring. The distributive lattice generated by this theory (called in this paper the Zariski lattice of R) is isomorphic to the lattice of compact open of the Zariski spectrum of R, while the Boolean algebra generated by this theory is isomorphic to the algebra of the constructible sets.